Multiple Chemical Sensitivities
A Workshop

This workshop was conducted at the request of the Environmental Protection Agency. This volume contains the papers prepared and presented by individual workshop participants; the papers have not undergone peer review.

Board on Environmental Studies and Toxicology
Commission on Life Sciences

National Research Council

Advancing the Understanding of Multiple Chemical Sensitivity
Proceedings of the Association of Occupational and Environmental Clinics (AOEC) Workshop on Multiple Chemical Sensitivity

20-21 September 1991, Washington, D.C.

Edited by: Kathleen M. Rest, Occupational Health Program
Department of Family and Community Medicine
University of Massachusetts Medical Center

Sponsored by:
The Association of Occupational and Environmental Clinics (AOEC)
1010 Vermont Street, N.W., Suite 513, Washington, D.C. 20005

Supported by:
U.S. Department of Health and Human Services
Public Health Service
Agency for Toxic Substances and Disease Registry

Proceedings of the Conference on Low-Level Exposure to Chemicals and Neurobiologic Sensitivity

6–7 April, 1994, Baltimore, Maryland

Editor: Frank L. Mitchell, D.O., M.P.H. • *Co-Editor:* Patricia Price, D.O.
Agency for Toxic Substances and Disease Registry

Sponsored by:
U.S. Department of Health and Human Services
Public Health Services
Agency for Toxic Substances and Disease Registry

PREFACE

The development in some persons of sensitivity to a wide variety of substances presents difficult and complex medical and scientific issues. This condition, often called Multiple Chemical Sensitivity (MCS), has been the subject of a great deal of interest and controversy.

The Agency for Toxic Substances and Disease Registry (ATSDR) has maintained an interest in this topic for a number of years. The goals of the Agency have been to serve as a conduit of information on the subject and to encourage research that would help to clarify the phenomenon. Therefore, ATSDR has provided support to three national conferences on the topic of MCS. The first was a National Academy of Sciences meeting held in March, 1991. The second was a meeting sponsored by the Association of Occupational and Environmental Clinics (AOEC) in September, 1991. The third was a Conference on Low-Level Exposure to Chemicals and Neurobiologic Sensitivity, held in April, 1994. Each of these meetings made valuable contributions toward identifying key scientific, clinical, and public health issues concerning MCS.

In order to further the goal of making scientific information relating to MCS widely available, ATSDR has published the Proceedings from these three meetings in one volume. Each is presented as it was originally published, including pagination.

This format allows scientists and interested members of the public to more easily consider the various viewpoints, clinical approaches, and suggestions for research that are presented.

It is the hope of ATSDR that the insights presented in this volume will stimulate clinical and laboratory research. As results from this research begin to appear, ATSDR, in concert with other federal agencies, can then distribute that information and help clarify the many scientific issues related to this problem.

ISBN: 0-911131-53-1; Library of Congress Catalog Card Number: 94-074818

TABLE OF CONTENTS

PART I: MULTIPLE CHEMICAL SENSITIVITIES — A WORKSHOP

PART II: ADVANCING THE UNDERSTANDING OF MULTIPLE CHEMICAL SENSITIVITY — PROCEEDINGS OF THE ASSOCIATION OF OCCUPATIONAL AND ENVIRONMENTAL CLINICS (AOEC) WORKSHOP ON MULTIPLE CHEMICAL SENSITIVITY

PART III: PROCEEDINGS OF THE CONFERENCE ON LOW-LEVEL EXPOSURE TO CHEMICALS AND NEUROBIOLOGIC SENSITIVITY

Multiple Chemical Sensitivities

A Workshop

This workshop was conducted at the request of the Environmental Protection Agency. This volume contains the papers prepared and presented by individual workshop participants; the papers have not undergone peer review.

Board on Environmental Studies and Toxicology

Commission on Life Sciences

National Research Council

NATIONAL ACADEMY PRESS
Washington, D.C. 1992

TABLE OF CONTENTS

Introduction

Jonathan M. Samet and Devra Lee Davis

The diagnostic label of *multiple chemical sensitivity* is being applied increasingly, although definition of the phenomenon is elusive and its pathogenesis as a distinct entity is not confirmed. Multiple chemical sensitivity has become more widely known and increasingly controversial as more patients have received the label.

The emergence of multiple chemical sensitivity as a phenomenon that needs investigation coincides with recognition that myriad exposures to environmental agents are sustained in indoor and outdoor environments (National Research Council 1991; Samet, Marbury, and Spengler 1987, 1988). Regulatory activity has focused on outdoor air pollution, but children and adults in the United States and other developed countries spend most of their time indoors (National Research Council, 1991) and personal exposures to many air pollutants are thus determined largely by indoor pollutant concentrations. Monitoring studies conducted during the 1970s and 1980s showed that homes, public buildings, and nonindustrial workplaces can be contaminated by diverse gaseous and particulate pollutants that originate in unvented combustion, evaporation of volatile agents from various materials and solutions, grinding and abrasion, soil gas, and biologic sources (Samet, Marbury, and Spengler, 1987, 1988). Absorption of the pollutants can occur through the lungs, gastrointestinal tract and skin.

The concept that multiple chemical sensitivity is a distinct entity that is caused by responses to chemicals originated in the work of Randolph in the 1950s (American College of Physicians 1989, Ashford and Miller 1991). In the disease model proposed by Randolph, multiple chemical sensitivity consists of an inability to adapt to chemicals and the development of responsiveness to extremely low concentrations after sensitization (Randolph 1956); the model postulates multiple symptoms that reflect involvement of multiple organ systems. Randolph's pathogenic schema includes "adaptation." Symptoms can occur on exposure to chemicals or on withdrawal from exposure after an adaptive response has taken place. Randolph and others who apply this model of pathogenesis have used controlled exposures to establish the presence of multiple chemical sensitivity: patients are placed in environments judged to eliminate deleterious agents and then exposed to suspect chemicals.

Many of the physicians who apply that model are now referred to as clinical ecologists. Randolph and others founded the Society for Clinical Ecology in 1965 and it became the

American Academy of Environmental Medicine in 1984. This organization offers a definition of multiple chemical sensitivity which is also referred to as ecologic illness: "Ecologic illness is a poly-symptomatic, multi-system chronic disorder manifested by adverse reactions to environmental excitants, as they are modified by individual susceptibility in terms of specific adaptations. The excitants are present in air, water, drugs, and our habitats." Clinical ecology has been controversial, and committees of other specialty organizations have considered its diagnostic and therapeutic approaches to be inadequately supported by published studies (American Academy of Allergy and Immunology 1986; California Medical Association 1986; American College of Physicians 1989). Several recent reviews provide a more comprehensive background on multiple chemical sensitivity (Cullen 1987, Bascom 1989, Ashford and Miller 1991).

Although the literature addressing multiple chemical sensitivity is increasing, information on its frequency and natural history is lacking. Even defining multiple chemical sensitivity and developing criteria for diagnosis have proved difficult. Diverse pathogenic mechanisms have been postulated, but experimental models for testing them have not been established.

At the request of the Environmental Protection Agency, the National Research Council conducted a workshop to develop a research agenda to study the phenomenon of multiple chemical sensitivity. The workshop was convened with the overall objective of addressing the gaps in the scientific evidence on multiple chemical sensitivity. Workshop participants were multidisciplinary and included clinicians, immunologists, toxicologists, epidemiologists. psychiatrists, psychologists, and other involved in research or clinical activity relevant to the problem. The participants had an extensive range of experience and views on multiple chemical sensitivity. Three groups were formed to develop a research agenda. The subjects assigned were 1) case evaluation and criteria for diagnosis 2) mechanisms potentially underlying multiple chemical sensitivity and 3) epidemiologic approaches to investigating multiple chemical sensitivity.

This volume includes the papers prepared and presented by individual workshop participants; the papers have not undergone peer review. The papers offer a variety of views of the pathogenesis of multiple chemical sensitivity, its definition and diagnosis, and its management. They also illustrate the range of opinions as to the legitimacy of multiple chemical sensitivity as a unique clinical entity and the approaches used to treat it.

The paper by Lebowitz reviews key concepts of sensitization central to considering possible immunologic mechanisms of multiple chemical sensitivity. Lebowitz points to the distinction between sensitization (sensitized persons respond at lower doses than nonsensitized persons) and irritation. Burrell and Meggs also address the immune system and multiple chemical sensitivity; multiple chemical sensitivity is not seen as consistent with the specificity of immune responses, and authors point to immune mechanisms that might produce this phenomenon.

Bell emphasizes the need to use an integrated neuropsychiatric and biopsychosocial approach in considering multiple chemical sensitivity. Both she and Miller and Ashford suggest a possible role of the limbic system in triggering by odor. Experimental approaches to addressing mechanisms are described by Karol (an animal model) and by Meggs (a clinical research protocol).

Miller and Ashford address the difficult problem of defining multiple chemical sensitivity, offering this operational definition: "The patient with multiple chemical sensitivity can be discovered by removal from the suspected offending agents and by rechallenge, after an appropriate interval, under strictly controlled environmental conditions.

Causality is inferred by the clearing of symptoms with removal from the offending environment and recurrence of symptoms with specific challenge." In a separate paper, Miller and Ashford address the distinction between allergy and multiple chemical sensitivity.

Several papers describe patients in whom multiple chemical sensitivity has been diagnosed. Heuser and colleagues report on 135 patients, providing the results of serologic testing for various antibodies. Rea and colleagues review 16 years of experience with over 20,000 patients studied at the Environmental Health Center in Dallas. Fiedler and Kipen provide the results of a detailed neurobehavioral and psychologic assessment of a small number of patients who met rigid case criteria.

Welch describes aspects of occupational asthma, an entity caused by specific chemicals or exposures. DeHart summarizes the positions of four professional societies that question many of the concepts concerning multiple chemical sensitivity and clinical ecology.

Three working groups were formed and provided an approach to research that needs to be done in this area. The first working group provided lists of case criteria, candidate populations for investigation, and approaches for evaluating patients with multiple chemical sensitivity. The group advocated the use of an environmental control unit to test the adaptation-deadaptation hypothesis. The workshop participants agreed that the recommendations of this working group should have the highest priority in any research agenda.

The second group also proposed the use of controlled exposures of subjects in whom multiple chemical sensitivity is diagnosed and of control subjects. This group advocated the formulation of animal models, the investigation of tissues, and the development of a data base of chemicals, foods, and drugs reported to be associated with multiple chemical sensitivity.

The third group was concerned with epidemiologic approaches. A multicenter clinical assessment of patients was advocated for early attention. The results of the multicenter study would be used to develop methods for determining the prevalence of multiple chemical sensitivity. This group also suggested followup of populations with discrete and sudden chemical exposures.

This volume is published as an addendum to Biologic Markers in Immunotoxicology (NRC, 1992).

REFERENCES

American Academy of Allergy and Immunology, Executive Committee. 1986. Clinical ecology. J. Allergy Clin. Immunol. 78:269-270.

American College of Physicians. 1989. Clinical ecology. Ann. Intern. Med. 111:168-178.

Ashford, N.A., and C.S. Miller. 1991. Chemical Exposures. Low Levels and High Stakes. New York: Van Nostrand Reinhold. 214 pp.

Bascom, R. 1989. Chemical Hypersensitivity Syndrome Study. Baltimore: Department of Environment of the State of Maryland.

California Medical Association Scientific Board Task Force on Clinical Ecology. 1986. Clinical ecology—A critical appraisal. West. J. Med. 144:239-245.

Cullen, M.R. 1987. The worker with multiple chemical sensitivities: An overview. Occup. Med. State Art Rev. 2:655-661.

NRC (National Research Council). 1991. Human Exposure Assessment for Airborne Pollutants: Advances and Opportunities. Washington, D.C.: National Academy Press. 321 pp.

NRC (National Research Council). 1992. Biologic Markers in Immunotoxicology. Washington, D.C.: National Academy Press.

Randolph, T.G. 1956. The specific adaptation syndrome. J. Lab. Clin. Med. 48:934.

Samet, J.M., M.C. Marbury, and J.D. Spengler. 1987. Health effects and sources of indoor air pollution. Part I. Am. Rev. Respir. Dis. 136:1486-1508.

Samet, J.M., M.C. Marbury, and J.D. Spengler. 1988. Health effects and sources of indoor air pollution. Part II. Am. Rev. Respir. Dis. 137:221-242.

OVERVIEW

To develop a research agenda on multiple chemical sensitivity syndrome, participants were assigned to three multi-disciplinary groups. The groups were asked to focus on: 1) the design of a research protocol for clinical evaluation; 2) studies to evaluate relevant exposures and mechanisms; and 3) epidemiologic approaches. The specific charges were:

- ***Development of a Working Definition for the Syndrome***
- ***Evaluation of All Potential Causes***
- ***Evaluation of Uncertainties Associated with the Data and Data Gaps***
- ***Preparation of Report on Research Recommendations and Protocol***

The reports of the three working groups follow. Because consensus was achieved within each group on the substance and language of the recommendations, the reports are presented unedited, as accepted at the workshop.

Research Protocol for Clinical Evaluation

WORKING GROUP I

The group agreed that patients have been identified with a condition of multiple and often diverse symptoms that have been attributed to chemical agents in the environment. These patients may have recognized disease syndromes. However, symptomatology related to multiple chemicals is a distinct feature of these patients that is not classifiable by existing criteria used in conventional medical practice for psychiatric or physical illness. Thus, this feature cannot be uniquely coded in either DSM-III-R or ICD-9.

CASE CRITERIA

Criteria for the selection of cases for evaluation of multiple chemical sensitivity were discussed, and the committee agreed on criteria for research purposes as discussed below (definitions for other purposes were not addressed by this group):

1. Sensitivity to chemicals. By sensitivity we mean symptoms or signs related to chemical exposures at levels tolerated by the population at large that is distinct from such well recognized hypersenstivity phenomena as IgE-mediated immediate hypersensitivity reactions, contact dermatitis, and hypersensitivty pneumonitis.
2. Sensitivity may be expressed as symptoms and signs in one or more organ systems.
3. Symptoms and signs wax and wane with exposures.

It is not necessary to identify a chemical exposure associated with the onset of the condition. Preexistant or concurrent conditions, e.g. asthma, arthritis, somatization disorder or depression, should not exclude patients from consideration.

CANDIDATE POPULATIONS

The selection of subjects for research protocols will depend on the specific hypotheses to be tested. Identifiable populations include but are not restricted to:

1. Symptom or sign based: Patients with reactivity to environmental chemicals, either through self-reporting or meeting case selection criteria.
2. Disease based: Patients with specific diseases that are suspected to be caused by or exacerbated by chemical exposures.
3. Exposure based: Groups characterized by a common exposure, such as workers at a specific factory, occupants of a particular building, or residents of a contaminated community.
4. Population based: Groups such as school children or a random community sample.

Appropriate comparison groups should be chosen in each case.

EVALUATION

Research design and methods should be chosen as appropriate for each hypothesis. However, a minimum uniform data base across these groups needs to be developed which shall include the items in Table One. Specialized evaluations need to be developed for investigating specific hypotheses as listed in Table Two. It is beyond the scope of this committee to develop these in detail.

A research priority will be the study of the adaptation-deadaptation hypothesis, and this study must be conducted at an early stage because the outcome of this investigation will influence future study design. An environmental control unit, a clinical research unit in which subjects are housed for the control of exposures to foods and chemicals, should be developed to study the adaptation-deadaptation hypothesis. Other possible uses of units would include control of exposures and challenging subjects in a well-defined environment. Use of this modality will be very complex, and its indications and use must be researched and tested in detail. Other modalities of testing such as challenge chamber studies will have utility, and also must be studied in detail.

Prospective longitudinal studies of exposure based events are very important and should be performed.

TABLE 1

Uniform Core Database
history of sensitivity to foods, chemicals, and drugs
age
gender
marital status
race
education
dietary history
socioeconomic status
standardized medical evaluation
occupational history
environmental and exposure history
smoking history
medication history
alcohol and drug use history
psychological screening
atopy

TABLE 2

Specialized Evaluations
Nutritional assessment
Metabolic phenotype
Psychological profile
Neuropsychological profile
Assessment of cell-mediated immunity
HLA haplotype
Provocative challenge
Bronchial hyperreactivity
Autonomic tone
Other immunological assessment
Other neurological assessment

Exposures and Mechanisms

WORKING GROUP II

There are clinical reports of a syndrome which has been described as follows: Chemicals (odorous and nonodorous VOC's, solvents, pesticides, etc.) cause a 'sensitization' (induction phase) in a subset of individuals. Upon subsequent exposure to lower concentrations, these individuals may respond in a polysymptomatic fashion (triggering). This 'sensitivity" may spread to other chemicals, foods or drugs. The underlying mechanism(s) is unknown at present but may include immunological, neurological, endocrinological, psychological, or other factors.

Currently, there is insufficient objective evidence for the entity called multiple chemical sensitivity caused by chemical exposures. Research is needed to test characteristics of the entity and define mechanisms that may be involved. Therefore the following research approaches are suggested.

HUMAN STUDIES

Following a comprehensive history, including environmental exposures, physical examination and appropriate laboratory testing, subjects diagnosed with MCS and control individuals, should be challenged by exposure to a mixture of offending agents determined historically. A double blind controlled procedure should be employed. The possible role of "adaptation" and "deadaptation" should be considered in the protocol. Endpoints for response should include immunologic, neurologic, endocrinologic, psychologic, social, etc. markers or measures. Dose-response relationships should be examined.

A second approach could be evaluation of individuals, over time, in their usual environment.

OTHER EXPERIMENTAL APPROACHES

Animal models should be developed that mimic the human syndrome. Exposures and parameters measured in the animal model should mimic those measured in humans wherever possible.

Tissues obtained by biopsy and necropsy from patients, animals and their controls should be evaluated for pathologic changes. In addition, functional and morphological changes in cells and tissues should be evaluated using *in vitro* techniques.

ADDITIONAL NEEDS

Attempts should be made to develop a data base of chemicals, foods and drugs, and associated signs and symptoms, which have been reported to be associated with MCS. Evaluation of this data base should be explored to identify possible linkages between exposure and mechanisms.

Epidemiology

WORKING GROUP III

The initial purpose and goal of epidemiologic research would be to establish the magnitude of the problem caused by the MCS phenomenon in the population and to characterize the cases sufficiently for further work. Full use of epidemiologic methods to study this issue is handicapped by the lack of a precise, agreed upon case definition. The lack of understanding of mechanisms and of the exposure patterns involved also restrict epidemiologic approaches at this time.

An important distinguishing feature of the syndrome is that the concentrations of contaminants to which the individual responds is reported to be orders of magnitude below the concentration to which the majority of people respond. Because the lack of an agreed upon case definition is a limiting factor in making progress, studies aimed at improving this definition should receive emphasis.

An early priority should be a multi-center clinical case-comparison study in occupational/environmental medicine clinics and other appropriate facilities. Patients who report responding with signs and symptoms to concentrations of environmental chemicals orders of magnitude below where the normal public responds would be enrolled. The study would take place at multiple sites using an agreed upon set of criteria. Clinical histories, batteries of tests and structured interviews should be used to define case characteristics and natural histories. The efficacy, specificity and sensitivity of different tests and instruments should be evaluated and validated between centers. The tests should, among others, assess psychologic-psychiatric conditions, immune function, neurotoxic reactions, etc. The study subjects would be compared with clinic controls (e.g., low back injuries) for the purpose of understanding case definition, the descriptive epidemiology of cases and general cross-sectional study.

The information from the multi-center study can then be used to construct a population-based study to determine the prevalence of the MCS entity and associated conditions. The provisional nature of the case definition suggests that it would be important for a broad set of symptom prevalences to be determined to allow flexible construction of a variety of definitions. Such descriptive studies could be started with review of the Health Interview and NHANES surveys conducted by the National Center for Health Statistics. It is possible that one or more of the population-based cohorts such as the Framingham,

Alameda County, Tecumseh, or Washington County cohorts could have an addition to follow-up surveys to probe for prevalence of conditions such as MCS.

Population based methods should also be used to determine the basic descriptive epidemiology of certain multi-organ conditions linked by some to MCS, such as SLE, scleroderma, MS and somatization disorder.

Epidemiology has an important role to play in supporting methods evaluation and development for test instruments used in clinical studies as well. Determination of normal ranges for new test modalities, establishing the sensitivity and specificity of screens and biomarkers, and the construction of survey instruments are examples where epidemiologic input is important and we recommend the inclusion of such expertise in these areas. The committee felt that the follow-up of a defined population subjected to a discrete and sudden chemical exposure (such as a spill or pesticide misapplication) would also be useful to assess the initiation of hypersensitivity to environmental chemicals and its natural history. It recommends that a protocol for evaluation and follow-up be developed and pre-tested and that resources be set aside for designated centers (e.g., environmental/occupational clinics) to respond rapidly in the event of an incident.

Key Concepts: Chemical Sensitization

Michael D. Lebowitz

INTRODUCTION

Sensitization to chemicals can be defined as changes in the organism, usually the immunochemical system, by exposure to a chemical such that further chemical exposure leads to recognition by the organism. Such recognition will lead to a response that is marked by a greater reaction at lower doses than what would be observed in non-sensitized individuals. This is usually called hypersensitivity (Turner-Warwick, 1978). Inhalation of the antigen/allergen in an individual previously sensitized leads to an allergic reaction, such as rhinitis or conjunctivitis. If the skin is sensitized, as in allergic contact dermatitis, then contact will cause an oedematous response and/or a rash. Pulmonary (airway) sensitization manifests itself through bronchial constriction or obstruction (Davies and Blainey, 1983; Hetzel and Clark, 1983; Ramsdale et al., 1985). Some chemicals can produce different types of "allergy". There are various known and hypothesized mechanisms for sensitization. There are also host susceptibility factors, including genetic predisposition, which will play a role in sensitization and in disease manifestation (Turner-Warwick, 1978; Gregg, 1983). There must be some differentiation also between irritation and sensitization (Burge et al., 1979; Newman-Taylor and Davies, 1981). Irritants do not have the immunochemical recognition signal. Irritation in a non-sensitized individual usually leads to a slower, less serious response, at higher doses. This does not prevent irritants from producing or enhancing inflammation, which can be prolonged, and which can produce hypersensitivity reactions (e.g., bronchial hyperreactivity) when stimulated (discussed below). There are about 10-20% of almost any population which show greater "sensitivity" to irritants, which is usually organ (e.g., eye) specific, by responding at lower doses (Weber, 1984; EPA, 1986). (Acute infectious diseases may produce some of the same symptoms also, which would require other means of differentiation.) Chemicals have different potencies as sensitizers, not entirely dependent on the mechanism of sensitization. For instance, platinum salts have a very high potency (i.e., up to 95% properly exposed become sensitized) while formaldehyde has a low potency (i.e., 1-5% properly exposed become sensitized) (Brooks, 1982; Newman-Taylor and Davies, 1981.).

MECHANISMS OF SENSITIZATION

Classic Sensitization

The B cell IgE mediates various immediate hypersensitivity states, including typical allergy to biological aero-allergens. IgE mediation is involved heavily in the pathogenesis of asthma (Hetzel and Clark, 1983; Turner-Warwick, 1978). Several chemicals, such as TDI, TMA, platinum salts, form protein bound haptens that act as antigens (Weill and Turner-Warwick, 1981; Brooks, 1982; Chan-Yeung, 1990; Newman-Taylor and Tee, 1990). Basically, IgE forms antibodies to certain antigens/allergens that are inhaled, absorbed, and/or ingested. This is the recognition (and sensitization) phenomena. Specific IgE then persists in the sensitized; it is an outcome that can be assayed. There are IgE receptor sites on basophils and mast cells. When an IgE antibody-antigen binding occurs, the IgE can attach to these receptors. This usually leads to degranulation of the cells and subsequent release of histamine, chemotactic factors for platelets and other active cells (e.g., eosinophils, neutrophils), and other active mediators (e.g., PG's/LTB). This starts a chain reaction (or cascade) of immuno-chemical phenomena which leads to the hypersensitivity response, and can lead to inflammation. (Ibid.) Anti-IgG can lead to basophil activation as well, as can the cells and some of the chemotactic factors (e.g., PAF) directly (Marone, 1989). (The mechanisms are discussed in more detail in the Immunology and Animal Models section.) IgE is genetically linked, as are other immunoglobulins (Turner-Warwick, 1978; Gregg, 1983; Lebowitz et al., 1984; McGue et al., 1989). IgE also has an amplification-control cycle that probably involves some subsets of IgG and secretory IgA. IgG itself is important in sensitization and hypersensitivity (e.g., in TMA hypersensitivity). IgE mediated activity is associated also with T-lymphocyte activity in the lung, as found from broncho-alveolar lavage (BAL) studies (Gerblich et al., 1991). There can also be IgG blocking antibody that somehow prevents IgE mediated reactions. (Secretory IgA is the most abundant immunoglobulin in the lung, but its role in amplification-control of IgE is poorly understood. Turner-Warwick, 1978). Sensitization can occur in the lower respiratory tract, usually called hypersensitivity pneumonitis (or extrinsic allergic alveolitis), such as isocyanate disease. The nature of the protein-hapten deserves attention; T- and B-cells appear to have specific roles (Ibid.). Mediators (such as lymphokines) and receptors, and of other factors require further elucidation. Secretory IgA, along with IgG, and T-cells (especially suppressor/cytotoxic CD8 cells) are usually involved in hypersensitivity pneumonitis (Schlueter, 1982; Patterson et al., 1990; Semenzato, 1991). Some of the other known IgE-mediated chemical hypersensitivities are platinum and nickel salts, cobalt, other isocyanates (in addition to TDI), other anhydrides (in addition to TMA, such as phthalic acid anhydride), and organic acids (such as plicatic acid from red cedar) which are essentially all low molecular weight compounds which form protein haptens (Brooks, 1982; Marone, 1989; Allegra et al., 1989; Nemery, 1990; Chan-Yeung, 1990). Based on skin reactions, chromium salts, and colophony flux probably have IgE mediated immediate hyper-sensitivity. Based on this, and sensitization through ingestion, there are other chemicals, such as formaldehyde and ethylene oxide, which may turn out to have some IgE mediated hypersensitivity (Rockel et al., 1989). The various metals/metal salts appear to have IgG involvement also (as shown by precipitin formation), as do isocyanates and anhydrides (Adams et al., 1988; Thurmond and Dean, 1988; Patterson et al., 1990). Some of the haptens formed will produce late phase reactions as well, related to influxes of eosinophils, PMN's, inflammatory mediators (op cit.). Non-atopics will produce IgE antibodies if antigen is administered with certain adjuvants (Marsh et al., 1972), though these responses are short-lived. This implies an IgE suppressive mechanism, which appears to be T cell mediated, and this suppression is probably due to CD8+ T cells (Diaz-Sanchez and Kemeny, 1990). Certain antigens can sensitize non-atopics as well (e.g., platinum salts, castor bean dust), which implies that they have IgE-potentiating properties. Such potentiation appears related to increased CD4+ T cells (Ibid.). Certain common chemicals (e.g., ozone) or mixes (e.g., diesel exhaust, including with HCHO) appear to be appropriate adjuvants (Meggs). Sometimes, elevated IgE and eosinophilia are associated with smoking (Burrows et al., 1989), and with acute and persistent respiratory disease, as in rubber workers involved in a thermoinjection process, where the suggested sensitization has not been related to one of the component chemicals (Bascom et al., 1990). Lymphocyte factors related to immunoglobulin activity also activate human basophils (Marone, 1989). One has to consider the triggering mechanism in predisposed individuals, such as specific triggers of inflammation (e.g., HCHO). Asthmatics and those with BHR, or those with such predispositions may be such susceptibles. It is claimed that those with autoimmune disorders (e.g., Lupus/SLE, rheumatoid arthritis-RA) may be predisposed also (Thrasher et al., 1989; Ashford). SLE and RA are associated with circulating immune complexes, which may play a role (discussed elsewhere) (Turner-Warwick, 1978). One also has to consider factors that affect interleukins which have such an impact in the chain of events after presentation with the antigen, including activation of basophils (e.g., by IL-3). IL-1 is a basic mediator of intercellular activity within and between immune system functions, including having a role in tissue homeostasis; IL-1 inhibitory activity (IHA) is reduced in interstitial lung disease (Nagai et al., 1991), and IL-1 is related to cell-mediated immunity as well.

Inflammation

Whatever the mechanism, inflammation (in airways, skin, possibly other organs) appears to be a key trait in many sensitization conditions (Reed, 1988; Bonini et al., 1989; Hogg et al., 1991). (Although it is discussed elsewhere as well, it is worth mentioning some major components of inflammation.) Epithelial release of mediators, including histamine and chemotactic factors, appears to be a major component of inflammation and a mechanism of importance in sensitization and subequent triggering by chemicals. The increased activity of mononuclear cells (including PAM's), polymorphonuclear cells (including neutrophils) and eosinophils are part of the basic mechanisms involving inflammation. The release of oxidants, such as hydrogen peroxide (H2O2), as with exposure to formaldehyde, and superoxide anion from these cells is a key contributor to damage and responses, occurs more in sensitized organisms, and can occur with im- mune complexes (Adams et al., 1988; Rossi et al., 1989; Allegra et al., 1989; Cerasoli et al., 1991). The lack of sufficient, effective antioxidants could be a mechanism of sensitivity (as in those with genetic defects, such as those who have little or no ceruloplasmin). It is interesting to note that only scarce airway inflammatory cell activity has been reported in at least some asthmatics between attacks (Saetta et al., 1989). Airway inflammation is probably the main cause of bronchial hyperreactivity by itself or by contributing to bronchial obstruction (although obstruction can occur by other mechanisms). The increase of airway or skin permeability, permitting processes that lead to mast cell or basophil degranulation, leads to mediator release. Mucosal edema, an aspect of inflammation, can increase epithelial permeability, alter smooth muscle dynamics, and stimulate neural pathways (Bucca and Rolla, 1989). The occurrence of immune complexes in situ is important also (as discussed above); basophils can be activated by C3a and C5a (Marone, 1989). Macrophages are major cells in acute and chronic inflammation, immune responses including the complement system and immune deficiencies, are involved in hematopoesis and coagulation, and promote proliferation of many other cells. They present antigen to T cells. They release plasminogen activator and classes of toxins such as neutral proteases. Exposure to lymphokines is part of macrophage activation. Derived factors can activate basophils directly (Turner-Warwick, 1978; Adams et al., 1988; Thurmond and Dean, 1988).

Other Mechanisms of Sensitization

There are several hypothesized mechanisms, each with preliminary experimental verification of its potential role. Epithelial cells in the mucosa of bronchi and intestine can express Ia antigens. Ia-bearing epithelium can present antigens to specific T-cells, to initiate the classic immune response (Allegra et al., 1989). "It is possible that epithelial cells, like all antigen- presenting cells, act as accessory cells in immune responses by producing IL-1, which is responsible for T-cell migration and proliferation." (Ibid.) Also, HLA-DR antigens from airway epithelial cells are seen in those with interstitial lung disease (Ibid.). HLA-DR is part of the MHC (major histocompatibility) locus, and MHC-peptide linkages have been shown to be specific for some compounds (op cit.). Further, there may be epithelial cell-surface receptors with potential protein (i.e., hapten) specificity, probably involving IgG binding and myosin, which would induce intracellular changes (op cit; DeMarzo et al., 1989). Mast cells have receptors for a variety of antigens and histamine-releasing factors, in addition to anti-IgE and -IgG (Marone, 1989). It is unknown whether the "antigen" receptor would accept specific chemical-protein haptens. IgG, which has such high affinity for binding proteins, may play a role in this function. In rhinitis, and probably in asthma, there are specific chemical sensitive mast cells in the epithelium, such as formaldehyde-sensitive mast cells in the nasal epithelium, which are related to increased numbers of mast cell progenitors in airway mucosa (Otsuka et al., 1985, 1986, 1987; Barret and Metcalfe 1987). These epithelia can produce cytokines which promote mast cell growth in the presence of lymphocytes (Ibid.; Denburg et al., 1987). Local cells synthesized close to progenitors influences which type of mast cell develops, such that "derangements in the airway epithelium may result in a generation of factors which can induce a selective expansion of a mast cell sub-population." (Allegra et al., 1989). These mast cells, as found in bronchial secretions, BAL's and bronchial epithelium of asthmatics, are leaky and have the characteristics of the specific chemical-sensitive mast cells (as described above) (Allegra et al., 1989). The increased numbers of these specific mast cells can be activated in the airway, by provocation and possibly by other triggers, and cause bronchoconstriction only in those with these characteristics (Ibid.). The same exposure -epithelial mast cell response would be likely in the intestine as well, and possibly other locations where there are mast cells with their superficial location in epithelium. Epithelial cells are capable of activating mast cells also thru the release of specific mediators (e.g.,[15] HETE) (Marone, 1989).

The possibility of neuro-sensitizers has been mentioned as a possibility, involving the adrenergic, cholinergic and/or NANC (neither) system(s). Mast cell-nerve interactions are biologically significant as reviewed recently (Bienenstock et al., 1991). Endogenous biochemical compounds are important in the regulation of smooth muscle tone, which is more evident in sensitized individuals (such as asthmatics) (Barnes et al., 1984; Randem et al., 1987). The autonomic system appears to be involved as well (Nadel and Barnes, 1984; Postma et al.,1985). The adrenergic system may be more important in this regard (Barnes et al., 1984). Central parasympathetic efferent signals are important for the circadian rhythm of bronchial tone (Dreher and Koller, 1990). The endogenous factors may only differentiate sensitized individuals; this requires further exploration. Although there is no knowledge of the specificity of muscarinic receptors (in the cholinergic system) for specific haptens, changes in the patterns of muscarinic receptors in sensitized individuals requires further exploration (Paggiaro et al., 1989). Neuropepities are important in asthma (Barnes, 1991). Threshold doses of PGD2 and takykinins (neuropeptides in the NANC system) to produce acetycholine induced bronchospasm (a cholinergic phenomena) were modified in animals presensitized (with ovalbumen) (Omini et al., 1989), implying a different set and time-course of reactions in sensitized organisms. Psychological conditioning may cause most cell degranulation in the rat (Bienenstock et al., 1991). Other ANS involvement, including that related to the limbic system, and hypothalmic responses is discussed elsewhere (C.Miller, I.Bell).

Different Pathways

The hypersensitive also may use a different pathway that essentially does not exist in the general population (as stated also by Burrell, and partly discussed above), which may be genetically determined. This may relate to inflammation directly, or to BHR (which appears to be, at least as far as entrainment, an independent genetic trait). (Figure 1) Asthmatics have a similar distribution, but different immunoreactivity of myosin heavy chains in bronchial smooth muscle compared to normal subjects (De Marzo et al., 1989).

Irritant Mechanisms

Pulmonary alveolar macrophages are engaged in response to irritants. They can release chemotactic factors (including PAF, ECT, NCT) and other mediators after stimulation by irritants (op cit.). Lysosomal enzymes are released, specifically after phagocytosis and cell lysis. Other mononuclear cells may play a role also. Irritants also stimulate PMN's (neutrophils), which release proteases. Eosinophils release MBP endotoxin, and all release free radicals (see above). These cells are the hallmarks of inflammation and the biochemicals released increase it, even when the stimulant is an irritant. Irritants can also produce endothelial cell damage which results in arachadonic acid metabolism (and circulation of prostaglan- dins, leukotrienes). Increased permeability leads to exposure of mast cells or basophils, which can then be activated to release other mediators (Thurmond and Dean, 1988). However, the type and level of mediator activity is less under irritation stimuli. Stimulation of C-fibers lead to release of Substance P and other neuropeptides (NKA, NKB), including vasoactive intestinal peptide (VIP); these are part of the NANC (nonadrenergic noncholinergic) system. Such stimulation may lead to epithelial release of mediators, and mast cell degranulation (Marone, 1989). Stimulation of cholinergic mechanisms (the vagal afferent pathway) may be important, though there is no indication that muscarinic receptors would have a "memory" for specific irritants (see above). Thus, chemical irritant contact may produce symptoms of disease that mimic

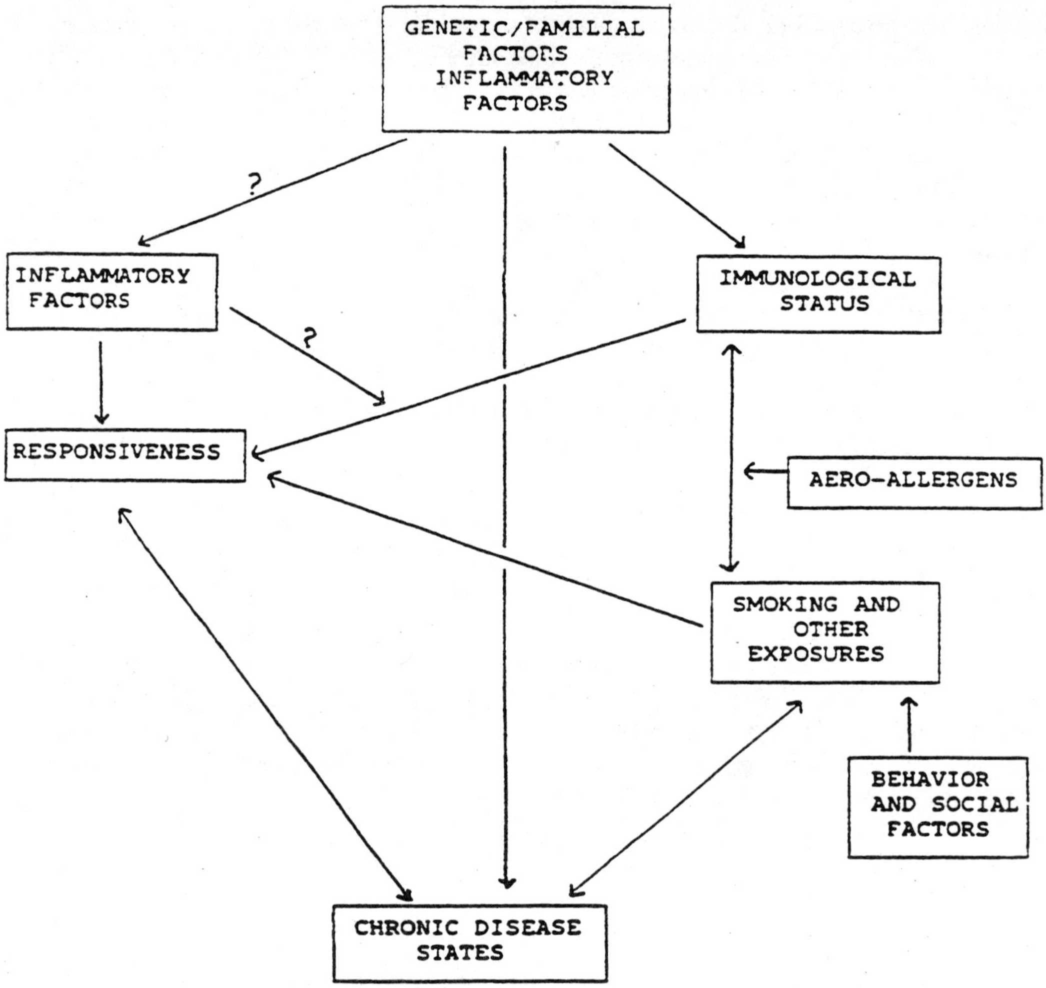

FIGURE 1 Hypothesized relationships.

immunological syndromes, but have a lack of antigen (Brooks, 1988; Thurmond and Dean, 1988). Protective responses, such as sneezing, coughing, and eye tearing, occur with stimulation of the epithelial irritant receptors. Irritants may even produce direct injury to epithelium. The "pseudoallery" or "reactive airway disease syndrome" usually requires exposure to relatively large amounts of the agent, usually occurs without a latency period after exposure (i.e., no "sensitization" period), and usually don't show reactivation of symptoms upon challenges with smaller amounts of the irritant or similar irritant (Ibid.). The irritant mechanisms last not much longer than due to the presence of the stimulus. Chronic inflammation can occur as well, leading to persistent symptoms, with high exposures to irritants; this may lead to reactive airway disease (Tarlo and Broder, 1989; Brooks, 1988). This form of chronic inflammation can result in increased susceptibility to other agents and to infection; implicated agents include chlorine, ozone, particles, SOx, NO2 (Gearhart and Schlesinger, 1986; Bylin et al., 1988; Wegman and Eisen, 1990). Examples of exposure agents that may produce such disease include acids, chlorine, phosgene, paints, calcium oxide, sulfur dioxide (Tarlo and Broder, 1989). Irritant stimuli may lead to adaptation or tolerance as well; ozone, NO2, H2S, phosgene and calcium oxide are agents that lead to reduced/attenuated responses (Wegman and Eisen, 1990). Irritant responses must be differentiated also from odor responses (Cain et al., 1987). The mechanisms of such tolerance or adaptation are not known. Certain "sensitizers" are pulmonary, dermal and sensory irritants as well, such as isocyanates, anhydrides, formaldehyde, ethylene diamine and paraphenylenediamine (Cain et al., 1987). Differentiation of responses is not always easy initially. After sensitization, stimulation can become non-specific as well, such that these and other irritants may produce similar responses. Even odors can produce non-specific responses in asthmatics (Shim and Williams, 1986).

FORMALDEHYDE

Formaldehyde is ubiquitous, especially in the indoor environment (Table 1). Formaldehyde is regarded as a upper respiratory irritant because it has a high solubility in water and is therefore captured by the wet mucus membrane. Retention of inhaled formaldehyde is shown to be almost 100% at the inhaled site in dogs (Egle, 1972). Malorney et al., (1965) showed that during an immediate following intravenous (IV) infusion of formaldehyde, the erythrocytes rapidly pick it up, leading to a very short-term increase in plama befor the erythrocytes oxidize it to formic acid. This is partly (1/6th) secruted in urine; the rest is oxidized to carbon dioxide and water. Formaldehyde (HCHO) irritant effects are presumed to be due to a nonspecific bronchial reflex response, which can increase flow resistance (Coffin and Stokinger, 1977). The effects of formaldehyde may be aggravated by interactions of other chemicals (Goldsmith and Friberg, 1977). The detection of the odor of formaldehyde in the air starts at about 0.05 ppm (0.06 mg/m3) (WHO/EURO, 1987) and is recognizable through most concentrations of exposure (Table 2). Exposure of a half hour to an hour is all that is needed to produce primary irritation; there is a dose response (Loomis, 1979: NRC 1981a). The lowest concentration causing pharyingitis is said to be 0.5 ppm (Stokinger and Coffin, 1968). Exposure to concentrations of formaldehyde gas of 0.1 ppm causes general irritation of the upper respiratory tract and other symptoms in children (Burdach and Wechselberg, 1980; NRC, 1981a). Concentrations from 1-5 ppm provoke coughing, constriction in the chest, and other symptoms (Loomis, 1979; NRC, 1981; Bardana, 1980; and Walker, 1964). Tolerance to eye and upper airway irritation may occur after 1 to 2 hours of exposure (NRC, 1981a; Blejer and Miller, 1966; Kerfoot and Mooney, 1975; Shipkovitz, 1968). However, even if tolerance develops, the irritation symptoms can return after 1 to 2 hour interruptions of exposure (Ibid; CPSC Ad Hoc Task Force, 1979). Ingestion usually leads injury to the larynx and trachea, gross effects on the GI tract, pneumonia, hardening of the lungs, hyperemia and edema of the lungs, and systemic damage. Increased mucolytic (decreased ciliary) activity has been shown related to formaldehyde exposure in animals in various studies (Carson et al., 1966; Cralley, 1942; Dalhamn, 1956; Hoffmann and Wynder, 1977; Kensler and Battista, 1963; Kotin, 1966). Decreased ciliary activity reduces the host defense mechanism to chemicals and micro-organisms, and may lead to increased cell wall permeability. Acute respiratory infections have been associated with formaldehyde exposure in children (Burdach and Wechselberg, 1980; NRC, 1981a; Tuthill, 1984). Trinkler (1968) found inflammatory changes in bronchial tracts of subjects who used formalin solution as a histological fixative.

Orringer and Mattern (1976) and Rockel et al., (1989) saw an increase frequence of acute and chronic bronchitis in dialyzed patients. There was a reactive, eosinophilic component.

TABLE 1

Contribution of Various Atmospheric Environments to Average Exposure[1]

Source	mg/d
1. Air	0.02
Ambient (10% 0f time)	
Indoor	
Home (65% of time)	
Conventional	0.5-2
Prefabricated (particle board)	1-10[3]
Workplace (25% of time)	
No occupational exposure[2]	0.2-0.8
With 1 mg/m^3 occupational exposure	4-5
ETS	0.1-1
2. Smoking (20 cigarettes/d)	1

[1] Contributions of food and water is low, so they are ignored here.
[2] Assuming the normal formaldehyde concentration is in conventional buildings
[3] Currently unusual.

Source: WHO/EURO 1987 Air Quality Guidelines for Europe (EURO Series #23) Copenhagen, 1987, pp 91-104.

Potential Mechanisms of Formaldehyde Induced Respiratory Sensitivity

Formaldehyde possesses the capacity to bind and alter protein constituents, such as: amino acids, proteins, nucleic acids, nucleosides, and nucleoproteins (Morin et al., 1964; Semin et al. 1974.; Goh el al., 1979; Bardana, 1980). Therefore, it is quite possible that this reactive chemical combines with respiratory tract proteins to form an immunoreactive, hapten-protein complex. Formaldehyde reacts directly with free amino groups, altering the

TABLE 2

Effects of Formaldehyde on Humans After Short-Term Exposure

Concentration of formaldehyde mg/m^3 Estimated median	Reported range	Effect
0.1	0.06 - 1.2	Odor threshold in 50 % of people (including repeated exposure)
0.5	0.1 - 1.9	Eye irritation threshold
0.6	0.1 - 3.1	Throat irritation threshold
3.1	2.5 - 3.7	Biting sensation in nose, eyes
5.6	5 - 6.2	Tolerable for 30 min (tearing)
17.8	12 - 25	Strong flow of tears, lasting for 1 hr
37.5	37 - 60	Danger to life, edema, inflammation, pneumonia
125	60 - 125	Death

Source: WHO/EURO 1987 Air Quality Guidelines for Europe (EURO Series #23) Copenhagen, 1987, pp 91-104.

character of the proteins, which could make the chemically altered constituents antigenic in themselves or can result in the formation of macro-molecular formaldehyde protein complexes and thus produce hyper- sensitivity in some persons (Bardana,1989; Loomis, 1979; Stokinger and Coffin, 1968). Supposedly, an IgE ISO RAST (Pharmacia, Sweden) was developed for formaldehyde and was positive for exposed dialysis patients (Rockel et al., 1989); others have evidently used formaldehyde conjugated to human serum albumen for immunoglobulin determinations (see below). Anti-N-like antibodies were found in dialysis patients using HCHO-treated cells (Sandler et al., 1979), which is evidence of type II auto-allergy (Consensus Workshop, 1984). Complement activation (via the alternate pathway) is another possible mechanism (Hakin et al., 1984; Rockel et al., 1989). C3a and C5a can activate basophils (Marone, 1989). More than one mechanism may be at work (Ibid.). Thrasher et al. (1990) report increased IL2 receptor cells and total B Cells (using monoclonal antibodies to LEU10) in those with long-term HCHO exposures (by report) and various presenting multiple organ system symptoms, compared to controls (with short-term exposure as chiropractic students. They also found increased titres (1:8 or greater) to B-cell

(IgG, IgM and IgE) isotypes to formaldehyde conjugated to human serum albumen; they indicate that these isotypes have proven positive in other investigations; increases in both IgG and IgE are hard to explain, and IgM increases haven't been found related to such immune disorders before (see above section, Turner-Warwick, 1978). Helper/Suppressor T-cell ratios don't appear different. They also reported increased titres (1:20) to auto antibodies (ASS, APC, ABB, AMIT and ANA). The meaning of significant increases in ASS only in mobil home residents, of APC and AMIT in variable percents in different groups, of ABB in office workers, and of ANA in both mobile home and office occupants are difficult to interpret. Increases in ANA hadn't been found in occupational asthma (Newman-Taylor and Davies, 1981; Newman-Taylor and Tee, 1990; op cit.). Other possible mechanisms have been discussed previously. It is quite likely that eosinophilia may be part of the immunological picture in formaldehyde-induced sensitivity (Popa et al., 1969; Hoy and Cestero, 1979; Nagornyi et al., 1979; Orringer and Mattern, 1976; Rockel et al., 1989). It has been shown that aldehyde pyrolysates from thermal degradation of polyethylene wrap is characterized by eosinophilia as well as by nocturnal attacks of dyspnea, wheezing and cough (Skerfving et al., 1980). There are an increasing number of studies which suggests formaldehyde can induce an inflammatory bronchitis reactivity airway disease. Particulates will carry formaldehyde gas (LaBelle et al., 1955; Patterson et al., 1990; Takhirov, 1974). Thus formaldehyde might have its affects in tracheobronchial or alveolar regions. Thus, particulates or aerosols have synergistic effects (Krzyzanowski et al., 1990), or the respiratory response is potentiated with simultaneous exposure to particulates as well as to other irritative chemicals (NRC, 1981b), or the aerosolized HCHO produces both sensitization and hyperresponsiveness (Hendrick and Lane, 1975, 1977; Frigas et al., 1982; Burge et al., 1985). Ostapovich (1975) studied sensitization of formaldehyde (2, 7, 15 mg/m3) in rats and guinea pigs injected with Freund's adjuvant prior to priming in order to stimulate antibody formation. Intermittent exposure, to concentration of 7 mg/m3 were given. Sensitization was induced by unequivocal increases in leukocyte agglutination, agar gel precipitation and passive hemaglutination, decreases in all but cholinesterase activity, from concentrations of 2 and 7 mg/m3. The time required for leukocytosis was shorter in rats than in guinea pigs. Zaeva (1968) summarized similar findings in animals. Nagornyi et al. (1979) found similar results, and eosinophilia, in guinea pigs with chronic exposures to 0.5 mg/m3. Sensitization was enhanced by intermittent inhalation. One mechanism of sensitization measured by leukocyte activity was formulated by Jerne, whose modified technique was utilized in this above study. This mechanism states that somatic generation of large number of genes expressed in leukocytes initially are directed against the animal's own histocompatibility antigens, which in the process of reacting with the cells that can form antibodies against them, cause these cells to mutate early, giving rise to many clones of antibody-forming cells with new and different specificities (Rose et al., 1973). Other in vitro and animal studies (Adams et al., 1988; Damiani, 1989) have shown that formaldehyde (like SO2 and nickel salts) increase macrophage phagocytosis and increase hydrogen peroxide levels. The latter promotes lipid metabolism, increases inflammation, and decreases host resistance; these studies have shown little direct effect of HCHO on B cells/immunoglobulins. Activated macrophages are associated with increased IL-1 and subsequent enhanced inflammatory processes (also by secretion of lytic enzymatic chemo attractants). (Increased IL-1 also stimulates T cell migration.) Phagolysomes degrade foreign molecules and resulting peptides associated with class I or II major histocompability complex (MHC) glycoproteins (which secondarily are presented as antigens to T cells (Damiani, 1989). Allegra et al. (1989) indicate that HCHO-sensitized most cells in the epithelium are associated with increased progenitors in airway mucosa and thus symptom severity. Mast cell promotion and differentiation in vitro occur by cytokines produced by such individual's epitherlium (with or without the presence of lymphokines) (op cit.)

Evidence of Asthma in Humans

The first case of formalin asthma was in an occupational setting and was published by Vaughan in 1939. Large numbers of European case studies reported on formaldehyde-induced asthma, claiming that an allergic mechanism was responsible, but immunological data is not presented usually (Vaughan, 1939; Coulant, 1961; Gervais, 1966; Tara, 1966; Nova and Touraine, 1956; Paliard et al., 1949; Sakula, 1975). Baur and Fruhmann (1979) showed an IgG increase in one case. In formaldehyde resin makers and other formaldehyde workers, formaldehyde, including formalin (40% formaldhyde by volume, an aqueous solution of formaldehyde) will produce primary irritational or allergic hypersensitivity. Further, formaldehyde reaction products, for example, dimethylurea, can be sensitizers (Milby et al., 1964). These and other studies (Hendrick and Lane, 1975, 1977; Popa et al., 1969; Kuzmenski et al., 1975; Kratchovil, 1971; Spassowski, 1976; led an NAS Committee (1981b) to conclude that formaldehyde has been shown to cause bronchial asthma in humans. The Committee stated that asthmatic attacks related to exposures to formaldehyde at low concentrations are due specifically to formaldehyde sensitization in some cases; controlled inhalation studies with formaldehyde are positive in these instances (Hendrick and Lane, 1975, 1977; Frigas et al., 1982; Burge et al., 1985). Nordman et al. (1985) found that only 12 of 230 persons with explicit suspicion of formaldehyde asthma actually reacted to formaldhyde gas broncho-provocation tests. They had significant dose-response reactions between 0.4 - 2.5 mg/m3 HCHO. Of 218 who didn't, 71 reacted to histamine broncho-provocation. Cockroft et al. (1982) reported two cases of occupational asthma caused by urea formaldehyde particle board, with positive histamine bronchial challenge tests, but no increased RAST specific IgE antibodies directed against formaldehyde human serum albumin conjugate. A third asthmatic didn't have a response with exposure, indicating (to them) a specific sensitization to the agent. Formaldehyde also seems to act as a direct airway irritant in persons who have bronchial asthmatic attacks from other causes (op cit.) Hakim et al. (1984) also reported acute asthma-like symptoms of chest tightness and dyspnea in dialysis patients when formaldehyde-treated membranes were used. The symptoms were correlated with complement activation (C3a). Asthma-like symptoms and bronchospasm also were reported with similar exposure by Rockel et al.(1989), associated with increased eosinophilia, IgE, elevations, and complement activation (C3a, C5a). HCHO conjugated specific IgE (RAST) elevations were found in some; other agents were found to have similar effects (ethylene oxide, phthalates, isocyanate). Airway Obstructive Disease due to formaldehyde has been shown in various studies also (Yefremov, 1970; Kerfoot and Mooney, 1975; Schoenberg and Mitchell, 1975; Kilburn et al., 1985; Krzyzanowski et al., 1990). Most recently, a community population study (Krzyzanowski et al., 1990) showed higher rates of asthma and chronic bronchitis in children related to monitored formaldehyde exposure, with a synergistic contribution of ETS. Formaldehyde-induced pesistent bronchial responsiveness by itself occurred primarily in the lower SES group (Quackenboss et al., 1989a). The monitoring used a modified LBL method with no temperature or humidity interferences (Quackenboss et al., 1989b). They also found acute changes in children, especially asthmatics, who were specifically exposed overnight in their bedrooms; morning PEF had decreased significantly with a demonstrable exposure-response relationship.

Skin Sensitization

Skin contact with formaldehyde has been reported to cause a variety of acutaneoud problems in humans, including irritation, allergic contact dermatitis (Type I allergy), and urticaria (Type IV allergy) (Odom and Maibach, 1977; Roth, 1969; Sneddon, 1968; NRC, 1981a; Consenses Workshop, 1984). Dermal exposure to formaldehyde may cause an acute and inflammatory reaction of the affected region concentrations of 1 to 3 ppm over minutes to hours produces irritant effects which may be tolerated by normal adults, but produces noticeable skin irritation from direct exposure (Pirila and Kilipo, 1949; Roth, 1969).

Current Levels of Knowledge

It is commonly accepted there is reasonable good knowledge of expsoure assessment for formaldehyde, including: sources, monitoring instrumentation, and distribution of exposures (WHO/EURO, 1986, 1987, 1991) as represnted by Tables 1, 3 and 4. The suggestions of the Consensus Workshop (1984) for future research are pertinent still. Knowledge levels are reasonable for certain effects: odor, sensory and mucosal irritation. More knowledge is needed about airway effects, including sensitization (Ibid.) and neurological effects (Kilburn et al., 1985). It is estimated that between 1 and 6% of individuals in the US are probably sensitive to low concentrations of formaldehyde (Moschandreas et al., 1980). Means of control have included technical and regulatory procedures (WHO/EURO, 1991); educational control methods have been recommended also.

TABLE 3

Formaldehyde Sources With Usual Range of Related Concentrations Indoors

Primary: Particle/ Press Board,	0.1 -2.0 mg/m3
Plywood (resin), Fabric	
Secondary: ETS	5- 50 Mg/ m^3
Wood Burning	
Infiltration: Vehicles in garages	?
From outdoors	?*

* Expected to increase with oxygenated fuels, especially methanol.

TABLE 4

Concentrations, Distribution of Exposures and Exposure-Response Relationships

Concentration	People with exposure		Levels of concern	
(mg/m^3)				
Range (mg/m^3)	Low	High	Limited	Definite
0-20-0.1-2.0	most	few	<0.03	0.10

Source: WHO/EURO 1991 (in press)
The European Short-Term Air Quality Guideline (WHO/EURO, 1987)

REFERENCES

Adams, D. O., J. G. Lewis, and J. H. Dean. 1988. Activation of mononuclear phagocytes by xenobiotics of environmental concern: analysis and host effects. Pp. 351-369 in Toxicology of the Lung, D. E. Gardner, J. D. Crapo, and E. J. Massaro, eds. New York: Raven Press.

Allegra, L., L. M. Fabbri, G. Picotti, and S. Mattoli. 1989. Bronchial epithelium and asthma. Eur Respir J 2(6 Suppl.):460S-468S.

Bardana, Jr., E. J. 1980. Formaldehyde: hypersensitivity and irritant eactions at work and in the home. Immunol and Allergy Practice 2(3):11-23.

Barnes, P. J. 1991. Neuropeptides and asthma. Am Rev Respir Dis 143:S28-S32.

Barnes, P. J., G. A. Fitzgerald, M. Brown, and C. T. Dollery. 1984. Nocturnal asthma and changes in circulating epinephrine, histamine and cortisol. N Engl J Med 303:263-267.

Barrett, K. E., and D. D. Metcalfe. 1987. Heterogeneity of mast cells in the tissues of the respiratory tract and other organ systems. Am Rev Respir Dis 135:1190-1195.

Bascom, R., M. E. Baser, R. J. Thomas, J. F. Fisher, W. N. Yang, and J. H. Baker. 1990. Elevated serum IgE, eosinophilia, and lung function in rubber workers. 45(1):15.

Baur, X., and G. Fruhmann. 1979. Bronchial asthma of allergic or irritative origin as an occupational disease. Prax Klin Pneumol 1(Suppl. 33):317-22.

Bienenstock, J., G. Macqueen, P. Sestini, J. S. Marshall, R. H. Stead, and M. H. Perdue. 1991. Inflammatory cell mechanisms. Mast cell/nerve interactions in vitro and in vivo. Am Rev Respir Dis 143:S55-S58.

Blejer, H. P., and B. H. Miller. 1966. Occupational health report of formaldehyde concentrations and effects on workers at the Bayly Manufacturing Co., Visalia. Pp. 6 in Study Report No. S-1806. Los Angeles: State of California Health and Welfare Agency, Dept. of Public Health, Bureau of Occupational Health.

Bonini, Se., E. Adriani, St. Bonini, G. De Petrillo, G. G. Loggi, A. Adabbo, M. Pani, and F. Balsano. 1989. Allergy and asthma: distinguishing the causes from the triggers. An eye model for the study of inflammatory events following allergen challenge. Eur Respir J 2(6 Suppl.):497S-501S.

Brooks, S.M. 1982. The evaluation of occupational airways disease in the laboratory and workplace. J Allergy Clin Immunol 70(1):56-66.

Bucca, C., and G. Rolla. 1989. Mucosal oedema and airway hyperreactivity. Eur Respir J 2(6 Suppl.):520S-522S.

Burdach, S. and K. Weschselberg. 1980. Damage to health at school. Complaints due to the use of materials which emit formaldehyde in school buildings [German]. Forstschritte der Medizin 98(11):379-384.

Burge, P. S., M. G. Harries, W. K. Lam, I. M. O'Brien, and P. A. Patchet. 1985. Occupational asthma due to formaldehyde. Thorax 40:255-260.

Burge, P. S., I. M. O'Brien, and M. G. Haries. 1979. Peak flow records in the diagnosis of occupational asthma due to colophony. Thorax 34:308-312.

Burrows, B., F. D. Martinez, M. Halonen, R. A. Barbee, and M. G. Cline. 1989. Association of
asthma with serum IgE levels and skin-test reactivity to allergens. New Eng J Med 320:271-277.

Bylin, G., G. Hedenstierna, T. Lindvall, and B. Sundin. 1988. Ambient nitrogen dioxide concentrations increase bronchial responsiveness in subjects with mild asthma. Eur Respir J 1:606-612.

Cain, W. S., T. Tosun, L. C. See, and B. Leaderer. 1987. Environmental tobacco smoke: sensory reactions of occupants. Atmospheric Environment 21(2):347-353. Carson, S., R. Goldhamer, and R. Carpenter. 1966. Mucus transport in the respiratory tract. Am Rev Respir Dis 93 (Supl):86-92.

Cerasoli, Jr., F., J. Tocker, and W. M. Selig. 1991. Airway eosinophils from actively sensitized guinea pigs exhibit enhanced superoxide anion release in response to antigen challenge. Am J Respir Cell Mol Biol 4:355-363.

Chan-Yeung, M. 1990. Occupational asthma. Chest 98(5 Suppl):148S-161S.

Coffin, D. L., and H. E. Stokinger. 1977. Biological effects of air pollutants. Pp. 231-360 in Air Pollution. Vol. II The Effects of Air Pollution, A. C. Stern, ed. New York: Academic Press, Inc. Consensus Workshop. 1984. Report on the Consensus Workshop on Formaldehyde. Environ Health Perspect 58:323-381.

Coulant, M. P., and G. Lopes. 1961. L'allergie au formal et a ses derives. Archives des maladies professionelles, de medicine du travail et de securite sociale (Paris) 22:769.

Cralley, L. B. 1942. The effect of irritant gases upon the rate of ciliary activity. J Ind Hyg Tox 24:193-198.

Cockcroft, D. W., V. H. Hoeppner, and J.Dolovich. 1982. Occupational asthma caused by cedar urea formaldehyde particle board. Chest 1:49-53.

CPSC Ad Hoc Task Force. 1979. Formaldehyde. Washington, D.C.

Dalhamn, T. 1956. Mucous flow and ciliary activity in the trachea of healthy rats and rats exposed to respiraory irritant gases. Acta Physiologica Scandinavica 36(Suppl. 123):5-161.

Damiani, G. 1989. The role of macrophages and lymphocytes in bronchial asthma:

perspectives for new research models. Eur Respir J 6 (Suppl.):469S-472S.

Davies, R. J., and A. D. Blainey. 1983. Occupational Asthma. Pp. 205-213; 228-230 in Asthma, T. J. H. Clark and S. Godfred, eds. London: Chapman and Hall.

De Marzo, N., P. Di Blasi, P. Boschetto, C. E. Mapp, S. Sartore, G. Picotti, and L. M. Fabbri. 1989. Airway smooth muscle biochemistry and asthma. Eur Respir J 2(5 Suppl.):473S-476S.

Denburg, J. A., M. Ohnisi, J. Ruhno, J. Bienenstock, and J.Dolovich. 1987. The effects of basophil/mast and eosinophil cell colony stimulating activities derived from human nasal polyp epithelial scrapings are T cell-dependent. J Allergy Clin Immunol 79(Suppl.), 167A.

Diaz-Sanchez, D., and D. M. Kemeny. 1990. The sensitivity of rat CD8+ and CD4+ T cells to ricin in vivo and in vitro and their relationship to IgE regulation. Immunology 69:71-77.

Dreher, D., and E. A. Koller. 1990. Circadian rhythms of specific airway conductance and bronchial reactivity to histamine: the effects of parasympathetic blockade. Eur Respir J 3:414-420.

Egle, J. L. 1972. Retention of inhaled formaldehyde, propionaldehyde and acreolein in the lung. Arch Environ Health 25:119-124.

EPA. August 1986. Evaluation of Integrated Health Effects Data for Ozone and Photochemical Oxidants. Chapter 13 in Air Quality Criteria Document for Ozone and Other Photchemical Oxidants, EPA-600/8--84-020eF, Washington, D.C.

Frigas, E., W. V. Filley, and C. E. Reed. 1982. UFFI dust. Non-specific irritant only? Chest 82:511-512.

Gearhart, J. M. and R. B. Schlesinger. 1986. Sulfuric acid-induced airway hyper-responsiveness. Fundamental and Applied Toxicology 7:681-689.

Gerblich, A. A., H. Salik, and M. R. Schuyler. 1991. Dynamic T-cell changes in peripheral blood and bronchoalveolar lavage after antigen bronchoprovocation in asthmatics. Am Rev Respir Dis 143:533-537.

Gervais, P. 1966. L'asthme professionel dans l'industrie des matieres plastiques. Poumon et le Coeur 22:2199.

Goh, K., and R. V. M. Cestero. 1979. Chromosomal abnormalities in maintenance hemodialysis patients. J Med 10:16/7.

Goldsmith, J. R., and L. T. Friberg. 1977. Effects of air pollution on human health. Pp. 457-610 in Air Pollution. Vol. II The Effects of Air Pollution, A. C. Stern, ed. New York: Academic 59 Press, Inc.

Gregg, I. 1983. Epidemiological Aspects. Pp. 242-294 in Asthma, T. J. Clarke and S. Godfrey, eds., 2nd ed., London: Chapman and Hall.

Hakin, R. M., J. Breillatt, J. M. Lazarus, and F. K. Port. 1984. Complement activation and hypersensitivity reactions to dialysis membranes. New Eng J Med 311(14):878-882.

Hetzel, M. R., and T. J. H. Clark. 1983. Adult Asthma. Pp. 473-481 in Asthma, T. J. H. Clark and S. Godfred, eds. London: Chapman and Hall.
Hendrick, D. J., and D. J. Lane. 1975. Formalin asthma in hospital staff. Brit Med J 1:607-608. 60.

Hendrick, D. J., and D. J. Lane. 1977. Occupational formalin asthma. Brit J Ind Med 34:11-18.

Hoffmann, D. J., and E. L. Wynder. 1977. Organic particulate pollutants - chemical analysis and bioassays for carcinogenicity. Pp. 361-455 in Air Pollution. Vol. II The Effects of Air Pollution, A. C. Stern, ed. New York: Academic Press, Inc.

Hogg, J. C., A. L. James, and P. D. Pare. 1991. Evidence for inflammation in asthma. Am Rev Respir Dis 143:S39-S42.

Hoy, W. E., and R. V. M. Cestero. 1979. Eosinophilia in maintenance hemodialysis patients. J Dialysis 3(1):73-79.

Kensler, C. J., and S. P. Battista. 1963. Components of cigarette smoke with ciliary-depressant activity. New Eng J Med 269:1161-61 1166.
Kerfoot, E. J., and T. F. Mooney. 1975. Formaldehyde and paraformaldehyde study in funeral homes. Am Ind Hyg Assoc J 36:533-537.

Kilburn, K. H., R. Warshaw, C. T. Boylen, S. J. S. Johnson, B. Seidman, R. Sinclair, and T. Takaro. 1985. Pulmonary and neurobehavioural effects of formaldehyde exposure. Arch Environ Health 40:254-260.

Kotin, P. 1966. (need title) Can Cancer Conf 6:475. Kratochvil, I. 1971. Effects of formaldehyde on the health of workers in the crease-resistant clothing industry. Pracovni Lekarstvi 23(1):374-375.

Krzyzanowski, M., J. J. Quackenboss, and M. D. Lebowitz. 1990. Chronic respiratory effects of indoor formaldehyde exposure. Environ Res 52:117-125.

Kuz'menko, N. M., et al. 1975. Study of the sensitizing action of formaldehyde under the conditions of plastic manufacture. Vrachebnoe Del 6:131-134.

LaBelle, C. W., J. E. Long, and E. E. Christofano. 1955. Synergistic effects of aerosols. AMA Arch Ind Health 11:297-304.

Lebowitz, M. D., R. Barbee, and B. Burrows. 1984. Family concordance of IgE, atopy and disease. J Allergy Clin Immunol 73:259-266.

Loomis, T. A. 1979. Formaldehyde toxicity. Arch Pathol Lab Med 103:321-324.

Malorney, R., N. Reitbrock, and M. Schneider. 1965. Die oxydation des formaldehyds zu ameisensaure im blut, ein beitrag zum stoffwechsel des formaldehyds. Naunyn-Schmiedebergs Arch Pharmacol 250:419-436.

Marone, G. 1989. The role of mast cell and basophil activation in human allergic reactions. Eur Respir J 2(6 Suppl.):446S-455S.

Marsh, D. G., L. M. Lichtenstein, and P. S. Norman. 1972. Induction of IgE-mediated immediate hypersensitivity to group 1 rye grass pollen allergen and allergoids in non-allergic man. Immunology 22:1013.

McGue, M., J. W. Gerrand, M. D. Lebowitz, and D. C. Rao. 1989. Commingling in the distributions of immunologulin levels. Human Heredity 39(4):196-201.

Milby, T. H., M. M. Key, R. L. Gibson, et al. 1964. Chemical Hazards. Pp. 63-242 in Occupational Diseases, WM. M. Gafafer, ed. U.S. Department of Health, Education, and Welfare, Publication No. 1097, Washington,D.C., U.S. Government Printing Office.

Morin, N. R., P. E. Zeldin, Z. O. Kubinski, P. K. Bhattacharya, and H. Kubinski. 1977. Macromolecular complexes produced by chemical carcinogens and ultra-violet radiation. Cancer Res 37:3802-3814.

Moschandreas, D. J., R. H. Moyer, J. R. Ward, H. E. Rector, J. T. Schakenback, et al. 1980. An evaluation of formaldehyde problems in residental mobile homes. Pp. 1-146. Final Task Report I, GEOMET Report No.. ESF-797, April, for the Department of Housing and Urban Development. Washington, D.C.: EPA. 65

Nadel, J. A., and P. J. Barnes. 1984. Autonomic regulation of the airways. Am Rev Med 35:451-467. Nagai, S., H. Aung, M. Takeuchi, K. Kusume, and T. Izumi. 1991. IL-1 and IL-1 inhibitory activity in the culture supernatants of alveolar macrophages from patients with interstitial lung diseases. Chest 99:674-680.

Nagornyi, P. A., Z. A. Sudakova, and S. M. Schablenko. 1979. General toxic and allergenic action of formaldehyde. Gigiena Truda i professional'nye Zabolevaniya 1:27-30. National Research Council (NRC). 1981a. Formaldehyde and other aldehydes. National Academy Press, Washington, D.C.

National Research Council (NRC). 1981b. Indoor Pollutions. National Academy Press, Washington, D.C. Nemery, B. 1990. Metal toxicity and the respiratory tract. Eur Respir J 3:202-219. Newman-Taylor, A. J., and R. J. Davies. 1981. Inhalation Challenge Testing. Pp. 143-168, in Occupational Lung Diseses, H. Weill and M. Turner-Warwick, eds. New York:
Marcel Dekker. Newman-Taylor, A., and R. D. Tee. 1990. Environmental and occupational asthma. Exposure assessment. Chest 98(5 Suppl.):209S-215S.

Nordman, H., H. Keskinen, and M. Tuppurainen. 1985. Formaldehyde asthma - rare or overlooked? J Allergy Clin Immunol 75:81-99. Nova, M. M. H., and R. G. Touraine. 1956. Asthme au formol, Pp. 293-294 in Societe De Medecine Du Travail, Lyon.

Odom, R. G., and H. I. Maiback. 1977. 1977. Contact urticaria: a different contact dermatitis. Pp. 441-452 in Advances in Modern Toxicology, Vol. 4, Dermatotoxicology and Pharmacology, F. N. Marzulli and H. I. Maibach, eds. Washington, DC: Hemisphere Publishing.

Omini, C., G. Brunelli, A. Hernandez, and L. Daffonchio. 1989. Role of the mediators in pulmonary hyperreactivity: the cocktail interaction hypothesis. Eur Respir J 2(6 Suppl.):493S-496S. Orringer, E. P., and W. D. Mattern. 1976. Formaldehyde-induced hemolysis during chronic hemodialysis. New Eng J Med 294(26):1416-1420.

Ostapovich, I. K. 1975. Conditions of respiratory route exposure to sulfur dioxide and formaldehyde and subsequent sensitization. Gigyena I Sanitariya 2:9-13.

Otsuka, H., J. A. Denburg, and J. Dolovich. 1985. Heterogeneity of metachromatic cells in human nose: significance of mucosal mast cells. J Allergy Clin Immunol 76:695-702.

Otsuka, K., J. Dolovich, A. D. Befus, et al. 1986. Basophilic cell progenitors, nasal metachromatic cells and peripheral blood basophils in ragweed allergic rhinitis. J Allergy Clin Immunol 78:365-371.

Otsuka, H., J. Dolovich, M. Richardson, J. Bienenstock, and J. A. Denburg. 1987. Metachromatic cell progenitors and specific growth and differentiation factors in human nasal mucosa and polyps. Am Rev Respir Dis 136:710-717.

Paggiaro, P. L., E. Bacci, N. Pulera, P. Bernard, F. L. Dente, D. Talini, and C. Giuntini. 1989. Vagal reflexes and asthma. Eur Respir J 2(6 Suppl.):502S-507S.

Patterson, R., L. C. Grammer, C. R. Zeiss, K. E. Harris, and M. A. Shaughnessy. 1990. Use of immunologic technology in the diagnosis of environmental and occupational immunologic lung disease. Chest 98(5 Suppl.):206S-208S.

Paliard, F., L. Roche, C. Exbrayat, and E. Sprunk. 1949. Chronic asthma due to formaldehyde. Arch Mal Prof 10:528.

Pirila, V., and O.Kilpio. 1949. On dermatitis caused by formaldehyde 70 and its compounds. Am Med Intern Fenn 38:38-51.

Popa, V, D. Teculescu, D. Stanescu, and N. Gavrilescu. 1969. Bronchial asthma and asthmatic bronchitis determined by simple chemicals. Dis Chest 56(5):395-404.

Postma, D. S., J. J. Keyzer, G. H. Koeter, H. J. Sluiter, and K. DeVries. 1985. Influence of the parasympathetic and sympathetic nervous system on nocturnal bronchial obstruction. Clin Sci 69:251-258.

Quackenboss, J. J., M. D. Lebowitz, C. Hayes, and C. L. Young. 1989. Respiratory responses to indoor/outdoor air pollutants: combustion products, formaldehyde, and particulate matter. Pp. 280-293 in Combustion Processes and the Quality of the Indoor Air Environment, J. Harper, ed. Pittsburgh: Air Pollution Control Association.

Quackenboss, J. J., M. D. Lebowitz, and C. Hayes. 1989a. Epidemiological study of respiratory response to indoor/outdoor air quality. Environ Int 15:493-502.

Ramsdale, E. H., M. M. Morris, R. S. Roberts, and F. E. Hargreave. 1985. Asymptomatic bronchial hyperresponsiveness in rhinitis. J. Allergy Clin. Immunol. 95(5):573-577.

Randem, B., M. H. Smolensky, B. Hsi, D. Albright, and S. Burge. 1987. Field survey of circadian rhythm in PEF of electronics workers suffering from colophony-induced asthma. Chronobiology International 4(2):263-271.

Reed, C. E. 1988. Basic mechanisms of asthma. Role of inflammation. Chest 94(1):175-177 Rockel, A., B. Klinke , J. Hertel, X. Baur, C. Thiel, S. Abdelhamid, P. Fiegel, and D. Walb. 1989. Allergy to dialysis materials. Nephrol Dial Transplant 4:646-652.

Rose, N. R., F. Milgram, and C. J. van Oss. 1973. Principles of Immunology. New York: MacMillan Publishin Co.,Inc.

Rossi, G. A., O. Sacco, F. Vassallo, S. Lantero, A. Morelli, U. Benatti, and G. Damiani. 1989. Oxidative metabolism of human peripheral blood eosinophils and neutrophils: H2O production after stimulation with phorbol myristate acetate and immune complexes. Eur Respir J 2(6 Suppl.):435S-440S.

Roth, W. G. 1969. Tylotic palmer and planter eczema caused by steam ironing clothes containing formaldehyde. Berufs-Dermatosen 17:263-267.

Saetta, M., L. M. Fabbri, D. Danieli, G. Picotti, and L. Allegra. 1989. Pathology of bronchial asthma and animal models of asthma. Eur Respir J 2(6 Suppl.):477S-482S.

Sakula, A. 1975. Formalin asthma in hospital laboratory staff. Lancet 2:816.

Sandler, S. G., R. Sharon, M. Bush, M. Stroup, and B. Sabo. 1979. Formaldehyde-related antibodies in hemodialysis patients. Transfusion 19(6):682-687.

Semenzato, G. 1991. Immunology of interstitial lung diseases: cellular events taking place in the lung of sarcoidosis, hypersensitivity pneumonitis and HIV infection. Eur Respir J 4:94-102.

Semin, Y.A., E. N. Kolomyitseva, and A. M. Poverennyi. 1974. Effects of products of the reaction of formaldehyde and amino acids on nucleotides and DNA. Molekulyarnaya Biologiya 8:276.

Schlueter, D.P. 1982. Infiltrative lung disease hypersensitivity pneumonitis. J Allergy Clin Immunol 70(1):50-55.

Schoenberg, J. B., and C. A. Mitchell. 1975. Airway disease caused by phenolic (Phenol-formaldehyde) resin exposure. Arch Environ Health 30:574-577.

Shim, C., M. H. Williams, Jr. 1986. Effect of odors in asthma. Amer J Med 80:18-22.

Shipkovitz, H. D. 1968. Formaldehyde vapor emissions in the permanent-press fabrics industry. Report No. TR-52. U.S. Public Health Service, Cincinnati, Ohio.

Skerfving, S., B. Akesson, and B. G. Simonsson. 1980. "Meatwrappers' asthma" caused by thermal degradation products of polythylene. Lancet 1:211.

Sneddon, I. B. 1968. Dermatitis in an intermittent haemodialysis unit. Brit Med J 1:183-184.

Spassovski, M. 1976. Health hazards in the production and processing of some fibers, resins, and plastics in Bulgaria. Environ Health Perspec 17:199-202.

Stokinger, H. E., and D. L. Coffin. 1968. A formaldehyde and its homologs, Pp. 483-484 in Volume I Air Pollution, A. C. Stern, ed. New York: Academic Press.

Tara, M. S. 1966. Intolerance? Sensitization? Formaldehyde-induced pulmonary allergy? Arch Mal 17:66-68.

Tarlo, S. M., and I. Broder. 1989. Irritant-induced occupational asthma. Chest 96(2):297-300.

Tathirov, M. T. 1974. Experimental study of the combined effect of six atmospheric pollutants on the human body. Gig Sanit 5:100-102.

Thrasher, J. D., R. Madison, A. Broughthon, and Z. Gard. 1989. Building-related illness and antibodies to albumin conjugates of formaldehyde, toluene diisocyanate and trimellitic anhydride. Am J Ind Med 15:187-195.

Thrasher, J. D., A. Broughton, and R. Madison. 1990. Immune activation and autoantibodies in humans with long-term inhalation exposure to formaldehyde. Archives Environ Hlth 45(4):217-223.

Thurmond, L. M., and J. H. Dean. 1988. Immunological responses following inhalation exposure to chemical hazards. Pp. 375-392 in Toxicology of the Lung, D. E. Gardner, J. D. Crapo, and E. J. Massaro, eds. New York: Raven Press.

Trinkler, H. 1968. Working with formaldehyde. Medizinische Laboratorium (Stuttgart) 21:283. Turner-Warwick, M., ed. 1978. Immunology of the Lung. London: E. Arnold (Publishers) Ltd.

Tuthill, R. W. 1984. Woodstoves, formaldehyde, and respiratory disease. Amer J Epidemiol 120(6):952-955.

Vaughn, W. T. 1939. The Practice of Allergy. P. 677. St. Louis: C. V. Mosby. Walker, J. F. 1964. Formaldehyde. Pp. 484. New York: Rinehold Publishing Corp.

Weber, A. 1984. Acute effects of environmental tobacco smoke. Eur J Resp Dis 68(Suppl.133):98-108.

Wegman, D. H., and E. A. Eisen. 1990. Acute irritants. More than a nuisance. Chest 97(4):773-775.

Weill, H., and M. Turner-Warwick, eds. 1981. Occupational Lung Diseases: Research Approaches and Methods. New York:Marcel Dekker, Inc.

WHO/EURO. 1986. Indoor Air Quality: Radon and Formaldehyde. World Health Organization Regional Office for Europe, Environ Health

WHO/EURO. 1987. Air Quality Guidelines. World Health Organization Regional Office for Europe, Vol. 2, Chap. 5, Copenhagen/Geneva.

WHO/EURO. 1991. Indoor Air Quality: Combustion Products, Copenhagen, in press. Yefremov, G. G. 1970. The state of the upper respiratory tract in formaldehyde production employees. Zh Ushn Nos Gorl Bolezn 30:11-15.

Zaeva, G. N., et al. 1968. Materials for the revision of the maximum 80 permissible concentration of formaldehyde in the interior atmosphere of industrial premises. Gigiena Truda Professional'nye Zab 12(7):16-20.

Multiple Chemical Sensitivity-What is It?

Roy L. DeHart

"If you don't know where you are going any road will get you there."

-Anonymous

This workshop on multiple chemical sensitivity accomplishes many objectives; perhaps most significant was gathering many professionals with widely divergent opinions regarding MCS together. One important objective unfortunately must await another time: the definition of MCS.

The inability to reach concensus of such an important aspect of this phenomenon speaks volumes on the controversy that is MCS. This controversy is neither new nor unstated. Several disciplines in organized medicine have addressed the issues and contributed to the controversy. Four of these positions are presented for review.

The California Medical Association established a Task Force on Clinical Ecology under its Scientific Board. The report of the Task Force was published in February, 1986 which in part states:

The task force collected material as for any subject review and included all information supplied by individual clinical ecologists and by their professional organizations. There was extensive description of the basic hypotheses of clinical ecology, and an ample and varied collection of anecdotal reports and individual patient testimonials. In contrast, there was a surprising paucity of published studies to prove or disprove clinical ecology hypotheses. Critical analyses of patients and cohorts, detailed data collection, validation and confirming laboratory assays were not provided.

No convincing evidence was found that patients treated by clinical ecologists have unique, recognizable syndromes, that the diagnostic tests employed are efficacious and reliable or that the treatments used are effective. Even though clinical ecology has existed for approximately 50 years, only a few studies have been conducted that are scientifically sound. Most have such serious methodological flaws as to make their conclusions unacceptable. Those few studies that used scientifically sound methods have provided evidence that the effectiveness of certain treatment methods used by clinical ecologists is based principally on placebo response.

Undoubtedly, some patients suffer from illnesses that cannot be readily diagnosed and for which only supportive treatments exist. It may even be true that some or all of the hypotheses and treatments proposed by clinical ecologists are valid but we found no evidence to support them. These hypotheses and treatments should be subjected to modern, scientific methods of evaluation. We think that this can be done provided genuine interest exists.

The task force is concerned that unproved diagnostic tests are being widely used by clinical ecologists in what may be incorrect or inappropriate applications. Decisions made on the basis of these tests can lead to misdiagnosis, resulting in patients being denied other supportive treatments and becoming psychologically dependent, believing themselves seriously and chronically impaired. This possibility underscores the need for more adequate scientific studies to prove or disprove the value of clinical ecology tests and treatments. To consider the current practice of clinical ecology experimental is misleading, however. It can only be considered experimental when its practitioners adhere to scientifically sound research protocols and inform their patients about the investigative nature of their practice.[1]

The same year the Executive Committee of the American Academy of Allergy and Immunology issued a position statement on Clinical Ecology. The critique follows:

The environment is very important in the lives of every human being. Environmental factors, such as chemicals and pollutants, have been demonstrated to influence health. The idea that the environment is responsible for a multitude of human health problems is most appealing. However, to present such ideas as facts, conclusions, or even likely mechanisms without adequate support, is poor medical practice.

The theoretical basis for ecologic illness in the present context has not been established as factual, nor is there satisfactory evidence to support the actual existence of "immune system disregulation" or maladaptation. There is no clear evidence that many of the symptoms noted above are related to allergy, sensitivity, toxicity, or any other type of reaction from foods, water, chemicals, pollutants, viruses, and bacteria in the context presented. Properly controlled studies defining objective parameters of illness, properly controlled evaluation of the treatment modalities, and appropriate patient assessment have not been done. Anecdotal articles do not constitute sufficient evidence of a cause-and-effect relationship between symptoms and environmental exposure. The major techniques used by the clinical ecologists are controversial and unproven. The American Academy of Allergy and Immunology has previously published position statements concerning subcutaneous and sublingual provocation neutralization procedures and found them to be unproven. More recent review of new data submitted by a number of clinical ecologists to the Practice Standards Committee of the Academy has not changed that recommendation. There are no adequate studies of the cyclic diets, elimination diets, injection therapy with chemicals, or even the environmentally controlled units to substantiate their use. Many of the patients are reported to have a normal physical examination and normal laboratory tests.

There are no immunologic data to support the dogma of the clinical ecologists. To suggest that these patients lack suppressor T cell function has not been supported by published data. The suggestion that neutralization therapy can provide rapid relief within minutes or hours cannot be supported by controlled clinical studies or immunologic data.[2]

The American College of Physicians published a position paper on Clinical Ecology in 1989. The summary from that publication is quoted:

Clinical ecologists propose the existence of a unique illness in which multiple environmental chemicals, foods, drugs, and endogenous *C. albicans* have a toxic effect on the immune system, thereby adversely affecting other bodily functions. The proposal uses some concepts that superficially resemble those that apply to clinical allergy and toxicology and others that are novel.

Review of the clinical ecology literature provides inadequate support for the beliefs and practices of clinical ecology. The existence of an environmental illness as presented in clinical ecology theory must be questioned because of the lack of a clinical definition. Diagnoses and treatments involve procedures of no proven efficacy.

Case reports by clinical ecologists and evaluation of these patients by other physicians indicate that this diagnosis is applied most frequently to persons with symptoms of physiologic (somatic) or psychologic dysfunction, or both. Proof of cause-effect relations between environmental factors and symptoms of "environmental illness" is particularly difficult because clinical ecologists implicate such a broad range of agents, including chemicals, foods, hormones, and microorganisms. Most patients are believed to react to multiple environmental substances by any route of exposure, and some are said to be intolerant to the entire environment, the so-called "total allergy syndrome."

The principal method of proof cited by clinical ecologists for the existence of "environmental illness" is the symptom-provocation test used in diagnosis of individual cases after the condition is suspected because of a history of symptoms and suspected causes. Published studies on the provocation test employed widely different subject-selection methods and outcome-measurement criteria. All were seriously flawed by the absence of matched patient-control groups, absence or inadequacy of placebo, and failure to achieve or document randomness of trials. Not surprisingly, therefore, the conclusions from these studies are conflicting.

Those studies reporting results of immunologic tests are insufficient to address theories of environmental illness; the number of cases is small and selection criteria are not clear. Enumeration of lymphocyte subsets and quantitation of serum immunoglobulin and complement levels in patients with "environmental illness" have not yielded clear-cut evidence of immunologic abnormality.

Clinical ecologists use a treatment program that includes avoidance of environmental chemicals, rotation of foods in the diet, and neutralization of symptoms with injected or sublingual extracts. Except for small-dose oral nystatin, which is used for treatment of patients with the candida hypersensitivity syndrome, drug therapy is intentionally avoided, although some clinical ecologists recommend mineral salts, oxygen, vitamins, minerals, and antioxidants for relief of symptoms. There are only two controlled studies on neutralization therapy. One evaluated only eight patients for 20 days; the other was a study of dust allergy in patients with allergic rhinitis, not "environmental illness." To date, there have been no controlled studies on diet therapy, environmental elimination therapy, or nystatin in patients with candida hypersensitivity. Thus there is no body of evidence that clinical ecology treatment measures, singly or in combination, are effective.

Others who have reviewed the diagnostic and therapeutic methods of clinical ecology have also arrived at the conclusion that these methods are of unproven value. Potential adverse effects from these procedures and the costs of clinical ecology diagnosis and treatment have not been evaluated.[3] More recently, on April 28, 1991, the American College of Occupational Medicine issued the following position statement on multiple chemical hypersensitivity syndrome:

Multiple Chemical Hypersensitivity Syndrome is one of over 20 names or descriptions (20th century disease, Environmental Illness, Total Allergy Syndrome, Chemical AIDS, etc.) given to a clinical complex characterized by recurrent polysystem symptoms often without observable physical findings or laboratory abnormalities believed to be associated with repeated exposure to specific biological, chemical or physical agents. Practitioners involved in the diagnosis and treatment of this phenomenon may identify themselves as clinical ecologists or environmental medicine specialists.

The pathophysiological mechanism described by these practitioners do not, in general, conform to what is currently known of human biological functions. To explain the phenomenon, these practitioners draw on new or modified mechanisms such as "total body load, spreading, and switching." The scientific foundation for managing patients with this syndrome has yet to be established by traditional clinical investigative activities that withstand critical peer review.

The following medical associations have carefully reviewed the available scientific evidence regarding this phenomenon:
The American Academy of Allergy and Immunology
The California Medical Association
The American College of Physicians

Their conclusions have been similar and state that the scientific and clinical evidence supporting the pathophysiological mechanisms and treatment regimens as articulated by these practitioners is lacking. It is the position of the American College of Occupational Medicine that the Multiple Chemical Hypersensitivity Syndrome is presently an unproven hypothesis and current treatment methods represent an experimental methodology. The College supports scientific research into the phenomenon to help explain and better describe its pathophysiological features and define appropriate clinical interventions. This research should adhere to established principles of scientific inquiry and the results submitted for publication in recognized peer-reviewed journals.[4]

In each of these position statements there is one integrating theme-the clear need for rigorous scientific placebo-controlled, double-blind investigations that are subject to exacting peer review. Although the working group was unable to characterize the patient with multiple chemical sensitivity, the investigators and scientists must provide a definition as part of protocol development. Case definition and criteria for a diagnosis must be explicit if the study population is to be appropriately identified. Further, if one is to define the mechanism of MCS or to conduct epidemiological research focused on MCS, case definition is not an option, but a necessity.

If the question cannot be answered as to what MCS is, how can there be approval of research protocols or acceptance of investigative results? In order to appropriately address the controversies surrounding this phenomenon we must know where we're going!

REFERENCES

California Medical Association. *Clinical Ecology-A Critical Appraisal.* Western Journal of Medicine; 144:2,239-49, 1986.

American Academy of Allergy and Immunology. *Clinical Ecology.* Journal of Allergy and Clinical Immunology 78:2,269-71; 1986.

American College of Physicians. *Clinical Ecology.* Annals of Internal Medicine 111:2,168-78; 1989.

American College of Occupational Medicine. *Multiple Chemical Hypersensitivity Syndrome.* Board minutes, April 28, 1991.

Case Definitions for Multiple Chemical Sensitivity[1]

Nicholas A. Ashford and Claudia S. Miller

TERMINOLOGY

A wide array of names has been applied to the syndromes suffered by patients with heightened reactivity to chemicals (Table 1). Each name has specific implications regarding the underlying cause, mechanism, or manifestations of the disease, and they overlap. A major hindrance in achieving scientific respectability has been the difficulty in agreeing upon a definition for this condition (or conditions). It should be realized that no single case definition, even if agreed upon in one context--e.g., for diagnosis purposes--will suffice for use in other contexts. It is important to distinguish case definitions for use in diagnosis, epidemiological studies, research on the nature of the condition, regulatory standard-setting, compensation awards, and situations requiring alternative employment or housing. The social and political consequences of the use of a particular definition in a specific context requires the case definition to be carefully constructed.

Cullen (1987b) has emphasized the importance of establishing a uniform case definition before meaningful epidemiologic studies can be undertaken, but cautions, "However constructed, the goal of descriptive studies must be refinement of the diagnostic criteria, in particular the very tentative boundaries with other diagnostic entities such as allergic, anxiety, panic and post-traumatic stress disorders, and physiologic sequelae of central nervous system (CNS) intoxication or injury, especially by organic solvents." He acknowledges possible overlap among these entities and offers the following case definition:

Multiple chemical sensitivities (MCS) is an acquired disorder characterized by recurrent symptoms, referable to multiple organ systems, occurring in response to demonstrable exposure to many chemically unrelated compounds at doses far below those established in the general population to cause harmful effects. No single widely accepted test of physiologic function can be shown to correlate with symptoms. (Cullen 1987a)

This case definition, intended for epidemiological use, is intentionally narrow. Cullen

[1] Excerpted in part from N.A. Ashford and C.S. Miller, Chemical Exposures: Low Levels and High Stakes, Van Nostrand Reinhold, New York 1991.

TABLE 1

Cause	Mechanism	Effect
Environmentally induced illness	Immunologic illness	Multiple chemical sensitivities (MCS)
	Immunotoxicity	
Chemically induced (or acquired) hypersusceptibility	Immune dysfunction	Multiple chemical sensitivity syndrome
Chemically acquired immune deficiency syndrome (chemical AIDS)	Immune dysregulation	Chemical hypersensitivity syndrome
	Conditioned odor response	
		Universal allergy
The petro-chemical problem	Fear/anxiety	
		20th-century illness
	Mass psychogenic illness	
		Total allergy syndrome
	Various psychiatric disorders	
		Environmental allergy or illness
		Cerebral allergy
		Environmental maladaptation syndrome
		Food and chemical sensitivity

This case definition, intended for epidemiological use, is intentionally narrow. Cullen excludes persons who react to substances no one else in aware of on the basis that such individuals may be delusional and excludes persons who have bronchospasm, vasospasm, seizures, or "any other reversible lesion" that can be identified and specifically treated. Clinical ecologists, however, would argue that persons with bronchospasm, vasospasm, seizures, and other illnesses excluded by Cullen may well have the chemical sensitivity problem. Each issue of the clinical ecologists' journal, Clinical Ecology, contains the following definition:

Ecologic illness is a chronic multi-system disorder, usually polysymptomatic, caused by adverse reactions to environmental incitants, modified by individual susceptibility and specific adaptation. The incitants are present in air, water, food, drugs and our habitat.

Although the patients the clinical ecologists and Cullen see are demographically divergent, the definitions of their illnesses are remarkably alike. Both describe the chemically sensitive patient in similar terms. [See Miller and Ashford, "Allergy and Multiple Chemical Sensitivity (MCS) Distinguished" in this report for a discussion of sensitive populations.]

However, what is sorely needed is an objective test that can be applied in each individual case to determine, incontrovertibly, whether a particular person has multiple chemical sensitivities.

Given the multitude of environmental exposures (both chemical and food) that allegedly can result in a seemingly endless array of physical and mental syndromes and the frequent absence of findings on routine physical examination, the practitioner who sees these patients with their divergent and unfamiliar litany of complaints is at great disadvantage in trying to diagnose the condition.

To circumvent this problem, we propose the following operational definition of multiple chemical sensitivity for diagnostic purposes, a definition that is based upon environmental testing:

The patient with multiple chemical sensitivities can be discovered by removal from the suspected offending agents and by rechallenge, after an appropriate interval, under strictly controlled environmental conditions. Causality is inferred by the clearing of symptoms with removal from the offending environment and recurrence of symptoms with specific challenge.

Challenges conducted for research purposes should be performed in a double-blind, placebo-controlled manner.

This operational definition is essential to resolving, once and for all, the debate about whether an individual's symptoms are or are not environmentally induced. An environmental unit is necessary for scientific validation of the concept of chemical sensitivity. Because of the expense and time required by patients and physicians alike, we are not arguing that the unit be used for all patients. Such stringent measures are not necessary for most patients. For severe cases, however, no alternative is available at present, and only from firsthand observation of hospitalized patients can physicians have the opportunity to understand this illness better. In time, as more clinical data on these patients accumulate, physicians may be able to diagnose this disorder on the basis of the patient's history and a few key laboratory tests. For now, reliance must be placed on rigorous study in an environmental unit. Ultimately a phenomenological definition may emerge that allows physicians to diagnose, at least tentatively, chemical sensitivity based on a history of a specific sensitizing event (such as a pesticide exposure) followed by evidence of chemical and food sensitivities, multisystem effects, improvement after avoidance of exposure, and similar experiences of persons with like histories.

Criteria for the selection of cases *for research purposes* are:

- Sensitivity to chemicals, i.e., symptoms or signs related to chemical exposures at levels tolerated by the population at large that is distinct from such well recognized hypersensitivity phenomena as IgE-mediated immediate hypersensitivity reactions, contact dermatitis, and hypersensitivity pneumonitis.
- Sensitivity, expressed as symptoms and signs, in one or more organ systems.
- Symptoms and signs that wax and wane with exposures.

It is not necessary to identify a chemical exposure associated with the onset of the condition. Preexistent or concurrent conditions, e.g. asthma, arthritis, somatization disorder or depression, should not exclude patients from consideration.

The selection of subjects for research protocols should depend on the specific hypotheses to be tested. Identifiable populations should include but not be restricted to:

- Symptom or sign based: Patients with reactivity to environmental chemicals, either through self-reporting or meeting case selection criteria.
- Disease based: Patients with specific diseases that are suspected to be caused by or exacerbated by chemical exposures.
- Exposure based: Groups characterized by a common exposure, such as workers at a specific factory, occupants of a particular building, or residents of a contaminated community.
- Population based: Groups such as school children or a random community sample.

Appropriate comparison groups should be chosen in each case.

Whatever case definitions emerge for diagnostic or research purposes, decisions concerning "safe levels" of chemical exposure for regulatory standard-setting, criteria of eligibility for disability awards under workers' compensation or social security, and criteria of eligibility for alternative employment or housing ultimately need to be based on suitable definitions. These definitions should be expected to differ in exactness and stringency, and may be in flux until more research results are obtained.

REFERENCES

Cullen, M., "The Worker with Multiple Chemical Sensitivities: An Overview." In: Cullen, M. (ed.) Workers with Multiple Chemical Sensitivities, Occupational Medicine: State of the Art Reviews (1987a), Hanley and Belfus, Philadelphia, 2(4):655-662.

Cullen.M., "Multiple Chemical Sensitivities: Summary and Directions for Future Investigators." In: Cullen, M. (ed.) Workers with Multiple Chemical Sensitivities, Occupational Medicine: State of the Art Reviews (1987b), Hanley and Belfus, Philadelphia, 2(4):801-804.

Allergy and Multiple Chemical Sensitivities Distinguished[1]

Claudia S. Miller and Nicholas A. Ashford

When von Pirquet coined the term "allergy" in 1906, he defined it as "altered reactivity". Thus the word "allergy" as originally conceived encompassed both immunity and hypersensitivity. In 1925 European allergists influenced their American colleagues to redefine "allergy" in the context of antibodies and antigens, effectively excluding hypersensitivity on any other basis.

Forty years ago, Theron Randolph, a classically trained allergist, noted that a cosmetic saleswoman he had been seeing for rhinitis, asthma, headache, fatigue, irritability, depression, marked weight fluctuation and intermittent episodic loss of consciousness developed her symptoms following exposure to gas, oil, coal and their combustion products (Randolph, 1987, pp. 73-76). Randolph developed a diagnostic-therapeutic maneuver which consisted of removing the patient from all suspected environmental exposures and subsequently reintroducing single elements of the environment, one at a time, while observing for changes in the patient's condition (Randolph, 1960). Although what he observed in his patients appeared to be some kind of hypersensitivity, it was not allergy. Subsequently, Randolph and other physicians known as clinical ecologists published clinical descriptions of patients with polysomatic complaints, frequently including mood and cognition difficulties, triggered by a wide variety of chemical exposures, but especially petrochemical exposures, and often with concomitant food and drug intolerances. These clinical descriptions bear striking resemblance to today's MCS patients, many of whom have never heard of clinical ecology.

A review of the literature on exposure to low levels of chemicals reveals four groups or clusters of people among whom individuals with heightened reactivity have been reported (Ashford and Miller, 1989):

1. Industrial workers
2. Occupants of "tight buildings," including office workers and school-children

---

[1]Excerpted from Ashford, N. A. and Miller, C. S. 1991. Chemical Exposures: Low Levels and High Stakes. New York: Van Nostrand Reinhold.

3. Residents of communities whose air or water has been contaminated by chemicals
4. Individuals who have had personal and unique exposures to various chemicals in domestic indoor air, pesticides, drugs, and consumer products

These four groups are listed for comparison in Table 1. Note that they differ in professional and educational attainment, age and sex, and the mix and levels of chemicals to which they are exposed, but that all have multiple symptoms involving multiple organ systems with marked variability in the type and degree of those symptoms. Symptoms are often "subjective". For example, central nervous system (CNS) symptoms such as difficulty concentrating or irritability are common, and physical examinations are frequently unremarkable for individuals in each category.

TABLE 1

Chemically Sensitive Groups

Group	Nature of Exposure	Demographics
Industrial workers	Acute and chronic exposure to industrial chemicals	Primarily males: blue collar; 20 to 65 years old
Tight-building occupants	Off-gassing from construction materials, office equipment or supplies; tobacco smoke; inadequate ventilation	Females more than males; white collar office workers and professionals; 20 to 65 years old; schoolchildren
Contaminated communities	Toxic waste sites; aerial pesticide spraying; ground water conatmination; air contamination by nearby industry and other community exposures	All ages, male and female; children or infants may be affected first or most; pregnant women with possible effects on fetuses; middle to lower class
Individuals	Heterogeneous; indoor air (domestic), consumer products, drugs, and pesticides	70-80% females; 30 to 50 years old (Johnson and Rea 1989); white middle to upper middle class and professionals

Many affected individuals report a major precipitating (inducing or "sensitizing") exposure which marked the onset of their chemical sensitivities. In one survey of 6,800 persons claiming to be chemically sensitive, 80 percent asserted that they knew "when, where, with what, and how they were made ill" (National Foundation for the Chemically Hypersensitive, 1989). Of the 80 percent, 60 percent (that is, almost half of those who replied) blamed pesticides. The respondents to the survey were self selected, and the result must be interpreted with caution. Nevertheless, the results suggest that future surveys of persons with different exposure histories and symptoms might contribute to an understanding of underlying mechanisms and causes.

In some chemically sensitive patients no single, identifiable, "high-level" exposure seems to have been associated with the onset of their difficulties. Exposures could have occurred but were not recognized or remembered. Some observers suggest that repetitive or cumulative lower level exposure events may lead to the development of sensitivities. Still others implicate genetic predisposition, pregnancy, major surgery with anesthesia, physical trauma, or major psychological stress as contributors to the illness. Based upon the increasing number of outbreaks of sick building syndrome, increased reporting of symptoms in contaminated communities to state health departments, increased recognition of problems in the industrial workplace, and the increasing numbers of physicians treating chemically related sensitivities, the existing evidence does suggest that chemical sensitivity is on the rise and could become a large problem with significant economic consequences related to the disablement of productive members of society.

The fact that the demographically diverse groups listed in Table 1 share similar patterns of illness (that is, onset after a major chemical exposure; subsequent hyperreactivity to low levels of a variety of chemicals commonly encountered in the environment such as cigarette smoke, perfume, and traffic exhaust; and multisystem complaints with frequent mood, memory and concentration difficulties) suggests that the problem may be real.

In addition, the temporal cohesiveness of onset of illness within groups of individuals sharing a recognized, major chemical exposure event, for example, the development of symptoms in several family members, co-workers or community members exposed simultaneously, help point to the problem as potentially real in those circumstances.

Although a definitive and accurate picture is yet to come, at this time these pieces - viewed collectively - provide sufficient evidence to conclude that chemical sensitivity does exist as a serious health and environmental problem.

The different meanings of the term "sensitivity" are at least in part responsible for the confusion surrounding chemical sensitivity. In the classical, toxicological use of the word "sensitivity", those individuals who require relatively lower doses to induce a particular response are said to be more sensitive than those who would require relatively higher doses before experiencing the same response. A hypothetical distribution of sensitivities, that is, the minimum doses necessary to cause individuals in a population to exhibit a harmful effect, is shown in curve A in Figure 1-1. (If we plot the cumulative number of individuals who exhibit a particular response as a function of dose, we generate a population dose-response curve; see curve A in Figure 1-2). This distribution describes the traditional toxicological concept of sensitivity. Curve A in Figure 1-1 illustrates that health effects of classical diseases are seen in a significant portion of the normal population at a certain dose; the sensitive and resilient populations are found in the tails of the distribution.

A second meaning of the word "sensitivity" appears in the context of classical IgE-mediated allergy (atopy). IgE is one of five classes of antibodies made by the body, and is, from the perspective of classically allergic individuals, the most important antibody. Atopic individuals have IgE directed against specific environmental incitants, such as ragweed or bee venom. Positive skin tests in these individuals correlate with a rapid onset of symptoms when they are actually exposed to those allergens. The atopic individual exhibits a reaction whereas non-allergic individuals do not, even at the highest doses normally found in the environment. A hypothetical sensitivity distribution for an atopic effect is shown in curve B of Figure 1-1, and the dose-response curve derived from that distribution is found in curve B

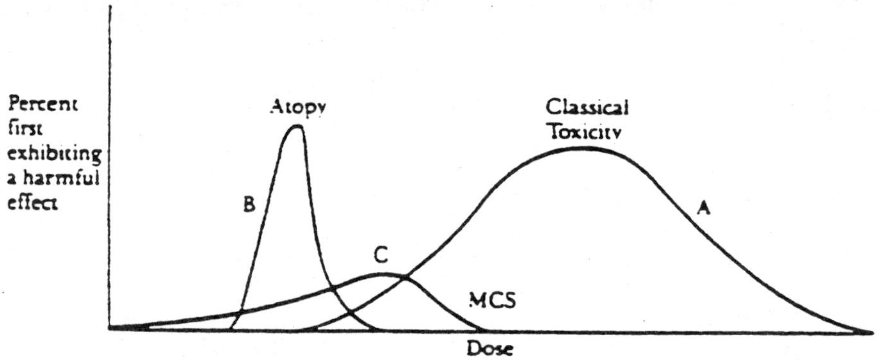

FIGURE 1-1 Hypothetical distribution of different types of sensitivities as a function of dose. Curve A is a sensitivity distribution for classical toxicity, e.g., to lead or a solvent. Sensitive individuals are found in the left-hand tail of the distribution. Curve B is a sensitivity distribution of atopic or allergic individuals in the population who are sensitive to an allergen, e.g., ragweed or bee venom. Curve C is a sensitivity distribution for individuals with multiple chemical sensitivities who, because they are already sensitized, subsequently respond to particular incitants, e.g., formaldehyde or phenol.

of Figure 1-2. Allergists include in the term "allergy" well-characterized immune responses that result from industrial exposure to certain chemicals, such as nickel or toluene diisocyanate (TDI). Most allergists refer to such responses as "chemical sensitivity", and reserve this term for responses that have a distinct immunological basis, preferring to use a term such as "chemical intolerance" for nonimmunological responses to chemicals.

Patients suffering from multiple chemical sensitivities (MCS) may be exhibiting a third and entirely different type of sensitivity. Their health problems often, but not always, appear to originate with some acute or traumatic exposure, after which the triggering of symptoms and observed sensitivities occur at very low levels of chemical exposure. The inducing chemical or substance may or may not be the same as the substances that thereafter provoke or "trigger" responses. (Sometimes the inducing substance is described as "sensitizing" and the individual affected as a "sensitized" person). Reactions may sometimes be observed at incitant levels similar to those to which classically sensitive and atopic patients respond. Unlike classical toxicity, however, here the effects of low-level exposures are not simply those effects observed in normal populations at higher doses. The fact that normal persons - for example, most doctors - do not experience even at higher levels of exposure those symptoms that chemically sensitive patients describe at much lower levels of exposure probably helps to explain the reluctance of some physicians to believe that the problems are physical in nature. (Although this also describes atopy, in this case the sensitivity is not IgE mediated). To compound the problem of physician acceptance of this illness, multiple organ systems may be affected, and multiple substances may trigger the effects. Over time, sensitivities seem to spread, in terms of both the types of triggering substances and the systems affected (Randolph, 1962, pp. 98 and 119). Avoidance of the offending substances is usually effective but much more difficult to achieve for these patients than for classically sensitive patients because symptoms may occur at extremely low doses

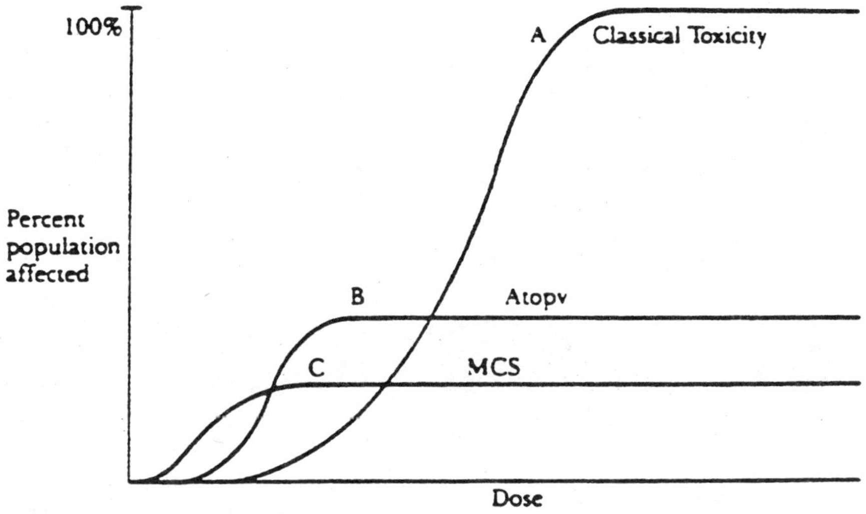

FIGURE 1-2 Hypothetical population dose-response curves for different effects. Curve A is a cumulative dose-response cur e for classical toxicity, e.g., to lead or a solvent. Curve B is a cumulative dose-response curve for atopic or allergic individuals in the population who are sensitive to an allergen, e.g., ragweed or bee venom. Curve C is a cumulative dose-response curve for individuals with multiple chemical sensitivities who, because they are already sensitized, subsequently respond to particular incitants, e.g., formaldehyde or phenol.

and the exposures are manifold and ubiquitous. Adaptation to chronic low-level exposure with consequent "masking" of symptoms (discussed more fully later) may make it exceedingly difficult to discover these sensitivities and unravel the multifactorial triggering of symptoms.

A hypothetical sensitivity distribution for a single symptom for the already chemically sensitive person in response to a single substance trigger is shown in curve C of Figure 1-1, and the corresponding dose-response curve is shown in curve C of Figure 1-2. It should be emphasized, however, that individuals who become chemically sensitive may have been exposed to an initial priming event that was toxic, as classically defined, and which was the cause for their having developed chemical sensitivities in the first place.

Conceivably, exposure to certain substances, such as formaldehyde, might elicit all three types of sensitivities.

The fact that sensitivity means something quite different to toxicologists, allergists, and clinical ecologists reflects the different disease paradigms under which each operates. Neither traditional allergists nor toxicologists fully appreciate the two-step process of induction and triggering that seems to characterize multiple chemical sensitivities.

Those clinical ecologists who reference the literature on classical chemical toxicity to buttress their case for chemical sensitivity may be adding to the confusion and contributing to others' reluctance to accept their ideas. Likewise, allergists who dismiss chemical sensitivity on the grounds that it is not consistent with a recognized immunologic mechanism may be overlooking another kind of sensitivity in their patients. Although chemicals may act in some manner (via a toxic mechanism, for instance), to predispose or cause the body to be reactive to subsequent low-level chemical exposures, the resulting hyperreactivity to low levels of chemically diverse and unrelated substances is not toxicity as classically defined or understood at this time. Some allergists maintain that the term "chemical sensitivity" should not be used in the context we have used it here, but should be reserved only for those responses having an immunological basis. However, the term sensitivity has broader applicability. A parallel might be the word "resistance", which is widely understood whether one is talking about electricity, psychiatry, or an infectious disease. Similarly, "sensitivity" is easily understood when used in any of the three contexts illustrated in this section; it is not the exclusive property of the allergist.

Although chemically sensitive patients were first described by Randolph in the 1950's, the problem seems more prominent in the past decade or so. There are some historical developments which may have contributed to a recent increase in chemical sensitivity.

Americans today spend many more hours per day indoors at work and at home, in schools, shopping malls, and other buildings than preceding generations (Environmental Protection Agency, 1989). On the average, we spend 90 percent of our time indoors. With the concern for energy conservation following the oil embargo of the 1970's, homes and office buildings in the United States were constructed more tightly and make-up air (fresh air intake) was cut to a minimum. Similarly, homeowners and new home builders caulked and sealed, installed storm windows and extra insulation, and effectively reduced fresh air infiltration. On the average, newer homes have half as much fresh air infiltration as older homes (0.5 versus 0.9 air changes per hour) (Mage and Gammage, 1985).

Over the past decade, EPA has conducted TEAM (Total Exposure Assessment Methodology) studies on a variety of volatile organics (1980-1987), carbon monoxide (1982-1983), pesticides (1986-1989), and particulates (1987-present). Samples of 20 volatile organic compounds in the personal (indoor) air, outdoor air, drinking water, and breath of approximately 400 residents of New Jersey, North Carolina, and North Dakota were collected (Wallace, 1987).

Levels of indoor air contaminants were often many times higher than outdoor levels, and sometimes orders of magnitude higher than outdoor levels. Breath levels for most chemicals measured were 30-40 percent of indoor air levels, but measured up to 90 percent of indoor air levels in some cases - tetrachloroethylene, for example. A study of non-occupational pesticide exposure also showed dramatically higher concentrations of pesticides inside homes than out of doors (Immerman, 1990).

Remarkably, the sources of pollutants that were identified by the EPA in homes are the same ones individuals with multiple chemical sensitivities identify as provoking their often vague and seemingly inexplicable symptoms, for example, room air deodorizers, attached garages, hot showers and spas, dry cleaned clothing, cleaning agents and disinfectants.

Before World War II, U.S. production of synthetic organic chemicals totaled fewer than a billion pounds per year. By 1976, production had soared to 163 billion pounds annually (Odell, 1980). Increased sources of indoor air pollution, coupled with decreased fresh make-up air, have transformed the indoor environment. Community exposures to toxic chemicals, industrial and office exposures, and other episodic exposures of individuals also increased, reflecting the rise in production of coal- and oil-derived chemicals and synthetics.

These changes in chemical production, consumer products, and building design have been accompanied by an increasing number of people who appear to react to low levels of environmental pollutants. Interestingly, since World War II certain illnesses, such as asthma

(Sly, 1983) and depression (Klerman and Weissman, 1989), seem to have shown upsurges. It is easy to imagine that asthma could be related to chemical exposures. In the case of depression, it is recognized that solvent-exposed workers experience more depression and cognitive difficulties. Further, the majority of indoor air contaminants are solvents, albeit concentrations are generally much lower than those found in an industrial setting. Randolph often referred to chemical sensitivities as the "petrochemical problem" because the increase in the incidence of this illness seems to parallel the growth of the petrochemical industry and the increased use of synthetic materials such as particleboard, pesticides, synthetic textiles, plastics, and food additives by consumers since World War II.

Randolph, who had been hospitalizing patients and testing them for their food sensitivities, found a critical element in many of his patients' recoveries was avoidance of environmental chemical exposures in their jobs and homes while in the hospital. He developed "comprehensive environmental control", a diagnostic approach in which patients avoid exposure to synthetic chemicals in order to facilitate diagnosis of chemical sensitivity.

Briefly, this technique involves placing the patient in a specially constructed environment devoid of materials that off-gas; avoiding the use of drugs, cosmetics, perfumes, synthetic fabrics, pesticides, and similar substances; and having the patient fast for a period of days until symptoms resolve. This initial period of avoidance and fasting requires approximately 4 to 7 days on the average. During this time, the patient exhibits withdrawal symptoms such as headache, malaise, irritability, or depression. By the end of this time, the patient's symptoms, if environmentally related, should clear, provided that end-organ damage has not occurred. At the end of this avoidance phase, the patient reportedly has a markedly lower pulse rate and an increased sense of well-being, as well as resolution of symptoms. Drinking waters from a variety of sources also are tested to find one most compatible with the patient. Next, individual foods are reintroduced, one per meal, over a two- to three-week period. Following this, the patient is placed on a rotating diet of "safe" foods (i.e., foods that did not provoke symptoms for that particular patient). Finally, the patient is challenged with very low levels (levels routinely encountered in daily living) of common chemicals. Those exposures, both food and chemical, that induce symptoms are to be avoided.

A description of comprehensive environmental control and its role in diagnosis and therapy first appeared 30 years ago in Clinical Physiology (Randolph, 1960) and again in the Annals of Allergy in 1965 (Randolph, 1965).

The detailed design of an environmental unit is beyond the limits of this discussion, however, some of the essentials are noted here. Although conceptually simple and scientifically elegant, achieving a well controlled environment within the average hospital is technically difficult.

First, by employing construction materials, furnishings, and clothing that are less likely to off-gas, very low levels of volatile organic compounds (for example, from synthetics) are maintained inside the unit. To create and operate a unit that is as free as possible from chemical pollution requires knowledge, precision, and vigilance while working with architects, ventilation engineers, contractors and their suppliers, nurses, dieticians, food and water suppliers, and maintenance and custodial staffs.

Several units have been operated by the clinical ecologists and one, which was patterned after those of the ecologists, was operated by John Selner, a Denver allergist. Currently none are in operation, although all of the physicians who have been involved with these units have found them to be a valuable tool for the evaluation of certain patients.

The clinical ecologists' environmental units and Selner's unit shared many of the same design and operational parameters (Table 2). Studies from ecologists' units leave much to be desired in terms of study design. Unfortunately, no studies were ever published from the allergists' unit in Denver.

TABLE 2

Features of Environmental Units[a]

Characteristics/Practices	Allergists' Unit[b]	Clinical Ecologists' Units[c]
Construction using materials that do not off-gas (primarily glass, stell, ceramic; cotton bedding and clothing). Avoidance of synthetic materials. No perfumes, cosmetics, odorous cleaners/ soaps,etc.	Yes	Yes
Air supply filtered; patients' rooms under positive pressure to reduce contamination from adjacent areas; airlocks	Yes	Yes
Patients' medications discontinued insofar as possible; gradual withdrawal from steroids, etc.	Yes	Yes
Patients fasted for 4 to 8 days to clear symptoms.	Yes, if symptoms do not clear after several days in unit	Yes at time of admission to unit
Organic foods used for food testing; commercial foods tested also	Yes	Yes
Patients tested for acceptable water	Yes	Yes
Challenges performed using single foods and chemicals after period of avoidance (to eliminate masking)	Yes	Yes

[a]None of the units described in this table is currently in operation.
[b] Selner in Denver (Selner and Staudenmayer, 1985).
[c]Randolph in Chicago and Rea in Dallas.

By isolating his patients from their usual environments and then re-exposing them to various foods and chemicals one by one, Randolph observed that adaptation seemed to play an important role in his patients' responses to many common substances they ate, drank or inhaled. Adaptation is known in other contexts as "acclimation" or "acclimatization", "habituation", "developing tolerance" and even "addiction". Randolph used the terms "adaptation" and "addiction" most often. However, reference to one of the other words may make it easier to grasp the concept. "Acclimatization" is a widely used term in occupational health that refers to workers gradually becoming accustomed to exposures on the job, for example, heat stress. Understanding adaptation is important for two reasons: (1) adaptation may interfere with the discovery of the effects of a particular exposure on the body and (2) chemical exposures may adversely impact adaptation mechanisms and thus lead to illness.

That human beings respond to chronic exposure to environmental challenges by adapting, acclimating, acclimatizing, or even becoming addicted is widely recognized for a variety of substances. Most would agree that the use of narcotics, alcohol, nicotine, and even caffeine can be addicting. For example, the first cigarette ever smoked might be associated with eye and throat irritation, but over time, with more cigarettes, most individuals adapt, and primarily the pleasurable effects of nicotine on the brain are experienced. After months or years, more cigarettes (or alcohol or caffeine or drugs) may be required for the same amount of lift. The individuals may exhibit addictive behavior, seeking cigarettes more frequently. Subsequently, quitting cigarettes (or alcohol, caffeine, or drugs) may lead to withdrawal symptoms including irritability, drowsiness, fatigue, moodiness, and headache. The reformed smoker may become extremely intolerant of the smoke of others, even in tiny amounts. Suddenly recalled are the irritation and unpleasant feelings associated with the first cigarette ever smoked. Over time the individual had "adapted" to those effects. Adaptation, which on the surface would seem good for the organism, may in fact be a two-edged sword. Developing tolerance for the noxious properties of the exposure may allow the individual to remain in the exposure more comfortably while other harmful consequences of the exposure continue. Thus the heavy smoker who is "adapted" to tobacco smoke is at increased risk for developing emphysema, lung cancer or vascular disease. While often occurring at much lower levels of exposure than the above examples, food and chemical adaptation and addiction have been observed by some physicians in their patients. In the case of MCS patients, *multiple* incitants, not only tobacco smoke, may be involved and *all* may need be avoided simultaneously for improvement to occur. Thus, frequent exposure to a substance results in adaptation (irritation and other warning signals may disappear). Continued exposure may lead to addiction. Reduction or cessation of exposure generally results in withdrawal symptoms.

What may confuse patients and practitioners is that the symptoms for which the individual is most likely to seek a physician's help are those that occur during withdrawal when the person is no longer exposed (or is less exposed) to the offending agent. Thus headaches may occur when the individual smokes fewer cigarettes than usual or drinks less caffeine. Indeed, these unpleasant withdrawal symptoms may be forestalled by smoking another cigarette or taking another drink of coffee, thus perpetuating addiction. Patients may report that smoking a cigarette or drinking a cup of coffee in the morning (after 8 or so hours without) relieves their headache (a withdrawal symptom) and they feel better, not suspecting that the cigarette or coffee might also be the cause of their headache.

Occupational health presents many examples in which acclimatization, inurement, or tolerance to a substance is known to develop, for example, exposure to ozone, nitroglycerin, and solvents. Note that the incitants mentioned thus far are all quite different from one another: some are ingestants, others inhalants; some are solid, others liquid or gaseous in form; some are simple molecules, whereas others are complex mixtures. The point is that the human body appears able to adapt to an endless array of substances.

By isolating MCS patients from their usual environments and then re-exposing them to various foods and chemicals one by one, physicians have observed that many common substances patients eat, drink or inhale seem to provoke symptoms.

A biphasic response to some of these substances (Figure 2) has been reported. Initially the individual might experience a stimulatory effect (adapted response; tolerance develops) lasting varying periods of time depending upon the incitant. However, this "up" phase was generally followed by a withdrawal phase (maladapted response; loss of tolerance). Upon beginning to experience unpleasant withdrawal symptoms, the individual would seek, consciously or unconsciously, more of the same substance. These ups and downs follow a sort of sinusoidal (biphasic) pattern, as depicted in Figure 2. On the graph, beginning at zero, the patient is free of symptoms and at baseline health status. Following a one-time or occasional exposure to a provoking substance, stimulatory effects result; after a period of time (minutes to hours to days, depending upon the nature of the incitant), the stimulatory effects subside and give way to withdrawal symptoms. The frequency of these up and down reactions depends upon the frequency of exposures, and the amplitude of the stimulatory and withdrawal portions of the reaction depend upon the substance and the individual's susceptibility (degree of adaptation or addiction) to it. The particularly sensitive person exhibits larger amplitudes than the normals. The key to understanding multiple

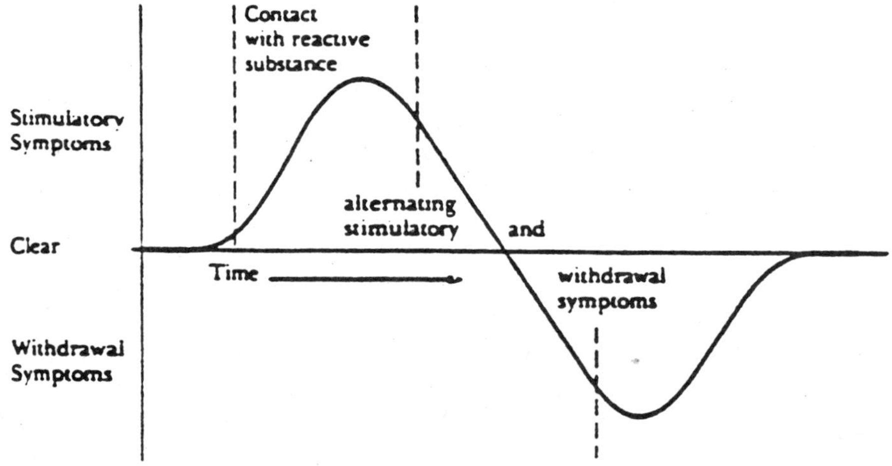

FIGURE 2 Symptom progression of a single reaction to an incitant. During the early phases of exposure to a particular substance, stimulatory symptoms predominate ("up," "hyper," "jittery"). As exposure to the offending agent continues, adaptation occurs and fewer of these symptoms are experienced. With removal from (or discontinuance of) exposure, the individual experiences withdrawal symptoms ranging in intensity from mild to severe. (From O'Banion, D.R., Ecological and Nutritional Treatment of Health Disorders, 1981, p. 68. Courtesy of Charles C. Thomas, Publisher, Springfield, Illinois.)

chemical sensitivity may lie in recognizing these ups and downs that appear to occur after exposure to many different substances. The amplitude of a reaction varies from person to person and incitant to incitant, but the pattern is reported to be quite constant.

After long-term exposure to a given incitant (for instance, alcohol), especially in certain sensitive individuals, the degree and duration of stimulation may become less and less while the withdrawal or depressed phase becomes deeper and more prolonged. At face value, this sinusoidal reaction to a substance might seem a somewhat artificial construct, but Randolph asserts it is not.

Chemical sensitivities may be difficult to assess while a patient remains at home or even in most hospitals because these places generally contain background low levels of natural gas, disinfectants, perfumes, cleaners, tobacco smoke, paints, varnishes, adhesives, and other substances. The patient's symptoms may be masked by the presence of these contaminants.

Under normal living circumstances, the stimulatory and withdrawal levels for foods and chemicals overlap each other (Figure 3) so that in real life - outside an environmental unit - at any given moment what the organism may be feeling is a summation of all effects, whether stimulatory or depressive, of all substances recently inhaled, contacted, or ingested. Figure 3 illustrates that attempts to identify the effects of single substances would be frustrated by the overlapping responses. Only by placing the individual in an environment devoid of chemical and food incitants would one be able to determine whether the illness is alleviated. Assuming the patient improves (which occurs in the majority of cases, according to ecologists), the next step would be to reexpose the person to individual substances in order to avoid overlapping responses, and then to observe the result. If all possible food and chemical contributors are not removed, an effect may be missed. Hence, in order to rule out environmental illness definitively, an environmental unit would be required. Conceivably environmental illness could be ruled in on an outpatient basis, but not ruled out.

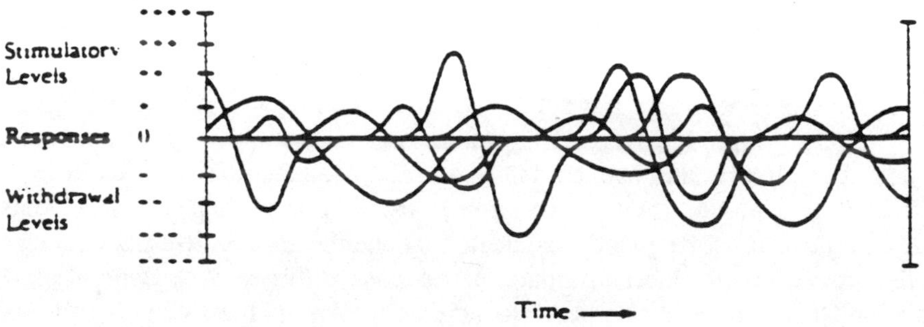

FIGURE 3 Overlapping of responses to food and chemical incitants in an individual with multiple exposures and multiple chemical sensitivities.

In real life, stimulatory and withdrawal reactions are observed but often not understood. For example, an asthmatic might feel well after spending a week on a Carribean island, breathing relatively uncontaminated air and eating a diet devoid of usual foods, only to have a severe, life-threatening asthmatic response to exhaust from the engine of a boat taking the individual home. Once back home in a metropolis, the asthmatic readapts, acclimatizes to auto exhaust, combustion products and other air pollutants in the area, and experiences only chronic wheezing. Thus, following deadaptation (removal from incitants), the individual exhibits a more acute and convincing reaction upon reexposure. This appears to be what occurs in an environmental unit during testing. So acute and convincing are some of these reactions that patients themselves erroneously (at least in the eyes of some) surmise they must have an "allergy" to a particular substance. However, if the patient is not deadapted (unmasked) when tested, a reaction may not occur, convincing the physician that the "allergy" was all in the patient's mind.

Occupational health has several widely recognized examples of adaptation that are analogous (Ashford and Miller 1991). They, too, fit a biphasic pattern. Industrial hygienists and occupation health physicians know that one of the most valuable clues to work-related illness is a history of intense symptoms following return to work after a vacation or weekend (leading to withdrawal and deadaptation).

Ozone, an air pollutant of special concern to residents of Los Angeles and other cities, has been the focus of considerable research relevant to adaptation. Intrigued by how little respiratory illness and death occurred relative to the high levels of ozone in very polluted cities and suspecting adaptation might play a protective role, Hackney and associates (Hackney et al., 1977a) compared the responses of four Canadians (not adapted) and four Californians (adapted) to ozone challenges. Although reactivity varied greatly from individual to individual, Californians were only minimally reactive to levels that for the Canadians caused coughing, substernal discomfort and airway irritation, pulmonary function test decrements, and increased red blood cell fragility.

In another experiment, six volunteers with respiratory hyperreactivity were placed in an environmental chamber with ozone at 0.5 ppm (parts per million), typical of high ambient levels, for 4 days (Hackney et al., 1977b). Five of six had decreased pulmonary function during days 1 to 3, but gradually improved almost to baseline by day 4, suggesting adaptation had occurred. The authors note that not all adverse effects of ozone may be prevented by adaptation; for example, increased red blood cell fragility may persist. Therefore, adaptation or masking of some symptoms may occur while other physiological alterations continue.

Individuals' abilities to adapt to ozone appear to depend upon their initial sensitivity to it. More sensitive persons adapt more slowly and cannot maintain the adaptation as long; they usually remain adapted less than 7 days following cessation of exposure (Horvath, 1981). While nitroglycerin and ozone adaptation (and deadaptation) may differ in certain respects from the adaptation (and deadaptation) described in MCS patients, solvents are among the chemicals most frequently implicated by chemically sensitive patients who attribute the onset of their illness to a particular exposure (Terr, 1989; Cone et al., 1987) and adaptation to solvents has also been documented. Vapors from various solvents are the most prevalent of indoor air contaminants (Molhave, 1982). The volatile organic compounds (VOCs) associated with sick building syndrome are in large part solvent vapors. The sensory irritation, headache, drowsiness, and other symptoms noted by occupants of tight buildings are consistent with known effects of solvent vapors, albeit at much higher concentrations.

Those who have painted or used solvents to any major extent are well aware of the olfactory fatigue (nasal adaptation) that occurs and may have experienced the stimulatory and depressive properties of solvents. Alcoholic beverages contain the solvent ethanol, which has related and familiar stimulatory and withdrawal effects.

Studies of xylene, one of the most prevalent solvents in indoor air, demonstrate that its effects are attenuated as exposure continues, presumably due to adaptation (Riihimaki and

Savolainen, 1980). Riihimaki and Savolainen exposed healthy male volunteers to constant (100 or 200 ppm) and varying (200 or 400 ppm hourly peak) concentrations of xylene, adjusting baseline concentrations in the latter case so that a mean concentration of 100 or 200 ppm was maintained. Exposures occurred over a six-hour period (with a one-hour break at noon) for five days, followed by a two-day weekend and one to three more days of active exposure to xylene. A variety of psychophysiologic parameters were measured, including reaction time, body balance, manual dexterity, and nystagmus.

Of particular interest, Riihimaki and Savolainen (1980) observed that most of the adverse effects of xylene upon their normal subjects "tended to disappear after a few succeeding days of exposure." However, "after the weekend away from exposure, the effects were again discernible." They conclude: "This phenomenon suggests that tolerance had developed over a few days with regard to psychophysiological effects by xylene."

With regard to patients with chemical sensitivities who also develop dietary intolerances, Bell notes that "foods are not only sources of nutrients, but also complex mixtures of organic chemicals. For instance, it is the unique pattern of chemical constituents that make a tomato a tomato rather than an apple" (Bell, 1982, pp. 35-36). Interestingly, limonene and pinene which are present in oranges also are constituents of room air deodorizers which provoke symptoms in some chemically sensitive patients. Like airborne pollutants, foods contain a wide range of chemical constituents and are in intimate contact with the organism for long periods of time. The surface area of the gastrointestinal tract is enormous, and the chemical load, in terms of both quantity and diversity of exposure, is huge.

We have mentioned a number of exposures that are recognized as involving adaptation. What is clear is that individuals with or without multiple chemical sensitivities undergo adaptation to a wide variety of substances in their environment. What is not clear is the specific role adaptation plays in the dramatic responses patients with food and chemical sensitivities have to low-level exposures that do not overtly affect others. These concepts are familiar to occupational health practitioners and industrial hygienists because they observe such effects firsthand among workers exposed to chemicals. Randolph states that most physicians see patients long after adaptation has occurred and at the time when end organ damage is setting in: "It is much as if the physician arrived at the theatre sometime during the last scene of the second act of a three act play--puzzled by what may have happened previously to the principal actor, his patient" (Randolph, 1962, p. 7). Through comprehensive environmental control (that is, an environmental unit), one may be able to overcome the masking effect of adaptation and back up or reverse the exposure to allow monitoring of toxicity in progress. The environmental unit may represent a kind of *dynamic toxicology*; traditional medical approaches provide only a snapshot of what is happening to the patient.

There are several reasons why de-adapting patients is critical to the study and diagnosis of MCS:

1. People are often exposed to dozens of different incitants simultaneously (such as volatile organic compounds in a tight home or building) and literally hundreds of different incitants over the course of a single day, so that health effects of these exposures may overlap, making it difficult to discern cause-and-effect relationships.
2. With continuous or frequent exposure to the same substance or chemically-related substances (such as xanthines in coffee, tea, chocolate and colas), individuals adapt or, in other words, develop tolerance to those exposures. Acute symptoms gradually may give way to chronic symptoms that bear no apparent

relationship to any particular exposure. Exposures may never stop long enough for the patient to reach baseline.

3. Exposures that are initially pleasant or stimulating (such as alcohol, solvents, or nicotine) generally also have withdrawal effects such as headache, depression or irritability, associated with them. Such withdrawal symptoms may occur hours to a few days after cessation of, or reduction in, exposure, greatly confounding attempts by patients and physicians to relate symptoms to a particular incitant.

Comprehensive environmental control, that is, use of an environmental unit, can overcome the masking effect of adaptation and the problems of overlapping exposures that result in overlapping responses to multiple agents. The environmental unit can back up or reverse the experience of adaptation and allow the investigator to monitor toxicity in progress. Figure 4 graphically depicts the changes in symptoms that might occur in a patient after entering an environmental unit. The advantages dynamic toxicology of this nature has over conventional methods for determining toxicity include facilitating detection of subclinical, prepathological effects of chemicals and providing more than just a snapshot of an individual's response to substances. Removing the person from interacting, time-dependent stimuli in this way allows the unraveling of multiple causes. The environmental unit is an essential tool. Many carefully conducted studies of chemical effects that have had negative or equivocal outcomes in the past may have been flawed by their failure to take adaptive mechanisms into account. The potential consequences of such an oversight are major.

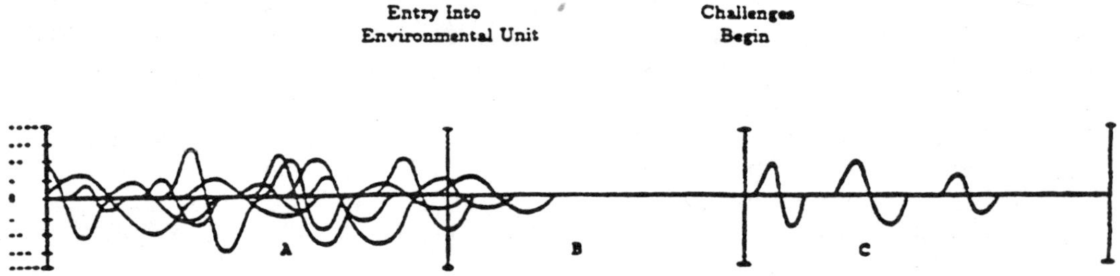

FIGURE 4 Graphical representation of an individuals symptoms before and after entering an environmental unit. In time period A an individual is responding to multiple incitants encountered in normal daily living (chemicals and/or foods), with stimulatory and withdrawal effects that overlap in time. At any particular time, how the person feels is determined not only by ongoing exposures, but by previous exposures whose effects may still be waning. In time period B, the individual enters an environmentally controlled facility, fasting. With cessation of contributory exposures, withdrawal effects occur, for example, headache, fatigue, and myalgias. Symptoms continue for some time (typically for 4-7 days) until the individual reaches "0" baseline. In time period C, single challenges to suspected incitants are administered. Symptoms, often robust, develop soon after challenges, allowing patient and physician to begin to observe the cause-and-effect relationship between exposures and symptoms for that individual.

Important questions that must be addressed in future studies of chemical sensitivity include:

1. Are subjects in a deadapted state prior to challenges so that extraneous

exposures during and prior to a challenge (up to several days before) do not interfere with testing?

2. Are open challenges performed first to confirm that the placebo (clean air or a masking odor such as peppermint) is in fact a placebo and that the "active" challenge is something to which the patient has had demonstrable reactions?
3. What is the recency and latency of the patient's exposure to the substance being tested? In other words, has enough time elapsed (about a week or so) that the person is no longer adapted or reacting to the last exposure but not so much time that the sensitivity has waned? Recency of exposure is recognized as a crucial variable in conducting challenges in patients with occupational asthma, for example.

The rift between allergy and clinical ecology has been fueled by the difficulty inherent in communicating these complex observations concerning adaptation, with unfortunate consequences for patients. An ancient proverb observes "When elephants fight, it is the grass that suffers". When physicians are embattled, it is the patient who suffers. Carefully designed studies of deadapted patients in an environmental unit, using double-blind placebo-controlled challenges, are an essential first step for helping resolve current professional antagonisms and placing research in this field on scientific footing.

SUMMARY OF ADAPTATION HYPOTHESES

Symptoms of exposure to many chemicals, whether inhaled or ingested, appear to follow a biphasic pattern. Adaptation is characterized by acclimatization (habituation, tolerance) with repeated exposures that result in a masking of symptoms. Withdrawal occurs when exposure is discontinued. Once a person has adapted, then the experimental consequences are that further exposures have very little additional effect and therefore may not be observed. The observer may not be able to witness the stimulatory or reactive event because a kind of "saturation" effect has set in.

Adaptation and withdrawal occur for a wide variety of organic and inorganic substances in many physical forms, including various dusts and fumes, solvents, nitroglycerin, ozone, drugs and foods.

An individual is exposed to a variety of substances at different times with varying frequency, duration, and intensity of exposure for each of these substances and with varying frequency and duration of reduction in or cessation of exposure for each substance. The individual may be in different stages (stimulatory or withdrawal) simultaneously for different substances. These stages may overlap and interfere with attempts to observe cause-and-effect relationships.

Adaptation may mask some symptoms or effects while other physiological alterations may continue.

Comprehensive environmental control, that is, use of an environmental unit, can overcome the masking effect of adaptation and the problems of overlapping exposures that result in overlapping responses to multiple agents. The environmental unit can back up or reverse the experience of adaptation and allow the investigator to monitor toxicity in progress. The advantages dynamic toxicology of this nature has over conventional methods for determining toxicity include facilitating detection of subclinical, prepathological effects of chemicals and providing more than just a snapshot of an individual's response to substances. Removing the person from interacting, time-dependent stimuli in this way allows the unraveling of multiple causes.

REFERENCES

Ashford, N. A. and Miller, C. S. 1991. Chemical Exposures: Low Levels and High Stakes. New York: Van Nostrand Reinhold.

Bell, I. R. 1982. Clinical Ecology. Bolinas: Common Knowledge Press.

Cone, J. E., Harrison, R., and Reiter, R. 1987. Patients with multiple chemical sensitivities: clinical diagnostic subsets among an occupational health clinic population. Philadelphia: Hanley & Belfus.

Environmental Protection Agency, Office of Air and Radiation. 1989. Report to Congress on Indoor Air Quality.

Hackney, J. D., Karuza S. K., Linn, W. S. 1977a. Effects of ozone exposure in Canadians and southern Californians, evidence for adaptation? Arch. Environ. Health. 32:110-116.

Hackney, J. D., Linn, W. S., Mohler, J. G., and Collier, C. R. 1977b. Adaptation to short-term respiratory effects of ozone in men exposed repeatedly. J. Appl. Psychol. 43:82-85.

Horvath, S. M., Gliner, J. A. and Folinsbee, L. J. 1981. Adaptation to ozone: duration of effect. Am. Rev. Respir. Dis. 123:496-499.

Johnson, A. and Rea, W. J. 1989. Review of 200 cases in the environmental control unit, Dallas. Presented at the Seventh International Symposium on Man and His Environment in Health and Disease, February 25-26, Dallas, Texas.

Immerman, F. and Schaum, J. 1990. Final Report of the Nonoccupational Pesticide Exposure Study. U.S. EPA, Research Triangle Park.

Klerman, G. L., and Weissman, M. M. 1989. Increasing rates of depression. J. Amer. Med. Assoc. 261: 2229-2235.

Mage, D., and Gammage, R. B. 1985. Evaluation of changes in indoor air quality occurring over the past several decades. In Gammage, R. B. and Kaye, S. (editors). Indoor Air and Human Health. Chelsea: Lewis Publishers.

Molhave, L. 1982. Indoor air pollution due to organic gases and vapors of solvents in building materials. Environ. Internat. 8:117-127.

National Foundation for the Chemically Hypersensitive. 1989. Cheers. 1:6.

O'Banion, D. R. 1981. Ecological and Nutritional Treatments of Health Disorders. Springfield, IL: Charles C. Thomas.

Odell, R., Environmental Awakening. 1980. Cambridge, MA: Ballinger.

Randolph, T. G. 1960. A third dimension of the medical investigation. Clin. Physiology. 2(1):42-47.

Randolph, T. G. 1962. Human Ecology and Susceptibility to the Chemical Environment. Springfield, Illinois: Charles C. Thomas.

Randolph, T. G. 1965. Ecologic orientation in medicine: comprehensive environmental control in diagnosis and therapy. Ann. Allergy 23:7-22.

Randolph, T. G. 1987. Environmental Medicine: Beginnings and Bibliographies of Clinical Ecology. Fort Collins, Colorado: Ecology Publications.

Randolph, T. G. and Moss, R. W. 1989. An Alternative Approach to Allergies. New York: Harper & Row.

Riihimaki, V. and Savolainen, H. 1980. Human exposure to m-Xylene. Kinetics and acute effects on the central nervous system. Ann. Occup. Hyg. 23:411-432.

Selner, J. C. and Staudenmayer, H. 1985. The relationship of the environment and food to allergic and psychiatric illness. Pp. 102-145 in Psychobiology of Allergic Disorders. Young, S. and Rubin, J. editors. New York: Praeger.

Sly, R. M. 1988. Mortality from asthma. J. Allergy Clin. Immunol. 82(5):705-717.

Terr, A. I. 1989. Clinical ecology in the workplace. J. Occupat. Med. 31(3):257-261.

Wallace, L. A., Pellizari, E. D., Hartwell, T. D., Sparacino, C., Whitmore, R., Sheldon, L., Zelon, H. and Perritt, R. 1987. The TEAM study: personal exposures to toxic substances in air, drinking water, and breath of 400 residents of New Jersey, North Carolina, and North Dakota. Environ. Res. 43:290-307.

Design of Animal Models to Probe The Mechanisms of Multiple Chemical Sensitivity

Meryl H. Karol

INTRODUCTION

Recent years have witnessed the emergence of a syndrome called multiple chemical sensitivity (MCS) characterized by recurrent symptoms related to multiple organ systems (1). Symptoms associated with the disorder include: wheeze, cough, shortness of breath, as well as headache, fatigue, depression, confusion, muscle ache, weakness, and gastro-intestinal upset (2). The ailment has been associated with diverse environmental exposures, most frequently to volatile organic solvents (2).

As with many newly recognized medical syndromes, development of an animal model has been suggested as an approach for gaining increased understanding of the pathogenesis of the disorder. Critical design of an animal model is essential to permit both acquisition of meaningful information relative to disease onset and pathogenesis, and to allow scientific acceptance of the model. With MCS, however, the broad range of symptoms, together with their subjective nature, makes it difficult to design an appropriate animal model.

The alternative approach frequently used for development of animal models, simulation of causal exposure conditions, is also fraught with difficulty. Exposures associated with onset of the syndrome are poorly defined and typically involve mixtures of chemicals, glues, solvents, and pesticides (2).

Design of an animal model for pulmonary disease resulting from environmental exposures will be discussed and emphasis placed on features of the model which have application to MCS.

ANIMAL MODELS OF ENVIRONMENTAL LUNG DISEASES

Animal models have been developed for numerous environmental lung diseases including: allergic hypersensitivity (3), acute (4) and chronic byssinosis (5), organic disease toxic syndrome (6), sensory irritation (7), as well as for chronic bronchitis, emphysema and lung cancer (8). Many have proved beneficial in elucidation of: disease pathogenesis (8), progression (5) or in identification of biomakers associated with chemical exposures (9).The characteristics of an ideal animal model have been summarized (Table 1). The use of small animals wherever possible is both economically beneficial and fosters use of the lowest possible phylogenetic species. Exposure of the model should resemble as closely as possible that of the human, to enable distribution of the chemical xenobiotic to appropriate target organs. Lastly, endpoints of toxicity must be carefully selected to resemble, or have relevance to, those observed in humans. Endpoints should occur with low spontaneous incidence.

TABLE 1

Features of an Ideal Animal Model for Environmental Disease

- Use of a small animal species
- Exposure through a natural route
- Similarity in site of lesion development in humans
- Similarity in cell type affected
- Low incidence of toxicity end point occurrence in controls
- Similar biological handling of toxicant (absorption, metabolism, storage, excretion) to humans

Modified from ref. 8.

Attention to design in development of an animal model for pulmonary sensitivity (10) has allowed its use in study of several environmentally related lung disorders. The basic features allowing for exposures and monitoring of responses are presented in Figure 1 and listed in Table 2. Guinea pigs are used since they are readily accessible, small rodents known to demonstrate strong allergic anaphylactic lung responses. Exposure is by inhalation of airborne atmospheres of chemical xenobiotics. Chemical atmospheres are generated in the mixing chamber. The concentration of xenobiotic, as well as the size of aerosol particles is established from samples taken at the ports in the mixing chamber. The desired chemical concentrations are achieved by adjusting the generator feed and airflow exhaust from this chamber.

Animals are exposed to this atmosphere by connecting individual plethysmographs to the mixing chamber using polytetrafluoroethylene (PTFE) tubing. The system is designed to accommodate 4 plethysmographs simultaneously.

Although exposure is predominantly via the inhalation route, dermal contact with the xenobiotic also occurs. Further, through grooming maneuvers and operation of the respiratory escalator, ingestion of the xenobiotic also occurs. Additionally, deliberate exposure through the ingestion route can be provided in this model by incorporating xenobiotics into the food supply placed within each plethysmograph.

During exposure, animals are neither sedated nor restrained within plethysmographs. The benefit derived from this design is the maintenance of a normal body position and therefore an appropriate breathing pattern of animals. This results in the deposition of inhaled particles in the appropriate regions of the respiratory tract and thereby allows

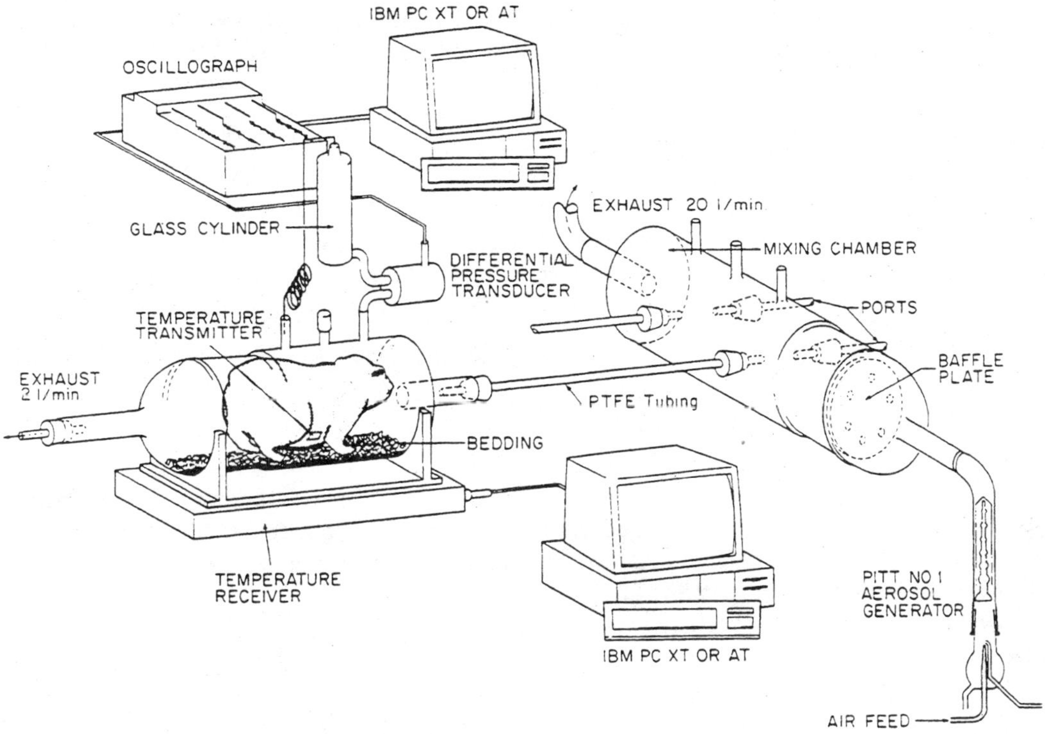

Figure 1 Schematic presentation of the exposure system and equipment for monitoring of pulmonary and febrile responses to airborne xenobiotics. (From ref. 11 with permission).

TABLE 2

Features of an Animal Model for Pulmonary Sensitivity
— Use of guinea pigs (250-400g)
— Exposure by inhalation of airborne chemicals
— Use of unsedated and unrestrained animals
— Avoidance of immunologic adjuvants to achieve sensitization
— Passive recording of responses without disturbance of animals
— Continuous monitoring of pulmonary and febrile responses for 24 hours

suitable biologic handling of xenobiotics (see Table 1).

A second feature of the model is the avoidance of adjuvants to induce immunologic sensitization and of physical restraint within the chamber.

Exposure duration may extend from seconds to hours since atmospheres are continuously created using an aerosol or dust (11) generator.

Monitoring of Responses

An essential feature of this model is the provision for passive monitoring of responses. Continuous recording of breathing pattern is achieved through the use of pressure transducers. Core temperature is monitored using radio frequency transmitters and receivers (see Figure 1). Fever has been associated with several pulmonary syndromes including hypersensitivity pneumonitis and organic dust toxic syndrome.

All readings are collected electronically without disturbance of animals (12), and recorded continuously for up to 24 hrs. The latter feature permits detection of reactions, such as breathing difficulty and fever, which have a variable time of onset. This feature was paramount to identification of late-onset airway hypersensitivity responses to ovalbumin (13) and tuberculin protein (14). The model is applicable for study of illness thought to result from environmental contact.

EVALUATION OF ANIMAL MODELS

Once developed, an animal model must be evaluated. Ideally, the syndrome elicited in the model will closely resemble that reported for the human experience. Such situations generate confidence that mechanisms of pathogenesis in the animal system will have relevance for extrapolation to human disease pathogenesis. The guinea pig model for environmental lung disease has been evaluated for three clinical syndromes as described below.

Respiratory Hypersensitivity

Respiratory hypersensitivity to aeroallergens is characterized by occurrence of immediate, or late-onset airway disturbance (15). Evaluation of the model has been performed by comparison of responses in the guinea pig with those reported for clinical sensitization. Characteristic features of clinical sensitivity are occurrence of early and late onset falls in FEV1 and eosinophilic inflammation of the lung (15). The guinea pig model of pulmonary sensitivity demonstrated each of these responses (12). Animals responded to inhalation provocation challenge with an immediate airway constrictive response (see Figure 2). Recovery was complete within one hour. Continuous monitoring of the breathing pattern revealed a second response with maximum intensity occurring about 5 hours following provocation challenge. This late-onset response resolved over several hours.

At the height of response, histopathologic examination of pulmonary tissue revealed eosinophilic infiltration of the submucosal and epithelial regions of both the large and small airways (12, 13). Since eosinophilic inflammation is a primary characteristic of allergic pulmonary sensitivity, the animal system demonstrated the major physiological and histopathological features of the human syndrome it was designed to model.

The animal model of allergic sensitivity has been additionally evaluated by comparison of immunologic findings in the guinea pigs with those from clinical patients. In experimental studies of TDI hypersensitivity, antibodies produced in guinea pigs were shown to have specificity towards 2,4 and 2,6 TDI isomers (16). Further, the antibody titer was found to vary directly with the concentration of TDI to which animals had been exposed

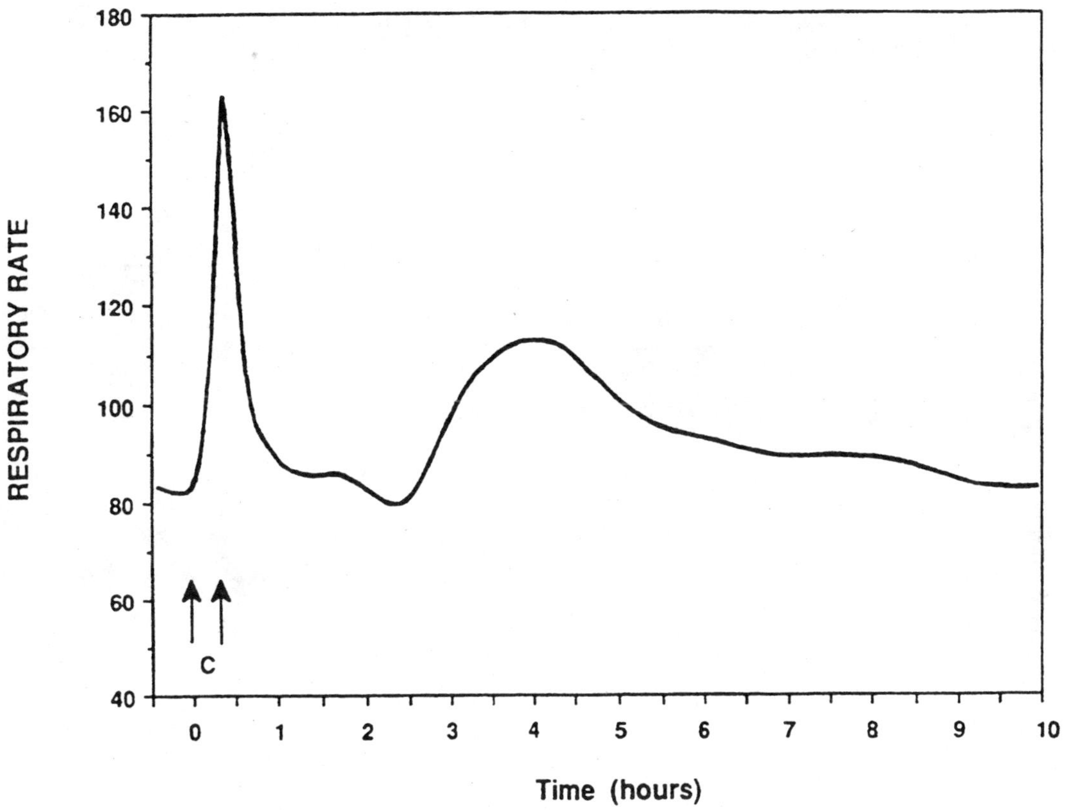

Figure 2 Immediate- and late-onset pulmonary responses of a sensitized guinea pig to ovalbumin. Arrows indicate the period of inhalation challenge. The immediate response occurred within 21 minutes of inhalation challenge. The late-onset response was observed 3.5-4.5 from the start of the challenge exposure. (from ref. 12 with permission).

(17). Both the antibody specificity and its concentration-dependent production were found in sera from individuals occupationally exposed to TDI (16,18).

Guinea Pig Model of Airway Hyperreactivity

One of the cardinal features of asthma is the presence of airway hyperreactivity (AHR) (19). This condition is most frequently assessed as a response to a lower than normal amount of inhaled histamine or methacholine. Study of the pathogenesis of asthma requires methodology for detecting occurrence of AHR. The basic guinea pig model for environmental lung disease was applied to detection of AHR. Methodology was developed to quantify airway reactivity in animals based upon their airway response to increasing concentrations of inhaled histamine (20).

To validate the model, it was necessary to assess its sensitivity in detecting transient AHR associated with an immunologic hypersensitivity response. Accordingly, a study was undertaken to determine whether the newly developed methodology could detect AHR which had been established using conventional methodology (21). Data presented in Figure 3 indicate the occurrence of AHR in each of the experimental animals following the hypersensitivity episode (22).

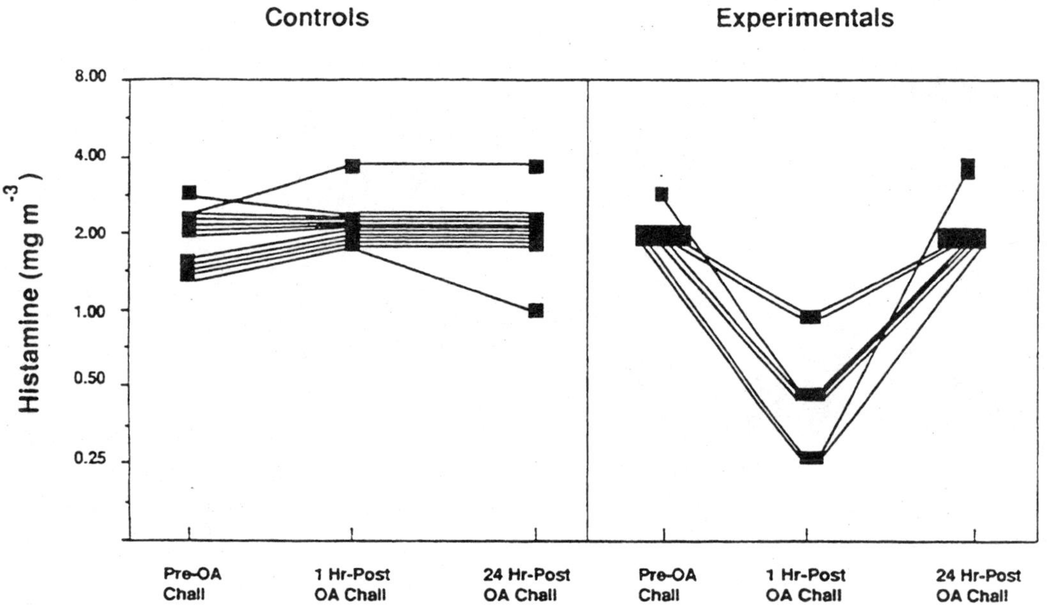

Figure 3 Airway reactivity of control and ovalbumin (OA)- guinea pigs prior to and following inhalation challenge with antigen. Hyperreactivity of experimental animals was apparent 1 hr post OA challenge by the response to lower than normal concentrations of histamine. (from ref. 22 with permission).

The model was further evaluated by comparing its sensitivity for detecting AHR with that of the previously employed methodology. The sensitivity of the two methods appeared comparable. From measurement of dynamic lung compliance and tidal volume (21), reactivity increased from 2.5 to 0.75 mgm-3 histamine required to evoke response. Using the non-invasive methodology (22), reactivity increased from 2.10 mgm-3 histamine to 0.5 mgm-3 required for response. The non-invasive methodology had the additional advantage of accommodating repeated measurements over a prolonged period of time as would be necessary for study of a chronic disease such as asthma.

ANIMAL MODEL FOR ORGANIC DUST TOXIC SYNDROME

A syndrome characterized by fever and neutrophilic alveolitis has been described (23). The disease is associated with environmental exposure to organic dusts and accordingly has been named Organic Dust Toxic Syndrome (ODTS). In order to investigate both the conditions conducive to development of the syndrome, and mechanisms underlying the neutrophilic inflammation, an animal model was sought.

The basic guinea pig inhalation model had features desirable for application to ODTS (Table 2). These features included: capability to provide prolonged exposures via the inhalation route, and to monitor pulmonary and febrile responses which might develop over several hours. In addition, bronchoalveolar lavage could be readily performed in guinea pigs for the evaluation of neutrophilic pulmonary inflammation.

The model was adapted to study of ODTS by exposing animals for 6 hr to atmospheres of cotton dust obtained from a cotton mill in Memphis, TN (6). Monitoring of animals revealed development of the cardinal features of the syndrome, ie., fever, and neutrophilia (Figures 4,5).

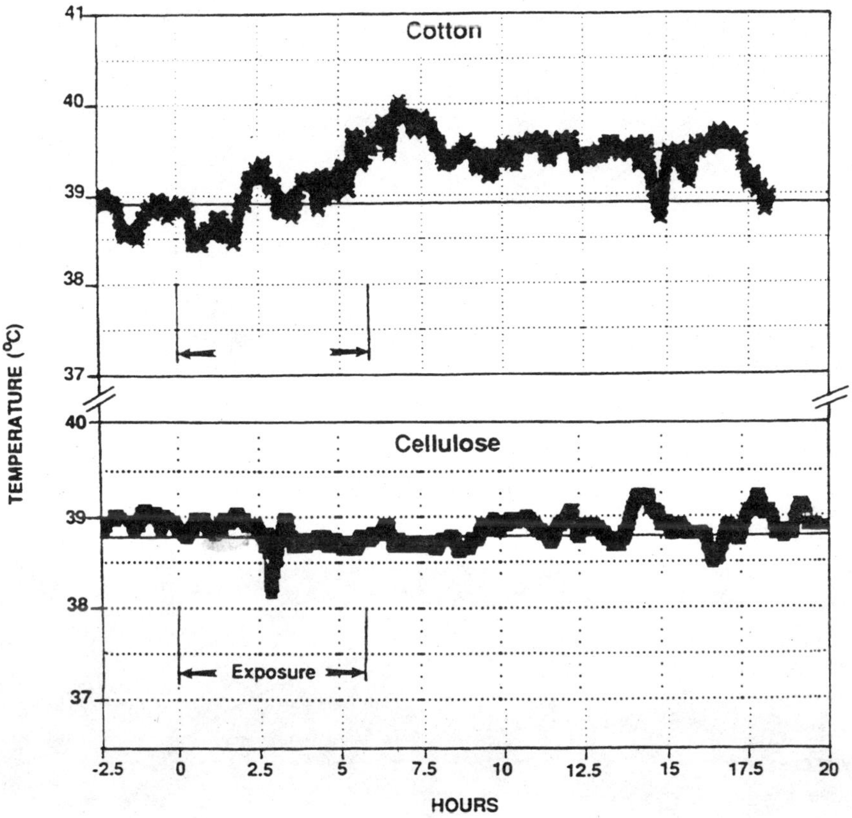

Figure 4 Febrile response of a guinea pig upon inhalation of cotton dust (upper) and lack of response to cellulose dust. Exposure period is indicated by arrows. The febrile response (1.1 oC rise in temperature) was sustained for several hours. (from ref. 6 with permission).

Assessment of the model was achieved by further experiments using "control" dusts. Frequently, in developing animal models, high dosages (or concentrations) of xenobiotics are employed. The rationale for this procedure is based on the fundamental dose-response

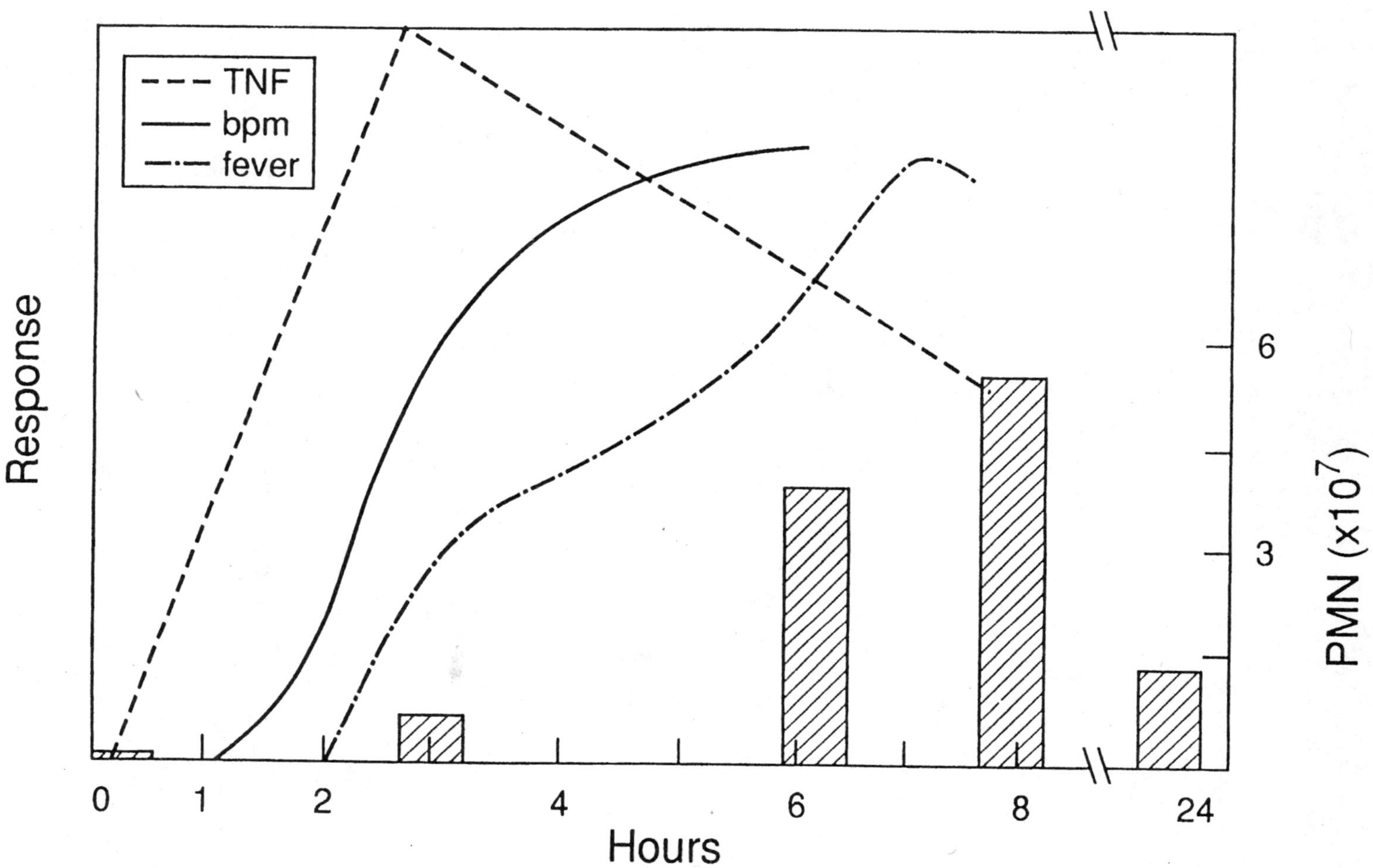

Figure 5 Physiological changes in guinea pigs resulting from inhalation of cotton dust. Hatched bars indicate polymorphonuclear (PMN) leukocyte count (right axis). Left axis: tumor necrosis factor (TNF) activity, breathing frequency, and fever response all shown relative to corresponding pre-exposure values. (from ref. 6 with permission).

principle of toxicology, ie., that if an agent is responsible for an effect, high concentrations of agent will produce sizable responses which can be readily measured. However, high concentrations of agents may create unnatural effects, such as those resulting from overload of biologic systems. For example, inhalation of high concentrations of particulates produced dust overload in rats, resulting in reduced alveolar clearance and altered particle retention (24). The possible occurrence of such effects must be avoided in design of animal models. Experiments must always be performed however, to test for their existence.

Validation of the model for ODTS was achieved by assessing the response of guinea pigs to a high concentration of cellulose dust (22), a dust without reported adverse health effects. Cellulose was selected since it represents the structural matrix of cotton but is devoid of the chemical and biological contaminants usually associated with cotton dust. The absence of both a febrile response, (see Figure 4) and evidence of alveolar inflammation in animals exposed to 41 mgm-3 cellulose, indicated that the response previously detected with cotton dust was not a result of dust overload in the lungs. Rather, it represented pulmonary toxicity resulting from the cotton dust.

AN ANIMAL MODEL OF MCS

Acceptance of an animal model is dependent upon how well the model reproduces the symptom complex associated with the particular disorder. With MCS, the symptoms are highly variable, involving many organ systems. Acceptance of a model for MCS will likely be difficult since there are no objective criteria which are accepted as defining the disorder. Further, the "adaptation" phenomenon, ie., unresponsiveness to acute chemical exposure, implies that negative findings are to be expected and occurrence of symptoms will be unpredictable.

The design of the animal model described above is compatible with study of chronic disease since animals are not instrumented and measurements (such as respiratory patterns and core temperature) are obtained passively and continuously. In addition, blood samples can be obtained readily for immunoglobulin and lymphocyte assessments. RAST (radioallergosorbent) testing can be done to ascertain food intolerance. The model also allows assessment of hypersensitivity, either by skin testing for cutaneous sensitivity, or through provocative inhalation challenge with aeroallergens.

Validation of a model for MCS will be difficult. Which symptoms will be accepted as representing the disorder? Will the absence of immunologic or pulmonary function changes signify a valid consequence of exposure, or an inadequacy of the animal model? Thus, although the animal model described here has many features which make it attractive for use in study of MCS, the absence of responses in animals would be hard to interpret. "Adaptation" could be invoked as a possible reason for unresponsiveness.

A more productive approach for development of an animal model for MCS would appear to be experimental reproduction of exposure conditions associated with development of the disorder. Continuous assessment could be made of physiologic changes (for example, changes in respiratory, neurologic, behavioral or immunologic function). The changes could then be assessed in control animals treated identically to those exposed, but without xenobiotic presence.

It should be noted that reproduction of causative exposure conditions of MCS presents a considerable difficulty since exposures associated with development of the disorder are not well defined. Concentrations of chemicals have not been reported. Moreover, implicated atmospheres are typically variable and complex. One novel solution to this problem would be to introduce animals into the actual environment associated with MCS and to monitor reactions of the animals. However, the uncharacterized and unpredictable nature of exposures associated with development of MCS makes this a difficult undertaking. Simulation of "contaminated" environments appears a more practical yet improbable alternative.

SUMMARY

MCS is a newly recognized disorder associated with exposure to environmental chemicals. Symptoms associated with it are highly diverse and involve many organ systems. Development of an animal model to study the pathogenesis of the disorder is a laudable goal but standard approaches appear implausible. The difficulty lies both with the absence of a clear definition of the disorder and with lack of diagnostic criteria. Further, phenomena associated with the syndrome, such as adaptation, implies that absence of symptoms may be interpreted as a representative "response".

Approaches to animal model development typically focus on either reproduction of responses or of exposures associated with the particular disorder. For MCS, both approaches present difficulties since there is an absence of agreement on diagnostic criteria and little information on agents, lengths of exposure, or number of repeated exposures necessary for development of the disorder.

Animal models have been developed for many environmentally related disorders and have provided information essential to elucidation of disease pathogenesis, diagnosis and treatment. It is hoped that this approach can be applied to study of MCS since much remains to be learned about this very puzzling but increasingly reported disorder.

ACKNOWLEDGEMENT

Support from NIEHS 01532 is gratefully acknowledged.

REFERENCES

1. Cullen, M.R. The worker with multiple chemical sensitivity: an overview. In: Workers with Multiple Chemical Sensitivity. State of the Art Reviews, Occupational Medicine (M.R. Cullen, Ed.) Hanley and Belfus, Inc., Philadelphia, PA, pp. 655-662 (1987).

2. Cone, J.E., Harrison, R. and Reiter, R. Patients with multiple chemical sensitivities: Clinical diagnostic subsets among an occupational health clinic population . In: Workers with Multiple Chemical Sensitivity. State of the Art Reviews, Occupational Medicine (M.R. Cullen, Ed.) Hanley and Belfus, Inc., Philadelphia, PA, pp. 721-738 (1987).

3. Thorne, P.S., Hillebrand, J., Magreni, C., Riley, E.J. and Karol, M.H. Experimental sensitization to subtilisin. I. Production of immediate- and late-onset pulmonary responses. Toxicol. Appl. Pharmacol. 86: 112-123 (1986).

4. Ellakkani, M.A., Alarie, Y.C., Weyel, D.A., Mazumdar, S. and Karol, M.H. Pulmonary reactions to inhaled cotton dust: An animal model for byssinosis. Toxicol. Appl. Pharmacol. 74: 267-284 (1984).

5. Ellakkani, M.A., Alarie, Y., Weyel, D. and Karol, M.H. Chronic pulmonary effects in guinea pigs from prolonged inhalation of cotton dust. Toxicol. Appl. Pharmacol. 88: 354-369 (1987).

6. Griffiths-Johnson, D., Ryan, L. and Karol, M.H. Development of an animal model for organic dust toxic syndrome. Inhal. Toxicol. (1991) (in press).

7. Alarie, Y., Kane, L. and Barrow, C. Sensory irritation: the use of an animal model to establish acceptable exposure to airborne chemical irritants. In: Toxicology: Principles and Practices, Vol. 1. (A.L. Reeves, Ed.) John Wiley and Sons, Inc., New York, NY, 1980, pp. 48-92.

8. Hakkinen, P.J. and Witschi, H.P. Animal Models. In: Toxicology of Inhaled Materials. (H.P. Witschi and J.D. Brain, Eds.) Handbook of Experimental Pharmacology Vol. 75, Springer-Verlag, Berlin Heidelberg, 1985, pp. 95-114.

9. Karol, M.H., Thorne, P.S. and Hillebrand, J.A. The immune response as a biological indicator of exposure. In: Occupational and Environmental Chemical Hazards (V. Foa, E.A. Emmett, M. Maroni, A. Colombi, Eds.) Ellis Horwood Ltd., Chichester, 1987, pp. 86-90.

10. Karol, M.H., Dixon, C., Brady, M. and Alarie, Y. Immunologic sensitization and pulmonary hypersensitivity by repeated inhalation of aromatic isocyanates. Toxicol. Appl. Pharmacol. 53: 260-270 (1980).

11. Weyel, D.A., Ellakkani, M., Alarie, Y. and Karol, M. An aerosol generator for the resuspension of cotton dust. Toxicol. Appl. Pharmacol. 76: 544-547 (1984).

12. Karol, M.H., Hillebrand, J.A. and Thorne, P.S. Characteristics of weekly pulmonary hypersensitivity responses elicited in the guinea pig by inhalation of ovalbumin aerosols. Toxicol. Appl. Pharmacol. 100: 234-246 (1989).

13. Thorne, P.S. and Karol, M.H. Association of fever with late-onset pulmonary hypersensitivity responses in the guinea pig. Toxicol. Appl. Pharmacol. 100: 247-258 (1989).

14. Stadler, J. and Karol, M.H. Experimental delayed-onset pulmonary hypersensitivity: Identification of retest reactions in the lung. Toxicol. Appl. Pharmacol. 65: 323-328 (1982).

15. Frigas, E. and Gleich, G.J. The eosinophil and the pathophysiology of asthma. J. Allergy Clin. Immunol. 77: 527-537 (1986).

16. Jin, R. and Karol, M.H. Specificity of antibodies to toluene diisocyanate identified in workers and induced in an animal model. Am. Rev. Respir. Dis. 143: A439 (1991).

17. Karol, M.H. Concentration-dependent immunologic response to toluene diisocyanate (TDI) following inhalation exposure. Toxicol. Appl. Pharmacol. 68: 229-241 (1983).

18. Karol, M.H. Survey of industrial workers for antibodies to toluene diisocyanate. J. Occup. Med. 23: 741-747 (1981).

19. Armour, C.L., Black, J.L. and Johnson, P.R.A. A role for inflammatory mediators in airway responsiveness. In: Mechanisms In Asthma: Pharmacology, Physiology and Management (C.L. Armour, J.L. Black, Eds.) Alan R. Liss, Inc., New York, NY, 1988, pp. 99-110.

20. Thorne, P.S. and Karol, M.H. Assessment of airway reactivity in guinea pigs: Comparison of methods employing whole body plethysmography. Toxicology 52: 141-163 (1988).

21. Popa, V., Douglas, J.S. and Bouhuys, A. Airway responses to histamine, acetylcholine, and propanolol in anaphylactic hypersensitivity in guinea pigs. J. Allergy Clin. Immunol. 51: 345-356 (1973).

22. Griffiths-Johnson, D. and Karol, M.H. Validation of a non-invasive technique to assess development of airway hyperreactivity in an animal model of immunologic pulmonary hypersensitivity. Toxicology 65: 283-294 (1991).

23. Rask-Andersen, A. Organic dust toxic syndrome among farmers. Brit. J. Ind. Med. 46: 233-238 (1989).

24. Stober, W., Morrow, P.E. and Morawietz, G. Alveolar retention and clearance of insoluble particles in rats simulated by a new physiology-oriented compartmental kinetics model. Fund. Appl. Toxicol. 15: 329-349 (1990).

Antigen Specific and Antigen Non-Specific Immunization

Robert Burrell

The word "hypersensitivity" in the multiple chemical hypersensitivity epithet means quite different things to those depending on whether they be toxicologists or immunologists. The term unfortunately prejudices immunologists because hypersensitivity has very precise connotations based on several decades of molecularly oriented studies which concluded that classical immunological reactions are the result of highly specific molecular interactions between precisely defined prosthetic groups on antigens (epitopes) reacting intimately with either specific immunoglobulins induced by prior contact with that antigen or with T lymphocyte antigen specific receptors (TCR). Although a certain degree of cross reactivity among different molecules can be expected, such may be explained on the basis of molecular structures having closely identical configurational fits. To suggest that chemically, quite-unrelated molecules cross react in the classical immunological sense is utterly foreign to those steeped in the lore of the enormous number of studies of such pioneers as Landsteiner, Heidleberger, Haurowitz, Campbell, and many other immunochemists.
The studies of Landsteiner (1945) on the specificity of serological reactions spanned nearly four decades and one of his innumerable studies is summarized in Fig. 1. His method of diazotizing defined haptens on carrier molecules to obtain antisera and then reacting these antisera with homologous and closely related haptens forms one of the basic foundations of classical immunology.

Later, such reactions were placed on a more quantitative basis by people like Eisen (1990) who used the techniques of physical chemists and enzyme kineticists to study these same kinds of hapten-specific interactions. Through the use of equilibrium dialysis, one can obtain such information as antibody valence and degree of heterogeneity in an antibody population and even the affinity constant of an antigen-antibody reaction (Fig. 2).
From these kinds of studies, it soon became evident that antigens and antibodies did not react by forming covalent bonds, but rather were held together by a number of physical forces. These forces (hydrogen bonding, electrostatic, Van der Waals, etc.) in themselves are non-specific and ordinarily are relatively weak, but become quite strong if the two molecules are very close together. Van der Waals forces for example decrease inversely with the seventh power of the distance separating two molecules. But what allows them to approximate themselves to each other so closely is the fine degree of complementary, struc-

GIVEN: An antiserum to **Meta-Aminobenzene Sulfonic Acid**

Reactive Haptens

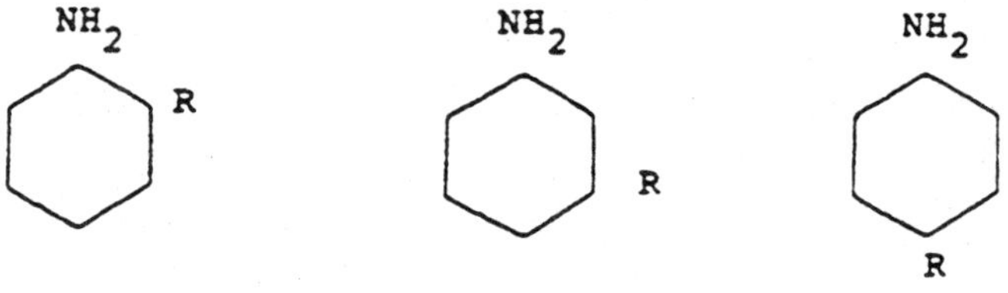

	ORTHO-	META-	PARA-
<u>Rx Antigen</u>:			
Homologous	+	+++	±
Aminobenzene As	-	+	-
Aminobenzoate	-	±	-

If R is large, cross reaction is even less because such a molecule can not make a close fit.

Fig. 1 One of the clasical experiments of Landsteiner illustrating the use of antibodies directed to synthetic haptens. In this case, antibodies were prepared against the hapten, meta-aminobenzene sulfonic acid conjugated to an appropriate carrier. Maximum serologic reactivity was obtained only when this antiserum was reacted with the hapten used in its production and not with molecules that were closely rleated chemically. Further, the specificity even extended to the position on the benzene ring on which the sulfonic group was attached.

tural configuration in the respective molecules. If a closely-related molecule has an additional conformational change, even moving a prosthetic group from a para- to a meta-position (Fig. 1), then that close fit is greatly compromised and there is either no reaction or a very much weaker one.

The availability of myeloma immunoglobulins, for which antigenic reactivity is known, afforded the immunochemist with a homogenous population of antibodies whose light and heavy chain variable domains could be sequenced. Now the exact molecular structure of the antibody as well as the antigen could be known. The amino acid sequence of a polypeptide determines how that peptide will fold three-dimensionally in space and that degree of folding is the antibody characteristic which allows the complementary fit of its homologous antigen. Fig. 3 shows the key amino acids in light and heavy chains that are in most intimate contact with a phosphorylcholine hapten. Essentially, the antigen fits in a very tight cleft formed by the folding of the light and heavy immunoglobulin chains. The lock-and-key analogy used for so many decades to explaining antigen-antibody fit has proven to be remarkably accurate.

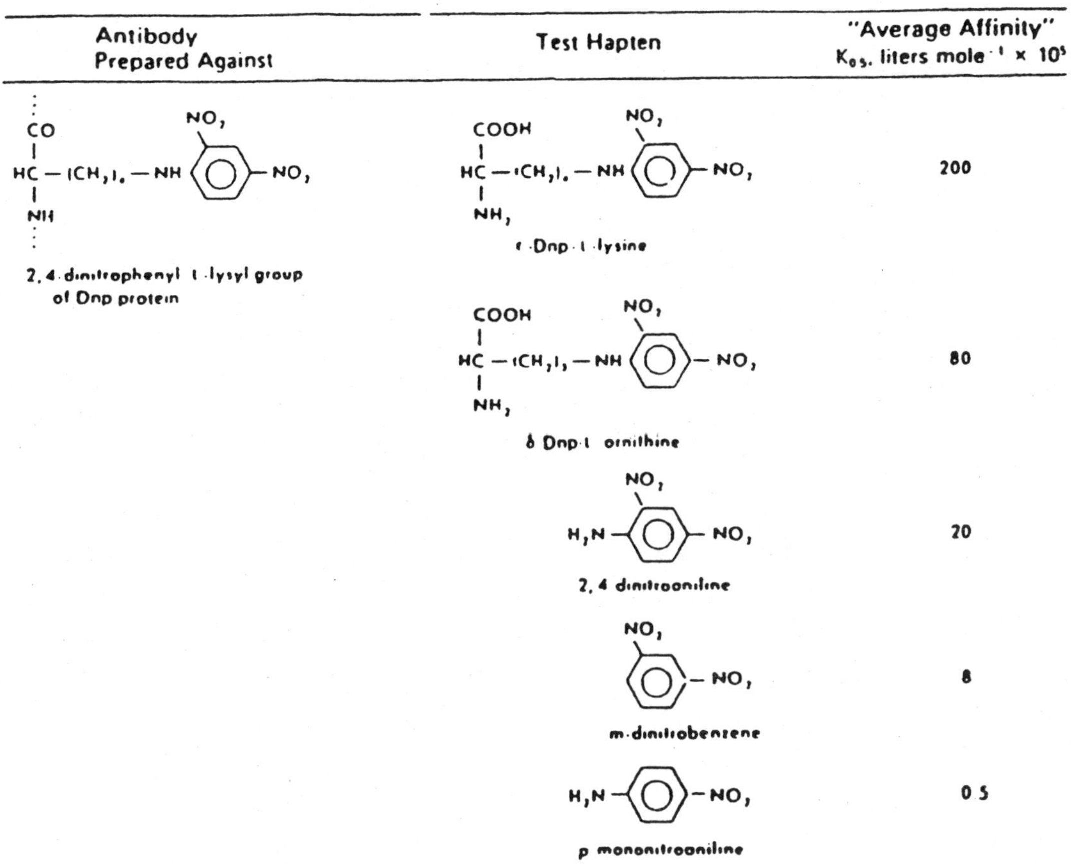

Fig. 2 When serological studies are extended quantitatively, affinity constants can be obtained using equilibrium dialysis methods. Note that the maximum affinity constant occurs only when the antiserum is reacted against the hapten used in its production and that this constant rapidly declines, the further removed the chemically substituted molecules are that are used in reacting against the same antiserum. From H.N. Eisen, General Immunology, 3rd Edition, 1990, Lippincott, Philadelphia, p. 17.

Contemporary molecular immunology is now focussed on the cellular aspects of the immune response, particularly the afferent induction pathways. The pathways involve first contacting the antigen, then processing the antigen so that peptide fragments of the native antigen are presented on the cell membrane surfaces of antigen presenting cells in context with major histocompatibility (MHC) molecules. This presentation is made to T helper lymphocytes who express receptors (TCR) for the presented antigen within this MHC context. Representative TCRs have also been sequenced and their structures can be considered analogous to antibody molecules (Fig. 4). Processed antigen molecules fit within the TCR cleft in a manner analogous with the way antigen fits in an imunoglobulin cleft. Thus even antigen-lymphocyte interactions are governed by the same kinds of intermolecular constraints that antigens and antibodies are subject to.

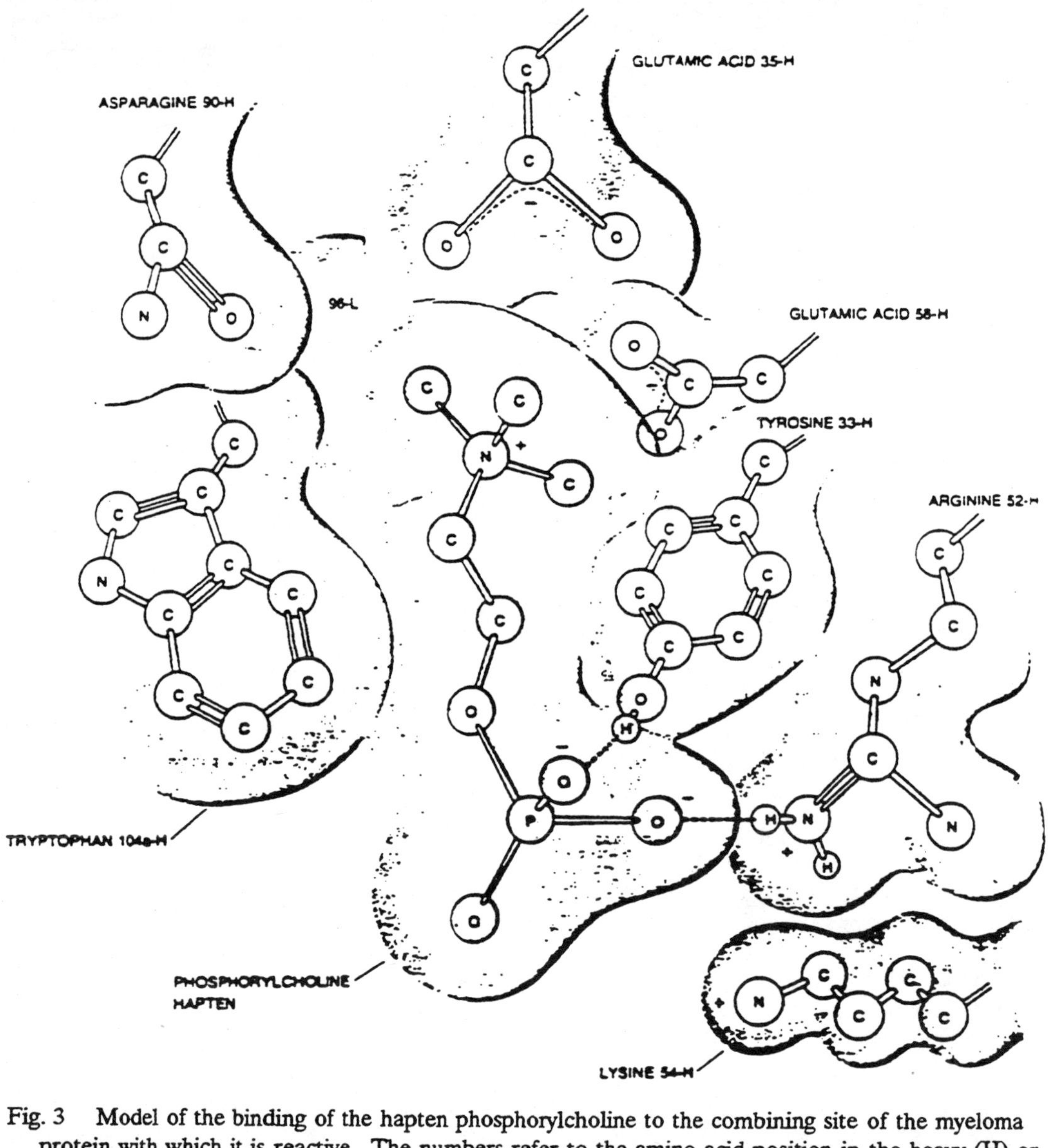

Fig. 3 Model of the binding of the hapten phosphorylcholine to the combining site of the myeloma protein with which it is reactive. The numbers refer to the amino acid position in the heavy (H) or light (L) chain respectively which is in closest approximation with the key polar forces on the haptenic molecule. The way in which the peptide chains of the antibody fold provide for a three-dimensional cleft in which the hapten may be attracted in a very close, conformational fit. From Scientific American, 1977, 236(1):58.

But with the tremendous increase in modern knowledge of immunity has come the realizations that 1) the immune system is far more complicated than intimate reactions between closely-fitting, complementary molecules and 2) that the very same immune system

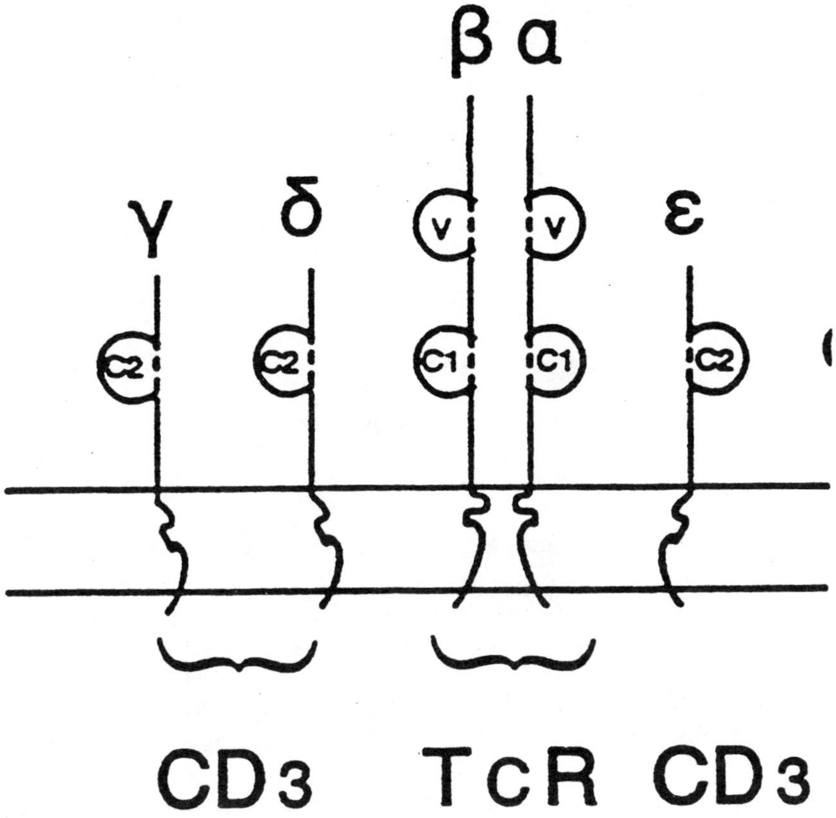

Fig. 4 A stylized model of the T cell receptor (TCR) molecule showing the resemblance to an immunglobulin molecule. It is composed of two chains embedded in the lymphocyte membrane. These chains also have variable and constant domains in a manner analogous to immunoglobulins. The CD3 chains are associated in proximity to the TCR and participate in stabilization of the molecular interactions with specific antigen and antigen- presentation markers.

that recognizes specific antigens also can be stimulated non-specifically to carry out its effects independently of antigens. Knowledge emanating from studies of such matters as the plethora of cytokines required for immunological reactions, the many forces involved in priming and translating immune recognition into inflammation, the pharmacology of immune reactions, the ability of some very important immunological cells to recognize foreignness in an antigen-independent manner (e.g., NK cells), and the growing awareness of how the central nervous system and the immune system are inter-connected are a few examples that suggest that the immune system can play an important role in effecting antigen non-specific reactions as well.

Table 1 compares classical, antigen specific immunity with that of antigen non-specific reactions from several aspects. Non-antigen specific reactions are triggered by many kinds of agents, e.g., polysaccharides that activate the alternative complement pathway in the absence of antibody or tumor cells that activate NK cells. Many types of immuno-

TABLE 1

Comparison of Antigen Specific and Antigen Non-specific Immune Reaction

Feature	Antigen Specific	Antigen Non-specific
Stimulus	Antigen	Immunomodulator
Response measured	Antibodies T_c cells	Mediator production NK cells Pathology
Secondary	Anamnestic	Enhanced, Ag non-spedcific
Longevity	Long, Ag-dependent	Temporary

modulators or substances that mimic allergies can fall into this category. Although one usually measures humoral antibody or T cytotoxic cell responses from antigen-driven reactions, increases in mediator production or responsiveness of target tissues are the parameters usually assessed in non-antigen specific reactions. One of the earmarks of classical immune reactions is the anamnestic response, but increased reactivity can also be brought about by a number of non-antigen specific mechanisms as well; the Shwartzman reaction is a good example. Longevity of antigen specific reactions is mostly dependent on dose and presence of antigen as well as the induction of memory cells. Non-antigen specific reactions are often temporary, that is enhanced responsiveness is only present within a short time range at such time when two independent stimuli have been received by the target tissue. At other times, increased hyperreactivity may be more permanent as a result of a major event such as an infection which then renders the subject hyperresponsive at other times years later, especially when another infection or irritation may take place. Table 2 lists several examples of antigen non-specific reactions.

Classical immunology of course is concerned with the effects of antigen on antigen-presenting cells, all classes of lymphocytes, antigen processing, effector regulation and the like, but these same processes are either closely allied or one and the same as the inflammatory mechanisms generated by the induction of immune responses. Even for antigen specific responses, it has been pointed out by Warren and Chedid (1987) that if immunology is defined as the body's reaction to foreign insult, then one must also consider such non-antigen specific, induced responses as the involvement of endothelial cells, activation of the complement and procoagulation cascades, vascular changes, cytokine generation, lysosomal enzyme release, the generation of toxic oxygen radicals, arachidonic acid metabolism, even the effect of interleukin 1 (IL-1) on the hypothalamus, among many others (Table 3).

The most important factor in trying to explain any hypersensitivity reaction is the matter of dealing with why, of all of those exposed to a given agent, only a small subset respond in such a deleterious manner. Classical allergists at least have an explanation largely in the form of antigen-specific IgE or lymphocytes that have been produced.

Explaining individual idiosyncrasies that are not allergy-based is difficult owing to the

TABLE 2

Examples of Antigen Non-Specific Reactions
Activation of alternative C pathway
Peptigoglycans
Endotoxin
Thrombin
NK cell recognition
Tumor cells
Virus-infected cells?
Acute phase reactants
C-reactive proteins
Interaction with bacterial polysaccharides
Haptaglobin
Anti-inflammatory factor
Low density lipoprotein
Partial detoxification of bacterial endotoxin

dearth of studies on other possible explanations. Pharmacologists recognize idiosyncratic reactions to drugs and account for them by a number of physiologic or pharmacologic host factors that determine reactivity to such agents. Those who manifest hyperresponsiveness to low dose levels have a genetic component that is responsible for either how the drug is metabolized or how that person's cellular/molecular mechanisms respond. These pharmacogenetic bases are further complicated by the numerous types of environmental factors that affect gene expression or the occurrence of the clinical manifestation. An individual with an abnormal gene product would never be noticed unless that person becomes exposed to an agent in which there is an interaction with that product. In these instances, only a certain subset of the population is going to metabolize the agent in such a way as to result in toxic by-products. When an environmental stimulus is additionally required to express the abnormal gene action, as in the case of cutaneous photosensitization in certain porphyrias, clinical manifestations might be erroneously considered immune in origin, but the interesting thing here is the requirement for two separate signals. In addition to genetic factors, other conditions which are known to alter pharmacologic reactivity to substances are age, presence of concurrent disease, nutritional status, hormone levels, diurnal variation in drug disposition or sensitivity, environmental factors that induce or inhibit drug metabolism, and efficiency of repair mechanisms (Nebert and Weber, 1990).

Although a fair amount is known about pharmacogenetics, much less is know about genetic factors that affect physiologic responsiveness. Some of the host factors that definitely have bases contributing to the way individuals hyperrespond to environmental agents are airway hyperresponsiveness to methyl choline challenge, increased propensity of basophils to release histamine, dermatographism, and hyperreactivity to histamine.

In MCS, not only are we faced with the problem of explaining hyperresponsiveness, we

TABLE 3

Examples of Host Responses that Must Also Be Considered Immunological
Endothelial cell reactions
Activation of C cascades in the absence of antibody
Activation of procoagulant activity
Acute phase reactant responses
Aracidonic acid metabolism
NK cell enhancement
Lysosomal enzyme release
Generation of toxic oxygen radicals
IL-1 effect on the hypothalamus
Beta-endorphin activity
Generation of neuropeptides

have an additional problem of explaining how this hyperresponsiveness is directed towards numerous agents with no obvious molecular relationships in a given individual. One of the environmental factors that has been studied more concerning the subsequent development of hyperresponsiveness has been that of association of certain types of infections with triggering development of bronchial asthma particularly in children and to some extent in adults. But even in these instances, other factors are additionally required, e.g., the type of virus causing the infection, the age at which the infection first takes place, the presence of certain constitutional symptoms at the time of infection, genetic predisposition, pre-existing airway hyperreactivity, and sex. Busse (1990) has recently reviewed the possible mechanisms (summarized in Table 4) by which this altered reactivity might occur and among them proposed are that certain viruses, notably respiratory syncitial virus, may enhance IgE synthesis and mediator release from basophils either directly or through the action of interferon from virus-stimulated lymphocytes. The viruses may also interfere with beta-adrenergic function either by directly inhibiting such function, by passively allowing the mast cell to promote more mediator release, or by enhancing cholinergic stimulation of airways smooth muscle. Finally, the virus may desquamate epithelium thus exposing afferent neurons to stimulation and decreasing the amount of an endopeptidase that normally degrades the amount of the peptidergic smooth muscle contractant, substance P. Thus, extrinsic asthma may require 1) previously altered tissue growth/damage from an infectious agent or an irritant, 2) genetic predisposition, and 3) a subsequent trigger later in

TABLE 4

Possible Mechanisms to Explain Asthma Predisposing Factors of Respiratory Infections

1. Viruses induce polyclonic IgE antibody responses
   Genetic predisposition
   Associated with particular viruses, e.g. RSV
   Elevated airway histamine
   Histamine levels correlate with hypoxia

2. Viruses enhance basophil mediator release/chemotaxis
   Can substitute interferon for virus

3. Viruses may diminish beta-adrenergic function
   Could establish a permissive situation promoting mast cell mediator release

4. Viruses promote cholinergic-dependent responses
   Not in smooth muscle directly, but in inflammatory cell mediator release

5. Viruses affect peptidergic responses
   Damage airway epithelium
   Affect bronchodilator effects of LPO products
   Unopposed constriction of Substance P
   Infection inhibits SP degrading enzyme

HYPOTHETICAL SCENARIO

Inflammatory granulocytes, altered by infection
Recruited to the airways
Enhanced mediator release *in situ*
Leads to airway hyperresponsiveness
Must take place in the generally-predisposed

life. We might ask with regard to multiple chemical sensitivity, what the role might be of a previous infection or toxic event at a critical point of development in a genetically-predisposed individual such that neurologically controlled effector mechanisms become subsequently more hyperreactive. But again, in trying to explain MCS, one confronts the observations that its signs and symptoms are not referrable to specific shock organs such as bronchial smooth muscle in asthma. Instead, many of the symptoms are so vague, subjective, and rarely subject to measurement.

Another reason certain individuals might be excessively responsive <u>at a given time</u> might be because primary, incidental stimuli may serve to prime inflammatory mediators that temporarily render that individual more responsive to a second, unrelated stimulus. The coincidental exposure to two or more stimuli within a given time may induce synergistically-altered responses whereas if these same two unrelated stimuli were received at separate times further apart, there would be no abnormal reaction. Much of what we know about this kind of exaggerated responsiveness has been learned from the study of host response to bacterial endotoxin and the prototype of this type of response is the Shwartzman reaction. The endotoxin-induced Shwartzman reaction can be manifest either in a dermal or systemic form--both lead to increased target organ hyperreactivity following exposure to the second stimulus and thus meets the definition of altered reactivity. Neither is antigen specific reaction because the two doses do not need to be antigenically related, the time between sensitization and provocation is too short to induce antigen specific reactivity, and the increased sensitivity does not last more than a few days.

TABLE 5

Examples of "Second Signal" Mediator Responses
Shwartzman reaction
Adult respiratory distress syndrome
Endotoxin enhanced responsiveness
Infection-asthma
Receptors-cAMP/arachidonic acid metabolites
Immunomodulator/adjuvants
Interferon stimulation
Pharmacogenetic hyperresponsiveness

It is now felt that this type of temporal, increased reactivity may be due to several mediator systems being turned on to such an extent by various types of priming stimuli that occur prior to exposure to a further stimulant. Much attention has been directed to the idea of endotoxin-initiated "second signals" wherein endotoxin primes target cells for subsequent enhanced reactivity. This is an exceedingly complex area and could involve several different types of systems including calcium ion channel triggering, activation of cyclic nucleotide synthesis, increased toxic oxygen radical formation, and enhanced cytokine production, e.g., TNF and IFN- (Heremans et al., 1990), among others (see Table 5 for examples). This may relate to multiple chemical sensitivity in that if an individual encounters an immunomodulating or pharmacologically-altering agent at near the same time he encounters a triggering substance, the combined effects of mediators from the two stimuli may synergistically cause an exaggerated reaction.

REFERENCES

Busse, W.W. 1990. Respiratory infections: Their role in airway responsiveness and the pathogenesis of diseases and asthma. J. Allergy Clin. Immunol. 85:671-683.

Eisen, H.N. 1990. General Immunology, pp. 11-18. Philadelphia:J.B. Lippincott.

Heremans, H., J. Vandamme, C. Dillen, R. Dijkmans, and A. Billiau 1990. Interferon-gamma, a mediator of lethal lipopolysaccharide- induced Shwartzman-like shock reactions in mice. J. Exp. Med. 171:1853-1869.

Landsteiner, K. 1945. The Specificity of Serological Reactions. 450 pp. Boston:Harvard University Press.

Nebert, D.W., and W.W. Weber 1990. Pharmacogenetics. In Principles of Drug Action, eds.

W.B. Pratt, and P. Taylor, pp. 469-531. New York:Churchill Livingstone.

Warren, H.S. and Chedid, L.A. 1987. Strategies for the treatment of endotoxemia: Significance of the acute-phase response. Rev. Infect. Dis., 9 (Supp 5), S630-S638.

Neuropsychiatric & Biopsychosocial Mechanisms In Multiple Chemical Sensitivity: An Olfactory-Limbic System Model

Iris R. Bell

INTRODUCTION

The purpose of this paper is to review the clinical and research psychiatric and psychophysiologic literature on multiple chemical sensitivity patients (MCS) and to develop an integrative neuropsychiatric and biopsychosocial systems model of possible mechanisms. Much of the controversy over MCS has focused on hypotheses that the clinical syndromes must derive exclusively either from physiological or from psychogenic sources. However, such a dichotomous view of MCS is overly simplistic. A large body of data from both human and animal studies supports a more complex biopsychosocial approach involving an interplay of multiple influences in the expression of all human illnesses (Engel, 1977; Schwartz, 1982), including the clinical phenomenology of chemical sensitivities (Bell, 1987). This model means that MCS would involve a continuum and interaction of mechanisms in terms of the relative contributions of biological, psychological, and social factors in a given patient. Genetically-based neurochemical and/or receptor vulnerabilities in the central and autonomic nervous systems would make certain subsets of the population more likely to experience adverse effects of low dose chemical exposures.

Furthermore, even psychological and social "stress" can express health effects via biological mechanisms (Shavit et al., 1984; Sapolsky et al., 1990). Most of the major psychiatric disorders have genetic and neurochemical components (e.g. monozygotic twins have a 65-75% concordance vs dizygotic twins with a 14-19% concordance for major depression - Talbott et al., 1988). Thus, evidence for one type of factor (e.g. psychological) cannot rule out the contribution of any other type of factor (e.g. biological) to a clinical syndrome. Appropriate research in MCS must test plausible hypotheses by which a particular factor could causally contribute to or ultimately produce the clinical picture. Phenomenological labels alone fail to meet the latter requirement.

A logical site to examine the potential convergence of biological, psychological, and social factors in MCS is the central nervous system (CNS), especially the interconnected olfactory and limbic systems (Bell, 1982; Ashford and Miller, 1991). The CNS receives input from and sends output to other subsystems of the body, including behavioral, endocrine, immune, and autonomic functions, each of which may be disturbed in certain MCS patients.

The CNS is an integrative center that transduces biological and psychosocial experiences into changes in neural activity; in turn, these send biological, psychological, and social behaviors back out into the environment. Many pesticides and solvents, which are putative triggers for MCS, have effects on central nervous system function (Isaacson et al., 1990; Llorens et al, 1990; Tham et al., 1990) that could interact with individual vulnerabilities.

The olfactory system is the usual pathway by which airborne chemicals from the external environment interact with the brain. The limbic system is the phylogenetically older part of the CNS that receives extensive input from the olfactory system and regulates numerous functions including mood and social behaviors, cognitive function, and eating/drinking/reproductive behaviors (Cain, 1974). Dysfunctions of the limbic system may play a key role in unipolar and bipolar depression, panic and other anxiety disorders, and schizophrenia. Damage to the hippocampus in the limbic system is an important feature of dementia, especially of the Alzheimer type (Sapolsky, 1990).

CONTROLLED STUDIES ON MULTIPLE CHEMICAL SENSITIVITY PATIENTS

At the outset, a model for MCS must be able to account for the clinical phenomenology. Patients with MCS report multiple symptoms in multiple systems. Clinical observations also include initial sensitization with acute high dose exposure, chronic low-dose sensitivity, spreading of number of sensitivities from one to many chemicals, adaptation to chronic exposures, and concomitant multiple food sensitivities (MFS). Little rigorous data are available to characterize the MCS population and their responses to chemical challenges (Tables 1a and 1b).

TABLE 1a.

Multiple Chemical and Food Sensitivity Studies: Patient Differences from Controls

Study	Controls (n)	Patients (n)	Current or past Depression	Cognitive Problems
Doty (1988)	normals (18)	MCS (18)	yes	Problems
Staudenmayer (1990)	normals (55)	MCS (58)	N/A	Not tested
Simon (1990)	chem-exposed normals (23)	MCS (13)	Yes	Not tested
Black (1990)	normals (33)	MCS (26)	Yes	Not tested
Bell (1990)	normals (23)	MFS with MCS (25)	Yes	Yes

TABLE 1b.

Multiple Chemical and Food Sensitivity Studies: Patient Differences from Controls

Study	Controls (n)	Patients (n)	ANS Sx	Other
Doty (1988)	normals (18)	MCS (18)	↑ resp. rate ↓ nasal patency = HR, = BP	N/A
Staudenmayer (1990)	normals (55)	MCS (58)	= skin temp = skin resist	↑ scalp EMG ↑ spectral EEG beta activity
Simon (1990)	normals (23)	MCS (13)	N/A	↑ SCL-90 somatization subscale ↓ Barsky Amplification Scale
Black (1990)	normals (33)	MCS (26)	N/A	↑ anxiety disorders ↑ somatization disorder
Bell (1990)	normals (23)	MFS with MCS (25)	↓ skin temperature in poor cognitive performers	+ tinnitus + daytime sleepiness

Despite extensive debate, only one controlled study has examined the effects of inhaled chemicals on individuals who report the MCS syndrome. Doty et al. (1988) compared the olfactory thresholds and selected autonomic responses to inhalation of phenyl ethyl alcohol and methyl ethyl ketone of 18 MCS patients with a matched control group of 18 normals. They found that the MCS group did not differ from normals on olfactory thresholds or cardiovascular responses to the two test chemicals, but that the MCS patients showed increased respiration rate and decreased nasal airway patency during chemical challenges. This study presented the chemicals in an environmentally controlled test chamber, but did not mask the identity of the chemical exposures or manipulate possible adaptation to the test items. Thus, these data confirm the presence of autonomic differences in MCS patients in terms of respiration and nasal airway function to specific chemical challenges, but do not address the larger clinical syndrome under consideration.

The remainder of the controlled data describes nonspecific phenomenology in subsets of

MCS patients. Most of these studies did not document tests of whether or not chemicals trigger illness in MCS patients. For example, Simon et al (1990) compared cohorts of plastics workers who did and did not develop chronic disabling MCS from low dose exposures to phenol, formaldehyde, and methyl ethyl ketone. They found that 54% of the 13 workers who developed MCS had a prior history of anxiety or depressive disorders in comparison with 4% of control workers. Staudenmayer and Selner (1990) observed higher levels of scalp electromyographic activity in MCS patients and similar patterns of electroencephalographic beta activity in MCS and mixed psychological outpatients in comparison with normal controls; this group reported performing negative controlled chemical challenges on the MCS subjects, but did not present their specific dependent measures or objective findings on this key point. Black's group (Black et al., 1990) reported that 65% of 23 MCS patients had histories of current or past mood, anxiety, or somatoform disorder in comparison with 28% of matched community controls. Although Black suggested that the psychiatric diagnoses could "explain some or all of their symptoms," they did not challenge these patients with chemicals to rule out chemically-induced illness. By consensual agreement within the American Psychiatric Association (Diagnostic and Statistical Manual III-revised, 1987), psychiatric diagnoses are descriptive labels only for phenomenology, not etiological or mechanistic explanations for syndromes. Thus, a psychiatric diagnosis labels a pattern of signs and symptoms, but offers no hypothesis concerning the mechanism(s) of the clinical phenomena (Davidoff et al., 1991). Finally, studies that have found psychiatric diagnoses in MCS patients have used non-random samples of the potentially chemically-sensitive population. Subject selection of MCS patients for research (Simon et al., 1990) and clinical series (Terr, 1986) has often derived from disabled subsamples with worker's compensation and other litigation claims, those self-identified with the field of clinical ecology (Doty et al., 1988; Black et al., 1990), or even those seen in psychiatric clinics (Black et al., 1990). Even in more recognized clinical syndromes involving chronic pain or traumatic injury, those individuals with associated legal and monetary cases represent a unique subset of any given chronically ill population. The discovery of psychiatric problems in a sample drawn in part from those seen in a psychiatric clinic (Black et al., 1990) is clearly a skewed representation of the true prevalence of psychiatric problems in the overall MCS population. Proper epidemiological research on characteristics of MCS will require case identification with more representative sampling from the general, chemically-exposed population.

Moreover, none of the studies that found increased depression, anxiety, and/or somatoform disorder histories compared MCS patients with chronically-ill control groups as opposed to healthy normals. Since chronic illnesses of all types, including cancer, heart disease, and autoimmune disorders, have a current or past prevalence of depression and anxiety disorders ranging from 5% to as high as 45% (general average, 20%) (Katon and Sullivan, 1990), the finding of depression and anxiety would be expected in chronically ill MCS patients and would not establish a specific causal link between the affective disturbance and the chronic illness. In addition, somatic concerns, which constitute clinical criteria for depression and somatoform disorders, are often excluded in research studies of medically ill populations in order to avoid false positive diagnoses of depression on largely somatic symptomatology. In contrast with other investigations on geriatric and medically-ill populations - which use modified interview and/or self-rating scales free of somatic bias or adjust statistically for medically-related somatic complaints, no study of the psychiatric syndromes associated with subsets of MCS patients has ever made any such methodological adjustments.

OLFACTORY-LIMBIC SYSTEM MODEL AND AFFECTIVE SPECTRUM DISORDERS

Affective Spectrum Disorders

Ironically, the finding of psychiatric dysfunctions may nonetheless offer clues to the biology of the disorder in a subset of MCS cases. That is, the mechanisms of depression and anxiety, as well as other psychiatric symptoms, involve putative dysregulation of brain chemistry and neurotransmitter receptors in specific neural pathways (Reiman et al., 1986; Talbott et al., 1988). Thus, one might expect individuals vulnerable to major psychiatric disorders be among the most susceptible to low doses of those environmental chemicals that could worsen their inherent dysfunctions in brain chemistry, either by direct action or by activation of endogenous mediators (Bell, 1982).

For example, animal models of depression include drug-induced elevation of brain acetylcholine, drug-induced depletion of brain serotonin and/or norepinephrine, and destruction of the olfactory bulbs (Jesberger and Richardson, 1985, 1988; van Riezen and Leonard, 1990). Psychiatric investigators have already raised the possibility that the presence of major depression in certain patients indicates an increased neurochemical vulnerability to certain environmental chemicals such as organophosphate pesticides (Rosenthal and Cameron, 1991). The latter agents raise brain acetylcholine levels, an effect which can induce depression in susceptible persons (Dilsaver, 1986).

The overlap between the biology of depression and certain MCS and MFS syndromes is further supported by converging evidence that many of the same disorders that some clinicians have claimed to be triggered by adverse food and chemical reactions also improve during chronic treatment with antidepressant medications. The disorders within medicine that share responsivity to these drugs include major depression, panic disorder, bulimia, obsessive-compulsive disorder, attention-deficit disorder with hyperactivity, migraine headache, irritable bowel syndrome, and cataplexy (Hudson and Pope, 1990). Notably, 87% of patients with the somatization disorder known as Briquet's syndrome (hysteria) also experience major depression during the course of their illness; and 76% of such patients report multiple food intolerances as a criterion for diagnosis (Purtell et al., 1951), another overlap with MCS/MFS phenomenology. In addition, Bell et al. (1990) have recorded tinnitus in 48% of one sample of community-recruited MFS/MCS patients, a condition also known to be responsive to antidepressant medications (Sullivan, et al., 1989). In terms of environmental reactivity, it is striking that a number of studies have found persons with a history of major depression to have an inordinately high prevalence of atopic allergy histories in comparison with both normals and other types of psychiatric patients (Nasr et al., 1981; Sugerman et al., 1982; Bartko and Kasper, 1989; Marshall, 1989; Bell et al., 1991). Such data contrast with the current lack of evidence for atopic mechanisms in MCS or MFS, despite the depression histories. More thorough epidemiological investigation of this point is needed to clarify the prevalence of atopy in depressed versus nondepressed MCS/MFS patients. Interestingly, however, in addition to other mechanisms, several tricyclic antidepressants exert also significant antihistamine effects at both H1 and H2 receptors (Schatzberg and Cole, 1991).

From the above list, controlled studies have indicated that adverse food and ingested chemical reactions may worsen certain cases of ADD with hyperactivity (Egger et al., 1985; Kaplan et al., 1990; Swanson and Kinsbourne, 1980), migraine headache (Egger et al., 1983; Egger et al, 1989), and irritable bowel (Jones, et al., 1982), though some researchers have had negative findings in the latter disorder (Bentley et al., 1983). Thus, it is possible that some brain neurochemical dysfunction susceptible to worsening by adverse food and chemical reactions and to improving with antidepressant drugs is common across a range of disorders, not all purely "psychogenic" in nature. This hypothesis raises two possibilities in identification of possible MCS cases: (a) choosing subjects with personal and family histories loaded for more than one of the above disorders may optimize chances of finding true cases; for example, Egger's group observed that the MFS dietary responders were those with histories of both migraine and seizures, but not with seizures alone (Egger, et al., 1989); (b) screening subjects for those who improve from 4-6 week trials with antidepressant medications, regardless of clinical diagnosis, should assist selection of biologically more homogeneous patients.

Olfactory Bulbectomy Model of Depression

It is also striking that one of the best animal models of depression currently used by drug companies to identify new antidepressant medications is that of olfactory bulbectomy (Jesberger and Richardson, 1988; van Riezen and Leonard, 1990). As in human depression, animals with olfactory bulb damage demonstrate neuroanatomic dysregulation of the limbic-hypothalamic axis, neurochemical imbalances of serotonin, norepinephrine, gamma-amino-butyric acid, and acetylcholine, and improvement from chronic but not acute treatment with antidepressant drugs (Jesberger and Richardson, 1988). Furthermore, olfactory bulbectomy increases the vulnerability of another limbic structure, the amygdala, to kindling (see below). The amygdala is especially important in regulation of affect (e.g. fear-avoidance; rage), drive, and related autonomic/endocrine/immune functions (Mesulam, 1985). Given the role of the olfactory bulb in transmitting sensory and nonsensory information concerning airborne chemicals to the rest of the brain (Cain, 1974), it is reasonable to hypothesize that individuals with dysfunctional olfactory bulb pathways secondary to inherent neurochemical and/or receptor alterations may be the population most sensitive to developing MCS.

Anxiety Disorders and Environmental Chemicals

Persons prone to anxiety may also have inherent neurochemical defects that make them particularly susceptible to certain environmental chemicals. For example, drugs and certain pesticides such as lindane that may antagonize receptors for the inhibitory neurotransmitter gamma-aminobutyric acid (GABA) have apparent anxiogenic properties in animals and humans (Llorens et al., 1990), opposite to those of antianxiety drugs such as the benzodiazepines. Thus, persons with benzodiazepine-responsive anxiety disorders that place them closer to threshold for anxiety due to relatively impaired GABAergic function at baseline might be expected to experience increased anxiety on a biological basis from lindane or related exposure. Another affective spectrum anxiety disorder that a small subset of MCS patients may experience is panic disorder (Dager et al., 1987). Panic disorder is a chronic disorder associated with a broad range of symptoms, especially recurrent acute attacks of difficulty breathing and hyperventilation, palpitations, dizziness, fatigue, and an impending sense of doom. Chemical agents such as low concentrations of carbon dioxide (e.g. 5.5%) in room air and as excess lactate trigger panic attacks in panic patients much more often than in normals under open and blind conditions (Woods, et al., 1986; Roy-Byrne, et al., 1988). In view of the olfactory-limbic model for MCS, it is notable that panic disorder patients have hemispheric asymmetry of blood flow in parahippocampal brain regions on positron emission tomography scans (Reiman et al, 1986). The parahippocampus is a limbic site that receives input from the main olfactory bulb, a major way-station for input from the nose and olfactory nerves. Furthermore, Dager et al. (1987) have hypothesized that occupational solvent exposures could induce panic disorder, possibly by activating a kindling process in the limbic system. Thus, a plausible hypothesis for some cases of environmentally-activated panic disorder is a biologically-based dysfunction of limbic pathways in response to olfactory system input, triggered by certain environmental chemicals.

Kindling and Long-Term Potentiation in the Limbic System: Possible Role in the Mechanisms of MCS

Kindling occurs when repeated subthreshold stimuli summate to trigger seizure activity in brain cells with previously normal activity. The limbic system is especially susceptible to kindling, which may also play a role in the developing chronicity and treatment resistance of certain bipolar affective disorders (Post et al., 1984). In view of the hypothesized responsiveness of affective spectrum disorders to antidepressants, it is notable that these drugs reportedly raise the threshold for or even prevent kindling (cf. Jesberger and Richardson, 1985). Therefore, another, more mechanism-oriented label for the spectrum of disorders that is common to both affective spectrum and MFS/MCS might be kindling-related syndromes potentially responsive to antikindling medications.

Another property of limbic brain cells perhaps related to kindling is that of long-term potentiation (LTP). LTP involves persisting enhancement of synaptic response initiated by brief high-frequency stimulation of excitatory pathways at subictal levels (Racine et al., 1983; Stripling et al., 1988; Jung et al., 1990; Kanter and Haberly, 1990). LTP is hypothesized to subserve memory formation and retrieval in the hippocampal region of the limbic system, but also occurs in olfactory cortex following high-frequency stimulation of the olfactory bulb (Stripling et al., 1988).

Notably, the nature of kindling - and perhaps long-term potentiation - suggests that these biological mechanisms might participate in development and/or expression of MCS - that is, by either initial acute high doses (cf. high doses=high frequency stimuli) or chronic summated low doses (cf. low doses=low frequency stimuli) of environmental chemicals stimulating the firing, long-term potentiation, and/or kindling of olfactory pathways - and from there, many other regions of the limbic system in genetically-predisposed individuals. Indeed, Bokina et al. (1976) reported a rabbit model of low dose environmental pollutant intolerance that matches the latter hypothesis. Their group noted that low concentrations of pollutants combined with a functional load (rhythmic flashing light) triggered abnormal paroxysmal activity in the olfactory bulb and corticomedical amygdala after an initial exposure to high concentrations of the same pollutants. Furthermore, Bokina reported alterations in visual evoked potentials during chronic exposure to low concentrations of formaldehyde and ozone.

In summary, olfactory pathway kindling and LTP might participate in registering information concerning past high dose and/or cumulative low dose chemical exposures that would increase the likelihood of limbic responsivity to subsequent low dose exposures. Resultant dysfunctions in numerous behavioral, autonomic, and endocrine subsystems under limbic regulation might then lead to multiple symptoms through a range of end-organ mechanisms. Finally, it is also possible that chemical exposures which kindle certain limbic pathways could damage capacity for long-term potentiation in the hippocampus (Racine et al., 1983; Armstrong et al., 1990). Such events might in turn disrupt capacity for normal memory formation. These hypotheses offer the possibility of developing animal models for MCS at the level of the central nervous system that would permit greater experimental control and eliminate concerns regarding subject expectation and bias.

Brain Electrical Activity and Chemical Exposures

Although we have no further EEG data on MCS patients or chemically-exposed animals, Lorig and colleagues (Lorig and Schwartz, 1988; Lorig et al., 1988; Lorig, 1989; Lorig et al., 1990a; Lorig et al., 1990b) have performed a series of studies on the effects of low concentration odors on electroencephalographic responses, mood, and cognitive activity in normal human subjects (Lorig et al., 1990a; 1990b). Their findings suggest that chemical odors below the olfactory threshold (i.e. that are consciously undetectable) nonetheless produce distinct EEG responses, worsened mood, and poorer performance on a visual search task (Lorig et al., 1990b). They also examined evoked potentials during a common test of auditory attention and observed increased midline P200 wave amplitude even at an undetected low concentration of a common perfume constituent (galaxolide)(Lorig et al., 1990b). The P200 findings indicated that even undetected airborne chemicals may have a disruptive influence on attention processes. Moreover, in another investigation, they reported that EEG alpha frequency activity in the left hemisphere was lower and EEG beta activity showed greater spatial diversity between posterior and anterior regions during nose breathing than during mouth breathing of unfiltered indoor room air (Lorig et al., 1988). They concluded that undetected odors in indoor air inhaled via the nose exert a distinct effect on brain function outside conscious awareness or changes due to sensory perception (Lorig et al., 1988). Therefore, despite the subcortical location of the olfactory system, its broad connections into the forebrain provide the neuroanatomical and neurophysiological substrate for broad effects of low levels of environmental chemicals on EEG and behaviors regulated by the limbic system.

If such effects are measurable in normals, the olfactory-limbic system model for MCS would predict that the EEG, evoked potential, and cognitive findings would be even more pronounced in MCS patients. Preliminary evidence does support the possibility of abnormalities of attention and concentration in MCS. For example, Bell et al. (1990; submitted for publication, 1991) noted significantly slower performance on a timed mental arithmetic task in outpatients with multiple food and chemical sensitivities than in normal controls. Over 70% of this same sample of patients reported the symptom of difficulty concentrating in comparison with 13% of the controls ($p < 0.008$). In addition, Feidler and Kippen (this volume) have reported evidence of poorer overall performance by MCS patients on the California Verbal Learning Test, a measure of immediate, delayed, and recognition memory also sensitive to attention processes.

In brief, the limbic-olfactory model suggests that even unperceived low concentrations of airborne chemicals can induce changes in brain electrical activity and resultant behaviors and physiology in both animals and human subjects. In certain individuals with particular neurophysiological and/or neurochemical predisposition to limbic system dysfunction (e.g. depression or anxiety disorders), such chemical exposures would activate clinical syndromes described as MCS. The environmental chemicals enter the brain via the olfactory system (Cain, 1974; Shipley, 1985) and would mobilize the abnormal brain processes in the limbic brain with or without sensory awareness via kindling and/or long-term potentiation with initial high dose exposures or with repeated low concentration exposures. One important marker of exogenous chemical olfactory-limbic effects may be abnormalities of attention processes, which could be quantified with neuropsychological and EEG evoked potential tests during undetected chemical as well as sham exposures. Animal models may permit depth electrode recordings with behavioral learning tests, while MCS patients may provide surface computerized EEG and evoked potential data with attention and memory tests to evaluate the above hypotheses (cf. James et al., 1987). Time factors in terms of sequencing high to low dose exposures and of giving repeated low dose exposures may be necessary to elicit the proposed phenomena.

PSYCHOSOCIAL MODULATORS OF BIOLOGICAL EVENTS

Within its descriptive labels, modern psychiatry has embraced a biopsychosocial perspective for diagnosis. The DSM-IIIR (American Psychiatric Association, 1987), for example, employs a multiaxial system to maximize appreciation of the multidimensionality of psychiatric disorders. Thus, axis I covers clinical syndromes, axis II personality disorders, axis III medical disorders, axis IV severity of psychosocial stressors, and axis V global assessment of functioning. Obviously, patients with psychotic delusions of chemical sensitivity, those with only personality disorders, or those with completely incorrect, though nonpsychotic, misattribution of symptoms to chemical reactions may exist within the continuum of patients claiming to have MCS. At least in the case of psychosis, however, none of the studies that assessed psychiatric status reported finding any psychotic type disorders in certain MCS samples (Doty et al., 1988; Black et al., 1990; Simon et al., 1990; Staudenmayer and Selner, 1990). More refined and objective measures for identification of true cases will assist in screening out such individuals as well as any cases of nondelusional misattribution from MCS study populations.

Nevertheless, illnesses generally have psychosocial modulators, which interact with underlying biological vulnerabilities. These modulators include: premorbid personality and coping style (McCrae and Costa, 1986; Jamner et al., 1988); perceived social support (Cohen and Syme, 1985); perceived familiarity of the surrounding environment (Siegel et al., 1982); sense of control over the illness (Rodin, 1986; Sanderson et al., 1989); and classical conditioning of responses to both triggering and treating agents (Eikelboom and Stewart, 1982; Bolla-Wilson, et al., 1988; MacQueen et al., 1989). As emphasized above, identification of these factors in patients with MCS would be expected. For example, MCS patients have organized local and national support groups, which may have both positive benefits in terms of providing buffering effects from social support networks and negative effects in terms of reinforcing illness behaviors (Bell, 1987). They merit specific study to optimize clinical care and to understand the MCS syndromes, but demonstration of these psychosocial phenomena would not by itself "disprove" a role for biological causes of the syndromes.

At the same time, coincidental or deliberate manipulation of these factors would be likely to exert a marked influence on course and outcome in various individuals. Even in the case of terminal cancer, for example, a defiant and demanding versus submissive personality style (Levy, 1985) and participation in a cancer patient support group therapy versus no group therapy (Spiegel et al., 1989) are associated with longer survival times. Providing nursing home residents with a sense of control over their immediate physical surroundings and daily activities substantially reduces subsequent mortality rates over those of residents without such sense of control (Rodin, 1986). In the case of affective spectrum panic disorder, providing patients with an illusion of control over exposure to carbon dioxide will attenuate the severity of the carbon dioxide-induced anxiety (Sanderson et al., 1989), even though in other blinded studies the CO2 clearly can trigger panic over and above that set off by sham inhalation

(Woods et al., 1986). At the same time, perceived stress leads to immune dysfunctions in human (Kiecolt-Glaser and Glaser, 1987) and animal (Shavit et al., 1984) subjects. The administration of opiate drugs to opiate-tolerant animals in a strange cage versus a familiar cage doubles the mortality from the same high dose (64% versus 32%; whereas 96% of drug-naive animals die) (Siegel et al., 1982). The mediating mechanisms of these types of effects may involve hormonal, central nervous, and autonomic nervous system elements. Without accounting for the entire clinical situation in MCS, many similar factors might still contribute to the day-to-day course of the syndrome.

Possible Mediators in MCS

In modern health psychology and psychosomatic medicine, it is becoming apparent that the mind/brain and body have a two-way communication system that utilizes biological messengers to implement the effects of "stress" on the body and to change mood and behavior. These include common hormones and other mediators such as cortisol (Sapolsky et al., 1990), interleukins (Sapolsky et al., 1987; Spadaro and Dunn, 1990), vasopressin (Weingartner et al., 1981), substance P, vasoactive intestinal peptide, prostaglandins, kinins, histamine, and opiate-like peptides (enkephalins, endorphins) (Sloviter and Nilaver, 1987; Wiedermann, 1987). For instance, prostaglandins are elevated in major depression (Calabrese et al., 1986; Ohishi et al., 1988), interleukin 1 in allergy and infection (Walter et al., 1989), and opiate-like peptides in sleep apnea (Gislason et al., 1989) and depressive disorders (Agren et al., 1982). PGE2 induces wakefulness (Matsumura et al., 1989); IL-1 produces somnolence and reduces exploratory behavior (Walter et al., 1989; Spadaro and Dunn, 1990); opiate-like peptides facilitate antigen-induced histamine release and lower natural killer cell activity (Shavit et al., 1984; Mediratta et al., 1988).

Many of these same agents participate in more purely physiological events (e.g. inflammation, infection, atopy). However, multiple pathways that originate with biological as well as psychosocial triggers can set off release of these mediators. Once they act at their end-organs, the clinical effects appear similar, regardless of the nature of the original trigger. It is reasonable to hypothesize that many of these mediators contribute to symptom production in MCS from multiple initiating pathways. Clinical symptom pictures in MCS patients (e.g. MFS/MCS daytime somnolence - see Bell et al., 1990 versus somnolence-inducing properties of prostaglandin D2 and interleukin 1) and data on mediators in syndromes overlapping MCS can guide the selection of possible endogenous agents for study.

Classical Conditioning Hypothesis

The hypothesis of classical conditioning of symptom flares to olfactory stimuli represents a potential intersection between psychological and biological mechanisms. For example, antigen exposure in a sensitized animal will mobilize histamine release on a physiological basis. If this event is paired in time with a previously neutral event (e.g. a physiologically-inactive odor), subsequent release of histamine in response to the classically conditioned odor without antigen will occur (Russell et al., 1984; MacQueen et al., 1989). Numerous drug responses including those to opiates and insulin can also be classically conditioned, although the direction of the conditioned response is often opposite from that to the direct biological stimulus (Eikelboom and Stewart, 1982).

In all classical conditioning, it is important to note that the phenomenon requires an initially biologically active stimulus to pair with the previously inactive stimulus. As stated earlier, the existence of classically conditioned responses does not rule out the co-existence of more directly biologically-initiated responses in the same individual. Thus, while it is important to explore classical conditioning of chemical responses in MCS patients and normals (cf. Kirk-Smith et al., 1983; Lorig and Roberts, 1990), the first priority of MCS research must be the biological effects of chemicals. Once the parameters of the biological effects are clarified, then possible psychosocial modifiers and their associated biological mediating mechanisms should be examined experimentally.

METHODOLOGICAL ISSUES

Methodological issues in the study of MCS are critically important. While the scope of this paper is limited to multiple chemical sensitivities, the body of controlled literature on multiple food sensitivities offers valuable clues to the proper design of studies on chemicals and chemically-sensitive patients (Pearson et al., 1983; Egger et al., 1985; Kahn et al., 1988, 1989; Kaplan et al., 1989). Table 2 summarizes the concerns regarding design, subject prescreening, sample size, placebo controls, blinding, duration and timing of avoidance and/or challenges, number of independent measures, and sensitivity of dependent measures. Importantly, the intense debate on the testing and treatment technique known as provocation-neutralization (King, 1988) should be held separately from the more fundamental questions concerning the nature and mechanisms of adverse chemical reactions. For the present, research on MCS should emphasize delivery of real-life-like concentrations via customary routes of exposure (i.e. nose inhalation, dietary ingestion) rather than shifting focus to any particular testing technique or unusual route of administration.

TABLE 2

Methodological Issues

Issue	Less Sensitive	More Sensitive
design	no control condition	between or within subjects
subject prescreening	all subjects with diagnosis	comparison
# subjects	smaller	subset with diagnosis
placebo control	too complex - active	larger
blinding	no blind or single blind	adequate - inactive
duration of study	single challenge	double blind
# independent measures	one chemical or food	weeks of challenges
sensitivity of dependent	dichotomous global + or -	multiple items
measures	for reaction	continuous variables

From the MFS literature, we have learned that it is necessary to prescreen subjects for current - not merely historical - reactivity to the proposed test agents. Under experimental conditions, researchers have generally not detected effects when studying whole populations - but have found effects when studying subsets of prescreened populations. The latter point raises statistical power considerations in the experimental design (King, 1985). If the effects are limited to a subset of a population, the less certainty we have in identifying these cases on entry into the study necessitates the use of larger sample sizes to avoid Type II error (missing an effect that is actually present by failing to choose enough responders as subjects).

It is also clear that blinding and selection of placebo conditions are difficult issues in a highly sensitive sample of subjects. It would be unfortunately easy to generate a placebo that is physiologically active, though otherwise indistinguishable from the intended active agent. Lorig's research on normal human subjects suggests a possible strategy, which involves a two-step procedure: (i) determination of olfactory thresholds for each test substance in the study subjects, (ii) experiments using sub-threshold concentrations of the test agent (Lorig et al., 1990a,b). This approach avoids sensory awareness and expectations, as well as the complication of possible biological reactivity to added masking odors intended to be inactive while hiding the actual active test agent.

With the background noise of chronic symptomatology and chronic everyday chemical exposures, the experimental approach in MFS also suggests ways to maximize emergence of test reactions under research conditions. That is, investigators need (a) to use repeated, multiple challenges over extended periods of time rather than a single challenge on one day (Kahn et al., 1988, 1989; Kaplan et al., 1989); and (b) to employ multiple rather than single chemical agents for tests (Kaplan et al., 1989). Patients who are in fact chemically sensitive to various substances may nonetheless hold incorrect beliefs about their reactivity to any specific chemical on a single day for a wide range of reasons, including the reality that no chemical exposure in real life is experienced separate from thousands of others in ambient air. Thus, Molhave's approach of starting investigation of tight building subjects and normals with mixtures of volatile organic compounds similar to those found in indoor air may be the most appropriate first step in MCS research as well (Molhave et al., 1986; Molhave 1990; Otto et al., 1990). MCS studies performed in environmentally controlled units with measurably low levels of ambient chemical noise may encounter fewer of these issues of background exposures in their design, especially with subjects resident for weeks at a time.

Finally, it is extremely important to select appropriate and sensitive dependent measures. It is wasteful to perform a controlled study, only to ask subjects whether or not they believe they are having adverse reactions (Pearson et al., 1983). Such dichotomous measures are notoriously insensitive, lower statistical power, and are likely to create Type II error of missing an actual effect of the chemical challenges. In addition to objective laboratory tests such as EEG, evoked potentials (Lorig, 1989), polysomnographic sleep studies (Kahn et al., 1988), blood and cerebrospinal fluid changes in putative mediators and their receptors, and neuropsychological cognition tests (Swanson and Kinsbourne, 1980; Molhave et al., 1986; Lorig et al., 1990b), researchers have a host of standardized self-report and observer-rated scales from medicine and psychology to grade changes in physical and emotional symptoms, behaviors, and mood along a continuum. All such measures offer the possibility of more sensitive continuous rather than dichotomous dependent variables for study.

CONCLUSIONS

In view of the obvious role of the olfactory system in communicating information about the chemical environment to the brain and thereby to the rest of the body, the olfactory-limbic model outlined above is a plausible alternative to the previous emphasis on putative immunological mechanisms for MCS. As indicated, neurons of the limbic brain have excitability properties that could provide the basis for amplification and spreading of adverse reactions to low dose chemical exposures. Furthermore, numerous psychosocial modifiers could participate in the overall clinical course of the illness via their transduced effects on the brain and the bodily functions it regulates. Overall, as is common in the clinical practice of biopsychosocial patient care, the first priority should be to study the direct biological effects of low dose chemical exposures on health in the subset of individuals vulnerable to multiple chemical sensitivity. The next priority is to examine interactions between the biological and psychosocial dimensions of MCS. Eventually, well-done investigations will permit the objective evaluation of the relative proportions of the variance in MCS symptomatology to which biological, psychological, and social factors contribute.

Despite the controversies surrounding multiple chemical sensitivities, it is possible to develop experimentally testable hypotheses concerning the nature and mechanisms of MCS. This work requires synthesis of approaches and information from multiple fields, including occupational medicine and toxicology, internal medicine, health psychology, basic and applied neurosciences, and pharmacology. In this area, it is ultimately insufficient to frame questions in terms of exclusively biological versus psychosocial causes or to show a fit or lack of fit between known clinical entities and MCS. The problems raised by chemical sensitivities demand scientific thought beyond the boundaries of existing fields. The solutions are likely to expand our knowledge base in novel ways toward understanding interactions between human beings and their environment.

REFERENCES

Agren, H., L. Terenius, A. Wahlstrom 1982. Depressive phenomenology and levels of cerebrospinal fluid endorphins. Annals N.Y. Acad. Sci. :388-398.

American Psychiatric Association 1987. Diagnostic and Statistical Manual of Mental Disorders. Third Edition, revised. Washington, D.C.: American Psychiatric Association.

Armstrong, D.L., J.L. Polan-Curtain, T. Osaka, M.R. Murphy, S.Z. Kerenyi, S.A. Miller, S.L. Hartgraves 1990. Central nervous system protection against effects of long-term low-dose nerve agents. Report for USAF School of Aerospace Medicine, Human Systems Division (AFSC), Brooks Air Force Base, Texas (USAFSAM-TR-90-31), U.S. Naval Medical Research and Development Command.

Ashford, N.A., C.S. Miller 1991. Chemical Exposures. Low Levels and High Stakes. N.Y.: Van Nostrand Reinhold.

Bartko, J.J. and S.Kasper 1989. Seasonal changes in mood and behavior. A cluster analytic approach. Psychiatry Res. 28:227-239.

Bell, I.R. 1982. Clinical Ecology: A New Medical Approach to Environmental Illness. Bolinas, CA: Common Knowledge Press.

Bell, I.R. 1987 Environmental illness and health: the controversy and challenge of clinical ecology for mind-body health. Advances 4(3):45-55.

Bell, I.R., Markley, E. J., B. Zumalt, D.S. King 1990. Psychological and somatic symptom and illness patterns in individuals with self-reported adverse food reactions. Biol. Psychiatry 27:168A.

Bell, I.R., M.L. Jasnoski, J. Kagan, D.S. King 1991. Depression and allergies: survey of a nonclinical population. Psychother. Psychosom. 51, in press.

Bentley, S.J., D.J. Pearson, K.J.B. Rix 1983. Food hypersensitivity in irritable bowel syndrome. Lancet 2:295-297.

Black, D.W., A. Rathe, and R.B. Goldstein 1990. Environmental illness. A controlled study of 26 subjects with '20th century disease'. J. Amer. Med. Assoc. 264:3166-3170.

Bokina, A.I., N.D. Eksler, A.D. Semenenko, and R.V. Merkur'yeva 1976. Investigation of the mechanism of action of atmospheric pollutants on the central nervous system an comparative evaluation of methods of study. Environ. Health Perspectives 13:37-42.

Bolla-Wilson, K., R.J. Wilson, M.L. Bleecher 1988. Conditioning of physical symptoms after neurotoxic exposure. J. Occupational Med. 30:684-686.

Cain, D.P. 1974. The role of the olfactory bulb in limbic mechanisms. Psychological Bull.

81:654-671.

Calabrese, J.R., R.G. Skwerer, B. Barna, A.D. Gulledge, R. Valenzuela, A. Butkus, S.Subichin, N.E. Krupp 1986. Depression, immunocompetence, and prostaglandins of the E series. Psychiatry Research 17:41-47.

Cohen, S. and S.L. Syme, eds. 1985. Social Support and Health. New York: Academic Press, Inc.

Dager, S.R., J.P. Holland, D.S. Cowley, D.L. Dunner 1987. Panic disorder precipitated by exposure to organic solvents in the work place. Am. J. Psychiatry 144:1056-1058.

Davidoff L.L., T.J. Callender, L.A. Morrow, G. Ziem 1991. Letter to the Editor. J. Psychosom. Res. 35(4):1-3.

Dilsaver, S.C. 1986. Cholinergic theory of depression. Brain Res. Rev. 11:285-316.

Doty, R.L., D.A. Deems, R.E. Frye, R. Pelberg, A. Shapiro 1988. Olfactory sensitivity, nasal resistance, and autonomic function in patients with multiple chemical sensitivities. Arch. Otolaryngol. - Head & Neck Surgery 114:1422-1427.

Egger, J. C.M. Carter, J. Wilson, M.W. Turner, J.F. Soothill 1983. Is migraine food allergy? A double-blind controlled trial of oligoantigenic diet treatment. Lancet 2:865-869.

Egger, J., C.M. Cater, J.F. Soothill, J. Wilson 1989. Oligoantigenic diet treatment of children with epilepsy and migraine. J. Pediatr. 114:51-58.

Eikelboom, R. and J. Stewart 1982. Conditioning of drug-induced physiological responses. Psychological Rev. 89:507-528.

Engel, G.L. 1977. The need for a new medical model: a challenge for biomedicine. Science 196:129-136.

Gislason, T., M. Almquist, G. Boman, C. Lindholm, L. Terenius 1989. Increased CSF opioid activity in sleep apnea syndrome. Regression after successful treatment. Chest 96:250-254.

Hudson, J.I. and H.G. Pope 1990. Affective spectrum disorder: does antidepressant response identify a family of disorders with a common pathophysiology? Am. J. Psychiatry 147:552-564.

Isaacson, L.G., S.A. Spohler, D.H. Taylor 1990. Trichloroethylene affects learning and decreases myelin in the rat hippocampus. Neurotoxicology Teratology 12:375-381.

James, L., A. Singer, Y. Zurynski, E. Gordon, C. Kraiuhin, A. Harris, A. Howson, R. Meares 1987. Evoked response potentials and regional cerebral blood flow in somatization disorder. Psychother. Psychosom. 47:190-196.

Jamner, L.D., G.E. Schwartz, H. Leigh 1988. The relationship between repressive and defensive coping styles and monocyte, eosinophile, and serum glucose levels: support for the opioid peptide hypothesis of repression. Psychosom. Med. 50:567-575.

Jesberger, J.A. and J.S. Richardson 1985. Animal models of depression: parallels and correlates to severe depression in humans. Biol. Psychiatry 20:764-784.

Jesberger, J.A. and J.S. Richardson 1988. Brain output dysregulation induced by olfactory bulbectomy: an approximation in the rat of major depressive disorder in humans. Intern. J. Neuroscience 38:241-265.

Jones, V.A., P. McLaughlan, M. Shorthouse, E. Workman, J.O. Hunter 1985. Food intolerance: a major factor in the pathogenesis of irritable bowel syndrome. Lancet 2:1115-1117.

Jung, M.W., J. Larson, G. Lynch 1990. Long-term potentiation of monosynaptic EPSPs in rat piriform cortex in vitro. Synapse 6:279-283.

Kahn, A., G. Francois, M. Sottiaux, E. Rebuffat, M. Nduwinana, M.J. Mozin, J. Levitt 1988. Sleep characteristics in milk-intolerant infants. Sleep 11:291-297.

Kahn, A., M.J. Mozin, E. Rebuffat, M. Sottiaux, M.F. Muller 1989. Milk intolerance in children with persistent sleeplessness: a prospective double-blind crossover evaluation. Pediatrics 84:595-603.

Kanter, E.D., L.B. Haberly 1990. NMDA-dependent induction of long-term potentiation in afferent and association fiber systems of piriform cortex in vitro. Brain Research 525:175-179.

Kaplan, B.J., J. McNicol, R.A. Conte, H.K. Moghadam 1989. Dietary replacement in preschool-aged hyperactive boys. Pediatrics 83:7-17.

Katon, W, M.D. Sullivan 1990. Depression and chronic medical illness. J. Clin. Psychiatry 51[6, suppl]:3-11.

Kiecolt-Glaser, J.K. and R. Glaser 1987. Psychosocial moderators of immune function. Annals Behavioral Medicine 9:16-20.

King D.S. 1985. Statistical power of the controlled research on wheat gluten and schizophrenia. Biol. Psychiatry 20:785-787.

King D.S. 1987. The reliability and validity of provocative food testing: a critical review. Med. Hypotheses 25:7-16.

Kirk-Smith, M., C. Van Toller, G. Dodd 1983. Unconscious odor conditioning in human subjects. Biological Psychology 17:221-231.

Levy, S.M. 1985. Behavior and Cancer. San Francisco: Jossey-Bass.

Liskow, B., E. Othmer., E.C. Penick, C. DeSouza, W. Gabrielli 1986. Is Briquet's syndrome a heterogeneous disorder? Am. J. Psychiatry 143:626-629.

Llorens, J., J.M. Tusell, C. Sunol, E. Rodriguez-Farre 1990. On the effects of lindane on the plus-maze model of anxiety. Neurobiology Teratology 12:643-647.

Lorig, T.S. 1989. Human EEG and odor response. Progress in Neurobiology 33:387-398.

Lorig, T.S. and G.E. Schwartz 1988. Brain and odor: I. Alteration of human EEG by odor administration. Psychobiology 16:281-284.

Lorig, T.S., G.E. Schwartz, K.B. Herman 1988. Brain and odor: II. EEG activity during nose and mouth breathing. Psychobiology 16:285-287.

Lorig, T.S., K.B. Herman, G.E. Schwartz 1990a. EEG activity during administration of low-concentration odors. Bulletin Psychonomic Soc. 28:405-408.

Lorig, T.S., E.Huffman, A. DeMartino, J. DeMarco 1990b. The effects of low concentration odors on EEG activity and behavior. J. Psychophysiology 5, in press.

Lorig, T.S., M. Roberts 1990. Odor and cognitive alteration of the contingent negative variation. Chemical Senses 15:537-545.

MacQueen, G., J. Marshall, M. Perdue, S. Siegel, J. Bienenstock 1989. Pavlovian conditioning of rat mucosal mast cells to secrete rat mast cell protease II. Science 243:83-85.

Marshall, P. 1989. Attention deficit disorder and allergy: a neurochemical model of the relation between the illnesses. Psychol. Bulletin 106:434-446.

Matsumura, H., K. Honda, W.S. Choi, et al. 1989. Evidence that brain prostaglandin E2 is involved in physiological sleep-wake regulation in rats. Proc. Natl. Acad. Sci. 86:5666-5669.

McCrae, R.R. and P.T. Costa 1986. Personality, coping, and coping effectiveness in an adult sample. J. Personality 54:385-405.

Mediratta, P.K., N. Das, V.S. Gupta, P. Sen 1988: Modulation of humoral immune responses by endogenous opioids. J. Allergy Clin. Immunol. 81:27-32.

Mesulam, M.M. 1985. Principles of Behavioral Neurology. Philadelphia: F.A. Davis Co.

Molhave, L. 1990. Volatile organic compounds. Indoor air quality and health. Indoor Air 5:15-33.

Molhave, L., B. Bach, and O.F. Pedersen 1986. Human reactions to low concentrations of volatile organic compounds. Environment International 12:167-175.

Nasr, S., E.G. Altman, H.Y. Meltzer 1981. Concordance of atopic and affective disorders. J. Affective Disord. 3:291-296.

Ohishi, K., R. Ueno, S. Nishino, T. Sakai, O. Hayaishi 1988. Increased level of salivary prostaglandins in patients with major depression. Biol. Psychiatry 23:326-334.

Otto, D., L. Molhave, G. Rose, H.K. Hudnell, D. House 1990. Neurobehavioral and sensory irritant effects of controlled exposure to a complex mixture of volatile organic compounds. Neurotoxicology Teratology 12:649-652.

Pearson, D.J., K.J.B. Rix, S.J. Bentley 1983. Food allergy: how much in the mind? A clinical and psychiatric study of suspected food hypersensitivity. Lancet 1:1259-1261.

Post, R.M., D.R. Rubinow, J.C. Ballenger 1984. Conditioning, sensitization, and kindling: implications for the course of affective illness. Pp. 432-466 in Neurobiology of Mood Disorders, R.M. Post and J.C. Ballenger, eds. Baltimore: Williams and Wilkins.

Purtell, J.J., E. Robins, M.E. Cohen 1951. Observations on clinical aspects of hysteria. A quantitative study of 50 hysteria patients and 156 control subjects. J. Am. Med. Assoc. 146:902-909.

Racine, R.J. N.W. Milgram, S. Hafner 1983. Long-term potentiation in the rat limbic forebrain. Brain Research 260:217-234.

Reiman, E.M., M.E. Raichle, E. Robins, F.K. Butler, P. Herscovitch, P. Fox, J. Perlmutter 1986. The application of positron emission tomography to the study of panic disorder. Am. J. Psychiatry 143:469-477.

Rodin, J. 1986. Aging and health: effects of the sense of control. Science 233:1271-1275.

Rosenthal, N.E., Cameron, C.L. 1991. Exaggerated sensitivity to an organophosphate pesticide. Am. J. Psychiatry 148:270.

Roy-Bryne, P.P. and T.W. Uhde 1988. Exogenous factors in panic disorder: clinical and research implications. J. Clin. Psychiatry 49:56-61.

Russell, M., K.A. Dark, R.W. Cummins, et al. 1984. Learned histamine release. Science 216:436-437.

Sanderson, W.C., R.M. Rapee, D.H. Barlow 1989. The influence of an illusion of control on panic attacks induced via inhalation of 5.5% carbon dioxide-enriched air. Arch. Gen. Psychiatry 46:157-162.

Sapolsky, R., C. Rivier, G. Yamamoto, P. Plotsky, W. Vale 1987. Interleukin-1 stimulates the secretion of hypothalamic corticotropin-relasing factor. Science 238:522-524.

Sapolsky, R.M., H. Uno, C.S. Rebert, C.E. Finch 1990. Hippocampal damage associated with prolonged glucocorticoid exposure in primates. J. Neuroscience 10:2897-2902.

Schatzberg, A.F., J.O. Cole 1991. Manual of Clinical Psychopharmacology, 2nd ed. Washington, D.C.: American Psychiatric Press, Inc.

Schwartz, G.E. 1982. Testing the biopsychosocial model: the ultimate challenge facing behavioral medicine? J. Consult. Clin. Psychol. 50:1040-1053.

Shavit Y., J.W. Lewis, G.W. Terman, R.P. Gale, J.C. Liebeskind 1984. Opioid peptides mediate the suppressive effects of stress on natural killer cell cytotoxicity. Science 223:188-190.

Shipley, M.T. 1985. Transport of molecules form nose to brain: transneuronal anterograde and retrograde labeling in the rat olfactory system by wheat germ agglutinin-horseradish peroxidase applied to the nasal epithelium. Brain Res. Bull. 15:129-142.

Siegel, S., R.E. Hinson, M.D. Krank, J. McCully 1982. Heroin "overdose" death: contribution of drug-associated environmental cues. Science 225:733-734.

Simon, G.E., W.J. Katon, and P.J. Sparks 1990. Allergic to life: psychological factors in environmental illness. Am. J. Psychiatry 147:901-906.

Sloviter, R.S., G. Nilaver 1987. Immunocytochemical localization of GABA-, cholecystokinin- vasoactive intestinal peptide-, and somatostatin-like immunoreactivity in the area dentata and hippocampus of the rat. J. Compar. Neurol. 256:42-60.

Spiegel, D., J.R. Bloom, H.C. Kraemer, E.Gottheil 1989. Effect of psychosocial treatment on survival of patients with metastatic breat cancer. Lancet 2(8673):888-91.

Spadaro, F. A.J. Dunn (1990). Intracerebroventricular administration of interleukin-1 to mice alter investigation of stimuli in a novel environment. Brain, Behavior, and Immunity 4:308-322.

Staudenmayer, H. and J.C. Selner 1990. Neuropsychophysiology during relaxation in generalized, universal 'allergic' reactivity to the environment. A comparison study. J. Psychosomatic Res. 34:259-270.

Stripling, J.S., D.K. Patneau, C.A. Gramlich 1988. Selective long-term potentiation in the pyriform cortex. Brain Research 441:281-291.

Sugerman, A.A., D.L. Southern, J.F. Curran 1982. A study of antibody levels in alcoholic, depressive, and schizophrenic patients. Annals Allergy 48:166-171.

Sullivan, M.D., R.A. Dobie, C.S. Sakai, W.J. Katon 1989. Treatment of depressed tinnitus patients with nortriptyline. Ann. Otol. Rhinol. Laryngol. 98:867-872.

Swanson, J.M., M. Kinsbourne 1980. Food dyes impair performance of hyperactive children on a laboratory learning test. Science 207:1485-1487.

Talbott, J.A., R.E. Hales, S.C. Yudofsky, eds. 1988. The American Psychiatric Press Textbook of Psychiatry. Washington, D.C.: American Psychiatric Press, Inc.

Terr, A.I. 1986. Environmental illness: a clinical review of 50 cases. Arch. Intern. Med.

146:145-149.

Tham, R., B. Larsby, B. Eriksson, M. Niklasson 1990. The effect of toluene on the vestibulo- and opto-oculomotor system in rats, pretreated with GABAergic drugs. Neurotoxicology Teratology 12:307-311.

van Riezen H. and B.E. Leonard 1990. Effects of psychotropic drugs on the behavior and neurochemistry of olfactory bulbectomized rats. Pharmac. Ther. 47:21-34.

Walter, J.S., P. Meyers, J.M. Krueger 1989. Microinjection of interleukin-1 into brain: separation of sleep and fever responses. Physiol. & Behavior 45:169-176.

Weingartner, H., P. Gold, J.C. Ballenger, S.A. Smallberg, R. Summers, D.R. Rubinow, R.M. Post, F.K. Goodwin 1981. Effects of vasopressin on human memory functions. Science 211:601-603.

Weidermann, C.J. 1987. Shared recognition molecules in the brain and lymphoid tissues: the polypeptide mediator network of psychoneuroimmunology. Immunol. Lett. 16(3-4):371-378.

Woods, S.W., D.S. Charney, J. Loke, W.K. Goodman, D.E. Redmond, G.R. Heninger 1986. Carbon dioxide sensitivity in panic anxiety. Arch. Gen. Psychiatry 443:900-909.

Neurobehavioral and Psychosocial Aspects of Multiple Chemical Sensitivity

Nancy Fiedler and Howard Kipen

Over the past decade an increasing number of patients have presented with a symptom complex that has been labeled multiple chemical sensitivities or environmental illness. These patients have been characterized to have persistent or recurrent somatic and psychological symptoms attributed to chemical exposures. The symptoms reported and their duration, however, are not consistent with the accepted toxicological properties of the chemicals. Schottenfeld and Cullen (1986), using DSM-III-R diagnostic criteria, initially described such patients to have one of three psychiatric disorders: 1) typical post-traumatic stress disorder; 2) atypical post-traumatic stress disorder; or 3) somatoform disorder. Commensurate with interest expressed in the lay and professional literature, further attempts have been made to characterize these individuals. Cullen (1987) edited an Occupational Medicine State of the Art Review in which he proposed the following relatively rigorous definition: 1) symptoms were acquired in relation to some initial identifiable environmental exposure(s); 2) the symptoms involve more than one organ system (e.g. respiratory and nervous system); 3) the symptoms recur and abate in response to predictable stimuli; 4) the symptoms are elicited by exposures to chemicals of diverse structural classes and toxicologic modes of action; 5) the exposures that elicit symptoms are very low, i.e. many standard deviations below the "TLV" and at levels not known to cause adverse human responses; 6) no single widely available test of organ system function can explain symptoms. This definition was proposed to differentiate MCS patients from patients with post-traumatic stress disorder and somatoform disorder.

Up to this point, subjects reported in the literature have not fully met this more rigorous definition (Brodsky, 1983; Terr, 1986; Bolla-Wilson, et al., 1988; Black et al., 1990;Rosenberg et al., 1990; Simon et al., 1990). For example, Simon et al. (1990) labeled a group of subjects who were currently in workers compensation litigation as chemically sensitive based solely on self-reported lifestyle modifications. Black et al. (1990) reported on psychological profiles of a group of subjects diagnosed as chemically sensitive by physicians practicing clinical ecology. The triggering event for these patients was not uniformly environmental (e.g. stress) nor was the associated diagnosis always chemical sensitivity. For example, some subjects were diagnosed as having chronic candidiasis. Because of the significant regulatory implications of "chemical" sensitivities a need exists to determine if a subset of patients can be found who meet the more rigorous definition proposed by Cullen (1987). If such patients can be identified, it is also important to develop a standardized method for evaluating these patients. Therefore, the purpose of the present paper is to document that this relatively narrow group of patients can be identified and to describe the standardized battery of tests we have chosen to characterize these patients.

CHARACTERISTICS OF MCS PATIENTS

Over the past two years approximately 25 patients have presented at our Environmental and Occupational Health Clinical Center with symptoms suggestive of chemical sensitivity. That is, all patients reported becoming symptomatic to low levels of multiple chemicals/substances encountered in their daily lives. Each patient underwent a comprehensive medical examination by an occupational physician. This consisted of a lifetime and current medical history, review of previous medical records, brief psychiatric history, physical examination, and routine hematology, blood chemistry, thyroid and urine studies. Further testing such as spirometry or chest x-ray were done as needed to document organic conditions which might reasonably explain the symptomatology. While previous organic conditions were often present (e.g. asthma) patients were not excluded if these conditions did not explain the current symptoms.

With regard to psychiatric history, patients who reported the following psychiatric diagnoses on their medical questionnaire or during a medical history were not included in the more narrowly defined group of MCS patients:

1. Evidence of current psychosis, organic brain syndrome, mania, or major depression.
2. History of treatment, within 10 years of onset of symptoms attributed to toxic exposure, for psychosis, organic brain syndrome, hypomania, major depression, somatoform disorder, dissociative disorder, phobia, panic disorder, post-traumatic stress disorder, obsessive compulsive disorder, or personality disorder.

Also, individuals were excluded from the rigorously defined subset of chemically sensitive patients if the following were present: 1) in litigation at the time of the evaluation; 2) in treatment with a clinical ecologist. While such patients may have chemical sensitivity, for various reasons these factors may confound the interpretion of the evaluation measures.

For example, patients involved in litigation or being treated by a clinical ecologist may have altered symptom patterns based on these factors. That is, patients in litigation may have a vested interest in appearing more symptomatic while the treatments or information provided by a clinical ecologist could potentially change a patient's symptom profile.

Based on the above criteria, eleven of the twenty-five patients qualified as presenting symptoms consistent with chemical sensitivity while not having other current or previous conditions or psychosocial factors that could account for their symptoms.

Among those patients who did not qualify, eight did not strictly meet the Cullen (1987) criteria. That is, on careful consideration either they did not have symptoms that wax and wane with exposure or they could not identify a specific point in time or exposure event after which their symptoms began.

Among the remaining six patients, two were involved in litigation, and one had a previous psychiatric history of major depression with electroconvulsive treatment. Three others were not available for further study due to logistic reasons, e.g. moved out of state. Thus, eleven patients met the full criteria proposed by Cullen (1987) and did not have the additional factors outlined previously which could confound an understanding of chemical sensitivity.

Chemically sensitive patients were three men and eight women ranging in age from 28 to 57 with a mean age of 42 for men and 43 for women. Educational level ranged from 12 to 16 years with a mean of 15 years for men and 14 years for women. Two male patients had previously worked in the chemical industry while the third was currently working as a draftsman. Two of the eight women were currently working as secretaries and two were administrative assistants, while the remainder worked as a teacher, bus driver, health educator, and sales clerk. All patients were employed at the time their symptoms began. All of the male patients worked with significant chemical or solvent exposures and first experienced symptoms in response to a change in ventilation or in a process at work, although none experienced a particular episode associated with accidental release or acute intoxication. The identified precipitating exposures and subsequent symptoms for five of the eight women occurred at work. For the remaining female patients, precipitating exposures began outside of work. The nature of the identified precipitating exposures included laying new carpet (N=4), indoor air quality complaints in an office environment (N=2), an adverse reaction to a prescribed drug (N=1), and a home pesticide application (N=1).

PSYCHOSOCIAL AND PSYCHIATRIC EVALUATION

Neurobehavioral Assessment Measures:

The primary neurobehavioral symptoms reported by MCS patients include impaired memory, concentration, and visuomotor speed and coordination. Standardized tests of these functions were selected to assess MCS patients and to compare their performance to a normative group of the same age and sex.

Tests of Memory:

A. California Verbal Learning Test (Delis, et al., 1987) - short and delayed recall of common words with an interference trial:
This test assesses the strategies an individual employs to encode information as well as immediate recall. Therefore, it is sensitive not only to deficits in the amount remembered but also to the way in which material is remembered. It also provides an indicator of short- and long-term memory as well as memory following interference from other tasks. Thus, this test more closely approximates the kind of memory required in day-to-day functioning and the kind of memory deficits often cited among chemically exposed subjects.

B. Digit Span (Wechsler, 1981) - short-term recall of an increasing string of digits:
This test is a part of the World Health Organization's (WHO) core battery of tests recommended for assessment of neurobehavioral effects of toxic chemicals. It has age specific normative data and has been used widely in the psychological and exposure assessment literature. It provides an evaluation of immediate memory for digits which may

require different skills from the California Verbal Learning Test which requires memory for common words.

C. Wechsler Memory Scale, Visual Reproduction I & II (Wechsler, 1987) short- and long-term memory for designs:

This task requires drawing visual designs from memory and, therefore, assesses the ability to reproduce rather than simply recognize designs.

Tests of Concentration:

A. Digit Span test to assess concentration:

In addition to memory, this test also provides an assessment of the ability to concentrate.

B. Stroop Color and Word Test (Trenerry et al. 1989) - determines the ability to concentrate with interference from outside stimuli:
This task is more complex assessing the ability to concentrate when interfering stimuli are presented. As such, it provides a more realistic assessment of concentration. It has been successful in distinguishing patients with brain dysfunction and, therefore, should provide a useful addition to our assessment of dysfunction due to exposure.

Visuomotor Coordination:

A. Grooved Pegboard Test (Trites, 1981) - determines fine motor coordination in a timed situation:
This task is used frequently as an indicator of fine motor coordination. Savage et al. (1988) as well as other investigators have found significant deficits on this task among exposed subjects.
B. Digit Symbol (Wechsler, 1981) - requires sustained attention and coordination between eye and hand to record symbols associated with specific digits;
This task is one of the World Health Organization's recommended tasks for the detection of brain dysfunction due to toxic exposure. It is particularly sensitive to brain dysfunction at a minimal level (Lezak, 1983).

Performance on the above tests of neurobehavioral function can be confounded by differences in age, sex, race, and intelligence (Valciukas, 1985). That is, performance deficits may be due to these variables rather than due to exposures or trauma from exposures. Normative information for the tests cited above has taken into account differences in performance due to age and sex. However, the normative information available generally does not take into account differences in performance due to general intellectual ability. Therefore, a test of verbal ability is also administered as an estimate of overall intellectual ability. It is generally assumed that basic verbal abilities such as vocabulary and well-rehearsed general information will not be affected by acute or transient exposures to toxic substances (Hartman, 1988). Significant discrepancies in performance between a test of reading or vocabulary and those of concentration, memory, and visuomotor skills may help validate deficits reported by patients clinically. The reading portion of the Wide Range Achievement Test - Revised (Jastak and Wilkinson, 1984) was chosen to assess general word knowledge and as an estimate of overall intellectual ability.

Psychiatric Assessment:

Since many of the symptoms reported by MCS patients may also reflect psychological distress, several standard measures were selected to assess psychiatric symptomatology. The Structured Clinical Interview for the Diagnostic and Statistical Manual of Mental Disorders - Third Edition - Revised (SCID-III-R - DSM-III-R) (Spitzer et al., 1990) is administered to assess current and previous psychiatric symptomatology and diagnoses.

Questions are based on the criteria set forth in the DSM-III-R for the diagnosis of psychotic, mood, substance use, anxiety, somatoform, eating, and adjustment disorders. Evaluation of psychopathology prior to the development of MCS is also important since previous psychiatric status is often cited as a precursor to the development of MCS (e.g. Simon et al., 1990). To serve this purpose, the SCID-III-R interview was modified so that patients are specifically asked if their symptoms had existed prior to the development of chemical sensitivities. Where the SCID-III-R covers only current episodes, it was modified by asking patients if they had any occurrence of the symptoms prior to the development of their current chemically related symptomatology.

Since previous history of somatization disorder is particularly important to consider among these patients, the Diagnostic Interview Schedule-III-A (DIS-III-A) section for somatization disorder is also separately administered to assess lifetime prevalence of somatization disorder (Robins & Helzer, 1985).

The interview format for the DIS-III-A follows a strict set of procedures to ascertain whether physical symptoms reported could be explained medically or by the use of drugs or alcohol. Therefore, this interview format was chosen to be administered by an occupational health nurse with psychiatric experience.

A physician then reviewed all medical explanations given by the patients to ascertain their medical plausibility. Based on this procedure, the number of medically explained and unexplained somatic symptoms, occurring before and after the development of MCS, can be determined.

The Minnesota Multiphasic Personality Inventory (MMPI) (Hathaway & McKinley, 1967) was given to assess current psychiatric symptoms. This self-report inventory has been in use for the past 47 years with a variety of clinical and non-clinical populations. Results from this inventory include validity scales to assess whether the patient responded with a particular response set such as under- or over-reporting of symptoms. If the validity scales are within acceptable limits, then one may proceed to interpret the clinical scales with some evidence that they truly represent the patient.

Psychosocial Functioning:

MCS has also been reported to have a major impact on patient's functioning in daily life (Cullen, 1987). Therefore, several standardized measures of social functioning were also selected. The Psychosocial Adjustment to Illness Survey Self-Report (PAIS-SR) (Derogatis

& Lopez, 1983) was selected to assess the impact that MCS has on the following dimensions of a patient's life: health care orientation, vocational environment, domestic environment, sexual relationships, extended family relationships, social environment, and psychological distress. The scores of MCS patients may be compared to medically ill patients (i.e. diabetics) to contrast the level of social and life functioning. Other chronically ill groups are also available for comparison such as patients with severe coronary artery disease.

The Social Adjustment Scale-Self Report (SAS-SR) (Weissman & Bothwell, 1976) was selected to assess social functioning compared to other psychiatric patients (e.g. depressed or alcoholic patients). This scale gives an overall adjustment score which is based on items covering social adjustment as it pertains to work in or out of home, social and leisure activities, family and marital relationships and economic issues. It has been shown to differentiate depressed patients from normals.

RESULTS AND CONCLUSION

Neurobehavioral Performance:

Chemically sensitive patients did not reveal deficits in concentration or visuomotor skills although some memory tests suggested impairment.

Psychosocial Function:

Psychosocial functioning of MCS patients was impaired relative to diabetics on the PAIS-SR. That is, they had significant disruptions in their social, home and work lives as well as significant disillusionment with the medical care system. On the SAS-SR, they also revealed disruptions in their home and work functioning. These disruptions were comparable to depressed and alcoholic patients.

CONCLUSION

A subset of MCS patients who meet more rigid criteria can be identified and characterized using standardized neurobehavioral, psychiatric, and psychosocial measures of functioning. To make studies comparable, it is important to develop standard selection criteria for patient selection and standardized batteries for patient characterization.

REFERENCES

Black D.W., Rathe A., Goldstein R. Environmental Illness - A Controlled Study of Twenty-six Subjects with 20th Century Diseases. JAMA 1990;264: 3166-3170;#24.

Bolla-Wilson K., Wilson R.J., Bleecher M.L. Conditioning of Physical Symptoms After Neurotoxic Exposure. J. Occup Med. 1988;30:684-686.

Brodsky C.M. Psychological Factors Contributing to Somatoform Diseases Attributed to the Workplace. J Occup Med. 1983;25:459-464.

Cullen M.R. The Worker With Multiple Chemical Sensitivities: An Overview. In: Cullen M, ed. Workers With Multiple Chemical Sensitivities. Philadelphia: Hanley & Belfus Inc; 1987:655-662.

Delis D.C., Framer J.H., Kaplan E., Ober B.A. California Verbal Learning Test Manual. San Antonio: The Psychological Corporation; 1987.

Derogatis L.R., Lopez M.C. The Psychosocial Adjustment to Illness Scale (PAIS and PAIS-SR) Administration, Scoring and Procedures Manual-I., 1983.

Hathaway S.R., McKinley J.C. Minnesota Multiphasic Personality Inventory Manual - Revised. New York: The Psychological Corporation, 1967.

Hartman D.E. Neuropsychological Toxicology - Identification and Assessment of Human Neurotoxic Syndromes. New York: Pergamon Press; 1988.

Jastak S., Wilkinson G.S. Wide Range Achievement Test Administration Manual. Wilmington: Jastak Associates Inc; 1984.

Lezak M.D. Neuropsychological Assessment. New York: Oxford University Press; 1983.

Robins L.N., Helzer J.E. Diagnostic Interview Schedule (DIS) Version III-A. St. Louis: Washington Univ. School of Medicine; 1985.

Rosenberg S.J., Freedman M.R., Schmaling K.B., Rose C. Personality Styles of Patients Asserting Environmental Illness. J Occup Med. 1990; 32:678-681.

Savage A.P., Keefe T.J., Mounce L.M., Heaton R.K., Lewis J.A., Burcar P.J. Chronic Neurological Sequelae of Acute Organophosphate Pesticide Poisoning. Arch of Env Health 1988;43:38-45.

Schottenfeld R.S., Cullen M.R. Recognition of Occupational-Induced Post Traumatic Stress Disorders. J. Occup Med. 1986;28:365-369.

Simon G.E., Katon W.J., Sparks P.J. Allergic to Life: Psychological Factors in Environmental Illness. Am J Psychiatry. 1990;147:901-906.

Spitzer R.L., Williams J.B.W., Gibbon M., First M.B. User's Guide for the Structured Clinical Interview for DSM-III-R. Washington, DC: American Psychiatric Press, Inc.; 1990.

Terr A.I. Environmental Illness: A Clinical Review of 50 Cases. Arch Int Med. 1986;146:145-149.

Trenerry M.R., Crosson B., DeBoe J., Leber W.R. Stroop Neuropsychological Screening Test Manual. Odessa FL: Psychological Assessment Resources, Inc; 1989.

Trites, R.L. Neuropsychological Test Manual. Montreal, Que: Technolab, 1981.

Valciukas, J.A., Control of Confounding Demographic Factors in the Analysis of Neurotoxicological Data. In Environmental Health Document 3. Neurobehavioral Methods in Occupational and Environmental Health. Copenhagen: World Health Organization, 1985.

Wechsler D. Wechsler Adult Intelligence Scale - Revised. New York: The Psychological Corporation; 1981.

Wechsler D. Wechsler Memory Scale - Revised Manual. San Antonio: The Psychological Corporation; 1987.

Weissman M.M., Bothwell S. Assessment of Social Adjustment by Patient Self-Report. Arch Gen Psychia 1976;33:1111-1115.

Acknowledgments

This project was supported by a grant from the Hazardous Substance Management Research Center, A National Science Foundation Industry/University Cooperative Center and A New Jersey Commission on Science and Technology Advanced Technology Center. We wish to acknowledge the contributions of Kathie Kelly- McNeil and Carol Natarelli, who have both been instrumental in providing the evaluation and care of our patients. We also appreciate the consultation of Bernard Goldstein, M.D. and Michael Gochfeld, M.D., Ph.D. in formulating this project. Finally, we appreciate the support services of Joyce Kosmoski and Patricia Hutty in preparing this and other documents for this project.

Diagnostic Markers Of Multiple Chemical Sensitivity

G. Heuser, A. Wojdani, and S. Heuser

One hundred thirty five patients (75% females) were evaluated for complaints of often disabling sensitivity to small concentrations of multiple chemicals after chemical exposure in the recent or distant past.

A comprehensive evaluation of all subjectively involved systems showed a high yield of abnormal objective findings on random (unrelated to time of exposure) testing. When properly timed, certain immune function tests (TA1 cells and chemical antibody levels) became abnormal, or more abnormal, after unintentional self-reported acute exposure and were thus shown to be potential markers of multiple chemical sensitivity (MCS).

We suggest that appropriate tests of the central nervous system, peripheral nervous system, nose and sinuses, pulmonary function, T-cell subsets, chemical antibodies and autoimmunity be performed. If four of these seven systems show abnormality, the diagnosis of MCS is supported. If certain functions become abnormal or more abnormal after unintentional significant exposure, the diagnosis is confirmed.

INTRODUCTION

The senior author has directed a headache and chronic pain clinic for more than ten years. As clinical histories were taken in increasing detail over the years, it became apparent that in many patients, headaches were triggered by chemicals in the environment. What appeared to be very small concentrations of perfumes, fumes, smoke and mist, would trigger headaches in some patients but not in friends and family members who were present in the same environment. Eventually, the senior author realized that small amounts of chemicals cannot only cause headaches but also a multitude of other complaints in some patients. This realization became a starting point for his increasing interest in chemical sensitivity.

One of the co-authors (A.V.) is also chemically sensitive. This is why, as an immunologist, he became interested in developing appropriate tests for chemical sensitivity. He made chemical antibody testing commercially available to our patients.

The other co-author (S.H.) specializes in computer assisted medical information retrieval. She works one day a week in a one story office which was built more than ten years ago. Furniture, equipment, carpets, and paint are also about ten years old. Never-the-less, she recently developed headaches and nose bleeds. These regularly developed the morning after she had worked in the office. An investigation showed that the manually set air intake valve for that office building had fallen shut. As a result, all air in the office was now being recirculated. No fresh air could enter anymore. When this situation was corrected, both headaches and nose bleeds disappeared. No other person in the office developed headaches and nose bleeds. The conclusion was reached that S.H. is chemically sensitive.

The three of us combined forces in order to develop a team approach to multiple chemical sensitivity (MCS). Consultants in varying specialties are complementing the efforts of our core group.

Initially, we became interested in complaints of chemical sensitivity in our headache patients. Now, we attract an increasing number of environmentally ill patients ("EIs"). Many of them claim disability and request help so that they can get State or Social Security. Others find themselves in litigation (personal injury, workers' compensation). Therefore, an objective evaluation of all patients became mandatory.

Patients with MCS present to their physicians with a startling spectrum of complaints. These complaints are often presented in a manner highly suggestive of a psychiatric disorder. None of them easily lend themselves to measurement. General malaise and weakness, fatigue, headaches, irritability, depression, memory problems, itching, numbness, "creepy and crawly" sensations, burning sensations in the nose, sore throat, hoarseness, shortness of breath, cough, abdominal distress, are all "soft" complaints and therefore not easily documented.

Patients may arrive in the office equipped with oxygen tank, air filter, and mask. Typically, patients describe either one single or repeated exposures to one or more chemicals in the past and the development of chemical sensitivity thereafter. It is striking how similar the history is, regardless of education or walk of life or background. This consistency from one patient to another certainly suggests that we are dealing with a disease entity.

On physical examination there is striking paucity of abnormal findings. Small erythematous lesions over the skin areas exposed to chemicals come and go. On chest examination, wheezing is occasionally found. Soft signs may or may not be present on neurological examination.

Routine laboratory tests (CBC, blood chemistry) and EKGs are also usually normal. We quickly learned that chemically sensitive patients almost regularly refuse to deliberately expose themselves to chemicals. We thus abandoned the idea of exposing our patients to chemicals in a controlled environmental chamber. Instead, we waited for unintentional exposure to occur and suggested testing within a given interval thereafter. The results were then compared with base line values obtained either before or a longer time after acute exposure.

Our results led us to conclude that patients who claim MCS could and should be used as their own controls. Obviously, whenever research funds are available, this patient population should be compared to non-exposed and non-symptomatic matched controls.

A patient with multiple complaints is difficult to evaluate. Most physicians have learned that such a patient usually turns out to suffer from "psychosomatic illness". This is why even a well trained physician is easily persuaded to assign a psychiatric diagnosis to patients with multiple chemical sensitivity. By contrast, it became our goal to look for objective markers of multiple chemical sensitivity. We believe that this paper shows initial steps in this direction.

MATERIALS AND METHODS

Control Population. One hundred sixty healthy volunteers were examined. Their ages ranged from 22 - 55 years. They had no known disease and denied drug use and smoking. 60% were caucasian, 20% african-americans, 20% were hispanics. 62% were females, 38% were males. Their bloods were examined and normal ranges were established for all immune function tests reported in this paper.

Experimental subjects. One hundred thirty five patients (75% females) were selected from our private patient population. All complained of sensitivity to small amounts of chemicals in their work, home, or other (commuting, shopping, hobbies) environments. This sensitivity had diminished the quality of life in most patients many of whom claimed disability. In the extreme they now lived in self-imposed isolation from society. Some had gone to extraordinary efforts to live in an environmentally "clean" home. Some lived in tents on the beach, some in remote locations in California or out of state. Some had held well paying jobs, had made significant contributions to society but had to give it all up because of their environmental illness.

All patients were personally examined by the senior author. Their work or home environments could not be examined by the authors who had to rely on the patients' reports. Material safety data sheets (MSDS) were examined whenever available from the work site.

Table I lists the settings in which patients claimed to have been exposed. Please note that most exposures were due to groups of chemicals rather than a single chemical.

TABLE I

Group of 135 Patients Exposed to Chemicals

Pesticides	39	Electron Microscopy	1
Sick homes/Buildings	55	Printing	2
Solvents/Fuels	15	Plating	2
Chlorine Gas	1	Art/Wood	3
Copper Compounds	2	Others	15

All laboratory tests were initially obtained at random i.e. regardless of time of chemical exposure or severity of symptoms. Whenever possible, tests were thereupon specifically timed so as to be obtained a given number of days after exposure and resultant symptoms. Patients were thus instructed to wait for follow-up tests until they were involuntarily exposed and developed symptoms as a result.

Whenever symptoms suggested impairment of neurological or psychological function, we suggested an electroencephalogram (EEG) and evoked response studies. MRI of the brain was usually ordered only if these electrophysiological studies were borderline or abnormal. More recently, we started ordering brain mapping and SPECT studies in a select group of patients.

Whenever symptoms suggested a peripheral neuropathy, we ordered nerve conduction and/or neurometric (perception threshold) studies.

If patients specifically complained of nasal or throat symptoms, consultation was obtained from a board certified ENT-specialist. Studies then frequently included sinus x-rays.

Shortness of breath and related complaints led to pulmonary function testing (PFT) and if indicated, chest x-rays.

Patients underwent immunological studies whenever possible.

We routinely obtained CBC and blood chemistry. As indicated, additional studies (viral, bacterial, endocrine and others) were done.

Neurological testing. Most EEGs were performed in the office of G.H. and included spontaneous sleep, hyperventilation, and photic stimulation. A 16 channel Beckman instrument was used. EEGs had at times been performed elsewhere.

Most MRI brain scans were performed at Medical Imaging Center of Southern California and read by a neuroradiologist.

Perception thresholds were tested with a Neurometer and evaluated by a neurologist with special experience in the use of this instrument. The results are indicators of peripheral nerve (A to C fibers) function.

Evoked response and conduction studies were done and evaluated by board certified neurologists.

SPECT scanning was done at Harbor General Hospital (UCLA affiliated).

Nasal and sinus evaluation. All symptomatic patients were referred to a board certified specialist.

Pulmonary evaluation. Pulmonary function tests (PFT) were done with a Brentwood Spirometer in our office. Some patients had undergone testing elsewhere.

Immune testing. Tests were performed at Immunosciences Laboratories under the direction of A. Wojdani, Ph.D.. In view of the importance of these tests in the context of this evaluation, the methodology is described in detail.

Preparation of formaldehyde-human serum albumin (F-HSA) and Formaldehyde-bovine serum albumin (F-BSA) conjugates:

F-HSA and F-BSA was prepared by the method of Patterson et. al. (1985). Briefly, one mg. of HSA or BSA (Biocell, Carson, Ca.) in PBS (Phosphate Buffered Saline) PH 7.4, each separately, were exposed to 270 of formaldehyde (Fisher Scientific, Fairlawn, N.J.). The mixture was incubated for 30 minutes at 37 C and was then extensively dialyzed against PBS. The F-HSA or F-BSA was sterilized with a 0.2 m filter (Milliport Corp., Bedford, Mass.).

Electrophoretic and immunoelectrophoretic comparison of HSA, BSA with F-HSA and F-BSA was performed to determine conjugation occurrence. Conjugation was evidenced by altered mobility of F-HSA, F-BSA when it was compared with HSA or BSA respectively. Moreover, the number of free amino acid groups present in the F-HSA or F-BSA was determined by the method of Snyder and Sobocinski (1975) and was used to assess the amount of substitution. The number of amino groups bound to formaldehyde was 26 for HSA and 31 for BSA. In this calculation, the formation of intermolecular cross-linking was considered.

Preparation of toluene diisocyanate-human serum albumin (TDI-HSA) and Bovine serum albumin (TDI-BSA) conjugates:

This preparation was similar to the methods of Dewar and Baur (1982). According to this method, 1g HSA or 1g BSA was dissolved in 100 ml of a buffer solution containing potassium chloride (0.05 mol/l), sodium borate (0.05 mol/l), PH 9.4 and cooled to 4 C. Dioxane (10 ml) containing 0.15 ml of toluene diisocyanate was then added dropwise while stirring over a period of 3 hours, followed by addition of 2 ml of ethanolamine, centrifugation, dialysis filtration and lyophilization. Similar to F-HSA and F-BSA, conjugation was confirmed by electrophoresis and determination of free amino groups present in the conjugate. The number of amino groups bound to TDI was 37 for HSA and 43 for BSA. In addition, spectrographic analysis of the conjugate was undertaken according to Zeiss et. al. (1980). There was a marked increase in absorption from 230 to 260 nm which indicated that TDI had become covalently linked to the protein carrier. This increase in absorption did agree with NH_2 group determination only 76% for HSA and 81% for BSA

Preparation of trimellitic anhydride-human serum albumin (TMA-HSA) and Trimellitic anhydride bovine serum albumin (TMA-BSA):

To prepare these conjugates 25 mg. of TMA was dissolved in 0.5 ml of dioxane and added dropwise either to 25 mg of HSA or BSA dissolved in 5 ml of cold 7% NaHCo3 in water. After stirring for 60 minutes at 4 C the conjugates were dialyzed against four changes of 0.1 M NaHCO3 and one change of buffer. Finally the conjugates were filtered and kept at -20 C until used. OD analyses of TMA-HSA, and TMA-BSA were done to determine the number of TMA residues linked to the corresponding carrier protein. The concentration of the carrier protein was converted to molar concentration with the molecular weight of HSA and BSA. From the ratio of the molar concentration of the TMA ligand and the protein carrier, the ratio of TMA residues per molecules of carrier was calculated. TMA-HSA was estimated to contain 5 TMA residues per HSA molecules and for TMA-BSA seven residues per albumin molecule (Pien et al., 1988).

Preparation of phthalic anhydride-human serum albumin (PA-HSA) and Phthalic anhydride bovine serum albumin (PA-BSA) conjugates:

These hapten-conjugates were prepared by adding 75 mg of PA to a cooled solution of 300 mg of HSA or BSA in 100 ml of H_2O. The reaction mixture was stirred overnight, dialyzed against 0.1 M PBS using tubings with a cutoff of 8000 dalton. Using the method of

Zeiss et. al. (1977) the molar ratios were calculated. The molar ratios were found to be 22/28 for PA/HSA and 25/30 for PA/BSA.

Preparation of benzene ring HSA (B-HSA) and Benzene ring BSA (B-BSA) conjugates:

For these preparations, 40 mg. of P-aminobenzoic acid was dissolved in 2 ml of 1 N HCL and cooled by immersion in an ice bath. A cold solution of 14 mg/ml was added dropwise. After each addition, the mixture was stirred for 30 seconds. In parallel, one gram of HSA or BSA was dissolved in boric acid 0.16 M sodium chloride (0-15 M) buffer PH 9.0 (PH was raised with NaOH). The beakers containing the solutions of albumins were surrounded by an ice bath on magnetic stirrer. The solution of diazonium salt was added dropwise, with rapid stirring, to the cold protein solution. After addition of each drop the PH is readjusted to 9.0 to 9.5 with one normal NaOH. When all the solution had been added, the reaction was allowed to continue with slow stirring for at least an hour with further additions of NaOH solution and maintaining the PH at the range of 9.0 to 9.5. Unreacted small molecules were removed by extensive dialysis or by passage through a column of sephadex G-25 in the cold room, with an isotonic salt solution as the eluting buffer. OD analyses of orange color development of B-HSA and B-BSA were done to determine the number of B residues linked to the corresponding carrier protein. The amount of B substitution for HSA were approximately 41 and for BSA 53 (Migrdichian, 1957).

Antibody Determinations:

Specific antibodies against F-HSA, TDI-HSA, TMA-HSA, PA-HSA and B-HSA were analyzed by a noncompetitive ELISA assay. Wells of microtiter plates (Dynatech, Alexandria, VA) were coated with 100 l of antigen solutions (100 g/ml) in 0.1 M PBS PH 7.2 overnight at 4 C. Plates were washed 4 times with 0.1 M PBS containing 0.05% tween 20 between each step. Free absorption sites were blocked with 2% protease free bovine serum albumin at room temperature for 4 hours and stored at -20 C until used.

Analytical Procedure:

The procedure included the following: (1) washing four times, (2) addition of 100 l of diluted serum (1:2 for IgE and 1:100 for IgM and IgG) in PBS tween-20 with 1% BSA (3) incubation for 4 hours at 20 C, followed by washing 4 times, (4) addition of 100 l of an optimal dilution of alkaline phosphatase labeled affinity purified goat anti-human IgE ( ) (1:200), IgM (1:500) and anti IgG (1:1000), purchased from KPI (Maryland) (5) incubation for 120 minutes at 20 C, (6) washing 6 times, (7) addition of 100 l of P-nitrophenyl phosphate (Sigma Chemical Co.) (8) incubation for 60 minutes at 20 C (9) addition of 50 l of 3 N sodium hydroxide solution, and (10) duplicate reading. The results were calculated based on absorbances of duplicate samples of 405 nm using microtiter reader. All samples were read against an HSA antigen as a control of
binding not specific to F-HSA, TDI-HSA, TMA-HSA, PA-HSA and B-HSA. Results were

expressed as titer. Titer is being the last dilution of serum giving absorbance twice of HSA control.

Specificity and Cross-inhibition Studies:

For determination of antibody specificity a cross-inhibition study was undertaken. Positive sera for each hapten-protein conjugate were run after appropriate incubation and precipitation with tenfold increasing increments of hapten bound HSA or BSA as inhibitors to cover the range of antibody to antigen excess. This range was between 50 g to 1000 g for hapten-BSA and 80 g to 1000 g for hapten-RSA. After incubation at 37 C and removal of precipitate by centrifugation, the samples from before and after cross-inhibition study were then placed on plates with wells coated with the specific conjugate. The subsequent steps were followed as described above for the ELISA study.

IgG and IgM antibody binding to different conjugates was inhibited by hapten-HSA or hapten-BSA from 36-85%. At a given concentration, both hapten-HSA and hapten-BSA inhibited the antibody level in similar manners.

Partial inhibition of IgE antibody binding to different hapten conjugates was observed when serum was pre-incubated with hapten-HSA or hapten-BSA. This incomplete observation of inhibition of IgE antibody was mainly related to nonavailability of serum with high IgE titers against different chemicals in our laboratory.

Determination of Normal Levels of Antibodies (Controls):

Based on the above procedures, 160 blood donor samples of healthy individuals, of both sexes, between the ages of 22-55, were examined for antibody levels against F-HSA, TD-HSA, TMA-HSA, PA-HSA, and B-HSA. The average titer was 1:800 400 for IgG, 1:3200 1600 for IgM and 1:8 4 for IgE. Thus, in our laboratory titers greater than 1:1600 for IgG, 1:6400 for IgM and 1:16 for IgE are considered positive.

In a given patient, rises or falls in antibody titers by more than one dilution were considered significant (see Tables).

Lymphocyte Subset Enumeration:

A single laser flow cytometer (Epics Profile: Coulter Epics, Inc., Hialeah, FL) which discriminates forward and right angle light scatter, as well as two colors, was used with a software package (Quad Stat: Coulter). Mononuclear cell populations were determined by two-color direct immunofluorescence by using a whole-blood staining technique with the appropriate monoclonal antibody and flow cytometry (Fletcher et al., 1989) The following pairs of fluorescein isothiocyanate (FITC), or phycoerythrin (PE)-conjugated monoclonal antibodies (Coulter immunology) were selected: T11-RDI/B4-FITC, T4-RDI/T8-FITC, T3-FITC/NKH-1-RDI and T11-FITC/Ta1-PE for determination of T-cell/B cell, T-helper/T-suppressor, NKHT3+/NKHY3- and for alternate pathway of lymphocyte activation respectively.

To monitor lymphocyte markers, bit maps were set on the lymphocyte population of the forward-angle light scatter versus a 90 light scatter histogram. The percentage of positively stained cells for each marker pair, as well as the percentage of doubly stained cells was determined. Estimates of absolute numbers of lymphocytes positive for the respective surface markers were determined by multiplying peripheral lymphocyte cell count by the percentage of positively stained cells for each marker pair. Also, the percentage of doubly stained cells was determined. Estimates of absolute numbers of lymphocytes positive for the respective surface markers were determined by multiplying peripheral lymphocyte cell count by the percentage of positive cells for each surface marker.

Measurement of anti-myelin basic protein antibodies:

Human myelin basic protein (HMBP) was prepared by the method of Diebler et al. (1972) and checked for purity by polyacrylamine gel electrophoresis. Antiserum to HMBP was induced in rabbits by repeated injection of HMBP in complete Freund's adjuvant. Antibody activity in the rabbit sera and patient's samples was detected by adding different dilutions (1:100 to 1:10,000) of sera to wells of a microtiter plate previously coated with HMBP as follows: HMBP 250 g/ml was dissolved in carbonate buffer, PH 9.6 and 200 l of this solution were added to each well. After incubation, washing and blocking as above, 200 l of either diluted rabbit or human serum were added to the wells. After incubation for 1 hour at 37 C the sera were shaken out of the wells and then were washed 5 times with wash solution. 200 l of peroxidase-conjugated goat anti-rabbit or goat anti human IgG, IgM or IgA (optimal dilution) were added to the appropriate well. After incubation and repeated washing 200 l of ABTS substrate were added to each well. Plates were incubated for one hour at room temperature and read in a microtiter reader at 405 nm wavelength. Using rabbit antisera, a titration curve was plotted and patient's sera were compared to this standard curve. Based on more than 200 controls and patients' sample determinations, titers greater than 1:2000 for IgA, 1:5000 for IgM, and 1:8000 for IgG were considered positive.

RESULTS

Table II lists the neurological tests done in patients who had complaints of headaches, irritability, memory loss, depression, numbness, tingling, crawling sensation etc.
Some studies were done in sufficient number to be suggestive of a significant trend. EEGs were abnormal in 45% of tested patients. They showed mild rather than severe abnormalities, with mostly unilateral (at times bilateral) intermittent slowing, dysrhythmia, and occasional single sharp waves and spikes in the temporal and adjacent leads. MRI scans were also abnormal in a high percentage of cases. In some scans there was a definite impression of atrophy (13%) or demyelinating disease (7%). Others (8%) had more ill-defined non-diagnostic lesions. Visual evoked (VER) and brain stem auditory evoked (BAER) responses were also abnormal in a high percentage of cases.

The number of patients who underwent single photon emission computerized tomography (SPECT) studies of brain perfusion and metabolism as well as computerized analysis of their EEG activity (BEAM) was small but initial results suggest further studies in a greater number of patients.

TABLE II

Neurological Tests in Patients with Multiple Chemical Sensitivity

Test	% Abnormal	# of Patients
Spect	75	4
Conduction	62	13
Neurometer	47	7
EEG	45	76
BEAM	43	7
BAER	33	18
MRI	28	54
VER	25	32

Current perception threshold studies by Neurometer were also performed in only a small number of patients. Never-the-less, they hold promise as potential markers of peripheral neuropathy, just as conduction studies do.

TABLE III

Examination of Nose, Sinuses and Pulmonary Function in Patients with Multiple Chemical Sensitivity

TEST	% ABNORMAL	# of PATIENTS
ENT Specialist	100	19
Sinus x-rays	52	25
PFT	62	78
Chest x-rays	16	32

Table III shows that a thorough ENT exam will show abnormalities in a high percentage of patients. The consistent findings were atrophic rhinitis in patients with severe nasal complaints. Sinusitis or at least thickened mucous membranes were found on sinus examination.

The typical abnormality on PFT was a decrease of FEF 25 - 75% to below 70% of predicted value, indicating small airway disease.

The tests done in table II and III were obtained at random i.e. unrelated to time of exposure.

Table IV shows that, again on random testing, increase in TA1 cell count and percentage is the most frequent abnormality in patients with MCS. Helper/suppressor (H/S) ratios can be increased (50%), unchanged, or decreased upon random testing. Suppressor cells were decreased in 27% of 110 patients. Whether their continued decrease leads to auto-immune disease is not yet apparent from our initial data. Mitogenesis was abnormal in 42% of 12 patients. Normal ranges in our control group were as follows: TA1 Cells 0-432/mm3 or 0 - 10%; H/S ratio 1 - 2.2; Helper Cells 336 - 2,376/mm3 or 35 - 55%; Supressor Cells 192 - 1598/mm3 or 20 - 37%; Lymphocytes 960 - 4,320/mm3 or 20 - 40%; B Cells 48 - 648/mm3 or 5 - 15%.

TABLE IV

Immune Cell Population in Patients with Multiple Chemical Sensitivity

TEST	% ABNORMAL	# OF Patients
TA1	66	92
H/S	50	109
Helper	43	109
Lymphs	27	110
B-Cells	25	104

For normal ranges, see text.

Table V shows abnormal levels of chemical antibodies in a high percentage of patients with MCS. By contrast "normal" (patients unaware of symptoms from chemical exposure) individuals remained in the normal range. (see Materials and Methods section).

Table VI illustrates elevated levels of chemical antibodies in a symptomatic patient. There was a significant change in IgG (benzene ring) and IgM (isocyanate) levels which decreased after exposure ceased and the patient became "asymptomatic" (while however still chemically sensitive).

We also examined for IgE chemical antibodies which were not elevated in any of our studies.

Table VII shows a significant decrease to normal levels of chemical antibodies in a patient who traveled out of state and stayed in a "non-contaminated" environment where she slowly became "asymptomatic". TA1 cells were not a good marker of MCS in this case.

TABLE V

Elevation of Antibodies to Chemicals in Patients with Multiple Chemical Sensitivity

TESTS	% ELEVATED	# of PATIENTS
All/Any	64	111
TMA	41	111
Benzene	28	74
Isocyanate	30	110
Formaldehyde	30	111
Phthalic Anhydride	6	81

For normal ranges see TABLE VI.

TABLE VI

	SYMPTOMATIC 3-30-90 IgG/IgM	ASYMPTOMATIC 6-22-90 IgG/IgM	NORMAL RANGE IgG/IgM
FO	3,200/1,600	1,600/800	1,600/6,400
ISO	3,200/3,200	1,600/800	1,600/6,400
TMA	6,400/1,600	3,200/800	1,600/6,400
PHTH A	1,600/3,200	800/1,600	1,600/6,400
BENZ	6,400/3,200	1,600/3,200	1,600/6,400

K. W., 42, female, was symptomatic from exposures to auto mechanic repair shop in 3/90. No more exposures after early April 1990.

Table VIII illustrates a case in which chemical antibodies were a poor marker of chemical sensitivity. By contrast, cells of the immune system were significantly abnormal and slowly approached normal as the patient stayed away from her home for several months. While the patient was originally bed ridden from exposure at her home, she was ambulating and much improved after four months away from her home. By contrast, her

TABLE VII

	Before IgG/IgM	+ 4 Mos. IgG/IgM	Normal Range IgG/IgM
FO	6,400/12,800	800/800	1,600/6,400
ISO	6,400/3,200	800/800	1,600/6,400
TMA	3,200/3,200	1,600/3,200	1,600/6,400
TA_1 #	207	168	0-432
TA_1 %	7.6	7.5	0-10

M. D., 53, female, with MCS before and after leaving "toxic" environment at her home.

husband denied chemical sensitivity (he actually had some minor symptoms on detailed questioning) and could only be persuaded to be tested after he was away from his home for four months. It should be noted that the couple re-entered their home every few weeks to fetch some of their belongings. This led to intermittent exposure and may account for the slow recovery of the wife. While the patient claimed exposure to malathion, her home had possibly become contaminated by other pesticides.

Table IX illustrates rapid increase in TA1 cells and decrease in immunocompetent natural killer cells (NKHT3) in a student with MCS who entered an anatomy laboratory for sufficient length of time to become severely symptomatic. By contrast, T-cells and helper-suppressor ratios did not change within that same time interval.

Table X illustrates increase in IgM antibodies to TMA, phthalic anhydride and compounds with a benzene ring, and also in IgG antibodies to the benzene ring, approximately two weeks after significant exposure. By contrast changes in TA1 cells were seen within only one day! Note that other parameters were unchanged when studied during this short time interval.

Table XI illustrates that antibodies to TMA can be the only antibodies elevated after exposure. In table XII only two antibodies (isocyanate and chemicals containing the benzene ring) are elevated after significant exposure. Table XIII illustrates increase in benzene related antibodies in the wife who showed more evidence of chemical sensitivity than the husband. Note that other antibodies were not elevated. Results in tables XI, XII and XIII suggest specificity in the respective chemical antibody tests.

Table XIV shows increased antibodies in husband and wife. Both claimed total disability from severe MCS with the wife showing more symptoms. Testing was done at random and years after the acute exposure. Table XV also illustrates the long term effects of exposure in a mobile home which the couple had moved out of three years earlier. They both claimed MCS and were symptomatic when seen in our office. Note elevated antibodies to formaldehyde and TMA together with high TA1 counts in both husband and wife.

Table XVI depicts the presence of auto-antibodies in our patient population.

TABLE VIII

	Wife + 4 Wks.	Wife + 6 Wks.	Wife + 4 mos.	Husband + 4 mos.	Normal Range
H/S	0.7	0.8	1.1	2.1	1 - 2.2
$NKHT_3$ +#	471	396	77	27	14 - 216
$NKHT_3$ +%	15.1	9.2	3.4	1.7	1,5 - 5.0
TA_1 #	624	689	452	192	0 - 432
TA_1 %	22	16	20	12.4	0 - 10
FO IgG/IgM	800 / 800	----	800 / 800	800 /800	1,600/ 6,400
ISO IgG/IgM	800 / 800	----	800 / 800	800 / 800	1,600/ 6,400
TMA IgG/IgM	800 / 1,600	----	800 / 800	800 / 800	1,600/ 6,400
Phth IgG/IgM	800 / 800	----	1,600 / 1,600	1,600 / 6,400	1,600/ 6,400

Married couple, (male, 45 and female, 56) exposed to malathion with symptoms far more severe in the wife.

Not shown in table XVI are additional results with respect to autoimmunity: ANA titers were positive in 17% of 96 patients. The highest percentage of elevated antibodies was seen when anti-myelin antibodies were studied. They were positive in 80% of 50 patients studied. Typically the elevation was in the IgM and/or IgA rather than IgG antibodies.

In contrast to our findings, "normals" have been reported to have a very low incidence of positive auto-antibodies: ANA 3-4%, parietal approximately 2%, smooth muscle approximately 3%, mitochondrial approximately 1%.

DISCUSSION

We embarked on our studies with the hope that patients with a claim of MCS could eventually be objectively evaluated. We felt that an approach should be found which could accommodate a great number of patients and could be supervised by a primary physician.

Our studies were done by specialists and with equipment available in most cities in the US. The immunological studies described in this paper are sophisticated but can be executed on blood which is mailed over-night to an appropriate laboratory.

The growing number of patients claiming MCS will in our opinion make it impossible

TABLE IX

	BASELINE 7 - 21 - 90	+ 2 DAYS 8 - 14 - 90	NORMAL RANGE
T #	2,975	2,005	701 - 3,758
T %	85.8	85.7	73 - 87
H/S	2.3	2.2	1 - 2.2
TA_1 #	523.7	985	0 - 432
TA_1 %	15.1	42.1	0 - 10
$NKHT_3$ + #	10.4	4.7	14 - 216
$NKHT_3$ + %	0.3	0.2	1.5 - 5

Patient L. W., 34, female with MCS, reacting to exposure in an anatomy laboratory on 8-12-90.

to study them all in environmental chambers under controlled conditions. These chambers will however be needed for further research.

Table XVII shows a list of chemicals brought to the senior author by a patient with MCS. She was exposed to some or all of these chemicals at work on an everyday basis. This table illustrates the dilemma for both patient and physician in trying to attempt to disentangle a complex problem such as this. An environmental chamber approach might not be practical in such a case as it would take too long and would be too expensive to study this matter in great detail.

While exacting research requires well selected controls, these are not easily found in our polluted urban environment. For example, most people are more or less aware of and effected by pollution in Los Angeles. Thus, there seems to be a whole spectrum of sensitivity, with our patient population being at one extreme.

Nevertheless we, have accumulated in excess of one hundred patients who did not complain of MCS and had no elevation of chemical antibodies (also see Material and Methods section for discussion of controls) nor of TA_1 cells.

In view of the above we feel that a patient should serve as his/her own control and that therefore all studies should be longitudinal. Our results show that properly timed studies can bring about significant changes in certain parameters after self-reported exposure. We are not certain at this time when TA1 cell counts reach a peak after exposure. We now know however, that these cells are elevated one to two days after exposure. We are also uncertain when chemical antibody levels reach a peak value. All we have shown so far is an elevation about two weeks after exposure.

Chemical exposure and its effects on the immune system has recently become the subject of discussion by leading allergists (Salvaggio, 1990). Changes in immune cell

TABLE X

Chemical Antibodies and Subpopulations Before and After Exposure

TESTS	9-14-90	10-19-90	TESTS	9-14-90	9-28-90
FO-IgG	800	1,600	WBC	7,300	8,800
-IgM	800	1,600	LYMPH %	27	37
ISO-IgG	800	1,600	T%	76	79
-IgM	800	1,600	H%	47	45
TMA-IgG	1,600	3,200	S%	27	29
-IgM	800	3,200	H/S	1.7	1.6
Phth-IgG	800	800	B%	21	17
-IgM	800	3,200	TA_1%	4.4	16.1
Benz-IgG	800	3,200	TA_1 #	86	512
-IgM	800	6,400			

A. G., Female, 52, MCS from work in sick bldg. Re-entry into same bldg. after severl months on 9-27-90.

populations, specifically TA1 cells, after exposure to chemicals were recently described by another group (Thrasher et al., 1989; Thrasher et al., 1990).

Chemical antibody measurements in chemically exposed patients recently became commercially available. Appropriate immunological procedures were originally developed by Dr. Wojdani and used in patients of the senior author (Thrasher et al., 1987). The original procedures were then expanded by Dr. Wojdani to include additional chemicals (this paper) and adopted and verified by another laboratory (Antibody Assay Laboratories) where additional research was done and published (Thrasher et al., 1989; Thrasher et al., 1990).

At this time, it should be noted that benzene is not per se antigenic. However, our data suggests that some chemical compounds containing the benzene ring are antigenic. Further studies are needed to determine which of these compounds cause antibody formation.

Our data suggest that chemical exposure can push some patients in the direction of autoimmune disease. Multiple sclerosis is an example. A number of our patients are suspected of having that disease on the basis of not only their clinical presentation but also abnormal MRI and evoked response studies together with high anti-myelin antibodies. This was previously discussed (Gard and Heuser, 1990).

Studies other than immune tests should also be done in a longitudinal fashion. Cost containment, a lack of research funding and other factors made this impossible in our patient population. However, PFT were at times studied immediately after exposure and became abnormal. It is possible that some neurological parameters (EEG, BEAM, and SPECT) may also show some significant changes.

TABLE XI

	BASELINE IgG / IgM	+ 2 WEEKS IgG / IgM
FO	1,600 / 1,800	800 / 800
ISO	800 / 800	800 / 800
TMA	1,600 / 3,200	3,200 / 12,800
BENZ	6,400 / 3,200	3,200 / 6,400

Patient, L. W., 34, female, with MCS, reacting to exposure in a "Sick Building".

TABLE XII

	8-14-90 IgG / IgM	10-9-90 IgG / IgM	NORMAL RANGE IgG / IgM
FO	800 / 800	1,600 / 1,600	1,600 / 6,400
ISO	800 / 1,600	3,200 / 1,600	1,600 / 6,400
TMA	800 / 800	1,600 / 800	1,600 / 6,400
PHTH A	800 / 800	1,600 / 800	1,600 / 6,400
BENZ	800 / 800	3,200 / 3,200	1,600 / 6,400

Patient L. C., 45, male, with MCS before and after exposure to "noxious" environment on 9-25-90.

While we have come to expect a high percentage of abnormal immune function tests, we were surprised at the high percentage of abnormal neurological tests. This indicates that the "psychiatric" presentation by many of these patients may well have a neurological basis.

The high number of abnormal test results in our patient population is probably explained by the fact that many patients were disabled with MCS and therefore quite sick.

Our EEG and SPECT studies point toward the limbic system as being involved in MCS. This system's possible role is aptly discussed in this conference by Dr. Miller. It, together with the role of the olfactory system deserves further study (also see Dr. Bell's presentation at this conference). Early studies by Russian authors (Bokina et al., 1976) pointed in this same direction.

TABLE XIII

	WIFE IgG / IgM	HUSBAND IgG / IgM	NORMAL RANGE IgG / IgM
FO	800 / 1,600	800 / 800	1,600 / 6,400
ISO	800 / 1,600	800 / 1,600	1,600 / 6,400
TMA	800 / 1,600	800 / 800	1,600 / 6,400
PHTH A	800 / 3,200	800 / 800	1,600 / 6,400
BENZ	1,600 / 12,800	800 / 6,400	1,600 / 6,400

Married couple (male, 43 and female, 35) with MCS after intermittent exposure (ongoing) to gasoline fumes.

TABLE XIV

	WIFE IgG / IgM	HUSBAND IgG / IgM	NORMAL RANGE IgG / IgM
FO	6,400 / 3,200	6,400 / 1,600	1,600 / 6,400
ISO	12,800 / 1,600	6,400 / 1,600	1,600 / 6,400
TMA	6,400 / 1,600	6,400 / 1,600	1,600 / 6,400
PHTH A	1,600 / 3,200	1,600 / 1,600	1,600 / 6,400

Married couple (male, 38 amd female, 34) with MCS after exposure to roofing materials approximately 2 years earlier.

While EEG studies showed mostly mild abnormalities in the temporal and adjacent leads (see results), our youngest patients (sister and brother, ages two and four respectively) developed actual clinical seizures with grossly abnormal EEGs about three weeks after moving into a new home and playing on the brand new carpet. For several months thereafter the mother observed MCS in both her children. Seizures and MCS slowly abated after the family moved out of the new home.

Diagnostic criteria for MCS. The diagnosis of MCS should be suspected if a patient reports

TABLE XV

	WIFE	HUSBAND	NORMAL RANGE
H/S	2.2	2.0	1.0 - 2.2
TA_1 #	923	1,579	0 - 432
TA_1 %	45	47	0 - 10
FO (IgG / IgM)	6,400 / 3,200	3,200 / 3,200	1,600 / 6,400
ISO (IgG / IgM)	1,600 / 1,600	800 / 1,600	1,600 / 6,400
TMA (IgG / IgM)	8,000 / 12,800	4,000 / 3,200	1,600 / 6,400

Married couple (male, 60 and female, 58) with MCS after exposure to Formaldehyde in 1986. Tests done in 1989.

TABLE XVI

Autoimmunity in Patients with Multiple Chemical Sensitivity (N = 92)

PARIETAL CELL	24%	MITOCHONDRIAL	4%
SMOOTH MUSCLE	17%	BRUSH BORDER	4%
MICROSOMAL	15%	RETICULIN	3%
THYROGLOBULIN	7%		

impaired well-being whenever exposed to more than one chemical in concentrations which do not effect the general population. In the extreme, the concentrations are very low and the patient is very sick and claims disability.

A comprehensive evaluation of seven systems should then be undertaken. It should be understood that not all seven systems (central nervous system, peripheral nervous system, nose and sinuses, PFT, T-cells subsets, chemical antibodies, autoimmune panel) are always affected. However, we suggest that abnormalities in four out of these seven systems strengthen the suspicion of MCS. If parameters become abnormal or become more abnormal following self-reported acute exposure, the diagnosis is basically confirmed. Studies in environmentally controlled chambers will be necessary to further advance the field.

TABLE XVII

Chemicals Used at Work
Polychlorinated Biphenyls (PCBs)
Pesticides
1,1,1 Trichloroethane
[methyl] Isocyanates
Xylene
N Pentane
Cyclopentane
Hexane Isomers
N-hexane
Formaldehyde
Perchloroethylene
Monochlorodifluoromethane/Chloropentafluoroethane
Barryman blend
Tectyl 802A
PSKD-NF/ZP-9 Developer
SKL-HF/S Spotcheck Penetrant
Nitrocellulose lacquer [Sinclair]
Krylon Enamel
Tapmatic No. 1 cutting fluid
Industrial lacquer
Chevron thinner

The suggestion to use four out of seven criteria is taken from the diagnostic criteria for systemic lupus erythematosus (SLE), where four out of eleven criteria have to be present to make the diagnosis (see table XVIII). It should be noted that psychiatric features are seen in a significant number of patients with SLE. Thus, the same should not be unexpected in patients with MCS.

CONCLUSIONS

Patients who present with complaints of MCS deserve a comprehensive objective evaluation. If this is performed, a high percentage will be shown to have abnormal test results. This is true if the central and peripheral nervous systems as well as pulmonary and immune functions are tested. Also, anatomical changes are frequently found in the nasal passages on close inspection. By contrast, CBC and blood chemistry are usually within normal limits. So are findings on general physical examination.

Whenever possible, longitudinal studies should be performed in which the patient is used as his/her own control. Increases in TA1 cells and chemical antibodies can then be seen following self-reported unintentional exposure and are therefore suggested as markers of MCS.

Our results suggest diagnostic criteria for MCS. These are sorely needed as the number of patients who claim disability as a result of MCS is growing. Millions of dollars are potentially at stake as claims increase. Patients who are truly sick deserve attention and help from industry, housing authorities and government agencies as well as physicians. Patients who make unjustified claims should be quickly identified.

Patients and industry and government are all in need of a practical approach to the diagnosis of MCS. We believe that our findings are pointing the way to such an approach.

Acknowledgements: We thank Dr. D. Alessi for ENT evaluations, Dr. R. Holgate (neuro-radiology) for evaluation of MRI brain scans, Dr. R. Lawrence (neurology) for interpreting the neurometer studies and Dr. I. Mena (nuclear medicine) for evaluation of SPECT brain scans.

REFERENCES

Bokina, A.I., N.D. Eksler, A.D. Semenenko, and R.V. Merkur'yeva. 1976. Investigation of the mechanism of action of atmospheric pollutants on the central nervous system and comparative evaluation of methods of study. Environ. Health Perspect. 13:37-42.

Dewar, M.A., and X. Baur. 1982. Studies on antigens useful for detection of IgE antibodies in isocyanate-sensitized workers. J.Clin. Chem. Clin. Biochem. 20:337-340.

Diebler, G.E., R.E. Martenson, and M.W. Kies. 1972. Large scale preparation of myelin basic protein from central nervous tissue of several mamalian species, Prep. Biochem. 2:139-165.

Fletcher, M.A., S. Azen, B. Adelsberg, G. Gjerset, J. Hassett, J. Kaplan, J. Niland, T. Maryhon, J. Parker, D. Stites, and J. Mosley. 1989. Immunophenotyping in a multicenter study. The transfusion safety study experience. Clin. Immunol. Immunopathol. 52:38-47.

Gard, Z. and G. Heuser. 1990. Re: Multiple sclerosis, solvents and pets. Letter to the Editor. Arch. Neurol. 47:138.

Migrdichin, V. 1957. Conjugated and synthetic antigens in organic synthesis. Reinhold, New York.

Patterson, R., K.E. Harris, and L.C. Grammer. 1985. Canine antibodies against formaldehyde-dog serum albumin conjugates: induction, measurement, and specificity. J. Lab. Clin. Med. 106:93-100.

Pien, L.C., C.R. Zeiss, C.L. Leach, N.S. Hatoum, D. Levitz, P.J. Garvin, and R.Patterson. 1988. Antibody response to trimellitic hemoglobin in trimellitic anhydride induced lung injury. J. Allergy Clin. Immunol. 82:1098-1103.

Salvaggio, J.E. 1990. The impact of allergy and immunology on our expanding industrial environment. J. Allergy Clinical Immunol. 85:689-699.

Snyder, S.L. and P.Z. Sobocinski. 1975. An improved 2,4,6-Trinitrobenzosulfonic acid method for the determination of amines. Ann. Biochem. 64:285-288.

Thrasher, J.D., A. Wojdani, G. Heuser, and G. Cheung. 1987. Evidence for formaldehyde antibodies and altered cellular immunity in subjects exposed to formaldehyde in mobile homes--a brief communication. Arch. of Environ. Health. 42:347-350.

Thrasher, J.D., R. Madison, A. Broughton, and Z. Gard. 1989. Building-related illness and antibodies to albumin conjugates of formaldehyde, toluene diisocyanate, and trimellitic anhydride.
Am. J. Industrial Med. 15:187-195.

Thrasher, J.D., A. Broughton, and R. Madison. 1990. Immune activation and autoantibodies in humans with long-term inhalation exposure to formaldehyde. Arch. Environ. Health. 45:217-223.

Zeiss, C.R., R. Patterson, R., J.J. Pruzansky, M.M. Miller, M. Rosenberg, and D. Levitz. 1977. Trimellitic anhydride-induced airway syndromes. Clinical and immunologic studies. J. Allergy Clin. Immunol. 60:96-103.

Zeiss, C.R., T.M. Kanellakes, J.D. Bellone, D. Levitz, J.J. Pruzansky, and R. Patterson. 1980. Immunoglobulin E-mediated asthma and hypersensitivity pneumonitis with precipitating anti-hapten antibodies due to diphenylmethane diisocyanate exposure. J. Allergy Clin. Immunol. 65:346-352.

Possible Mechanisms For Multiple Chemical Sensitivity: The Limbic System and Others [1]

Claudia S. Miller and Nicholas A. Ashford

The limited data available at this time suggest that any mechanism or model that would purport to explain the syndrome of multiple chemical sensitivities would need to address the features most closely associated with this illness (Ashford and Miller, 1991):

1. Symptoms involving virtually any system in the body or several systems simultaneously, and frequently the central nervous system (particularly mood, memory and concentration difficulties).
2. Differing symptoms and severity in different individuals, even among those having the same exposure.
3. Induction by a wide range of environmental agents, including pesticides and solvents.
4. Subsequent triggering by lower levels of exposure than those involved in initial induction of the illness.
5. "Spreading" of sensitivity to other, often chemically dissimilar substances; each substance may trigger a different, but reproducible, constellation of symptoms.
6. Concomitant food and drug intolerances, estimated to occur in a sizeable percentage of those with chemical sensitivities.
7. Adaptation (masking), that is, acclimatization or tolerance to environmental incitants, both chemical and food, with continued exposure; loss of this tolerance with removal from the incitant(s); and augmented response with reexposure after an appropriate interval (for example, 4 to 7 days).
8. An apparent threshold effect referred to by some as the patient's *total load*, a theoretical construct that has been invoked to help explain why an individual develops this syndrome at a particular time. Illness is said to occur when the total load of biological, chemical, physical, and psychological stressors exceeds some threshold for the patient. This concept has emerged from clinical

[1]Excerpted from Ashford, N. A. and Miller, C. S. 1991. Chemical Exposures: Low Levels and High Stakes. New York: Van Nostrand Reinhold.

observations; no direct experiments have been done to test its validity in humans; however, animal models do exist. The concept aligns with Selye's work on the general adaptation syndrome (Selye, 1946).

Items 2 and 3 above confound epidemiological investigation of this problem.

Items 3 and 4 suggest a two step process - (1) induction or sensitization resulting from an initial "major" exposure and (2) triggering of symptoms by subsequent lower level exposures to many different incitants.

Although knowledge of the mechanism of a disease may be useful for developing better therapies, such knowledge is not a prerequisite for intervention. Preventing the development of multiple chemical sensitivities in those not yet afflicted may be possible by controlling environmental exposures that cause the initial sensitization.

The most frequently cited physiological theories to explain chemical sensitivity involve the nervous system, the immune system, or the interaction between them because these two systems most clearly link the external environment and the internal milieu.[2] The rapid responsiveness of these systems also makes them attractive candidates because symptoms of food or chemical sensitivity have been reported to develop within seconds of exposure. Many chemicals, such as polybrominated biphenyls (PBBs) and trichloroethylene, affect both the nervous system and the immune system. Until 1980, the idea of a possible direct communication between the nervous and immune systems was widely debated. Subsequently, the existence of a neuroimmunoendocrine axis has been increasingly realized. Several discoveries have helped to confirm the presence of two-way communication between the nervous and immune systems (Payan et al., 1986).

Kilburn proposes that the human nervous system, because it is so highly evolved, may be most susceptible to environmental agents (Kilburn, 1989):

> "Sensitivity may be its undoing. The intuitive hypothesis is advanced that the nervous system is the most liable of the body's systems to damage from environmental toxins. Appreciation of damage may be masked because subtle dysfunction is concealed by the nervous system's remarkable redundancy and substitution of functions, or it is overlooked in clinical evaluations which are usually only qualitative."

MECHANISMS INVOLVING THE LIMBIC SYSTEM

The hypothalamus (part of the limbic system) has attracted considerable attention because it is the focal point in the brain where the immune, nervous, and endocrine systems interact (Bell, 1982). Bell notes that assuming a direct cause-and-effect relationship would be premature, but that the hypothalamus could mediate food and chemical addictions in patients with multiple chemical sensitivities. The olfactory system has known links to the hypothalamus and other parts of the limbic system, which has led Bell to speculate that "the olfactory system, hypothalamus and limbic system pathways would provide the neural circuitry by which adverse food and chemical reactions could trigger certain neural, psychological and psychiatric abnormalities." Patients with chemical sensitivities have

---

[2]Discussions of possible immune system mechanisms and psychological hypotheses for chemical sensitivity appear in Ashford and Miller, 1991.

reported food cravings, binges, violence, or hypersexual activity following chemical exposures. A model involving the hypothalamus could help to explain such behavioral changes.

Some authors have alleged that psychological conditioning to odors is responsible for patients' reactions to chemicals. Of course, odor conditioning may occur in selected cases. However, physiological mechanisms involving the limbic system may also occur. A direct pathway from the oropharynx to the brain and hypothalamic and limbic region has been demonstrated in rats (Kare, 1968; Maller et al., 1967). Substances placed in the oropharynx migrated to the brain in minutes via a pathway other than the blood stream and in higher concentrations than if administered via the gastrointestinal tract, suggesting a direct route from mouth (or nose) to brain. Similarly, Shipley showed that inhaled substances that contact the nasal epithelium may cross into the brain and be distributed widely via transneuronal transport (Shipley, 1985). Thus, molecules that are inhaled and contact the olfactory apparatus could influence functions in other parts of the brain.

Ryan and associates studied 17 workers who attributed changes in thought processes, particularly memory and concentration difficulties, or changes in mood to their exposure to solvents (Ryan et al., 1988). Those workers with "cacosmia" (a heightened sensitivity to odors resembling that reported by chemically sensitive individuals) performed most poorly on neurobehavioral tests requiring verbal learning or visual memory. The authors felt their findings supported a hypothesis that chronic solvent exposure may affect the "rhinencephalic structures" (the primitive "smell" brain), the evolutionary precursor of the limbic system.

This phylogenetically ancient part of the brain is present in all mammals. It influences the organism's interaction with its environment in many subtle ways essential for preservation of the individual, its offspring, and the species. *Limbus* (Latin for "margin" or "rim") refers to its appearing like a rim around the edge of the cerebral hemispheres. Figure 1 shows its component parts. Note the close anatomical relationship to the olfactory bulb. The olfactory nerves are the brain's most direct contact with the external environment. Each nerve consists of bipolar neurons, which have one end in the upper part of the nose and the other in the brain (olfactory bulb). Strong odors and even milder ones cause electrical activity in the amygdala and hippocampal areas of the limbic system (Monroe, 1986). Subsensory exposure to chemicals can cause protracted, if not permanent, alterations in the electrical activity of the limbic region, beginning first with the most sensitive structures, particularly that portion of the amygdala that analyzes odors (Bokina, 1976). All parts of the limbic system are intimately interconnected. Interestingly, ablation of the olfactory bulb in laboratory animals serves as a model for depression which investigators have used for testing the efficacy of various antidepressants (Jesberger and Richardson, 1988).

The amygdala is involved in feelings and activities related to self-preservation, such as searching for food, feeding, fighting, and self-protection (MacLean, 1986). The cingulate gyrus appears to influence maternal care and nursing, separation cries between mother and offspring, and playful behavior, including wit and humor (MacLean, 1986). The septum involves feeling and expression relating to procreation. Lesions in the septal area may cause hyperresponsiveness to physical stimuli (such as touching, sounds, or temperature changes), hyperemotionality, loss of motivation, excessive sugar and water intake, and fear of unfamiliar situations (Isaacson, 1982), phenomena reported by some chemically sensitive individuals.

The hippocampus appears important for laying down new memories and thus is essential for learning (Gilman, 1982). Hippocampal lesions may cause difficulty in retaining

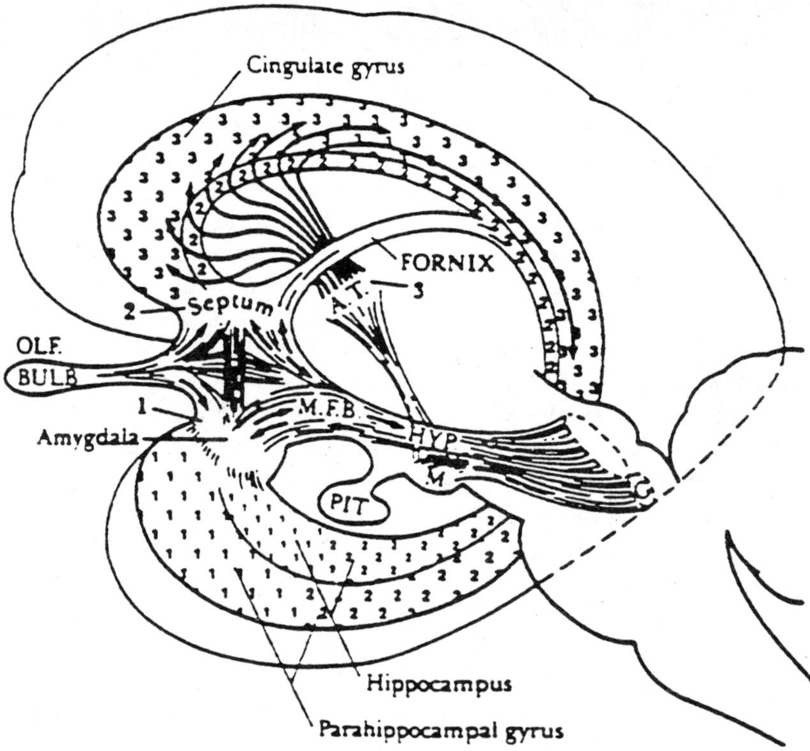

FIGURE 1 Three major subdivisions of the limbic system. The small numerals 1, 2, and 3 overlie, respectively, the amygdaloid, septal, and thalamocingulate divisions. The corresponding large numerals identify connecting nuclei in the amygdala, septum, and anterior thalamus. Abbreviations: AT, anterior thalamic nuclei; G, tegmental nuclei of Gudden; HYP, hypothalamus; M, mammillary bodies; MFB, median forebrain bundle; PIT, pituitary; OLF, olfactory. Source: MacLean, P.D., "A Triune Concept of the Brain and Behavior," in Boag, T., and Campbell, D., The Hincks Memorial Lectures (1973), University of Toronto Press, Toronto, Ontario, p. 15.

recent memories (Isaacson, 1982). The hippocampus, at the intersection of numerous neural pathways and in a critical position to affect the transfer of information from one brain region to another, acts as an information switching center. **Comment:** Learning and memory decrements are a frequent consequence of exposure to toxic substances, and some researchers view the hippocampus as a prime target for such toxins (Office of Technology Assessment, 1990; Walsh, 1988). Damage to the hippocampus itself, or to nerves leading to or from it, may adversely affect the synthesis, storage, release, or inactivation of the excitatory and inhibitory amino acids that serve as neurotransmitters in this region of the brain. Toxins may disrupt the delicate balance of these amino acids, perhaps leading to the release of a flood of excitatory neurotransmitters that damage neighboring cells, a phenomenon that has been called "excitotoxicity" (Office of Technology Assessment, 1990). Remarkably small perturbations of hippocampal function may have large and long-lasting effects upon behavior and cognition (Walsh, 1988).

The most vital component of the limbic system, the hypothalamus, governs: (1) body temperature via vasoconstriction, shivering, vasodilation, sweating, fever, and behaviors such as moving to a cooler or warmer environment or putting on or taking off clothing; (2) reproductive physiology and behavior; (3) feeding, drinking, digestive, and metabolic activities, including water balance, addictive eating leading to obesity, and complete refusal of food and water leading to death; (4) aggressive behavior, including such physical manifestations of emotion such as increased heart rate, elevated blood pressure, dry mouth, and gastrointestinal responses (Gilman, 1982).

The hypothalamus is also the locus at which sympathetic and parasympathetic nervous systems converge. Many symptoms experienced by patients with food and chemical sensitivities relate to the autonomic (sympathetic and parasympathetic) nervous systems; for example, altered smooth muscle tone produces Raynaud's phenomenon, diarrhea, constipation, and other symptoms reported by these individuals.

The hypothalamus also appears to influence anaphylaxis and other aspects of immunity (Stein, 1981). Conversely, antigens may affect electrical activity in the hypothalamus (Besedovsky, 1977).

It is important to recognize that thoughts arising in the cerebral cortex that have strong emotional overtones also can trigger hypothalamic responses and recreate the physical effects associated with intense anger, fear, and other feelings. To implement its effects, the hypothalamus not only has a direct electrical output to the nervous system but also produces its own hormones, many of which stimulate or inhibit the pituitary's production of hormones (Gilman, 1982). Of interest in this regard, is that a disproportionate number of chemically sensitive individuals seem to have been treated for thyroid hormone deficiency at some time in their lives.

Most of the neural input to the hypothalamus comes from the nearby limbic and olfactory areas (Isaacson, 1982). Lesions in the limbic region may be associated with irrational fears, feelings of strangeness or unreality, wishing to be alone, and sadness (MacLean, 1962). A feeling of being out of touch with or out of control of one's feelings and thoughts, not unlike that described by many patients with chemical sensitivity, may be perceived. Some report feeling "spacey" or that "the camera isn't on" unless they make an enormous effort to focus their attention.

Doane describes potential difficulties for patients with limbic dysfunction (Doane, 1986):

> Activity controlled by the limbic system may seem largely irrational and often is not perceived within one's self in ways that are easily understood or communicated in verbal language.

The dynamic involvement of the hypothalamus and limbic system in virtually every aspect of human physiology and behavior makes injury to these structures an intriguing hypothesis to explain the development of chemical sensitivity with its diverse manifestations. Rich neural connections lie between the olfactory system and the limbic and temporal regions of the brain. Surgical or epileptic patients with damage to the limbic or medial temporal portions of the brain may experience persistent alterations in odor perception (for example, an unusual smell that characteristically precedes seizure activity) as well as learning and memory difficulties (Ryan, 1988).

Bell hypothesizes that chemically sensitive patients may have olfactory-limbic-temporal pathways that are more easily "kindled" (Bell, 1990). In other words, a small signal or insult would more readily trigger firing of nerve cells in brain regions where kindling was present.

Kindling might be enhanced by genetic endowment, prior environmental exposures, psychological stress, hormonal variations, or other factors. Unlike surgical ablation, which destroys a brain area, kindling is a kind of stimulatory lesion (Girgis, 1986).

Kindling has been described previously in the context of seizures. The amygdala, for example, which is particularly susceptible to electrical discharge following either electrical (Girgis, 1986) or chemical provocation (Bokina, 1976), is subject to long-lasting alteration when strong or repetitive stimuli are administered. Very potent or repeated stimuli, whether electrical or chemical, may permanently augment the tendency for neurons to fire in the presence of future stimuli, even when challenged with much lower levels than those originally involved.

Girgis reports a decrease in acetylcholinesterase (AChE), an enzyme that breaks down the neurotransmitter acetylcholine in junctions between nerve cells, that parallels the increase in supersensitivity to stimuli (Girgis, 1986). The limbic system is especially rich in AChE, which is strongly bound to the nerve cell membranes and very stable. The AChE may play a protective role by enzymatically maintaining acetylcholine concentrations at nerve junctions within safe bounds and protecting susceptible cells in the limbic system from developing "bizarre sensitivity" (Girgis, 1986). Interestingly, physicians who treat patients with multiple chemical sensitivities have noted some of the most severe and debilitating exposures for these patients have involved organophosphate pesticides, which inhibit AChE.

Bokina found impaired speed of execution and coordination of execution and complex motor processes in humans repeatedly exposed to carbon disulfide for 10- to 15-minute intervals at subsensory levels (Bokina, 1976). Animals primed by high or chronic lower concentrations of various chemicals, such as formaldehyde and ozone, and subsequently reexposed to even lower concentrations of the same chemicals showed an increased tendency toward paroxysmal electrical discharge in the amygdala (Bokina, 1976). Bokina observed that although the chemicals he used to sensitize the animals were different in terms of their structure and physical and chemical properties, their effects upon the limbic system were remarkably similar.

Kindling could help to explain the apparent loss of adaptive capacity and spreading of sensitivities to chemically unrelated substances reported in multiple chemical sensitivity. Formerly well-tolerated low-level exposures to, for example, tobacco smoke or perfume might trigger symptoms in individuals whose limbic areas have been kindled by a prior pesticide or solvent exposure.

One intriguing aspect of the limbic system as a mechanism for multiple chemical sensitivities is the system's responsiveness to both chemical and cortical stimuli. Therefore, conscious thought processes and emotional states influence limbic activity just as chemical or physical stimuli can. The former may be under more or less conscious control of the individual, whereas the latter are almost entirely unconscious and automatic. However, conscious efforts that play into the delicate circuitry of the limbic system may be able to alter or suppress concurrent electrical activity evoked by environmental agents. Some patients with chemical sensitivities report being able to "will" their way out of a mild reaction to a food or chemical and attempt to control their symptoms in this manner. Most say such efforts do not work for their most problematic incitants. In fact, the ability to exercise any conscious effort, even that of simply getting away from the exposure, may be lost during a reaction. Monroe reported the case of a man for whom exposure to the odor of stale beer caused greatly increased electrical activity in the limbic system (amygdala and hippocampal areas). Various memories, some associated with beer, also increased electrical activity in the same region. However, simple arithmetic computations would immediately stop such activity. Therefore, conscious thought processes could alter some electrical activity in the limbic system (Monroe, 1986).

An intriguing example of the competing effects of exposure and psychological state has been reported (Sanderson, 1989). Carbon dioxide levels greater than 5 percent in the air have been shown to induce panic attacks ("fight or flight" responses depend upon limbic activity) in patients suffering from panic disorder. The fact that patients in this study who believed that they had control over the carbon dioxide level to which they were exposed had fewer and less intense panic disorder symptoms suggests that psychological factors (that is, the illusion of control), could mitigate the biological response to an environmental stressor.

Thus, experimental evidence suggests a delicate interplay occurring in the limbic region. Conceivably, chemicals contacting olfactory nerve projections in the nose could either be transported into or relay electrical signals to the limbic region, leading to a vast array of symptoms. Likewise, thought processes and mood states trigger limbic activity or may, in some cases, interrupt preexisting limbic activity. At present, however, there is no evidence to suggest that limbic activity triggered by environmental exposures can be entirely overcome by psychological interventions.

One important ramification of a limbic hypothesis, if true, is that there may be no convenient biological marker for multiple chemical sensitivity.

Detection of chemical stimuli in the nose is not limited to the olfactory nerve, but involves the trigeminal nerve and its afferents which may also play a role in this condition. Trigeminal free nerve endings in the nose and mouth detect noxious chemicals and reflexively initiate protective responses including cessation of breathing, constriction or dilatation of the airways, reduction in heart rate and cardiac output, constriction of most blood vessels (except capillaries in the head), increased epinephrine release, changes in blood pressure and efforts to withdraw (Silver and Maruniak, 1981). This is a powerful reflex which serves an obvious protective role. Most familiar is the trigeminal reflex response to smelling salts. Apparently trigeminal responses can occur with non-irritating stimuli. Thus far, no odor has been found that stimulates the trigeminal nerve alone or the olfactory nerve alone (Silver and Maruniak, 1981) making study of either system in isolation difficult. Potentially both play a role in chemical sensitivity.

BIOCHEMICAL MECHANISMS

Rea and other ecologists who see chemically sensitive patients have noted vitamin and mineral abnormalities in many of them. Others argue that these patients are often sick, debilitated, and malnourished, and therefore, such findings are not surprising.

Individuals who have defective enzyme detoxification systems could be more susceptible to low level exposures. Conceivably, chemically sensitive individuals could have defective detoxification pathways, and be more affected while others in the same environment tolerate the same exposures without symptoms (Rogers, 1990). Rea has noted that many of his chemically sensitive patients have decreased levels of detoxifying enzymes, such as glutathione peroxidase. This possibility is particularly intriguing because such enzyme systems are inducible (that is, can be stimulated) and thus might conform to an adaptation hypothesis. Scadding and associates noted poor sulfoxidation ability in 58 of 74 patients with well-defined reactions to foods versus 67 of 200 normal controls $p < 0.005$) (Scadding et al., 1988). Similarly, Reidenberg reported the case of a laboratory technician who developed a lupuslike disease in response to hydrazine (Reidenberg et al., 1988). She was genetically a slow acetylator, which may might have predisposed her to developing a lupuslike disorder after sufficient exposure to the chemical. A deficiency of one or more particular enzymes could help to explain why some patients are more susceptible to foods and chemicals than others. Further, damage by a toxin might compromise detoxification pathways so that other substances formerly metabolized by this pathway could not be degraded properly and thus might provoke symptoms at low exposure levels, a hypothetical basis for the spreading phenomenon.

Levine has proposed that environmental sensitivities are the result of toxic chemicals reacting with cell constituents to create free radicals (which are formed when a molecule loses an electron) (Levine, 1983). He hypothesizes that if an antioxidant molecule (such as vitamin A, C, E, or selenium) is not present nearby to supply the missing electron, then an electron may be removed from an unsaturated lipid in a cell membrane, leading to membrane damage, release of prostaglandins and other inflammatory mediators, and formation of antibodies to chemically altered tissue macromolecules.

VASCULAR MECHANISMS

Rea hypothesizes that blood vessel constriction, inflammation, or leakage in multiple organ systems could explain the bizarre combinations of symptoms in these patients. In his view, particular complaints could mirror the site and size of affected blood vessels. Spasms in large-caliber arteries, either acutely or chronically, could reduce blood supply to an organ or limb and result in dysfunction, pain, or even necrosis (Rea, 1975). Chemical injury to the fragile walls of smaller vessels, however, would be more likely to cause hemorrhage (resulting in petechiae and bruises) or edema (Rea, 1979). The walls of blood vessels contain smooth muscle. Rea notes that other tissues containing smooth muscle such as the respiratory, gastrointestinal, and genitourinary systems are frequently involved in these patients (Rea, 1977). Impaired blood vessel or altered smooth muscle function might explain the diverse and seemingly unrelated symptoms occurring in patients with multiple chemical sensitivities. In the case of either blood vessel or smooth muscle dysfunction, clearly, neurological and immune alterations or changes in membrane permability or receptors (for example, denervation - supersensitivity) could play primary roles. A vascular hypothesis might also explain why some patients experience increased pain or other symptoms at the site of an earlier injury or surgery, where blood flow may be relatively compromised.

Perhaps the mechanism for multiple chemical sensitivities is not identifiable; that is, after all avenues of biochemical and immunological inquiry have been exhausted, no single explanation for this disorder is forthcoming. The theory of substance-specific adaptation is based upon observations of the responses of patients in a deadapted state who are worked up in an environmental unit. Adaptation is *only* an observation at this time, not a mechanism. However, biological *limits* might regulate how much an organism can adapt, limits that could be highly individual and vary by orders of magnitude. Certainly adaptation occurs at all levels of biological systems, from enzyme systems to cells, tissues, organs, and even behavior (Fregly, 1969). Theoretically, a major insult or the accumulation of lower-level injuries within these systems could lead to a kind of "overload" or "saturation" effect with respect to adaptive capacity that would cause an individual to have environmental responses, which, instead of being flexible and fluid, are now fragile and overly responsive. Many patients report that even after years and in some cases decades following the onset of their problems they have recovered only a portion of their former energies and tolerance for their environment. Their descriptions seem to suggest the loss of an intangible capacity to adapt, parts of which may be temporary and recoverable and other parts of which may not.

REFERENCES

Ashford, N. A. and Miller, C. S. 1991. Chemical Exposures: Low Levels and High Stakes. New York: Van Nostrand Reinhold.

Bell, I. R., Clinical Ecology. 1982. Common Knowledge Press: Bolinas, California.

Bell, I. R. 1990. The biopersonality of allergies and environmental illness. Paper presented at the Eighth Annual International Symposium on Man and His Environment in Health and Disease. Dallas, Texas.

Besedovsky, H., Sorkin, E., Felix, D. and Haas, H. 1977. Hypothalamic changes during the immune response. Eur. J. Immunol. 7:323-325.

Bokina, A. I. Eksler, N. D., Semenenko, A. D., and Merkuryeva, R. V. 1976. Investigation of the mechanism of action of atmospheric pollutants on the central nervous system and comparative evaluation of methods of study. Environ. Health. Perspect. 13:37-42.

Doane, B. K. and Livingston, K. E., editors. 1986. Clinical psychiatry and the physiodynamics of the limbic system. Pp. 285-315 in The Limbic System: Functional Organization and Clinical Disorders. New York: Raven Press.

Fregly, M. J. 1969. Comments on cross-adaptation. Environ. Res. 2:435-441.

Gilman, S., and Winans, S. W. 1982. Manter and Gatz's Essentials of Clinical Neuroanatomy and Neurophysiology. Philadelphia: F. A. Davis Company.

Girgis, M. 1986. Biochemical patterns in limbic system circuitry: biochemical-electrophysiological interactions displayed by chemitrode techniques. Pp. 55-65 in Limbic System: Functional Organization and Clinical Disorders. New York: Raven Press.

Isaacson, R. L. 1982. The Limbic System. New York: Plenum Press.

Jesberger, J. A. and Richardson, J. S. 1988. Brain output dysregulation induced by olfactory bulbectomy: an approximation for the rat of major depressive disorders in humans. Intern. J. Neurosci. 38:241-265.

Kare, M. 1986. Direct pathways to the brain. Science. 163:952-953.

Kilburn, K. H. 1989. Is the human nervous system most sensitive to environmental toxins? Arch. Environ. Health. 44(6):343-344.

Levine, S. A. and Reinhardt, J. H. 1983. Biochemical pathology initiated by free radicals, oxidant chemicals, and therapeutic drugs in the etiology of chemical hypersensitivity disease J Orthomolecular Psychiatry. 12:166-183.

MacLean, P. D. 1967. The brain in relation to empathy and medical education. J Nerv

Ment Dis. 144(5):374-382.

Maclean, P. D., Boag, T. and Campbell, D. 1973. A Triune Concept of the Brain and Behavior: The Hincks Memorial Lectures. Pp. 6-66. Toronto, Ontario. University of Toronto Press.

Maclean, P. D. 1986. Culminating developments in the evolution of the limbic system: the thalamocingulate division. In Doane, B. K. and Livingston, K. E., editors. Pp. 1-28. Organization and Clinical Disorders. New York: Raven Press.

Maller, O., Kare, M. R., Welt, M. and Behrman, H. 1967. Movement of glucose and sodium chloride from the oropharyngeal cavity to the brain. Nature 213(2):713-714.

Monroe, R. R. 1986. Episodic behavioral disorders and limbic ictus. Pp. 251-266 in The Limbic System: Functional Organization and Clinical Disorders. Doane, B. K., and Livingston, K. E., editors. New York: Raven Press.

Office of Technology Assessment, Congress of the United States. 1990. Identifying and Controlling Poisons of the Nervous System. Washington, D. C.: United States Government Printing Office.

Payan, D. G., McGillis, J. P. and Goetzl, E. J. 1986. Neuroimmunology. Adv. Immunology. 39:299-323.

Rea, W. J. 1977. Environmentally triggered small vessel vasculitis. Ann. Allergy. 38:245-51.

Rea, W. J. 1979. The environmental aspects of ear, nose and throat disease. Part I. Oto-Rhino-Laryngology & Allergy Digest 41(7):41-56.

Rea, W. J., Bell, I. R. and Smiley, R. E. 1975. Environmentally triggered large-vessel vasculitis. Pp. 185-198 in Allergy and Medical Treatment. Johnson, J. and Spencer, J. T., editors. Chicago: Symposia Specialists.

Reidenberg, M. 1983. Lupus erythematous-like disease due to hydrazine. Am. J. Med. 75:363-370.

Rogers, S. A. 1990. A practical approach to the person with suspected indoor air quality problems. Indoor Air. Fifth International Conference on Indoor Air Quality and Climate, Toronto, Ontario 5:345-349.

Ryan, C. M., Morrow, L. A. and Hodgson, M. 1988. Cacosmia and neurobehavioral dysfunction associated with occupational exposure to mixtures of organic solvents. Am. J. Psychiatry 145(11):1442-1445.

Sanderson, W. 1989. The influence of an illusion of control on panic attacks induced via inhalation of 5.5 percent carbon dioxide-enriched air. Arch. Gen. Psychiatry 46:157-162.

Scadding, L. 1988. Poor sulphoxidation ability in patients with food sensitivity. Br. Med. J. 297(6641):105-107.

Selye, H. 1946. The general adaptation syndrome and the diseases of adaptation. J. of Allergy. 17:231-247, 289-323, 358-398.

Shipley, M., 1985. Transport of molecules from nose to brain: transneuronal antegrade and retrograde labeling in the rat olfactory system by wheat germ agglutinnin-horseradish peroxidase applied to nasal epithelium. Brain Res. Bulletin. 15:129-142.

Silver, W. L. and Maruniak, J. A. 1981. Trigeminal chemoreception in the nasal and oral cavities. Chemical Senses 6(4):295-305.

Stein, M., Keller, S. and Schleifer, S. 1981. The hypothalamus and the immune response. Pp. 45-65. In Weiner, H.,Hofer, M. A. and Stunkard, A. J., editors. Brain, Behavior and Bodily Disease. New York: Raven Press 45-65.

Walsh, T. J., and Emerich, D. F. 1988. The hippocampus as a common target of neurotoxic agents. Toxicology. 49:137-140.

Developing Clinical Research Protocols for Studying Chemical Sensitivities

William J. Meggs

The decade of the eighties brought progress in identifying and quantifying the low level exposures to volatile organic compounds found in the indoor air of homes, schools, and offices. In addition, some data is now available on both endogenous and added low molecular weight organic chemicals in food and water. Claims of adverse health effects from these exposures are far-ranging, and the challenge is now upon the clinical investigator to scientifically determine if there are adverse health effects associated with these exposures. There are specific difficulties associated with the development of research protocols in this area, and issues that must be addressed include: (a) Determination of patient populations to be studied, (b) the determination of proper control groups, (c) proper techniques for blinding challenges with odorous substances, (d) assessing psychiatric problems, (e) allowing for "adaptative" phenomena, (g) designing studies which will simultaneously establish if stimulus-response relationships exist while investigating possible underlying mechanisms and perhaps most difficult, (f) designing studies that will be definitive in a highly controversial and politicized area of investigation.

There are three approaches to defining patient populations to be studied. In the first circumstance, patients are recruited on the basis of reactivity to chemicals, either by meeting a case definition for multiple chemical sensitivity syndrome or claimed reactivity to multiple chemicals, with the focus on the etiological agents which may or may not reproduce the syndrome. In the second approach patients are recruited that have specific diseases such as asthma, rheumatoid arthritis, or depression, with effects of withdrawing and reexposing patients to specific chemicals determined. In the third design, a given population such as a group of workers in a specific factory are sampled for response. In all three approaches, studies must be designed employing objective parameters to measure disease activity. Studies on patient populations with polysomatic complaints and no objective findings is not likely to yield definitive data. Patients with both objective illness and alleged chemical reactivity are available and are preferred patients for study.

Diseases are potentially induced or exacerbated by environmental chemicals if: (a) the disease incidence has increased in recent years, (b) occupational exposure induces or exacerbates the disease, (c) there are case reports or anecdotal data relating the disease to chemical exposures, (d) the disease activity is linked to pharmaceuticals, as in drug-induced autoimmune hemolytic anemia, and (e) if animal models of the disease show induction by chemical exposure. Using these criteria, a group of diseases that need to be studied is given in Table 1. We do not claim that all of these diseases rigorously meet each of the above criteria. Common features of these diseases are a waxing and waning course, localized tissue inflammation, and ultimately tissue destruction. They are candidates for Stage 2 and Stage 3 illnesses of the Hypothetical Chemical Stress Syndrome, discussed elsewhere in these proceedings.

TABLE 1

Some of the Diseases Whose Relationship to Environmental Chemicals Needs Clarification
Asthma
Autoimmune hemolytic anemia
Crohn's Disease
Depression
Manic depressive illness
Multiple sclerosis
Rheumatoid Arthritis
Schizophrenia
Systemic Lupus Erythematosis
Ulcerative Colitis

Clinical research protocols must be designed to incorporate suitable control groups. Investigators must recognize that the entire population is exposed to low-level volatile organic chemicals (VOC's), and there is data suggesting that normal volunteers exposed in challenge chambers to mixtures of VOC's can become symptomatic. Some effects of chronic low-level VOC exposure may be seen in both normal individuals and patients.

Clinical studies of adverse effects of chemicals must be double-blinded, but unique difficulties arise in the case of odorous inhalants. Odor masking by exposing the subject to a high dose chemical considered innocuous and a low dose chemical to be tested presents unique problems in this setting and will require systematic study. Anesthetizing the olfactory bulb may mask olfactory mediated physiologic reactions. The experience of blinding techniques in the study of food allergy and food intolerance should be heeded, and these investigators have found that blinding techniques should not be accepted a priori, but should be subjected to scientific scrutiny. Exposures of patients during sleep may be the most effective technique, but will need controlled investigation due to altered physiology during sleep.

Clinical studies must allow for psychiatric illnesses in the patient population. It is recognized that there are behavioral aspects to this illness. In the original description of the multiple chemical sensitivity syndrome, it was claimed that psychiatric symptoms including depression, hallucinations, and manic states can be induced in some patients by chemical exposure. A contrary view is that this population consists of psychiatric patients with somatization and chemical phobias. Clinical studies can address this controversy by assessing the psychiatric state of patients in the presence and absence of chemical exposures. Great care must be employed in the design of double-blinded psychiatric evaluations.

In the original clinical description of the multiple chemical sensitivity syndrome, it was claimed that chemically sensitive patients chronically exposed to chemical inhalants do not react acutely to chemical challenges and go through a withdrawal phase lasting approximately five days when removed from the chemical environment. Chronic symptoms are said to resolve after withdrawal, but the patients develop acute symptoms when rechallenged with chemicals. Clinical studies must allow for this adaptation mechanism in order to be accepted. Considerations of possible adaptative phenomena will require residential clinical research units in which indoor air can be controlled. A further consideration during the period of study includes monitoring for reactivity to both endogenous and added organic chemicals in food, and contaminants of water.

Study design should include specific laboratory studies to objectively address defined questions. Though the pathogenesis of this syndrome has not been established, proposed mechanisms should be considered and used as a basis to design a laboratory panel to be followed as patients enter the protocol, undergo withdrawal from the chemical environment in a clinical research unit, and then are reexposed to specific agents. For instance, Are there changes in helper and suppressor lymphocyte numbers, or lymphocyte activation, during this process? Following acute challenge, are substances such as histamine or metabolites of neurotransmitters released into the blood or urine?

In summary, scientific understanding in this area of clinical medicine will only develop with the establishment of clinical research units designed to minimize volatile organic chemicals in the air, staffed with multidisciplinary teams of specialists, and employing well-designed clinical protocols. Patients should be studied in defined populations such as those characterized by quantifiable, objective diseases. Disease activity and laboratory parameters should be followed from entry through withdrawal from organic chemicals, and then upon reexposure. Trials must be double-blinded, with careful assessment for changes in disease activity and in laboratory parameters. Normal volunteers should undergo identical evaluations. Costs of this research and potential sources of funding will be discussed.

Immunological Mechanisms of Disease and the Multiple Chemical Sensitivity Syndrome

William J. Meggs

INTRODUCTION

The Multiple Chemical Sensitivity Syndrome (MCS) is a highly controversial illness which afflicts an unknown number of people. The very existence of this syndrome is questioned, but it is clear from the experience of physicians from several specialties that a population of patients exists that goes to great lengths to avoid a group of odorous organic chemicals at levels that are known to occur in contemporary homes, schools, and offices. These chemicals include the products of combustion, cleaning products, fragrances, pesticides, outgassing from synthetic materials used for clothing, furnishing, and building products and are commonly found in the indoor air.

This type of sensitivity is colloquially called "allergy" but differs from traditional allergic reactions. At the present time the mechanism of this intolerance of chemicals is as controversial as the syndrome itself, but it is important to consider hypothetical mechanisms at this early stage of scientific scrutiny. Some say there is not scientific evidence that this syndrome exists, but the syndrome exists because the patient population exists, and in extreme cases these people are disabled and no longer able to function productively in society. What is expressed by disclaimers is a belief that the mechanism of the syndrome is a mental disturbance such as somatization, and at this stage the somatization hypothesis is one of several to be considered by investigators designing clinical research protocols.

This article will discuss proposed and hypothetical immunological mechanisms for the Multiple Chemical Sensitivity Syndrome. In the next section, descriptions of this syndrome will be discussed, and the syndrome will be described in terms of a chemical stress syndrome. Known immunological mechanisms for the pathophysiology of established diseases will then be discussed and compared with MCS to see if any are feasible explanations. Finally, the hypothesis that the mechanism of MCS involves neuroimmunology, the interplay between the immune and nervous systems, will be presented.

DESCRIPTIONS OF THE MULTIPLE CHEMICAL SENSITIVITY SYNDROME

An *opinion* that exposure to environmental organic chemicals could lead to illnesses was expressed in 1962 in a short book, *Human Ecology and Susceptibility to the Chemical Environment*, written by a Chicago allergist (Randolph, 1962). The characterization of this illness is presented in Table 1, and this type of sensitivity was presented as being distinct from classical allergic sensitization. Notable features of this description include an adaptation phenomenon, with: (a) sensitive individuals having chronic symptoms while living in the chemical environment which resolve several days after removal from the chemicals, (b) withdrawal symptoms upon removal, and (c) "shock reactions" when these individuals are reexposed to chemicals. Patients affected by this disorder have a variety of physical and *mental* illnesses which can be managed by avoiding incriminated chemicals. A spreading phenomenon was described, in which affected individuals become progressively sensitive to more and more chemicals found in indoor air, water, and foods. A set of chemicals which were found in clinical practice to cause illness in some patients is given in Table 2. This work was presented as an opinion based upon the clinical practice of having patients avoid chemicals and foods using trial and error, and no scientific data such as double-blinded studies were presented to substantiate this point of view.

TABLE 1

Randolph's Description of Chemical Sensitivity Disease
Acquired, often after chronic insidious or acute exposure to a petrochemical.
Chemical exposures trigger physical (arthritis, asthma, colitis, etc.) or mental (depression, difficulty with concentration, mania, psychosis, etc.) symptoms.
Specific Adaptation Syndrome. Adaptation to specific chemicals with chronic exposure is followed by maladaptation and chronic illness, withdrawal symptoms when removed from the chemical environment, and "shock reactions" on reexposure.
Spreading phenomena. As an individual becomes maladapted to the chemical environment, intolerance to increasing numbers of environmental chemicals develops.
Avoidance. By avoiding the chemical environment, chronic illnesses may resolve.

A case definition of the Multiple chemical Sensitivity Syndrome based on the experience of occupational medicine specialists around the country has been presented (Cullen, 1989). Table 3 gives the seven criteria presented in this case definition. Comparison of Tables 1 and 3 shows considerable overlap, and a discussion of the relationship between the two will be given below.

An operational definition of MCS has been given (Ashford and Miller, 1989, 1991),

TABLE 2

Chemicals Alleged to Cause Reactions in chemically Sensitive Individuals (From Randolph, 1962)

Indoor air contaminants
- utility gas
- coal smoke
- fumes from fresh paint, turpentine, mineral spirits, detergents
- fragrances from toiletries and perfumes
- cleaning products such as bleach, ammonia, disinfectants
- inspect sprays and repellants
- odorous synthetic carpets, pads, adhesives, and building materials

Outdoor air contaminants
- automobile and diesel exhaust
- industrial air pollutants
- paint manufacturing and sulfur-processing fumes
- fumes from roof tar and roads

Chemical additives and contaminants of food and water
- insecticides and fumigant residues
- some chemical preservatives
- sulfur residues
- chemical flavoring and sweetening agents
- plastic containers and lined tins
- chloride in water

Synthetic drugs, cosmetics, and miscellaneous chemicals
- medications, including aspirin, sulfonamides, synthetic vitamins
- allergic extracts or other biologic materials that contain phenol
- cosmetics
- synthetic textiles
- bed linens washed with detergents, dried in gas driers, or impregnated with synthetic starch or sizing

that "The patient with multiple chemical sensitivities can be discovered by removal from the suspected offending agents and by rechallenge, after an appropriate interval, under strictly controlled environmental conditions. Causality is inferred by the clearing of symptoms with removal from the offending environment and recurrence of symptoms with specific challenge."

A description of this syndrome is best given in the context of what we call the "Hypothetical Chemical Stress Syndrome", a concept which is not original and draws heavily on the work of others (Selye, 1946; Randolph, 1962). The word "hypothetical" is used because at the present time there is insufficient data to establish a consensus about the role

TABLE 3

Examples of Host Responses that Must Also Be Considered Immunological
Endothelial cell reactions
Activation of C cascades in the absence of antibody
Activation of procoagulant activity
Acute phase reactant responses
Aracidonic acid metabolism
NK cell enhancement
Lysosomal enzyme release
Generation of toxic oxygen radicals
IL-1 effect on the hypothalamus
Beta-endorphin activity
Generation of neuropeptides

of this syndrome in human disease. The word "chemical" is used because our focus is on the stress of exposure to environmental chemicals, though some would argue that emotional and other stresses can "cross react" with chemical exposures.

The stages of the Hypothetical Chemical Stress Syndrome (HCSS) are given in Figure One. Individuals in Stage 0 tolerate chemical exposures without difficulty. Progression from Stage 0 to Stage 1 occurs when an individual is chemically stressed either by an acute high-dose chemical exposure, or by a chronic insidious exposure. Individuals in Stage 1 have symptoms on exposure to chemicals, but no physical findings on physical examination. Hence, Stage 1 is the "-algia" stage, with -algia being the Greek suffix used in medicine to denotes pain; e.g., arthralgia denotes joint pain while myalgia denotes muscle pain.

Stage 2 is reached when an individual develops foci of inflammation from chemical exposures. It is at this stage that both findings on physical examination appear and a medical diagnosis can be given. The specific medical diagnoses which might be amenable to this approach are discussed elsewhere in these proceedings (Meggs, 1991). This stage is referred to as the "-itis" stage, with -itis being the Greek medical suffix denoting inflammation. Arthritis is inflammation of joints, myositis is inflammation of muscle, and so forth. The progression from Stage 1 to Stage 2 again follows increasing chemical exposures, and if tissue damage has not occurred, the inflammation can be reversed by removal of the chemical stimuli. That is, progression between Stages 1 and 2 is a two way process, with progression from Stage 2 to Stage 1 being possible if chemicals are avoided. The inflammation of Stage 2 can be reduced by medications such as corticosteroids and the non-steroidal anti-inflammatory agents, but these medications are not curative. If the chemical stimuli are not removed, there is immediate relapse of inflammation. Further, these medications do not prevent the progression from Stage 2 to Stage 3.

Stage 3 occurs if an individual suffers damage to tissue, so this stage can be

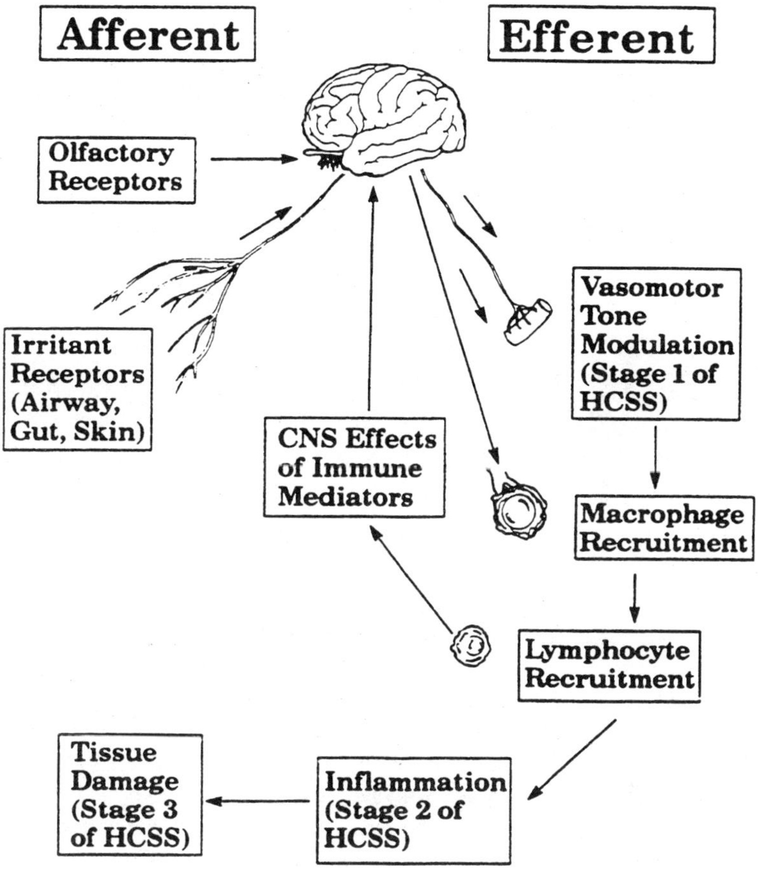

FIGURE 1 Hypothetical chemical stress syndrome.

characterized by "-osis". In medicine, the Greek suffix -osis denotes a change in living tissue; e.g., necrosis means death of tissue and fibrosis means the replacement of functional tissue by scar tissue. Unfortunately, once tissue is damaged there is little hope in current medical practice for a reversal, and organ function is lost.

Generally, the progression into Stages 2 and 3 occurs in one organ system, such as the skin, joints, or lungs. Later, a second organ system may become involved. It may still be important to remove an individual who has reached the irreversible tissue damage of Stage 3 from the chemical environment because it may be that inflammation can occur in other tissues.

What many clinicians call the Multiple Chemical Sensitivity Syndrome is Stage 1 of the Hypothetical Chemical Stress Syndrome. These patients have multiple complaints which they attribute to chemical exposures, but little if any findings on physical examination. Hence, it is easy to regard these patients as having a somatization disorder with no real medical illness. It is important that studies of adverse reactions to environmental chemicals include patients with diagnosable medical conditions, that is Stage 2 and 3 patients, and that the hypothesized progression between these stages be studied. In this regard, longitudinal studies will be very important.

The concept of a stress syndrome was originally introduced as the "Generalized

Adaptation Syndrome" to explain observations in animal models of disease (Selye, 1946). In this syndrome, laboratory animals exposed to stresses progress through stages of adaptation similar to the stages described above. The syndrome was initially discovered while exposing animals to chemical stresses, and the word "generalized" was introduced in the name of this syndrome because other stresses, for example the stress of confinement, was observed to lead to similar results.

The "Specific Adaptation Syndrome" has been described in humans to differ from the "Generalized Adaptation Syndrome" in that maladaptation can occur to one specific stress, in particular a single chemical (Randolph, 1962). The extent to which these stress syndromes are equivalent can be debated, but in any event they are very similar. The purpose of formulating the Hypothetical Chemical Stress Syndrome is to develop a specific hypothesis for studying adverse reactions to environmental chemicals such as those found in the indoor air.

IMMUNOPATHOGENESIS OF DISEASE AND CHEMICAL SENSITIVITIES

It is well known that the interaction of the immune system with environmental chemicals can cause disease. In this section, the known immunological mechanisms of disease will be reviewed, and their possible relationship to the multiple chemical sensitivity syndrome will be discussed. The function of the immune system is to protect the organism from invasion by pathogens such as bacteria, fungi, viruses, and parasites. It is known that the immune system has different arms, or branches, as listed in Table 4, which are recruited to deal with different types of pathogens. We know that these branches are distinct and recent investigations suggest that activation of one branch can actually deactivate another (Moore et al., 1990).

TABLE 4

Branches of the Immune System and Their Function

Branch	Function
IgE-eosinophil system	Destruction of tissue parasites
IgG-neutrophil/macrophage system	Destruction of encapsulated bacteria
lymphocyte system	Destruction of fungi and viruses
IgA-secretory system	Destruction of pathogens at mucosal surfaces

Further, activation of the immune system can lead to inflammation, disease, and tissue damage. IgE mediated reactions arise when foreign proteins or low molecular weight compounds such as drugs bound to foreign proteins interact with IgE bound to the surface of mast cells to release histamine and other immune mediators, leading to the classical allergic reactions to foods, aeroallergens, pharmaceuticals, and insect venoms. Clinical manifestations of these reactions are asthma, urticaria, rhinitis, anaphylactic shock, or gastrointestinal problems. Some of the mediators released in allergic reactions are chemotactic factors, that is, they attract white blood cells into the tissues. Some patients with MCS report flushing and itching after chemical exposures, and a minority of patients report that they developed classical allergic disease after the onset of MCS. These anecdotes do not change the fact that MCS is distinct from classical allergy, as originally stated by Randolph.

Immune complexes are high molecular weight aggregates of antigen and antibody which can deposit in vessel walls, activate complement, and lead to inflammation and tissue damage, with serum sickness being the classical and experimental form of immune complex disease. The clinical picture of serum sickness is distinct, with fever, urticaria, fatigue and malaise, arthritis, and sometime renal failure. Immune complexes have been associated with autoimmune diseases and a host of other clinical conditions (Cochrane, 1988). While they do not play a role in MCS, they may play a role in stage 2 and 3 illnesses of fig. 1 related to chemicals.

Delayed type hypersensitivity, also sometimes referred to as cellular immunity and mediated by T-lymphocytes, can lead to inflammatory changes in tissues after a sensitizing exposure followed by reexposure. The initial exposure leads to antigen-specific T-cells which become activated on reexposure and release cytokines. Experimentally this is most commonly demonstrated in the skin, and contact dermatitis in humans arising from organic chemicals is mediated by cellular immunity. Of the known types of immune reactivity, cellular immunity is the one most likely to play a direct role in adverse reactions to environmental chemicals. The delayed or withdrawal nature of the inflammatory symptoms described in these patients is similar to the delay seen in cellular inflammation. Two of the interleukins (IL-1 and IL-2) are known to have effects on the central nervous system (Obal et al., 1990 and De Sarro et al., 1990). Several authors have claimed abnormalities of cellular immunity in these patients, such as alterations of helper and suppressor ratios and the presence of activated lyumphocytes in peripheral blood, as discussed below.

There are at least theoretical analogies in immunology to the spreading phenomena discussed for patients with multiple chemical sensitivities. The concept of an environmental adjuvant has been introduced to describe substances which induce an immune response to other substances. In immunology, an adjuvant is a substance which potentiates an immune response to other substances. An example of an adjuvant is alum, for an animal injected with alum mixed with a protein will begin producing IgE specific for the protein. Examples of inhaled environmental adjuvants in experimental models include sulfur dioxide, which can induce allergic asthma to co-administered aeroallergens in a guinea pig model (Riedel et al, 1988), and ozone, which has been shown to potentiate platinum sensitivity in monkey (Biagini et al., 1986).

The mechanism of environmental adjuvancy may be that inflamed tissue processes antigen. Substances like sulfur dioxide and ozone produce tissue irritation, which then leads to inflammation with the recruitment of macrophages to the area. Macrophages then present antigens in the area to lymphocytes to produce an immune response to the antigen. Subsequent exposure to either the adjuvant or the antigen could produce further inflammation and processing of other antigens. The process may be initiated by the interaction of chemical irritants with sensory nerves.The repertoire of triggers of inflammation would grow, with increased risk of more substances becoming triggers. Such an immunological spreading phenomena has not been demonstrated to my knowledge but is theoretically possible.

REVIEW OF SELECTED PUBLISHED DATA ON IMMUNE ABNORMALITIES IN CHEMICALLY SENSITIVE PATIENTS

Studies have appeared which attempt to establish an immunological basis for patients with multiple chemical sensitivities, as described in Table 5. There are studies depicting increases in T4/T8 ratios in chemically sensitive patients, decreases in this ratio, and no abnormalities in this ratio. It is important to note, however, that the patient populations differ in these studies. One study described three groups of chemically sensitive patients, those with vascular dysfunction (vasculitis), asthma, and rheumatoid arthritis (Rea et al., - +1986). Chemical sensitivity was verified by chemical challenges in these study patients. Rea's patients would be considered stage 2 and 3 patients in the Hypothetical Chemical Stress Syndrome discussed in section I above. Statistically significant decreases in T8 (suppressor) cells and increases in the T4/T8 (helper:suppressor ratio) were found in patients with vasculitis and rheumatoid arthritis but not in asthmatics.

Levin and Byer (1989) have looked at populations of patients who have had documented environmental exposures to chemicals such as trichlorethylene exposure, PCB's, or a mixture of industrial dyes, solvents, and pesticides. These patients were said to be chemically sensitive following these exposures, and may represent Stage 1 patients in the HCSS. This study found decreased T4:T8 ratios in these patients.

Terr examined the medical records of 50 patients referred by the California Worker's Compensation Board of Appeals who claimed to have difficulties with workplace chemicals and found no consistent pattern of immunological abnormality. In contrast to the patients studied by Rea and his collaborators, these patients had a variety of medical diagnoses and no attempt was made to challenge them with chemical exposures to see if they were sensitive to chemicals. Rather, it was concluded that, since another medical diagnosis could be given, they did not have chemical sensitivities and instead had a somatization disorder.

These three studies obtain conflicting results on the T4:T8 ratio in chemically sensitive populations, but the patient populations differ dramatically. Rea's group studied patients with known inflammatory diseases which improved when isolated from the chemical environment, Levin and Byer studied populations environmentally exposed to organic chemicals, and Terr studied patients claiming chemical sensitivity who had had a claim for worker's compensation rejected. These studies illustrate the importance of using well-designed clinical research protocols to assess chemically sensitive patients, using well-defined patient populations. It can be argued that the expression multiple chemically sensitivity is best used as an etiology, not as a medical diagnosis. The hypothesis presented in Section II is that a number of inflammatory conditions can arise from the chronic stress of exposure to environmental chemicals, and it may be that the various conditions developing from chemical exposures may have different immunological profiles reflecting the medical diagnosis rather than the chemical etiology.

One group of investigators has proposed that antibodies to formaldehyde coupled to

TABLE 5

Selected Published Data on Immune Abnormalities in Chemically Exposed Patients

Reference	Patient Group	Number	Abnormality
Rea et al., 1986	Vasculitis	70	T8 decreased[a], T4/T8 increased[a]
	Asthma	27	T4% decreased[b]
	Rheumatoid arthritis	7	T8 decreased[a], T4/T8 increased[c]
Terr, 1986	Worker's comp. appeals	50	No immunological abnormality
Broughton & Thrasher, 1988	Formaldehyde exposure	67	Antibodies to formaldehyde-albumin Increased CD26 cells
Levin & Byers, 1989	TCE exposure	25	T4/T8 decreased[d]
	Industrial Chemical Exposure	10	T4/T8 decreased[e]
			T4/T8 decreased[e]
	PCB exposure	21	T4/T8 decreased[e]
	Electronic workers	78	
Patterson et al., 1989	Chemically sensitive		No antibodies to formaldehyde-albumin
	Dialysis patients		Antibodies to formaldehyde-albumin
McConnachie & Zahalsky, 1991	Chlordane/heptachlor	27	Increased CD26, decreased CD45R/T4 Decreased ConA mitogen response Decreased mixed lymphocyte response

a $p<0.0001$
b $p<0.05$
c $p<0.001$
d $p<10^{-8}$
e $p<10^{-6}$

human serum album (f-HSA) may be a good marker for chemical sensitivity (Broughton and Thrasher, 1988; Thrasher et al., 1988; Thrasher et al., 1989). These investigators found low level IgG, IgE, and IgM titers to f-HSA, in the 1:4 to 1:16 range, in chemically sensitive patients. Well-established associations between diseases and antibody titers, such as the association of systemic lupus erythematosis with anti-DNA titers, are characterized by extremely high titers of 1:1,000 or higher. Non-specific binding at the high serum concentrations found in low titer wells make low-titer associations with disease of dubious clinical significance. Another group has found no association between formaldehyde conjugates and chemical sensitivity, but has found titers to be a possible marker of exposure to formaldehyde in renal dialysis patients (Patterson et al., 1989).

An increased frequency of the CD26 lymphocyte marker in chemical sensitivity individuals has been reported (Broughton and Thrasher, 1988). This marker has also been reported with greater frequency in chlordane/heptachlor exposed individuals (McConnachie and Zahalskyu, 1991), though the issue of chemical sensitivity is not addressed in this study. Since CD26 is a lymphocyte activation marker, the notion that chronic activation of cellular immunity by chemical exposures is an attractive one which deserves further study.

NEUROIMMUNOLOGY AND CHEMICAL SENSITIVITIES

Over the past decade there has been an increased appreciation of the numerous interactions between the nervous system and the immune system. It is tempting to speculation that an understanding of chemical sensitivity may be found in the study of neuroimmunology, given the neurological and behaviorally abnormalities in this patient population and the altered reactivity to normally harmless levels of environmental chemicals reminiscent of immune mediated phenomena. In this section we propose a hypothetical model of disease based on the interactions of the immune and nervous systems, as depicted in Fig. 2. Odorous chemicals excite the nervous system in two ways, through olfactory nerves and irritant nerve fibers in the airway. According to the hypothetical tenet of neuroimmunology, given in Table 6, stimulation of irritant fibers in the airway by volatile organic chemicals leads to inflammation. Normal volunteers exposed for four hours to a mixture of volatile organic chemicals were found to have increased numbers of polymorphonuclear cells in nasal lavage washings eighteen hours (Koren et al., 1990), suggesting an association between irritant chemicals and inflammation.

TABLE 6

The Hypothetical Tenets of Neuroimmunology
Tenet 1. Irritancy leads to inflammation and vasomotor abnormalities.
Tenet 2. Inflammation provokes an immune response, that is inflamed tissue processes antigen.
Tenet 3. Immune mediators modulate the central nervous system.

As the airway becomes inflamed, macrophages are recruited to the area and begin to process antigens, including organic chemicals bound to tissue proteins, for presentation to lymphocytes. We propose that this process leads to sensitized lymphocytes specific for the chemicals which were present in the airway. Subsequent exposure to these chemicals can

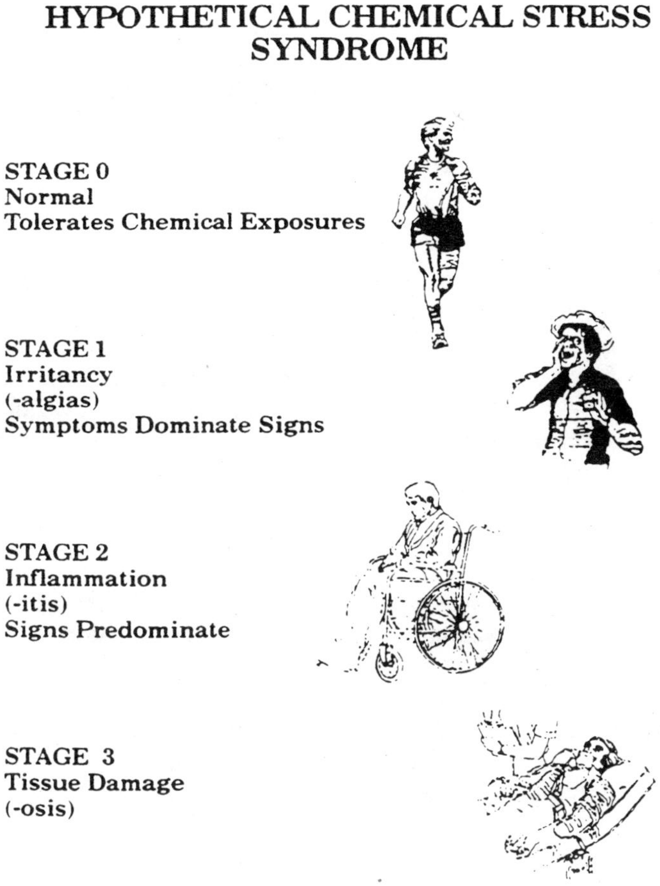

FIGURE 2

lead to the production of cytokines which can have systemic effects and may effect the central nervous system. Two interleukins, IL-1 (Obal et al., 1990) and IL-2 (De Sarro et al., 1990) are known to have activity on the central nervous system. Perhaps the difficulty with concentration that is reported by chemically sensitive individuals following exposures is related to the release of interleukins with central nervous system activity.

Blood flow in vessels throughout the body is modulated by the contraction and relaxation of smooth muscles in the blood vessel walls, or vasomotor tone, and the tone of these muscles is controlled by the nervous system. Tenet Number 1 in Table 6 proposes that stimulation of irritant receptors by chemicals can alter blood flow by dilating or constricting vessels. Clinical manifestations would include headache, nasal congestion, and rhinorrhea. It is interesting to speculate if neuronal pathways can be acquired so that stimulation by chemicals in the airway might lead to vasomotor phenomena in other organs. One would then have the possibility of protean manifestations of chemical exposures depending on one's "learned" or acquired responses.

It is instructive to discuss the Hypothetical Chemical Stress Syndrome of fig. 1 in terms of the neuroimmunological mechanisms proposed in fig. 2. Stage 1 patients in the HCSS have an exaggerated vasomotor response to the stimulation of irritant receptors in the airway, with vasomotor tone alterations in organs other than the airway. Stage 2 patients are those who have developed recurrent foci of inflammation resulting from chemical exposures. Stage 3 arises when tissue is destroyed by the inflammatory products including free radicals released by macrophages and neutrophils.

SUMMARY

We have presented a speculative but comprehensive and mechanistic model to explain the alleged adverse reactions to environmental chemicals. The multiple chemical sensitivity syndrome must be discussed in the broader context of stress syndromes. The patient with multiple complaints related to chemicals but few objective signs of disease manifests an early stage of chemical induced disease. We believe that the stimulation of irritant receptors by chemicals, primarily in the airway, leads to modulations of vasomotor tone. In the chemically sensitive patient, this response is exaggerated and can occur various organs through acquired neuronal pathways.

As the disease progresses, foci on inflammation develop, with infiltrations of macrophages, polymorphonuclear cells, and perhaps lymphocytes into the affected tissue. At this point, the patient develops a diagnosable disease and is traditionally treated with corticosteroids, non-steroidal anti-inflammatory agents, bronchodilators, or other pharmaceuticals as appropriate. The inflammation can be reversed by avoiding the causative chemicals if tissue damage has not occurred.

The inflammatory response leads to processing antigen and specific immunological responses to chemicals. Immune mediators then cause further symptomatology including effects on the central nervous system. Tissue destruction can be a final outcome of chemical sensitivity.

This speculative model must be subjected to several tests. Do patients with chemical sensitivities, including those with diagnosable disorders such as rheumatoid arthritis and multiple sclerosis, progress through the stages of the stress syndrome when their chemical environments are manipulated? This question can be answered by placing patients with candidate diseases in an Environmental Control Unit and studying clinical and laboratory changes in their health.

Are the mechanisms proposed above operative in chemically sensitive patients? This question can be addressed by measuring immune parameters, neurological parameters, and mediators in blood and urine while patients' chemical environments are manipulated. An Environmental Control Unit will be necessary for this type of work. Animal models may be helpful in studying the details of the response to chemical exposures depicted in figure 2.

REFERENCES

Ashford, N. A., and Miller, C. S. 1990. Multiple chemical Sensitivities. Report to the New Jersey Department of the Environment.

Ashford, N. A., and Miller, C. S. 1991. Chemical Exposures: Low Levels and high Stakes. Van Nostrand Reinhold, New York.

Biagini, R. E., W. J. Moorman, T. R. Lewis, and I. L. Bernstein. 1986. Ozone enhancement of platinum asthma in a primate model. Am. Rev. Respir. Dis. 134:719-725.

Broughton, A., and J. D. Thrasher. 1988. Antibodies and altered cell mediated immunity in formaldehyde exposed humans. Clinical Toxicology 2, 155-174, 1988.

Cochran, C. G. 1988. Immune Complex Diseases, In J. B. Wyngaarden and L. H. Smith, Jr., ed., Cecil Textbook of Medicine, 18th edition, W. B. Saunders, Philadelphia. chapter 424, p 1960.

Cullen, M. 1989. "The Worker with Multiple Chemical Sensitivities: An Overview." In, Cullen, M. (ed.) Workers with Multiple Chemical Sensitivities. Occupational Medicine: State of the Art Reviews. 2(4):655-662. Hanley & Belfus, Philadelphia.

De Sarro, G. B., Y. Masuda, C. Ascioti, M. G. Audino, G. Nistico. 1990. Behavioral and ECoG spectrum changes induced by intracerebral infusion of interferons and interleukin 2 in rats are antagonized by naloxone. Neuropharmacology 29:167-79.

Koren, H. S., R. B. Devlin, D. Hourse, S. Steingold, and D. E. Graham. The inflammatory response to the human upper airways to volatile organic compounds. 1990. Proceedings of the 5th International conference on indoor air quality and climate. Ottawa, p. 325.

Levin, A. S. & V. S. Byers. 1989. Environmental illness: a disorder of immune regulation. In M. R. Cullen, ed., Workers with multiple chemical sensitivities. State of the art reviews: Occupational Medicine. 2(4)669-681.

Meggs, W. J. "Development of a Clinical Research Protocol for the Multiple Chemical Sensitivity Syndrome."

Moore K. W., P. Vieira, D.F. Fiorentino, M. L. Trounstine, T. A. Khan, T. R. Mosmann. 1990.Homology of cytokine synthesis inhibitory factor (IL-10) to the Epstein-Barr virus gene BCRFI. Science 248:1230-4.

Obal, F. Jr., M. Opp, A. B. Cady, L Johannsen, A. E. Postlethwaite, H. M. Poppleton, J. M. Seyer, and J. M. Krueger. 1990. Interleukin 1 alpha and an interleukin 1 beta fragment are somnogenic. Amer Jour Physiology 259:R439-46.

Patterson, R. et al. 1989. IgG antibodies against formaldehyde human serum proteins: a comparison of other IgG antibodies against inhalant proteins and reactive chemicals J

Allergy Clin Immunol 84:359-366.

Randolph, T. G. 1962. Human Ecology and Susceptibility to the Chemical Environment. Charles C Thomas, Springfield, IL.

Rea, W. J., A. R. Johnson, S. Youdim, et al., 1986. T&B lymphocyte parameters measured in chemically sensitive patients and controls. Clinical Ecology 4, 11.

Riedel, , F., M. Kramer, C. Scheibenbogen, and C. H. Rieger. !988. Effects of SO2 exposure on allergic sensitization in the guinea pig. Jour Allergy Clin. Immunol. 82:527-534.

Terr, A. I. 1986. Environmental Illness: a clinical review of 50 cases. Arch Int Med 146

Considerations for the Diagnosis of Chemical Sensitivity

William J. Rea, Alfred R. Johnson, Gerald H. Ross, Joel R. Butler, Ervin J. Fenyves, Bertie Griffiths, and John Laseter

INTRODUCTION

The study of the effects of the environment upon the individual is now feasible due to new technology developed in the construction of environmental units.[1,2,3]. Our observations reveal that individual or multiple organs may be involved. The brain is the target organ in only a subset of chemically sensitive patients, and its involvement should not be confused with psychosomatic disease.

Over the last 16 years physicians and scientists at the Environmental Health Center in Dallas have had an opportunity to observe over 20,000 patients who had chemical sensitivity problems. These patients were studied under various degrees of environmental control. This experience is unique in the world and has resulted in numerous peer-reviewed scientific articles, chapters in books, and books on this subject.

Studies have resulted in over 32,000 challenge tests by inhalation, oral, or injection methods, of which 16,000 are double-blind. Blood chemical levels and fat biopsies for organic hydrocarbons number over 2,000, while the measurement of immune parameters are over 5,000 tests. Objective brain function tests have been accomplished in over 5,000 patients. Other objective tests, like computerized balance studies, depollutant enzyme levels, and autonomic nervous system changes as measured by the Iriscorder, number near 1,000.

We wish to share our findings with the participants of the National Academy of Sciences Committee for the study of chemical sensitivity.

DEFINITION AND PRINCIPLES

Chemical sensitivity is defined as an adverse reaction to ambient doses of toxic chemicals in our air, food, and water at levels which are generally accepted as subtoxic. Manifestation of adverse reactions depend on: (1) the tissue or organ involved; (2) the chemical and pharmacologic nature of the toxin; (3) the individual susceptibility of the exposed person (genetic make-up, nutritional state, and total load at the time of exposure); (4) the length of time of the exposure; (5) amount and variety of other body stressors (total load) and synergism at the time of reaction. (6) the derangement of metabolism that may occur from the initial insults.

To demonstrate cause-and-effect proof of environmental influence on an individual's health, one must understand several important principles and facts. These principles involve those
of *total body load* (burden), *adaptation* (masking, acute toxicological tolerance), *bipolarity*, *biochemical individuality*. Each principle will be discussed separately.

TOTAL BODY LOAD (BURDEN)

This is the patient's total pollutant load of whatever source (usually from air, food and water or surroundings [1,2,4]. The body must cope with this total burden; usually it must be utilized, expelled or compartmentalized. Total body load includes: (1) *physical factors* (e.g. hot, cold, weather changes, positive ions.[5] electromagnetic phenomena.[6] radon); (2) *toxic chemicals* (e.g. inorganics: Pb, Cd, Hg, Al, Br, etc.; organics: pesticides, formaldehydes, phenols, car exhausts, etc.).[7-21] (3) *biological* (bacteria, virus, parasites, molds,[22] food.)[23,24] (4) *psychological or emotional factors* also significantly affect the patient, confirmed by recent work in psychoneuroimmunology, linking the psyche and the neuroendocrine and immune systems.[25,26,27,28] Failure to reduce the total body load prior to pollutant challenge will frequently yield inaccurate results. Accordingly, we believe it is essential to conduct investigative procedures in controlled environmental circumstances with the total load reduced.

ADAPTATION
(MASKING, ACUTE TOXICOLOGICAL TOLERANCE)

Induced by the internal or external environment, this is a change in the homeostasis (steady state), of body function with adjustment to a new "set point".[29,30,31,32] Adaptation is an acute survival mechanism in which the individual "gets used to" a constant toxic exposure in order to survive, at the same time suffering a long-term decrease in efficient functioning and perhaps longevity. Selye was among the first to describe this compensatory mechanism.[33] Because of adaptation or tolerance, the patient's total body load may increase undetected because the *perception* of a cause-and-effect relationship is lost. With no apparent correlated symptoms, repeated exposures may continue to damage his immune and enzyme detoxification systems.[34,35] The eventual result of continued toxic exposure over a period of days, weeks, months to years is end-organ failure.[11] Withdrawal or avoidance of an offending substance for at least four days will aid in reducing the total body load, after which a controlled re-exposure challenge will reproduce cause-and-effect reactions. In these deadapted individuals, there is high reproduciblity of these evoked reactions permitting the physician to acquire sound scientific information.[36]

BIPOLARITY

After an exposure, the body initially develops a bipolar response of a stimulatory phase followed by a depressive phase,[37,29,1] usually with induction of immune and enzyme detoxification systems.[38] If the incitant is strong enough, or if substantial size or duration of exposure occurs, the induced enzyme and immune detoxification systems are depleted or depressed by overstimulation and overutilization. An individual may also initially experience a stimulatory reaction in the brain, perceiving the inciting substance not as being harmful, but as actually producing an energizing "high". Therefore, he continues to acquire more exposures. After a period of time, however, be it minutes, months, or years, his body's defenses are adversely overstimulated and he develops disabling depression-exhaustion symptoms.[31] This stimulation and depression-exhaustion pattern has been observed with many pollutant exposures, including ozone.[30,12] When studying the effects of pollutants upon adapted individuals, the stimulatory phase is often missed or misinterpreted as being normal, thus giving faulty data. Studies in the controlled environment, involving 16,000 challenges in 2,000 deadapted patients, has proven this bipolarity phenomenon repeatedly.

BIOCHEMICAL INDIVIDUALITY

Another principle necessary to understand environmental aspects of health and disease, and especially chemical sensitivity, is that of biochemical individuality. Biochemical individuality of response is the individual's uniqueness.[39] This uniqueness of response depends on the differing quantities of carbohydrates, fats, proteins, enzymes, vitamins, minerals, immune and enzyme detoxification parameters with which an individual is equipped to handle pollutant insults. These variations determine an individual's ability to process the noxious substances he encounters. They further contribute to the intensity of his reaction to toxic exposures and to his susceptibility to chemical sensitivity. Thus, a group of individuals may be exposed to the same pollutant. One person may develop arthritis, one sinusitis, one diarrhea, one cystitis, one asthma, and one may remain apparently unaffected.

We have differing quantities and interactions of carbohydrates, fats, proteins, enzymes, vitamins, minerals, and immune parameters with which to respond to environmental factors. One simple example is the noted relationship between low serum magnesium levels and the HLA B35 genotype.[40] This biochemical individuality allows us to either clear the body of noxious substances, or to collect them and contribute to the body burden. Biochemical individuality is dependent on at least three factors: genetic endowment, the state of the fetus's nutritional health and toxic body burden during pregnancy, and the individual's present toxic body burden and nutritional state at the time of exposure.

Some individuals, for example, are born with significantly lower quantities of specific enzymes (it may be 75%, 50% or even 25% of the norms). Their response to environmental stimuli is often considerably weaker than those born with 100% of the normal detoxifying enzymes and immune parameters. Examples are the babies with phenylketonuria or the individuals with transferase deficiency, who do well until exposed to their environmental triggers, and then damage sets in. There are over 2,000 genetically-transmitted metabolic errors, suggesting that most of the population will have at least one abnormality.[41] Toxic volatile organic chemicals have been shown by Laseter to bioconcentrate in the fetus, increasing the acquired burden in some babies.[42]

It is well known that some individuals acquire their toxic load at work or around their homes.[42] This changes with different seasons and weather conditions thus giving variable effects and responses over time. Extreme care must be taken in evaluation of each patient, who may exhibit unique clinical responses due to his specific biochemical individuality. As an example, it is well known that not all patients will exhibit every reported symptom associated with systemic lupus erythematosus (SLE). Similarly, each patient exposed to the same environmental pollutant will react with his or her unique complex of symptoms. Because this vital fact is misunderstood, many studies are flawed when the wrong signs and symptoms are assessed for that individual.

SPREADING PHENOMENA

Spreading is a secondary response to pollutants that can involve new incitants or new target organs. Spreading that involves new incitants occurs when the body has developed increased sensitivity to increasing numbers of biological inhalants, toxic chemicals, and foods at increasingly smaller doses. At this time, overload becomes so taxing that a minute toxic exposure of any substance may be sufficient to trigger a response or autonomous triggering my occur. For example, a person initially may be damaged by a pesticide and then eventually have his disease process triggered by exposure to a myriad of toxic chemicals and foods, such as phenol, formaldehyde, perfume, beef, lettuce, etc.

Spreading may occur for many reasons. It may be due to a failure of the detoxification mechanisms--oxidation, reduction, degradation and conjugation--brought about by toxic overloading, or it may occur because of depletion of the enzyme or coenzyme's nutrient fuels, such as zinc, magnesium, all B vitamins, amino acid, or fatty acid. This depletion may account for an increasing inability to detoxify and respond appropriately. The blood brain barrier or peripheral cellular membranes of the skin, lung, nasal mucosa, gastrointestinal or genitourinary systems may be damaged allowing previously excluded toxic and nontoxic substances to penetrate to areas that increase the risk of harm. Immune or pharmacological releasing mechanisms may become so damaged that they are triggered by many substances toxic, then non-toxic (such as food) in addition to the specific one to which they were intended to respond. It is well substantiated that antigen recognition sites may be disturbed or destroyed by pollutant overload. Hormone deregulation (feedback mechanisms) may occur allowing for still greater sensitivity.

In contrast to patients who experience increased sensitivity to multiple triggering agents, some chemically sensitive patients may have one isolated organ involved in their disease process for years, only to have dysfunction spread to other organs as their resistance mechanisms break down. This kind of spreading from one to another or multiple end-organs enables the progression of hypersensitivity and the eventual onset of fixed named disease.

SWITCH PHENOMENON

The switch phenomenon is the changing of one end-organ response to another. This usually occurs acutely, but may occur over a much longer period of time. This phenomenon was first described by Savage in the 1700s. He observed that when mental patients were at their worst, they usually had a remission of their asthma or sinusitis. When they were better mentally and they were seen in the outpatient clinic, they had a much higher incidence of sinus and asthma problems. Randolph and most other environmentally oriented physicians have also observed this phenomenon. At the EHC-Dallas, we have observed similar occurrences in our patients and, in fact, take cognizance of this phenomenon when evaluating therapy outcome.

In observing thousands of controlled challenges in the environmental unit, we have seen the target organ responses of many of our patients switch to several different ones during a long (e.g., 24-hour) reaction. Often we have seen, for example, transient brain dysfunction followed by arthralgia, followed by diarrhea, followed by arrhythmia.

Therapy can appear to have been effective, even when a pollutant has not been totally eliminated. In this case, a new set of symptoms may begin indicating that a pollutant response has simply switched to another end-organ. This phenomenon occurs frequently when symptom-suppressing medication therapy or inadequate environmental manipulation is used. For example, a patient may have his sinusitis cleared by medication (e.g., cortisone). but later, since the cause has not been eliminated, he may develop arthralgia and eventually arthritis, or his colitis may have cleared only later to have cystitis develop.

This life-long progression of disease does not have to occur if the switch phenomenon is recognized during initial evaluation. To prevent it, individuals and physicians need simply to be cognizant of seemingly unrelated events. For example, statements are often made to the effect that a child will outgrow a problem when, in reality, one symptom complex dissipates only to be replaced by a new set of symptoms. For example, a child may have recurrent ear infections. Eventually, these may stop, but bedwetting may ensue. Over time the bedwetting may cease, but the child may then develop asthma. These changes in health may appear to be totally unrelated; in this instance, however, they are switch phenomena. The situation of an adult who sprays pesticides in his home, and then visits a neurologist with complaints of headaches and a rheumatologist with symptoms of arthritis is similar. Both the physician and the patient frequently fail to recognize the relevancy of these seemingly disparate symptoms as being part of a larger pattern needing further investigation.

POLLUTION FACTS

In order to accomplish concise studies of the chemical sensitivity phenomena, one must understand some facts about environmental pollutants.

Modern technology's rapidly accelerating rate of growth has produced a wide variety of chemical products, that contribute to the total chemical environment. Recent studies show that nearly 50% of the global atmospheric pollutants are generated by man, (either isolated from natural products or synthesized), and the ubiquitous nature of the toxic chemical agents is widely appreciated.[8,13,14] It has been estimated that more than 2,000 new chemical compounds are introduced annually, and that over 60,000 different organic chemicals are used commercially today.

The widespread presence of hazardous chemicals has rendered critical the environmental sensitivity problems described by Randolph almost 40 years ago.[43] While celebrated instances of gross contamination have long been the object of professional attention, only recently have literally thousands of synthetic chemical products, heretofore believed innocuous, been incriminated as agents of homeostatic dysfunction[14,11].

Current data affirm the view that standard methods for the determination of chemical incitants may no longer be effective.[7,8,36,1] With the finding that sensitivities can occur from subthreshold and picomolar quantities of chemicals, has come the discovery that standard procedures, such as skin prick or scratch tests, often fail to demonstrate positive reactions which are otherwise verifiable.

Recent literature confirms the harmful effects of chemical incitants, like formaldehyde,[44,45] phenol,[45,46] some pesticides,[7] chlorine,[47] and petroleum alcohol,[48]

Commonly encountered chemicals like glycine,[9,49] DDT, toluene and turpentine,[50,51,52] and drugs such as hydralazine have been found to induce advanced-staged disease process.[53]

A number of familiar metals have also been incriminated, among them nickel, cobalt, chromium,[54] aluminum,[55] mercury,[56] and platinum.[57] Other common environmental chemical incitants include xylene,[58] various acrylates,[59] and acrylated prepolymers,[60] benzyl peroxide, carbon tetrachloride,[61] sulfates,[62] dithiocarbamates,[63] and diisocyanates.[64]

WATER POLLUTION

Water has an important role in delivering contaminant minerals, toxic organic and inorganic chemicals, particulate matter and radiation to the human organism. In developed nations, the incidence of many chronic diseases, particularly cardiovascular disease, is associated with water characteristics, like purity and mineral content.[13] Hardness, or the lack thereof, is involved in heart disease, hypertension, and stroke.[13] Among the theorized protective agents found in hard water are calcium, magnesium, vanadium, lithium, chromium and manganese.[13] Certainly, once cardiovascular pathology is induced, waters with high sodium content may be harmful. Other adverse agents include the metals cadmium, lead, copper, and zinc, which tend to be found in higher concentrations in soft water. Nitrates in water (usually from fertilizer) pose immediate threats to children under three months of age due to production of methemoglobin,[65] and sulfur can also cause reactions in susceptible patients.

City water, much of it secondhand, often contains from 100 to 10,000 times as many synthetic compounds as natural spring water.[66] This, coupled with the rapid growth in the use of synthetic chemicals, has focused concern on the chemical quality of drinking water.[13] Although microbes are important, attention is now being drawn to the microchemical contaminants. Advances in analytic chemistry has been able to reveal chemical contaminants in the parts-per-billion or parts-per-trillion range. It is a serious mistake to assume that extensive contamination of drinking water with "low" levels of synthetic pollutants is "normal." These chemicals are widespread, and we should not be lulled into assuming these contaminants are innocuous. Examination of our ground water has revealed many hundreds of toxic chemicals in these ranges.[67,68]

Many examples of water contamination exist and have been documented, including Times Beach, Missouri with winter floods flushing dioxin-contaminated oil used 20 years ago, Niagara's Love Canal area, Waterbury, Connecticut, and Middleboro, Kentucky.[69]

In many cases, deadly materials have been accumulating for years in dumps and landfills. In the United States, some 80,000 pits and toxic waste lagoons hold chemicals ranging from carbon tetrachloride to discarded mustard-gas bombs.[68] Slowly escaping from their burial sites, these leftovers directly contaminate our ground water. Polluted ground water exists at 347 of the nations 418 worst chemical dumps, and probably is occurring in the rest.[68] Laseter[7] and others[70] have shown that a virtual organic chemistry laboratory exists in most drinking water.

In the early 1980s, California, New York, New Jersey, Arizona, Nova Scotia, and Pennsylvania condemned dozens of public water supply wells due to trichloroethylene or tetrachloroethylene pollution.[71] Leaking fuel tanks contaminated nine Kansas public water supplies in 1981.[71] Officials in New Mexico identified 25 cities where hydrocarbons and solvents contaminated the ground water.[71] Analysis of New Orleans drinking water alone revealed the presence of 13 halogenated hydrocarbons.

Sources of water pollution fall into three major categories: (1) municipal sewage; (2) agricultural wastes; and (3) industrial wastes. Approximately 55% of the water treated in municipal plants is from homes, while another 45% is from industry. Agricultural wastes include those from livestock and toxic chemicals (pesticides, herbicides, fertilizers), and farm runoff collects in rivers, lakes, and ground water. Industrial wastes, however, contain some of our more toxic substances. Over one-half of the total volume of industrial wastes come from paper mills, organic chemical manufacturing plants, petroleum companies, and steel manufacturing. The major pollutants are chemical byproducts, oil, grease, radioactive waste and heat. Other sources of contamination are drinking water disinfectants and byproducts;[68] it should be remembered that chlorine, interacting with organic material, produces toxic trihalomethanes and other organochlorines. Alternatives to treating water with chlorine include ozone, chloramines, ultra-violet irradiation, iodination, or home reverse-osmosis and charcoal filtration.[68]

Chloride, added at many sewage treatment plants, can also react with organic matter in the water to form chlorinated hydrocarbons, many of which are also known to cause cancer. Copper sulfate, aluminum sulfate and fluorine are other major contaminants which may add to the total body burden.[68]

Over a thousand different toxic chemicals have been found in public water supplies including pesticides, herbicides, industrial solvents, and polychlorinated biphenyls, just to name a few.

Inorganic pollutants include arsenic, cadmium, chromium, copper, manganese, mercury, silver, and selenium.[13] The use of inorganic arsenic insecticides has lead to high arsenic levels in some water supplies.[13] Barium (greater than 1 mg/L) has toxic effects on the heart, blood vessels, and nerves,[68] while cadmium at levels greater than .Olmg/L has adverse arterial effects. At levels greater than lmg/L or one ppm, the following metals found as water contaminants have produced severe chronic toxicity: antimony,[72] beryllium,[73] cobalt,[73] gold,[73] iodine,[73] lithium,[73] mercury,[74] and vanadium.[73] In Minamata, Japan, between 1953 and 1960, various plastic companies dumped methyl mercury chloride into the water, producing 50 to 85 ppm of mercury in nearby fish. Four hundred and six people died after ingesting these mercury-contaminated fish, and the adverse toxicological effects on developing children are continuing to be measured.[75]

A recently completed study found that skin absorption contributed from 29 to 91% of the total body dose of pollutants (from water), with an average of about 64%.[76] This is even more important when one looks at the large number of volatile organic compounds found in our drinking and bath water.

Radiation occurs in some waters in the form of radon, a naturally occurring radionuclide that seeps from rocks and may be concentrated in airtight homes, especially the basements. At this stage, more information is needed to fully assess its effects. It probably, however, can increase the total body load.

In 1965, a serious drinking-water problem was seen in 40 percent of patients hospitalized for a program of comprehensive environmental contro1.[1,77,78] Today it is up to 80%. We have found that patients susceptible to water contaminants virtually always exhibit multiple sensitivities, with advanced and severe environmental reactions, especially to airborne chemicals.[1] Interestingly, water sensitivity in children was found to increase on a circadian and seasonal basis.[79] Increased severity was seen during June and July or in September and October, when grass, pollen, and mold counts were also high.[79] Some ECU patients had difficulty with waters containing high levels of sodium, others with calcium, and still others with high bicarbonate waters. A few individuals tolerated distilled water, even though it may contain some hydrocarbon residuals. Hundreds of outpatients have shown symptoms in reaction to both chlorinated and nonchlorinated waters, including numerous spring, charcoal-filtered, and distilled waters. If these water-induced symptoms remain undiscovered, food and chemical testing, may be distorted. It is vital to test and find safe water before proceeding with other testing in these severely sensitive individuals.

CHEMICAL CONTAMINATION IN FOODS

The contamination of our urban food supplies is the result of widespread use of food additives, preservatives, and dyes in growth, manufacturing and processing. Virtually all commercially grown and prepared foods have pesticides and herbicides in them.[79,9]

The literature abounds with reports of chemical sensitivities to many additives.[80,81] Contaminant reactions complicate the study of food sensitivity, forcing one to define more clearly the nature of the incitant, not only as it is encountered in foods, but in the air and water as well. Be11 has reported urticarial reactions and immunological changes to exposures to a number of food additives.[82] Condemi[83] and Bell both suggest that food dyes may trigger reactions in sensitive individuals; including conditions commonly thought to be psychogenic, or certain forms of hyperactivity.[28,84,85,86,87,88,89] Lindemayer has associated urticarial reactions with several additives such as p-hydroxybenzoic acid propylester, benzoic acid, sodium benzoate, ponceau rouge, and indigo carmine.[90] Monroe's data indicate a casual role played by tartrazine azo dyes and salicylates in the provocation of vascular alterations.[36] Other additives, including sodium nitrite and sodium glutamate, have been found to trigger migraine phenomena in susceptible patients.[91]

Sulfur dioxide[16] and sodium salicylate can provoke asthmatic reactions,[92] while aspirin-like food contaminants and dyes may trigger urticaria, angioedema, bronchoconstriction and purpura.[93] An even wider variety of symptoms, including severe gastro-intestinal disorders, has been associated with sensitivities to aniline, commonly found in rapeseed oi1.[94]

In our experience, natural toxic components of foods, such as alkaloids, phenols, lectins, etc. must also be accounted for when studying the secondary food sensitivity which occurs from pollutant overload in the chemically sensitive. Therefore, three factors must be considered when evaluating the total food load. These are man-made pollutant contamination, natural toxic effects of foods, and food sensitivity. Failure to consider all three in the chemically sensitive patient may color or negate otherwise a clearly defined case of chemical sensitivity.

CHEMICAL INCITANTS IN THE HOME ENVIRONMENT

Indoor air pollution in the home environment has produced a multitude of sensitivities to chemicals.[8,95] Time and space limitations allow only a cursory review of the numerous commercial hygienic products which can be noxious for chemically susceptible individuals. Among these are a wide variety of cosmetics.[96,97] particularly those containing glycerin, propylene glycol, or butylene glyco1,[98] perfumes,[99] and hair products such as dyes,[100,101] creams,[102] sprays,[103] and shampoo.[104] Moreover, sensitivities have been demonstrated to occur in association with lip salve,[105] fingernail preparations,[106] soaps,[107] sanitary napkins,[108] mouthwash,[109] antiperspirants,[110] contact lenses.[111] contact lens solutions,[112] and suntan lotions.[113]

Reports are widespread of sensitivities to chemicals in textiles, including synthetic acrylic fibers,[114] polyester spin finishes,[115] the epoxy resins, and synthetic clothing.[116] Products such as fabric spray starch may also be considered toxic for the chemically sensitive individual[117] for whom even the metallic buttons on blue jeans may trigger reactions to nickel.[118] Formaldehyde[44] on synthetics or tetrachloroethylene, from dry-cleaned clothing can also produce problems.

Household cleaning products, particularly those containing formaldehyde, phenols and chlorine are hazardous for many patients. Several laundry products and detergents may be identified as household incitants[119], as well as a number of products used to clean and polish furniture[120].

The very construction of many homes may prove dangerous for the chemically sensitive patient. Data suggests that chemicals contained in wood preservatives (e.g., pentachlorophenols) are environmental incitants capable of triggering a variety of symptoms.[121,122,123] Others report problems with reactions to formaldehyde-containing pressboard, carpets, plywood and petrochemical contaminants.[124]

Current data confirm earlier findings regarding the hazards of pesticides[125] such as 2,4,DNP and fungicides[126]. Moreover, research increasingly suggests the possibility of sensitivities to apparently innocuous items such as rubber bands,[127] coins,[128] epoxy,[129] and countless paper products.[130,131] Pesticides, along with oil, gas or coal are major offenders for sensitive individuals.

Research shows house plants[132,133] and common insects[134] can now be viewed as environmental incitants or causes of homeostatic dysfunction. In addition, sensitivities to cold and heat,[36] and to contaminants in household water supplies have been associated with symptoms ranging from urticaria to severe respiratory distress.

Natural gas heat and stoves, and routine insecticiding or termite proofing of homes can be prime offenders in chemical sensitivity. One must consider these potential sources of contaminants when developing studies on chemical sensitivity. In our experience, failure to evaluate building and home environments before challenge testing will often make challenge studies invalid for the diagnosis of chemical sensitivity.

MECHANISMS

The mechanisms involved in chemical sensitivity are becoming clearer, one of which has pollutant injury occurring to the lungs or liver, with resultant free radical generation[135]. Disturbances then occurs at the cellular, subcellular and molecular levels, producing injury either immunologically, or nonimmunologically through enzyme detoxification systems. Vascular or autonomic nervous system dysfunction will then occur with one or a myriad of end-organ responses.

IMMUNOLOGICAL

Type I hypersensitivity is usually mediated through the IgE mechanism on the vessel wall. Classic examples are angioedema urticaria, and anaphylaxis due to sensitivity to pollen, dust, mold, or food,[136] or some chemicals such as toluene diisocyanate. Ten percent of the patients with immunological involvement with chemical sensitivity seen at the ECH-Dallas seem to fall within this category.

Type II cytotoxic damages may occur with direct injury to the cell. A clinical example of this is seen in patients exposed to mercury[137]. A group in Minimata, Japan developed neurological disease from eating fish exposed to toxic methyl mercury chloride. Mercurial pesticides fall into this category. Twenty percent of the patients with immunological involvement seen at the EHC-Dallas seem to fall into this Type II category.

Type III shows immune complexes of complement and gamma globulin damaging the vessel wall. A clinical example of this is lupus vasculitis. Numerous chemicals, including procainamide and chlorothaizide, are known to trigger the autoantibody reaction of lupus-like reactions. Many other toxic chemicals can also trigger the autoimmune response[138]. Other chemicals, such as vinyl chloride, will produce microaneurysms of small digital arterioles, probably due to this mechanism[139,51].

Type IV (cell-mediated) immunity occurs with triggering of the T-lymphocyte. Numerous chemicals such as phenol, pesticides, organohalides, and some metals will also alter immune responses, triggering lymphokines, and producing the Type IV reactions[138]. Clinical examples are polyarteritis nodosa, hypersensitivity angiitis, Henoch-Schonlein purpura, and Wegener's granulomatosis [1,139]. A recent study done at the Environmental Health Center - Dallas on 104 proven chemically-sensitive individuals (70 vascular, 27 asthmatic, and 7 rheumatoid), comparing them with 60 normal controls, showed that those manifesting a chemical sensitivity through their vascular tree had suppression of the suppressor T-cells (greater than 4 S.D.)[47]. Clearly the larger portion of our patients with immunological involvement fall into the Type III and IV categories.

NON-IMMUNE ENZYME DETOXIFICATION

Non-immune triggering of the cell and vessel wall may occur. Complement may be triggered directly by molds, foods, or toxic chemicals[139], and mediators like kinins, postaglandins, etc. may also be directly triggered. These reactions then cause vascular spasm, with resultant hypoxia and release of lysozymes, which further produces more spasm, hypoxia, etc. Eventually end organ failure will occur.

Triggering of the enzyme detoxification, mostly in the systems liver and respiratory mucosa, plays an important role in clearing of pollutants. It occurs, however, to a lesser extent in all systems. Foreign compound biotransformations have considerable variability, depending on genetic factors, age, sex, nutrition, health status, and the size of the dose.

The metabolism of foreign compounds usually occurs in the microsomal fraction (smooth endoplasmic reticulum) of liver cells. A few biotransformations are non-microsomal (redox reactions involving alcohols, aldehydes and ketones). There are basically four biotransformation categories -- oxidation, reduction, degradation, and conjugation.

The first three biotransformation pathways for xenobiotics are the same pathways that the body uses to process food and nutrients. If these enzyme systems are over-utilized by competing foreign pollutants, inadequate handling of food proteins can result, with the subsequent induction of food sensitivities. However, because these detoxification pathways are dependant on nutrient and mineral cofactors, these systems are *inducible* by appropriate oral or systemic supplementation. Such supplementation serves as an important factor in stabilizing and treating patients with chemical sensitivity. The fourth category of biotransformation, that of conjugation, is almost exclusively for handling foreign compounds. Conjugation appears to be uniquely utilized for the catabolism of foreign compounds, using amino acids and their derivatives with peptide bonds, and carbohydrates and their derivatives with glucide bonds. Simpler compounds like sulfate and acetate are also involved in conjugation with linkage of ester bonds. Activated conjugated compounds plus specific enzymes are often detoxified by coupling with co-enzymes, Examples: co-enzyme A with acetate, and other short-chain fatty acids adenosine or phosphoadenosine phosphate is detoxified with a methyl group from sulfate methionine, or the ethyl group from ethionine. Similarly, uridine and phosphate with glucose and glucuronic acids[140,141].

There are generally five major categories of foreign-compound conjugative processes[140]. These are: 1) *acetylation* through co-enzyme A, for detoxifying aromatic amines and sulfur amides; 2) *peptide conjugation* with glycine and aromatic carboxylic acids to hippuric acid; 3) *sulfonation* with glutathione (containing cysteine) or PAPS, and microsomal enzyme conjugation for multi-ring systems such as naphthalene, anthracene, and pheno-anthracene, which eventually results in benign mercaptic acids or alternatively benign sulfate esters; 4) *alkylations* by methionine of amines, phenols, thiols, noradrenalin, histamine, serotonin, pyridine, pyrogallol, ethylmucaptin sulfites, selenites and tellurites; 5) *Glucuronation*. Glucuronides detoxify pesticides, alcohols, phenols, enols, carboxylic acid, amino hydroxamines, carbamides, sulfonamide and thiol[140,141]. All of these processes are dependent upon nutrient fuels to keep these processes running efficiently. Toxic chemicals disturb the supply of the nutrient fuels by 1) producing poor quality food, 2) reducing intake, 3) reducing normal absorption, 4) setting up competitive absorption in the gut with nutrients, 5) imbalancing intestinal flora, 6) disturbing transport mechanisms, 7) disturbing proper decomposition and metabolism, 8) causing renal leaks, and 9) directly damaging nutrients. If nutrient inadequacy occurs, normal metabolism is overloaded and disturbed, resulting in selective changes in the pools of nutrients such as vitamins, minerals, amino acids, enzymes, lipids, and carbohydrates. Once this occurs, there is a viscous cycle of dysmetabolism, often with production or worsening of chemical sensitivity. These detoxification and metabolic defects are often measurable and have been accomplished in over 2,000 chemically sensitive patients.

DIAGNOSIS

The diagnosis of chemical sensitivity can now be made with a combination of the following history, physical examination, immune tests including IgE, IgG, complements, T & B lymphocyte subsets, blood levels of pesticides, organic compounds, heavy metals (intracellular), and occasionally objective brain function tests. Antipollutant enzymes, such as superoxide dismutase, glutathione, peroxidase, and catalase have been found to be suppressed in the chemically sensitive. Vitamin deficiencies, mineral deficiencies and excess, amino acid deficiency and disturbed lipid and carbohydrate metabolism has been observed.

Challenge tests are the cornerstone of confirmatory diagnosis. These may be accomplished through oral, inhaled, or intradermal challenges. Care should be taken to rule out inhalant problems with pollen, dust, and molds. Food sensitivity occurs in approximately 80% of the people with chemical sensitivity; and must be evaluated. When diagnosing chemical sensitivity, one must investigate water contaminant sensitivities, as 90% of people with chemical sensitivity have water contaminant reactions[4]. This can be checked by placing the patient on less chemically-contaminated, charcoal filtered, distilled, or glass-bottled spring water for four days, with subsequent rechallenge of the patient's regular drinking water. This procedure will often elicit a reaction to the water pollutants in the sensitive individual.

Patients frequently know where and when the onset of their problems occurred, e.g., sudden exposure to pesticides, working around printing machines and factory machines, etc. They usually develop increased odor perception to gasoline, perfumes, new paints, car exhausts, gas stoves, fabrics, clothing or carpeting stores, chlorine and chlorox, and cigarette smoke. Not only will they find these smells offensive, but may have marked reactions to them as well. Other symptoms can range from the almost universally-seen fatigue, to classic end-organ failures. Physical findings frequently are vascular in nature, with edema, petechiae, spontaneous bruising, purpura, or peripheral arterial spasm. Frequently there is flushing, adult-onset acne, and a yellowness of the skin without jaundice. Chronic, recurring nonspecific inflammation is usually a significant sign, e.g., colitis, cystitis, vasculitis, etc. Laboratory findings are often non-specific, e.g., sedimentation rates may increase or liver profile may be mildly off. Fifteen percent of environmentally sensitive patients have positive C-reactive proteins. Twenty-five percent show abnormal serum complement parameters. Fifty percent of the chemically sensitive patients have depressed T cells. Twenty-five percent have impaired blastogenesis, and twenty-five percent have impaired delayed hypersensitivity, as evidenced by cell-mediated immunity skin tests. Of the patients with T-cell abnormalities, the depletion of the suppressor cells is seen, by over four standard deviations from a control group of normals[141]. Ten percent of these patients have elevated IgE or IgG. Patients with recurring infections have impaired phagocytosis and killing capacity. Very accurate blood measurements are now available for the chlorinated pesticides as well. The following were found in over 200 chemically sensitive patients:

PESTICIDE IN BLOOD	% DISTRIBUTION IN 200 PATIENTS
DDT and DDE	62.0%
Hexachlorobenzene	57.5
Heptachlor Epoxide	54.0
beta-BHC	34.0
Endosulfan I	34.0
Dieldrin	24.0
gamma-Chlordane	20.0
Heptachlor	12.5
gamma-BHC (Lindane)	9.0
Endrin	5.5
delta-BHC	4.0
alpha-BHE	3.5
Mirex	2.0
Endosulfan II	1.5

Organophosphate levels are only positive within 24 hours after exposure, and are not much help. Lab tests for pentachlorophenols and organic solvents like hexane and pentane, are also now available, as are herbicide levels. General volatile organic hydrocarbons are found in a large portion of chemically sensitive patients. Their presence indicates either recent exposure, or a failure in the enzyme detoxification system. Those found in over 500 chemically sensitive patients include benzene, toluene, trimethylbenzene, xylene, styrenes, ethylbenzene, chloroform dichloromethane, 1,1,1,-trichloroethane, trichloroethylene, tetrachloroethylene, dichlorobenzenes. Metals including lead, mercury, cadmium, and aluminum are sometimes found in the intracellular contents of some chemically sensitive patients. These again are found in 10% of the patients.

Fat biopsies have been preferred on many patients with over 100 different compounds studied. Often there is more in the fat than blood in some cases such as organochlorine insecticides and more in the blood than fat such as seen with such substances as 2-methylpentane and 3-methylpentane.

Skin biopsies of bruising and petechiae reveal perivascular lymphocyte infiltrates around the vessel wall in chemically sensitive patients.

Challenge tests can be done by the sublingual or intradermal route. The efficiency of these tests is now well established as numerous studies, (several double-blind), have now been done[4,47,142,24,143,144,145]. These need to be done since 80% of the chemically sensitive are food sensitive. Blind intradermal challenge for chemicals can now be done with terpenes, petroleum derived ethanol, glycerine, formaldehyde, phenol, perfume, and newsprint, whereby production of symptoms will help establish the patient's chemical sensitivity.

Over 200,000 intradermal challenges of chemicals have been done under environmentally-controlled conditions at the EHC-Dallas. These are clearly reliable, especially as they meet the positive criteria of sign and symptom reproduction, wheal growth and negative placebo response.

Inhalation challenge is another method for the diagnosis of chemical sensitivity, done under varying degrees of environmentally controlled conditions. For best results, one uses an anodized aluminum and glass booth to do ambient dose challenge of any toxic chemical in a hospitalized, environmentally controlled setting. Some studies done in our center, under strictly controlled conditions in an environmental unit, showed significant findings (4 S.D.) of the chemical reactors over the controls when using less than .20 ppm formaldehyde, less than .0025 ppm phenol, less than .33 ppm chlorine, less than .50 ppm petroleum derived ethanol, less than .034 ppm of the pesticide, 2,4,DNP, along with 3 placebos. These tests have been used in over 3,000 patients with over 99% accuracy. Similar studies can be done in the office setting, although controls are much more difficult and one finds many more placebo reactions. This is because environmentally-controlled conditions are generally much more difficult to achieve and patients are often studied in the masked or adapted state, wherein symptoms may not be perceived. With the inhaled challenges, one can measure and plot blood levels, immune parameters, metabolic changes as well as sign and symptom scores.

Vitamin and intracellular mineral levels are needed to completely evaluate the chemically-sensitive individual. In our Center, analysis of over 300 chemically sensitive patients showed the following vitamin deficiencies: 64% with B6 deficiency, 30% with B2, 29% with Bl, 27% with folic acid, 24% with vitamin D, 19% with B3, 6% with vitamin C, 3% with vitamin B12. Out of 190 chemically sensitive patients with mineral deficiencies, 88% had chromium deficiency, 12% selenium, 8% zinc, 40% magnesium and 35% sulfur. Many had mineral excess in their blood cells.

TREATMENT

The cornerstone of treatment for chemical sensitivity is avoidance[146]. This will decrease total body burden, allowing recovery of the overtaxed detoxification systems. Less chemically contaminated water (including spring, distilled, and charcoal filtered), may be used, but only in glass or steel containers. Water will leach a variety of contaminants from the walls of synthetic plastic containers. A rotary diet of less chemically contaminated food, should also be used to reduce load and keep the patient in the unmasked state. Remove as many household incitants as possible, including petroleum-derived heat, insecticides, synthetic carpets and mattresses, and formaldehyde-containing substances such as pressboard and plywood. Toxic exposures can be monitored by the general volatile organic hydrocarbon blood tests. Some job changes may be needed, while occasionally the most severely affected patients have to leave badly polluted areas. Techniques should be developed for follow-up and monitoring of these modalities.

Injection therapy for inhalants, foods, and some chemicals will also help this problem[24,144,145,147,148,149,150]. Low-dose sublingual therapy in patients with allergic rhinitis was effective[151]. These treatments can be done daily, but usually every four to seven days. In our opinion, a properly balanced rotary diet is essential in treating the patient with food sensitivity, whether or not it may be induced by chemical overload. Vitamin and mineral supplementation is often necessary to replace the deficiencies that occur from the direct toxic damage, exhausted enzymatic detoxification pathways, and from the direct competition absorption. In rare cases, nutritional replacement with intravenous hyperalimentation is needed for severely debilitated patients. Techniques should be developed for monitoring and evaluating the outcome.

CONCLUSION

The philosophy and techniques of environmental medicine developed over the last 25 years offers a means to scientifically investigate and treat patients affected by pollutants. This approach gives the physician valuable, accurate information, in the pursuit of optimum health for these environmentally-sensitive patients.

REFERENCES

1. Randolph, T.G., Human Ecology and Susceptibility to the Chemical Environment. Springfield, Illinois: Charles C. Thomas, 1962 (sixth printing 1978).

2. Dickey, L.D. (Ed.), Clinical Ecology. Springfield, Illinois: Charles C. Thomas, 1976.

3. Bell, I.R., Clinical Ecology. Bolinas, California: Common Knowledge Press, 1982.

4. Rea, W.J. and Mitchell, M.J. Chemical sensitivity and the environment. Immun Allerg Prac, Sept/Oct: 21-31, 1982.

5. Wordan, J.I. The effect of air in concentrations and polarity on the C02 capacity of mammalium plasma. Fed Proc 1954; 13:557.

6. Choy, R.V.S., Monro, J.A., Smith, C.W., Electrical Sensitivities in Allergy Patients. Clin Ecol Vol IV No. 3. pp. 93-102.

7. Laseter, J.L.; DeLeon, I.R.; Rea, W.J.; Butler, J.R. Chlorinated hydrocarbon pesticides in environmentally sensitive patients. Clin Ecol, 2(1):3-12, 1983.

8. National Research Council. Indoor Pollutants. Washington, D.C.: Natl. Academy Press, 1981.

9. Gilpin, A. Air Pollution, 2nd Ed. St. Lucia, Queensland: University of Queensland Press, 1978.

10. Wark, K. and Warner, C. Air Pollution: Its Origin and Control. New York: Harper & Row, 1961.

11. Adelman, R.C. Loss of adaptive mechanisms during aging. Fed. Proc. 38:1968-1971, 1979.

12. Stokinger, H.E. Ozone toxicology: A review of research and industrial experience, 1954-1962. Arch Environ Health, 110:719, 1965.

13. The National Research Council. Water hardness and health. In Drinking Water and Health. New York. National Academy of Science, 1977, 439-447.

14. Winslow, S.G. The Effects of Environmental Chemicals on the Immune System: Selected Bibliography With Abstracts. Oak Ridge, TN: Toxicology Information Response Center, Oak Ridge National Laboratory, 1981.

15. Calabrese, E.J. Pollutants and high-risk groups. The Biological Basis of Increased Human Susceptibility to Environmental and Occupational Pollutants. NY: Wiley -

Interscience, 1977.

16. Freedman, B.J. Sulphur dioxide in foods and beverages: its use as a preservative and its effect on asthma. Br J Dis Chest, 74(2):128-134, 1980.

17. Davidson, M. and Fienleib, M. Disulfide poisoning: A review. Am Heart J, 83:100, 1972.

18. Innami, S., Tojo, H., Utsuja, S.; Nakamura, A.; and Nagazama,S. PCB Toxicity and Nutrition I. PCB Toxicity and Vitamin A. Jap J Nutr, 32:58-666, 1974.

19. Villeneune, D.C., Grant, D.L., Phillipos, W.E.J.; Clark, M. L.; and Clegg, D. J. Effects of PCB administration on microsomal enzyme activity in pregnant rabbits. Bull Environ Contam Toxicol, 6:120, 1971.

20. Rea, W.J.; Johnson, A.R.; Smiley, R.E.; Maynard, B.; and Dawkins-Brown, 0. Magnesium deficiency in patients with chemical sensitivity. Clinical Ecology, 4(1): 17-20, 1986.

21. Pan,Y., Johnson, A.J., Rea, W.J.: Aliphatic Hydrocarbon Solvents in Chemically Sensitive Patients. Clin Ecol 5(3):126-131,1987-88.

22. Olson, K.R., et al: Amanita phalloides-type mushroom poisoning. West. J. Med. (37:282, 1982).

23. Brostoff, Jonathan, and Challacombe Stephen; Food Allergy and intolerance, London, Bailliere Tindall/W.B. Saunders, 1987.

24. Rapp, O.J. Double-blind confirmation and treatment of milk sensitivity. Med J Aust. 1:571-572, 1978.

25. Rea, W.J.; Pan,Y. et al: Toxic Volatile Organic Hydrocarbons in Chemically Sensitive Patients, Clin Ecol 5(2):70-74,1987.

26. Hill, Johanna Peptides and Their Receptors as the Biochemicals of Emotion, proceedings of the American Acad. Envir. Med. Annual Mtg., Nashville, TN, Oct. 29, 1987.

27. Ader, Robert Behavioral Influences of Immunity, proceedings of Amer. Acad. Envir. Med. Annual Mtg., Nashville, TN, Oct. 29, 1987.

28. King, D.S. Can allergic exposure provoke psychological symptoms? A double-blind test. Biol Psychiat. 16:3-19, 1981.

29. Randolph, T.G. Specific Adaptation. Ann Allergy. 40:333-45, 1978.

30. Mustafa, M.G. and Tierney, D.F. Biochemical and metabolic changes in the lung with oxygen, ozone, and nitrogen dioxide toxicity. Am Rev Resp Dis, 118(6), 1978.

31. Stokinger, H.E. and Coffin, D.L. Biological effects of air pollutants. In A.C. Stern (Ed.), Air Pollution. New York:
American Press, 1968.

32. Rinkel, H.J.; Randolph, T.G.; and Zeller, M. Food Allergy. Norwalk, Connecticut: New England Foundation of Allergic an Environmental Disease, reprint of 1951 text.

33. Selye, H. The general adaptation syndrome and the diseases of adaptation. J. Allergy, 17:23, 1946.

34. Jollow, D.J., et al. 1977 Biological Reactive Intermediates: Formation, Toxicity, and Innetivation. New York.

35. Wand, A.M., et al. 1976 Immunological mechanisms in the pathogenesis of vinyl chloride disease. Brit. Med. J.1, 6015:936-938.

36. Rea, W.J.; Peters, D.W.; Smiley, R.E. et al. Recurrent environmentally triggered thrombophlebitis. Ann Allergy, 47:338-344, 1981.

37. Savage, G.H. Insanity and Allied Neuroses. Philadelphia: Lea and Febiger, 1884.

38. Rowe, A.H. Allergic toxemia and migraine due to food allergy. Calif West Med. 33:785-792, 1930.

39. Williams, R.J. Biochemical Individuality. New York: Wiley, 1963.

40. Noordhout, B., et al. "Latent Tetany, magnesium and HLA tissue antigens." Magnesium-Bulletin 3 (offizielles organ Gesellshaft fur magnesium forsehung) 9, 3, 1987.

41. Stanbury, J.B., et al. The Metabolic Basis of Inherited Disease, 5th printing New York, McGraw-Hill Book Co., 1983.

42. Nadal,A. and Lee, L.Y. Airway hyperirritability induced ozone. In S.D. Lee (Ed.), Biochemical Effects of Environmental Pollutants. Michigan: Ann Arbor Science Publishers, 1977.

43. Randolph, T.G. Sensitivity to petroleum. Including the derivatives and antecedents. a. Lab Clin Med 40:931-932 Dec. 1952.

44. Fisher, S.A.: Dermatitis due to the presence of formaldehyde in certain sodium lauryl sulfate (SLS) solutions. Curtis 27(4):360-2, 366, Apr, 1981.

45. Fregert, S.: Irritant dermatitis from phenol-formaldehyde resin powder. Contact Dermatitis 6(7):493, Dec., 1980.

46. U. S. Environmental Protection Agency: Unfinished Business. A Comparative Assessment of Environmental Problems.

47. Rea, W.J., Johnson, A.R., Youdim, S., Fenyves, E.J., Samadi, N. T & B Lymphocytes. Parameters Measured in Chemically Sensitive Patients and Controls.

48. Mohave, L., Moller, J: The Atmospheric Environment in Modern Danish Dewllings -- Measurements in 39 Flats. Institute of Hygiene, University of Aarhus, Denmark. Paper presented at International Indoor Climate Symposium. Copenhagen, 1978.

49. Suhonen, R.: Contact allergy to dodecyl-di-(aminoethyl) glycine (Dedsimex i). Contact Dermatitis 6(4):290-1, Jun., 1980.

50. McMillian, R.: Environmentally-induced thrombocytopenic purpura, JAMA 242(22):2434-5, Nov. 30, 1979.

51. Rea, W.J.; Bell, I.R.; and Smiley, R.E., Environmentally triggered large-vessel vasculitis. In F. Johnson and J.T. Spence (Eds.), Allergy: Immunology and Medical Treatment. Chicago: Symposia Specialist, 1975.

52. Rea, W.J. Review of cardiovascular disease in allergy. In C.A. Frazier (Ed.), Biannual Review of Allergy. Springfield, Illinois, 1979-80.

53. Peacock, Al: Case reports, hydralazine-induced necrotizing vasculitis, Br Med J 282(6270):1121-2, April 4, 1981.

54. Dooms-Goossens, A., Ceuterick, A., Vanmaele, N., Degreef, H.: Follow-up study of patients with contact dermatitis caused by chromates, nickel, and cobalt. Dermatologica 160(4):249-60, 1980.

55. Clemmensen, 0., Knudsen, H.E.: Contact sensitivity to alumi- num in a patient hyposensitized with aluminum precipitated grass pollen. Contact Dermatitis 6(2):305-8, Aug., 1980.

56. Vermeiden, Oranje, A.P., Vuzevski, V.0.: Stolz, E. Mercury exanthem as occupational dermatitis. Contact Dermatitis, 6(2):88-90, Jan., 1980.

57. Hughes, E.G.: Medical surveillance of platinum refinery workers. Soc Occup Med 30(1):27-30, Jan., 1980.

58. Bell, I., King, D.: Psychological and physiological research relevant to clinical ecology: an overview of the recent literature, Clin Ecol, 1982.

59. Bjorkner, B., Dahlquist, I., Fregert,S.: Allergic contact dermatitis from acrylates in ultraviolet curing inks. Contact Dermatitis 6(6):405-9, Oct. 1980.

60. Bjorkner, B.: Sensitization capacity of acrylated prepolymers in ultraviolet curing inks tested in the guinea pig. Acta Derm Venercol (Stockh), 61(1):7-10, 1981.

61. Romaguera, Gimalt, Sensitization to benzoyl peroxide, retinoic acid and carbon

tetrachloride. Contact Dermatitis, 6(6):442, Oct., 1980.

62. Satish, J. Importance and measurement of sulfates in floating dust. Soz Praeventivmed 25(4):201-2, Sep., 1980.

63. Kleibl, K., Rackova, M.: Cutaneous allergic reactions to dithiocarbamates. Contact Dermatitis 6(5):348-9, Aug., 1980.

64. Gallagher, J.S., Tse, C.S., Brooks, S.M., Bernsten, I.I.: Diverse profiles of immunoreactivity in toluene diisocyanate (TOI) asthma. JOM 23(9):610-6, Sep., 1981.

65. Wilson C.W.M. Hyperactivity to Maine tap water in children: its clinical features and treatment. Nutri and Health 2: 51-63, 1983.

66. Glaze, W.E., North Texas State University, Environmental Division, Denton, Texas, Personal communication.

67. Cherry, R. and Cherry, F.: What's in the water we drink? The New York Times Magazine, Dec. 8, 1974.

68. Environmental Protection Agency (EPA): Is your drinking water safe? (Booklet)

69. Trimble, J. U.S. News and World Report, Feb. 28, 1984, pp.27-32.

70. Spalding, R.F.; Junk, G.A. Richard, J.J. Water: Pesticides in ground water beneath irrigated farmland in Nebraska, Pesticides Monitoring Journal 1(2): 70-73, 1980.

71. Junk, G.A.; Spalding, R.F. and Richard, J.J. Aureal, Vertical, and temporal differences in ground water chemistry. II. Organic constituents, 1983.

72. Corwin, A.H. Heavy metals in air, water, and our habitat In Dickey, L.D. (Ed) Clinical Ecology, Springfield, Il: Charles C. Thomas, 1976, pp. 292-305.

73. Moran, J.M.; Morgan, M.O. Wiersma J.N Introduction to Environmental Science 2nd Ed. New York: W.H. Freeman and company, 1986, pp. 177-209.

74. Skiba, N.S. Role of river discharge of solids in migration of mercury, Zap Kirgizsk Otd Vses Mineralog Obschchestva 4:31, 1963; Chem Abst, 61:5357e, 1964.

75. Takenchi, T. Morikawa, N.; Matsumoto, H.; and Shiraishi, Y. Apathological study of Minamata disease in Japan, Acta Neuropathol 2:40, 1962.

76. Brown, H.S. Bishop, D.R.; Rowan, C.A. The role of skin absorption as a route of exposure for volatile organic compounds in drinking water, Am J Pub Health 74: 479-484, 1984.

77. Randolph, T.G. Ecologic orientation in medicine: Comprehensive environmental control in diagnosis and therapy. Ann Allergy. 23:7-22, 1965.

78. Randolph, T.G. The ecologic unit. Hospital Management, Part I, 97:45, 1965; Part II, 97:46, 1965.

79. Wilson, C.W.M. Hypersensitivity to Maine tap water in children: its clinical features and treatment. Nutrition and Health 2:51-63, 1983.

80. Juhlin, L.: Incidence of intolerance to food additives. Int J Dermatol 19(10):548-51, Dec., 1980.

81. Fries, J.H.: Food allergy: Current concerns. Ann Allergy 46(5):260-3, May, 1981.

82. Bell, I, King, D.: Psychological and physiological research relevant to clinical ecology: An overview of the recent literature. Clin Ecol, 1982.

83. Condemi, J.J.: Aspirin and food dye reactions.Bull N.Y. Acad Med 57(7):600-7, Sep., 1981.

84. Rea, W.J. Environmentally triggered small vessel vasculitis. Annals of Allergy, 38:245-51, 1977.

85. Potkin, S.G.; Weinberger, D.; Kleinman, J. et al. Wheat gluten challenge in schizophrenic patients. Amer J. Psychiat.138:1208-1211, 1981.

86. Hanninen, H.; Eskelinen, L.; Husman, K.; and Nurminen, M. Behavioral effects of long-term exposure to a mixture of organic solvents. Scand. J. Work Environ. Health. 4:240-55, 1976.

87. Randolph, T.G. The history of ecologic mental illness. In C.A. Frazier (Ed.), Annual Rev Allergy 1973. Flushing, NY: Med Exam Pub Co, 1974.

88. Savage, G.H. Insanity and Allied Neuroses Philadelphia: Lea and Febiger, 1884.

89. Rea, W.J.; Butler, J.R.; Laseter, J.L.; and DeLeon, I.R. Pesticides and brain function changes in a controlled environment. Clinical Ecology, 2(3):145-50, 1984.

90. Lindemayr, H. Schmidt, J.: Intolerance to acetylsalicylic acid and food additives in patients suffering from recurrent urticaria (author's transl.). Wien Klin Wochenschr 91(24):817-22, Dec., 21, 1979.

91. Hanington, E.: Diet and migraine. J. Hum. Nutr. 34:175-80,1980.

92. Dahl, R.: Sodium salicylate and aspirin disease. Allergy 35(2):155-6, Mar., 1980.

93. Wuthrich, B., Fabro, L.: Acetylsalicylic acid and food additive intolerance in urticaria, bronchial asthma and rhinopathy. Schweiz. Med. Wochenchr. 111(39):1445-50, Sep. 26, 1981.

94. Tabuenca, J.M.: Toxic-allergic syndrome caused by ingestion of rapeseed oil

denatured with aniline. Lancet 12:2(8246):567-8, Sep. 1981.

95. Cullen M. ed. Workers with Multiple Chemical Sensitivities. Occup. Med. State of the Art Reviews. 2(4),1987.

96. Thune, P: Photosensitivity and allergy to cosmetics. Contact Dermatitis 7(1):54-5, Jan., 1981.

97. Zaynoun, S.T., Aftimos, B.A., Tenekjian, K.K., Kurban, A.K.: Berloque dermatitis a continuing cosmetic problem. Contact Dermatitis 7(2):111-6, Mar., 1981.

98. Fisher, A.A.: Reactions to popular cosmetic humectants, Part III. Glycerin, propylene glycol, and butylene glycol. Curtis 26(3):243-4, 269, Sep., 1980.

99. Guin, J.D., Berry, V.K.: Perfume sensitivity in adult females. J Am Acad Dermatol 3(3):299-302, Sep., 1980.

100. Foussereau, J., Reuter, G., Petitjean, J.: Is hair dyed with PPD-like dyes allergenic? Contact Dermatitis 6(2):143, Jan., 1980.

101. Klein, A.D. 3d, Rodman, O.G.: Allergic contact dermatitis to paraphenylenediamine in hair dye. Milit Med 146(1):46-7, Jan., 1981.

102. Ramsay, C.A.: Skin responses to ultraviolet radiation in contact photodermatitis due to fentichlor. J Invest Dermatol 72(2):99-102, Feb., 1979.

103. Schlueter, L.E., Soto, R.J., Baretta, E.D., Herrmann, A.A., Ostrander, L.E., Stewart, R.D.: Airway response to air spray in normal subjects and subjects with hyperreactive airways. Chest 75(5):544-8, May, 1979.

104. Stoll, D., King, L.E., Jr.: Disulfiram-alcohol skin reaction to beer-containing shampoo. JAMA 244(18):2045, Nov., 1980.

105. Hindson, Phenyl salicylate (Salol) in a lip salve. Contact Dermatitis 6(3):216, Apr., 1980.

106. Fisher, A.A.: Permanent loss of finger nails from sensitization and reaction to acrylic in a preparation designed to make artificial nails. J Dermatol Surg Oncol 6(1):70-1, Jan., 1980.

107. Jordan, W.P., Jr.: Contact dermatitis from D & C yellow 11 dye in a toilet bar soap. J Am Acad Oermatol 4(5):613-4, May, 1981.

108. Larson, W.G.: Sanitary napkin dermatitis due to the perfume. Arch Dermatol 115(3):363, Mar., 1981.

109. Mathias, C.G., Chappler, R.R., Maibach, H.I.: Contact urticaria from cinnamic aldehyde. Arch Dermatol 116(1):74-6, Jan., 1980.

110. Aust, L.B., Maibach, H.I.: Incidence of human skin sensitization to isostearyl alcohol in two separate groups of panelists. Contact Dermatitis 6(4):269-71, Jun., 1980.

111. Sertoli, A., Di Fonzo, E., Spallanzani, P., Panconesi, E.: Allergic contact dermatitis from thimerosal in a soft contact lens wearer. Contact Dermatitis 6(4):292-3, Jun., 1980.

112. Zeigen, S.R., Jacobs, I.H., Weinberger, G.I.: Delayed hyper-sensitivity to thimerosal in contact lens solutions. J Med Soc NJ 78(5):362-4, May, 1981.

113. Jackson, R.T., Nesbitt, L.T., Jr., DeLeo, V.A.:6-Methylcoumarin photocontact dermatitis. J Am Acad Dermatol 2(2):124-7, Feb., 1980.

114. ElSaad El-Rifaie, M.: Perioral dermatitis with epithelioid cell granulomas in a woman: a possible new etiology. Acta Derm Venereal (Stockh) 60(4):359-60, 1980.

115. Burrows, D., Campbell, H.S.: Contact dermatitis to polyesters in finishes. Contact Dermatitis 6(5):362-3, Aug., 1980.

116. Imbeau, S.A., Reed, C.E.: Nylon stocking dermatitis. An unusual example. Contact Dermatitis 5(3):163-4, May, 1979.

117. McDaniel, W.H., Marks, J.G., Jr.: Contact urticaria due to sensitivity to spray starch. Arch Dermatol 115(5):628, May, 1979.

118. Larson, F.S., Brandrup, F.: Nickel release from metallic buttons in blue jeans. Contact Dermatitis 6(4):298, Jun, 1980.

119. Rudzki, E., Kozlowska, A.: Causes of chromate dermatitis in Poland. Contact Dermatitis (63):191-6, Apr., 1980.

120. Kleinhans, Days. Formaldehyde contact allergy. Derm Beruf Umwelt 28(4):101-3, 1980.

121. Klaschka, F., von Nieding,F., Walter, R. Skin sensitization to wood preservatives containing pentachlorophenol (PGP). Zentralbl Arbeitsmed Arbeitsschutz Prophyl 29(6):150-4, Jun., 1979.

122. Bach, B., Pedersen, N.B.: Contact dermatitis from a wood preservative containing tetrachloroisophthalonitriles. Contact Dermatitis 6(2):142, Jan., 1980.

123. Spindeldreier, A., Deichmann, B.: Contact dermatitis against a wood preservative with a new fungicidal agent. Derm Beruf Umwelt 28(3):88-90, 1980.

124. Frigas, E., Filley, W.V., Reed, C.E.: Asthma induced by dust from urea-formaldehyde foam insulating material. Chest 79(6):706-7, Jun., 1981.

125. Matsushita, T., Nomura, S., Wakatsuki, T.: Epidemiology of contact dermatitis from

pesticides in Japan. Contact Dermatitis 6(4):255-9, Jun., 1980.

126. Rudzki, Mapiorkowska, Dermatitis caused by the Polish fungicide Sadoplon 75. Contact Dermatitis 6(4):300-1, Jun, 1980.

127. Weinstein, L.H., Fellner, M.J.: Rubber band dermatitis. Int J Dermatol 18(7):558, Sep., 1979.

128. Conde-Salazar, L., Romero, L., Guimaraens, D.: Metal content in Spanish coins. Contact Dermatitis 7(3):166, May, 1981.

129. Fregert, S.: Epoxy dermatitis from the nonworking environment. Br J Dermatol 105 Sppl 21:63-4, Sep., 1981.

130. Guin, J.D.: Contact dermatitis to perfume in paper products letter. J Am Acad Dermatol 4(6):733-4, June, 1981.

131. Marks, J.G., Jr.: Allergic contact dermatitis from carbonless copy paper. JAMA 245(22):2331-2. June 12, 1981.

132. Calnan, C.D.: Dermatitis from ivy (Hedera canariensis variegata). Contact Dermatitis 7(2):124-5, Mar., 1981.

133. Sugai, T., Takahashi, Y., Okuno, F.: Chrysanthemum dermatitis in Japan. Contact dermatitis 6(2):155, Jan., 1981.

134. Fuchs, E. Insects as inhalant allergens. Allergol Immunopathol (Madr) 7(3):227-30, May-June, 1979.

135. Levine, S. and Kidd, P. Antioxidants Adaptations, BioDivision, Allergy Research Group, San Leando, California, 1985.

136. Theorell, H.; Blambock, M; Kockum, C. Demonstration of reactivity to airborne and food antigen in cutaneous vasculitis by variation in fibrino6peptide and others, blood coagulation, fibrinolysis, and complement parameters, Thrombo Haemo Sts (Stattz), 36: 593, 1976.

137. Gaworoki, C.L. and Sharma, R.P. The effects of heavy metals on (3H) thymidine uptake in lymphocytes, Toxicol App Pharmacol, 46(2): 305-13, 1978.

138. Winslow, S.G. The effects of environmental chemicals on the immune system: a selected bibliography with abstracts, Oak Ridge, Tennessee: Toxicology Information Response Center, Oak Ridge National Laboratory, 1981, pp. 1-36.

139. Rea, W.J. and Suits, C.W. Cardiovascular disease triggered by foods and chemicals In Gerrard, J.W. (Ed.) Food allergy: new perspectives, Springfield,Illinois: 1980. Charles C. Thomas, 1980.

140. Jakoby, W.B. (Ed.) Enzymatic Basis of Detoxication, Vol. I 1983, Academic Press.

141. Reeves, A.L. (Ed.) Toxicology: Principles and Practice Vol. 1981, John Wiley & Sons, New York.

142. O'Shea, J.; Porter, A. Double-blind study of children with hyperkinetic syndrome treated with multi-allergen extract sublingually, J Learning Disabil 14: 189-91, 1981.

143. Forman, R.A. Critiques of evaluation studies of sublingual and intracutaneous provocation tests for food allergy, Med hypotheses 7: 1019-27, 1981.

144. Miller, J.B. A double-blind study of food extract injection therapy, Ann Allergy 38(3): 185-91, 1977.

145. Boris, Shiff, Weindorf, S. Bronchoprovocation blocked by neutralization therapy, J Allergy Clin Immunol 7: 92, 1983.

146. Rea. W.J., Pan,Y, Johnson,A.R. Clearing of Toic Volatile Hydrocarbons from Humans. Clin Ecol 5(4)"166-170, 1987/88.

147. Boris M., Schiff M., Weindorf S. Antigen-Induced Asthma Attenuated by Neutralization Therapy. Clin Ecol, 1985; 3:59-62.

148. Boris M, Schiff M, Weindorf S. Injection of Low-dose Antigen Attenuates the Response to Subsequent Bronchoprovocative Challenge. Otolaryngol-Head Neck Surg, 1988; 98:538-545.

149. Miller JB, Intradermal Provocative/Neutralizing Food Testing and Subcutaneous Food Extract Injection Therapy. In Brostoff J, Challacombe S, (Eds): Food allergy and Intolerance, London, Bailliere Tindall Publishers, 1987: 932-946.

150. Rea WJ, et al: Elimination of Oral Food Challenge Reaction by Injection of Food Extracts, Arch Otolaryngoll, 1984; 110:248-252.

151. Scadding G. Brostoff, J.: Low-dose Sublingual Therapy in Patient With Allergic Rhinitis Due to House Dust Mite, Clin Allergy, 1986; 16: 483-491.

Occupational Asthma

Laura S. Welch

Asthma consists of reversible airway obstruction. There is no uniform definition of occupational asthma, but most experts would call it a new state of bronchial hyper-reactivity secondary to some agent at work [1], or reversible obstruction of the airways caused by inhalation of a substance or material used by a worker or present at his work [2]. Asthma can be a pre-existing disease that is exacerbated by exposures at work, rather than having these exposures as the primary cause; this is not defined as occupational asthma but rather as exacerbation of pre-existing disease. Exacerbation of pre-existing disease is work-related, but I will reserve the term occupational asthma for use as defined above.

Three percent of all Americans have asthma [3], and 5 to 15% of all asthma is estimated to be occupational [4,5]. The talk will present identified causes of occupational asthma, and working case definitions of the disease.

Even though there has been a great deal of research, not everyone will agree on the case definition of occupational asthma. Definitions vary by the study and the instrument used. Questionnaire has been useful for screening, but are less sensitive than methacholine challenge in detecting cases [6]. If pulmonary function is used, as usual definition is a drop of greater than 10% in FEV1 over a work shift, or a greater than 20% drop in peak flow over a work day. It may take repeated measurements in any one worker to detect such a drop, for some react only at the end of a work week.

Much has been written on the mechanism of occupational asthma, and both Dr. Liebowitz and Dr. Karol have discussed this in their presentations; I will not review that in detail here. As an overview, occupational asthma can be allergic or non-allergic. Asthma may develop as a specific sensitivity to one agent (or a family of related agents), or as a response to exposure to non-specific irritants.

There is certainly a major role of the immune system in the development of occupational asthma specific to one agent. In studies of asthma secondary to Western red cedar dust, considered a prototype of asthma from low-molecular weight compounds, investigators have found specific IgE in 40% of cases (but note that it is not detectable in 60%) [4]. Studies of isocyanate-induced asthma have also found IgE in some cases but not all. Many investigators think there is a role for direct histamine release in the lung. But in addition, there is much emphasis on the role of inflammation in the initiation and continuation of asthma.

The clinical presentation of occupational asthma can consist of an immediate reaction, a late-phase reaction, or commonly a mixture of both types [2,4]. In Western red cedar cases, for example, late asthmatic reactions alone occurred in 44%, both in 49% and immediate reactions alone in only 7%. There was no correlation between development of red cedar asthma and previous atopic status. These findings suggest that the allergic reaction is more complex than IgE alone. In addition, this same study reported that nonspecific bronchial hyper-reactivity (as measured with methacholine) correlated strongly with specific reactivity to plicatic acid, and decreased as patients became asymptomatic.

The second syndrome, a response to what we think is non-specific irritation rather than a specific allergy, was described by Brooks [7], and called RADS (reactive airway dysfunction syndrome); characteristics of this group of patients was also described by Tarlo and Broder [8]. Tarlo reported that this syndrome made up 10% of patients coming for evaluation to two centers. In reviewing 15 cases, these individuals has a low incidence of atopic disease, and were more likely to be smokers. These findings need further investigation.

What we know about occupational asthma can help us in beginning research on multiple chemical sensitivity. Study of occupational asthma first and foremost illustrates both the difficulty of developing a case definition, and the important of having one, as well as the importance of standardized measurement. These are basic research concepts, but it has taken many scientists to develop what tests and clinical signs are appropriate for the study of occupational asthma.

An approach for epidemiologic study that has worked with asthma is to use a questionnaire to define cases, and then proceed to a more detailed case investigation. We know that we do not capture all occupational asthma cases with a questionnaire, but this approach is efficient, acceptable, and yields very useful information of the subgroups of asthmatics that have a set of symptoms.

Occupational asthma is one disease for which exposure is part of the case definition. A particular set of symptoms of clinical findings develops in relationship to exposure. I expect that a case definition of MCS will include such a response to exposure as well, and we can use the lessons learned from occupational asthma to guide us here.

Occupational asthma has many causative agents, and more than one mechanism that results in the same clinical presentation. In asthma, there is clearly a role for both the immune system and direct inflammation.

Of particular importance, studies of occupational asthma have taught us about the role of controlled provocative challenges, and the importance of these challenges being performed in a standard fashion in a few specialized centers [9]. The diagnosis of occupational asthma usually does not need a controlled challenge, and such studies serve as a research toll primarily.

Finally, we have learned a great deal about the role of testing for specific antibodies or other tests of the immune system. Research in occupational asthma has shown that great attention must be paid to quality assurance and quality control, and that analyses must be standardized between laboratories [10].

REFERENCES

1. Hendrick, D.J., Fabbri, L.: Compensating occupational asthma. Thorax 1981; 36: 881-884.
2. Brooks, Stuart M.: Bronchial asthma of occupational origin. Scand J Work Environ Health 1977; 3: 53-72.
3. Occupational Disease Surveillance: occupational asthma. MMWR 1990; 39: 119-123.
4. Chan-Yeung, M., Lam, S.: Occupational asthma. Am Rev Respir Dis 1986; 133: 686-703.
5. Blanc, P.D.: Occupational asthma in a national disability survey. Chest 1987; 92: 613-617.
6. Bernstein, D.I., Cohn, J.R.: Guidelines for the diagnosis and evaluation of occupational immunologic lung disease: preface. J Allergy Clin Immunol 19; 5: 791-793.
7. Brooks, Stuart M., Weiss, Mark A., Bernstein, I.L.: Reactive airways dysfunction syndrome--case reports of persistent airways hyperreactivity following high-level irritant exposures. J Occup Med 1985; 27: 473-476.
8. Tarlo, S.M., Broder, I.: Irritant-induced occupational asthma. Chest 1989; 96(2): 297-300.
9. Brooks, Stuart M.: The evaluation of occupational airways disease in the laboratory and workplace. J Allergy Clin Immunol 1982; 70: 56-66.
10. Grammer, L.C., Patterson, R., Zeiss, C.R.: Guidelines for the immunologic evaluation of occupational lung diease. J Allergy Clin Immunol 1989; 84: 805-813.

Workshop To Develop Research Protocols for Multiple Chemical Sensitivity

Contributors

Nicolas A. Ashford
Massachusetts Institute of Technology

Iris R. Bell
Department of Psychiatry, University of Arizona

Robert Burrell
School of Medicine, West Virginia University

Joel B. Butler
Clinical and Behavior Psychology, Dallas

Devra Lee Davis
National Academy of Sciences

Roy L. DeHart
University of Oklahoma Health Sciences Center

Ervin J. Fenyves
University of Texas at Dallas

Nancy Fiedler
Department of Environmental and Community Medicine, UMDNJ

Bertie Griffiths
Environmental Health Center, Dallas

Gunnar Heuser
School of Medicine, UCLA

S. Heuser
Director, Environmental Medical Research and Information Center, Los Angeles

Alfred R. Johnson
Environmental Health Center, Dallas

Howard Kipen
Department of Environmental and Community Medicine, UMDNJ

John Laseter
AccuChem Laboratories, Richardson, Texas

Michael Lebowitz
College of Medicine, University of Arizona

Claudia S. Miller
The University of Texas Health Science Center at San Antonio

William J. Meggs
Department of Medicine, East Carolina University

William J. Rea
Environmental Health Center, Dallas

Gerald H. Ross
Environmental Health Center, Dallas

Jonathan Samet
University of New Mexico

Laura S. Welch
School of Medicine, George Washington University

A. Wojdani
Director Immunosciences Laboratory, Los Angeles

Workshop To Develop Research Protocols for Multiple Chemical Sensitivity

Participants

Jonathan Samet, *Chairman*
Pulmonary Division
UNMH Tumor Registry
2211 Lomas N.E.
Albuquerque, NM 87131

Roy E. Albert
Department of Environmental Health
University of Cincinnati
3223 Eden Avenue
Cincinnati, OH 45267-0056

Joseph F. Albright
Basic Immunology Branch
National Institute of Allergy
and Infectious Diseases
National Institutes of Health
Room 757, Westwood Building
Bethesda, MD 20892

Nicholas A. Ashford
Associate Professor of
Technology and Policy
Massachusetts Institute of Technology
77 Massachusetts Avenue
Cambridge, MA 02139

Robert B. Axelrad
Indoor Air
Environmental Protection Agency
401 M Street S.W. (ANR-445W)
Washington, DC 20460

Rebecca Bascom
Assistant Professor of Medicine
University of Maryland
22 South Greene Street
Baltimore, MD 21201

Iris R. Bell
Department of Psychiatry
University of Arizona
Health Sciences Center
1501 N. Campbell Avenue
Tucson, AZ 85724

Eula Bingham
University of Cincinnati
College of Medicine
Kettering Lab, ML-56
Cincinnati, OH 45267

John H. Boyles, Jr.
7076 Corporate Way
Centerville, OH 45455

Robert Burrell
Department of Microbiology
West Virginia University
2082 HSN
Morgantown, WVA 26506

James E. Cone
University of California
San Francisco General Hospital
Building 9, Room 109
San Francisco, CA 94110

Jeff Davidson
Environmental Protection Agency
401 M Street SW (PM-273)
Washington, DC 20460

Earon S. Davis
Environmental Health Consultant
2530 Crawford Avenue, Room 115
Evanston, IL 60201

Roy DeHart
Division of Occupational Medicine
University of Oklahoma
School of Medicine
800 N.E. 15th Street, Room 218
Oklahoma City, OK 73104

Dan Ein
4636 Kenmore Drive, N.W.
Washington, DC 20007-1924

Nancy Fiedler
Environmental & Community Medicine
Robert Wood Johnson Medical School
675 Hoes Lane
Piscataway, NJ 08854

Jordan Fink
Allergy Department
Medical College of Wisconsin
8700 West Wisconsin Avenue
Milwaukee, WI 53226

Elizabeth Gresch
The Dow Chemical Co.
2030 Dow Center
Midland, MI 48674

Judy Guerriero
University of California
San Francisco General Hospital
Building 9, Room 109
San Francisco, CA 94110

Gunnar Heuser
NeuroMed and NeuroTox Associates
323 S. Moorpark Road
Thousand Oaks, CA 91361

William J. Hirzy
Environmental Protection Agency
401 M Street, SW (TS-778)
Washington, DC 20460

Ken Hudnell
Office of Research & Development
Environmental Protection Agency
Neurophysiology Branch
Mail Drop 74-B
Research Triangle Park, NC 27711

Pauline Johnston
Indoor Air
Environmental Protection Agency
401 M Street SW (ANR-445W)
Washington, DC 20460

Meryl Karol
Graduate School of Public Health
Environmental and Occupational Health
University of Pittsburg
130 DeSoto Street
Pittsburg, PA 15261

Howard R. Kehrl
Physiology Section
Environmental Protection Agency
Human Studies Division
Clinical Reserach Branch, HERL/ORD
Building C-224-H
Chapel Hill, NC 27514

Howard Kipen
Environmental & Community Medicine
Robert Wood Johnson Medical School
675 Hoes Lane
Piscataway, NJ 08854

Loren D. Koler
College of Veterinary Medicine
Oregon State University
Corvallis, OR 97330

Michael D. Lebowitz
Professor of Internal Medicine
University of Arizona College
of Medicine
1501 N. Campbell Avenue
Tucson, AZ 85724

Max Lum
Director, Health Education
ATSDR
1600 Clifton Road
MS E-33
Altanta, GA 30333

Roberta Madison
Department of Health Sciences
California State University
18111 Nordhoff Street
Northridge, CA 91330

William J. Meggs
Assistant Professor of Medicine
Allergy/Immunology
East Carolina University School
of Medicine
Brody Building
Greenville, NC 27858-4354

Mark Mendel
EETB, California State Department
of Health
5900 Hollis, Suite E
Emeryville, CA 94608

Claudia S. Miller
Allergy and Immunology
University of Texas
Health Science Center
7703 Floyd Curl
San Antonio, TX 78284

Frank Mitchell
Senior Medical Officer
ATSDR
1600 Clifton Road
MS E-33
Atlanta, GA 30333

Lars Molhave
John B. Pierce Foundation Laboratory
Department of Epidemiology
& Public Health
Yale University School of Medicine
290 Congress Avenue
New Haven, CT 06519

David Ozonoff
Boston University
School of Public Health
80 East Concord Street
Boston, MA 02118

Andrew Pope
Institute of Medicine
National Research Council
2101 Constitution Avenue, NW
IOM-2137
Washington, DC 20418

Patricia H. Price
ATSDR
1600 Clifton Road
MS E-33
Atlanta, GA 30333

William J. Rea
8345 Walnut Hill Lane, Suite 205
Dallas, TX 75231-4262

Kathleen E. Rodgers
Livingstone Laboratories
1312 N. Mission Road
Los Angeles, CA 90033

Anne Sassaman
Division Director, Extramural Program
DERT, NIEHS
North Campus., Building 3, Room 301A
Research Triangle Park, NC 27709

Jan Stolwijk
John B. Pierce Foundation Laboratory
Department of Epidemiology
& Public Health
Yale University School of Medicine
290 Congress Avenue
New Haven, CT 06519

Virginia Sublet
ATSDR
1600 Clifton Road
Atlanta, GA 30333

Abba I. Terr
450 Sutter Street
San Francisco, CA 94108

Jack D. Thrasher
11330 Quail Creek Road
Northridge, CA 91326

William J. Waddell
Professor and Chairman
Department of Pharmacology
& Toxicology
University of Louisville
School of Medicine
Louisville, KY 40292

Laura Welch
Associate Professor of Medicine
George Washington University
School of Medicine
2150 Pennsylvania Venue, NW
Washington, DC 20037

James E. Woods
W. E. Jamerson Professor of Building
Construction, College of Architecture
and Urban Studies
Virginia Polytechnic Institute & State
University
202 Cowgill Hall
Blacksburg, VA 24061-0205

Grace Ziem
Occupational and Environmental Health
Research Office
1722 Linden Avenue
Baltimore, MD 21217

STAFF:
Robert P. Beliles
National Research Council, Room HA
354
2101 Constitution Avenue, NW
Washington, DC 20418

Devra Lee Davis
National Research Council, Room NAS
341b
2101 Constitution Avenue, NW
Washington, DC 20418

Richard Thomas
National Research Council, Room HA
354
2101 Constitution Avenue, NW
Washington, DC 20418

Sharon Holtzmann
National Research Council, Room NAS
359
2101 Constitution Avenue, NW
Washington, DC 20418

Toxicology and Industrial Health

AN INTERNATIONAL JOURNAL

ADVANCING THE UNDERSTANDING OF MULTIPLE CHEMICAL SENSITIVITY

PROCEEDINGS OF THE ASSOCIATION OF OCCUPATIONAL AND ENVIRONMENTAL CLINICS (AOEC) WORKSHOP ON MULTIPLE CHEMICAL SENSITIVITY

20-21 September 1991
Washington, D.C.

Edited by:
Kathleen M. Rest
Occupational Health Program
Department of Family and Community Medicine
University of Massachusetts Medical Center

Sponsored by:
The Association of Occupational and Environmental Clinics (AOEC)
1010 Vermont Street, N.W., Suite 513
Washington, D.C. 20005

Supported by:
U.S. Department of Health and Human Services
Public Health Service
Agency for Toxic Substances and Disease Registry

Published by:
Princeton Scientific Publishing Co., Inc.
Princeton, New Jersey

Toxicology and Industrial Health
Vol. 8, No. 4

PROCEEDINGS OF THE AOEC WORKSHOP ON MULTIPLE CHEMICAL SENSITIVITY

TABLE OF CONTENTS

DESCRIPTIONS OF THE SPONSORS

AGENCY FOR TOXIC SUBSTANCES AND DISEASE REGISTRY

The Agency for Toxic Substances and Disease Registry (ATSDR) is a federal public health agency located in Atlanta, Georgia. It is part of the Public Health Service within the U.S. Department of Health and Human Services. Created by Superfund legislation in 1980, ATSDR has a mission to prevent or mitigate adverse human health effects and diminished quality of life resulting from exposure to hazardous substances in the environment. To carry out its mission and to serve the needs of the American public, ATSDR conducts activities in the following areas: public health assessments, health investigations, exposure and disease registries, emergency response, toxicological profiles, physician and community education, and applied research.

ASSOCIATION OF OCCUPATIONAL AND ENVIRONMENTAL CLINICS

The Association of Occupational and Environmental Clinics (AOEC) is a national network of 40 clinical facilities, most affiliated with academic institutions, and 250 individual members. The organization is dedicated to higher standards of patient-centered, multidisciplinary occupational and environmental health care. AOEC emphasizes prevention and total health through information sharing, quality service, and collaborative research.

AOEC members are diverse in their training, expertise, and research interests, but are drawn together by their orientation and commitment to a philosophy of care embodied in the AOEC Patient Bill of Rights and clinic membership criteria.

PREFACE

The Association of Occupational and Environmental Clinics (AOEC) sponsored a workshop for its members on the emerging syndrome known as multiple chemical sensitivity (MCS). The workshop was supported in whole by funds from the Comprehensive Environmental Response, Compensation, and Liability Act trust fund through an interagency agreement with the Agency for Toxic Substances and Disease Registry, U.S. Public Health Service. The workshop was held in Washington, D.C., on 20-21 September 1991. This was the first time the AOEC convened a workshop on a substantive clinical issue, and it was well-attended by its members. This volume contains the papers prepared by invited speakers as well as the recommendations made by workshop participants.

These proceedings have been edited only for technical content and consistency of style; the content of each article is the sole responsibility of its author(s). Mention of products or company names is for identification only and does not imply any endorsement by the sponsors of the workshop.

The sponsors and workshop coordinators wish to express their appreciation to the program planning committee for their hard work in putting together an exciting program, to the authors who presented papers and submitted them for the proceedings, to the AOEC members who attended and enthusiastically participated in the workshop, to Jeanne A. Bucsela at ATSDR for her superb editorial assistance, and to the AOEC staff for their untiring logistical and administrative efforts.

ADVANCING THE UNDERSTANDING OF MULTIPLE CHEMICAL SENSITIVITY (MCS): OVERVIEW AND RECOMMENDATIONS FROM AN AOEC WORKSHOP

KATHLEEN M. REST, PhD, MPA
Occupational Health Program, Department of Family and Community Medicine
University of Massachusetts Medical Center
Worcester, Massachusetts

INTRODUCTION

Mysterious clinical entities that both puzzle and polarize the medical community are not unusual. In trying to understand any emerging problem, health professionals question and probe for margins of the new, so-called disease within the framework of their previous knowledge, experience, and belief systems. That framework varies for different observers, and this may help explain the conflict that sometimes develops among and between professionals and patients. Health professionals ask several questions: What exactly is this problem? Is it new? Is it real? Is it the same as something else? What causes it? Is it serious? Who gets it? How widespread is it? What can be done about it? What works? While clinicians (and patients) wait for research and an expanding body of clinical experience to provide answers to even the most basic questions, they must make diagnostic and therapeutic decisions to help their affected patients.

The syndrome called multiple chemical sensitivity (MCS) is the latest mystery illness to challenge the medical and scientific communities. Even its name evokes controversy. Not surprisingly, the struggle to understand this problem occurs in an emotionally charged and sometimes acrimonious environment. These proceedings reflect an attempt by one organization—the Association of Occupational and Environmental Clinics (AOEC)—to provide a forum for its members to share their experiences and learn from experts with diverse views in an intellectually challenging, honest, and open environment.

AOEC involvement in a workshop on MCS followed logically from the interests of its members—experienced occupational and environmental health professionals. AOEC members have assumed and often enjoyed the challenge of investigating exposure and disease relationships, and some had recently discussed MCS in the literature and other public forums. In 1987, an AOEC member edited the first comprehensive collection of articles on MCS (Cullen, 1987a). Other members participated in National Research Council workshops on the subject and in special task forces convened by other groups, such as the American Public Health Association. By 1990, it was clear that many AOEC members were seeing patients with MCS-like symptoms and were puzzled. As both practitioners and researchers, they needed a forum for discussion and for developing ideas to enhance their understanding of the syndrome.

Address all correspondence to: Dr. Kathleen M. Rest, Occupational Health Program, University of Massachusetts Medical Center, 55 Lake Avenue North, Worcester, MA 01655.

A workshop was timely. Activity concerning MCS had been increasing in many sectors. Professional medical organizations had issued position papers on MCS (California Medical Association, 1986; American Academy of Allergy and Immunology, 1986 and 1990; American College of Physicians, 1989; American College of Occupational Medicine, 1991). Debates and disagreements had occurred at professional meetings. A comprehensive and somewhat controversial book that reviewed MCS literature was published (Ashford and Miller, 1991). The government of Canada sponsored a workshop on environmental sensitivities[1] – and the National Research Council convened a workshop to develop a research agenda to study the MCS phenomenon (National Research Council, 1992). At the same time, MCS activists were becoming increasingly visible and vocal nationwide (Perlman, 1990).

With support from the Agency for Toxic Substances and Disease Registry (ATSDR), AOEC planned a workshop that would allow its members to 1) share clinical experiences as a basis for understanding the spectrum of MCS; 2) examine current research findings; 3) explore possible mechanisms and models of MCS; and 4) consider a variety of approaches for treatment and follow-up. The workshop was convened to develop recommendations for a research agenda that could be pursued by government agencies, research institutes, the academic community, and others interested in furthering the scientific understanding of MCS.

Workshop participants included health care providers (primarily physicians) from AOEC member clinics and invited speakers from a variety of disciplines, including allergy, immunology, clinical ecology, psychiatry, psychology, family medicine, pulmonology, orthopedics, and occupational medicine. Both invited speakers and AOEC members had a diversity of views, experiences, and perspectives on MCS. The workshop format included plenary sessions with formal paper presentations and four small groups that focused on research needs in characterizing patients, characterizing events, developing treatment strategies, and exploring possible mechanisms.

Although the program planners tried to provide clear direction to persons making presentations, they had no control over the presentations themselves. It seems that nearly everyone wanted to provide some socio-medical commentary, advance some theory, or provide some clinical pearls, and the speakers often covered myriad subjects. Thus, the papers in this volume cannot be divided neatly into topic areas. Moreover, they were not subject to traditional peer review, although they did receive technical review for logic and some editorial assistance. Thus, the papers may not be as well documented as readers might like. Indeed, they may even describe practices and assert opinions that readers dislike or find objectionable. The papers are printed here in a spirit of open inquiry, with no attempt to censor non-traditional or minority views. The reader is urged to pocket any pre-conceived notions while reading these papers and to give every author a fair hearing. Perhaps, when taken together, these papers will help shed some light on this perplexing problem.

---

[1] Ministry of Health and Welfare Canada. Environmental Sensitivities Workshop. May 24, 1990, Ottawa, Canada.

SUMMARY OF PRESENTATIONS

Clinical Experience

The workshop began with an interdisciplinary panel of physicians who discussed their clinical experiences with individual patients and with exposed groups. Redlich described a population of MCS patients evaluated by the Yale Occupational and Environmental Medicine Program over a five-year period. Ross presented information on 200 MCS patients seen at the Environmental Health Center in Dallas. Lenoir, whose paper is not included in these proceedings, described a subpopulation of MCS patients in his allergy practice. Although these physicians were from different types of clinical practices—occupational medicine, clinical ecology, and allergy—their MCS patients shared some common characteristics. The majority were women; many were in their 30s and 40s; and many had white- or pink- collar jobs.

Other panelists described their work with specific exposure episodes or outbreaks among groups of workers. Cone investigated an illness outbreak in a hotel casino; Simon studied a group of workers in an aerospace manufacturing firm; and Welch described an episode of sick building syndrome in a federal office facility. These papers report interesting physical and psychologic findings and suggest how exposure episodes could be used to further the scientific understanding of MCS.

Diagnosis and Treatment

The controversy generally associated with MCS surfaced during the second panel, which examined various approaches to diagnosis, treatment, and management of MCS patients. Rest presented results of a survey of AOEC physicians, which found many common approaches to assessment and treatment and some interesting differences in physicians' attitudes towards MCS. Subsequent presentations were provocative and revealed significant areas of disagreement. For example, should MCS patients avoid or be increasingly exposed to the substances that evoke their symptoms? Simon discussed the value of behavioral interventions that seek to desensitize patients by gradually increasing their exposure to and tolerance of the offending agents. In sharp contrast, Ziem called for avoidance and control of chemical exposures. Ross also recommended avoidance and described several controversial evaluation and treatment methods that have been used with MCS patients. Levin, a clinician and immunologist, also called for avoidance and described his clinical experience and treatment methods with MCS patients. His suggestion that MCS may be the prodromal phase of other serious diseases was especially provocative. Jewett, an orthopedic surgeon, discussed issues of diagnosis and treatment during a presentation on research. (For these proceedings, he divided his presentation into two separate papers.) He recommended a counseling approach to treatment and hypothesized that success is a function of the practitioner's beliefs. Participants who discussed these presentations were concerned about the mutually exclusive nature of the recommendations. The small group dealing with treatment methods made recommendations summarized later in this article.

Research Studies

A third panel focused on research studies and strategies. Daniel, whose paper is not included here, reported preliminary findings of a case-control study of immunologic and psychologic parameters in MCS patients seen in allergy practices. Kipen presented preliminary immunologic findings from a larger study of the medical, psychiatric, neuropsychologic, and immunologic status of a series of MCS patients.

King discussed methodologic issues involved in research on food sensitivities and presented results of a study of behavioral effects of food challenges in children and a study of beliefs about food and behavior in a population of university students.

Selner presented results of a study that used neuropsychologic methods to compare baseline characteristics of MCS patients with those of controls. He also reported on a study that evaluated the use of symptom self-report inventories in differential diagnosis. Selner also shared his clinical experience on the frequency of a history of childhood sexual and physical abuse in his MCS patients.

Buchwald reviewed the literature on chronic fatigue syndrome (CFS), a condition often suggested as overlapping MCS. She also presented results of several studies that evaluated the immunologic, virologic, and psychiatric dysfunction in patients at a chronic fatigue clinic. One study compared differences in characteristics and symptoms among patients with MCS, fibromyalgia, and CFS.

Finally, Jewett discussed the difficulties of doing research on MCS and proposed a careful review of a variety of topics before decisions are made about research strategies and methods.

Mechanisms

One of the most troublesome aspects of understanding MCS is its apparent lack of explanatory mechanisms. Until the syndrome can be explained mechanistically, clinicians may favor psychiatric or psychologic explanations of etiology and resist any suggestion that MCS is physical in origin. The final panel proposed several mechanisms deserving of research.

Miller discussed the possible role of the limbic system in MCS and described several conceptual issues, such as "adaptation," and their importance for the design of future research studies.

Two speakers discussed immunologic mechanisms. Meggs provided an overview of the immune system, the mechanisms of immunopathology, and selected published data on immune abnormalities in chemically exposed patients. He suggested that a neuroimmunologic hypothesis might explain the MCS phenomenon. Urging scientific rigor, Albright suggested approaches and several testable hypotheses that might help elucidate any immunologic component of MCS.

Bascom presented a third hypothesis in examining the relationship between low-level exposure to environmental irritants and altered function of the respiratory mucosa. She also explored the basis of biologic variability in the acute response to irritants.

Finally, Davidoff described two psychologic models that have served as default etiologic explanations of MCS—the somatization disorder model and the conditioning model—as well as a chronic disease model. Using interview data obtained from an ongoing study of MCS patients, she examined the explanatory power of the three models in relation to MCS.

All presentations evoked lively discussion, quiet mumblings, and a fair amount of head shaking in both directions. The outcomes and recommendations that follow reflect attempts to reach consensus in small group discussions that paralleled the broad areas covered by the speakers.

SMALL GROUP OUTCOMES AND RECOMMENDATIONS

Characterizing Patients

This group identified the need for a case definition that could be used in descriptive epidemiologic research and in challenge studies. Additional definitions might be needed and appropriate for other purposes, such as for treatment and compensation. Their proposed research definition, which is described as a modification of the definition first offered by Cullen (1987b), is shown in Table 1.

TABLE 1. A Proposed Research Definition for MCS

- A change in health status identified by the patient
- Symptoms triggered regularly by multiple stimuli
- Symptoms experienced for at least six months
- A defined set of symptoms reported by patients
- Symptoms that occur in three or more organ systems
- Exclusion of patients with other medical conditions
  (Psychiatric conditions are not considered exclusionary.)

The first criterion requires that the patient be able to identify a change in health status. This criterion rejects the notion that MCS must be associated with a single event. Rather, it permits patients to identify some time period in which they felt well and a subsequent time period in which they did not. The timing of this change may vary with the patient; for some, it may have occurred years ago, and for others, it may be a recent event.

The second criterion requires that the patient's symptoms be triggered regularly by multiple stimuli. "Regularly" is defined as "most of the time," that is, the patient need not always react to a specific substance.

The third criterion requires that the patient's symptoms be of at least six-months duration. Although this six-month period was acknowledged as somewhat arbitrary, it is the same time period used in the case definition of chronic fatigue syndrome.

The fourth criterion requires that patients report specific symptoms that would be listed in a defined set of symptoms to be derived from existing and ongoing research. This yet-to-be-developed symptom checklist could be used to categorize and stratify patients for research and to help ensure validity in any multi-center research study.

The fifth criterion requires that the symptoms involve three or more organ systems, i.e., they be multi-systemic in nature. The group arbitrarily chose three as the organ system cut-off point.

The final criterion requires that patients with other medical conditions be excluded from research to help avoid misclassification. The group did not develop a list of these conditions but suggested it include diagnoses that might otherwise explain the symptoms, for example, reactive airways disease. After considerable discussion, the group agreed that psychiatric conditions should not be excluded because concurrent psychiatric morbidity is an endpoint of interest in descriptive or analytic epidemiologic studies. Many members of the group believed that this criterion should be based entirely on history provided by the patient and not on any medical or laboratory tests.

Discussion. These definitional criteria generated considerable discussion. One clinician suggested that requiring the involvement of three or more organ systems and a defined number of symptoms might bias any research done on treatment methods. Such criteria would result in selecting sicker patients and excluding patients who are more mildly affected, and, in this clinician's experience, sicker MCS patients respond less well to treatment than do patients who are less sick. The group spokesperson explained that members wanted to establish narrower rather than broader criteria for research studies.

There was discussion about the requirement of multiple stimuli. Could one chemical that produces different symptoms in different organ systems qualify or must a certain number of chemicals be involved? Although the group did not discuss this point specifically, they gave more weight to reported symptoms than to the number, type, and nature of exposures.

Some discussion centered on symptom severity and functional status. Should a criterion on functional impairment be included in its research definition, for example, 50% impairment of activities over a given period? The group did not reach consensus on this point, so no functional criteria were included in the definition. One workshop participant found it curious that psychiatric conditions—many of which are based on functional abilities—were included, but functional criteria themselves were excluded from the proposed definition. It was noted that the definition could be used to identify subpopulations for testing specific hypotheses and that functional status criteria could be added for any subgroup at a later time.

Participants of the recent National Research Council workshop on MCS also grappled with a case definition. Their deliberations resulted in a broader, less exclusionary definition, which did not require symptoms in more than one organ system and did not necessarily exclude patients with standard medical diagnoses from research studies (National Research Council, 1992).

Characterizing Events

This small group recommended questions and priorities relating to epidemiologic and population-based research and how event- or exposure-specific precipitants of MCS might be characterized. They identified the need to assess incidence and prevalence of MCS. Community-based studies and existing surveys, such as the National Health and Nutrition Examination Survey and the National Health Interview Survey, could be used to obtain data on incidence and prevalence.

Prevalence surveys in specific occupational cohorts might also help enhance understanding of patient demographics and the role of high-versus low-exposure levels. Cross-cultural studies of "naive" populations might also be valuable. Such populations might include pesticide-exposed persons in the third world and exposed cohorts in countries outside the United States where MCS has not been described.

Existing "natural experiments" might provide fertile ground for research. For example, new apprentice pesticide applicators could be examined at baseline and followed over time to assess the frequency of developing symptoms consistent with MCS. Populations in other exposure situations, such as sick buildings, could also be followed in this way. Such longitudinal studies might also provide opportunities for applying experimental study designs.

The group noted a lack of knowledge about the natural history of MCS and identified a need for and the challenges of designing studies to elucidate its natural history. The group offered several ideas about how such studies might be done but did not necessarily endorse them. Case registries could be established to follow MCS patients. A registry could be used for descriptive as well as serologic studies of panels of MCS patients to determine how many convert to certain rheumatologic conditions, compared with control groups. Registry data could also be linked to cancer incidence data, if such a relationship comes into question.

The group identified a need for more formal diagnostic assessment of MCS patients and recommended use of a double-blind, placebo-control, cross-over research design. They emphasized the need to develop methods for demonstrating that research subjects can be truly blinded to exposure before any challenge studies are done. Finally, the group articulated the need to identify reliable and objective outcome measures for research.

Discussion. The issue of demonstrating blinding in exposure studies provoked some discussion. It may be difficult to demonstrate that exposures are indeed subsensory in studies that involve subsensory exposures. Researchers in Sweden had difficulty shamming exposure to electrical energy in a double-blind, crossover study of "electrical allergy." Unanticipated subtleties, such as noise, raised methodologic challenges for the researchers. Blinding chemical exposures may be even more difficult. Several speakers from the audience urged that research using challenge studies be broad enough to include stimuli that people report as causing their response, that is, that studies not be limited to assessing the effects of subsensory exposures. Although odor conditioning and the olfactory system may be important factors in MCS and should be studied, research will inevitably be criticized if it eliminates the exposure levels that seem to cause problems. The group supported studies of a continuum of exposure levels—from subsensory to sensory—to help elucidate dose-response relationships. They did not, however, try to develop a research protocol.

The group did not define the types of events that should be studied. Patients seem to report a variety of precipitating events—from pesticide exposure to automobile accidents. Several group members believed that efforts to define events for research would presume etiologies or mechanisms that simply are not known at this time and, thus, that such studies would be premature. Others believed that event- or exposure-driven research would provide valuable insight into understanding the natural history of MCS. The proposed case definition does not require identification of a specific event, nor does it exclude event-driven research, which might be useful if properly designed and conducted. For comparison, participants of the National Research Council workshop on MCS endorsed exposure- and event-driven research (National Research Council, 1992).

Treatment Methods

For research concerning treatment methods, the group presumed that 1) a general case definition is needed that can be subcategorized for targeted studies; 2) quantifiable biologic outcome data are scarce; 3) knowledge of the syndrome's natural history is inadequate; 4) no population-based estimates of incidence and prevalence or secular trends are available; 5) no primary care-based studies have been done; all cases reported to date come from specialists who have their own selection and practice biases; and 6) no primary prevention studies have been done. The latter might include early intervention with an exposed population, such as occupants of a sick building. Critical incident counseling might be one possible intervention. The group questioned whether intervention studies might not yield the biggest cost-benefit. Treatment should emphasize the patient's reintegration into society with optimal function. Although avoidance therapy may be one intervention to achieve this end, the group discouraged the "radical separatist avoidance approach." Outcome measures should be long term because what might appear as a successful outcome after six months may disappear at one year or more.

Specifically, the group recommended use of randomized, controlled trials whenever possible. Treatment studies should be multi-center and collaborative and should collect enough data to identify and analyze subgroups. A broad case definition will be needed to conduct such studies, and exclusion criteria for major diseases, such as systemic lupus erythematosus, are needed.

Treatment research should begin with therapies that appear to have some reasonable theoretical basis. These include avoidance, biofeedback, behavioral desensitization, provocation and neutralization, olfactory blockade, and functional rehabilitation. Therapies without adequate theoretical bases, such as rotation diet and antifungal and antibiotic therapies, should be placed at the end of the research priority list. The group did not, however, present a priority ranking.

Treatment outcomes to be measured should be functional in nature. Standardized and validated instruments, such as the Sickness Impact Profile, should be used whenever possible. Laboratory outcomes do not currently appear to be adequately validated. Subjective outcomes, including belief structures of both patients and providers, should also be studied. Treatment studies will be affected by numerous confounders that can be handled statistically or as part of the exclusion criteria. Research designs must demonstrate an attempt to deal with these confounders which include, but are not limited to, age, gender, and other demographics; psychologic disease; laboratory markers, such as anti-nuclear antibodies; chemical exposures (a dependent variable but also a possible confounder); substance abuse; family history; allergy status; litigation status at entry and during the study; and clinician enthusiasm. The latter would present a measurement challenge.

Discussion. The audience added several other possible outcomes and confounders. One participant suggested that change in patients beliefs about etiology would be an interesting outcome to study. Another suggested that the patient's perceived relationship with the provider might be an important confounder.

The role of food sensitivity in MCS and the group's comment that rotational diets not be studied at this time generated considerable discussion. Should avoidance studies include avoidance of food? Many MCS patients report sensitivity to certain foods, and the pharmacologic properties of food have been recognized. Several participants noted that diet control has been an important therapeutic modality in many diseases, including diabetes, migraine, arthritis, and cancer. One participant claimed that almost 50 different specific enzyme syndromes, which can present as specific diseases, have been identified and are related to food ingestion. Eliminating certain foods from a diet can greatly mitigate symptoms of many diseases. It was noted that food intolerance, food sensitivity, and food allergy are not synonymous.

The group consensus was that moderate food avoidance practices constitute a sensible and testable therapy for study, but that complicated and expensive trials of elaborate rotation diets are not warranted at this time. One outcome of a recent study of MCS sponsored by the government of Ontario was funding of research on food sensitivities at the University of Toronto (Thompson, 1985).

Finally, the small group emphasized that it did not endorse the use of provocation and neutralization for diagnosis or treatment but only endorsed its inclusion in research studies of the therapeutic efficacy of various treatment modalities.

Mechanisms

This small group discussed a variety of specific research areas and questions, as well as appropriate control groups for MCS studies. Challenge studies, primarily inhalation studies, including but not limited to chamber studies, were considered a fruitful area for research. Such studies might address the following research questions. Are symptoms reproducible and to what extent? What is the role of sensation? What is the dose-response? What is the physiology of the acute response? How are physiologic responses correlated with symptom response? What exactly is the exposure, for example, what constituents of diesel exhaust provoke the symptoms? Researchers doing chamber studies should not limit outcome measures to symptom self-reports or global measures, but include a thorough assessment of all types of symptoms, as well as physiologic measures.

Key issues in the design of challenge studies include the level of control and the extent of subject preparation and trade-offs that must be made in the face of limited resources. Studies can be performed in relatively uncontrolled environments, such as physicians' offices, or in environmentally controlled exposure chambers or environmental units. Highly controlled studies will be more expensive and may necessarily involve fewer subjects. Similarly, the degree of subject preparation will affect both design and expense. If subjects are to be in a "de-adapted" state before participating in a challenge study, the expense in time and money may limit the number of subjects that can be studied. The pursuit of basic, mechanistic studies was enthusiastically endorsed. The role of the nasal-olfactory-limbic pathway in the genesis of MCS was of considerable interest. Both morphologic and functional studies are needed. Studies of olfactory function should include both clinical and neurophysiologic assessments. Imaging studies, such as PET and SPECT scans, were also of interest. As an area of research, the use of peripheral blood markers to study immunologic function generated little enthusiasm.

Research on the role of psychologic factors in the etiology of MCS was of interest, but the group agreed that further cross-sectional studies would not be informative or productive at this time. The existence of psychologic symptoms in MCS patients, although common, reveals little about their etiology. Indeed, psychologic symptoms represent a common pathway for many different etiologies. Prospective studies might be more fruitful and would help address issues of recall bias. However, such studies would require large numbers of people because the prevalence of MCS, although unknown, appears to be limited. Prospective studies might be feasible for following cohorts of persons sensitive to chemicals; however, these persons may not be the same as those categorized as MCS patients by any research definition. It was noted that psychologic factors may be powerful effect modifiers even if they are not primarily etiologic.

Pharmacologic probes, specifically psychopharmacologic agents directed at certain brain areas or neuroreceptive pathways, might enhance understanding of the syndrome. Although such probes might help identify which chemicals are involved in the brain, they may reveal nothing about etiology.

Family studies were also mentioned as an area for research because of interest in the types of morbidity, both medical and psychologic, that occur in the families of patients with MCS.

The issue of appropriate control groups was raised. Because MCS appears to be a function of individual variability, exposure-matched controls might be appropriate in some cases. Studies of individuals exposed at the same level might help identify the determinants of individuals' variable responses to exposure. Groups that might appropriately be employed as *comparison* groups include patients with chronic fatigue syndrome; patients with more well-defined immunologic syndromes, such as rheumatoid arthritis and systemic lupus erythematosus; patients with atopic or allergic disease, especially allergic rhinitis; and patients with more well-defined psychiatric illnesses, such as panic disorder and depressive illness. Such groups would function as comparison (versus control) groups because the overlap between these disorders and MCS is unknown.

Discussion. It was suggested that chamber studies address the issue of adaptation even if it adds to the expense. Studies that do not address adaptation will be criticized and will not lead to consensus. There was also a concern about using patients with autoimmune disease as control or comparison populations, because there is some evidence that inhaled chemicals can cause autoimmune disease.

It was noted that research on biochemical mechanisms requires a subpopulation of patients with reproducible symptoms. This points to the need for chamber studies, which can validate the reproducibility of symptoms and identify a subpopulation for further research.

GENERAL DISCUSSION

Following discussion of the small group presentations, members of the audience participated in a general discussion period.

Noting the logistic and financial difficulties of studying groups of exposed workers or community residents after an outbreak or an exposure event, one participant suggested that the National Institute for Occupational Safety and Health (NIOSH) consider collaborating on and using a research protocol when it is asked to do health hazard evaluations involving building-related illnesses or other mass outbreaks. A NIOSH representative expressed interest in discussing the matter further.

Another participant commented on the remarkable amount of agreement he saw among the workshop attendees on the clinical presentation of MCS—a homogeneity he found lacking in discussions of other complicated disease processes.

Several participants asked that health professionals, especially clinical ecologists, be more receptive to the positive role that psychologic counseling can play in treating MCS patients. However, they noted that it should not simply be used to refer patients elsewhere. Similarly, one participant urged health professionals to recognize that the judicious use of appropriate medication may also be beneficial in some cases. However, MCS patients are especially difficult to convince that medication may provide some relief of symptoms because of their prior experience with adverse effects.

Several government agencies have expressed interest in MCS or are already involved in areas related to MCS, for example, the Environmental Protection Agency (EPA) in indoor air quality; NIOSH in occupationally-exposed groups; NIH in research; and ATSDR in its health assessments of Superfund sites and its demonstrated support of the workshop. These interests may spawn the submission of grant proposals and applications for research funds that may be reviewed by panels composed of persons who attended this workshop. The workshop, which has endorsed broad-based inquiry and the investigation of a wide spectrum of hypotheses and research questions, will be of little value if persons use the review process to rule out certain research or to prevent the investigation of minority views and new ideas. Everyone, especially persons who are out of the mainstream or who have minority points of view, must be diligent and scrupulous in guarding the scientific integrity and rigor of their research proposals, protocols, data, and presentation of findings.

Finally, a show of hands revealed that while most attendees indicated they had heard something new at the workshop, fewer believed they had learned anything, and still fewer admitted having changed their minds on an important issue. It is not clear what this means for furthering the understanding of MCS. Whether clinicians and researchers will really support open inquiry on this syndrome remains to be seen.

REFERENCES

AMERICAN ACADEMY OF ALLERGY AND IMMUNOLOGY. Executive Committee of the American Academy of Allergy and Immunology. (1986). Position Statements - Clinical Ecology. J. Allergy Clin. Immunol. 78(8):269-271.

AMERICAN ACADEMY OF ALLERGY AND IMMUNOLOGY. Practice Standards Committee Subcommittee on Environmental Medicine. (1990). Position Statement: Management of the patient with "environmental illness." 4:8, 15.

AMERICAN COLLEGE OF OCCUPATIONAL MEDICINE. (May 1991). Position Statement: Multiple chemical hypersensitivity syndrome.

AMERICAN COLLEGE OF PHYSICIANS. (1989). Position paper: Clinical Ecology. Ann. Intern. Med. 111(2):168-177.

ASHFORD, N.A. and MILLER, C.S. (1991). Chemical Exposures: Low Levels and High Stakes. Van Nostrand Reinhold, New York.

CALIFORNIA MEDICAL ASSOCIATION TASK FORCE ON CLINICAL ECOLOGY. (1986). Clinical ecology: a critical appraisal. West. J. Med. 144(2):239-245.

CULLEN, M. (ed). (1987a). Workers With Multiple Chemical Sensitivities. Occup. Med.: State Art Rev. 2(4):655-806.

CULLEN, M. The worker with multiple chemical sensitivities: An overview. (1987b). In: Workers with Multiple Chemical Sensitivities. (M.R. Cullen, ed.). Occup. Med.: State Art Rev. 2(4): 655-66.

NATIONAL RESEARCH COUNCIL (1992). Multiple Chemical Sensitivities: Addendum to Biologic Markers in Immunotoxicology. National Academy Press, Washington, D.C.

PERLMAN, D. (1990). Protesters disrupt allergy conference. (Nov. 10, 1990). San Francisco Chronicle. A4.

THOMPSON, G. (1985). Report of the Ad Hoc Committee on Environmental Hypersensitivity Disorders. Report to the Minister of Health, Toronto, Ontario.

THE EXPERIENCE OF THE YALE OCCUPATIONAL AND ENVIRONMENTAL MEDICINE CLINICS WITH MULTIPLE CHEMICAL SENSITIVITIES, 1986-1991

MARK R. CULLEN, MD, PATRICIA E. PACE, MPH, AND CARRIE A. REDLICH, MD, MPH
Yale Occupational and Environmental Medicine Program
Yale University School of Medicine
New Haven, Connecticut

INTRODUCTION

The occupational and environmental medicine clinics at Yale have been providing diagnostic services to patients referred from throughout Connecticut since 1979. More than 5,300 patients have been seen in 12 years, and basic information regarding demographics, environmental exposures, and diagnoses have been recorded in an ongoing database. Results of the first eight years of experience have been published and summarized (Cullen, 1989).

Since the development in 1986 of a monograph on the problem of multiple chemical sensitivities (MCS) (Cullen, 1987a), the Yale clinics have been a referral center for this newly recognized clinical problem. Although patients with MCS had been seen since the inception of the clinic and formed the basis for parts of the monograph, a reasonably uniform approach to diagnosis and coding was not initiated until after 1985. Since that time, the following general strategy has been undertaken:

1. Referrals are accepted for consultation from only the Connecticut region and immediate surroundings to prevent overloading of clinic resources. Patients with suspect or established MCS who live outside the Connecticut area are discouraged from coming for either medico-legal consultation or treatment.

2. All newly referred patients with suspect MCS or those for whom consideration of MCS arises *de novo* during evaluation at Yale are assessed clinically for conformity to the six-point case definition published in the monograph (Cullen, 1987b). No specific clinical or laboratory tests are performed routinely; only those tests deemed necessary to exclude alternative diagnoses suggested by the history and examination are performed if they have not been done previously.

3. No specific tests to elucidate possible theories about the etiology of MCS are performed because of the conviction that these are likely sources of diagnostic and therapeutic confusion.

Address all correspondence to: Dr. Mark Cullen, 333 Cedar Street, New Haven, CT 06510.

4. A diagnosis of MCS is entered into the database if a patient meets the published criteria (Cullen, 1987b) for the disorder or if no other alternative diagnosis appears more secure in the judgment of the treating clinician after discussing the case in clinic conference session.

This paper describes a study of the Yale MCS population since 1986, using data in the clinic database. Our aim was to describe our MCS population and contrast it with the larger clinic population as a contribution to the descriptive epidemiology of MCS. No attempt at etiologic inference or assessment of the strategy used for evaluation or diagnosis was intended nor could any such be reasonably drawn given the nature of the (selected) population studied and the methods of analysis employed.

METHODS

A search was made through the database for all patients coded for MCS (ICDA 9 code 478.8 by convention at our center) between 1986 and the present. All other patients seen at the clinic during this period were used as a comparison population.

The following data were individually abstracted for each case: age, gender, referral source, toxin associated with onset of symptoms (initially), and whether the source of the initial problem occurred in a job or in an ambient environmental setting. With patients whose initial exposure(s) was (were) associated with a job, the first digit of the SIC code of the employer was listed, along with occupational title as given by the patient in text format.

Because of incompleteness or ambiguity of data in about one-fourth of these records, the author (MRC) pulled the actual charts of the patients to fill in any missing data and validate ambiguous codes. In no case was the diagnosis reassessed or changed.

For the control group (all other patients during the period 1986-91), the variables noted above, except for job title, were summarized. No attempt was made to list any of the individual data for this large comparison group, and therefore no chart review was undertaken as for the cases noted above.

Analysis was limited to the use of a t-test for the single continuous variable—age, and differences in proportions were evaluated by chi-squares for trend.

RESULTS

A total of 49 patients coded as MCS were identified in the database, about 1 case per month over the 53 months for which coding was largely completed. The comparison group consisted of 2710 other patients seen during the same time period.

The 49 patients ranged in age from 22 to 66 years, with a mean of 45.3 years, standard deviation 9.9. By comparison, the clinic population was slightly older with a mean age of 50.2 years and standard deviation of 14.2. This difference was significant at the 0.003 level.

The gender distribution of the MCS patients was markedly different from the clinic population as a whole. Sixteen (32%) were men compared with 82.3% in the general clinic population, $p<0.0001$.

Referral sources were also quite different, with 49% of the MCS patients being physician referred compared with only 23% of the larger group, $p<0.001$. On the other hand, 18% of the MCS patients were self-referred compared with 11% of the clinic as a whole, $p<0.01$.

Another difference was the source of initial exposure for the MCS patients. Eight (16%) involved "ambient" exposures occurring outside a work setting compared with only 5.3% of the overall clinic population, $p<0.01$. Among the MCS patients who were initially exposed on the job (N=41, 83%), the distribution of industries was markedly different from that of the population as a whole. Only 2 (5%) of the MCS cases came from the construction industry (SIC 1000 series) compared with almost 29% of the clinic population, $p<0.0001$. Although nine (22%) MCS patients came from manufacturing, 51% of the clinic population came from this sector (SIC 2000 and 3000 series), $p<0.001$. By contrast, 19 (46%) of the MCS patients came from the service industry—almost entirely education and health care (SIC 8000 series)—compared with only 5% of the comparison group, $p<0.0001$.

Although we could not make a direct comparison with the clinic population as a whole because of our strategy for coding exposures, we analyzed the environmental agent associated with the onset of MCS for the MCS patients. Seventeen (34%) were coded as indoor air pollution, a designation given to office air settings where a particular source of exposure cannot be identified further. Another 20 (40%) involved organic hydrocarbons—including formaldehyde—used as solvents, oils, or plastic systems. Only three involved a pesticide exposure, and the remaining nine were variously coded as "chemical dusts" or "other."

Because the parameters involved are not uniformly coded, we were not able to analyze the MCS population according to symptom patterns, exposures eliciting symptoms, or findings on examination or laboratory tests. Similarly, functional and work status for the group were not analyzed or compared with the larger population.

DISCUSSION

The results of this study tend to confirm the impressions that we had formulated previously about MCS (Mooser, 1987) and that are held generally. MCS patients would appear to be younger than other patients with occupational or environmental health problems (although not so young!) and are predominantly female. Although it could not be tested directly, the MCS patients appear to come from higher socio-economic backgrounds judging by their occupations.

An important observation from this data set is the continued low rate of cases occurring in those industrial sectors that are responsible for most other chemical and physical injuries—SIC codes 1000-3999. Although this may represent a bias of referral or diagnosis, it seems extremely unlikely in the context of the Yale clinics. Referral bias is unlikely because these clinics are the overwhelming major referral site for all suspect occupational and environmental illness in the region. It seems improbable that MCS patients coming from industry would be systematically referred less often than patients with suspect lead poisonings or asbestosis. Conceivably, workers who have heavier industrial exposures to chemicals may be seen at the clinic but diagnosed as having something other than MCS, but this also seems quite unlikely because of the striking clinical characteristics of MCS and the awareness of our clinicians to it as a result of their experience and ongoing research in this arena. A third consideration may be that industrial workers are less likely than workers in the service sector to seek medical attention for symptoms typical of MCS. However, this too seems implausible in view of the high rate of dysfunction in patients who develop MCS.

We conclude, therefore, as previously suspected, that exposure backgrounds with low levels of chemical exposure appear, paradoxically, more likely to be associated with MCS than do high-exposure settings. The value this observation may have for studying the etiology of MCS remains to be elucidated, assuming it is, in fact, confirmed by others.

Finally, we were very struck in this simple study by the very small number of patients seen in our clinics who met the basic definition for MCS (Cullen, 1987b). Although some degree of underdiagnosis is theoretically possible relative to other clinics that use slightly less restrictive criteria, the clinicians at Yale tend to apply our published criteria liberally when alternative diagnoses are not apparent. Furthermore, a cursory search through potential competing diagnoses—such as upper respiratory irritation, solvent or pesticide poisoning or psychological syndromes—revealed an upper limit during this 5-year period of at most 2-3 times the number of diagnosed cases (no more than 150) that could possibly be considered MCS by any published set of criteria.

Although the Yale experience is admittedly small and we have discouraged specific referrals for MCS from outside our area, it seems clear that MCS, at least in a form prominent enough to cause referral to an occupational and environmental medicine clinic, is quite rare. Suggestions that this disorder may affect 1% or more of the entire population, as previously published and discussed (Mooser, 1987), seem to be substantial overestimates judging by our experience. Given this broad discrepancy, some more direct, population estimates of the size of the MCS problem should be developed as soon as possible.

REFERENCES

CULLEN, M.R. and CHERNIACK, M.G. (1989). Spectrum of occupational disease in an academic hospital-based referral center in Connecticut from 1979 to 1987. Arch. Intern. Med. 149:1621-1626.

CULLEN, M.R. (ed.). (1987a). Workers with Multiple Chemical Sensitivities. Occup. Med.: State Art Rev. 2(4). Hanley and Belfus, Philadelphia, PA.

CULLEN, M.R. (1987b). The worker with multiple chemical sensitivities: An overview. In: Workers with Multiple Chemical Sensitivities (M.R. Cullen, ed.). Occup. Med.: State Art Rev. 2(4):655-661.

MOOSER, S.B. (1987). The epidemiology of multiple chemical sensitivities. In: Workers with Multiple Chemical Sensitivities (M.R. Cullen, ed.). Occup. Med.: State Art Rev. 2(4):663-668.

HISTORY AND CLINICAL PRESENTATION OF THE CHEMICALLY SENSITIVE PATIENT

GERALD H. ROSS, MD, CCFP, DIBEM, FAAEM
Environmental Health Center
Dallas, Texas

INTRODUCTION

Early in 1991, an advertisement, illustrating the growing public awareness of potential problems with sensitivities to common substances around us, appeared on the back cover of Good Housekeeping magazine. It showed a little girl's party dress on a hanger, obviously not being used, with the title, "How can such a nice dress be so irritating?" The ad explained the benefits of a new laundry product made without perfumes or dyes and implied the new laundry liquid would not leave chemical residues in the fabric to subsequently irritate little girls' sensitive skin. It is doubtful that such an ad would have appeared ten years ago, expressing concern about the chemical residues we are all exposed to in modern society. This advertisement may be a sign of the times.

CHEMICAL SENSITIVITY

Chemical sensitivity is an abnormal state of health characterized by intensified and adverse responses to components found in food, air, water, or physical surroundings of the patient's environment (Hileman, 1991) The provoked symptoms or signs may be chronic, relapsing, or multisystem and are triggered by levels of exposure that are generally well tolerated by most people. The diagnosis is suspected on the basis of history and physical examination, and the condition may be confirmed by removing the offending agents and rechallenging patients under properly controlled conditions. The clearing or improvement of symptoms or signs by removing the suspected substances and the recurrence or worsening after specific, low-level challenge is highly indicative of environmental hypersensitivity and, in this context, chemical sensitivity. This definition is very similar to that proposed by Ashford and Miller (1989) in their report to the State of New Jersey and in their subsequent book (Ashford and Miller, 1991)

If we consider a patient's total environment as a complex interplay of genetic endowment; nutritional status; emotional stress; and exposure to foods, chemicals, microbes, inhalants, and even electromagnetic fields, then the expression of health is a function of the interplay of those factors and their effect on the patient (Randolph, 1962). Also necessary and important is

Address all correspondence to: Dr. Gerald Ross, Environmental Health Center-Dallas, 8345 Walnut Hill Lane, #205, Dallas, TX 75231.

the concept frequently used in environmental medicine of biochemical individuality, which is essentially the metabolic uniqueness that characterizes each person. No two individuals are identical in how they respond to a stressor (Calabrese, 1984). Although basic common sense and the lessons learned in medical school point to the concept of metabolic uniqueness, it is often overlooked in medical practice. For example, probably 10% of the population are slow metabolizers of the drug debrisoquine (Brostoff, 1987). The genetic polymorphism that determines fast or slow metabolism may be an important factor in expressing susceptibility to certain diseases or reactions to toxic environmental chemicals. Thus, the genetic variation (predisposition), phenylketonuria (PKU), sets the stage for severe problems if too much phenylalanine (stressor) is taken in the diet. Similarly, food-sensitive migraine patients frequently have phenylsulfotransferase deficiencies that may contribute to the onset of headaches when certain foods are consumed (Brostoff, 1987). Individually, many persons may have inborn variations of metabolism that predispose them to potential problems with metabolism or detoxication of substances in the chemical environment (Jacoby, 1980).

Randolph (1980), in his excellent book, *An Alternate Approach to Allergies*, described the chemically sensitive patient: "typically, however, patients have been polysymptomatic; that is, they have a long history of many problems, physical and mental, which have left them in a general state of misery. The more symptoms they accumulated, the less the doctors believe their complaints." All too often, once the physician reaches the end of the diagnostic repertoire for classifying a patient's complaints, the patient tends to be classified (by default) as having a functional or psychologic origin for confusing symptoms. This difficulty with diagnosis involving complex symptoms does not necessarily indicate a problem patient but perhaps a physician who lacks the proper knowledge to make a correct diagnosis.

SURVEY OF CHEMICALLY SENSITIVE PATIENTS

Future discussions of patients with multiple chemical sensitivity (MCS) may be aided by statistical data from a large survey, conducted by the staff at the Environmental Health Center in Dallas, Texas, of more than 200 consecutive patients being investigated and treated in the Environmental Control Unit (Sprague, 1987). The Environmental Health Center is a multidisciplinary facility of physicians, psychologists, research scientists, and support staff who have extensive experience with patients referred from around the country with symptoms of chemical sensitivity. This clinic is specially constructed with porcelain-on-steel walls and ceilings, terrazzo tile floors, extensive air-filtration systems, and other controls to minimize indoor air pollution and maximize the likelihood of patients being in a baseline state before appropriate investigation and treatment of allergies or chemical sensitivities.

Age distribution of the patients is shown in Figure 1. Consistent with findings of other studies, about three-quarters of these patients are women. Most present for evaluation in their 20s, 30s, or 40s but report that symptoms attributable to environmental problems first started at a much earlier age, even in the teenage years and earlier (Figure 2). Although disturbing in some respects, this information is encouraging because it raises the possibility of early intervention to prevent progression of this condition.

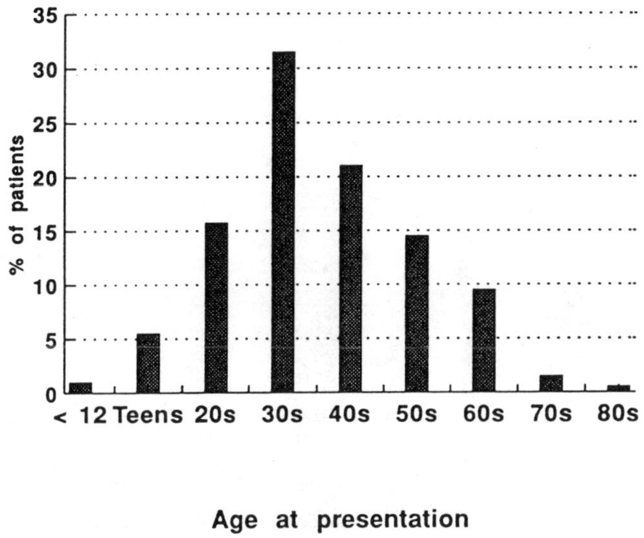

FIGURE 1. Age distribution of 200 chemically sensitive patients of Environmental Health Center-Dallas.

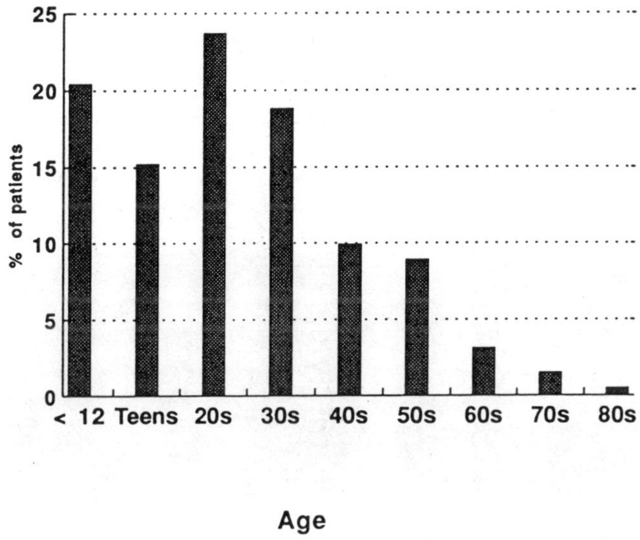

FIGURE 2. Age of symptom onset for 200 chemically sensitive patients of Environmental Health Center-Dallas.

The number of physicians previously seen by these patients varied, but it was not unusual for a patient to have consulted as many as 20 doctors in trying to find answers for their unusual health problems (Figure 3). The education level in this group tended to be rather high; many were college educated (Figure 4). These patients represented a wide spectrum of occupations and trades, including business, sales, labor, law, teaching, technical trades, other professionals, clergy, physicians, other medical personnel, clerical workers, homemakers, artists, students, and engineers (Figures 5a and 5b).

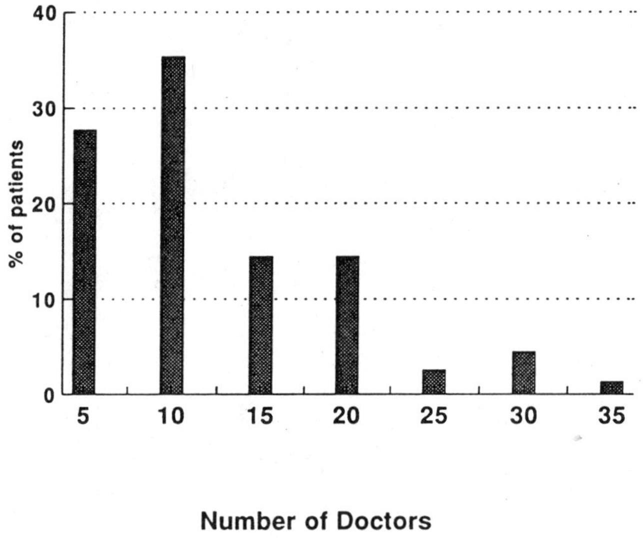

FIGURE 3. Number of doctors seen previously by 200 chemically sensitive patients of Environmental Health Center-Dallas.

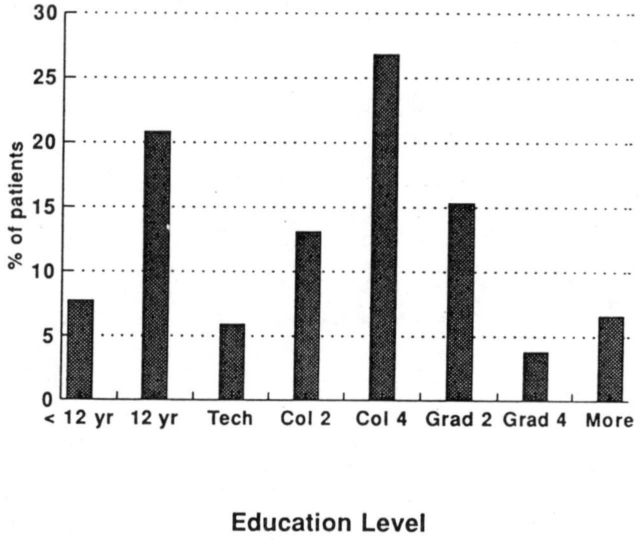

FIGURE 4. Education level attained by 200 chemically sensitive patients of Environmental Health Center-Dallas.

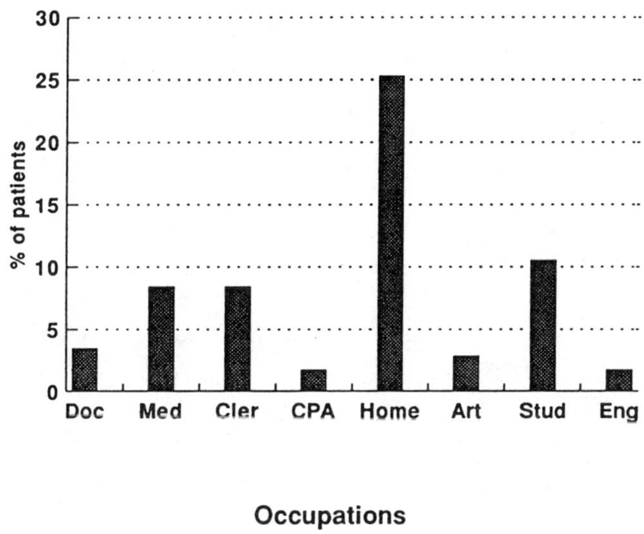

FIGURE 5(a). Occupations of 200 chemically sensitive patients of Environmental Health Center-Dallas.

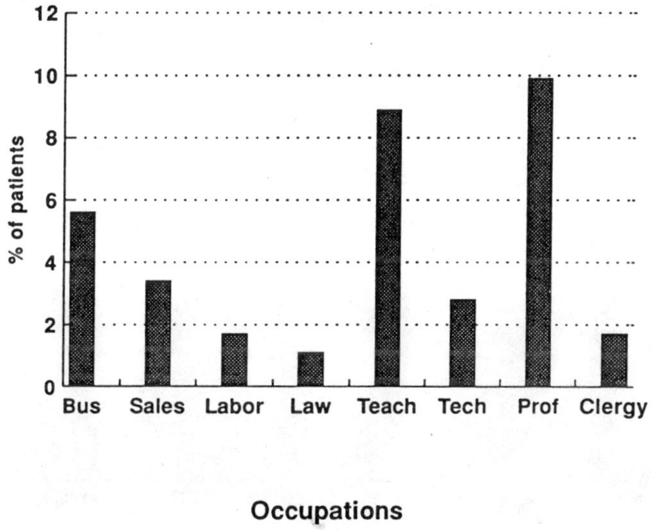

FIGURE 5(b). Occupations of 200 chemically sensitive patients of Environmental Health Center-Dallas.

Typically, the family history in many of these patients was positive for classical allergy, and the likelihood of migraine headaches, alcoholism, thyroid dysfunction, collagen vascular diseases, or psychiatric diagnoses in the family appeared high. A careful history of these patients might reveal a significant deterioration in their health after some identifiable event, such as redecorating the house, the arrival of a new baby, prolonged recovery from the flu, moving into a new building, or exposure to pesticides.

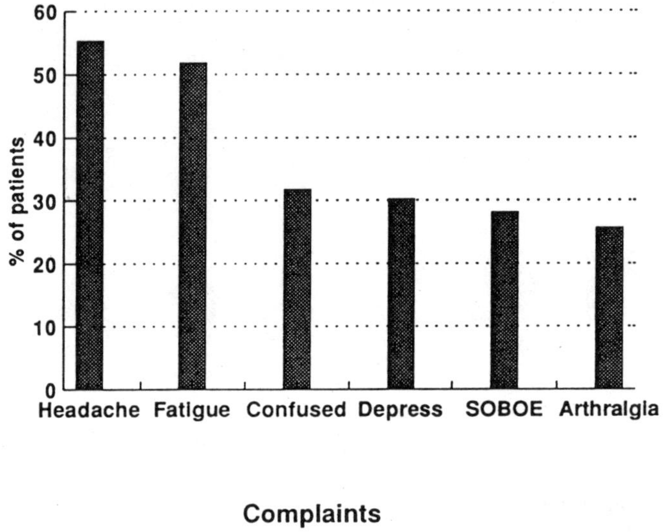

FIGURE 6(a). Most frequent complaints of 200 chemically sensitive patients of Environmental Health Center-Dallas.

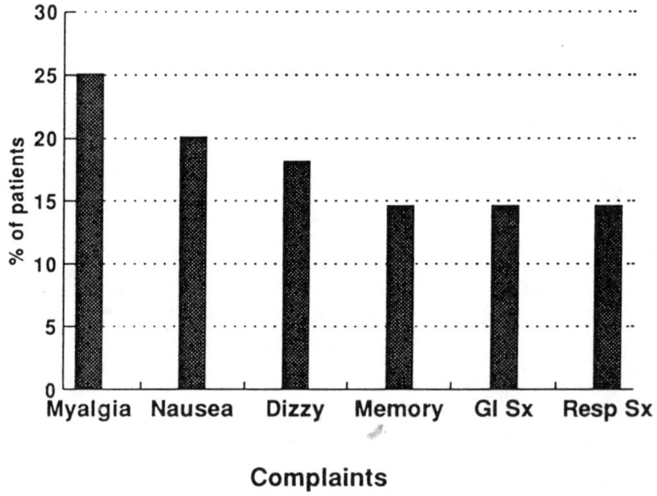

FIGURE 6(b). Most frequent complaints of 200 chemically sensitive patients of Environmental Health Center-Dallas.

The most frequent complaints in this survey of patients ranged from headache, fatigue, confusion, depression, shortness of breath, and arthralgia, to myalgia, nausea, dizziness, memory problems, gastrointestinal symptoms, or respiratory symptoms (Figures 6a and 6b). Neurologic symptoms, including migraine, poor memory, confusional states or "brain fog", tremor, weakness, and numbness, seemed to top the list. Patients frequently reported anxiety, depression, irritability, and dizziness, and some also reported that noise particularly irritated them. On occasion, they reported being tearful, with a higher-than-usual incidence of addiction. A frustrating aspect for these patients (and their doctors) was a lack of consistent and pathognomonic physical indicators of MCS. Patients sometimes said, "I feel so awful, that if one more person tells me how good I look, I'll scream!"

COMMON CHARACTERISTICS

Despite the lack of universally consistent physical findings in chemically sensitive patients, certain characteristics are seen more often than not. Although anecdotal, it may help physicians interested in MCS to know that chemically sensitive patients are more likely to have unstable physical balance, to be easily confused, and to have poor memory that can be measured objectively on neurologic evaluation (Rea, 1984). Frequently, they complain of cold hands and feet, possibly from autonomic instability and vascular spasms. They are much more likely to be underweight than overweight. The women may have a history of fibrocystic breast disease or thyroid dysfunction, and many have coated or geographic tongues and develop aphthous ulcers after eating certain foods. Children have a high incidence of learning disabilities or hyperactivity, often get very sleepy after lunch, and may be quite erratic in their school work. Patients with chemical sensitivities often have food cravings, histories of food and other addictions, and may experience withdrawal symptoms when they go without certain foods. Children especially may exhibit Jekell-and-Hyde-type sudden personality or mood changes (Rapp, 1991).

Chemically sensitive patients may be diagnosed objectively, however, and this is perhaps best demonstrated on double-blind, placebo-controlled, low-dose chemical challenges, under environmentally controlled conditions (Rea, 1990a).

SUMMARY

Although no one has all the answers to the mystery of chemical sensitivity, the reality of this condition, most recently called multiple chemical sensitivities, is not in doubt. Evidence is increasing of its possible physiologic mechanisms, which will be discussed later in this volume. From the evidence and from personal and professional experience, the author believes that chemical sensitivity is not a diagnosis of exclusion, and that fixed-name diseases may have environmental triggers or complicating factors (Rea, 1990b). With appropriate preparation and environmental controls, MCS can be investigated and diagnosed in a scientific and reproducible manner.

REFERENCES

ASHFORD, N.A. and MILLER, C.S. (1989). Chemical Sensitivity. A Report to the New Jersey State Department of Health. December 1989.

ASHFORD, N.A. and MILLER C.S. (1991). Chemical Exposures—Low Levels and High Stakes. Van Nostrand Reinhold, New York.

BROSTOFF, J. (1987). Mechanisms. In: Food Allergy and Intolerance. (J. Brostoff and S. Challacombe, eds.). Bailliere Tindall/W.B. Saunders, Philadelphia, PA. pp. 443-455.

CALABRESE, E.J. (1984). Ecogenetics: Genetic Variation in Susceptibility to Environmental Agents. Wiley, New York.

HILEMAN, B. (1991). Multiple chemical sensitivities. Chem. Engin. News. 69(20):26-42.

JACOBY, W.B. (1980). Detoxication Enzymes. In: Enzymatic Basis of Detoxication, (W.B. Jacoby, ed.). Vol 1. Academic Press, New York. pp.1-6.

RANDOLPH, T.G. (1980). An Alternate Approach to Allergies (revised ed.). Harper and Row, New York.

RAPP, D. (1991). Is This Your Child? William Morrow and Company, New York.

REA, W.J., BUTLER, J.R., LASETER, J.L., and DeLEON, I.R. (1984). Pesticides and brain function changes in a controlled environment. Clin. Ecol. 2(3):145-150.

REA, W.J., ROSS, G.H., JOHNSON, A.R. et al. (1990a). Confirmation of chemical sensitivity by means of double-blind inhalant challenge of toxic volatile chemicals. Clin. Ecol. 6(3):113-118.

REA, W.J., ROSS, G.H., JOHNSON, A.R., SMILEY, R.E., and FENYVES, E.J. (1990b). Chemical Sensitivity in Physicians. Clin. Ecol. 6(4):135-141.

SPRAGUE, D. (1987). The concept of an environmental unit. In: Food Allergy and Intolerance. (J. Brostoff and S. Challacombe, eds.). Bailliere Tindall/W.B. Saunders, Philadelphia, PA. pp. 947-960.

ACQUIRED INTOLERANCE TO SOLVENTS FOLLOWING PESTICIDE/SOLVENT EXPOSURE IN A BUILDING: A NEW GROUP OF WORKERS AT RISK FOR MULTIPLE CHEMICAL SENSITIVITIES?

JAMES E. CONE, MD, MPH
University of California, San Francisco

THOMAS A. SULT, MD
South Lake Tahoe, California

INTRODUCTION

Four segments of the population who may have multiple chemical sensitivities have been identified (Ashford and Miller, 1990):

1. Industrial workers
2. Occupants of 'tight' buildings, including office workers and school children
3. Residents of communities whose air or water is contaminated by chemicals
4. Individuals who have had personal and unique exposure to various chemicals in domestic indoor air, including pesticides, drugs, and consumer products

We evaluated workers from one building who had symptoms most likely associated with pesticide applications to control cockroach infestation. Many of these workers subsequently developed acquired intolerances to pesticides, solvents, or other agents, suggesting multiple chemical sensitivities. This may represent a new group of workers potentially at risk for the development of multiple chemical sensitivities.

BACKGROUND

The following discussion is excerpted from the local Occupational Safety and Health Administration (OSHA) inspector's report.[1]

A 'mystery illness' at a major casino resort hotel was reported in a local newspaper (Evans, 1990). Symptoms included dizziness, weakness, nausea, and sore throat, and employees repeatedly reported a "Raid[2]-like smell." The building manager, who reported that 17 employees had been ill on March 24 and 25, 1990, called the local OSHA office. Three workers were admitted to the hospital because of the severity of their symptoms, including

Address all correspondence to: Dr. James E. Cone, 2241 Channing Way, Berkeley, CA 94704.

---

[1] Meier, R., personal communication: OSHA Report (no date given).
[2] Use of trade names is for identification only and does not imply endorsement by the Public Health Service.

syncopal episodes while at work. Blood gases and carboxyhemoglobin levels were determined for a few workers. One worker showed respiratory alkalosis consistent with hyperventilation.

The OSHA inspector interviewed workers, who complained of dizziness, nausea, sweating, headaches, racing heartbeats, weakness, shaking or trembling, and tingling of lips and face. Some workers had red splotchy rashes on their skin, particularly on the face and exposed areas. A crisis atmosphere was present, and many workers were shaken or afraid. Even those who were not sick were extremely apprehensive.

Several more workers became ill with similar symptoms on the following Sunday. The OSHA representative arrived on site at midday, just as several more workers were becoming ill. In the pit area, the OSHA industrial hygienist (IH) detected a sweet unusual smell that came and went and changed direction, much like an ant spray. The IH, too, experienced similar symptoms, including cramps, chills, nausea, and headache.

By April 5, more than 113 employees had reported symptoms. A resident nurse was on duty in the first aid room to provide on-site urgent care and triage. During the following months, more than 250 individuals from the building consulted the author (T.S.) or other local physicians in urgent-care centers and emergency departments because of work-related symptoms.

Evaluation and treatment of the illnesses associated with this outbreak is ongoing. The objective of this study was to determine whether the workers had a pattern of illness that could be identified, and then whether this pattern was consistent with the hypothesis that the illness was related to workplace exposures, particularly to pesticides and their organic solvent carriers.

We present this case series in the hope that others may provide ideas for the best ways to evaluate, treat, and prevent such outbreaks in the future.

METHODS

Industrial Hygiene

Both the OSHA and the company consultant industrial hygienist did sampling. Air samples were taken on March 26 and April 4. GCMS samples of air were taken directly on traps containing mixed absorbent media. Pesticide sampling for organochlorine and organophosphate pesticides was done in the air and at floor level. Samples were analyzed by NIOSH Methods 5502 and 5012. Glass fiber filter and XAD-2 media were used for samples obtained on March 26 and April 2, and specific testing for coumaphos and propoxur was first performed on April 5.

Medical

The building management engaged the services of a private medical consultant to interview workers on site. Individuals were referred to the company physician, who performed laboratory testing including red blood cell (RBC) and serum cholinesterase levels for 24 workers. Analysis was performed by the local state-certified forensic toxicology laboratory, using the

Ciba-Corning Enzymatic Cholinesterase Testing Kit and a Gilford Spectrophotometer. Normal values for these tests were provided by Ciba-Corning.

Approximately nine months after the initial incident, 19 workers were referred to the UCSF Occupational Health Clinic for interview and examination by attending physicians and residents. Tests included the Lanthony D-15 color vision test, performed monocularly with correction for visual acuity. A color confusion index was calculated using a computer program by Jaques Gresset[3]. Individuals were referred for full neuropsychological testing, nerve conduction testing, and neurologist consultation when indicated by results of the initial examination.

The charts for workers were divided according to the presence or absence of symptoms associated with the initial exposure episode and the subsequent medical course. A case-comparison analysis was performed to compare the characteristics of individuals who reported symptoms consistent with the syndrome of acquired intolerance to solvents (AIS), pesticides, or other odorants or chemical agents. This syndrome has been described in the literature previously in connection with the chronic sequelae of pesticide poisoning (Tabershaw, 1966) and following acute or chronic solvent intoxication (Gyntelberg, 1986). We identified this syndrome as one of three clinical diagnostic subsets in our previous case series of multiple chemical sensitivities among patients at the Occupational Health Clinics at UCSF.

RESULTS

Industrial Hygiene

The company IH reported initial air sampling results to OSHA as not statistically significant. Initially, positive results were reported for two pesticides, coumaphos (7.4 ppb on XAD-2 Media only, PEL of 8 ppb) and fensulfothion. The fensulfothion result was thought to be a laboratory error. Trace organic compounds—xylene, 1,1,1-trichlorethane, methylene chloride, acetone (maximum of 240 ppb in one sample) and petroleum distillates—were also detected in ppb ranges (Walker, 1990).

Laboratory analysis of a second set of samples did not confirm the continued presence of coumaphos or fensulfothion. Pesticide application logs provided by the building management identified the primary active ingredients as pyrethrins and carbamates.

The OSHA industrial hygienists made the following observations[4]

> Baygon (trade name for carbamate propoxur) was applied into walls of kitchen areas in the basement. Solvent carriers included in the formulation were trichloroethane, methylene chloride, and xylene. This mix was pressurized by carbon dioxide, which would drive vapors and mist into and through inner wall spaces for cockroach control.

---

[3] School of Optometry, University of Montreal, Quebec. The program was translated into English by James E. Cone and Blanche Grosswald.

[4] Meier, R., personal communication: OSHA Report (no date given).

Pyrethrins used were generally either wettable powder or emulsifiable concentrates that were spot applied. Of note were applications of both, which coincided neatly with major outbreaks of employee symptoms. During the initial walkaround, the pesticide applicator indicated pyrethrin was applied under some booths in the employee cafe on March 23. Booths in question were favorite booths where workers would leave half-eaten food on a shelf behind the booth. The employees from the affected area work a 60-minutes-on, 20-minutes-off cycle during an 8-hour shift, so, on average, they spend more time in the cafe and other breakrooms than other employee classes. Another breakroom used by the workers was located in the affected area of the building. Two of three walls for this basement breakroom are shared back-to-back with kitchens where pesticides were applied into walls and cracks.

Medical

Nineteen workers were referred to the UCSF Occupational Health Clinic, and were seen over a 6-month period, from 9 to 15 months after the initial episode.

Characteristics of Workers

The mean age of the workers examined was 40 years (range 31-65, Standard Deviation = 9 years). Seventeen were female, and two were male. All were either dealers or 'pit bosses', who are supervisory personnel working on the floor of the casino. The mean length of service was 9.3 years, with a range of 1-26 years. Seven (37%) never smoked, four (21%) were past smokers, and 8 (42%) were current smokers. Eight (42%) reported histories of allergies.

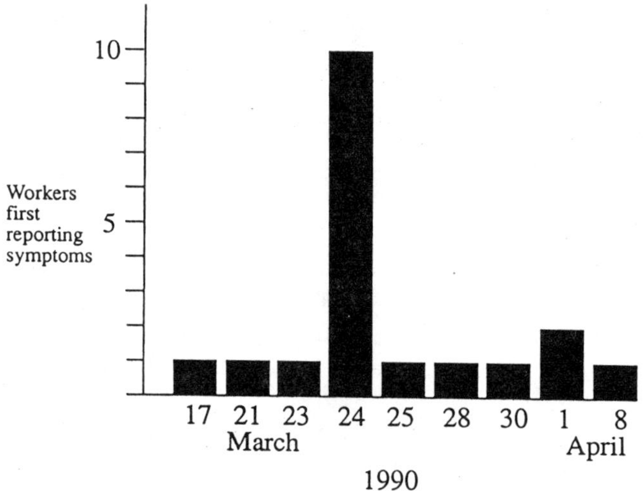

FIGURE 1. Onset of symptoms reported at a casino, 1990.

Reported Symptoms

Symptoms were initially reported on March 17, 1990 (Figure 1). The reported symptoms, organized by organ system, are listed in Tables 1-5. The predominant symptoms were headache, numbness, nausea, palpitations, memory loss, shakiness or tremors, dizziness, more fatigue than usual, acute intoxication symptoms at work, and depression (in more than 50% of the cases). Excessive sweating, syncope, visual symptoms, and mucous membrane irritation were also commonly reported.

TABLE 1. General Symptoms Reported by Casino Workers Exposed to Pesticides, 1990

SYMPTOM	NO. OF WORKERS	% OF WORKERS (EXCEPT WHERE NOTED*)
Abnormal menses	2	12*
Dysfunctional bleeding	5	29*
Excessive sweating	9	47
Joint pains	7	37
Metallic taste in mouth	3	16
Muscle cramps	3	16
Skin rash	9	47
Weight loss	8	42

* Menstrual symptom percentages are based on the total number of women (17) in the group examined.

TABLE 2. Eye, Ear, Nose, and Throat Symptoms Reported by Casino Workers Exposed to Pesticides, 1990

SYMPTOM	NO. OF WORKERS	% OF WORKERS
Blurred vision	8	42
Dizziness	14	74
Eye irritation	7	37
Nose irritation	6	32
Ringing in ears	5	26
Throat irritation	4	21
Tunnel vision	3	16

Thyroid disease was reported by 4/19 (21%). Tremors—mostly resting tremors—were confirmed in three cases on physical examination and were sometimes so severe that the person had difficulty walking or undressing for the examination. Test results for nerve-conduction velocity are pending for several workers but were normal in the two tested to date. Neurologic consultation has reported normal examinations for two workers but evidence of toxic encephalopathy ('painter's syndrome') and abnormal Trigeminal Evoked Potential for a third.

Symptoms reported in this case series were serious enough to lead to significant disability in 12/19 (63%) workers, and these workers were removed from further exposure to the workplace.

TABLE 3. Chest Symptoms Reported by Casino Workers Exposed to Pesticides, 1990

SYMPTOM	NO. OF WORKERS	% OF WORKERS
Chest pain	6	32
Cough	3	16
Palpitations	13	68
Shortness of breath	5	26

TABLE 4. Gastrointestinal Symptoms Reported by Casino Workers Exposed to Pesticides, 1990

SYMPTOM	NO. OF WORKERS	% OF WORKERS
Diarrhea	6	32
Nausea at work	13	68
Vomiting	6	32

TABLE 5. Neurologic and Neuropsychologic Symptoms Reported by Casino Workers Exposed to Pesticides, 1990

SYMPTOM	NO. OF WORKERS	% OF WORKERS
Acute intoxication symptoms at work	12	63
Anxiety	7	37
Depression	10	53
Difficulty concentrating on game	12	63
Fatigue more than usual	12	63
Headache	17	90
Memory loss, short term	11	58
Numbness of face or extremities	14	74
Shaking/tremors	10	53
Sleep disorders	1	5
Syncope	4	21

Color Vision Testing Results
A total of 11 cases of acquired blue-yellow color vision loss and three cases of complex color vision loss were diagnosed using the Lanthony D-15 Desaturated Color Vision test (Figure 2).

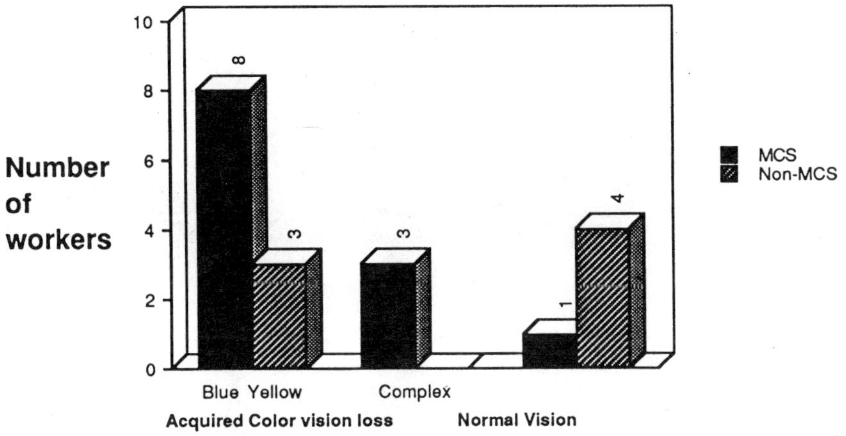

FIGURE 2. Acquired color vision loss of MCS vs. non-MCS workers at casino, 1990.

Acquired Intolerance to Solvents (AIS)
Of particular interest is the report by 12/19 (63%) workers of a new onset of 'sensitivity' or intolerance to perfumes, gasoline, newsprint, cleaning materials, pesticides and other solvent-containing materials. When the 12 individuals who developed AIS are compared with the 7 who did not, no statistically significant difference was noted in the proportion reporting symptoms related to the various organ systems. However, headache, a history of thyroid disease, and blurred vision tended to be reported more often by persons with AIS than by those without. Similarly, the color confusion index tended to be even higher among persons with AIS (mean of 1.83 in the right eye among those with AIS vs. a mean of 1.37 for those without; mean of 1.72 in the left eye with AIS, and 1.56 without), although this difference was not statistically significant. Nevertheless, these levels are very highly abnormal compared with unexposed populations.

Results of Cholinesterase Testing
In all, 24 workers reported to the company physician with symptoms and had blood drawn and analyzed for red blood cell (RBC) cholinesterase and serum cholinesterase. Figures 3 and 4 show results of RBC and serum cholinesterase tests. To determine whether these results differ significantly from expected values, a two-tailed t-test was performed to evaluate whether a difference exists between the sample means and population means (Zar, 1984), assuming the population means are normally distributed.

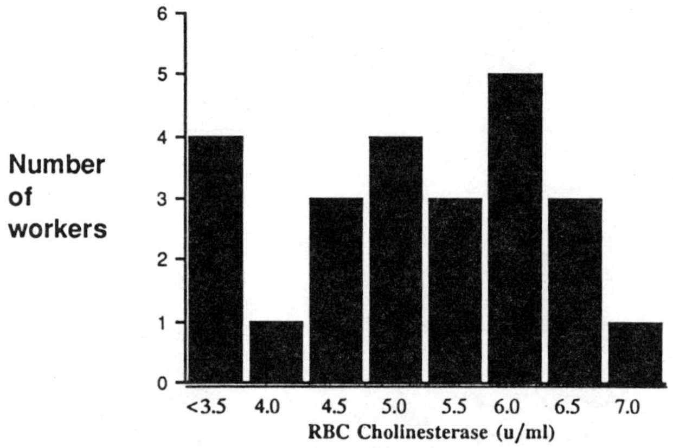

FIGURE 3. Range of RBC cholinesterase values for workers at casino, 1990.

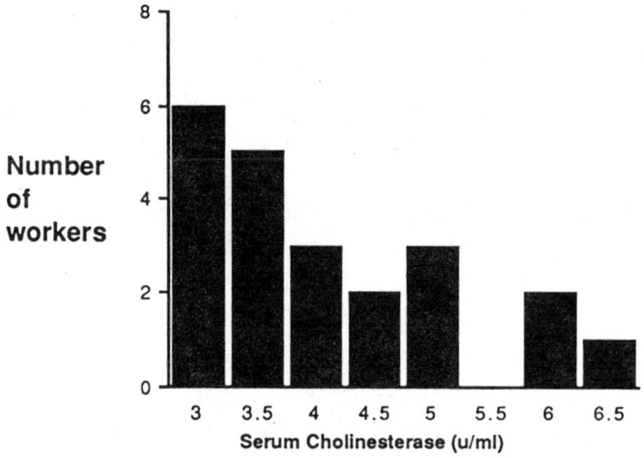

FIGURE 4. Range of serum cholinesterase values for workers at casino, 1990.

The difference in mean RBC cholinesterase for the workers tested (4.83 μ/ml) compared with the population mean (5.79 μ/ml) was statistically significantly (Table 6). On the other hand, the mean serum cholinesterase in the building occupants tested (4.317 μ/ml) was lower than the population mean (4.79 μ/ml), but the difference was not statistically significant (Table 7).

DISCUSSION

The company medical consultant met with management and advanced the theory that this event was mainly a "building-related mass psychogenic illness." The alternative hypothesis, however, is much more likely: that this outbreak represents acute and chronic poisoning with a cholinesterase-inhibiting pesticide (probably the carbamate pesticide, Baygon [Propoxur], mixed with organic solvent carriers).

TABLE 6. Statistical Testing of Sample Mean RBC Cholinesterase Testing Results for Casino Workers Compared with Laboratory-Expected Population Mean, 1990

H_0 : μ=5.79 u/ml
H_a: $\mu \neq$5.79 u/ml
alpha=0.05
n=24
population mean=x=5.79
sample mean = μ = 4.83
standard deviation = s = 1.1

s^2=1.21

$s_x = \sqrt{[s^2/n]} = \sqrt{[1.21/24]} = 0.224$

$$t = \frac{x-\mu}{s_x} = 4.286$$

degrees of freedom = v = n-1 = 24-1 = 23
thus, $t_{0.05(2), 23}$: p<0.001: **Reject H_0**

TABLE 7. Statistical Testing of Sample Mean Serum Cholinesterase Testing Results for Casino Workers Compared with Laboratory-Expected Population Mean, 1990

H_0 : μ=4.79 u/ml
H_a: $\mu \neq$4.79 u/ml

alpha=0.05
n=22
population mean=x=4.79
sample mean = μ = 4.317
standard deviation = s = 1.1

s^2=1.21

$s_x = \sqrt{[s^2/n]} = \sqrt{[1.21/22]} = 0.234$

$$t = \frac{x-\mu}{s_x} = 2.017$$

degrees of freedom = v = n-1 = 22-1 = 21
thus, $t_{0.05(2), 21}$: 0.05<p<0.1: **Fail to reject H_0**

The most important evidence to support this conclusion is the clear difference between the mean RBC cholinesterase in this sample, compared with the laboratory mean value, assuming the laboratory values represent a normal distribution. In addition, four workers had levels below the range of normal for this laboratory, which confirms the impression of clinical pesticide poisoning.

These results are more remarkable because carbamates do not cause as much depression of cholinesterase levels and because effects reverse more quickly. In organophosphate-associated pesticide poisoning, it is not unusual to see a more persistent depression in the RBC cholinesterase, compared with the shorter-duration depression of the serum cholinesterase, as was seen here. However, the apparent persistence of the cholinesterase depression (hours to days after the peak symptom episodes and the presumed exposures) is unusual with carbamate pesticides in general. This suggests either ongoing exposure or the presence of other cholinesterase-inhibiting pesticides as well.

The cholinesterase results support our primary hypothesis, that indeed the workers had suffered some significant cholinesterase inhibition. This would explain many of their symptoms and findings. Unfortunately, in this case, as in most non-agricultural settings, no preexposure cholinesterase levels were documented.

Reported Toxicology of the Carbamate, Baygon (Propoxur)

The insecticide, 2-(1-methylethoxy) phenyl methlcarbamate (Baygon or propoxur), is "characterized by fast knockdown, long residual and flushing effect" (Farm Chemicals Handbook). It is a white-to-tan crystalline solid, soluble in water (0.2% at 25°C) and soluble in most polar organic solvents. It is available as an emulsifiable concentrate, a wettable powder, and as bait and dust.

Toxicity testing shows that it has an acute oral LD50 (rat) of 100-200 mg/kg in males and 70-130 in females. Dermal LD50 is > 500 mg/kg in the male rabbit. An adult volunteer exposed to propoxur experienced a prompt fall in RBC cholinesterase activity that reached a low of 27% of the normal value in 15 minutes. This was followed by blurred vision, nausea, pallor, sweating, tachycardia, and vomiting. All symptoms abated, and enzyme levels were normal two hours later (Gosselin, 1984). Although propoxur inhibits cholinesterase, inhibition is rapidly reversible. Acute symptoms do occur following low-level exposures but are transitory in nature (American Conference of Governmental Industrial Hygienists, 1980). However, permanent behavioral aberrations were documented in rats that received propoxur repeatedly (Gosselin, 1984).

As the OSHA industrial hygienist pointed out, "sampling, specific for pyrethrins and carbamates, was not performed until 9-10 days after application. Active ingredients were present in the concentrate at very low levels (< 10%) before mixing. Mixing Baygon with water or CO2 adds more dilution. Both poisons, unlike organophosphates, are relatively short lived in the environment or break down readily. The presence of solvent carriers in air suggests that poisons either were once present or were present at levels that are not detectable. There are no PELs for the pyrethrins used, and PELs for pesticides in general are not universally accepted as providing adequate protection."

Multiple Chemical Sensitivities

Of particular note in this outbreak is the high frequency of reported cases resembling multiple chemical sensitivity or acquired intolerance to solvents or solvent-containing products, particularly perfumes. Before any inferences can be made about the incidence in the population of workers as a whole, a more adequate sample must be taken, but the rate was high compared with other similar exposure events, at least among those most severely affected who came to medical attention. The two previous building-associated pesticide poisoning outbreaks we evaluated have each involved similar numbers of ill workers, but only one or two cases of MCS-like symptoms were reported out of 20-30 workers affected.

The difference in this outbreak may be the ongoing nature of the exposure. The repeated applications of pesticides and carriers in the same environment contrast with the evacuation and cleanup that occurred in the other two building outbreaks.

In view of the trend of more severe color-vision abnormalities noted here in persons with MCS (even within a cohort exposed to the same neurotoxins), the assessment of acquired color-vision loss may be useful in future studies of individuals with MCS.

REFERENCES

AMERICAN CONFERENCE OF GOVERNMENTAL INDUSTRIAL HYGIENISTS (ACGIH). (1980). Documentation for TLVs. 4th ed. and supplement. ACGIH, Cincinnati, OH.

ASHFORD, N.A. and MILLER, C.S. (1990). Chemical Exposure: Low Levels and High Stakes. Van Nostrand Reinhold, New York.

EVANS, M. (1990). Mystery disease fells at least 50. The Gazette, March 25.

GOSSELIN, R.E. (1984). Clinical Toxicology of Commercial Products. 5th ed. Williams and Wilkins, Baltimore, MD.

GYNTELBERG, F., et. al. (1986). Acquired intolerance to organic solvents and results of vestibular testing. Am. J. Ind. Med. 9:363-370.

TABERSHAW, I.R. and COOPER, C. (1966). Sequelae of acute organic phosphate poisoning. J. Occup. Med. 8:5-20.

WALKER, T.J. (1990). Industrial Hygiene Report, June 5.

ZAR, J.H. (1984). Biostatistical Analysis. Prentice-Hall, Englewood Cliffs, NJ. pp. 97-101.

EPIDEMIC MULTIPLE CHEMICAL SENSITIVITY IN AN INDUSTRIAL SETTING

GREGORY E. SIMON, MD, MPH
Center for Health Studies, Group Health Cooperative
Department of Psychiatry and Behavioral Sciences
University of Washington, Seattle

INTRODUCTION

During late 1987, one of the work units at a large aerospace manufacturing firm introduced a new process for producing fire-retardant composite plastics. This process involved close work with solvents and irritants such as phenol and formaldehyde. More than 400 employees worked at this site, and approximately 200 worked directly with the composite plastic material.

By six months after the introduction of these materials, more than 50 workers had reported illness, which they attributed to worksite chemical exposure; 37 of them were ill enough to file workers' compensation claims. An Independent Medical Evaluation Board including the author and colleagues studied these workers. Portions of their findings were reported previously (Sparks et al., 1990; Simon et al., 1990).

REPORT OF SYMPTOMS

The 37 workers reported systemic symptoms (such as fatigue and weakness), symptoms of respiratory and mucous membrane irritation (such as cough, sore throat, eye irritation), cognitive symptoms (such as poor concentration, memory loss) and changes in mood or personality (such as irritability, anxiety, and mood instability). The majority of subjects reported a self-limited syndrome with symptoms limited to workplace exposure. Some had been transferred to other worksites and subsequently reported substantial improvement or complete resolution. A significant minority, however, developed a full multiple chemical sensitivity (MCS) syndrome. Symptoms persisted even after the workers left work or transferred to new locations. These subjects reported a generalized sensitivity to numerous common exposures so that the symptoms originally precipitated by workplace exposure could now be produced by synthetic fabrics, paint, or gasoline fumes. Many of these persons had left work and reported severe restriction of their daily activities to avoid chemical exposures that might provoke a recurrence of symptoms.

Note: Preparation of this presentation was supported in part by an appointment to the Agency for Toxic Substances and Disease Registry (ATSDR) Clinical Fellowship Program in Environmental Medicine administered by Oak Ridge Associated Universities through an interagency agreement between the U.S. Department of Energy and ATSDR.

Address all correspondence to: Dr. Gregory Simon, Center for Health Studies, 1730 Minor Ave., Seattle, WA 98101.

INVESTIGATION

Environment
The overall evaluation consisted of various components. Environmental measures included repeated air monitoring by the Washington State Division of Occupational Safety and Health and the National Institute for Occupational Safety and Health. All measured levels of phenol, formaldehyde, and total particulates were less than 10% of Occupational Safety and Health Administration Permissible Exposure Limits (PEL). All measured levels of styrene, acetone, methylene chloride, and antimony trioxide were less than 1% of the PEL.

Medical
Comprehensive medical examination revealed abnormalities in only a small number of subjects. The only findings on detailed physical examination included evidence of contact dermatitis in 4% and upper respiratory irritation in 13%. Laboratory examination revealed hematuria in 6% and elevation of serum thyroxine in 8%. An immunologic evaluation, including enumeration of lymphocyte subsets and measurement of antibodies to formaldehyde and trimetallic anhydrite, showed no immunologic differences between study subjects and healthy laboratory controls.

Social Conditions
Social conditions at this worksite appeared ripe for what I will call psychologic contagion. Many workers described a high level of perceived health threat. One described the disabling symptoms experienced by his co-workers as "a chemical kind of AIDS" and many believed they had suffered chemical poisoning or immunologic damage. Symptoms often began with highly visible attacks, which publicly emphasized the level of danger. Many subjects described seeing colleagues overcome by chemical fumes and removed from the production line by ambulance. Both labor and management at this plant were under high levels of stress. Production pressures had resulted in longer work hours and forced overtime at the time the new manufacturing process was introduced. Success of this new process was essential if management was to meet fast-approaching product delivery dates.

Psychiatric Evaluation
Results of psychiatric evaluations for this group of 37 subjects will illustrate the potential contribution of psychologic factors to this outbreak. The evaluation employed the following instruments:

> An abridged version of the Diagnostic Interview Schedule (DIS): a structured, psychiatric, diagnostic interview that assesses current and lifetime psychiatric symptoms (Robins et al., 1981). This interview allows assignment of diagnoses according to criteria of the American Psychiatric Association's diagnostic manual, the DSM-III. The version of the interview used in this study assesses symptoms of depression, panic attacks, and generalized anxiety and includes a section on somatization to assess physical symptoms or complaints that lack apparent medical

explanation. Interviews were scored in two ways. Initial scoring included all symptoms reported, even those associated with chemical sensitivity. Interviews were then edited to include only symptoms present before introduction of the new manufacturing process.

Selected subscales of the Hopkins Symptom Checklist-90 (SCL-90): a self-report measure of psychiatric symptoms (Derogatis, 1977).

The Whitely Index: a self-report questionnaire designed to measure hypochondriacal attitudes and beliefs (Pilowsky, 1967).

The Barsky amplification questionnaire: a self-report questionnaire designed to measure symptom amplification or the tendency to report a large number of physical symptoms (Barsky et al., 1988).

The prevalence of current psychologic morbidity was high among the 37 workers evaluated (Table 1). Sixty percent met diagnostic criteria for current major depression, 26% for current panic disorder, and 78% for one of the two. When considering only symptoms that predated chemical exposure, however, rates of pre-existing psychologic morbidity were not markedly elevated. The 23% lifetime prevalence of anxiety or depressive disorder is higher than community samples but similar to rates found among general medical patients. The entire group, therefore, showed a high level of psychiatric symptoms associated with this outbreak but did not appear to have a high rate of pre-existing psychiatric illness.

TABLE 1. Percentage of DIS Psychiatric Diagnosis Among Workers Filing Compensation Claims for Chemical-related Illness (n = 37)

	PERCENTAGE WITH DIAGNOSIS	
DIAGNOSIS	CURRENT	PRE-EXISTING
Major Depression	60	23
Panic Disorder	26	4
Either	78	23

SCL-90 scores showed a similar elevation of current psychologic morbidity, with mean T-scores for anxiety, depression, and somatization subscales all over 70 (2 standard deviations above the mean for general population samples). For each subject, all three scores were greater than one standard deviation above the community mean (T-score greater than 60).

EXAMINATION OF MCS SUBJECTS

To examine the subset of subjects with the full MCS syndrome more closely, we used four questions to operationalize our definition of MCS.

- Do you now need to follow any special diet because of chemical or food sensitivity?
- Do you now take special precautions in your home or home furnishings because of chemical sensitivity?
- Do you now need to wear particular clothes because of chemical sensitivity?
- Do you have trouble shopping in stores or eating in restaurants because of chemical sensitivity?

The frequency of positive responses to these four questions was bimodally distributed, with a large group having zero "yes" answers and a second group reporting three or more. We defined chemical sensitivity cases as those with two or more positive responses. By this standard, 13 subjects were classified as MCS cases and 23 as non-cases. One subject did not complete the questionnaire and could not be classified as a case or a non-case. The subject was excluded from this portion of the analysis. Varying the cut-point either way, to one or three positive responses, did not substantially affect the results that follow.

Both cases (92%) and non-cases (70%) showed similar high levels of current psychiatric morbidity (current anxiety or depressive disorder). SCL anxiety and depression scores were elevated in both groups and only slightly higher among cases.

The two groups did differ on prevalence of pre-existing morbidity. Structured psychiatric interview data were analyzed to include only symptoms occurring before initial exposure to the new manufacturing process. In this analysis, cases (54%) had much higher levels of prior anxiety or depressive disorder than did non-cases (4%), and a larger number of somatization symptoms (reported physical symptoms with no apparent medical explanation) in the somatization section of the DIS (6.2 vs 2.9).

Self-report measures of somatization were also much higher among the chemical sensitivity cases. Scores on the Whitely Index (measuring hypochondriasis), the Barsky Scale (measuring symptom amplification), and the SCL-90 Somatization scale were all significantly greater among persons developing MCS.

Anecdotal Impressions

Some anecdotal impressions left after the interviews were not well illustrated by the quantitative findings. Many subjects reported the sudden onset of symptoms during a period of high exposure (for example, working in a closed space). Often these acute attacks included panic or anxiety symptoms such as palpitations, dyspnea, sweating, dizziness, and paresthesias. A core group of the most affected and most vocal subjects appeared to have long prior histories of medically unexplained physical symptoms, high use of medical care, and long prior periods of disability. The secondary or less-severely affected subjects more often reported histories reminiscent of anxiety disorder: sudden onset of panic-like symptoms, avoidance of situations that might provoke recurrence, then gradual recovery. These findings suggest several psychologic mechanisms that might contribute to chemical sensitivity.

PSYCHOLOGIC MECHANISMS

The first of these is a behavioral sensitization model in which physical symptoms are viewed as conditioned responses to some sensory cue—often the odor of a solvent or other chemical. In this model, acute exposure to a solvent or toxin results in sensation to an unpleasant odor and produces acute physical symptoms. These symptoms may result from direct irritant effects of exposure (as with cough or sore throat), from mild intoxication (as with some cases of dizziness or confusion), or from fear-induced autonomic arousal (as with dyspnea or palpitations). The sensory cue (or odor) becomes paired with the physical symptoms so that a similar odor evokes a symptomatic response even in the absence of the original chemical exposure. Avoidance of situations likely to result in exposure to chemical odors reinforces the strength of the conditioned response.

Many of our subjects with limited symptoms appeared to fit this scenario. These were typically workers without extensive prior histories of somatization or psychiatric disorder. In response to a high level of perceived threat, they became acutely symptomatic. As the level of threat declined, they desensitized themselves and improved spontaneously.

A second model explains MCS as a form of somatization or symptom amplification. Persons exhibiting somatization or symptom amplification appear to have chronically high levels of psychologic distress associated with heightened perception of unpleasant physical sensations. This tendency to amplify symptoms can also be transiently increased by life stresses or social and cultural forces (such as workplace stress or a high level of perceived health threat). In this model, chemical sensitivity develops when a person with chronic symptom amplification develops the new belief that these symptoms are due to chemical exposure. This belief leads to avoidance of perceived exposure. Treatment based on this model would follow usual treatment for somatization patients-reducing symptom sensitivity and learning to attribute physical sensations to benign causes.

A final model focuses on the social environment and views MCS as an exaggeration or displacement of legitimate concerns about health risks. We are reminded almost daily that many significant environmental hazards are undetectable to our senses or individual experience. Radon gas, electromagnetic fields, benzene in drinking water have all been linked to fatal illness at levels far below those any individual can detect. This uncertainty about personal risk leads to increased vigilance and a tendency to attribute common symptoms to environmental hazards. We all sniff the air a little more closely when we worry about things we cannot smell. In addition, medical illness may seem the most valuable currency for presenting social or environmental concerns. The workers examined here would have received much less attention had they only complained that forced overtime was a strain on their families or that the odors involved in the new manufacturing process were irritating and unpleasant. In the specific case of chemical sensitivity, the term "allergy" is open to many interpretations. Allergy in the popular sense (as in "these chemicals make me feel sick") was somehow transformed into a more technical meaning (as in "these chemicals have affected my immune system").

These proposed models are certainly not mutually exclusive. All may contribute in any given case, and each may provide some clues about reducing chemical sensitivity and relieving patients' suffering.

REFERENCES

BARSKY, A.J., GOODSON, J.D., LANE, R.S., and CLEARY, P.D. (1988). The amplification of somatic symptoms. Psychosom. Med. 50:510-519.

DEROGATIS, L.R. (1977). SCL-90R: Administration, Scoring, and Procedures Manual. Johns Hopkins University Clinical Psychometrics Unit, Baltimore, MD.

PILOWSKY, I. (1967). Dimensions of hypochondriasis. Br. J. Psychiatry 113: 89-93.

ROBINS, L.N., HELZER, J.E., CROUGHAN, J., and RATCLIFF, K.S. (1981). The National Institute of Mental Health Diagnostic Interview Schedule: Its history, characteristics, and validity. Arch. Gen. Psychiatry 38:381-389.

SIMON, G.E., KATON, W.J., and SPARKS, P.J. (1990). Allergic to life: Psychological factors in environmental illness. Am. J. Psychiatry 147:901-906.

SPARKS, P.J., SIMON, G.E., KATON, W.J., ALTMAN, L.C., AYARS, G.H., and JOHNSON, R.L. (1990). An outbreak of illness among aerospace workers. West. J. Med. 153:28-33.

DEVELOPMENT OF MULTIPLE CHEMICAL SENSITIVITY AFTER AN OUTBREAK OF SICK-BUILDING SYNDROME

LAURA S. WELCH, MD AND ROSEMARY SOKAS, MD
Associate Professors of Medicine and Health Care Sciences
Division of Occupational and Environmental Medicine
George Washington University School of Medicine
Washington, DC

INTRODUCTION

This paper describes an outbreak of sick-building syndrome at a federal agency in Washington, D.C., and the subsequent evaluation of 20 persons involved in the outbreak.

The agency employs 5000 persons, 3700 of whom work in a building originally designed as a housing complex, with two towers and a mall between. The first floor of the mall is currently a shopping area; the upper floors have been divided into offices. This building was not likely designed for the level of occupancy it has had since it was built, and some occupants had complained of symptoms consistent with sick-building syndrome before the indoor renovation described below.

RENOVATION AND INDUSTRIAL HYGIENE INVESTIGATION

In October 1987, the agency undertook a major renovation project, beginning in the mall area between the two towers. The rate of employee complaints increased about a month or two after renovation started. The renovation consisted of installing new carpet, painting, and, in some cases, moving walls. At about the time the individual complaints were reported, the agency formed a task force and asked an industrial hygiene team from Research Triangle Park to perform some industrial hygiene sampling in the building.

Overall findings of the industrial hygiene sampling were consistent with those reported in previous investigations of sick-building syndrome. Some office areas in the mall where new carpet had been installed had as few as 0.2 air exchanges an hour. The air exchange in tower areas where some renovation had been done was significantly better. Carbon dioxide levels in the mall areas rose from 300 to 800 ppm over the course of the day. Individuals seen in the author's practice worked mostly in mall areas, so symptoms correlated very well with the location of ventilation problems in the building.

The agency had planned to limit solvent exposure during this building renovation even before any complaints were received; thus, carpet application was done without adhesives. Neverthe-

Address all correspondence to: Dr. Laura S. Welch, Director, Division of Occupational and Environmental Medicine, School of Medicine, George Washington University, 2150 Pennsylvania Avenue, N.W., Washington, DC 20037.

less, one compound, 4-phenylcyclohexane (4-PC), which had been reported to cause medical symptoms, was a matter of concern. It is a byproduct of the reaction between butadiene and styrene, which are constituents of the latex backing on carpet. Previous studies had suggested that 4-PC might be a factor in several indoor air outbreaks (Syracuse Research Corporation, 1988).

The agency requested some head space analysis on samples of carpet—both new pieces that had not been installed and pieces that had been installed two months earlier. Estimated levels of 4-PC on top of the new carpet were in the range of 50-90 ppb (EPA, 1988). Because the laboratory had to develop a method quickly to analyze for 4-PC, a large confidence interval might exist around these results. 4-PC was present in renovated areas of the building: at 1.5 ppb in the mall areas sampled and at 0.9 ppb in the tower areas with new carpet. Investigators estimated that levels of 4-PC could have been 5-15 ppb at the time of carpet installation. Newly carpeted areas and control areas showed no difference in levels of toluene, styrene, or xylene on initial sampling. No formaldehyde of any significance was detected.

The Environmental Protection Agency had commissioned some inhalation toxicology studies in animals to determine whether 4-PC has major irritant effects, that is, whether it causes changes in the nasal and the bronchial mucosa at levels in the range of one part per million (Syracuse Research Corporation, 1988). Because the studies found no effects in animals, they concluded that 4-PC is primarily an odorous compound rather than an upper-respiratory irritant. However, the Consumer Product Safety Commission has received carpet-related complaints (CPSC, 1989), and further investigation of this particular compound is expected. In addition, further air quality studies at this agency have been performed (EPA, 1990).

Currently, it is unclear whether 4-PC caused this outbreak. The complaints are similar to those reported in other outbreaks of sick-building syndrome and could certainly have occurred in the absence of 4-PC. However, the association of this compound with new carpet installation is worthy of follow-up.

CLINICAL INVESTIGATION

The 20 persons examined in connection with this outbreak were evaluated in a clinical fashion rather than with a standard research protocol. Thus, individuals with neurologic complaints were given more thorough neurologic evaluations; those without neurologic complaints received more limited neurologic assessments. The same tests were not performed on all patients.

Patients ranged in age from 24 to 64 years, with a median age of 40; 15 were 39 years or older. Nineteen patients were white, the majority were women (N=16), and more were members of the professional staff than of the secretarial staff at the agency. It is not known whether this age and sex distribution is representative of employment patterns in the agency or of the persons who had complaints during the outbreak.

Of the 20 persons examined, 18 reported upper respiratory symptoms and 14 reported lower respiratory symptoms. Lower respiratory symptoms were defined as chest complaints, including predominant chest tightness, wheezing, or a persistent cough. Thirteen persons reported neurologic complaints, including light headedness, paresthesias, and unsteady gait. Twelve reported an increased frequency of headache, and eight reported excessive fatigue. Half the persons examined had previous histories of allergic rhinitis, hay fever, or asthma. Fifty percent of the general population has some atopic disease, including asymptomatic disease and atopic skin disease. The rate of symptomatic respiratory allergies in the group reported here is higher than would be expected. Two persons reported previous chronic fatigue syndrome. One person had previous complaints of symptoms similar to those she experienced during this outbreak, but the symptoms were significantly worse during this period. Four had definite or probable depression as judged by the examining physician's impression at the time of initial evaluation. (Either they were currently being treated for depression or, at the time of the clinic visit, had characteristics of depressive disorder with mood disturbance, weight loss, and sleep disturbance.) In general, physical examinations of these 20 individuals were normal.

The laboratory tests performed were determined by the presenting complaints. Three individuals had reversible airway obstruction, and a third had a positive methacholine challenge test. Four persons who had not previously had allergy testing were demonstrated to be allergic to a mix of antigens. Two individuals were referred for full psychometric testing. Results for both were within the normal range overall, although one had a discrepancy of about 20 points between verbal and performance IQ subsets.

With the data obtained, specific diagnoses of new onset of asthma were made for three individuals. Eighteen persons met Molhave's definition of sick-building syndrome (Molhave, 1987), with a combination of some upper respiratory complaints and headache or eye irritation in association with some central nervous system complaints.

Over time, three of these twenty persons had ongoing symptoms consistent with a working definition of multiple chemical sensitivity (MCS). This definition was based on Cullen (1987), with the additional stipulation that the individual have some functional impairment from symptoms. Each of the three persons with MCS had a significant pre-existing condition. One had prior chronic fatigue syndrome. The second had a previous history of similar reactions to chemicals, although she had been able to stay at work. The current episode precipitated a significant worsening of her condition so that she was unable to continue working in the building and had been working at home. The third person, whose complaints were predominantly neurologic, had a previous history of significant neurotoxic injury from organic solvent exposure at a previous job. Some discrepancy existed between the verbal and performance IQ subsets on his neuropsychiatric test.

The 20 individuals examined here have not been seen in follow-up, and it is not known how many more building occupants at this federal agency had health complaints or might have developed MCS. Thus, there may be additional, as yet unidentified, cases of MCS. The physician who runs the agency's own health unit reported that approximately 60 persons sought attention over a 1-2 month period when this outbreak was occurring. Attempts were made to obtain complete medical records for all affected individuals to describe the overall outbreak, but issues of identification and patient confidentiality prevented the collection of a complete case series for those people. The patients described here represent approximately one-third of the persons who had enough complaints to report to their employer in the first two months of the outbreak.

SUMMARY

Investigation of this outbreak raises some important points for future research. Although for various reasons the case ascertainment for MCS was not complete, the three MCS patients described here all had preexisting conditions that may have put them at risk. In addition, one person among the 20 described had chronic fatigue syndrome but did not develop MCS.

Many of the persons described here continue to have ongoing complaints that are not MCS. Significant exacerbation of preexisting allergic disease and new onset of asthma occurred among those patients. As a group, they did not recover completely after the outbreak; several are no longer working in the building but in alternative work spaces. An important distinction should be made between individuals who met the definition used here for MCS and others who had significant exacerbation of some better-defined illness brought on by building conditions. New onset of MCS was a partial but not complete explanation of the clinical course for this group of 20 persons.

REFERENCES

CONSUMER PRODUCT SAFETY COMMISSION (CPSC). (1989). News from CPSC. August 24.

CULLEN, M. (1987). Multiple chemical sensitivities: An overview. In: Workers with Multiple Chemical Sensitivities. (M.R. Cullen, ed.). Occup. Med.: State Art Rev. 2(4):655-662.

MOLHAVE, L. (1987). The sick building—A subpopulation among the problem buildings. In: Indoor Air 1987. Proceedings of the IV International Conference on Indoor Air Quality and Climate. (B. Seifert, H. Esdorn, and M. Fischer, eds.). Institute for Water, Soil and Air Hygiene, Berlin. pp. 469-473.

SYRACUSE RESEARCH CORPORATION. (1988). 4-Phenylcyclohexane Screening: Draft Report, 8/19/88. Contract No. 68-02-4209, prepared for EPA Office of Toxic Substances, Washington, DC.

U.S. ENVIRONMENTAL PROTECTION AGENCY (EPA). (1988). Environmental Response Team, Report of Sampling, May. Unpublished report.

U.S.ENVIRONMENTAL PROTECTION AGENCY (EPA). (1990). Indoor Air Quality and Work Environment Study: EPA Headquarters Buildings—Volume 2: Results of Indoor Air Environmental Monitoring Study. Environmental Protection Agency, Atmospheric Research and Exposure Assessment Laboratory, Research Triangle Park, NC.

A SURVEY OF AOEC PHYSICIAN PRACTICES AND ATTITUDES REGARDING MULTIPLE CHEMICAL SENSITIVITY

KATHLEEN M. REST, PhD, MPA
Occupational Health Program, Department of Family and Community Medicine
University of Massachusetts Medical Center
Worcester, Massachusetts

INTRODUCTION

Multiple chemical sensitivity (MCS) is a poorly understood clinical syndrome characterized by recurrent symptoms, often in multiple organ systems. It develops in response to a variety of chemicals at exposure levels significantly lower than those that would cause reactions in the general population (Cullen, 1987; Ashford and Miller, 1991). The syndrome is cloaked in uncertainty and controversy, as health professionals struggle to describe and understand the disorder. At the same time, they are asked to diagnose, treat, and manage an increasing number of challenging patients.

Of all the medical specialties, occupational medicine is best designed to provide the training and clinical experience needed to manage patients whose health problems may be related to exposure to toxic substances. More than other physicians, occupational medicine specialists understand the potential health effects of a wide variety of toxic substances, know how to evaluate the work-relatedness of disease, are involved in screening and surveillance programs for groups of workers exposed to toxic substances, and generally are more willing to assume the burden and meet the challenge of diagnosing disease due to toxic chemical exposures. Because of this training and clinical experience, these same physicians are also better equipped to deal with illnesses associated with chemical exposures that occur outside the workplace—in the home or community environments.

Thus, one could logically look to occupational medicine specialists for help in furthering the scientific understanding of MCS. With their toxicologic expertise and their experience with chemically exposed workers, these physicians could provide valuable clinical insight and make a significant contribution toward unraveling the bewildering web of MCS.

One professional organization specifically founded to deal with the clinical aspects of occupational and environmental health is the Association of Occupational and Environmental Clinics (AOEC). AOEC is an organization of occupational and environmental health clinics

Address all correspondence to: Dr. Kathleen M. Rest, Occupational Health Program, University of Massachusetts Medical Center, 55 Lake Avenue North, Worcester, MA 01655.

and professionals, many of whom are associated with academic-based occupational and environmental medicine programs. The AOEC clinics provide a multi-disciplinary, patient-centered, prevention-oriented, public health approach to care and adhere to an organizational Patient Bill of Rights (AOEC, 1989). Occupational and environmental health professionals who do not work in AOEC member clinics can join and participate in the organization as individual members.

As AOEC prepared to sponsor a workshop on MCS, a survey of AOEC physicians seemed timely because no information was available on their current clinical practices and attitudes regarding MCS. A self-administered questionnaire was developed to collect basic information on diagnostic methods, treatment and management strategies, and general attitudes regarding MCS.

METHODS

Study Population

The study population for this survey consisted of all physician members of AOEC. These physicians became members by 1) working in clinics that formally applied for and received clinic membership status in the organization, or 2) joining AOEC as individuals. Physicians with individual membership may or may not be affiliated with a member clinic.

Data Collection

An anonymous, self-administered questionnaire was developed and sent to all AOEC member clinics (N=34) and to all physicians who were individual members of AOEC (N=108). The survey instrument consisted of closed-ended, five-point questions that inquired how often respondents performed specific activities when evaluating patients for and managing patients with MCS and how much they agreed or disagreed with certain attitudinal statements about MCS. The letter accompanying the questionnaire instructed the clinics to distribute the questionnaire to all attending physicians in the AOEC member clinic (residents were specifically excluded) and asked individual members to complete the survey themselves. To increase the response rate, anonymity was allowed. However, respondents were asked to indicate whether they worked in an AOEC member clinic and to identify that clinic by name.

The data were entered into the EpiInfo computer program, which was used for analysis. Two-tailed tests of significance (Chi square or Mehta's exact test) were used to compare differences between physicians who worked in AOEC member clinics and those who did not, and to compare differences between physicians who had evaluated ten or more patients for MCS and those who had evaluated fewer than ten patients over a two-year period.

RESULTS

One hundred twenty-two (122) physicians responded to the survey; 67 worked in member clinics, and 55 worked elsewhere but held individual membership in AOEC. No calculation of an individual response rate was possible because the degree of overlap between individual and clinic member physicians was unknown. However, the clinic response rate was 97%, as physicians in 33 of 34 member clinics responded to the survey. The number of respondents from each AOEC member clinic ranged from one to six, with 12 clinics having one respondent, 12 clinics having two respondents, 7 clinics having three respondents, 1 clinic having four respondents, and 1 clinic having six respondents.

The 55 respondents with individual memberships worked in a variety of settings (Table 1).

TABLE 1. Distribution of Physicians with Individual AOEC Membership, by Type of Practice

TYPE	NUMBER	PERCENT
Academic Medical Center	14	25
Private Practice	11	20
Hospital-Based Clinic	7	13
Group Practice	4	7
Industrial/Company Practice	4	7
Other	8	15
No Response	7	13
TOTAL	55	100

Most respondents (73%, N=89) reported that they had personally evaluated patients for MCS, and 71% reported actually making a diagnosis of MCS or an MCS-like syndrome. Physicians who worked in AOEC member clinics were more likely to have seen MCS patients than were those who worked elsewhere. Fully 88% (N=59) of physicians in member clinics reported that they had personally evaluated patients for MCS; 54% (N=30) of physicians who worked elsewhere reported personally evaluating such patients ($p < 0.00008$).

Physicians working outside member clinics seemed to have slightly more experience with MCS patients; 43% of individual member physicians reported they had evaluated more than ten patients over a two-year period compared with 32% of physicians in member clinics (Table 2). Two-thirds of all respondents had evaluated ten or fewer patients over the past two years.

The following sections describe the clinical evaluation, treatment, and management practices of the 89 respondents who had evaluated patients for MCS. The section on attitudes reflects responses from the entire study population of 122 physicians. Very few data were missing; most respondents replied to all questions.

Evaluating Patients for MCS

When asked about their clinical assessment methods, respondents reported thorough and comprehensive history-taking practices (Figure 1). Fully 95% of respondents reported that

TABLE 2. **Number (and Percent) of AOEC Physicians Evaluating Patients for MCS in a 2-year Period, by Membership Status**

NUMBER OF PATIENTS	CLINIC MEMBER	INDIVIDUAL MEMBER	TOTAL RESPONDING
1 - 5	26 (44)	7 (23)	33 (37)
6 - 10	14 (24)	10 (33)	24 (27)
11 - 15	6 (10)	5 (17)	11 (12)
16 - 20	5 (8)	1 (3)	6 (7)
21 - 30	3 (5)	2 (7)	5 (6)
31 - 40	2 (3)	1 (3)	3 (3)
41 - 50	1 (2)	0 (0)	1 (1)
50 +	1 (2)	4 (13)	5 (6)
No Response	1 (2)	0 (0)	1 (1)
TOTAL	59 (100)	30 (100)	89 (100)

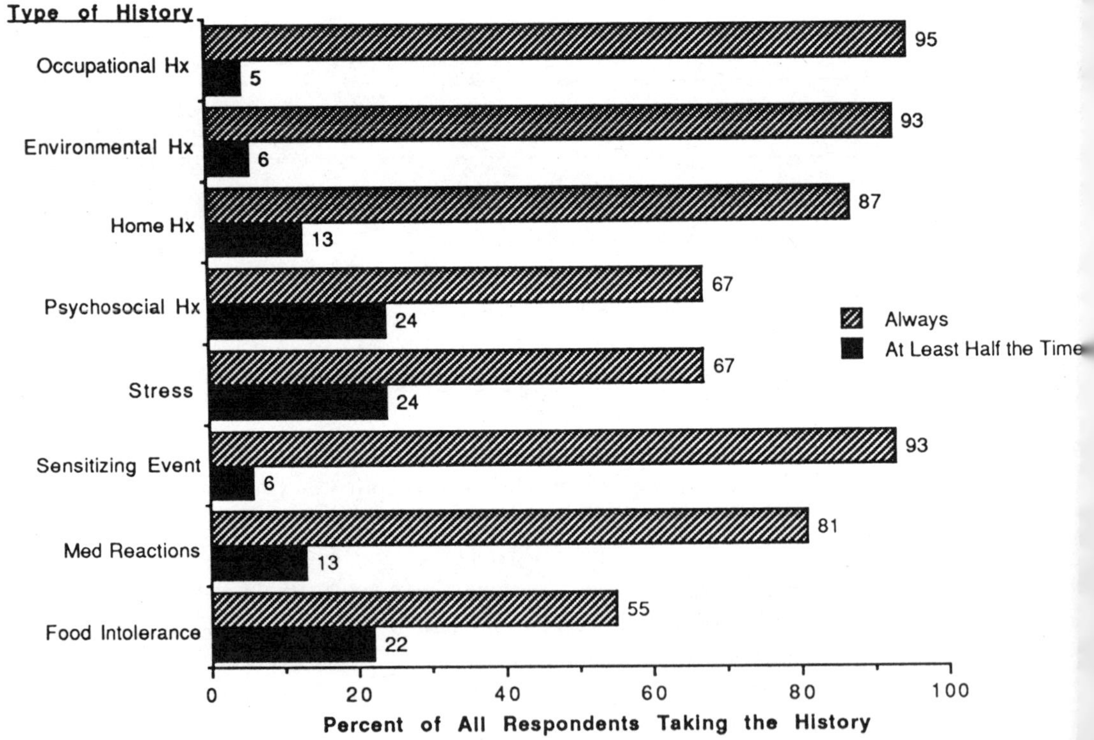

FIGURE 1. Respondents' history-taking practices in evaluating patients for MCS.

they always took a complete occupational history; 93% reported they always took a complete environmental exposure history. Similarly, 87% reported they always asked detailed questions about the home environment, and 93% reported they always tried to identify a specific "sensitizing" event. Figure 1 also illustrates a high frequency of inquiry about previous reactions to medications and drugs and about stress, as well as a high frequency of obtaining a psychosocial history. Few physicians reported inquiring about food intolerances.

Physicians who worked in AOEC member clinics and those who did not showed very few differences in history-taking practices. Physicians in member clinics (72%) were slightly more likely to always obtain a complete psychosocial history than were individual member physicians (57%) and to always ask detailed questions about stress (80% vs. 48%, $p < 0.007$).

Respondents who had evaluated more than ten patients for MCS over the past two years were slightly more likely (65%) to always inquire about food intolerances than were those who had evaluated fewer than ten patients (50%).

Respondents did not rely heavily on medical testing when evaluating patients for MCS (Figure 2). Although 71% reported obtaining standard blood chemistries at least half the time or always, far fewer ordered standard allergy tests (18%) or obtained special blood tests to detect immune system changes (17%) that often. Special brain scans, such as PET, SPECT, EEG, and BEAM, were hardly ever used. Neuropsychiatric assessments were performed more often; 61% of respondents performed them at least half the time or always.

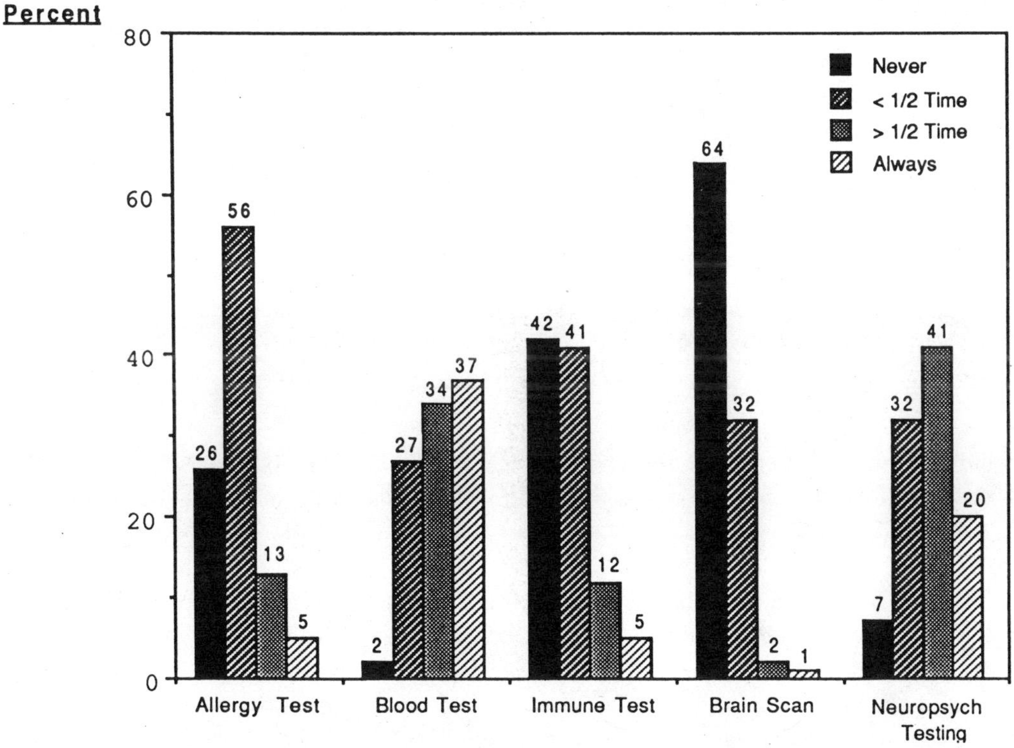

FIGURE 2. Respondents' test-ordering practices in evaluating patients for MCS.

TABLE 3. Number (and Percent) of Physicians Ordering Tests, by Membership Status and Number of Patients

TEST	BY PHYSICIAN MEMBERSHIP STATUS		BY MCS PATIENTS EVALUATED OVER A 2-YEAR PERIOD	
	Clinic Member	Indiv. Member	≤ 10 Patients	> 10 Patients
Allergy Tests				
Never	14 (25)	8 (29)	17 (31)	5 (18)
< half time	36 (63)	12 (43)	28 (51)	18 (64)
> half time	5 (9)	6 (21)	8 (14)	3 (11)
Always	2 (3)	2 (7)	2 (4)	2 (7)
Blood Tests				
Never	1 (2)	1 (3)	1 (2)	1 (3)
< half time	18 (31)	5 (17)	15 (27)	7 (24)
> half time	18 (31)	11 (38)	17 (31)	12 (41)
Always	20 (35)	12 (41)	22 (40)	9 (31)
Immune Test				
Never	30 (54)	6 (21)	27 (50)	8 (27)
< half time	20 (36)	15 (52)	19 (35)	15 (52)
> half time	4 (7)	6 (21)	6 (11)	4 (14)
Always	2 (3)	2 (7)	2 (4)	2 (7)
Brain Scan				
Never	42 (72)	14 (48)	41 (73)	14 (48)
< half time	15 (26)	13 (45)	14 (25)	13 (45)
> half time	1 (2)	1 (3)	1 (2)	1 (3)
Always	0 (0)	1 (3)	0 (0)	1 (3)
Neuropsych Test				
Never	4 (7)	2 (7)	5 (9)	1 (3)
< half time	14 (25)	13 (45)	16 (30)	10 (34)
> half time	28 (50)	7 (24)	23 (43)	11 (38)
Always	10 (18)	7 (24)	10 (18)	7 (24)

Table 3 illustrates several differences in medical testing practices between physicians who worked in AOEC member clinics and those who did not. Individual members appeared to use medical testing procedures somewhat more often than did their AOEC clinic member colleagues. For example, twice as many individual member physicians reported obtaining standard allergy tests at least half the time or always compared with clinic member physicians ($p<0.07$). Individual physicians also ordered special blood tests to detect immune system changes more often than their clinic member colleagues (28% vs. 10% obtained such tests at least half the time or always, $p<0.02$). However, fewer individual member physicians performed neuropsychiatric assessments than did physicians in AOEC clinics. Brain scans were seldom used, and more clinic member physicians reported that they never ordered them than did members who practiced elsewhere.

The number of patients seen by physicians seemed to have very little effect on their test-ordering practices. However, respondents who evaluated fewer than ten patients were less likely to order brain scans (73% reported they never ordered such scans) than were physicians who saw more than ten patients (48%). They were also less likely to order special immune tests.

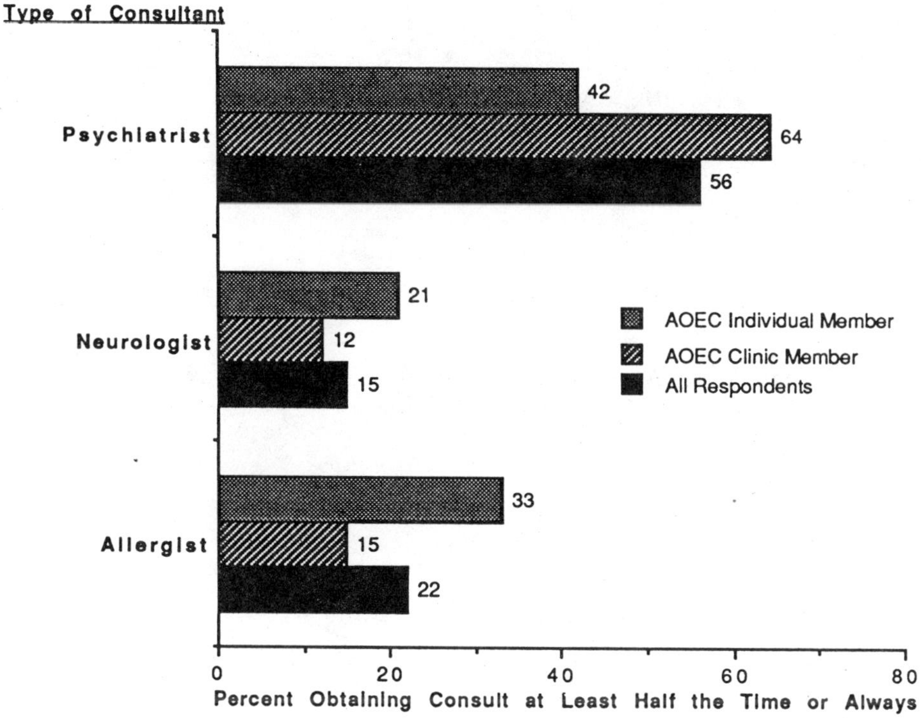

FIGURE 3. Respondents' reported use of consultants by membership status.

Respondents consulted psychiatrists or psychologists more often than they did other clinical specialists; 56% obtained such consultations at least half the time or always; 22% obtained consultations from allergists and 15% obtained consultations from neurologists as often (Figure 3). Use of consultants varied between physicians in AOEC member clinics and those who practiced elsewhere. Individual members consulted allergists more often than did AOEC clinic physicians ($p < 0.04$) but psychiatrists less often ($p < 0.04$) (Table 4). Respondents who evaluated fewer than ten patients were also more likely (26%) to consult allergists when evaluating patients for MCS than were respondents who evaluated more than ten patients (10%). Their use of psychiatric consultations varied little.

Respondents varied in their attempts to obtain environmental sampling data to help them evaluate MCS patients—20% always tried to obtain these data; 23% tried at least half the time; 43% tried less than half the time; and 14% never tried to obtain these data. Physicians who practiced in AOEC clinics differed little from those who did not in this regard, although more physicians in member clinics (17%) reported that they never try to obtain environmental sampling data (vs 7% individual physicians). Likewise, the number of patients seen made little difference, although respondents who saw fewer than ten patients were slightly more likely (47%) to try to get environmental sampling data than were those who saw more than ten patients (38%).

TABLE 4. Number (and Percent) of Physicians Using Consultants, by Membership Status and Number of Patients

CONSULTANT TYPE	BY PHYSICIAN MEMBERSHIP STATUS		BY MCS PATIENTS EVALUATED OVER A 2-YEAR PERIOD	
	Clinic Member	Indiv. Member	≤ 10 Patients	> 10 Patients
Allergist				
Never	13 (22)	7 (23)	13 (23)	6 (21)
< half time	36 (62)	13 (43)	29 (51)	20 (69)
> half time	9 (15)	7 (23)	13 (23)	2 (7)
Always	0 (0)	3 (10)	2 (3)	1 (3)
Neurologist				
Never	13 (22)	3 (10)	11 (20)	4 (14)
< half time	38 (65)	20 (69)	36 (64)	22 (76)
> half time	6 (10)	4 (14)	7 (12)	2 (7)
Always	1 (2)	2 (7)	2 (4)	1 (3)
Psychiatrist				
Never	1 (2)	1 (3)	2 (4)	0 (0)
< half time	20 (34)	16 (55)	24 (43)	12 (41)
> half time	32 (55)	7 (24)	24 (43)	13 (45)
Always	5 (9)	5 (17)	6 (11)	4 (14)

Only 16% of the respondents reported that they visited the home or worksite at least half the time when evaluating patients for MCS (no one reported always doing so). Physicians in AOEC member clinics were less likely (12%) to make such visits than were their colleagues who practiced elsewhere (24% made site visits at least half the time).

Treatment and Management of MCS Patients

Respondents appeared to use very simple and non-technical methods for treating and managing patients with MCS. Sixty-two percent reported that they always counseled patients to avoid exposure to the suspected offending agents on a trial basis (18% did so at least half the time); 39% reported that they always provided ongoing support to the patients via return visits (36% did so at least half the time).

A variety of other interventions were used but to lesser extents. In the behavioral sphere, 15% of respondents reported that they always suggested their patients obtain psychologic counseling, and 49% did so at least half the time. Few (15%) reported that they suggested biofeedback, although 43% suggested the use of other behavioral techniques at least half the time or always. Respondents who evaluated more than ten patients reported more use of biofeedback (23%) than did those who saw fewer patients (11% used biofeedback at least half the time or always). Both groups were similar in their use of other behavioral techniques. Few respondents ever prescribed anti-anxiety, anti-depressant, or sleep medications.

Overall, nutritional interventions were seldom used, although some differences existed between individual and clinic member physicians (Table 5). Fewer physicians in member clinics provided nutritional counseling ($p < 0.003$) or referred patients for such counseling ($p < 0.001$).

TABLE 5. Number (and Percent) of Physicians Using Nutritional Interventions, by Membership Status and Number of Patients

INTERVENTION	BY PHYSICIAN MEMBERSHIP STATUS		BY MCS PATIENTS EVALUATED OVER A 2-YEAR PERIOD	
	Clinic Member	Indiv. Member	≤ 10 Patients	> 10 Patients
Provide Nutritional Counseling				
Never	32 (55)	10 (36)	25 (46)	16 (53)
< half time	22 (38)	8 (29)	19 (35)	10 (33)
> half time	4 (7)	6 (21)	8 (15)	2 (7)
Always	0 (0)	4 (14)	2 (4)	2 (7)
Refer for Nutritional Counseling				
Never	42 (72)	10 (36)	34 (63)	17 (57)
< half time	16 (28)	13 (46)	18 (33)	11 (37)
> half time	0 (0)	3 (11)	1 (2)	1 (3)
Always	0 (0)	2 (7)	1 (2)	1 (3)
Suggest Food Rotation				
Never	42 (71)	18 (67)	40 (73)	18 (62)
< half time	15 (25)	6 (22)	11 (20)	10 (34)
> half time	2 (3)	2 (7)	4 (7)	0 (0)
Always	0 (0)	1 (4)	0 (0)	1 (3)

The respondents were relatively conservative in recommending environmental interventions; 37% recommended work restrictions at least half the time; 3% always did so. Thirty-four percent recommended a trial away from work at least half the time; 6% always did so. In addition, 37% recommended changes in the home environment at least half the time; 13% always did so.

Recommendations about environmental interventions varied among the respondents. Compared with individual member physicians, respondents in member clinics were more likely to never recommend work restrictions ($p < 0.05$) or a trial away from work ($p < 0.01$), although they were also slightly more likely to make such recommendations at least half the time (Table 6). Physicians in member clinics were also less likely to recommend changes in the home environment ($p < 0.01$). The number of patients evaluated for MCS also appeared to have some effect on respondents recommendations regarding environmental intervention. Respondents who evaluated more than ten patients over the past two years were somewhat more likely to recommend work restrictions, a trial away from work, and specific changes in the home environment than were respondents who reported evaluating less than ten patients over the same period. Only the difference regarding work restrictions approached statistical significance ($p < 0.06$).

TABLE 6. Number (and Percent) of Physicians Recommending Environmental Interventions, by Membership Status and Number of Patients

RECOMMENDED INTERVENTIONS	BY PHYSICIAN MEMBERSHIP STATUS		BY MCS PATIENTS EVALUATED OVER A 2-YEAR PERIOD	
	Clinic Member	Indiv. Member	≤ 10 Patients	> 10 Patients
Work Restrictions				
Never	8 (14)	0 (0)	5 (9)	2 (7)
< half time	26 (44)	18 (64)	31 (56)	12 (40)
> half time	24 (41)	8 (29)	19 (34)	13 (43)
Always	1 (1)	2 (7)	0 (0)	3 (10)
Trial Away from Work				
Never	8 (14)	0 (0)	5 (9)	2 (7)
< half time	25 (43)	19 (68)	29 (54)	15 (50)
> half time	23 (40)	6 (21)	18 (33)	10 (33)
Always	2 (3)	3 (11)	2 (4)	3 (10)
Change Home Environment				
Never	7 (12)	1 (4)	6 (11)	2 (7)
< half time	26 (44)	10 (36)	22 (40)	12 (4)
> half time	23 (39)	9 (32)	22 (40)	10 (33)
Always	3 (5)	8 (28)	5 (9)	6 (20)

Finally, respondents reported that they seldom employed alternative therapies or referred patients for such therapies as detoxification or provocation and neutralization. Indeed, 89% of respondents never used detoxification techniques and only 7% had ever referred a patient for detoxification. Similarly, 96% reported that they never used provocation and neutralization, and only 2% had ever referred a patient for such therapy.

In the behavioral arena, individual and clinic member physicians differed little except that respondents in member clinics reported more use of biofeedback (19%) than did their colleagues who practiced elsewhere (7% suggested biofeedback at least half the time or always).

Attitudes

Most respondents (83%) reported that they believe MCS has a combination of psychogenic and physical causes, although 62% agreed that MCS is more psychologic than organic in nature. Only 9% believed that MCS is predominantly physical in origin. However, approximately 25% of the respondents could neither agree nor disagree with statements about the predominance of physical versus psychologic origins of MCS.

Of respondents with an opinion (N=85), physicians in AOEC clinics were more likely (91%) than their non-AOEC clinic colleagues (81%, $p < 0.03$) to agree that MCS is more psychologic than organic in nature. However, physicians who evaluated more patients were less likely (72%) to point to psychologic causes than were those who saw fewer patients (94%, $p < 0.02$). Similarly, respondents with more clinical experience were also less likely to agree that most MCS patients have a psychiatric disorder that causes their symptoms ($p < 0.03$).

One-third of all respondents (N=122) had no opinion about the possible mechanisms for MCS—neither agreeing nor disagreeing with statements regarding the involvement of the immune and nervous systems. Of those with opinions, more believed that the nervous system is most involved with MCS (67%) than believed that the immune system is most involved (15%). Individual member physicians (82%) were more likely than clinic member physicians (56%) to agree that the nervous system is primarily involved in MCS ($p < 0.03$). No statistically significant differences in beliefs about mechanisms existed between physicians who evaluated more versus fewer patients for MCS.

Nearly a third of the respondents had no opinion about the role of odor conditioning in MCS, although 58% agreed that odor conditioning is a major factor. Of physicians with an opinion, fewer of the more experienced physicians (69%) believed that odor conditioning is a major factor in MCS than did their less experienced colleagues (92%, $p < 0.04$).

Although nearly one third of respondents had no opinion, 47% agreed that MCS is a symptom of a chemophobic society. Of those with an opinion, physicians with less MCS experience were slightly more likely (66%) to agree that chemophobia is a factor in MCS than were their colleagues with more clinical experience with MCS (53%).

Survey respondents had little use for clinical ecologists. Although 94% agreed that they could understand why patients with MCS complaints might seek help from alternative practitioners, 49% said that they would discourage a patient from seeing a clinical ecologist. Only 11% agreed that they might refer a patient to a clinical ecologist. Indeed, 23% agreed that clinical ecologists are responsible for the MCS controversy (42% disagreed, and 35% had no opinion). However, individual member physicians who had an opinion were more likely (50%) to believe that clinical ecologists are responsible for the MCS controversy than were their clinic member colleagues (20%, $p < 0.04$). Physicians who evaluated more patients for MCS differed little on this matter from those who evaluated fewer patients.

Although only 22% of respondents agreed that some treatments used by clinical ecologists could be of value in managing MCS patients (48% had no opinion), most (82%) agreed that these treatments should be studied in a scientific manner.

Finally, many respondents saw some overlap between MCS and other syndromes—77% agreed that there may be possible overlap between MCS and building-related illness, and 68% that chronic fatigue syndrome may overlap with MCS. Opinions between individual and clinic member physicians differed little, but physicians who had evaluated more than ten patients over the past two-year period were more likely to see an overlap with chronic fatigue syndrome (77%) and building-related illness (84%) than were the physicians who had evaluated fewer than ten patients (59% for chronic fatigue syndrome and 71% for building-related illness, respectively).

DISCUSSION

This survey found that AOEC physicians were eager to share information and opinions about MCS. Most respondents reported some clinical experience with MCS patients, but, for the majority, the experience was not extensive.

Physicians in AOEC member clinics were more likely to have evaluated MCS patients personally than were their colleagues who practiced elsewhere. AOEC member clinics may be easier for patients and referring physicians to identify because most have the words "occupational" or "environmental" in their names.

Because exposure histories are the mainstay of occupational medicine, it is not surprising that respondents used them extensively when evaluating patients for MCS. In addition, nearly all respondents always tried to identify a discrete "sensitizing" event through their history-taking practices.

Respondents obtained psychosocial information during their clinical assessment, although few reported that they always did so. Occupational medicine physicians may be more confident in their abilities to interpret information about the physical and chemical environments than in their abilities to obtain and use psychosocial data in clinical diagnosis.

Although some of the literature has suggested that food intolerances are sometimes reported by patients with MCS-like complaints (Rea, 1988; Bell, 1982; Randolph, 1962), respondents were somewhat less likely to inquire about such intolerances than they were to elicit information about exposures and stress. If food intolerance is a common complaint, it is not surprising that physicians with more experience in evaluating MCS patients were more likely to ask questions about food.

Respondents seldom used medical testing to help them evaluate patients for MCS. Only standard blood chemistries were used to any extent. Respondents were more likely to obtain consultations from other specialists than to order tests. Psychiatrists and psychologists were consulted significantly more often than were allergists and neurologists. This is consistent with respondents' reported beliefs in the predominantly psychologic origin and nature of MCS. Physicians with more MCS experience were less likely to consult allergists than were physicians who had more limited clinical experience with MCS. Because affected patients frequently associate their symptoms with specific exposures, it is not surprising that physicians with limited experience consult allergists when evaluating such patients. Physicians with more extensive experience may have found that allergists provide little help when patients' symptoms are not consistent with the traditional concepts of IgE-mediated sensitivity.

Two assessment tools commonly associated with occupational medicine practice—environmental sampling and site visits—were not used extensively in the evaluation process. Less than half the responders reported trying to obtain environmental data at least half the time or always when evaluating patients for MCS, and only 16% reported that they visit the home or workplace as often. Fewer clinic member physicians made such site visits even though they are likely to have more experience making them and to have better access to industrial hygienists for assistance. However, it may be that their experience suggests that site visits will not be informative in cases involving very low exposure levels.

When treating and managing patients with MCS, respondents appeared to rely primarily on advice, support, and behavioral interventions. Two-thirds reported that they suggested psychologic counseling at least half the time or always, and 42% suggested the use of behavioral techniques. Although 80% reported that they counseled patients to avoid exposure to the suspected offending agents at least half the time or always, respondents were somewhat reluctant to recommend specific changes in the home environment, work restrictions, or a trial away from work.

Because nutritional interventions are seldom used in occupational medicine, it is not surprising that their use by respondents was limited. Given the controversy about clinical ecology, it is also not surprising that alternative therapies, such as detoxification and provocation-neutralization, were very rarely employed.

Reflecting the lack of scientific studies about MCS, no consensus was found on the etiology or the mechanisms involved. Although almost two-thirds believed that MCS is more psychologic than physical in nature, 29% could neither agree nor disagree with that statement. Nearly half the respondents agreed that MCS is a symptom of a chemophobic society. Physicians who saw more than ten patients over a two-year period were less likely to agree that MCS is predominantly psychologic in nature than were physicians who evaluated fewer patients. This difference was consistent in two separate questions relating to organic versus psychologic etiologies.

Substantial and often acrimonious debate has arisen about clinical ecology and the therapies often associated with clinical ecologists (Ashford and Miller, 1991). AOEC physicians had strong opinions about clinical ecologists—very few would refer a patient to a clinical ecologist and half would discourage a patient from consulting one. Nearly one quarter of the respondents held clinical ecologists responsible for the MCS controversy. Although almost half the respondents had no opinion about the possible value of the treatments used by clinical ecologists, most agreed that such treatments should be studied in a scientific manner.

CONCLUSIONS

This survey provides baseline information about the attitudes and clinical assessment and management practices of one group of occupational physicians—members of the AOEC. It reveals a strong reliance on historical data; moderate use of psychiatric consultation; and very limited use of ancillary tests or procedures for assessment. Preferred treatment and management modalities are primarily supportive and behavioral.

AOEC physicians believe that MCS has a multifactorial etiology, probably more psychologic than physical, but have no consensus on possible mechanism. Attitudes about clinical ecologists and their treatments are strongly held and reflect the skepticism articulated by other professional groups of physicians (American Academy of Allergy and Immunology, 1986; American College of Physicians, 1989; American College of Occupational Medicine, 1990).

The survey data suggest that, although AOEC physicians may not understand MCS and may have strong opinions about it, as a group they have remarkably similar approaches to assessing and managing patients with MCS or MCS-like complaints. Their willingness to share experiences and their potential access to sizeable patient populations with MCS-like complaints make AOEC physicians a valuable resource in furthering the scientific understanding of this problem.

ACKNOWLEDGEMENTS

The author thanks Asim Razvi for his assistance and quick mastery of several computer programs as well as Eddy A. Bresnitz, MD, and David S. King, PhD, for their valuable criticism. The author also thanks the AOEC staff for the membership lists and mailing labels and the AOEC physicians whose rapid and enthusiastic response made this survey possible.

REFERENCES

AMERICAN ACADEMY OF ALLERGY AND IMMUNOLOGY. (1986). Position statements: Clinical ecology. J. Allerg. Clin. Immunol. 78(2): 269-271.

AMERICAN COLLEGE OF OCCUPATIONAL MEDICINE. (1990). ACOM adopts ACP's position on clinical ecology. In: ACOM Report (September).

AMERICAN COLLEGE OF PHYSICIANS. (1989). Position paper: Clinical ecology. Ann. Int. Med. 111(2): 168-178.

ASSOCIATION OF OCCUPATIONAL AND ENVIRONMENTAL CLINICS (AOEC). (1989). AOEC Patient Bill of Rights (unpublished).

ASHFORD, N.A. and MILLER, C.S. (1991). Chemical Exposures: Low Levels and High Stakes. Van Nostrand Reinhold, New York.

BELL, I. (1982). Clinical Ecology. Common Knowledge Press, Bolinas, CA.

CULLEN, M.R. (1987). The worker with multiple chemical sensitivities: An overview. In: Workers with Multiple Chemical Sensitivities. (M.R. Cullen, ed.). Occup. Med.: State Art Rev. 2(4): 655-661.

RANDOLPH, T.G. (1962). Human Ecology and Susceptibility to the Chemical Environment. Charles C. Thomas, Springfield, IL.

REA, W. (1988). Chemical hypersensitivity and the allergic response. Ear Nose Throat J. 67(1): 50-56.

PSYCHIATRIC TREATMENTS IN MULTIPLE CHEMICAL SENSITIVITY

GREGORY E. SIMON, MD, MPH
Center for Health Studies, Group Health Cooperative
Department of Psychiatry and Behavioral Sciences
University of Washington, Seattle

INTRODUCTION

Multiple chemical sensitivity (MCS) is a clinical syndrome defined by reactivity to common environmental exposures. Although a wide range of individual symptoms have been described, the most prominent symptoms are typically cardiorespiratory (dyspnea, cough, palpitations, chest pain), constitutional (fatigue, weakness, loss of appetite), and neuropsychologic (dizziness, confusion, memory loss, mood or personality changes). Individual exposures may precipitate acute symptomatic attacks that usually include cardiorespiratory and neurologic symptoms. Constitutional and psychologic manifestations often persist between acute attacks and may become chronic. No specific symptom pattern defines the MCS syndrome. Almost every conceivable symptom has been described in MCS cases, and no symptoms are said to be inconsistent with MCS.

No pathophysiologic mechanism for these symptoms has been established. Previous investigations have not revealed any clear pathologic or physiologic abnormality attributable to chemical exposure. Physical and conventional laboratory examinations typically show no abnormalities (American College of Physicians, 1989; Terr, 1986). Immunologic mechanisms have been most commonly proposed, but the data available do not support the immunologic hypothesis (American College of Physicians, 1989; Kahn and Letz, 1989; Terr, 1987; Task Force on Clinical Ecology of the California Medical Association, 1986; American Academy of Allergy, 1981).

Given the current lack of understanding about mechanism, MCS can be most clearly defined in terms of beliefs and behaviors. Patients with MCS believe that their symptoms result from low-level chemical exposure. Although some may object to use of the term "believe," no clear evidence exists that symptoms actually do correlate with chemical exposure. In addition, MCS patients respond to symptoms by avoiding situations of perceived exposure. Progressive withdrawal and the resulting disability are the hallmarks of severe chemical sensitivity.

Note: Preparation of this presentation was supported in part by an appointment to the Agency for Toxic Substances and Disease Registry (ATSDR) Clinical Fellowship Program in Environmental Medicine administered by Oak Ridge Associated Universities through an interagency agreement between the U.S. Department of Energy and ATSDR.

Address all correspondence to: Dr. Gregory Simon, Center for Health Studies, 1730 Minor Ave., Seattle, WA 98101.

The controversy surrounding the cause and treatment of MCS may seem unusual in medicine but is ubiquitous in psychiatry. All psychiatric diagnoses are actually loosely defined clinical syndromes without clear pathophysiology or cause. Psychiatric theories of etiology range from theories proposing genetically determined alterations in dopamine receptors to theories involving the long-term effects of childhood environment. Despite this array of seemingly incompatible ideas about cause, some consensus exists about effective treatment. The need to relieve patient's suffering has pushed clinicians past controversial and unresolvable questions about etiology to practical questions about what kind of treatment works. Given our current state of knowledge, although we may not know what's wrong, uncertainty should not stop us from finding what helps. If psychiatric treatment has anything to contribute to the care of patients with MCS, it is an attitude of practicality in the face of conflict and uncertainty.

The rationale for use of psychiatric treatments in MCS is not based on any argument that MCS is a psychiatric illness. Given the lack of other effective treatments and the prominence of psychologic symptoms in the MCS syndrome, application of psychiatric treatments has some promise. Even if a specific neurotoxic or immunologic mechanism for MCS were discovered, the argument for psychiatric treatment would not be diminished. Discoveries about cause rarely lead directly to cure, and clinicians are still left struggling with how to help.

TREATMENT

Establishing a Treatment Alliance

Building a working treatment alliance between physician and patient is an essential first step. Many MCS sufferers view the offer of psychiatric or psychologic treatment as an accusation that the symptoms are "all in your head." Because of the stigma attached to psychiatric illness, psychiatric referral may be perceived as a rejection. In some cases, this perception may be well-founded. Some physicians may hold stereotypes of psychologic symptoms as being "made up" and of psychiatric patients as being liars or malingerers. The success of psychiatric treatment (or any treatment) depends on building a sense of cooperation toward the goal of improving the patient's health.

Because the cause of MCS is unknown, treatment must be directed at the relief of symptoms. Success will require a crucial shift in emphasis from diagnosis to symptom management. This can be a difficult transition for physicians accustomed to the certainty of clear biomedical diagnoses and for patients accustomed to laboratory or diagnostic tests as the definitions of illness. Both patient and physician must be able to define the problem in terms of how the patient feels and functions. Relieving symptoms and improving function must be seen as the primary goals.

Focus on issues of disability, liability, and compensation may interfere with effective treatment. Given current knowledge about the pathophysiology of MCS, questions of cause and responsibility are difficult to resolve. Moreover, these questions often produce conflict and divert attention from helping patients to feel and function better. Some patients may be more interested in assigning responsibility than in relieving symptoms. Clinicians must accept this and understand that at some level these goals are incompatible. It is difficult to concentrate on assigning blame for an injury while simultaneously trying to minimize its effects.

In the treatment of MCS, rehabilitation is a more appropriate goal than cure. An unreasonable emphasis on cure means that success at symptom relief will still be viewed as failure. As with any chronic illness, good clinical care attempts to minimize symptoms and maximize function. Because some exposure to symptom-producing situations is inevitable, complete relief is an unrealistic goal. Instead, patient and physician must cooperate to make those inevitable exposures more tolerable.

The patient must set priorities for treatment. Rehabilitation is a much more collaborative process than is diagnosis. Although the physician may have expertise in applying various treatment methods, only patients are qualified to say which symptoms should be the focus of treatment. Treatment should focus on the specific symptoms patients identify as most distressing. Attempts to decrease withdrawal should focus on building tolerance for those activities patients would most like to regain. Patients will certainly be more motivated to work toward goals they have set for themselves.

Behavioral Treatments

Behavioral interventions are directed primarily at decreasing avoidance and reducing disability. This approach emphasizes that disability does not result from acute symptoms but from the resulting pattern of avoidance (Schottenfeld and Cullen, 1985; Bolla-Wilson et al., 1988). Withdrawal from activities should not be seen as the inevitable consequence of acute symptom attacks. In fact, accepting withdrawal reinforces disability and may increase the severity of subsequent attacks. Withdrawal is based on the assumption that symptoms cannot be controlled and that surrender is the only option. When initial avoidance does not relieve symptoms, patients will often withdraw even further. Ironically, the fact that avoidance is not effective becomes an argument to avoid more vigorously. Eventually, patients feel isolated, powerless, and discouraged.

Although exposure may provoke acute "attacks," no evidence exists that exposure worsens any underlying disease process. Proponents of avoidance may be too quick to equate symptoms with disease—assuming that exposures that produce symptoms also lead to some lasting pathologic damage. Even in MCS patients with long histories of severe symptoms and disability, no clear pathologic morbidity has been described. Without such evidence, it may be premature to assume that avoidance of symptom-producing situations will have long-term benefit.

Two clinical scenarios from other fields of medicine help illustrate this principle. First, in patients with asthma, acute exposures produce acute symptoms and repeated acute attacks can lead to chronic illness accompanied by pathologic damage. In more severe cases of asthma, avoidance may be the best medicine. Second, the relationship between exercise and ischemic heart disease illustrates an alternative scenario. Although acute over-exertion results in angina, sustained exercise can improve function and may help reverse the underlying coronary disease. Appropriate treatment for coronary disease often calls for exercise, not avoidance. Available evidence is inadequate to determine which of these relationships holds in the case of MCS. The current presumption that acute exposure causes greater injury can only lead to more withdrawal and disability.

The fundamental principle of behavioral treatment is desensitization by increasing exposure. Although overwhelming exposures may lead to intolerable symptoms and further withdrawal, gradual increases in exposure allow for accommodation and increasing tolerance. In this way, behavioral desensitization parallels immunologic desensitization. Through this gradual process, patients can eventually tolerate a level of exposure that would have been incapacitating initially.

Desensitization is best accomplished through an organized program. First, the patient must identify a specific situation or activity that he/she cannot now tolerate but would like to enjoy again. Then patient and physician work together to plan a series of small steps leading to eventual success. At each step, the patient would practice and persevere until a tolerance to that level of exposure is built. Only when one level of exposure is well tolerated should the patient move on to the next.

Various interventions can relieve symptoms and ease the process of desensitization. Relaxation training can reduce arousal and decrease the severity of autonomic symptoms, such as palpitations and tremor. Breath-control exercises will help to relieve respiratory symptoms, such as dyspnea and respiratory irritation. Visualizations and other hypnotic techniques can distract attention from other physical symptoms and serve as a reminder of positive goals.

A behavioral program should include plans for increasing the overall level of physical and social activity. Regular physical exercise will help improve stamina, increase the sense of well-being, and "desensitize" patients to autonomic changes. A return to regular activities of this type is essential for patients who have withdrawn from social or recreational activities to avoid chemical exposure. For example, a previously dedicated bingo player who now stays home because of reactions to cigarette smoke at the bingo hall should be encouraged to find other similar activities that are easier to tolerate.

The following case vignette will illustrate the use of behavioral treatments in a chemically sensitive patient:

> A 52-year-old engineer was referred with symptoms of chemical sensitivity that included dizziness, confusion, depressed mood, dyspnea, and palpitations. Although

these symptoms initially occurred following an acute exposure to solvent fumes, he had developed a disabling generalized sensitivity. He was rarely asymptomatic and had withdrawn from almost all his previous activities. Nearly all his sick-leave benefits were exhausted, and he was considering early retirement. Although quite skeptical of a behavioral approach, he agreed to work on a rehabilitation plan. His immediate goal was returning to work long enough to secure his pension benefits. He was instructed in relaxation techniques and breath-control exercises and told to practice these at least twice every day. He also agreed to begin a regular walking program to increase his physical strength. The plan for returning to work called for him to begin working two hours each day with regular breaks for relaxation and walks outdoors. He was strongly urged to persist despite any symptoms and was reassured that (with aid of his new skills) symptoms would gradually fade.

After two months, he was working approximately half time and was able to tolerate working indoors at his previous worksite for four hours at a time. His exercise tolerance had increased and he had resumed some of his previous recreational activities. Although his overall level of symptoms was not dramatically changed, his level of function was much higher.

Pharmacologic Treatments

Psychopharmacologic treatment is aimed at relief of the psychologic symptoms that commonly accompany chemical sensitivity: depression or mood swings, loss of energy or fatigue, insomnia, and increased tension or anxiety (Simon et al., 1990; Black et al., 1990; Stewart and Raskin, 1985). A recommendation for antidepressant or anti-anxiety medication does not imply that symptoms result from a primary psychiatric or psychologic illness. These treatments bring significant relief regardless of etiology. For example, depression associated with stroke or HIV encephalopathy responds to antidepressant medication as well as does depression occurring without any neurologic injury. Patients must understand that an antidepressant prescription is not meant to imply that they are "just depressed."

Use of medication should be guided by specific target symptoms. Symptoms most likely to respond to the use of antidepressant medication include depressed mood, insomnia, fatigue, and loss of interest or enthusiasm. Antidepressants are unlikely to benefit patients who do not have these symptoms. Effectiveness of treatment should be judged by how well these target symptoms are relieved. Anxiolytic medication may be effective in relieving subjective anxiety, tremors, and palpitations. Anxiolytics, however, should be viewed as short-term interventions, effective for up to four weeks. Long-term relief of anxiety symptoms will depend on relaxation exercises and other behavioral techniques.

Prescribing for MCS patients may involve special difficulties. Understandably, MCS sufferers are wary of all medications because of their concern about chemical exposure. Greater reassurance than usual may be required. MCS patients may also experience more difficulty with medication side effects. This may result from either a greater biologic sensitivity or a higher level of anticipation and arousal. In either case, initial doses should be low and dosage increases gradual.

Psychopharmacologic treatment should be considered a means of relieving specific psychologic symptoms, not as a cure for chemical sensitivity. MCS patients will be understandably skeptical about using a chemical to cure a problem brought on by chemical exposure. Therefore, physicians must explain the use of medication as a palliative treatment directed at specific target symptoms. This will help maintain the focus of treatment where it belongs—on rehabilitation and symptom relief.

SUMMARY

Despite the controversy and uncertainty surrounding the causes of the MCS syndrome, psychiatric treatments may provide some relief to MCS patients. These approaches may help patient and physician focus on the most important goals: relief of symptoms and improvement of function. Success requires that physician and patient establish a collaborative relationship and negotiate a treatment plan. Behavioral treatments are most effective at reducing disability and promoting a return to active life. When directed at appropriate target symptoms, psychopharmacologic treatments may reduce some of the most distressing symptoms of chemical sensitivity.

REFERENCES

AMERICAN ACADEMY OF ALLERGY. (1981). American Academy of Allergy position statements - controversial techniques. J. Allergy Clin. Immunol. 67(5):333-338.

AMERICAN COLLEGE OF PHYSICIANS. (1989). Position paper: Clinical ecology. Ann. Intern. Med. 111(2):168-177.

BLACK, D.W., RATHE, A., and GOLDSTEIN, R.B. (1990). Environmental illness: A controlled study of 26 subjects with '20th Century Disease'. J.A.M.A. 264(24):3166-3170.

BOLLA-WILSON, K., WILSON, R.J., BLEECKER, M.L. (1988). Conditioning of physical symptoms after neurotoxic exposure. J. Occup. Med. 30(9):684-686.

KAHN, E. and LETZ, G. (1989). Clinical ecology: Environmental medicine or unsubstantiated theory. Ann. Intern. Med. 111(2):104-105.

SCHOTTENFELD, R.S. and CULLEN, M.R. (1985). Occupation-induced posttraumatic stress disorder. Am. J. Psychiatry 142:198-202.

SIMON, G.E., KATON, W.J., and SPARKS, P.J. (1990). Allergic to life: Psychological factors in environmental illness. Am. J. Psychiatry 147:901-906.

STEWART, D.E. and RASKIN, J. (1985). Psychiatric assessment of patients with '20th Century Disease.' Can. Med. Assoc. J. 133:1001-1006.

TASK FORCE ON CLINICAL ECOLOGY OF THE CALIFORNIA MEDICAL ASSOCIATION. (1986). Clinical ecology: A critical appraisal. West. J. Med. 144(2):239-245.

TERR, A.I. (1986) Environmental illness: A review of 50 cases. Arch. Intern. Med. 146:145-149.

TERR, A.I. (1987). Multiple chemical sensitivities: Immunologic critique of clinical ecology theories and practice. Occup. Med.: State Art Rev 2(4):683-694.

MULTIPLE CHEMICAL SENSITIVITY: TREATMENT AND FOLLOWUP WITH AVOIDANCE AND CONTROL OF CHEMICAL EXPOSURES

GRACE E. ZIEM, MD, DrPH
Occupational and Environmental Medicine Research Office
Baltimore, Maryland

INTRODUCTION

Multiple chemical sensitivity (MCS) is defined as illness reactions associated with exposure to more than one chemical, at significantly lower exposure levels than would cause noticeable illness in the general population, not mediated solely by known immune mechanisms, often accompanied by unusual fatigue, involving symptoms in multiple organ systems, and usually including symptoms of the central nervous system.

Although initially skeptical that such illness reactions could occur following low-level chemical exposure, I became aware that existing chemical exposure limits were scientifically faulty (Castleman and Ziem, 1988; Ziem and Castleman, 1989; Roach and Rappaport, 1990) and that no-effect levels extrapolated from chronic animal studies were often orders of magnitude below current legal exposure limits (Cunningham, 1988). After evaluating a cluster of individuals overexposed to chlordane and with illness reactions fitting the above definition for chemical sensitivity, I became increasingly open to a new clinical entity, despite never having read or heard about it. Subsequently, after witnessing repeated illness reactions accompanied by central nervous system symptoms closely resembling an anesthetic-like effect, but without overt psychologic changes, in other chemically sensitive patients, I sought ways to reduce chemical exposures for such patients. The results have been gratifying.

This treatment of chemically sensitive patients has produced significant, although gradual, clinical improvement through control of chemical exposures that aggravate the illness. Chemicals associated with symptoms in each specific patient must be identified and then a plan developed to control these exposures as well as other potential exposures that have been associated with illness in other chemically sensitive patients. The goal is to gradually improve the underlying condition so that the patient can eventually return to a relatively normal life.

IDENTIFYING EXPOSURES

Chemical exposures can be divided into two groups: external exposures in the patient's environment and internal exposures resulting from body burdens of various chemicals from the

Address all correspondence to: Dr. Grace Ziem, Occupational and Environmental Medicine Research Office, 1722 Linden Avenue, Baltimore, MD 21217.

environment. The chemicals that tend to exacerbate or induce illness symptoms in patients with MCS, in my experience, have typically been pesticides, other petrochemicals, coal-derived chemicals, and combustion products. Chlorinated water containing organic debris can also induce illness reactions in these patients, possibly because organochlorines are formed in this situation. Heated surfaces containing chemical residues that volatilize with heating can also be problematic, e.g., electrical and electronic equipment.

The specific exposures that trigger symptoms vary from patient to patient. To identify problem exposures for a specific patient, a patient log of illness reactions is useful. The patient records the date of each illness reaction (symptom onset or, for chronically ill patients, significant worsening of symptoms). The accompanying symptoms are then listed, beginning with the most troublesome, e.g., most uncomfortable, most disabling in terms of ability to function normally. For each reaction, the chemical exposures are recorded for a period of hours prior to the reaction. A four-hour period before illness will probably catch most inhalation exposures, although some patients may have a significantly longer delay of symptom onset, perhaps 12-24 hours or even more. Ingestion exposures, such as foods containing pesticide residues, seem to have more delayed onset intervals than do inhalation exposures. Irritative symptoms may also be reported with low-level exposure and can occur within minutes. The latter are often accompanied by some systemic symptoms in chemically sensitive patients.

Symptoms the physician should look for in the patient log include unusual fatigue, difficulty thinking, impaired concentration, and short-term memory losses. The patient may use terms such as feeling spacey, but the doctor can inquire for more specificity. Headaches (e.g., in the sinus areas and sinusitis) may also occur, especially with pesticide and petrochemical exposures and not infrequently with irritants. Other symptoms may involve the upper or lower respiratory system where irritation symptoms are common. Generalized aching in muscles, joints, or connective tissue areas is not unusual. Symptoms can also involve internal organs, including gastrointestinal, genitourinary, and cardiac symptoms (e.g., palpitations, irregular heart beat). Unusual thirst is not uncommon.

For children with suspected chemical sensitivity, the physician should be alert that hyperactivity, attention deficit, and behavioral changes commonly follow chemical exposure; hyperactivity is much less common in chemically sensitive adults. The parent will need to inquire about exposures and maintain the log for younger children. A comparison of handwriting samples obtained during illness reactions and when the child feels well can be useful.

The physician should review the illness log during the patient visit for the type, duration, and severity of exposures that precede illness reactions. These exposures should be given serious attention for control measures, particularly when repeated patterns are observed, or reactions are accompanied by symptoms that interfere with the person's ability to function normally on the job, at school, or at home. Patients may not be aware of all their exposures. They should be encouraged to ask about possible exposures whenever they become ill outside the home, when new products have been brought into or used in the home, or when they have been around someone who may be using or wearing one or more products containing petrochemicals or coal-tar derivatives. Patients should include in their log any use of medications or beverages containing alcohol and the presence of smoke or other combustion products as exposures, because these can trigger or aggravate illness reactions.

Determining the exposures that trigger illness reactions can require some detective work by the physician. I repeatedly found that when clinical improvement did not occur, some exposure had not been identified. When the exposure was finally identified and controlled, significant improvement occurred in almost every clinical case involving my patients to date. The physician must be willing to evaluate carefully any exposures that may have been overlooked in patients who are not making clinical progress. More severely affected individuals are likely to improve more slowly, however.

Current or past employees and their physicians are legally entitled to material safety data sheets on all chemicals used in their workplace through 29 CFR 1910.1200 and 1910.20, the Hazard Communication standard (Federal Register, Nov. 25, 1983) and the Access to Medical Records rule (Federal Register, May 23, 1980) of the Occupational Safety and Health Administration (OSHA). The patient, or a physician with the patient's written release, can request information from the employer, including a safety data sheet on each chemical used in or near the patient's work area. The physician can often obtain information quickly by phone, but more complete information is best obtained by written request. Certified mail requesting a return receipt usually results in a faster written reply. When safety data sheets do not list all chemicals present, the physician, nurse, or industrial hygienist should contact the manufacturer directly, because OSHA regulation 1910.1200 allows health professionals access to trade-secret information when needed to care for an individual. For chemical sensitivity, the contents of products causing potential exposure must be carefully identified to avoid unnecessary illness. Health professionals must maintain trade-secret information confidentially to avoid inadvertent release. If a government agency has inspected the workplace, a copy of the inspection results may often be obtained through a freedom-of-information request or even informal inquiry.

Some chemically sensitive patients go through a period of reduced ability to detect chemical smells, and these patients are at a disadvantage in identifying early exposures. Conditioning or advising chemically sensitive patients to ignore chemical smells (and thus to continue exposure) can be a serious mistake (potential malpractice) unless the physician is absolutely certain that such exposures pose a zero risk to that person's illness. The clinical condition of several of my patients worsened significantly after they were told to ignore or not worry about chemical exposures. The detection of smell is also the detection of exposure (clean air does not smell) and is useful for reducing or avoiding exposures that can aggravate the underlying illness.

Physicians who too readily assign a psychologic origin to chemical sensitivity without a very comprehensive evaluation of both the patient's chemical environment and the onset and variation of illness in relation to that environment are also on ethical thin ice. Physicians who are less familiar with a patient are far more likely to assign psychologic causes. A small random survey of chemically sensitive patients in my practice revealed that, of the physicians consulted by those patients, only 3 of 27 (about 10%) who assigned a psychologic cause to the illness reactions had seen the patient three or more times in the five years preceding onset of the illness; whereas 11 of 21 physicians (about 50%) who considered chemicals to be the main problem had seen the person three times in the preceding five years.

REDUCING ILLNESS REACTIONS

Patients with MCS typically improve, albeit gradually over months or years, with adequate control of their chemical exposures. The focus should be on eliminating all unnecessary exposures to pesticides, petroleum and coal-derived chemicals, and combustion products. Combustion products from natural substances, such as wood and leaves, may also be a problem. Tobacco smoke represents combustion products from such things as tobacco leaves, additives, and pesticides.

Control efforts must include chemical exposures in the home, where most of the hours of exposure typically occur, and in the workplace, where exposures may be more intense. For chemically sensitive children and other students, the school environment must be controlled. Unfortunately, school classrooms and cafeterias still commonly have frequent and routine petrochemical pesticide applications regardless of the presence of pests. Community sources of exposures can also be a problem, e.g., gas stations; dry cleaning stores; landfills containing hazardous waste; industrial air or water emission sources; areas of heavy traffic; or nearby neighbors or commercial facilities using petrochemical pesticides for lawn, agricultural, or grounds treatment.

Patients who drink beverages containing alcohol would probably benefit from a trial period of complete avoidance for a month or longer. Then, if they wish to test their response to alcohol, they could have a single drink on a day without other known exposures. This may need to be repeated several times if the patient wishes to avoid unnecessary abstinence. Unusual fatigue within a day after the single drink or other symptoms unexpected from a single drink, such as significantly impaired thinking beyond what the patient would have experienced prior to the onset of chemical sensitivity, represents a probable positive reaction. Chemically sensitive patients often voluntarily reduce or eliminate their alcohol consumption because of symptoms, even before medical advice. Although ethanol is derived from grains or other foods, it is structurally related to petrochemicals. Some physicians believe that the grain or other food stuffs from which the alcohol is derived may be a factor if the patient has also developed increased sensitivity to specific foods.

Avoidance trials are helpful in evaluating whether a particular exposure aggravates illness in a particular patient. The exposure should be avoided for a minimum of two weeks (four weeks are more ideal) because individual illness reactions can persist for one to two weeks after removing the immediate offending agent, and increased fatigue is often the last symptom to clear. Additional time is needed for the patient to establish a baseline for comparison before and after the trial exposure.

Control of chemical exposures can be achieved most easily in the home, where the environment is more under the patient's control. Household exposure must be controlled for recovery because many patients spend 10-20 hours there daily. The chemically sensitive patient should list home-use products that may contain petroleum or coal-derived chemicals (and irritants like lauryl sulfate compounds, bleaching, and other cleaning agents, for patients also sensitive to irritants). The list should include brand names and manufacturers of products the patient wishes to continue using if they do not contain petrochemicals, coal-derived products, or irritants (e.g., cleaning agents, cosmetics and personal hygiene products [all the patient's personal use products and all aerosols used by family members], all pesticides, odorants and odor-masking agents [air fresheners], and other house, garden and garage and vehicle-maintenance supplies). The patient can then inquire from the Poison Control Center whether these products contain any petrochemicals, coal-derived substances, or irritants. If necessary, persons other than the patient can use vehicle-maintenance supplies containing petrochemicals if they are stored in a non-attached garage or shed and are not used on the vehicle interior. Use of petrochemicals that can enter the vehicle's interior should be limited whenever possible and avoided for several days before the patient will use the vehicle.

Petrochemicals typically evaporate, the volatile rapidly and the non-volatile more slowly. Thus, ongoing exposure from residues can occur for days, weeks, months, or even longer, until evaporation is complete and the residue gone. For example, chlordane residues persist in concentrations adequately effective against termites for 20 or more years; chemically sensitive patients may have difficulty with chlordane exposure even beyond this time. Thus, merely avoiding a room during or shortly after chemical use there does not avoid exposure and is insufficient to protect the chemically sensitive.

Most chemically sensitive patients seem to be adversely affected by petrochemical pesticides. Home and garden pests can usually be controlled adequately without the use of petrochemicals (Olkowski et al., 1991). Integrated pest management and nonpetrochemical pest-control alternatives are available for virtually every pest-control problem in commercial agriculture as well (Alternative Agriculture, 1989). It is necessary to eliminate petrochemical pesticide use from the patient's home, lawn, play areas, and other areas where any skin contact could occur or where vapor exposure occurs more than occasionally. Thus, pesticide alternatives are needed in the school and workplace. Pesticides in the diet should also be reduced or eliminated. Because the pesticide residue legally allowed on foods can be too high for chemically sensitive patients, foods grown without pesticides (often called "organic") are preferable. Thus, whenever possible, the physician should recommend such organic foods, particularly for patients who are sufficiently ill that their symptoms significantly interfere with work, school, and other activities needed for daily living. Animal fats, such as meats, dairy products containing fat, and shortening, can contain higher residues of fat-soluble petrochemicals, and their intake should be reduced whenever possible.

The use of petroleum or coal fuels for cooking (gas stoves), clothes drying, heating (oil, gas, or coal) and other purposes can cause illness in chemically sensitive patients. If any question exists about symptoms associated with cooking fuels, the patient should try using a hot plate and other electric appliances for cooking for several weeks, followed by a brief trial using gas to cook a meal, and should record any change in symptoms. Multiple agents can induce an illness reaction, and reactions can be missed in people with chronic illness or other exposures. Therefore, only when persons who are not chronically ill lack any detected reactions after multiple trials can the relative safety of gas cooking be suggested. Similar trials can be done for heating fuels by temporarily substituting electric space heaters (without coils that contain petrochemicals). In general, severely sensitive individuals show more symptoms with petroleum or coal-based fuels, and these persons may not recover adequately if the exposures are neglected.

Cosmetics and other personal-use products containing petroleum and coal-derived substances can also be a problem. Such substances applied to the skin often enter the body readily. Aerosols and substances that evaporate rapidly (hair spray, nail polish remover) can cause exposure when used by family members or others. Volatile substances on the skin or clothing of family members, friends, coworkers, or others can be a problem, especially for more sensitive patients. Many colognes and perfume agents are petrochemicals, and these can persist for hours at sufficient levels to cause illness reactions, especially in more sensitive patients. Fortunately, safer substitutes are available for many cosmetics. Some patients are sensitive to products containing lanolin from sheep treated with pesticides, because the pesticides are often fat-soluble and enter the lanolin readily.

As mentioned above, chlorinated water often contains chlorinated solvents, and exposure can result in increased body levels of chlorinated hydrocarbons (Aggazzotti et al., 1990). This often presents a problem for chemically sensitive patients who use chlorinated pools. Natural, nonchlorinated water is best for chemically sensitive patients who swim. Showering (and bathing) also causes exposures, which can be particularly problematic for more sensitive patients. Exposure will be directly proportional to the rate of flow, water temperature, and shower duration (Gilbert and Calase, 1988). Thus, long showers with high flow create much higher exposures than do short ones with reduced flow. Exposure can be reduced further by using an activated charcoal filter attached to the shower head or bath faucet. Water for drinking and cooking may also be a problem. Activated charcoal or reverse-osmosis-filtered water can minimize this exposure. Bottled spring water is likely to have lower petrochemical content than does most tap water, but the actual chemical content will vary depending on spring contamination. Water in plastic containers may have some contamination released from the container material, and this could be a problem for some patients.

Certain fabrics can pose a problem for chemically sensitive patients because of offgassing into the air, skin contact, and the ease with which petrochemicals enter the body through the skin. Fabrics with permanent-press treatment can contain formaldehyde or other petrochemical residues that bother many sensitive patients. These fabrics may present a problem during shopping in department stores and after purchase, until repeated washing reduces the levels of formaldehyde or other chemicals. Recently dry-cleaned garments often contain residues of perchloroethylene or other solvents at levels that may aggravate illness in sensitive patients. Although many sensitive patients prefer to avoid dry cleaning, recently dry-cleaned garments can be aired outside the home (not in a plastic bag) for several weeks and may then present less of a problem. Patients who are also bothered by irritants may find that residues of certain fabric softeners and some detergents on clothing and bedding are a problem. Less irritating substances are available as substitutes for home laundry in these cases. Some patients have difficulty with certain synthetic garments made from various petrochemicals. After repeated washings, symptoms may be reduced. Some wools may be a problem, both because pesticides were used on the sheep and because wools may have mothproofing treatment.

Chemically sensitive patients often develop symptoms from exposure to vehicle exhaust. This problem is more likely in heavy traffic, on freeways, in tunnels, and behind buses and trucks. The patient should not drive in such situations if reactions to these exposures include significant central nervous system symptoms, especially if symptoms come on fairly rapidly or without significant warning. Patients can often reduce such symptoms significantly by installing a charcoal filter device to treat air that comes into their vehicles; such devices can plug into the vehicle's cigarette lighter. Choosing less-congested routes and commuting schedules and avoiding tunnels is also helpful. Some patients may need personal, charcoal-filtered respirators for adequate symptom control while driving, but other controls usually suffice. If other frequent and intense chemical exposures are rigorously controlled, the illness eventually improves sufficiently for most patients so that respirator use is seldom needed indefinitely. It may still be needed on occasion, such as on long drives and in unavoidable congested areas.

Stickers for handicapped parking areas can help patients who experience illness with vehicle exhaust, particularly if they frequently park in large lots or garages and walk through exposed areas. Despite the above controls, some patients may have times in their illness when they are be too ill to drive safely. This can improve significantly or eventually resolve in most cases when other exposures are controlled.

Gasoline (or other fossil fuel) vapors can cause illness reactions in chemically sensitive patients. These patients should probably ask someone else to fill their tanks, particularly if they have noticed adverse reactions after being at service stations. Such patients should avoid gasoline service stations whenever possible. In practice, vapor-recovery devices do not often form adequate seals with the gas tank during filling, and exposure still occurs.

If certain plastics in the home (or elsewhere) offgas, they may be a problem. More sensitive patients are likely to be troubled by plastics, but any frequently or chronically ill patient should probably minimize such exposure. For example, microwaving with plastics and plastic wraps can cause offgassing and potential food contamination; glass is preferred. Plastic covers on pillows or mattresses can cause hours of daily offgassing close to the patient's breathing zone. Plastic shower curtains may also be a problem, especially in small rooms. Plastic offgassing in new vehicles is a common problem. In general, flexible plastics offgas more often than rigid ones, and new ones more often than old ones. Water is likely to be less contaminated when bottled in glass than in plastic containers, other factors being equal. While perhaps not a problem for most sensitive patients, some have reported repeated illness after using Teflon cookware.

Some lay references can help the patient reduce unnecessary exposures by suggesting cleaning and personal hygiene products, cosmetics, clothing, etc. that may be less likely to induce illness reactions (Dadd, 1984; Zamm, 1980; Olkowski et al., 1991). Another reference has useful avoidance tips, although discussion elsewhere in the book about clinical ecology is more controversial (Golas, 1979). Housing construction and indoor air pollution have also been addressed with useful suggestions for the chemically sensitive, albeit in a more technical manner (Small, 1983).

Chemically sensitive patients are often more sensitive to medications as well as chemicals. They seem more prone to unusual fatigue, other side effects, and perhaps more serious adverse reactions. They also seem more likely to develop significant illness reactions to dyes and other chemicals injected for diagnostic purposes. Thus, for pharmaceutical agents used for therapy or diagnosis, the physician will need to be unusually cautious or conservative and to weigh very carefully the expected benefit against increased risk of illness reactions. Non-pharmaceutical alternatives should be considered whenever possible, such as acupuncture for analgesia and anesthesia rather than novocaine-related agents and petroleum-derived anesthetics. Unfortunately, physicians may have difficulty selecting alternatives because medical education in many countries tends to ignore nonpharmaceutical remedies. When pharmaceuticals are required, beginning with lower dosages is strongly recommended whenever possible. Misdiagnosis can also lead to unnecessary medication. A common example is misinterpreting the very common sinus irritation seen in chemically sensitive patients as infection. The physician prescribes one antibiotic after another with no clinical improvement, but with increasing medication side effects.

Exposure limits legally permitted in the workplace are not designed to protect the chemically sensitive. In addition, occupational exposure limits were developed by a process in which chemical manufacturers served on the TLV Committee of the American Conference of

Governmental Industrial Hygienists and played major roles in developing proposed limits for their own products (Castleman and Ziem, 1988; Ziem and Castleman, 1989). Although these exposure limits show no statistical correlation with human health effects (Roach and Rappaport, 1990), many of the limits issued by this private organization were later adopted as legal OSHA workplace limits. However, there are alternatives. The federal government considers chemical sensitivity a handicap, and the Americans with Disabilities Act (Federal Register, July 26, 1991) requires that employers make reasonable accommodations for handicapped persons. Reasonable controls for employees, when medically recommended, can include reduced petrochemical use in the work area, advanced notice for building repairs and any pesticide or other petrochemical use, and activated charcoal filter devices in the work area, with replacement filters as needed. Integrated pest management that results in reduced pesticide use in the patient's building is not only reasonable but also effective and as economical as current pest control methods.

If exposures causing or aggravating chemical sensitivities are occupational, employers can reduce employee disability costs under the handicapped provision by controlling the workplace more strictly than OSHA requires. Physicians can help convince employers to do so. If an occupational exposure is suspected in part because symptoms began at work, become worse at work, or are better on weekends or vacations, a trial period of several weeks away from exposed work areas is advisable. An increase in symptoms after the employee returns to the regular work area tends to confirm a probable occupational component.

Room air filters with an activated charcoal filter can help patients in some situations. This works best when workplace petrochemical use has also been reduced. Area filters are more likely to help in small rooms and when the patient needs to spend significant time in the room. Potential candidates for area filters include an office that can be closed off from general work areas and the patient's bedroom and other living areas. Area filters specifically designed for the chemically sensitive are best because they usually avoid plastics, particle board, and other chemical components that can release contaminants. Some area filter companies will allow clients a 1-month trial period before purchase to evaluate symptom reduction.

Charcoal filters should be replaced periodically. Filters with plastic components should be avoided because of potential offgassing. The amount of petrochemicals in the air, air flow rates through the filter, and room size will determine how frequently filters must be replaced. By trial and error, the proper interval for replacing or renewing filters can be determined for each situation to avoid illness. At least one manufacturer (Allermed[1]) is developing a device that shows a signal light when filters need to be replaced. Activated charcoal filters should be stored in airtight containers (preferably metal canister or glass) when not in active use or they will continue to pick up chemicals.

---

[1] The use of company or product names is for identification only and does not imply endorsement by the Agency for Toxic Substances and Disease Registry.

Filter devices and replacement filters are necessary treatment for many chemically sensitive patients and should be regarded as medical devices for insurance reimbursement. Because they can reduce illness and disability, they are good investments for all parties. Activated charcoal devices (not ozone) are best for removing petrochemicals when patients are in the area, but particulate filters are needed where chemical particles (such as vehicle exhaust and fiberglass with formaldehyde-containing resin) are also involved. When quite small particles are present, HEPA or high efficiency particulate air filters may be needed with a prefilter to remove larger particles.

Building materials, such as particle board, fiberglass, and other insulation products containing formaldehyde resins, often offgas formaldehyde, particularly when new. Most chemically sensitive patients react adversely to low levels of formaldehyde. If these materials are present in the home or workplace, and were installed in the last few years, symptoms can increase. It may help if the patient can arrange to stay elsewhere for a few weeks, both to reduce symptoms and pinpoint their causes.

Patients with past exposure to petrochemicals that persist in the body for long periods (e.g., chlordane for termite treatment, PCBs, PBBs, DDT) have an internal exposure source stored in body fat that can be released gradually into the blood. More volatile petrochemicals (e.g., toluene, xylene, methylene chloride) are stored in body fat for substantial periods (Pyykko et al., 1977; Astrand, 1975; Engstrom and Bjurstrom, 1977 and 1978). Factors that mobilize body fat stores, such as weight loss, exercise, and chronic illness associated with poor appetite, can increase chemical release into the blood. This may explain why dieting increases symptoms in chemically sensitive patients, in my experience.

A body burden of petrochemicals can aggravate illness in chemically sensitive patients. A promising method for reducing such body burdens is sauna therapy shortly after exercise (Root et al., 1985; Schnare et al., 1984; Kilburn et al., 1989, Wisner et al.[2]) Exercise mobilizes the chemical from body fat, and sweating in the sauna increases excretion. If a medical workup does not reveal contraindications for exercise or sauna, I recommend a daily program (five to seven days a week) of exercise followed immediately by sauna. The duration and vigor of exercise will be determined partly by the fatigue and stamina of the patient. Because fatigue is common in chemical sensitivity, the patient initially may not be able to do long or vigorous exercise even in the absence of heart or lung problems. However, walking or less-strenuous aerobics are usually an adequate beginning for these patients. As control of both external and internal exposures produces clinical improvement, patients often find they can exercise more vigorously. Sauna temperatures of 140-160°F are usually suitable, although individual tolerances vary. Patients on diuretics or with cardiovascular disease may need somewhat lower temperatures. The patient stays in the sauna 15-20 minutes, leaving sooner if illness or

---

[2] Wisner, R.M., Shields, M.G., Curtis, D.L., and Beckman, S.L. Human contamination and detoxification: Medical response to an expanding global problem. (Unpublished paper describing three cases of body burdens of volatile chemicals substantially reduced with sauna.)

discomfort develops. Then, after a cool or tepid shower (in nonchlorinated water), the patient can re-enter the sauna for a second and final 15-20 minutes.

Home sauna units can be constructed economically, but building materials should be of minimal toxicity. Unfortunately, many commercially available saunas are not so constructed. Poplar wood for a bench; ceramic tiles on walls, ceiling and floor set in Portland cement without additives; and 100% clear silicon for caulking have worked well. Any sauna purchased for home use should have an accompanying statement that the sauna was designed for the chemically sensitive and contains no petrochemical products. A home sauna allows the patient to avoid use of commercial saunas, which may be cleaned with toxic agents and have other contaminants from maintenance and use. It also allows the patient to continue therapy without substantial added cost other than electricity. Such a small home sauna for treating the chemically sensitive should also be considered a medical device for insurance reimbursement.

PUBLIC POLICY

Public policy to reduce unnecessary exposures to petrochemicals and certain other chemicals could reduce illness in sensitive individuals as well as the number of new cases of chemical sensitivity, many of which occur after exposure to petrochemicals and certain other chemicals (Levine and Reinhardt, 1983). Although chemical sensitivity can begin without significant, documented, preceding chemical exposure (just as mesothelioma can occur without previous documented asbestos or fiber exposure), it commonly occurs following prolonged or too-intense exposure to pesticides, other petrochemicals, and coal-derived chemicals or combustion products. This is, in fact, typical of my patients. Thus, social efforts to avoid unnecessary exposure of workers and the public to these substances could have significant beneficial effects for primary prevention as well as for reducing illness in persons who are already chemically sensitive.

When exposures are found that aggravate or are associated with onset of chemical sensitivity, the physician can play an important role in prevention. With technical support, recommendations, and occasional legal referral, significant exposure reduction can often be achieved in work and school environments and sometimes also in community exposures. Legislative provisions for the handicapped can assist this endeavor. With chemical sensitivity increasingly being considered a handicap, legal protection for the handicapped is available, such as reasonable housing and employment accommodations for persons who are chemically sensitive. This can help facilitate changes in pesticide and other chemical use in apartment buildings, schools, other public buildings, and workplaces. Because one effect of current legal occupational exposure limits is discrimination (in the form of disproportionate illness) against the handicapped group of chemically sensitive persons, these exposure limits may come under increasing future challenge.

Further changes in chemical use policy are urgently needed. Public buildings, public facilities such as parks and golf courses, and public areas of multiple dwelling living units should be phased into integrated pest management where nonpetrochemicals are relied on primarily, and petrochemicals are only permitted when other measures have been used as directed but failed. In addition, whenever petrochemical pesticides are to be used in facilities or grounds, advance notice should be given for employees and the public to avoid more intense unnecessary exposures. Such notices should remain in place until residue levels are low enough to not cause illness in the majority of sensitive individuals. Obviously, a notice lasting 24-72 hours after pesticide use is insufficient. Repairs and renovations involving petroleum or coal-derived chemicals (e.g., paints, roofing agents, adhesives) in public facilities need similar policies: less toxic substitutes whenever possible, some advance notice, and notices remaining until residue levels are sufficiently low.

Chemicals used to mask odors are a significant problem. Area odorants and odor-masking agents used in bathrooms, vehicles, and other locations are typically petrochemicals with agreeable smells. One such agent, paradichlorobenzene, is a suspect cancer agent, but it and other petrochemical odorants can aggravate illness in chemically sensitive persons. The promotion of such products as air fresheners is misleading and harmful. When these chemicals are dispensed in areas used by employees or the public, the chemically sensitive cannot use rest rooms and other facilities without risk of illness. Chemical sensitivity can also be caused by using petrochemicals (e.g., toluene, xylene, and chlorinated solvents) to treat damaged furnishings and by insurance company use of these chemicals to reclaim homes or other building areas affected by unpleasant smells such as skunk or smoke odors. Such practices should be discontinued.

Pesticide residues on foods aggravate illness in the chemically sensitive, cause periodic acute illness in nonsensitive individuals, and could have additional unknown consequences for us all because the substances used do not undergo adequate toxicologic testing. Increased effort is needed to enhance knowledge and disseminate information about alternative means of agriculture. More candid labeling and other information for the consuming public will inform sensitive individuals and encourage a market shift to less contaminated foods. Wider production of such foods is likely to reduce their cost. In addition, reduced pesticide use in buildings and in agriculture will reduce water contamination, thus benefiting the general public, chemically sensitive or not, and other species (e.g., fish and other species in the human food chain).

Because formaldehyde exposure can cause formaldehyde allergy and chemical sensitivity, both with potentially serious consequences, building materials using formaldehyde, such as particle board, should be used for exterior construction only. Interior use may result in building contamination. Insulation material that releases formaldehyde (e.g., fiberglass with formaldehyde-containing resins, a common treatment of fiberglass) should be replaced with safer substances. Just as lead-based paint is no longer permitted for certain uses, formaldehyde products need to be labeled and the location of their use controlled.

SUMMARY

Reducing unnecessary chemical exposures, particularly pesticides and other petrochemicals, shows promise for reducing illness episodes in the chemically sensitive. Because similar types of exposures have been associated with the onset of chemical sensitivity, such precautions could have wider preventive value for the rest of society as well. Many uses of chemicals have dubious social benefits, and reduced use should be achievable.

The chemical industry will likely bitterly contest the reduced use of chemicals because it stands to lose substantial sales. Compensation and liability insurance carriers also stand to lose if the environment is found problematic, rather than individual psychology, for example. Professionals should also recognize conflicts of interest for the chemical and insurance industries by openly acknowledging funding sources for research.

The author believes that research on chemical sensitivity that blames the psyche of the victim rather than the chemical will more likely be funded by the insurance or chemical industry than will other research. Study designs should be developed in an atmosphere removed from financial conflicts of interest. This means a substantially larger role for government funding of research on chemical sensitivity to avoid biasing the knowledge base by financially interested parties. The time is critical for government funding of research on chemical sensitivity because the illness is being defined and characterized. If preliminary research is flawed by improper design and focus, our understanding of the problem could be delayed for years.

REFERENCES

ACCESS TO EMPLOYEE EXPOSURE AND MEDICAL RECORDS (1980). Occupational Safety and Health Administration. Federal Register, May 23, 1980.

AMERICANS WITH DISABILITIES (1991). Department of Justice. Federal Register. July 26, 1991.

AGGAZZOTTI, G. FANTUZZI, G., and TARTONI P.L. (1990). Plasma chloroform concentrations in swimmers using indoor swimming pools. Arch. Environ. Health 45:175-179.

ALTERNATIVE AGRICULTURE. (1989). National Academy Press, Washington, DC.

ASTRAND, I. (1975). Uptake of solvents in the blood and tissues of man: A review. Scand. J. Work Environ. Health 1:199-218.

CASTLEMAN, B.I. and ZIEM, G.E. (1988). Corporate influence on threshold limit values. Am. J. Ind. Med. 13:531-559.

CUNNINGHAM, K. (1988). A comparison of PELs and TLVs to health-based exposure limits derived from the IRIS data base (9/28/88). Submitted to Docket H-020, Occupational Safety and Health Administration, by R.T. Zagraniski, Assistant Commissioner, New Jersey Department of Health.

DADD, D.L. (1984). Nontoxic and Natural. Jeremy Tarcher, Inc. Distributed by St. Martins Press, New York.

ENGSTROM, J. and BJURSTROM, R. (1977). Exposure to methylene chloride: Content in subcutaneous adipose tissue. Scand. J. Work Environ. Health 3:215-224.

ENGSTROM, J. and BJURSTROM, R. (1978). Exposure to xylene and ethylbenzene II. Concentration in subcutaneous adipose tissue. Scand. J. Work Environ. Health 4:195-203.

GILBERT, G. and CALASE, E.J. (1988). Estimation of trichloroethylene inhalation exposure during the showering process. Northeast Reg. Env. Pub. Health Newsletter 2:2-4.

GOLAS, N. (1979). Coping With Your Allergies. Simon and Schuster, New York.

HAZARD COMMUNICATION STANDARD, FINAL RULE (1983). Occupational Safety and Health Administration. Federal Register. November 25, 1983. pp. 53280-53348.
KILBURN, K.H., WARSAW, R.H., and SHIELDS, M.G. (1989). Neurobehavioral dysfunction in firemen exposed to polychlorinated biphenyls (PCBs): Possible improvement after detoxification. Arch. Environ. Health 44:345-50.
LEVINE, S.A. and REINHARDT, J.H. (1983). Biochemical-pathology initiated by free radicals, oxidant chemicals, and therapeutic drugs in the etiology of chemical hypersensitivity disease. Orthomol. Psychiat.12:166-183.
OLKOWSKI, W., DAAR, S., and OLKOWSKI, H. (1991). Common Sense Pest Control. Taunton Press.
PYYKKO, K., TAHTI, H., and VAPAATALO, H. (1977). Toluene concentrations in various tissues of rats after inhalation and oral administration. Arch. Toxicol. 38:169-176.
ROACH, S.A. and RAPPAPORT, S.M. (1990). But they are not thresholds: A critical analysis of the documentation of threshold limit values. Am. J. Ind. Med. 17:727-753.
ROOT, D.E., KATZIN, D.B., and SCHNARE, D.W. (1985). Diagnosis and treatment of patients presenting subclinical signs and symptoms of exposure to chemicals which bioaccumulate in human tissue. Proceedings of the National Conference on Hazardous Wastes and Environmental Emergencies, May 14-16, 1985, Cincinnati, OH.
SCHNARE, D.W., BEN, M., and SCHIELDS, M.G. (1984). Body burden reductions of PCB, PBBs and chlorinated pesticides in human subjects. Ambio. 13:378-380.
SMALL, B.M. (1983). Indoor Air Pollution and Housing Technology. Prepared for the Planning Division of the Policy Development and Research Sector, Canada Mortgage and Housing Corporation. Available from Technology and Health Foundation, R.R.#1, Goodwood, Ontario, Canada LOC 1AO.
ZAMM, A.V. (1980). Why Your House May Endanger Your Health. Simon and Schuster, New York.
ZIEM, G.E. and CASTLEMAN, B.I. (1989). Threshold limit values: Historical perspectives and current practice. J. Occup. Med. 31:910-918.

TREATMENT OPTIONS IN MULTIPLE CHEMICAL SENSITIVITY

GERALD H. ROSS, MD, CCFP, DIBEM, FAAEM
Environmental Health Center
Dallas, Texas

INTRODUCTION

Multiple chemical sensitivities (MCS), a condition that probably has been present for hundreds of years in various manifestations, is gaining increasing recognition in medicine (Hileman, 1991). At the same time, the prevalence of classical allergy like asthma and hay fever is growing, and paralleling this is an increase of symptoms that some patients exhibit from exposure to commonly-encountered chemical substances. A simple definition of MCS is: a significant intolerance to low-dose components in the patient's personal environment, which most other people tolerate quite well.

Because of the complex nature of this condition, it is important to keep in mind the concept of total load or the total environment. This involves the complex interplay of genetic endowment, nutritional status, emotional stress, and exposure to foods, chemicals, microbes, inhalants, and even electromagnetic fields. How health is expressed is obviously a function of the cumulative effect of these factors on the patient.

Treatment options for MCS center around education, avoidance of allergens (including those in air, food, water), environmental clean-up procedures, rotational diets, enhancement of nutritional status, immunotherapy, immune stimulation, detoxification where appropriate, and addressing the patient's social and emotional health. In addition, the emerging awareness remains that electromagnetic fields may play a significant role in certain situations. (Fitzgerald, 1990). It is vital to establish a relationship whereby patients feel that physicians take their complaints (though perhaps myriad and long standing) seriously and in an atmosphere of mutual respect and trust. Within such a context, patients will more readily accept education about things they can do to potentially help themselves.

TREATMENT

Surely the basis of treatment for any allergic or sensitivity problem is to avoid the sensitizing agent. This can be done whether the primary problem is in air, food, water, or physical surroundings, and it may even apply to electromagnetic fields to a certain extent.

Avoidance

Because MCS patients frequently have wide varieties of allergies or sensitivities, they usually benefit from programs that minimize their contact with potentially sensitizing substances.

Address all correspondence to: Dr. Gerald H. Ross, Environmental Health Center-Dallas, 8345 Walnut Hill Lane, #205, Dallas, TX 75231.

Glass-bottled spring water is frequently helpful because it is much purer and does not contain the chlorine, fluorine, and other contaminants usually found in domestic drinking water supplies. Unfortunately, water bottled in plastic containers may pick up the phthalates, solvents, or plasticizers used to manufacture the plastic itself (Kailin and Brooks, 1963); for this reason the author recommends water bottled in glass.

One of the largest and best-recognized sources of contact we have with chemical contaminants is our food supply (Davies, 1986), which frequently contains pesticides and other chemical residues potentially harmful to everyone's health, and especially to persons with chemical sensitivity. Consequently, MCS patients often do much better clinically by consuming organically grown, less chemically-contaminated food products.

Because of the massive incidence of indoor air pollution, which is often 100 times that of outdoor air (Nero, 1988; NRC, 1981), air-filtration devices that remove both particulate and chemical substances from breathable indoor air usually benefit MCS patients considerably. In addition, steps to minimize indoor pollution by removing carpeting and combustion products from the home (such as natural gas, heating oil, or wood-burning stoves) usually bring about substantial improvements as well (Barron, 1990; Drerup et al., 1990).

Indoor air pollution is generally lower in homes that have hardwood or tile floors, with natural wood rather than fiber board, plywood, or other pressed-wood products that contain substantial amounts of glues, solvents, and other chemicals (including formaldehyde). The latter will off-gas and contaminate indoor environments for months or even years (Levin, 1989). Similarly, synthetic fabrics and other materials made from petroleum will slowly offgas volatile organic compounds (VOCs) and other chemical substances used in their manufacture, thus adding to the total load of indoor pollution. These latter substances frequently trigger symptoms and health deterioration in patients with MCS. An intelligent program of minimizing this kind of exposure usually has substantial therapeutic benefit for these patients.

Nutrition

Shandra (1980), a widely recognized international nutrition expert, has called undernutrition the most frequent cause of immunodeficiency. Because human enzymatic, metabolic, and detoxification pathways are largely dependant on adequate vitamins, minerals, and other essential co-factors, the nutritional status of patients with MCS is of obvious importance and concern (Jacoby, 1980). A survey of more than 330 patients at the Environmental Health Center-Dallas showed widespread vitamin deficiencies (including a B_6 deficiency in almost 58% of these patients) even in the presence of oral supplementation (Ross et al., 1990). A rotational diet is one way to minimize exposure to food substances to which the patient may be sensitive, and the diet can be readily designed by competent personnel within the context of nutritional adequacy (Radcliffe, 1987). A rotational diet does not repeat the same food or beverage item within the rotational cycle, and such diets can be constructed in several ways. To illustrate a 4-day, monorotational diet (one food per meal): On Day 1, the patient might consume eggs for breakfast and would not again consume any egg products until the morning of Day 5, thereby minimizing contact with potentially sensitizing foods. A rotational diet is relatively simple to institute, even before the foods to which the patient is actually sensitive are known. Such a diet minimizes over-exposure to any foods, and when properly constructed, can provide good nutritional adequacy and wide variety.

Patients with MCS are known to have a higher likelihood of nutritional imbalances (Ross et al., 1989), especially magnesium deficiency (Rea et al., 1987a). Because many patients with MCS seem to be nutrient deficient despite oral supplementation, intravenous infusions of certain vitamins and minerals (especially vitamin C, B vitamins, magnesium, and trace minerals) can have excellent therapeutic benefits. The exact mechanism of benefit is not known but may be through enhancing both immune function and detoxication pathways.

The Nova Scotia Environmental Medicine Clinic is currently engaged in a collaborative pilot study with the government of Nova Scotia in Canada on intravenous infusion of nutrients, including magnesium, for patients with chronic fatigue syndrome. In the author's experience, these patients also frequently have intracellular magnesium deficits.

Candida Sensitivity. One aspect of the total-load or the total-environment concept might include the possibility of candida sensitivity or overgrowth (Kroker, 1987). Although this issue is controversial, many of the author's patients with MCS have substantial elevations of antibodies against candida and may have had recurring vaginal or oral yeast infections. An anticandida program involving dietary modification, nutritional support, the replenishment of healthy gut flora, and possibly an antifungal medication, may prove very beneficial. Moreover, present knowledge about the potential adverse effects of candida on susceptible individuals appears far from complete. It seems reasonable to keep an open mind on this topic because the author has seen many patients who benefitted from such an approach, within a context of evaluating and lowering the total load.

Testing

Intradermal and Sublingual Testing. A vital question remains: how to evaluate and confirm chemical sensitivity. As in all medicine, initial reliance is primarily on history. Two main types of confirmatory testing, involving intradermal and inhalation challenge, are currently available. Most patients with a history of MCS have widespread sensitivities to chemicals, foods, and, to a lesser extent, inhalants (Rea and Ross, 1989). The testing antigens used for intradermal food testing at the Environmental Health Center-Dallas, are prepared from organically grown products, with no preservatives in the vaccines. This is in distinction to most testing antigens on the commercial market, which have either phenol or glycerine as stabilizers. Many patients with MCS cannot tolerate antigens that contain these substances for either testing or treatment. A variety of well-known, skin-testing techniques are available, including scratch or prick testing. Serial-dilution, end-point titration (SDET) and provocation-neutralization also give reliable data for use in a treatment program (King and King, 1990).

With provocation-neutralization, potential treatment vaccines may be drawn up for foods, chemicals, inhalants, microbial substances (like mixed respiratory vaccine and fluogen), and individual bacteria produced from autogenous vaccines; the latter are particularly helpful in patients with chronic rhinitis or sinusitis.

Sensitivity to a variety of chemicals may also be tested directly in the same way. Subsequent immunotherapy treatment with very low-dose, subcutaneous or sublingual vaccines can proceed in conjunction with avoidance of the offending agent (Scadding and Brostoff, 1986). Testing and treatment are also possible with neutralizing doses to neurotransmitters, like serotonin, dopamine, methacholine and others, which can produce substantial improvement in the patient's symptomatology. Intradermal testing involves not only evaluating the skin whealing response but also assessing symptoms and signs that may be produced from the testing itself. Sublingual testing, which relies almost exclusively on symptoms or signs, can nevertheless be a valuable testing technique.

Inhalation. One of the most scientifically rigorous evaluations of chemical sensitivity is specific challenge with low-dose chemical exposures. These take place inside in an enclosed glass and anodized-aluminum booth, under controlled circumstances (Rea et al., 1990). When done properly, these can help determine whether the patient is sensitive to a variety of chemical substances. Appropriate dilutions of chemicals are vaporized in the booth to achieve concentrations that are below odor thresholds and that approximate levels found in everyday settings (Rea et al., 1990). These tests are performed on a double-blind basis, using placebos of water or saline. The booth is evacuated and washed after each challenge to prevent any potential contamination. The absence of intake pipes or tubes that direct gas into the booth also prevents potential contamination and subsequent invalidation of subsequent tests at a later date. Chemical testing performed in such a booth must be done with the patients in the de-adapted state for reliable results (Rea et al., 1990).

The ideal location for this type of low dow-dose, chemical challenge is a hospital-based, environmental control unit (ECU) (Sprague, 1987) in which the patient is housed in a very clean environment for several days before testing begins. The principles of total load and de-adaption must be understood and followed with low-dose, inhaled, challenge testing, because incidental outpatient exposures to perfume, traffic exhaust, cigarette smoke, or other substances may trigger delayed responses that will cause inconsistent results.

Transfer Factor. Another potential treatment for patients with MCS is immune stimulation, especially in view of the frequent occurrence of immunosuppression with this illness. (Levin and Byers, 1987). The use of transfer factor may prove very helpful in these cases (Youdim et al., 1991). Transfer factor is an extract of white blood cells with known immune-stimulating properties. It is especially helpful in cases of leukopenia or in reduced leukocyte-killing capacity, which may contribute to recurring infections.

Pupilography. An additional method being developed to evaluate patients with MCS is the assessment of autonomic nervous function by pupilography (Shirakawa et al., 1992). Hamamatsu Photonics in Japan, in cooperation with the Department of Ophthalmology at Kitasato University, has developed a sophisticated optical device called the binocular iriscorder, that scans the eye surface in the infrared spectrum, and measures the pupilographic response to a specific light stimulus by very precise, computerized measurements. These are then compared with a standard reference range.

The change in pupil size, velocity of contraction and dilation, recovery time, and other parameters are measured in milliseconds following the light stimulus. These measurements are usually quite stable in each individual, with a slight diurnal variation. Patients with MCS have frequent deviations from baseline and exhibit considerable instability of autonomic function, usually in a sympatholytic pattern (Shirakawa et al., 1992). Many patients, whom the author has tested in blind, intradermal challenges with agents to which the patient is known to be sensitive, show distorted autonomic stability. Further research with this computerized analysis is pending.

Posturography. Another technique for evaluating patients with MCS is posturography testing, which essentially evaluates the neurologic integration of the nervous input from the eyes, ears, and peripheral nervous system by computerized methodology. Frequently, the author finds disorders of balance in patients with chemical sensitivity, and especially patients with histories of significant chemical exposures. When these patients undergo detoxification programs to enhance metabolism and the elimination of toxic chemical substances from their bodies, substantially improved measurements of these specific, neurologic balance functions usually result.

Adipose and Nerve Tissue

Toxic chemical substances may be stored for long periods in adipose and nerve tissue; many xenobiotics are direct neurotoxins (Seba et al., 1987). Indeed, DDT-type compounds and many others are often found in adipose tissue at much higher levels than would be found in blood under steady-state conditions. Certain individuals appear to have an impaired ability to effectively detoxify some of these exogenous substances. The chemicals may then build up over long periods following repeated exposures, especially in industrial settings (Seba et al., 1987). Levels of potentially toxic chemical substances at 10-40 times the population average, or even greater, are not unusual in MCS patients. With such toxic loads, it is perhaps not surprising that patients are unwell with various neurologic and other symptoms.

Detoxification. One method that is gaining considerable interest for reducing this total toxic load is a detoxification program using an integrated regimen of graduated exercise; nutrient replacement; and sauna-chamber, heat-depuration therapy (Schnare et al., 1982). The heat chamber at the Environmental Health Center-Dallas is built with ceramic tile on all inside surfaces and untreated wooden benches; it uses relatively low temperatures, in the range of

140°F. Patients first exercise and then spend time in the sauna to induce profuse perspiration. Under these conditions, stored xenobiotics appear to mobilize substantially (Schnare et al., 1982) and can then be metabolized, primarily through the liver and the gut, but also through the breath, urine, and perspiration. The clearance rate of stored xenobiotics from the body may be quite variable, and the author's experience shows that styrene and chloroform come out most easily. 1,1,1-Trichloroethane is ubiquitous and seems to be the most difficult to clear (Rea et al., 1987b). If, however, such a detoxification program proceeds too quickly, the patient may worsen for a time, usually in conjunction with a rise in liver transaminases, which suggests a significant strain on hepatic function. With caution however, these xenobiotics can be substantially reduced in MCS patients, usually with marked improvement in the patient's overall chemical sensitivity and general health. (Rea et al., 1987b).

Amalgams

Another area of growing interest and potential treatment is the controversial issue of amalgams and their adverse consequences on health. The television program "60 Minutes" reported on the hazards of silver/mercury tooth fillings and raised many questions about the safety of this heretofore well-accepted dental material. Although this subject is controversial and consensus is lacking, sensitization to mercury is a real phenomenon. Removing amalgams is an option that certainly requires investigation, especially for chemically sensitive patients.

SUMMARY

The essential features of treatment for chemical sensitivity are:

1) Encouraging the provision of clean air, food, water, and surroundings.
2) Identifying substances to which the patient is sensitive, with subsequent
   a) enhanced avoidance, or
   b) specific immunotherapy to reduce the patient's reactivity to those substances.
3) Assessing and enhancing the patient's nutritional status to maximize the body's ability to detoxify and to minimize the free-radical production and oxidative stress of xenobiotics.
4) Addressing concurrent problems such as infections, immunosuppression, and other medical conditions in an appropriate fashion.
5) Evaluating the patient's psychologic status and addressing any social and emotional problems in a compassionate manner.

The author believes that multiple chemical sensitivity is a real condition with documented physiologic abnormalities. It is not a functional or psychologic illness or a belief system of the patient. Second, this condition is diagnosable and treatable by various means. These treatment options not only make common sense but usually result in significant improvement for these unfortunate patients, who deserve the very best efforts of their health care providers.

REFERENCES

BARRON, S.R. (1990). Survey of the Medical Impact on Environmentally Hypersensitive People of a Change in Habitat. Canadian Mortgage and Housing Corporation. 30 April 1990.

DAVIES, K. (1986). Human Exposure to Selected Persistent Toxic Chemicals in the Great Lakes Basin: A Case Study of Toronto and Southern Ontario Region. City of Toronto Dept. of Public Health. December 1986.

DRERUP, O., MATLOCK, C., ROUSSEAU, D., and SALARIS, V. (1990). Housing for the Environmentally Hypersensitive (Survey and Examples of Clean Air Housing in Canada). Canadian Mortgage and Housing Corporation. July 1990.

FITZGERALD, K. (1990). Electromagnetic fields: The jury's still out. IEEE Spectrum, August 1990, pp 22-35.

HILEMAN, B. (1991). Multiple chemical sensitivities. Chem. and Eng. News. 69(20):26-42.

JACOBY, W.B. (1980). Detoxification enzymes. In: Enzymatic Basis of Detoxication. Vol 1. (W.B. Jacoby, ed.) Academic Press, New York. pp. 1-6.

KAILIN, E.W. and BROOKS, C.R. (1963). Systemic toxic reactions to soft plastic food containers: A double blind study. Med Annals of The District of Columbia 32:1-8.

KING, W.P., and KING, H.C. (1990). The evolution of otoloryngic allergy practices. Ear Nose Throat J. 69:11-26.

KROKER, G.F. (1987). Chronic candidiasis and allergy. In: Food Allergy and Intolerance. (J. Brostoff and S. Challacombe, eds). Bailliere Tindall/W.B. Saunders, Philadelphia, PA. pp. 850-872.

LEVIN, A.S. and BYERS, V.S. (1987). Environmental illness: A disorder of immune regulation. In: Workers with Multiple Chemical Sensitivities. (M.R. Cullen, ed.). Occup. Med: State Art Rev. 2(4):669-682.

LEVIN, H. (1989). Building materials and indoor air quality. In: Problem Buildings: Building Associated Illness and The Sick Building Syndrome. (J.F. Cone, ed.). Occup. Med: State Art Rev. 4(4):667-693.

NATIONAL RESEARCH COUNCIL (NRC). (1981). Indoor Pollutants. National Academy Press, Washington, DC.

NERO, A. (1988). Controlling indoor air pollution. Scientific American 258(5):42-48.

RADCLIFFE, M.J. (1984). Diagnostic use of dietary regimes. In: Food Allergy and Intolerance (J. Brostoff and S. Challacombe, eds). Bailliere Tindall/W.B. Saunders, Philadelphia, PA. pp. 806-822.

REA, W.J., JOHNSON, A.R., and SMILEY, R.E., et al. (1987a). Magnesium deficiency in patients with chemical sensitivity. Clin. Ecol. 5(1):17-20.

REA, W.J., PAN, Y., and JOHNSON, A.R. (1987b). Clearing of toxic volatile hydrocarbons from humans. Clin. Ecol. 5(4):166-170.

REA, W.J. and ROSS, G.H. (1989). Foods and chemicals as environmental incitants. Nurse Practitioner. 14(9):17-40.

REA, W.J., ROSS G.H., JOHNSON, A.R., et al. (1990). Confirmation of chemical sensitivity by means of double-blind inhalant challenge of toxic volatile chemicals. Clin. Ecol. 6(3):113-118.

ROSS, G.H., REA, W.J., and JOHNSON, A.R. (1989). A prospective cross sectional study comparing environmentally-sensitive patients and controls, measuring biochemical, hematological, immunological and toxicological parameters. Proceedings of the 7th Annual International Symposium on Man and His Environment in Health and Disease. Dallas, TX, February 23-26, 1989.

ROSS, G.H., REA, W.J., and JOHNSON, A.R. (1990). Evidence for vitamin deficiencies in environmentally sensitive patients. Clin. Ecol. 6:60-65.

SCADDING, G.K., and BROSTOFF, J. (1986). Low-dose sublingual therapy in allergic rhinitis due to house dust mite. Clin. Allergy 16:483-491.

SCHNARE, D.W. et al (1982). Evaluation of a detoxification regimen for fat storage xenobiotics. Med. Hypothesis 9:265-282.

SEBA, D.B., MILAM, M.J., and LASETER, J.L. (1987). Uptake, measurement and elimination of synthetic chemicals by man. In: Food Allergy and Intolerance, (J. Brostoff and R. Challacombe, eds.). Bailliere Tindal/W.B. Saunders, Philadelphia, PA. pp. 401-415.
SHANDRA, R.K. (1980). Immunology of nutritional disorders. In: Current Topics in Immunology. Series 12. Edward Arnold, London. pp. 1-7.
SHIRAKAWA S., REA, W.J., ISHIKAWA, S., and JOHNSON, A.R. (1992). The evaluation of the autonomic nervous system response by pupilographical study in the chemically sensitive patient. (In press).
SPRAGUE, D. (1987). The concept of an environmental unit. In: Food Allergy and Intolerance. (J. Brostoff and R. Challacombe, eds). Bailliere Tindal/W.B. Saunders, Philadelphia, PA. pp. 947-960.
YOUDIM, S., REA, W.J., and LIANG, H. (1991). Treatment of environmentally sensitive patients with transfer factor, Part 1: Immunologic studies. Clin. Ecol. 7(3):55-61.

MULTIPLE CHEMICAL SENSITIVITIES: A PRACTICING CLINICIAN'S POINT OF VIEW CLINICAL AND IMMUNOLOGIC RESEARCH FINDINGS

ALAN S. LEVIN, MD
VERA S. BYERS, MD, PhD
Immunology, Inc.
San Francisco, California

INTRODUCTION

Multiple chemical sensitivities (MCS), also known as environmental illness, is a widely recognized clinical syndrome. It is easily diagnosed by history, physical, and laboratory tests. In most cases, the physician's primary role is to rule out more serious disease. In the overwhelming majority of cases, MCS is a self-limiting process that requires little or no treatment. Treatment, when indicated, generally involves simple avoidance of the most recent inciting agent. In this paper, we will attempt to restate some of the basic tenets of an earlier discussion (Levin and Byers, 1987) and to update our research findings into this disorder.

Clinically, MCS is a syndrome characterized by neurologic, endocrine/metabolic, and immunologic symptoms. Like diabetes, systemic lupus erythematosus, atherosclerosis, and many other human diseases, it affects multiple organ systems and has a psychologic overlay. Unlike most other human diseases, its overlay provides a focus for most of its detractors. The most frequent error physicians make in evaluating psychiatric symptoms is to assume that a psychologic explanation is sufficient (Goldberg, 1983). A wide variety of physical diseases masquerade as psychiatric disorders. Clinicians may be distracted by a patient's chief complaint of anxiety, depression, or personality change and fail to pursue the possibility of organic disease (Martin, 1983).

Toxic chemicals interact with proteins and DNA in various biochemical reactions (Levin and Byers, 1987). Pathophysiologically, MCS appears to be an acquired or aggravated disorder of the neuro/endocrine/immune (biologic response modification system). Although many biochemical mechanisms can be implicated in this phenomenon, we believe the four major mechanisms are:

1) Specific interaction between a toxic chemical and a membrane receptor on the lymphoid target cell (such as the AH receptor in lymphocytes for aromatic halogenated hydrocarbons in mice) (Silkworth et al., 1986), which triggers a cascade of reactions involving cyclic AMP and calcium influx and which eventually ends in cell death through aproptosis. Loss of cell function can cause an alteration in immune regulation.

Address all correspondence to: Dr. Alan S. Levin, Immunolgy, Inc., 500 Sutter Street #512, San Francisco, CA 94102-1114.

2) Interaction of toxic chemicals with thymic epithelial cells, as seen in the mouse (Silkworth and Antrim, 1985). This interaction can produce sufficient damage to the epithelial stroma so that cells that would normally be negatively selected (autoreactive cells) are allowed to percolate through the organ and cause autoimmunity in the patient. The autoimmune disease would be more severe in children whose thymic function is most active and cause milder forms of disease in older adults whose thymic function is 1% of the child's level.

3) Interaction of toxic chemicals with critical DNA-rearranging enzyme proteins causing malfunction in immune reactive cells (Shuurman, 1991). This type of interaction can cause cell death or the generation of an oncogene, which in turn develops into an inheritable mutation and gradually into a malignancy.

4) The triggering of a previously quiescent oncogene and the gradual development of a malignancy.

DEFINITION

MCS is a syndrome characterized by symptoms referable to disorders in the neurologic, endocrine/metabolic, genitourinary, and immunologic systems. Symptoms include headaches, eye irritation, emotional irritability, short-term memory loss, and emotional depression. Other clinical manifestations may include acquired intolerance to environmental agents such as aerosolized petrochemicals and foods, and alteration in metabolic rate associated with increased body mass despite a stable or decreased caloric intake. Some patients describe abnormal menses, pain, and pelvic bloating. Patients commonly describe themselves as having been completely well until some finite time, usually punctuated by a flu-like syndrome that is worse during exposure to the polluted environment, such as the workplace, and that resolves on weekends. Patients who had a prior history of chronic disease, like diabetes or asthma, often describe significant worsening of these diseases with exposure. A more comprehensive description of the symptoms of MCS is provided elsewhere (Levin and Byers, 1987).

ETIOLOGY

As in all human diseases, the symptoms and severity of MCS are greatly dependent upon the genetic make up of the patients, their past medical history, their personal habits, and their environmental exposures. The acute onset of symptoms is most commonly associated with a viral-like illness and is often associated with significant exposure to toxic chemicals in the environment.

EPIDEMIOLOGY

Reports describing this syndrome have increased steadily in the United States over the past three decades. We believe this increased reporting has been coincident with several interdependent factors. These include the changing genetic pool of the population, the change in transmission of infectious diseases, the marked increase in industrial and domestic exposure to chemicals toxic to the human biologic response modification (neuro/endocrine/immunologic) systems, and jet travel.

Changing Genetic Pool

We now live among the first, second, and sometimes third generation of people who previously would not have survived without the use of antibiotics, insulin, sophisticated hospitalizations, and surgery. This dilutes the gene pool with individuals whose adaptive mechanisms are not as durable or as flexible as those of their ancestors.

Changing Transmission of Infectious Diseases

With the change in sexual practices, diseases that previously were unrecognized have now become epidemic. These diseases include the herpes infections, chlamydial infections, and disorders associated with retroviruses.

Increased Exposure to Toxic Chemicals

The domestic use of toxic agents, such as formaldehyde, organic solvents, halogenated hydrocarbons, organophosphates, and heavy metals, has increased dramatically since the industrial revolution. Exposure of the population to these agents has grown exponentially over the past several decades so that today as few as 5000 people in the entire world may not have detectable halogenated hydrocarbons in their bodies (Nader et al., 1981). A very large percentage of the chemicals in common use today have proven toxic to the biologic-response-modification system (neurologic/immunologic/endocrine) *in vitro* and *in vivo* in humans and animals.

Jet Transportation

With the advent of widespread world travel, we find ourselves trading pathogens with the world. In the past, any individual who contracted a disease while traveling in a distant foreign country would manifest the illness before having time to arrive back home. Today an individual could contract an endemic viral illness in Saudi Arabia and be in New York before becoming acutely ill. The average clinician practicing in the United States today must consider previously obscure etiologic agents in the differential diagnosis of patients.

CLINICAL MANIFESTATIONS

Like diabetes, systemic lupus, and syphilis, MCS has polymorphic clinical manifestations. The common thread is the finite time before which the patient reported being well and after which the above symptoms began. This time is usually punctuated by an acute flu-like syndrome characterized by headache, irritability, and emotional lability. If the toxic substance is airborne, the patient may describe eye and nasorespiratory irritation. If it is water borne, the patient may describe skin rashes and changes in bowel function. The physical examination may reveal anterior and posterior cervical, axillary, and inguinal lymphadenopathy. In our experience, fundoscopy may reveal vascular changes we feel are consistent with chronic toxic chemical poisoning. We call this "silver wiring" because the vessels appear similar to those seen in grade II to III Keith-Wagner-Baker retinopathy without the hemorrhages or exudates seen in hypertension. The thyroid may be enlarged. Examination of the cardiovascular system may reveal mitral-valve prolapse. Some patients will exhibit Raynauds phenomena. Hepatosplenomegaly is uncommon in this population. The balance of the physical examination is generally not remarkable.

LABORATORY FINDINGS

The laboratory tests we find most helpful in the diagnosis of MCS include complete blood counts (CBC), chemistry panels, erythrocyte sedimentation rate, immune profiles, and autoantibody titers.

In the CBC, megaloblastic changes are not uncommon. These may be due to anti-gastric parietal cell antibodies and may be harbingers of B12 or folate deficiency. Lymphopenia is a common finding in individuals with disorders of immune regulation due to chronic toxic chemical poisoning. Chemistry panels generally do not reveal significant abnormalities. Elevations in liver enzymes may, however, be noted in the acute phase of severe toxic chemical exposures.

The autoantibodies we use in assessing patients with MCS are:

Rheumatoid Factor (RF) - Found in patients with chronic autoimmune tissue damage. It is seen in a large proportion of adults with rheumatoid arthritis but is not diagnostic of that condition.

Anti-Nuclear Antibodies (ANA) - Indicative of chronic tissue damage. Positive in almost all systemic lupus erythematosus (SLE) patients and most other patients with chronic inflammatory disease.

Anti-Double Stranded DNA Antibodies (DSDNA) - Diagnostic of SLE.

Anti-SM Antigen Antibodies (SM antibodies) - Diagnostic of SLE.

SS Antibodies - Helpful in diagnosing Sjogren's syndrome.

Anti-Ribonuclear Proteins Antigens (RNP Antibodies) - Helpful in diagnosing SLE and other rheumatic disorders.

Serum Complement (C3) Tests - C3 is consumed in many autoimmune disorders. Because it is an acute-phase reactant, increased consumption often triggers overproduction of this protein. Thus, both high and low levels of C3 are diagnostic of a complement-consuming, immune-complex-mediated disorder.

Anti-Thyroid Antibodies (ATA): Microsomal and Thyroglobulin - We often find elevated ATAs in patients with multiple chemical sensitivities. An 11% incidence of ATAs has been reported in workers exposed to PCB and PBB oxides (Satran et al., 1987). They are indicative of chronic inflammation in the thyroid. Biopsy confirms the chronic autoimmune thyroiditis (Hashimoto's thyroiditis). The symptoms of thyroiditis far precede the reduced production of thyroid hormone (T4). Normal levels of thyroid hormone do not rule out autoimmune thyroiditis. A diagnostic trial of thyroid hormone therapy is indicated if autoimmune thyroid disease is suspected, even in the face of normal laboratory parameters.

Anti-Smooth Muscle Antibodies - Commonly found in patients with chronic active hepatitis. They are also seen in a very large percentage of patients with chronic, toxic, chemical poisoning.

Anti-Mitochondrial Antibodies - Found in patients with primary biliary cirrhosis (PBS). PBS can be triggered by exposure to toxic chemicals.

Anti-Gastric Parietal Antibodies - Found in patients who have atrophic gastritis. These patients tend to develop megaloblastic anemia due to B12 deficiency.

Anti-Adrenal Antibodies - Found in patients with autoimmune adrenal disease. Autoimmune adrenal disease can be triggered by toxic chemical exposure.

Anti-Ovarian Antibodies - Found in patients with autoimmune oophoritis. Autoimmune oophoritis can be triggered by exposure to toxic chemicals.

Anti-Testicular Antibodies - These are the male analog of anti- ovarian antibodies. This disorder can also be triggered by toxic chemical exposure.

Other Autoantibodies - These include anti-reticulin, anti-brush border, and anti-basement membrane antibodies and are usually found in patients with autoimmune renal disorders.

IMMUNE PROFILES

B and T Cells with T-Cell Subsets

This profile measures the different subpopulations of immune reactive cells. It gives a measure of the absolute levels of each population and the homeostatic relationship between these cells in the form of ratios. The most common ratio used is the CD4+/CD8+ (helper/suppressor) ratio (H/S). The immune profile and H/S are stable measurements in individuals with values in the normal range (Lifson et al., 1985). Instability or abnormal values are very unusual in the normal human population. We find that patients with severe MCS have unstable immune profiles. Early in the symptomatic portion of the disease, H/S often varies erratically. In individuals with severe chronic disease, the H/S stabilizes to a low value months or years after initial exposure. Individual patients with MCS sometimes have inverted H/S due to an excess of CD8+ cells. Multipatient studies show statistically significant increases in CD8+ cells or reduction in the H/S in cohort populations exposed to toxic levels of formaldehyde (Pross et al., 1987), pesticides (Fiore et al., 1986), and herbicides (Levin and Byers, 1987) compared with control populations. Appropriate treatment and recovery from the syndrome are marked by a reduction of the CD8+ cells and a normalization of the H/S. The B- and T-cell profiles are very helpful in diagnosing, treating, and monitoring MCS.

DISORDERS OF IMMUNE REGULATION

Upon initial evaluation, we find that MCS patients often have a form frust of SLE, rheumatoid arthritis, autoimmune thyroiditis, or scleroderma. The diagnosis of these disorders is made by careful and thorough history (with special attention to family history), physical examination, and laboratory testing. Occasionally, tissue biopsies are helpful in making a definitive diagnosis.

Less often, we find that patients who present with the symptoms of MCS have serious systemic illness such as cancer, autoimmune diseases, and endocrine/metabolic disorders. In these cases, the symptoms of MCS precede the discovery of the more serious disease.

In our experience, MCS patients, when followed for long periods, often develop cancer or severe autoimmune disease. Over the two decades we have followed these patients, it is our clinical experience that MCS patients with persistent symptoms often go on to develop cancer, systemic lupus, multiple sclerosis, or adult onset diabetes. In our opinion, these patients have a genetic propensity to a specific illness, which was triggered by an insult from an environmental agent. The first symptoms of this triggering were MCS.

TREATMENT

After diagnosis of MCS has been confirmed and the etiology has been assessed, a treatment protocol must be established. The first and most important treatment is avoidance. Individuals must be removed from the insulting environment. In our experience with over 1000 adult patients, we find that, with treatment and the passage of time, more than 85% of patients recover and resume normal lives within 18 months.

Our treatment protocol involves the integrated use of multiple therapeutic modalities. These include diet and environment control coupled, when indicated, with antiviral treatment, antifungal treatment, appropriate hormonal treatments, and immune modulator treatments.

Diet and Environment Control

Diet and environment control are major components in the treatment of all human illnesses. When the human immune system is damaged by any agent, the best and easiest treatment is to take the pressure off. This can be done by removing major offensive agents from the diet and environment. Diet control involves avoiding foods that may evoke adverse reactions. In the average American diet, the major offenders are foods that individuals eat day in and day out.

These include sucrose, cow's milk and cow's milk products, wheat, corn, and processed foods. For individuals who consume an oriental diet, the offending agents are often rice and soy. For Middle Easterners, the offending agent can be goats milk. In the case of MCS, environment control involves the avoidance of excessive exposure to toxic chemicals such as organic solvents, cigarette smoke, industrial petrochemicals, pesticides, and herbicides. More often than not, diet and environment control alone will successfully return the patient to normal function.

Antiviral Treatment

In our experience, we find that antiviral treatment appears to have a positive benefit in roughly 20% of patients who describe the symptoms of a chronic viral syndrome triggered by toxic chemical exposure. The three antiviral agents used in this population are acyclovir, amantidine, and gancyclovir.

Acyclovir (Zovirax) is a fallacious nucleotide analog originally approved for the treatment of Herpes Simplex II disease. It is also active against other Herpes viruses including CMV and Zoster. When used at 200 mg five times a day, we have found that it can effect a positive benefit in approximately 20% of patients who describe the symptoms of a chronic viral syndrome triggered by toxic chemical exposure. Rarely, a loading dose of 400-600 mg five times a day for one week produces dramatic improvement. The major side effect of this drug is nausea. Occasionally, serious cases warrant intravenous treatment, 5 mg/kg TID IV for 5 days followed up by 200 mg PO five times per day. At very high doses (4+ grams per day intravenous) renal toxicity is noted.

Amantidine (Symetril) is a symmetric amine that inhibits an early step in the replication of the influenza virus. It is used in a dose of 100 mg BID. Although rarely effective in patients who describe viral syndromes triggered by toxic chemicals, its safety and cost warrant a 48-72 hour trial treatment course.

Gancyclovir (DHPG) is an antiviral agent with proven effectiveness against CMV and other herpes viruses. Used at 5 mg/kg IV TID, it can completely eliminate CMV retinitis and pneumonia in immune compromised patients. Because of its severe myelotoxicity and the need for long-term intravenous access, it is used only sparingly in patients who describe viral syndromes triggered by toxic chemicals.

Antifungal Treatments

Mold allergy is a very common cause of chronic symptomatology (Kroker, 1987). The vast majority of molds to which the average patient is exposed are body molds (Rippon, 1982). Reduction in the population of body molds, primary *Candida albicans*, can effect a positive benefit in many patients whose immune systems appear to have been damaged by toxic chemicals. The first step in this process is appropriate diet control. Because the Candida organism feeds on sucrose, refined sugars should be totally eliminated from the diet. Antifungal drug therapy can then be instituted.

Nistatin (Nilstat) is an antifungal agent derived from the Streptomyces Narci. It is not absorbed from the gastrointestinal (GI) tract so it has few if any side effects. Used at 8 million International Units per day (4 pills qid or 1/2 tsp pure powder qid) with appropriate diet control, it can effect a positive response in 30-90 days. Diet control is the key to success in this treatment. Because this drug is fungolytic, Jarisch/Herxheimer reactions are common when this drug is used appropriately. This reaction usually starts on day 4 of treatment and resolves by day 10. The presence of a Jarish/Herxheimer reaction may be clinical confirmation of the disease. Symptoms that prevail into the third week of treatment may be those of an allergy to the Nilstat itself. Because it is the derivative of a mold, it can cause mild GI tract allergic reactions with prolonged use. The usual treatment course is 60-90 days. Although controversy has been raised in the medical literature regarding this diagnosis and treatment (Dismukes et al., 1990), we believe it should be tried as a first step in the treatment course of many patients with chronic dermatitis and GI symptomatology as well as chronic fatigue patients. This is because a large body of very credible practicing clinicians report positive clinical benefits, and because it is very safe and inexpensive (Bennett, 1990). We like to call it a simple solution to a complex question.

Amphotericin B (Fungazone) is an antifungal agent derived from the Streptomyces Orci. Like Nistatin, it is not absorbed from the GI tract so it has few if any side effects. Used in doses of 250 mg PO qid with appropriate diet control, it also can effect a positive response in 30-90 days. Because it, too, is fungolytic, Jarisch/Herxheimer reactions are common. The drug is manufactured in New Jersey and shipped to Europe and Asia where it has been used widely for decades. Because of a limited market, the manufacturer has elected not to spend the hundreds of millions of dollars necessary to seek United States Government Food and Drug Agency (US FDA) approval. It is therefore not approved by the US FDA and must be obtained from foreign pharmacies where it is readily available as an over-the-counter drug. Because it, too, is the derivative of a mold, it can cause mild GI-tract allergic reactions with prolonged use. The usual treatment course is 60-90 days.

Ketoconazole (Nizoral) is a potent systemic fungostatic agent. It is an imidizole drug similar to Metronidizole (Flagyl) and Levamizole. This drug is not only an antifungal agent but binds plasma cortisols, testosterone, estrogen, and progesterone. It is also a potent immunomodulating agent and is used as an anticancer drug in humans and animals. We find it the single most effective drug in the treatment of chronic viral and fatigue syndromes. Because it appears to effect a positive benefit long after the fungus population has been reduced, its therapeutic benefit may possibly be derived from its immunomodulatory activity or its anti-hormone effects. Of approximately 36 million persons who have taken the drug, 1 in 15,000 develop transient elevation in liver enzymes. Eight patients, one of whom was ours, have developed protracted elevations in liver enzymes between 5 weeks and 6 months after stopping the drug.

The literature cites three deaths from Nizoral, all from liver failure and all in post-menopausal females. None of these patients had liver-function tests monitored during therapy. Our clinic has hundreds of patients taking Nizoral with no problems. We require liver-function panels every two weeks during drug therapy. We discontinue the drug with any slight increase in enzymes and do not renew prescriptions unless the patient has current liver-function tests on file.

Fluconazole (Diflucan) is a new imidizole antifungal agent that is more potent than Nizoral in this respect. It does not bind to plasma sterols and is not known to be an immune modulator. Thus, it is effective only in those patients with relatively pure candida problems. The dose is 200 mg per day, and liver function should be followed every six weeks.

Hormonal Treatments

Thyroid hormone treatments often effect a positive benefit in patients whose immune/endocrine systems have been damaged by toxic chemical exposure. Because many of these patients have evidence of autoimmune thyroiditis (AIT), therapeutic trial with thyroid hormone is warranted. The available medications include synthetic and natural preparations. The synthetic preparations are often preferred by endocrinologists because their quality control is superior to natural preparations. One synthetic preparation, Synthroid (Flint Pharmaceuticals) has acacia gum as a filler. Because many California residents are allergic to the acacia tree, this drug must be used with some caution in California residents. The symptoms of AIT are of chronic inflammation and not of thyroid hormone deficiency. Thus, the treatment can be divided into two phases, supplemental and suppressive.

Supplemental treatment usually involves a dose of 0.05 mg per day. This will supplement the damaged gland and allow it to heal. We liken this type of treatment to putting an Ace Wrap on a sprained ankle. If after 4-6 weeks at that treatment dose, no improvement is noted, the dose is increased to 0.1 mg per day. At this dose the gland may shut down and the patient may experience increased fatigue. At this point, usually 4-6 weeks after increasing the dose, the dose may be increased to 0.125 or 0.15 mg per day. At this level the thyroid is suppressed. Because the gland can easily be brought back into function at any time, the temporary shut down of thyroid function is innocuous. The dose is then titrated between 0.125 and 0.3 mg per day with increased increments of 0.025 mg per day every 2-4 weeks until the ideal dose is reached. This suppressive therapy is likened to putting a cast on a sprained ankle. In most cases, the thyroiditis may be scarred down and quiescent after 18 months so that treatment can be discontinued. The treatment can be tapered down or cut off "cold turkey." In either case, the side effects of treatment discontinuation are minimal.

Sex Hormone Treatment

Females. Many female patients with toxic chemical damage to their immune/endocrine systems suffer from autoimmune polyendocrinopathy syndrome. In addition, recent research in osteoporosis confirms the hypothesis that all perimenopausal females should be treated with supplemental estrogen (Sowers et al., 1991) and probably progesterone (Gallagher et al., 1991; McNeely et al., 1991). Individuals have been shown to have aberrant secretion of estrogen from their ovaries as much as two years before menopause (McNeely et al., 1991). Much of the bone loss these individuals suffer occurs in the premenopausal period (Aloia et al., 1990; Rodin et al., 1990). Thus, many patients benefit from estrogen or progesterone replacement treatment. In selected patients, we use Estroderm patches at 0.05 mg or 0.1 mg throughout the month with provera at 10 mg per day for the last 10 days of the month. Some female patients benefit from occasional testosterone injections (50-100 mg IM every month for a maximum of three injections) or Buccal Android (5 mg per day). These anabolic steroids act as immune- and hematopoietic-system stimulants.

Males. Many male patients exposed to toxic chemicals suffer from the autoimmune polyendocrinopathy syndrome. Assessing testosterone and DHEA levels in these patients is often helpful. Frequently patients are found to have low or low-normal levels of these hormones. We have had considerable success using testosterone injections (150 mg IM q 2 wks) or Buccal Android (10 - 20 mg q d) as well as DHEA (50 mg qid) as a booster of the bone marrow, libido, and general energy levels. The side effects of these treatments are minimal. One study reported hepatic tumors in one strain of rats treated with high doses of human testosterone. To our knowledge, this finding has never been repeated in that strain of rats and has not been detected in other strains or species of animal. Weight gain and increased risk of atherosclerotic disease are considerations that can be obviated by diet control. In general, patients should not be kept on this type of therapy for more than 3 months.

Immunomodular Therapy

Intravenous Gammaglobulin. Gammaglobulin is that portion of the plasma that contains antibodies. In the early 1940s, the alcohol-fractionation methods Cohn developed for human plasma proteins allowed the isolation of gammaglobulin preparations for therapeutic use (Cohn et al., 1946). This early gammaglobulin preparation was an aggregated and fixed complement so that when given intravenously, it caused anaphylaxis. When given intramuscularly, however, it would be disaggregated by proteolytic enzymes released by the damaged muscle cells so that native gammaglobulin could slowly percolate into the blood stream with no significant side effects. Because the preparation was given intramuscularly, the dose was limited to a volume that could comfortably be given in muscle. Initially doses of 1-2 ml IM were the norm; now we give 10 ml IM routinely.

Because children with congenital agammaglobulinemia could be kept alive with regular gammaglobulin injections, and because intramuscular injections are painful and dose limited, pediatric immunologists implored the manufacturers of gammaglobulin to produce an intravenous (IV) preparation. Researchers developed a variety of techniques to disaggregate gammaglobulin, including reduction and alkylation (Fernandes and Lundblad, 1980), plasmin treatment (Sgouris, 1967), and sulfonation (Hofstaetter et al., 1983). Therapeutically effective

IV gammaglobulin finally became available in the late 1970s. Initially, gammaglobulin was thought to be useful only as a prophylactic treatment of infectious disease

IV gammaglobulin has been effective as a primary treatment for a variety of chronic infectious diseases and as augmentation to antibiotic treatment. Recent advances in immunology have changed our understanding of the role of gammaglobulin. Evidence is gathering that the primary role of gammaglobulin is not to protect the host from pathogens in the environment but to act as a regulatory molecule, a hormone. Gammaglobulins, through the immune network of idiotypes (Ab1), anti-idiotypes (AB2), and anti-anti-idiotypes (Ab3), serve as an interactive matrix of regulatory molecules that control the growth and differentiation of almost every organ derived from the mesodermal layer of the embryo. Thus, high doses of gammaglobulin have been safe and effective treatment for the control of the aberrant immune responses seen in disorders of immune regulation. Gammaglobulin has been a safe and effective treatment for most disorders of immune regulation. It is superior to steroids and cytotoxic agents for controlling most serious autoimmune disorders. Below is a partial list of the disorders of immune regulation amenable to IV gammaglobulin therapy.

Allergic Asthma
Aplastic Anemias
Autoimmune Adrenalitis
Autoimmune Arteritis (Kawasaki Syndrome)
Autoimmune Anemias
Autoimmune Oophoritis
Autoimmune Thyroiditis (Hashimoto's Disease)
Autoimmune Thrombocytopenia (ITP)
Crohn's Disease
Epilepsy
Human Adjuvant Disease (Silicosis from breast implants)
Immune Complex Mediated Complement Consuming Vasculitis
Irritable Bowel Syndrome
Multiple Sclerosis
Multiple Chemical Sensitivities
Rheumatoid Arthritis
Sarcoidosis
Scleroderma
Severe Food Allergies
Severe Inhalant Allergies
Sickle Cell Anemia
Systemic Lupus Erythematosus
Ulcerative Colitis
Vasculitis

The major side effects of gammaglobulin therapy are rare allergic reactions to the fragments of protein left behind in the disaggregation process. These reactions are self limited and easily treated. Occasionally patients describe headaches when the material is infused too rapidly. Although gammaglobulin is a blood product, it cannot transmit AIDS (Mirta et al., 1986; Kane, 1984). This has been shown by several experiments as well as a long history of lack of viral illness transmission (Greico, 1984). Rare anaphylactic reactions have been reported with gammaglobulin treatment (Buckley et al., 1991).

Thus, the therapeutic uses of IV gammaglobulin preparations are:

1) Antibody substitution in uses of generalized or partial antibody deficiency in immune-compromised patients. These include patients with congenital and acquired immunodeficiencies, agammaglobulinemia, hypogammaglobulinemia, immunoglobulin sub-class deficiencies, and specific immune incompetence (selective or idiotype deficiency). Acquired deficiencies may be induced by drugs, toxic chemicals, and viral immunosuppression.

2) Modulation of the immune system by immunoglobulin-immunoglobulin interactions (mainly idiotype-anti-idiotype interactions) and Ig-Fc-receptor interactions, as it is known from the RES blockade in the treatment of autoimmune thrombocytopenia, ITP (Bussel, 1989).

Transfer Factor

Transfer factor was first described in 1955 by Sherwood Lawrence at NYU (Lawrence, 1955). He demonstrated that one could passively transfer antigen-specific cellular immunity using a dialyzable extract of lymphocytes. Our group first used it in 1969 to successfully treat children with congenital immunodeficiencies (Levin et al., 1970) and cancer (Levin et al., 1975).

Transfer factor is an extract of T lymphocytes, which is a mixture of immune modulators with molecular weights of 15,000 Daltons and less. Thus, it contains fragments of helper cell (CD4) receptors, fragments of suppressor-cell receptors (CD8), non-specific suppressor substances, interleukin II (IL2), colony-stimulating factors (GMCSF, GCSF), fragments of interferon, and other unidentified immune modulators. When used in serious T cell deficiencies, it often restores immune functions as measured *in vivo* by skin tests and *in vitro* by blastogenesis assays (Levin, 1987).

Regular transfer factor therapy has been shown to reduce T-cell loss, morbidity, and mortality in HIV disease (Louie et al., 1987; Carey et al., 1987).[1] Transfer factor has proven safe and effective as prophylaxis and therapy for bacterial (Levin, 1987), viral (Steele et al., 1980), and

---

[1] Waites, L.A., Byers, V.S., and Levin, A.S. Transfer factor treatment in HIV disease (manuscript in preparation).

protozoal (Sharma et al., 1979) infections in multiple controlled studies published in prestigious medical journals.

The side effects of transfer factor therapy are enhancement of the immune response. Transfer factor cannot transmit AIDS and has never transmitted a viral illness. The material is repeatedly frozen and lyophilized and dialyzed to a molecular weight of 15,000 Daltons or less. The HIV virus is 250,000,000 Daltons. For an HIV virus to come though multiple dialysis membranes is akin to a whale coming through a fly screen. In addition, the material is derived from HIV- and hepatitis-screened blood. If, however, an individual is in peaceful co-existence with an organism somewhere in his body, such as the brain, heart, or liver, an enhanced immune response against that organism may cause a deleterious reaction in that organ. In the brain, it could cause a space-occupying lesion, in the liver it could increase hepatitis, in the lungs it could enhance pneumonitis, and in the muscles it could enhance myositis. These adverse reactions, however, are exceedingly rare. The cells enhanced by transfer factor are exquisitely sensitive to steroids so that one 60 mg. dose of prednisone orally has successfully "turned off" the adverse reactions in every case. Our group has given more than 25,000 doses of transfer factor and only had to "turn off" the transfer factor reaction five times. Each time was in patients with breast or renal cell carcinoma metastatic to the brain. Although the reactions were successfully stopped in all cases, the patients later succumbed to their cancers. All things considered, transfer factor is an exceedingly benign T-cell stimulator.

Thus, transfer factor is a low-molecular-weight immune modulator that has proven safe and effective treatment in the partial restoration of T-cell immune reactivity in disorders of immune regulation, malignant diseases, chronic viral illness, and chronic bacterial infections.

Low-Dose Oral Interferon

Interferon is an immune modulator shown to enhance immune reactivity in a variety of infectious and malignant disorders. It has several "windows of reactivity" so that very small doses have a greater immune modulatory effect than larger doses. Many reports have described spectacular benefits from therapy with very low doses of Interferon (2 units/kg) given orally. Although we have used this material and not seen such responses, some selected CFIDS patients have reported increased energy and reduction of apthus stomatitis lesions apparently attributable to low oral interferon therapy.

SUMMARY

The multiple chemical sensitivities syndrome is a symptom complex characterized by emotional depression, short-term memory loss, acquired intolerance to environmental agents such as aerosolized petrochemicals and foods, and alteration in metabolic rate associated with increased body mass. This syndrome can be caused by multiple etiologic agents. With careful evaluation, it is rare that a causal agent can not be identified in this syndrome. Our treatment regimens include combinations of diet and environment control as well as antiviral, antifungal, and immune modulator therapy. We find that the majority of patients can, through time and appropriate treatment, be restored to a normal and productive lives.

REFERENCES

ALOIA, J.F., VASWANI, A., ROSS, P., and COHN, S.H. (1990). Aging bone loss from the femur, spine, radius, and total skeleton. Metab. Clin. Exp. (USA) 39(11):1144-1150.

BENNETT, J.E. (1990). Searching for the yeast connection. N. Engl. J. Med. 323(25):1766.

BUCKLEY, R.H. and SCHIFF, R.I. Drug therapy: The use of intravenous gammaglobulin in immunodeficiency diseases. (1991) N. Eng. J. Med. 325(2):110-117.

BUSSEL, J.B. (1989). The use of intravenous gamma-globulin in idiopathic thrombocytopenic purpura. In: Autoimmunity and Immunointervention. (E.W. Gelfand, H.D. Ochs, and E.R. Stiehm, eds.), 53(2):S147-S155.

CAREY, J.T., LEDERMAN, M.M., and TOOSSI, Z. (1987). Augmentation of skin test reactivity and lymphocyte blastogenesis in patients with AIDS treated with transfer factor. J.A.M.A. 257(15):651-655.

COHN, J.E., STRONG, L.E., HUGHES, W.L., Jr., MULFORD, D.J., ASHWORTH, J.N., MELIN, M., and TAYLOR, H.L. (1946). Preparation and properties of serum and plasma proteins IV. A system for the separation into fractions of the protein and lipoprotein components of biological tissues and fluids. J. Am. Chem. Soc. 68:459-475.

DISMUKES, W.E., WADE, J.S., LEE, J.Y., DOCKERY, B.K., and HAIN, J.D. (1990). A randomized, double-blind trial of nystatin therapy for the Candidiasis hypersensitivity syndrome. N. Engl. J. Med. 323(25):1717-1722.

FERNANDES, P.M. and LUNDBLAD, J.L. (1980). Preparation of a stable intravenous gamma-globulin: Process design and scale-up. Vox Sang. 39:101-112.

FIORE, M.C., ANDERSON, H.A., HONG, R.A. et. al. (1986). Chronic exposure to aldicarb-contaminated groundwater and human immune function. Environ. Res. 41:633-645.

GALLAGHER, J.C., KABLE, W.T., and GOLDGAR, D. (1991). Effect of progestin therapy on cortical and trabecular bone comparison with estrogen. Am. J. Med. 90(2):171-178.

GOLDBERG, R.J. (1983). Psychiatric symptoms in cancer patients. Is the cause organic or psychologic. Postgrad Med 74(1):263-273.

GREICO, M.H.(1984). Risk of transferring AIDS or hepatitis via immune globulin. J.A.M.A. 24(31):256.

HOFSTAETTER, T., GRONSKI, P., KANZY, E.J., SCHORLEMMER, H.U., and SEILER, F.R. (1983). S-sulfonation: A reversible chemical modification of human immunoglobulins permitting intravenous application. II. Effect on Fc-mediated effector functions. Vox. Sang. 45:155-165.

KANE, M.A.(1984). Risk of transferring AIDS or hepatitis via immune globulin. J.A.M.A. 252(8):256.

KROKER, G.F.(1987). Chronic candidiasis and allergy. In: Food Allergy and Intolerance. (J. Brostoff and S.J. Challacombe, eds.). Bailliere Tindall, London. pp. 850-870.

LAWRENCE, H.S. (1955). The transfer in humans of delayed skin sensitivity to streptococcal M substance and to tuberculin with disrupted leucocytes. J. Clin. Invest. 34:219-225.

LEVIN, A.S. (1987). Transfer factor and allergies. In: Food Allergies, (J. Brostoff, and S.J. Challacombe, eds.). Bailliere Tindall, London. pp.995-1004.

LEVIN, A.S., BYERS, V.S., FUDENBERG, H.H., JOHNSTON, J.O., and WYBRAN, J. (1975). Osteogenic sarcoma: Immunologic parameters before and during therapy with tumor specific transfer factor. J. Clin. Invest. 55:487-494

LEVIN, A.S., SPITLER, L.E., STITES, D.P., and FUDENBERG, H.H. (1970). A genetically determined cellular immunologic deficiency: Clinical and laboratory responses to therapy with transfer factor. Proc. Natl. Acad. Sci. USA 67:821-829.

LEVIN, A.S. and BYERS, V.S. (1987). Environmental illness: A disorder of immune regulation. Occup. Med.: State Art Rev. 2:(4)669-681.

LIFSON, J.D., FINCH, S.L., SASAKI, D.T., and ENGLEMAN, E.G. (1985). Variables affecting T-lymphocyte subsets in a volunteer blood donor population. Clin. Immun. Immunopath. 36:151-160.

LOUIE, E., BORKOWSKY, W., and KLESIUS, P.H. (1987). Treatment of cryptosporidiosis with oral bovine transfer factor. Clin. Immunol. Immunopathol. (USA) 44(3):329-334.

MARTIN, M.J. (1983). A brief review of organic diseases masquerading as functional illness. Hosp. Commun. Psych. 34:(4)328-332.

McNEELY, S.G. Jr., SCHINEFELD, J.S., STOVALL, T.G., LING, F.W., and BUXTON, B.H. (1991). Prevention of osteoporosis by medroxyprogesterone acetate in postmenopausal women. Int. J. Gynecol. Obstet. 34(3):253-256

MIRTA, C., WONG, M.F., MOZEN, M.M., MCDOUGAL, J.S., and LEVY, J.A. (1986). Elimination of infectious retroviruses during preparation of immunoglobulins. Transfusion 26(4):394-397.

NADER, R., BROWNSTEIN, R., and RICHARD, J. (1981). Who's Poisoning America. Sierra Books, San Francisco.

PROSS, H.F., DAY, J.H., CLARK, R.H., and LEES, R.E.M. (1987). Immunologic studies of subjects with asthma exposed to formaldehyde and urea-formaldehyde foam insulation (UFFI). J. Allergy Clin. Immunol. 79:797-810.

RIPPON, J.W. (1982). Medical Mycology, 2nd ed. W.B. Saunders, Philadelphia, PA. pp. 507-508.

RODIN, A., MURBY, B., SMITH, M.A., CALEFFI, M., FENTIMAN, I., CHAPMAN, M.G., and FOGELMAN, I. (1990). Premenopausal bone loss in the lumbar spine and neck of femur: A study of 225 caucasian women. Bone (USA) 11(1):1-5.

SATRAN, M., PAUL., T.L., ROTI, E., and BRAVERMAN, L.E. (1987). Environmental factors affecting autoimmune thyroid disease. Endocrinol. and Metab. of No. America, pp. 327-338.

GOURIS, J.T. (1967). The preparation of plasmin-treated immune serum globulin for intravenous use. Vox. Sang. 13:71-84.

SHARMA, M., MOMTAZ, A., FIROUZ, R., and ALA, F. (1979). Transfer factor therapy of human cutaneous leishmania infection (CLI). A double blind clinical trial. In: Immune Regulators in Transfer Factor, (A. Kahn, C.H. Kirkpatrick, and N.O. Hill, eds.). Academic Press, New York. pp. 563-571.

SHUURMAN, H.(1991). Chemicals trophic for thymus and T cells: Risk for immunodeficiency and autoimmunity. Proceedings of 5th International Congress on Immunopharmacy. May 26-31.

SILKWORTH, J.B., ANTRIM, L., and SACK G. (1986). AH receptor mediated suppression of the antibody response in mice is primarily dependent on the AH phenotype of lymphoid tissue. Toxicol. Appl. Pharmacol. 86(3):380-390.

SILKWORTH, J.B. and ANTRIM, L. (1985). Relationship between AH receptor-mediated polychlorinated biphenyl-induced humoral immunosuppression and thymic atrophy. J. Pharmacol. Exp. Ther. 235(3):606-611.

SOWERS, M, CLARK, K., WALLACE, R., JANNAUSCH, M., and LEMKE, J. (1991). Prospective study of radial bone mineral density in a geographically defined population of postmenopausal caucasian women. Calcif. Tissue Int. 48(4):232-239.

STEELE, R.W., MYERS, M.G., and VINCENT, M.M. (1980). Transfer factor for the prevention of varicella-zoster infection in childhood leukemia. N. Engl. J. Med. 303:355-365.

DIAGNOSIS AND TREATMENT OF HYPERSENSITIVITY SYNDROME

DON L. JEWETT, MD, DPhil
Professor Emeritus, Department of Orthopaedic Surgery
University of California, San Francisco
Director of Research, Abratech Corporation, Mill Valley, California

INTRODUCTION

The interest in symptoms of hypersensitivity syndrome, as evidenced by this workshop on multiple chemical sensitivity, is heartening because this symptom complex deserves study as a significant factor in modern medical practice. Reasons for this statement and reasons for avoiding labels that imply chemical causation of these symptoms are given elsewhere in this volume (Jewett, 1992). Persons engaged in occupational health with an interest in patients who complain of adverse responses to chemicals may logically hypothesize that chemicals are the primary cause of the syndrome. But the long history of failed hypotheses regarding this clinical presentation (Layzer, 1991) will cause the careful investigator to be cautious about ascribing cause and interpreting the significance of success based on uncontrolled clinical experience.

Some observations that should alert the unwary investigator have been presented at this workshop. First, the typical multiple chemical sensitivity patient is an educated, white, middle-aged female—hardly a group likely to be differentiated from others in our society on the basis of chemical exposures. Second, despite decades of advocacy, proponents of a chemical cause for the symptoms of hypersensitivity syndrome have yet to produce adequate scientific evidence to support their claims, and a well-controlled, double-blind study of their own patients in their offices completely failed to confirm their method of detecting sensitivity to foods and chemicals (Jewett et al., 1990). Finally, the physicians in this workshop who presented papers on treating hypersensitivity syndrome patients each report success with a completely different therapeutic regimen! Ross (1992) and Ziem (1992) each report that avoidance of chemicals is needed; Simon (1992) suggests that exposure to chemicals should be gradually increased; Selner and Staudenmayer (1992) report that a large percentage of such patients were sexually abused as children (although repeated questioning over many months of treatment was needed before the alleged incident was recalled). This paper will report that success depends on the author's beliefs at the time of treatment (which must be one of the hypotheses considered for explaining such varied treatments of a supposed chemical disorder).

Address all correspondence to: Dr. Don L. Jewett, Abratech Corporation, 150 Shoreline Hwy., Bldg. E, Mill Valley, CA 94941-6610.

BACKGROUND

Some may find it unusual that an orthopaedic surgeon has had clinical and investigative experience with the hypersensitivity syndrome. My experience occurred in three ways: as a patient, as a treating physician, and as a scientific investigator. As a patient, I have had lifelong allergies, which became worse during an especially difficult period of my life. Traditional allergy shots produced some mild, but unsustained, improvement. I then started treatment under a clinical ecologist and ultimately spent five weeks in a Dallas environmental control unit. I ended with a diagnosis of "universal reactor," 70 different injection substances, and a change (but not a decrease) in symptoms (including hyperosmia).

After this I began seeing patients in a "sensitivity clinic" that I ran within the Orthopaedic Clinic at the University of California, San Francisco. I treated patients with diet changes (four-day rotation diet). Some patients seemed to experience significant positive effects (Jewett, 1983). Such results encouraged me to formulate a long-range plan to definitively prove the hypothesis that small amounts of substances to which we are commonly exposed can result in symptoms (at least in some patients at some times). Because the expected positive result was likely to have a significant impact on medical practice, I devised a scientifically rigorous study that both proponents and opponents agreed was a fair and appropriate double-blind test of symptom provocation by injections. However, the study showed provocation testing to be totally unreliable as a method for determining sensitivity (Jewett et al., 1990), and so I began to doubt the validity of this diagnostic and therapeutic approach. (Coincidentally, my own symptoms were still present and unsuccessfully treated—to the extent that the treatment actually seemed to perpetuate the problems.) Because of my reputation among the patient community, patients still came to me for diet therapy. Despite my internal doubts, I gave them the same written material for the diet that I had given previous patients, but the therapeutic effectiveness of the diets was noticeably less. Ultimately I abandoned diets altogether and used the methods described below. I saw an estimated 250 hypersensitivity syndrome patients over a ten-year period.

My role as an investigator is limited to the above-cited study because proponents immediately dropped association with me when the negative results were presented to a national meeting of the Society for Clinical Ecology. Over time, my own symptoms have lessened markedly and return as allergic reactions only during periods of particular stress. Of course, this personal experience, along with the scientific evidence, influences my therapeutic approach to hypersensitivity syndrome patients.

DIAGNOSIS

The diagnosis of hypersensitivity syndrome is made by excluding all other diagnoses but requires the presence of symptoms in multiple-organ systems that cannot be diagnosed otherwise. The history usually reveals one or more of the following characteristics (as seen in an orthopaedic practice).

1. No diagnosis has been established despite adequate study by competent specialists.
2. The cause of symptoms seems non-mechanical, and no single diagnosis accounts for the symptomatology.
3. The symptoms are migratory, either within the presenting complaint ("Some days it's my right leg that hurts, and some days it's my left") or over a longer time span ("I've had bursitis in both shoulders, my hips, and my knees at one time or another for years").
4. Even though symptoms may never be completely absent, they vary within a time frame of about 4-7 days ("Some days I wake up and my back feels fine, and then other times I have to stay in bed for several days").
5. The history of recovery is more rapid than would be expected from a mechanical disruption ("Two days ago my back went out and I had to crawl to bed, almost paralyzed, but today I can bend with only a little stiffness"). A variation may be a symptom that has been present for many days or weeks but disappears on the day of the office visit.
6. The original cause of the problem was relatively minor ("I reached out to pick up a can of beans and my back went out," or "I had a whiplash when the other car hit [$200 worth of damage], but I didn't notice anything wrong until the next day"). Note that the incident initiating the symptoms need not be a chemical exposure but can be traumatic events of various sorts (including events of social or psychological significance), and in some cases the symptoms start without an identifiable initiating factor.
7. Recovery has been prolonged past any reasonable healing phase ("My back hasn't healed from the initial back injury five years ago").
8. The patient has a general attitude of fragility or general distrust of his/her body ("I have to be careful of what I do").
9. The symptoms are aggravated by sustained body position ("I can't sit still because, if I do, I hurt," or "When I work in one position, my muscles ache").
10. The patient does not like to take drugs because they seem to be ineffectual or produce numerous adverse reactions.
11. The condition is relieved by injection of local anesthetic and cortisone but only for 24-36 hours.
12. The patient has visited four or more doctors without establishing a diagnosis or successful treatment. Many of these patients learn not to mention some of their symptoms because the remaining symptoms are then not taken seriously.
13. The patient has been previously treated with anti-inflammatory medicines for non-specific "___itis" or "___algia" (e.g., myositis, arthralgia).
14. The patient reports foods or chemicals that give adverse effects. He/she may report sensitivity to fumes not noticed by others, such as house gas, cigarette smoke, perfume, diesel fumes, the soap and cleanser aisle at the supermarket, or anti-static preparations for clothes dryers.
15. The patient reports stiffness that is worse in the morning and then clears up with activity, or pains that are generally worse after midnight and better upon arising.

16. The symptoms are aggravated by pre-menstrual swelling, or (less frequently) are coincident with ovulation.
17. The patient relates symptoms to swelling that is not immediately evident to the examining physician ("See how swollen my ankle is?" or "Sometimes my hands just bloat up to twice their size").
18. Intermittent swelling is often detected upon awakening by bags under the eyes or by the inability to form a tight fist. The degree of swelling may vary from day-to-day but usually diminishes when the patient is up and active.
19. The patient has had repeated weight gains and losses that are so rapid they could only be caused by water retention or loss, (e.g., a four-lb weight loss in two days associated with repeated urination).
20. Repeated infections occur when no other cause of reduced resistance is evident.
21. The patient has a history of allergies in childhood or over many years.
22. Complaints are restricted to, or are much greater on, one side of the body (so-called hemicorpalgia)—("My right side is the problem, both my leg and arm, and even the side of my face").
23. Symptoms are often aggravated with exercise ("When I start to run, I itch over most of my body," or "During warm-up, I have an embarrassing amount of rectal gas") that often disappear by the time perspiration begins.

If hypersensitivity syndrome is suspected, then the questionnaire in Appendix I is administered. The questionnaire is designed to indicate not only present symptoms (within the last six months) but also symptoms at any previous time, so that multiple-organ-system involvement (undiagnosed) over a longer period can be considered. The number of items checked on the questionnaire gives a rough indication of the degree to which these symptoms affect the patient's life. If less than 5% of the symptoms have been experienced, the patients usually consider themselves quite well and do not find the symptoms distressing enough to seek treatment. Patients who check 30%-50% of the symptoms are usually quite affected and have been searching for some medical solutions. Persons who check more than 75% of the symptoms are often disabled and unable to function effectively. Note that these are clinical impressions, and no study has been conducted to determine the redundancy of the items on the questionnaire.

Clearly a careful and thorough medical history is needed. It is important not to doubt the veracity of what the patient has experienced (the symptom), no matter how contrary the symptoms are to present medical knowledge. No signs are specific or diagnostic of hypersensitivity syndrome, but several may raise the index of suspicion because they are commonly seen in hypersensitivity syndrome patients.

1. Cold fingers and toes (sometimes measured at ambient room temperature).
2. Blue hands or feet, or patchy redness alternating with white areas of vasoconstriction.
3. Dark shadows under the eyes.

4. Dermatographia
5. Irregular, intermittent, muscular twitching, best seen in the forearm muscles and the orbicularis oculi beneath the eye.
6. Slow deliberate movements by the patient during examination to prevent eliciting spasms.
7. Painful spasms elicited by the examiner's too-rapid movements of the patient's limbs.
8. Patient avoidance of palpation of the symptomatic part by withdrawing or by grasping the examiner's hand.
9. Involuntary withdrawal from the examiner's light touch, even when the examination commences away from the symptomatic area.
10. Tinel's sign over multiple nerves, including paresthesias from applied pressure in the sciatic notch if the hip is flexed to 90 degrees.
11. Variable intermittent swelling of the hands detected by changes in tightness of rings worn on the fingers.
12. Easily elicited joint popping in various different locations.
13. Splinter hemorrhages at the tips of the nail bed (best seen with a low-power magnifying glass).

TREATMENT

If a lesson is to be learned from the apparent clinical successes with the disparate approaches mentioned previously, then it is that the physician's beliefs and the degree of enthusiasm for those beliefs must be critical variables in any "successful" treatment of this symptom complex. Furthermore, experience has led me to postulate that in our culture, such successful treatments need an aura of complexity, preferably with some technological aspect (an electrical gadget, many labeled bottles, or a computer print-out of test results) over which the practitioner can cogitate the implications for the patient's treatment. Bluntly stated, technological mumbo-jumbo is the best way to elicit a placebo response in a society in which technology is the mysterious domain of highly trained professionals. Note that this description is as applicable to mainstream medicine as it is to practitioners who treat hypersensitivity syndrome with unproven therapies. Clearly, the vast majority of ordinary medical practice is carried out without the basis of well-controlled, double-blind, clinical trials, and it is also subject to influence by placebo responses. (In an earlier era, it may be that the placebo response was best elicited by the kindness and caring of the now-idealized family doctor.)

My current treatments are based on the hypothesis that the pathophysiological consequences of prolonged stress can account for many characteristics of hypersensitivity syndrome. As the reader should understand by now, any success I have had with this treatment offers no basis whatsoever to judge whether this hypothesis is correct. My treatment begins during the initial office visit because I do not challenge the validity of any symptoms. This in itself is therapeutic because I accept the patient on her/his own terms, even though I may not believe what the patient believes about the causes of the symptoms. At the first visit, I satisfy myself that the symptoms have been adequately evaluated by appropriate specialists; otherwise, I make referrals and postpone further visits until appropriate studies are completed. The next several visits deal with the patient's perceptions of possible causes for the symptoms, and together we plan therapeutic trials appropriate to these ideas, so long as treatment is neither expensive nor dangerous. Examples include diet changes, lifestyle changes to reduce chemical exposure, oral Nystatin for supposed Candida albicans infection, or an exercise routine.

If these methods fail or lose efficacy and the patient seeks further relief, I suggest that prolonged stress can lead to some of the bodily effects the patient may be experiencing. I describe some of the most severe stress as caused by helplessness and hopelessness, and almost all my patients have nodded in agreement at this. I explain that planning to deal with problems is in itself stress-reducing, even before the problems themselves are solved. I then ask them to bring to the next visit a list of the three most important problems in their lives, and at that visit the patients clearly divide into two groups. The first group lists problems like: 1) finding fresh air, 2) reducing reactions to foods, and 3) creating a safe living environment. The second group lists problems like: 1) finding someone to love, 2) finding someone to love me, and 3) finding a satisfactory job. My therapeutic success with the first group is poor because they find it difficult or impossible to stop their compulsive search for a medical or technological solution to their condition. My success with the patients in the second group is much better; their symptoms slowly and gradually reduce as conditions in their lives gradually improve.

The methods for improvement vary greatly from patient to patient and may involve any of several approaches, including referral to support groups, to social workers and re-training programs, or to psychologists or psychiatrists; referral to a large, local bookstore is usually included. I suggest they peruse the psychology or personal-improvement sections to find a book that speaks to them. Many patients find such a book (which they later enthusiastically recommend that I mention to other patients) but rarely do two patients choose the same book. Two patients were relieved of their syndromes by "giving themselves to Jesus."

Sometimes, I have tried to help these patients by transmitting much of the information described above during the first visit (e.g., when the patient has traveled hundreds of miles and cannot likely make follow-up visits). In no case has this approach been successful (even when contact by phone was invited for follow-up). This leads me to believe that the patient must develop confidence in the physician for my approach to work. Essentially the patient hears the message: "To improve your health, you must overcome problems that you have been trying hard to solve for a long time, and failed." Then the patient must have confidence and receive gentle, persistent support to find new ways of approaching apparently unsolvable personal problems.

Thus, my treatment is probably best described as counseling. I am not a psychiatrist, but a few grey hairs help me recognize and empathize with persons who are temporarily losing life's battles; encourage a strategy that evaluates the entire campaign, rather than any one battle; and advise acceptance of an isolated defeat rather than straining endlessly in a hopeless cause. Of course, by the nature of medical practice and Darwinian selection, only patients who are likely to respond to my approach return for follow-up visits. This approach is time-consuming, sometimes frustrating, and definitely impractical economically. Perhaps the continuing economic pressure of medical practice to shorten the length of patient-physician contact prevents hypersensitivity syndrome patients from receiving treatment of this sort. Such a limitation may even be the basis for an increased incidence—if there has in fact been an increase. Such speculation can only be resolved when a basis for hypersensitivity syndrome has been established by appropriate scientific study and when treatment is founded on a more rational basis than clinical experience.

REFERENCES

JEWETT, D.L. (1983). Environmentally influenced hypersensitivity: Musculoskeletal manifestations. Clin. Ecol. 1:133-140.

JEWETT, D.L., FEIN, G., and GREENBERG, M.H. (1990). A double-blind study of symptom provocation to determine food sensitivity. N. Engl. J. Med. 323:429-433.

JEWETT, D.L. (1992). Research strategies for investigating "Multiple Chemical Sensitivity". Toxicol. Ind. Health (this issue).

LAYZER, R.B. (1991). Clinical symptoms in the Chronic Fatigue Syndrome. In: Chronic Fatigue Syndrome (G. Serratrice and J.L. Vilde, eds.). Springer-Verlag, Paris. pp. 5-9.

ROSS, G.H. (1992). Treatment options in multiple chemical sensitivity. Toxicol. Ind. Health (this issue).

SELNER, J.C. and STAUDENMAYER, H. (1992). Neuropsychologic observations in patients presenting with environmental illness. Toxicol. Ind. Health (this issue).

SIMON, G. (1992). Psychiatric treatments in multiple chemical sensitivity. Toxicol. Ind. Health (this issue).

ZIEM, G. (1992). Multiple chemical sensitivity: Treatment and follow-up with avoidance and control of chemical exposures. Toxicol. Ind. Health (this issue).

See also Appendix I on the following pages.

APPENDIX I

Instruction: Circle the number of any question to which you would answer "yes" for any time in your life. Add a check mark if you have had the condition or experience in the last six months.

Muscles

1. Have you had muscle spasms? Where?
2. Aching and soreness in the muscles of the neck?
3. Aching or soreness in the muscles of your lower back?
4. Aching or soreness in the muscle that runs from your neck to your shoulder?
5. Aching in other muscles? Where?
6. Sometimes wake up tired, even with a full night's sleep?
7. Times of utter fatigue, unable to do what you would like to do?
8. Stiffness or tightness in the muscles of your arms or legs?
9. Difficult to sit still without "restless legs?"
10. Tired sooner than you would like to be?
11. Avoid exercise?
12. Need to rest during prolonged physical activity?
13. Fatiguing to spend several hours shopping?
14. Toss and turn at night to get comfortable?
15. Twitching eyelids?
16. Muscles twitch in arms and legs?
17. Sometimes catch toes while walking?

Joints

18. Sometimes awaken with stiff hands?
19. Had swelling of the joints without accident or injury? Which joints?
20. Ankle swelling that comes and goes?
21. Had joints which are warm, red and painful without joint infection?
22. Joints stiff in the morning? Where?
23. Aching pains that come and go in the joint?
24. Do your joints pop? Which ones?

Extremities

25. Do you often have cold fingers or toes when others don't?
26. Have you seen "splinters" under your nails at the outer edge, even though there is no real splinter there?
27. Sometimes have sweating palms?
28. Tingling in your arms and hands, or legs and feet?
29. Arms or legs "fall asleep" from time to time?
30. Do you sometimes have white spots in your fingernails?

Skin

31. Itching skin more than once a month? Where?
32. Scratch skin or scalp even when it doesn't itch?
33. Moist or sweating skin without exercise?
34. Skin sometimes white or ghostly?
35. Sometimes have flushing or reddening of your skin?
36. Hands and fingers sometimes "bluish"?
37. Hives or eczema?
38. "Athlete's foot" or "jock strap itch" is a recurring problem?

Infections

39. Recurring ear infections?
40. More than one bladder infection?
41. Colds more frequently than every three months?
42. Repeated infections? Where?
43. Sinus infections?
44. "Tonsillitis" more than once?
45. Repeated skin pustules?

Nasal

46. Difficulty breathing through your nose?
47. Nasal discharge when you don't have a cold?
48. Frequent stuffy feeling in your nose and face?
49. Pains in your sinuses (above and below your eyes)?
50. "Sniffles" without a cold?
51. Itching and rubbing of the nose?
52. Sneezing or the urge to sneeze where it is not dusty?
53. Carry more than one handkerchief for blowing the nose?

Throat and Mouth

54. Do you have itching of the tongue or roof of the mouth?
55. Throat ever become sore, tight or swollen?
56. Difficulty in swallowing?
57. Clearing of the throat or coughing several times a day?
58. Swelling at the back of the throat or roof of the mouth?
59. Weakness of voice or hoarseness?
60. Excess saliva or mucus?
61. Do you cough or clear your throat at night or in your sleep?
62. Foods seems "dry" so it is "washed down" with something to drink?
63. Taste foods hours after eating?
64. Metallic taste in mouth?

65. Bad taste in mouth in morning?
66. Bad taste in mouth during day?
67. Numbness around mouth?

Ears

68. Do you have reddening or warmth of the ears?
69. Itching in the ear canal?
70. Feeling of fullness or blockage in the ear canal?
71. Ear-aches?
72. Hearing loss which varies?
73. Some sounds very annoying or painful?
74. High-pitched tone sensed in the middle of the head?
75. Ringing in the ear?
76. Pounding pulse in the ear?

Eyes

77. Do you have itching or burning of the eyes?
78. Rubbing of the eyes each day?
79. Watery eyes or tears?
80. Inflamed or reddened eyes?
81. Puffiness ("bags") under the eyes which vary?
82. Dark skin under the eyes or around the eyes?
83. Blurring of the vision at times?
84. Unable to see details at times, other times OK?
85. Spots sometimes seen?
86. "Floaters" (transparent lines or "worms") which move with the eyes?
87. Flashes in the vision at times?
88. Occasional darkness in the vision?
89. Sometimes see many swirls or "pinwheels" in vision?
90. Bothered by bright lights?
91. Have double vision at times?
92. Have difficulty reading, such as mixup of letters, letters becoming small or words moving around?
93. Heavy eyelids, hard to keep them up?

Lungs and Heart

94. Shortness or breath when not physically active?
95. Reduced air flow felt?
96. Times when your breathing is rapid?
97. Pains in your chest?
98. Rapid heart beat?
99. Forceful or pounding heartbeat?

100. Uneven heartbeat at times?
101. Heart "jump" at times?
102. Asthma?

Stomach - Intestines

103. Nausea or queasy stomach without the "flu?"
104. Burp or belch once a week or more?
105. Feeling of abdomen fullness or being bloated?
106. Tightness across the stomach?
107. Vomiting without getting the "flu?"
108. Pain or discomfort in the abdomen?
109. Cramps the come and go in the abdomen?
110. Rumbling or gurgling in the abdomen or stomach?
111. Rectal gas?
112. Soft bowel movement or diarrhea at times without the "flu"?
113. Sharp pains in the mid-stomach just below the ribs from "gall-bladder" or "ulcer"?
114. Constipation?
115. Acid stomach?
116. Take Alka Seltzer more than once every two months?
117. Strong hunger pangs?
118. Very thirsty at times?
119. Sometimes feel better after eating?
120. Can't resist some foods when available?
121. Eat repeated helpings of some foods? Which ones?
122. Eat some food each day routinely?
123. Some foods make you feel bad? Which ones?
124. Feel better after some foods? Which ones?
125. Bother you to skip one or two meals?
126. Think you have hypoglycemia?
127. Bowel movements float?
128. Hiccups more than once a year?

Kidney - Bladder

129. Have to urinate frequently at times?
130. Sometimes there is very little urine even though it was necessary to urinate?
131. Sometimes the urge to urinate is very strong?
132. Pain with urination?
133. Difficult to urinate at times?
134. Several times have seen the doctor for urinary problems, but there is no infection?
135. Vaginal discharge or yeast infection?
136. Itching of the genitals?
137. Breast swelling not connected with menstrual cycle?

138. Some days urinate more than usual amount of fluid?
139. Rapid weight gain or loss in a day or so?

Head - Headaches

140. Do you have migraines?
141. Do you have repeated headaches?
142. Aching or pressure in the head?
143. Itching of the scalp?
144. Tingling of the scalp?
145. Throbbing or stabbing pain in the head? One side or both?
146. Fainting spells?
147. Hallucinations?
148. "Blackouts"?

History

149. As a baby, were there any foods that you couldn't be fed?
150. Were you a "colicky" baby?
151. Did you "outgrow" childhood allergies?
152. Any dust or pollen allergies?
153. Any blood relatives with allergies, including parents and children?

General

154. Handwriting getting worse?
155. Sometimes misspell words by switching letters in the word?
156. Sometimes call your right side your left, or vice versa?
157. Insomnia?
158. Sleep more than 8 hours a day?
159. Drowsy at times during the day or evening?
160. Bothered by lack of concentration?
161. Poor memory?
162. Difficulty remembering words or names you should know?
163. Sometimes get words "mixed-up"?
164. Sometimes can't pronounce word you know?
165. Difficulty losing weight?
166. Difficulty gaining weight?

Moods

167. Life seems to repeatedly cycle from being happy to being discouraged?
168. Sometimes people comment how irritable you are?
169. Periods when you are more active and have more energy than most people?
170. Times when you are so discouraged or depressed you can't get anything done?
171. At times can't make even simple decisions?

172. Times when you really just don't care what happens?
173. Times when you are confused?

Miscellaneous

174. Any chemicals or smells which bother you? Which ones?
175. Has your condition worsened in the last 2-3 years?
176. Women: Any of above symptoms in this questionnaire worse just before menstrual period? Which? Other menstrual symptoms not included above?

IMMUNOLOGIC EVALUATION OF CHEMICALLY SENSITIVE PATIENTS

HOWARD KIPEN[1,4], NANCY FIEDLER[1,4], CLEMENT MACCIA[2], EDWARD YURKOW[3], JEANINE TODARO[3], AND DEBRA LASKIN[3,4]
[1]Department of Environmental and Community Medicine
[2]Department of Pediatrics
UMDNJ-Robert Wood Johnson Medical School
Piscataway, New Jersey
[3]Department of Pharmacology and Toxicology
Rutgers University, Piscataway, New Jersey
[4]Environmental and Occupational Health Sciences Institute
Piscataway, New Jersey

INTRODUCTION

There is a need for a rigorous definition and characterization of the symptom complex that has been labeled multiple chemical sensitivities (MCS) or environmental illness. Over the past decade, an increasing number of patients have presented with persistent or recurrent somatic and psychological symptoms attributed to chemical exposures. Prominent symptoms reported by these patients include, but are not limited to, central nervous system complaints (such as poor concentration and memory), arthralgias, headaches, and respiratory problems. The symptoms reported, their intensity, and their duration, however, are not consistent with the accepted toxicologic properties of the chemicals. In a 1987 review paper, Levin and Byers (1987) asserted that MCS was a disorder of immune regulation, frequently characterized by depression of the T-cell helper-to-suppressor ratio by an inordinately high number of suppressor cells.

Cullen (1987), in an occupational medicine, state-of-the-art review of MCS, proposed a relatively rigorous definition to characterize persons reporting chemical sensitivities and to differentiate them from patients with somatoform disorder and typical or atypical post-traumatic stress disorder. However, patients described in the peer-reviewed literature have not fully met this definition (Schottenfeld and Cullen, 1986; Bolla-Wilson et al., 1988; Simon et al. 1990; Brodsky, 1983; Rosenberg et al., 1990; Terr, 1986). For example, Simon et al. (1990) labeled a group of persons, who were in workers compensation litigation, as chemically sensitive based solely on self-reported lifestyle modifications. Black et al. (1990) described psychological profiles of a group of subjects diagnosed as chemically sensitive by physicians practicing clinical ecology. Criteria for the diagnosis were not specified. In addition, the initial identified exposure for these patients was not uniformly environmental

Address all correspondence to: Dr. Howard Kipen, UMDNJ-Robert Wood Johnson Medical School, Department of Environmental and Community Medicine, 675 Hoes Lane, Piscataway, NJ 08854.

(e.g., stress) nor was the associated diagnosis always chemical sensitivity. For example, some subjects were diagnosed with chronic candidiasis. Only the case series of Terr (1986) has presented data on immunologic parameters in a subset of patients. These data were described as substantially normal.

By contrast, the present project identifies persons whose chemical sensitivity had a specific onset associated with an environmental or other specific chemical exposure event and who subsequently reported multiple symptoms in response to multiple agents at low levels of exposure. Thus, this project is the first to characterize the medical, psychiatric, neuropsychological, and immunologic status of a series of patients who fit the specific diagnostic criteria for chemical sensitivities proposed by Cullen (1987). Preliminary reports of the psychiatric and neuropsychological findings have been presented[1] (Fiedler et al., 1991). This paper contains preliminary immunologic data on a subset of the persons tested and compares the results with other cohorts under study at our Environmental and Occupational Health Clinical Center.

METHODS

Subjects

Subjects were recruited from patients referred to our Clinical Center. Most patients were referred by community physicians who had questions about the putative relationship between chemical exposures and symptoms or illnesses presented by their patients. We have solicited such referrals for this research.

During the past two years, 35 patients have been referred for evaluation of chemical sensitivities. As of this writing, 16 met the following criteria for the diagnosis of MCS: 1) symptoms were acquired in relation to some initial identifiable environmental exposure(s); 2) the symptoms involve more than one organ system (e.g., respiratory and nervous systems); 3) the symptoms recur and abate in response to predictable stimuli; 4) the symptoms are elicited by exposures to chemicals of diverse structural classes and toxicologic modes of action; 5) the exposures that elicit symptoms are very low, i.e., many standard deviations below the threshold limit value (TLV) and at levels not known to cause adverse human responses; and 6) no single, widely available test of organ system function can explain the symptoms. Subjects had no previous physical or psychiatric history or current diagnosis that could explain the development, breadth, and severity of their symptoms. Information was obtained through a detailed medical history and examination and a self-reported medical questionnaire routinely administered to all clinic patients. Only a cursory overview of psychiatric history was included at this point, i.e., a history of psychiatric hospitalization or treatment by a mental health professional on an inpatient or outpatient basis. Furthermore, patients reported being in reasonably good health with productive home and work lives before

---

[1] Fiedler, N., Maccia, C., and Kipen, H.M. Medical and psychological evaluation of chemically sensitive individuals. Accepted, Journal of Occupational Medicine.

their initial environmental exposure. Information about the exposures and substances that initiated or elicited symptoms comes solely from the history provided by the subjects.

Fourteen patients were excluded for the following reasons: 1) met part but not all the above criteria (N=8); 2) referred specifically for litigation-related determinations (N=2); 3) had previous significant psychiatric history (N=1) or were employed as psychologist or psychiatrist (N=2); and 4) were unable to participate because moved out of state (N=1). These exclusions are mostly germane to the psychiatric aspects of the study but were applied to all subjects. Complete immunologic studies are not available for ten other subjects because of logistic or laboratory constraints in obtaining or handling specimens. The 11 subjects whose immunologic characterizations are described in this study are the first to be reported in the literature who specifically meet the above-listed selection criteria. Although our submitted report of the psychiatric characterization of such patients is also based on 11 subjects, only five subjects appear in both reports at this time because of the logistic constraints of collecting the immunologic data discussed above

EVALUATION MEASURES

Medical Evaluation

Each potential subject underwent a comprehensive medical examination at the Clinical Center under the supervision of a board-certified internist and occupational physician. In addition to routine laboratory evaluations, further testing, such as spirometry or chest x-ray, was performed as needed to document organic conditions that might reasonably explain the symptomatology. Patients were evaluated at a meeting of the investigators. Persons judged to have met the entrance criteria were then eligible for further studies as research subjects. Before participation, all subjects completed informed consent forms; no subject refused to participate.

Immunologic Evaluation

The medical aspects of the research protocol consist of an evaluation of immunocompetence. Humoral immunity was evaluated with tests of immunoglobulin (IgG, IgA, IgM, IgE) and complement components (CH50, C3, C4). Type I hypersensitivity was evaluated by positive histamine and negative glycerine controls and by skin-prick testing to a variety of common antigens. The antigens used were:

Dust Mite	Cat Hair - Epithelium
Alternaria	Dog Hair - Dander
Hormodendrum	Ragweed Mix
Timothy Grass	Oak Mix
Meadow Fescue Grass	Maple

Cell-mediated immunity was evaluated with an anergy battery of: PPD (5 TU); Trichophyton (1:100); Dermatophyton; and Candida (1:1000). Delayed-type patch tests for the following common North America antigens were also done.

Fragrance Mix	Parabens
Epoxy Resin	Quaternium-15
Thioram Mix	Wood Alcohol (lanolin)
P-tert-Butylphenol- formaldehyde	Mercapto Mix

All subjects received the same antigen panels, and no attempt was made to match antigens to the specific sensitivities reported by given subjects, except for excluding tuberculin testing of those subjects with a history of tuberculin reactions.

Immune system assays also included characterization of lymphocyte surface markers for each subject. Phlebotomy using 10 ml of peripheral venous blood was performed at approximately 12 noon when subjects arrived for their psychologic evaluation. Blood was drawn into vacutainers containing EDTA and transported to the laboratory for same-day labeling with fluorescently labeled antilymphocyte monoclonal antibodies, using a Coulter Q-Prep Immunology Work Station. Samples were then analyzed for antibody binding by flow cytometry with a Coulter Epics Profile flow cytometer. The percentage of cells binding the antibodies was calculated using Overton's cumulative subtraction routine from the Cytologic software program.

The following cell-surface antigens were assayed: CD2, CD3,CD4, CD8, CD14, CD25, CD26, CD45, CD45R, CD56, and I3 (HLA-DR). All antibodies to these antigens were obtained from Coulter Immunology (Hialeah, FL). Some antigens were analyzed as paired combinations enabling characterization of leukocyte subsets that express both markers simultaneously. Preliminary data for CD4 and CD8 will be presented here.

RESULTS

Characteristics of Patients

Patients included in this case series were recruited over a one-year period and were the first 11 persons who met the entrance criteria for MCS and had valid immunophenotypic data. The two men and nine women ranged in age from 28 to 57 years with a mean age of 35 for men and 44 for women. One male subject had previously worked in the chemical industry; the other was working as a draftsman at the time of evaluation. Three of the nine women were working as secretaries, two were analysts, one was a teacher, one a sales clerk, and two were housewives. All subjects except the two housewives were employed at the time their symptoms appeared. Both male subjects worked with significant chemical or solvent exposures and first experienced symptoms in response to a change in ventilation or in a process at work. Neither, however, experienced a particular episode associated with a medically treated acute intoxication or accidental release. The identified precipitating exposures and subsequent symptoms for five of the nine women occurred at work. For the remaining female subjects, precipitating exposures began outside work. The identified precipitating exposures included laying new carpet (N=4), indoor air quality complaints in an office environment (N=2), adverse reaction to a prescribed drug (N=1), and home pesticide application (N=2).

Data are available on two comparison groups for the immunophenotypic variables. The first comparison group consists of 11 male farmers with a mean age of 55 years. All were in good health except for frequent hypertension, which was treated with diuretics, beta blockers, or converting enzyme inhibitors in two cases. None of the subjects was asthmatic. The second comparison group consists of seven individuals with chronic fatigue syndrome (CFS), two males aged 34 and 50 years and five females with a mean age of 37 years. None was asthmatic, and none was receiving pharmacologic therapy, although all were under medical care.

Medical Evaluation

None of the MCS subjects had previous medical conditions or psychiatric history that would explain their symptoms, but five did have significant ongoing or previous medical conditions. One subject had a past history of recurrent pneumonia, although not for the last ten years. Four subjects had asthma, three with a longstanding, relatively mild condition; in one woman, it manifested concurrently with the MCS symptoms. Asthma therapy did not ameliorate the predominantly non-respiratory components of their symptomatology. No significant abnormalities of any of the hematology and chemistry tests were detected using laboratory standards.

Immunologic Function

Mean WBC count was 6.9, with a range of 5.5-8.3 (normal range 3.8-11.4). Mean total lymphocyte count was 2176 with a range of 1820 to 2945 (normal range 1049-3581) (Table 1). Data on serum complement and immunoglobulins are summarized in Table 2. One subject had an elevated IgE consistent with a history of atopy and asthma. Another individual had an elevated IgM of 418. All other values were within normal laboratory ranges.

TABLE 1. Summary of WBC Data for Patients with Multiple Chemical Sensitivities (n = 11)

WBC			
Mean	6.9	Group Range	5.5 - 8.3
Median	7.2	Normal Range	3.8 - 11.4
Lymphocytes			
Mean	2176	Group Range	1820 - 2945
Median	2153	Normal Range	1049 - 3581

TABLE 2. Summary of Complement and Immunological Data for Patients with Multiple Chemical Sensitivities

CH3, Serum			
Mean	82.0	Group Range	60.0 - 101.0
Median	82.0	Normal Range	50.0 - 120.0
CH 4, Serum			
Mean	30.0	Group Range	18.0 - 47.0
Median	28.0	Normal Range	15.0 - 50.0
IgE, Serum			
Mean	70.0	Group Range	15.0 - 192.0
Median	43.0	Normal Range	0.0 - 180.0
IgG, Serum			
Mean	1220.0	Group Range	714.0 - 164.8
Median	1240.0	Normal Range	700.0 - 1800.0
IgA, Serum			
Mean	241.0	Group Range	152.0 - 374.0
Median	234.0	Normal Range	70.0 - 440.0
IgM, Serum			
Mean	240.0	Group Range	69.0 - 418.0
Median	220.5	Normal Range	60.0 - 290.0
CH50, Total			
Mean	175.0	Group Range	106.0 - 280.0
Median	158.0	Normal Range	100.0 - 300.0

Evaluation of cellular immunocompetence revealed that two of ten subjects were anergic; a third was nonreactive to three antigens, with tuberculin being withheld because of a prior reaction history. Immediate hypersensitivity skin tests were negative except for the histamine control in six subjects; positive for two antigens in two subjects (with respective IgEs of 78 and 135); and negative to all, including the histamine control, in two subjects.

Patch testing was negative in eight subjects, positive in two who demonstrated positive responses to lanolin, and not performed in one.

Immunophenotypic Evaluation (Tables 3-5)

CD4: Normal range for our laboratory is approximately 35-55%. Two of eleven MCS subjects were out of range (20 and 20.9). Five of 11 farmers were clearly out of range (22.2, 23.0, 24.6, 28.2, and 29.8). None of the seven CFS subjects was out of range. Mean values for the three groups were 40.7 for the MCS subjects; 33.1 for the farmers; and 44.5 for the CFS subjects.

CD8: Normal range for our laboratory is approximately 20-36%. Five MCS subjects were out of range (11.5, 12.1, 6, 15, and 8). Six farmers were out of range

(14.1, 13.7, 10.1, 13.9, 9.3 and 10.1). One CFS subject was low (13.9). Mean values for the three groups were 18.0 for the MCS subjects; 18.0 for the farmers; and 22.4 for CFS.

CD4/CD8: The normal range for this ratio is frequently reported as 1-2 and can conservatively be expanded up to 2.5. Five of the eleven MCS subjects had elevated ratios (4.8, 3.3, 5.3, 2.6, 2.8) with a mean of 2.7 and a median of 2.4. Two of the farmers had elevated ratios (3.7 and 4.3) and one a depressed ratio (0.9) with a mean of 2.1 and a median of 1.8. One CFS subject had a high ratio (3.6), with a mean and median of 2.1.

TABLE 3. Immunophenotypic Evaluation for Multiple Chemical Sensitivity Subjects

	Sex	Age	CD4[1](%)	CD8[2](%)	CD4/CD8 Ratio[3]
1	F	48	55.6	11.5	4.8
2	M	28	39.8	12.1	3.3
3	F	44	42.6	19.0	2.2
4	F	22	38.9	20.3	1.9
5	F	39	48.0	25.9	1.9
6	M	42	32.0	6.0	5.3
7	F	40	20.0	15.0	1.3
8	F	50	20.9	8.0	2.6
9	F	52	46.7	30.1	1.6
10	F	51	59.7	21.7	2.8
11	F	43	44.0	27.9	1.6
Mean		42	40.7	18.0	2.7
Median		43	43.3	19.7	2.4

[1] Normal range 35-55%
[2] Normal range 20-36%
[3] Normal range 1-2.5

DISCUSSION

These data are preliminary because of the relatively small number of subjects tested and the lack of a normal group of control subjects for the immunophenotypic analyses. CFS subjects were selected as controls for the psychiatric and neuropsychologic aspects of the MCS protocol and were tested concurrently with the MCS subjects. Evaluation of appropriate asymptomatic controls is planned in the near future, concurrent with a larger number of MCS subjects. The farmers were evaluated as part of a separate study to assess the impact of chronic, high-level organophosphate exposure on immunophenotypic variables and is only presented here for illustration. Data on another 30 farmers and 6 unexposed, geographically matched controls are now being analyzed.

TABLE 4. Immunophenotypic Evaluation for Farmer Comparison Group

	Sex	Age	CD4[1](%)	CD8[2](%)	CD4/CD8 Ratio[3]
1	M	63	50.0	29.2	1.7
2	M	68	32.7	17.7	1.8
3	M	30	33.4	22.7	1.5
4	M	38	29.8	14.1	2.1
5	M	61	22.2	13.7	1.6
6	M	42	24.6	10.1	2.4
7	M	61	23.0	13.9	1.7
8	M	34	42.3	24.9	1.7
9	M	44	28.2	32.2	0.9
10	M	37	34.1	9.3	3.7
11	M	62	43.7	10.1	4.3
Mean		55	33.1	18	2.1
Median		42	33	15.9	1.8

[1] Normal range 35-55%
[2] Normal range 20-36%
[3] Normal range 1-2.5

TABLE 5. Immunophenotypic Evaluation for Chronic Fatigue Syndrome Comparison Group

	Sex	Age	CD41(%)	CD8[2](%)	CD4/CD8 Ratio[3]
1	F	51	40.9	21.7	1.9
2	F	31	41.1	20.0	2.1
3	F	33	43.2	21.9	2.0
4	F	32	47.9	29.2	1.6
5	M	50	42.5	19.7	2.2
6	M	34	45.4	30.6	1.5
7	F	42	50.3	13.9	3.6
Mean		39	44.5	22.4	2.1
Median		38	44.4	21.8	2.1

[1] Normal range 35-55%
[2] Normal range 20-36%
[3] Normal range 1-2.5

WBC counts, lymphocyte counts, immunoglobulin levels, and complement levels were unremarkable. The presence of two (and possibly three) delayed-hypersensitivity non-reactors is not clearly beyond what would be expected for our battery of four antigens. Only one of these non-reactors also had an elevated ratio of CD4 to CD8 cells.

Five of the eleven MCS subjects had substantially depressed percentages of CD8 positive (suppressor/cytotoxic) T cells. Five also had substantially elevated CD4/CD8 ratios, attributable to the low CD8 count in four and to a borderline high CD4 count (59.7) and borderline low CD8 count (21.7) in another. The farmers showed a relatively similar frequency of depressed CD8 percentages but only one elevated ratio because the CD4 counts were depressed in five of them. Their mean ratio was essentially normal. The CFS subject were substantially normal, with a single elevated ratio.

Only one study in the literature provides immunophenotyping data on individual MCS patients (Terr, 1986). This case series of 50 MCS subjects reported immunophenotyping data on ten subjects. Although individual percentages of CD4 and CD8 cells are not presented, five of ten had ratios of approximately 2.5 or greater based on the published figure. All CD8 absolute counts were reported to be within normal range.

The existing literature on immunophenotyping of subjects exposed to toxic substances provides only one example where CD8 cells were decreased and the CD4/CD8 ratio was elevated. This was reported in a group of asbestos-exposed individuals, 75% of whom had abnormal chest radiographs (Miller et al., 1983). Another study of automobile pattern makers did not report ratios but showed significantly depressed CD8 levels in the exposed group (Robins et al., 1991). Studies of women chronically exposed to aldicarb in drinking water reported increased CD8 levels and decreased ratios (Fiore et al., 1986; Mirkin et al., 1990). Individuals with TCE-contaminated drinking water also showed increased CD8 levels (Byers et al., 1988).

Two studies of CFS patients have reported increased CD4/CD8 ratios of approximately 2.6, although in both cases CD8 was normal and the increase apparently resulted from increased CD4 cells (Peter and Wallace, 1988; Strauss et al., 1985). Another study of CFS patients reported decreased ratios attributable to elevated CD8 counts (Klimas et al., 1990). A more recent study reported no difference in CD4, CD8, and ratios in 25 CFS patients compared with controls (Gold et al., 1990). Mean ratio was 1.6.

No definitive conclusions about the presence of immunophenotypic abnormalities in our MCS subjects are possible at this time because of the limited number of subjects and the lack of comparable normal controls in our laboratory. These preliminary findings do, however, suggest the possibility that individuals with MCS tend to have different CD4 and CD8 T-cell subsets compared with an age- and sex-matched group of patients chronically ill with CFS. The pathophysiologic implications of such differences, if substantiated, are not clear, in contrast with the devastating immunodeficient picture associated with chronically depressed CD4 counts and ratios.

The data appear comparable to the elevated CD4/CD8 ratios also reported in Terr's data (1986), although he did not conclude they were elevated. The lack of comparability to CFS patients does not support the idea that such immunophenotypic changes are associated with chronic symptomaticity itself. The CD8 values in MCS subjects are similar to those observed in pesticide-exposed farmers, although age and sex distribution are not comparable, and farmer ratios did not appear elevated because the CD4 counts were depressed concomitantly. Further investigation on these effects and their significance is ongoing in our Clinical Center, and these preliminary results should be considered more as a presentation of our method and approach than as confirmed findings. More firm conclusions may be possible when the analyses described above are completed.

ACKNOWLEDGMENTS

This project was supported by a grant from the Hazardous Substance Management Research Center, a National Science Foundation Industry/University Cooperative Center, and a New Jersey Commission on Science and Technology Advanced Technology Center, as well as by two seed money grants from the Environmental and Occupational Health Sciences Institute. The Institute is supported by NIEHS Center Grant ES 05022. We wish to acknowledge the contributions of Kathie Kelly-McNeil and Carol Natarelli, who have both helped provide the evaluation and care of our subjects. We also appreciate the consultation of Bernard Goldstein, M.D., and Michael Gochfeld, M.D., Ph.D., in formulating this project. Finally, we appreciate the support services of Joyce Kosmoski and Patricia Hutty in preparing this and other documents for the project.

REFERENCES

BLACK, D.W., RATHE, A., and GOLDSTEIN, R. (1990). Environmental illness - A controlled study of twenty-six subjects with 20th century diseases. JAMA 264:3166-3170.

BOLLA-WILSON, K., WILSON, R.J., and BLEECKER, M.L. (1988). Conditioning of physical symptoms after neurotoxic exposure. J. Occup. Med. 30:684-686.

BRODSKY, C.M. (1983) Psychological factors contributing to somatoform diseases attributed to the workplace. J. Occup. Med. 25:459-464.

BYERS, V.S., LEVIN, A.S., OZONOFF, D.M., and BALDWIN, R.W. (1988). Association between clinical symptoms and lymphocyte abnormalities in a population with chronic domestic exposure to industrial solvent-contaminated domestic water supply and a high incidence of leukaemia. Cancer Immunol. Immunother. 27:77-81.

CULLEN, M.R. (1987). The worker with multiple chemical sensitivities: An overview. In: Workers With Multiple Chemical Sensitivities. (M.R. Cullen, ed.). Occup. Med.: State Art Rev. 2(4): 655-662.

FIEDLER, N. and KIPEN, H.M. (1991). Neurobehavioral and Psychosocial Aspects of Multiple Chemical Sensitivity. Proceedings NAS Workshop to Develop Research Protocols for MCHS. National Research Council.

FIORE, M.C., ANDERSON, H.A., HONG, R., GOLUBJATNIKOV, R., SEISER, J.E., NORDSTROM, D., HANRAHAN, L., and BELLUCK. D. (1986). Chronic exposure to aldicarb-contaminated groundwater and human immune function. Environ. Res. 41:633-645.

GOLD, D., BOWDEN, R., SIXBEY, J. et al. (1990). Chronic fatigue: A prospective clinical and virologic study. J.A.M.A. 264:48-53.

HOLMES, G.J., KAPLAN, A., KOMAROTT, L. et al. (1988). Chronic fatigue syndrome: A working case definition. Ann. Intern. Med. 108:387-389.

KLIMAS, N.G., SALVATO, F.R., MORGAN, R., and FLETCHER, M.A. (1990). Immunologic abnormalities in chronic fatigue syndrome. J. Clin. Microbiol. 1403-1410.

LEVIN, A.S. and BYERS, V.S. (1987). Environmental illness: A disorder of immune regulation. In: Workers With Multiple Chemical Sensitivities. (M. R. Cullen, ed.). Occup. Med.: State of Art Rev. 2(4):669-681.

MILLER, L.G., SPARROW, D., and GINNS, L.C. (1983). Asbestos exposure correlates with alterations in circulating T cell subsets. Clin. Exp. Immunol. 51:110-116.

MIRKIN, I.R., ANDERSON, H.A., HANRAHAN, L., HONG, R., GOLUBJATNIKOV, R., and BELLUCK, D. (1990). Changes in T-lymphocyte distribution associated with ingestion of aldicarb-contaminated drinking water: A follow-up study. Environ. Res. 51:35-50.

PETER, J.B. and WALLACE, D.J. (1988). Abnormal immune regulation in fibromyalgia. Arthritis Rheum. 31.

ROBINS, T.G., WEINSTEIN, R.J., and DEMERS, R.Y. (1991). Lymphocytopenia, T-lymphocyte subsets, and colorectal polyps in automotive pattern and model makers. J. Occup. Med. 33:510-515.

ROSENBERG, S.J., FREEDMAN, M.R., SCHMALING, K.B., and ROSE, C. (1990). Personality styles of patients asserting environmental illness. J. Occup. Med. 32:678-681.

SCHOTTENFELD, R.S. and CULLEN M.R. (1986). Recognition of occupation-induced post traumatic stress disorders. J. Occup Med. 28:365-369.

SIMON, G.E., KATON, W.J., and SPARKS, P.J. (1990). Allergic to life: Psychological factors in environmental illness. Am. J. Psychiatry. 147:901-906.

STRAUS, S.E., TOSATO, G., ARMSTRONG, G., LAWLEY, T., PREBLE, O.T., HENLE, W., DAVEY, R., PEARSON, G., EPSTEIN, J., BRUS, I., and BLAESE, R.M. (1985). Persisting illness and fatigue in adults with evidence of Epstein-Barr virus infection. Ann. Intern. Med. 102:7-16.

TERR, A.I. (1986). Environmental illness: A clinical review of 50 cases. Arch. Intern. Med. 146:145-149.

BIOLOGIC AND SOCIAL FACTORS IN FOOD SENSITIVITIES

DAVID S. KING, PhD
Department of Psychiatry
University of California, Los Angeles

INTRODUCTION

The phenomenon of multiple chemical sensitivity bears similarities to that of food sensitivity, with many of the same questions unresolved. In fact, the two syndromes may overlap. Symptoms to food, including psychological and behavioral alterations, often play a prominent role in patients who suffer from multiple chemical sensitivities (Ashford and Miller, 1991). Some investigators have attributed these adverse behavioral effects from foods to biologic pathways, but others believe psychosocial mechanisms are responsible.

METHODOLOGIC ISSUES

Conceptual and methodologic problems have impeded scientific progress regarding food sensitivities and behavior. For example, no single biologic model of food sensitivity has emerged, although a variety of different hypotheses have been advocated. Different foods (e.g., food colorings, sugar, individualized food responsivity) and susceptible patient populations (e.g., hyperactive children, atopic individuals) have been implicated, but no agreement exists on exposure levels (dose) of a food necessary to provoke a measurable reaction, and the time course of reactions is unclear. Thus, definitive tests are difficult because it is not possible to study all combinations of such independent variables in a single experiment.

Although double-blind food challenges with placebo controls are essential for clear inference about responses to challenges, no optimal method for dose delivery has yet emerged. Compromise with the amount of food given, form of the challenge food (e.g., lyophilized vs. natural state), or adequacy of blinding is often necessary. For this reason, a *positive* challenge requires both a negative placebo challenge and adequate evidence of blinding for clear interpretation. A *negative* challenge requires demonstration that the dose and form of the food were adequate to provoke symptoms in individuals who are truly sensitive to it.

Under clinical conditions, however, alternative explanations for either result are still possible. Thus, a single *positive* test result in a patient could be due to baseline variability in symptom patterns or other factors despite the presence of a negative placebo challenge. More problematically, patients commonly report some level of symptoms on every trial because they are

Address all correspondence to: Dr. David S. King, Neuropsychology Assessment Laboratory, Neuropsychiatric Institute and Hospital, University of California, Los Angeles, 760 Westwood Plaza, Room C8-747, Los Angeles, CA 90024.

aware of on-going sensations in various parts of their bodies. How much increase in symptom reporting on a food challenge is necessary for it to be termed positive in comparison with a placebo dose? On the other hand, a *negative* challenge could be attributed to a decrease in the patient's sensitivity (possibly temporary) or the need for cumulative doses over time before symptoms emerge.

The latter possibility raises a common problem. A critical methodologic concern in research on food sensitivity is whether one is testing the test, testing the patient, or testing the hypothesis. For example, if negative experimental results are obtained with an unproven method of food testing, the findings may be uninterpretable without considerable ancillary data. That is, if a challenge method has never been shown to be adequately reliable and valid, or if it has low reliability and validity (cf. King, 1988), negative results (i.e., no overall difference between food and placebo challenges) could simply be due to an insensitive testing technique. Patients with true food sensitivity might fail to react under such conditions. To some degree, a large sample size could overcome the low reliability of a test by increasing statistical power (Cohen, 1977), but this would not help if the testing method was invalid.

A related potential stumbling block to clear inference is testing the patient rather than the hypothesis. Frequently, patient self-reports (or parental reports) of symptoms to food are essentially the sole criterion for admission to a study of food effects. When negative experimental results are then obtained under these conditions, it is unclear whether they suggest that such food effects do not actually exist or that self-report is not necessarily valid. In other words, it remains possible that some individuals who report symptoms related to foods are accurate in their self-observations (true responders), while others are not. If an experimental group consists primarily of the latter, the problem in interpretation of results is obvious.

Analogous problems occur when *positive* results are obtained under experimental conditions, with the primary threat to validity being inadequate blinding. Blinding methods could be validated, however, by formal pretesting or by including manipulation checks for awareness within the design. Such evidence should be included in the report because the simple assertion that preparations "were indistinguishable" is not convincing. Responses to placebo challenges (particularly at significantly greater than baseline levels) would be consistent with successful blinding.

We conducted two projects that bear on these issues regarding the behavioral effects related to food intake; one addressed biologic factors, the other social factors. The first investigated the effects of double-blind food challenges in children reported to be behaviorally responsive to selected foods.[1] The second explored the beliefs of college students about food effects on behavior.[2]

BEHAVIORAL EFFECTS OF FOOD CHALLENGES IN CHILDREN

The first project consisted of two parts: (1) developing and verifying a method for delivering blinded food challenges, and (2) using this method in an independent experimental test of the effects of food challenges in selected children. Because parental report was considered to be of potentially low validity, a positive screening test in addition to a history of behavioral response to a food or foods was required for admission to the experimental phase of the project.

In the first study, Hollister-Stier food extracts in a glycerol base were mixed with concentrated prune syrup as a disguising agent (de-ionized water was added to the syrup to reduce viscosity) for sublingual delivery. The placebo consisted of the diluent alone. Nineteen foods were prepared in this manner, including milk, wheat, combined cane/beet sugar, orange, and egg.

Pretest of the Blind

After being given identified samples of the placebo, 15 normal children (ages 6 1/2 to 12 1/2 years) with no known food sensitivities were administered food and placebo doses sublingually in a double-blind, randomized design. Random orders of paired food extracts and/or placebos were presented, and the children were asked to guess the identities of the doses (i.e., prune juice alone or prune juice mixed with another food). A second dependent measure was whether the pairs of doses (i.e. food/food, food/placebo, placebo/food, or placebo/placebo) tasted the same or different. The children performed at a level consistent with chance on both measures, indicating that they could not discriminate doses containing food extracts from the placebos and could not distinguish discrepant pairs (e.g., food/placebo) from identical ones (e.g., food/food).

Experimental Stage

These verified blind doses were then used to assess the behavioral and cardiac effects of food challenges in screened children. This process consisted of four steps. First, children who reportedly reacted behaviorally to a food were sought by advertisement and physician contact. Second, the parent or guardian was interviewed by telephone to eliminate children who were not appropriate for such reasons as having serious medical or psychiatric conditions or being unable to complete all parts of the study. A resultant pool of 122 children was admitted to the third step, the double-blind screening.

Screening. Each child was tested individually under double-blind conditions with as many as five food/placebo sublingual test pairs. That is, the test foods selected for each child (e.g., egg,

---

[1] King, D.S., Margen, S., Ogar, D.A., and Durkin, N. Double-blind food challenges affect selected children's behavior and heart rate. (Submitted for publication.)

[2] King, D.S. and Wisocki, P.A. A survey of student beliefs regarding behavioral effects of foods. (Submitted for publication.)

milk, wheat) were each paired with a placebo dose, and the sequence of food and placebo within each pair was randomized. On each trial, the child was given the coded dose, was asked to expectorate it after two minutes, and was then observed for ten minutes. After each trial, the child was rated on a ten-point scale for behavioral change, if any. If a greater change was rated on a food trial than on its paired placebo, the child was eligible for the subsequent experiment.

Experimental Phase Subjects. From the screening, 24 children (17 boys, 7 girls), ranging in age from 5 1/2 to 13 years (median = 8), were admitted to the experimental phase. Approximately one-third of the children were hyperactive by Conners' Abbreviated Parent Questionnaire (Conners, 1980), and none were on regular medication.

Method. In the balanced, repeated-measures design, each child was randomly assigned to one of six sequences of three test-food and three placebo challenges. The three test-food challenges for each child were identical, and consisted of the test food that evoked the greatest behavioral change relative to placebo during the screening tests. Dependent measures included heart rate, a five-minute continuous performance test, the finger tapping test (non-dominant hand), parent's blind ratings of behavioral change, and blind judges' ratings. The latter variable was obtained using videotapes of each trial. During the experimental trials, the child sat facing a one-way mirror through which the trials were videotaped and the parent observed.

For each trial of the experimental procedure, the child first sat quietly for a brief period to lower the heart rate, which was then measured for one minute. The sublingual dose was then given, held under the tongue for two minutes, and expectorated. A second one-minute heart rate was obtained while the child was asked to sit quietly. The continuous performance test was then administered, with the child required to press a button only when the letter X appeared on a screen after an A. The finger-tapping test was then presented, followed by the second post-dose heart rate (one minute). At this point, the parent completed the rating, and a ten-minute washout period followed. The latter was extended if symptoms were present and persisted. The videotapes were later viewed independently by two trained judges.

Results

The children's heart rates did not differ for test foods and placebos either before the doses or at the end of the trial after the doses. However, a significant elevation in heart rate occurred on the first post-dose heart rate measure for the test-food trials compared with the placebos ($p < 0.02$). The absolute change, although reliable across subjects, was small (1.6 beats increase per minute).

Heart rate variability also did not differ before or at the end of the trial. However, a significant substance- (test-food, placebo) by-position (1st, 2nd, 3rd dose) interaction ($p < 0.04$) suggested that the second test-food dose tended to suppress heart rate variability relative to the second placebo.

Analysis of the continuous performance test data also indicated a significant substance-by-position interaction ($p < 0.03$) for false positive responses. Significantly fewer false positives occurred on the third (and final) test food trial compared with the third placebo ($p < 0.006$).

Findings of the finger-tapping test were similar to those of the continuous performance test. The substance-by-position interaction was again significant ($p < 0.05$), with children tapping more after the third test-food trial than after the third placebo.

Parental ratings did not differ for test-food and placebo trials. Nor did they correlate with the ratings of the blind judges, whose ratings did correlate with each other ($r = 0.56$). The judges' ratings were averaged together, and a significant substance-by-position interaction was obtained ($p < 0.04$). These results were complex, however, and the reader is referred to King et al.[3], for additional details. Post-hoc analyses suggested that the judges may have detected cumulative dose effects.

Discussion

Taken together, these findings suggest that a wide variety of foods may be capable of evoking subtle cardiac and behavioral effects under double-blind conditions. The pattern of findings obtained is most consistent with an increase in arousal as a mediator, as suggested by research on stimulants. For example, stimulants tend to decrease false positives on the continuous performance test (Zahn et al., 1975). Although methylphenidate tends to increase heart rate, low doses of the drug (0.2-0.3 mg/kg) may have no effect on heart rate in hyperactive children (Hastings and Barkley, 1978), and it may also suppress heart rate variability (Porges et al., 1975).

Nevertheless, the effects obtained in this study, although statistically significant, were of no clinical importance because of their subtlety. That is, effects of the magnitude we observed cannot account for hyperactivity, emotional disturbance, or other clinically relevant symptoms sometimes attributed to foods. In addition, the small differences in ratings and other dependent measures between test-food and placebo challenges raise serious questions about the use of sublingual testing for individual diagnosis under conditions similar to ours.

A major limitation of this study was the necessity of using minute doses of foods via an unproven testing technique for the challenges. Ease of demonstrable blinding was a major consideration in our choice of this method, but generalization to more realistic doses of foods is difficult. Nevertheless, some studies employing dietary manipulations have obtained larger, more clinically relevant effects (Egger et al., 1983; Egger et al., 1985; Kaplan et al., 1989).

---

[3] King, D.S., Margen, S., Ogar, D.A., and Durkin, N. Double-blind food challenges affect selected children's behavior and heart rate. (Submitted for publication.)

BELIEFS ABOUT FOOD AND BEHAVIOR

As mentioned above, patients with multiple chemical sensitivities commonly report that they experience a wide spectrum of symptoms related to ingestion of a variety of foods. However, little is known about the prevalence of similar beliefs in the general population and the frequency among non-patients of beliefs in the behavioral effects of foods and in personal reactions to foods.

Method

To address this question, we developed the Food-Behavior Inventory,[4] , which consisted of three sections.

1. General beliefs about the effects of food on behavior (e.g., Hyperactivity can be caused or aggravated by sugar). Responses were indicated on a five-point Likert scale from 'strongly agree' to 'strongly disagree'.
2. Beliefs about the effects, if any, of sugar and other foods on the respondent. If the respondent indicated a belief that he or she was affected by sugar or other foods, a symptom check list was provided to indicate the effects perceived. The check list included both mental (e.g., depression) and physical (e.g., bloated stomach) symptoms.
3. Dietary alterations the respondent had made because of perceived personal effects of specific foods.

Using the Food-Behavior Inventory, we conducted a systematic mail survey of the student population at a large northeastern university [5]. Ninety-five of 232 questionnaires were returned, yielding a response rate of 41%.

Results

Results indicated that belief in the behavioral effects of foods is widespread in this population. For example, approximately 90% agreed that hyperactivity could be caused or aggravated by sugar, and large majorities reported that symptoms such as depression, irritability, or fatigue could be produced or aggravated by a food allergy or a food sensitivity.

These beliefs were not merely abstract, because 90% of the sample reported that they reacted with symptoms to sugar, at least on occasion. The most frequently indicated symptom following sugar ingestion was euphoria/high, followed by happiness, headaches, anxiety/tension, and fatigue. The students indicated other symptoms less commonly. In addition, more than a third of the students reported that some food other than sugar (or caffeine or alcohol) affected them behaviorally. Gastrointestinal symptoms were most frequently reported in this category, fol-

---

[4] King, D.S. and Wisocki, P. A survey of student beliefs regarding behavioral effects of foods. (Submitted for publication).

[5] King, D.S. and Wisocki, P. A survey of student beliefs regarding behavioral effects of foods. (Submitted for publication.)

lowed by fatigue, headaches, and irritability. Approximately two-thirds of the students reported that they had altered their diets because of food effects on them. The foods most commonly avoided were sweets and meats.

Demographic analyses indicated a direct relationship between years of education and the general belief in the behavioral effects of foods. The more educated the students, the more they tended to believe in such effects. This correlation may possibly reflect an "acculturation" process within the student population because gender, race, and social class of origin (parental) were not related to belief.

Discussion

Overall, the results indicate that the belief that foods are associated with symptoms, including behavioral symptoms, is virtually a cultural norm in this population. They suggest that, if directly questioned under clinical or research conditions, the majority of students may well ascribe personal symptoms to some food. Although items concerning chemicals were not included in this survey, the question arises whether or not similar belief structures would emerge regarding them.

In any case, our findings suggest that clinicians and researchers must view the self-report of symptoms to foods with caution, because nearly everyone in this population claimed to have experienced them. Although it is speculation, a more relevant distinction between patient and non-patient groups than the belief that one experiences symptoms to foods may be whether or not medical attention is actively pursued because of them.

These studies are consistent with the view that both biologic and psychosocial factors can contribute to food-related symptom perception and report. It remains to future research to determine the relative importance of each

REFERENCES

ASHFORD, N.A. and MILLER C.S. (1991). Chemical Exposures: Low Levels and High Stakes. Van Nostrand Reinhold, New York.

COHEN, J. (1977). Statistical Power Analysis for the Behavioral Sciences (revised edition). Academic Press, New York.

CONNERS, C.K. (1980). Food Additives and Hyperactive Children. Plenum Press, New York.

EGGER, J., CARTER, C.M., GRAHAM, P.J., GUMLEY, D., and SOOTHILL, J.F. (1985). Controlled trial of the oligoantigenic treatment in the hyperkinetic syndrome. Lancet i: 540-545.

EGGER, J., CARTER, C.M., WILSON, J., TURNER, M.W., and SOOTHILL, J.F. (1983). Is migraine food allergy? Lancet ii: 865-869.

KAPLAN, B.J., McNICHOL, J., CONTE, R.A., and MOGHADAM, H.K. (1989). Dietary replacement in preschool-aged hyperactive boys. Pediatrics 83:7-17.

HASTINGS, J.E. and BARKLEY, R.A. (1978). A review of psychophysiological research with hyperkinetic children. J. Abnorm. Child Psychol. 6:413-447.

KING, D.S. (1988). The reliability and validity of provocative food testing: A critical review. Med. Hypotheses 25:7-16.

PORGES, S.W., WALTER, G.F., KORB, R.J., and SPRAGUE, R.L. (1975). The influences of methylphenidate on heart and behavioral measures of attention in hyperactive children. Child Dev. 46:727-733.

ZAHN, T.P., ABATE, F., LITTLE, B.C., and WENDER, P.H. (1975). Minimal brain dysfunction, stimulant drugs, and autonomic nervous system activity. Arch. Gen. Psychiatry 32:381-387.

NEUROPSYCHOPHYSIOLOGIC OBSERVATIONS IN PATIENTS PRESENTING WITH ENVIRONMENTAL ILLNESS

JOHN C. SELNER, MD
HERMAN STAUDENMAYER, PhD
Allergy Respiratory Institute of Colorado
Denver, Colorado

BACKGROUND

In 1975, the authors recognized a clinical syndrome that varied widely among patients but seemed to have certain common characteristics. The syndrome typically involved multiple systems, expressed with innumerable symptoms and a general malaise that could not be explained by any single etiological factor. Patients typically exhibited no objective physical or laboratory evidence of disease (Terr, 1986, 1989). The two most common characteristics were fear of xenobiotics (chemicals or drugs, food or water) and a belief that symptoms were directly attributable to environmental agents (Brodsky, 1983). Many of these individuals had been treated by clinical ecologists and this appeared to be the origin or reinforcement of their beliefs. In the ecology literature (Randolph and Moss, 1979), these patients were referred to as universal reactors.

The authors' early attempts to validate the environmental beliefs about symptoms made use of open, noncontrolled, exposure challenges (Selner and Staudenmayer, 1985). With open challenges, patients consistently reported symptoms they had associated with ambient, workplace, or domestic exposure. However, when the suggestion was made that a patient would be exposed to an active substance but was in fact given only clean air or a placebo, the symptom response was often pronounced and dramatic. Likewise, when the suggestion was made that a placebo or clean air was being given when in fact a suspect agent was introduced, patients typically did not report symptoms. Thus, subjective reports seemed unreliable, and alternative methods of assessment were needed. It became necessary to:

1) Find some objective physiologic measures to assess change from an asymptomatic baseline, and

2) Mask the exposure and control it with sham or placebo challenges, relying on consistency of response across several challenges to determine the statistical probability of a reliable response not attributable to chance.

NEUROPSYCHOPHYSIOLOGY

The rationale for using psychophysiologic measurements of EEG, EMG, peripheral finger temperature, and galvanic skin response was twofold. First, the general malaise symptoms

Address all correspondence to: Dr. John C. Selner, Allergy Respiratory Institute of Colorado, 5800 E. Evans Avenue, Denver, CO 80222.

reported by universal reactors were very similar to those seen in psychophysiologic or stress disorders (Fann et al., 1982). In these disorders, psychophysiologic stress profiling for assessment and biofeedback for treatment had been reported to be effective. Second, clinical ecologists had used a version of Hans Selye's General Adaptation Syndrome stress model to describe the symptoms they exclusively attributed to environmental factors (Selye, 1946). Thus, stress physiology would seem to provide an assessment of objective changes during provocation challenge. Affected patients were presumably describing *state* change, and a more homeostatic physiology should be observed during relaxation (nonexposed) periods when symptoms were absent; deviations from this baseline physiology should be observed when patients were exposed.

For example, many of the universal reactors treated by the authors reported cognitive effects characteristic of attention disruption (e.g., difficulty concentrating, remembering, making decisions). These attention disruptions are characteristic of cognitive activation or scanning. This kind of state change in EEG had been reported in the study of an astronaut during a Gemini launch (Adey et al., 1967). Under conditions of relaxation, his EEG showed a homeostatic unimodal alpha frequency at 10 Hz. In the spacecraft shortly before lift-off, the spectral pattern showed a multimodal pattern of slow-wave theta (4-7 Hz) and several alpha frequency peaks in the 8-13 Hz range characteristic of changes in focus of attention in fractions of a second. When the astronaut reached orbit, the EEG normalized again. This response is a reflection of fear and anxiety in anticipation of lift-off.

Neuropsychophysiologic Methods

To explore the EEG correlates of attentional and emotional disorders and classify patients into distinct categories, the authors employed EEG neuromapping techniques, incorporating spectral analysis for slow EEG frequencies between 2-15 Hz characteristic of attentional processing and higher beta frequencies (15-25 Hz) associated with anxiety (Duffy, 1988). Peripheral physiology measurements of scalp EMG, galvanic skin response, and peripheral finger temperature were also employed with patients who showed CNS symptoms of depression and anxiety.

State vs. Trait in Neuropsychophysiology

The universal reactors the authors studied often did not show homeostatic physiology during the baseline relaxation period in which they were symptom free. A neuropsychophysiologic baseline is important because it defines the *trait* characteristics of the individual. The astronaut's trait EEG was homeostatic, and the alterations referred to as state changes associated with the anxiety of lift-off returned to normal when the astronaut reached orbit. Although EEG changes associated with cognitive activation or emotional responses are normal, the same states can become traits if they become chronic and persist under all circumstances including relaxation. Such trait states are well characterized in populations with chronic psychologic disorders (Itil, 1982). For that reason, this study used a population of patients with well-defined psychologic disorders as a control in assessing universal reactors.

The distinction between the assessment of EEG trait characteristics and state characteristics is the rationale for obtaining psychophysiologic measurements associated with provocation challenges. The psychologic state changes observed should correlate with the chemical exposure or the induced symptoms. For example, in toxicologic studies of the effects of caffeine on the central nervous system, reliable characteristic EEG changes in spectral displays have been demonstrated to be associated with dosing (Etevenon et al., 1988).

A Comparison Study of Baseline Trait Psychophysiology

Three groups of subjects were classified according to EEG spectral patterns for frequencies below 15 Hz, EEG beta level (15-25 Hz) and EMG scalp muscle activity during baseline relaxation when they were symptom free (Staudenmayer and Selner, 1990). Fifty-eight universal reactors, all of whom presented with environmental illness and beliefs reflecting multiple chemical sensitivities, constituted the experimental group. One control group consisted of 55 subjects who reported no psychologic symptoms during the past year. Either they had no symptoms at all or were under medical care for symptoms or disease without apparent primary or secondary psychologic disorders. The second control group consisted of 88 subjects who were outpatients presenting with primary psychologic disorders or secondary psychologic disorders that modulated physical conditions. The most common problems in this latter group were depression, anxiety, mood swings, phobia, panic, and insomnia.

These subjects did not identify elements of environmental illness to explain their problems and were seeking appropriate psychologic treatment.

Psychophysiologic data were recorded in a sound-attenuated, dimly lit room in which the subjects were supine and relaxing with eyes closed. The results showed

1) More universal reactors (43%) had high levels of scalp muscle tension than did controls (16.4%) and psychologic subjects (16.9%) ($p < 0.001$).

2) The proportions of universal reactors (43%) and psychologic subjects (38%) with high levels of EEG beta power were not significantly different. However, significantly fewer control subjects (22%) had high beta power ($p = 0.02$).

3) The distribution of subjects across the eight categories of slow-wave EEG spectral patterns below 15 Hz. were statistically different among the three groups ($p < 0.001$). Although the distribution of universal reactors was not different from the distribution of psychologic subjects ($p = 0.22$), both groups were significantly different from the distribution of control subjects ($p < 0.001$). Control subjects were more likely to be classified into the unimodal alpha categories, and subjects in the psychologic and universal reactors groups were more likely to be classified into the other spectral categories associated with attentional disorders.

Thus, this study suggested that universal reactors had *trait* psychophysiology patterns deviant from homeostatic levels but similar to patterns seen in patients with overt psychologic disorders.

Case Report

The following case illustrates how the examination of spectral patterns obtained by brain mapping can identify EEG correlates of symptoms of depression (insomnia, fatigue, and loss of concentration).

A 68 year old retired insurance agent presented with a history of recurrent asthma, rhinosinusitis, headaches, pharyngitis, gastritis, general malaise, insomnia, severe fatigue, mental confusion, loss of concentration, and memory loss. He had detected onset of these symptoms two years earlier after a fungicide fogging of a crawlspace at his home. An ecologist had used provocation chemical testing and diagnosed multiple chemical sensitivity (MCS). The patient was advised to leave his home for a motel, to discontinue driving his automobile, and to initiate detoxification strategies, including anti-oxidant supplements (Vitamin C).

Following this, an ecologic dentist removed his amalgams and reinforced the need for continued detoxification procedures. A second ecologist recommended similar therapy, but the patient did not improve significantly and continued to experience all his symptoms.

The patient's social history suggested occupational stress complicated by alcohol abuse and loss of his lifelong business. Psychologic assessment led to a working diagnosis of endogenous depression and xenophobia with iatrogenic reinforcement. EEG spectral patterns during a wakeful state, recorded from 19 sites according to the international 10-20 systems using a Neurosearch-24 system, Lexicor, Boulder, CO, were recorded at baseline before therapy was begun. The spectral patterns were uniform across sites and symmetric across hemispheres. There was significantly more slow wave delta activity (2-4.5 Hz) relative to alpha (7-12 Hz). Amitriptyline was initiated, and biofeedback-assisted attentional training and systematic desensitization strategies were introduced in which the patient was gradually exposed to increasing quantities of the suspect offending chemical odors. EEG trait characteristics associated with depression normalized when anti-depressant medication was introduced. Delta was reduced to homeostatic levels and alpha was the dominant EEG frequency. During the one-year course of Amitriptyline therapy, the patient experienced a cessation of all symptoms and was able to return to his home, wife, automobile, and the pursuit of his interests, including foreign travel, without recurrence of symptoms.

SYMPTOM SELF-REPORT INVENTORIES

The ability to discriminate verifiable medical disease from projections of psychosomatic illness is a fundamental diagnostic problem in this population. One of the most frequent questions clinicians ask is, "Is there a paper-and-pencil test to discriminate patients with psychologic problems from those with bonafide toxicologic problems?" In the authors' experience, universal reactors usually deny or do not report psychologic symptoms but instead emphasize somatic symptoms they attribute to biologic mechanisms set off by chemical exposure. The following study was designed to evaluate a short, symptom self-report instrument, the Deragotis Brief Symptom Inventory (Derogatis and Melisaratos, 1983), to see whether it could facilitate differential diagnosis.

Method

Subjects. A total of 202 subjects, aged 15-68 years, underwent medical and psychologic assessment by review of medical records. From this assessment, they were classified into three groups. These same subjects participated in the study of EEG characteristics described above (Staudenmayer and Selner, 1990).

Controls. Subjects were symptom-free during the past year or were under medical care for medical symptoms or disease without apparent primary or secondary psychologic disorders. The group consisted of 55 subjects, 24 men and 31 women, with a mean age of 38.9 (S.D., 11.4 years).

Psychologic patients. Subjects were outpatients who presented with primary psychologic disorders or had secondary psychologic disorders that modulated physical conditions. They received psychologic treatment and did not project their problems onto the environment. The group contained 88 subjects, 42 men and 46 women, with a mean age of 37.6 (S.D., 10.8 years).

Universal reactors. The subjects presented with environmental illness and a belief system reflecting multiple chemical sensitivities. The group comprised 58 subjects, 17 men and 41 women, with a mean age of 40.2 (S.D., 10.2 years). When challenge tested under controlled laboratory conditions, all subjects failed to show objective chemical sensitivity consistent with their reported experiences. Challenges included active chemical agents at concentrations consistent with exposure levels the subjects reported had induced adverse responses. These challenge tests were blinded, randomized, and placebo- (sham) controlled using olfaction maskers.

Procedure

The Brief Symptom Inventory (BSI) is a 53-item, self-report inventory—a short form of the SCL-90R (Derogatis, 1975). It has nine psychiatric dimensions and a general symptom index. Each item was rated on a five-point scale (not at all = 0, extremely = 4). Subjects were asked to rate the average amount of distress they experienced during the past year (or an appropriate period if onset of illness was less than one year ago for universal reactors). The score for each dimension was the average of all items in that dimension. Dimension scores were analyzed by one-way analysis of variance (ANOVA) with Tukey post-hoc group comparisons ($P < 0.05$).

TABLE 1. Brief Symptom Inventory Scores and Anova with Tukey Comparisons among Groups

	Group							
	CONTROL (C) (n=55)		UNIVERSAL REACTOR (UR) (n=58)		OUTPATIENT PSYCHOLOGIC (P) (n=88)		ANOVA	TUKEY Post Hoc
DIMENSIONS	M	SD	M	SD	M	SD	F P	COMPARISONS
SOMATIZATION	0.56	0.66	1.30	0.83	1.08	0.86	12.8<.001	C<P=UR
OBSESSIVE COMPULSIVE	0.90	0.59	1.42	0.91	1.79	0.91	19.6<.001	C<UR<P
INTERPERSONAL SENSITIVITY	0.75	0.63	0.94	0.86	1.60	1.05	17.9<.001	C=UR<P
DEPRESSION	0.67	0.61	0.94	0.90	1.63	0.99	23.4<.001	C=UR<P
ANXIETY	0.85	0.69	1.13	0.82	1.69	1.03	16.7<.001	C=UR<P
HOSTILITY	0.69	0.50	0.85	0.67	1.21	0.78	10.9<.001	C=UR<P
PHOBIC ANXIETY	0.26	0.43	0.41	0.63	0.69	0.75	7.9<.001	C<P
PARANOID IDEATION	0.70	0.60	0.85	0.84	1.10	0.87	4.5=.01	C<P
PSYCHOTICISM	0.35	0.42	0.52	0.64	1.17	0.85	28.9<.001	C=UR<P
GENERAL SYMPTOM INDEX	0.65	0.39	0.97	0.59	1.32	0.74	20.4<.001	C<UR<P

Results

Results for BSI dimensions are presented in Table 1. ANOVAs comparing the three groups were significant for all dimensions. The individual group comparisons showed different relations for the dimensions. The universal reactor group (UR) scored significantly higher on the general symptom index than the control group (C) but lower than the psychologic group (P) (C<UR<P). The same relation was found for the obsessive-compulsive dimension. For the phobic anxiety and paranoid ideation dimensions, UR was not statistically different from C or P, but C was significantly less than P (C<P). A pattern with no difference between UR and C, but with UR and C both significantly lower than P (C=UR<P) was found for the dimensions of interpersonal sensitivity, depression, anxiety, hostility, and psychoticism. Finally, UR scored highest on somatization and was not statistically different from P, but both UR and P were significantly higher than C (C<P=UR).

Discussion

A denial of psychologic symptoms, especially depression and anxiety, commonly seen in universal reactors is consistent with their belief about the cause of their problems. Simon et al., (1990) have reported similar results with symptom checklists. In the present study, symptom self-reports were not compatible with neuropsychophysiologic characteristics.

It is not surprising that the UR somatization score was high because this is consistent with their belief about environmental sensitivity. However, their elevated score on obsessive compulsiveness suggests a prevalence of perfectionism, a characteristic often seen in various psychologic disorders, including personality disorders.

Symptom checklists are of questionable value in diagnosis because they usually indicate how patients in this population present themselves. The Brief Symptom Inventory, like all self-report instruments, is difficult to interpret (Brewin, 1987) because the therapist does not know whether it represents:

1) How the individual experiences himself
2) How the individual would like to see himself, or
3) How the individual thinks the examiner would like to see him.

REPRESSED CHILDHOOD TRAUMA AND PERSONALITY DISORDERS

In treating universal reactors for the past 12 years, especially those with the most severe mobility, dietary, and social restrictions, the authors believe the number of revealing incidences of early childhood sexual and physical abuse was excessive (Selner and Staudenmayer, 1986). Many of these patients had suppressed or repressed memory of these events at the time of initial assessment and did not reveal or uncover them until trust was established in a therapeutic relationship. In a subset of 20 women, taken from the 41 universal reactor women in the study described above, who also underwent long term psychotherapy, the incidence of childhood sexual abuse defined as intercourse was 60%. These data suggest that somatization so commonly seen in universal reactors who are without objective evidence of disease may have a basis in the sequelae of childhood abuse. Furthermore, these personal and often traumatic memories are usually not shared during intake assessment.

These findings should not be misinterpreted as contending that many universal reactors are exclusively victims of sexual abuse. The more important question is, what is the developmental impact of childhood sexual and other abuse? The profound sequelae that such trauma have on personality development in the child have been well documented (Gelinas, 1983; Herman et al., 1986; Brown and Anderson, 1991). These sequelae can be characterized as leaving individuals without appropriate coping skills and with experiences of depression and anxiety. The child's reality is one of vulnerability in a family where the caretakers and loved ones are often abusers. Survival may demand that the psyche defend itself by repressing trauma. Projecting or attributing the effects of this trauma onto external factors is commonly seen in many forms of somatoform disorders. In the population of universal reactors the authors have studied, the specific projection identified is onto the environment (Staudenmayer and Camazine, 1989).

These patients are all too often dismissed as malingerers without the benefit of a rigorously pursued medical and psychologic evaluation. They should be recognized for the fragile psyches they possess and for how desperately they have attempted to survive in a world that they see as fundamentally unsafe.

CAUTIONARY NOTES AND DIRECTIONS FOR FUTURE RESEARCH

First, ambient workplace and domestic chemicals can cause definable physical disease and must be thoroughly evaluated in a medical workup. Second, the present claims about psychologic etiology have been restricted to a subpopulation called universal reactors. Categorically, the authors' studies, like those of others (Brodsky, 1983, 1987; Rix et al., 1984; Schottenfeld, 1987; Stewart, 1987), have shown that many universal reactors have a psychogenic basis for their symptomatology and belief. In some cases, their disability has been overcome with psychologic and psychiatric intervention, and they are relieved of extreme xenophobias and can return to a functional work and social existence. Third, to date the authors know of no operational definition for so-called multiple chemical sensitivity (MCS) as used by practitioners with a clinical ecologist orientation, Cullen's (1987) definition notwithstanding. However, they have not seen objectively verifiable disease that can be attributed to this mysterious diagnosis.

Psychophysiologic profiling and specifically EEG neuromapping are not definitive diagnostic tests in and of themselves. They are one tool in a complex armamentarium used in assessment. This tool like other behavioral measures, e.g., neurobehavioral or neuropsychologic testing, must be viewed cautiously in terms of attributing etiology and causality of effect (Matarazzo, 1990). Pre- and post-exposure data are one way of establishing a potential cause in observed change. Without existing data for pre-exposure baseline, causality data can only be interpreted as suggestive. Attributing causality to a particular variable in the absence of confirming data does not seem appropriate. Inferences are possible only when test results provide a scientific basis for establishing criteria of change with specific attribution to disease. Object disease resulting from single or multiple chemical exposure and involving multi-organ systems certainly exists, but in the nebulously defined syndrome called MCS, no such data exist.

Critique of Challenge Studies

Challenge studies may be questioned because of suboptimal dose or duration of exposure, just as irritant threshold levels may be affected by the presence or absence of co-existing factors (e.g., heat, humidity, sound, visual cues, or electromagnetic influence). The nutritional patterns of the patient, including specific foods or nutrient intake, may also be important to challenge outcomes. Furthermore, detoxification status is viewed in the clinical ecology literature as a determinant of response (Ashford and Miller, 1991), although no literature to date substantiates these critiques and contentions. However, they are potentially confounding issues to be evaluated in a parametric design under a scientific protocol.

Control Groups and the True Positive Dilemma

One of the fundamental problems in research involving MCS is to define an appropriate control group for this population. Subjects in a control group should have all the characteristics of the experimental group except the independent variable—in this case, MCS. Several kinds of populations have been suggested, each of which may have its own problems. If, for example, chronic pain and chronic fatigue populations are used as control groups and if patients with physical or laboratory evidence of objective disease are excluded, the remaining patients seem to have the same potential for psychologic aberrations as universal reactors do. Thus, they may not be appropriate control groups. The authors have tried to identify reproducible reactions to specific chemical exposures in a control group whose members have specific definable diseases such as asthma, chronic rhinitis, urticaria, and cardiovascular illness as well as functional disorders such as attentional deficit disorder and closed head injury. Gathering data on such control populations would allow for more conclusive interpretation of neuropsychophysiologic and neurobehavioral tests in a universal reactor group. The compilation of data on such groups is a direction for future research.

Interactive Effects of Toxicology and Psychologic Factors

The media have presented the polarization between clinical ecology on one side and allergy and psychiatry on the other as a dichotomy, that is, patients either have MCS or they have one form or another of a psychologic disorder. The authors do not share this position but believe several explanatory mechanistic options exist, including functional and anatomical neuropathology, other organ or multi-organ toxicologic response, immunotoxicologic response, metabolic dysfunction, and classical allergies—IgE mediated response, to name a few.

For study design, logic dictates four possible groups: a normal control group with neither psychologic or toxicologic disorders; a second group consisting of individuals with both psychologic and toxicologic problems; a third group comprising individuals with psychologic problems but without toxicologic disorders; and a fourth group of individuals with clear toxicologic problems but without psychologic disorders. In future research, subjects should be carefully screened with reliable and accepted assessment criteria for entry into each of the four groups.

CONCLUSION

The thrust of the authors' work over the last 15 years with patients complaining of environmental illness has been to develop methods for establishing and differentially diagnosing toxicologic and psychologic factors. This has been based on certain presuppositions.

1) In an assessment procedure, scientific method and protocol must be followed.

2) The absence of evidence does not constitute a basis for inferring causality.

3) Psychologic motivation of patients in terms of secondary gain (malingering) and primary gain (repressed or suppressed trauma) must be evaluated in a standard workup of environmental cases.

4) The well being and personal integrity of the patients must be respected. Environmentally ill patients, irrespective of motive or attributed etiology, should be recognized as people who are truly suffering. To arbitrarily dismiss them for one reason or another does them no service.

5) The indiscriminate use of controversial and unproven diagnostic and treatment techniques can reinforce inappropriate ideation about environmental exposure and deny patients access to both alternative explanations of symptom causality and possible avenues to wellness

REFERENCES

ADEY, W.R, KADO, B.S., and WALTER, D.O. (1967). Computer analysis of EEG data from Gemini Flight GT-7. Aerosp. Med. 38:345-359.

ASHFORD, N.A. and MILLER, C.S. (1991). Chemical Exposures: Low Levels and High Stakes. Van Nostrand Reinhold, New York.

BREWIN, C.R. (1987). Theories of motivation. In: Theoretical Foundations of Behavior Therapy. (H. Eysenck and I. Martin, eds.). Plenum, New York. p.282.

BRODSKY, C.M. (1983). Allergic to everything: A medical subculture. Psychomatics. 24:731-742.

BRODSKY, C.M. (1987). Multiple chemical sensitivities and other environmental illnesses: A psychiatrist's view." In: Workers with Multiple Chemical Sensitivities. (M.R. Cullen, ed.). Occup. Med.: State Art Rev. 2(4):695-704.

BROWN, G.R. and ANDERSON, B. (1991). Psychiatric morbidity in adult patients with childhood histories of sexual and physical abuse. Am. J. Psychiatry 148(1):55-61.

CULLEN, M.R. (1987). The worker with chemical sensitivity: An overview. In: Workers with Multiple Chemical Sensitivities. (M.R. Cullen, ed.). Occup. Med.: State Art Rev. 2(4):655-662.
DERAGOTIS, L.R. (1975). SCL-90 Manual. Deragotis, Baltimore, MD.
DEROGATIS, L.R. and MELISARATOS, N. (1983). The Brief Symptom Inventory: An introductory report. Pyschol. Med. 13:595-605.
DUFFY, F.H. (1988). Clinical decision making in quantified electroencephalographic analysis. In: Statistics and Topography in Quantitative EEG. (D. Samson-Dollfus, ed., J.D. Guieu, J. Gotman, P. Etevenon, co-eds.). Elsevier, Amsterdam. pp. 9-26.
ETEVENON, P., PERON-MAGNAN, P., GUILLOU, S., TOUSSAINT, M., GUEGUEN, B., DENIKER, P., LOO, H., and ZARIFIAN, E. (1988). A pharmacological model of "local cerebral activation": EEG cartography of caffeine effects in normals. In: Functional Brain Imaging. (G. Pfurtscheller and F.H. Lopes da Silva, eds.). Hans Huber, Toronto. pp. 171-180.
FANN, W.E., KARALAN, I., POKORNY, A.D., and WILLIAMS, R.L. (1982). Phenomenology and Treatment of Psychophysiological Disorders. Spectrum, New York.
GELINAS, D.J. (1983). The persisting negative effects of incest. Psychiatry 46:312-332.
HERMAN, J., RUSSELL, D., and TROCKI, K. (1986). Long-term effects of incestuous abuse in childhood. Am. J. Psychiatry 143:1293-1296.
ITIL, T.M. (1982). The use of electroencephalography in the practice of psychiatry. Psychosomatics 23:799-813.
MATARAZZO, J.D. (1990). Psychological assessment versus psychological testing: Validation from the Binet to the school, clinic and courtroom. American Psychologist 45:999-1017.
RANDOLPH, T.G. and MOSS, R.W. (1979). An Alternative Approach to Allergies. Lippencott and Crowell, New York.
RIX, K.J.B., PEARSON, D.J., and BENTLEY, S.J. (1984). A psychiatric study of patients with supposed food allergy. Br. J. Psychiatry 145:121-126.
SCHOTTENFELD, R.S. (1987). Workers with multiple chemical sensitivities - a psychiatric approach to diagnosis and treatment. In: Workers with Multiple Chemical Sensitivities. (M.R. Cullen, ed.). Occup. Med.: State of Art Rev. 2(4):739-754.
SELNER, J.C. and STAUDENMAYER, H. (1985). The practical approach to the evaluation of suspected environmental exposures: Chemical intolerance. Ann. Allergy. 55:665-673.
SELNER, J.C. and STAUDENMAYER, H. (1986). The relationship of the environment and food to allergic and psychiatric illness. In: Psychophysiological Aspects of Allergic Disorders. (S.H. Young, J.M. Rubin, H.R. Daman, eds.). Praeger, New York. pp. 102-146.
SELYE, H. (1946). The general adaptation syndrome and disease of adaptation. J. of Allergy 17:231-247.
SIMON, G.E., KATON, W.K., and SPARKS, P.J. (1990). Allergic to life: Psychological factors in environmental illness. Am. J. Psychiatry 147:901-6.
STAUDENMAYER, H. and CAMAZINE, M. (1989). Sensing type personality, projection and universal "allergic" reactivity. J. Psychological Type 18:59-62.
STAUDENMAYER, H. and SELNER, J.C. (1990). Neuropsychophysiology during relaxation in generalized, universal "allergic" reactivity to the environment: A comparison study. J. Psychosomatic Research 34:259-270.
STEWART, D.E. (1987). Total allergy syndrome: A critical review of the 20th century disease. Psychiatry in Canada 1:51-55.
TERR, A.I. (1986). Environmental illness: A clinical review of 50 cases. Arch. Intern. Med. 146:145-149.
TERR, A.I. (1989). Clinical ecology in the workplace. J. Occup. Med. 31:257-261.

CHRONIC FATIGUE SYNDROME

DEDRA BUCHWALD, MD[1], DEBORAH GARRITY, BS[4], RALPH PASCUALY, MD[6], PHALLA KITH, PA[5], RHODA L. ASHLEY, PhD[3], MARK H. WENER, MD[1,3], PAMELA G. KIDD, MD[3], WAYNE J. KATON, MD[2], AND JOAN E. RUSSO, PhD[2]

[1]Department of Medicine
[2]Psychiatry and Behavioral Sciences
[3]Laboratory Medicine/Pathology
[4]The School of Medicine
University of Washington
[5]Chronic Fatigue Clinic, Harborview Medical Center
[6]Sleep Disorders Clinic, Providence Medical Center
Seattle, Washington

OVERVIEW OF CHRONIC FATIGUE SYNDROME

During the past several years, much attention has been focused on an illness called chronic fatigue syndrome (CFS). CFS is a disabling systemic illness characterized by severe persistent fatigue often associated with fevers, pharyngitis, myalgias, headache, neurocognitive difficulties, sleep disturbances, and depression. It is a provisional diagnosis, to be considered only after all other potential causes of illness have been reasonably excluded. CFS frequently has a sudden onset following a flu-like, presumably viral, illness. Most commonly, CFS affects young, previously healthy adults and is more common in females. The cause of CFS remains obscure although findings from a growing body of literature suggest that immunologic abnormalities, infectious agents—particularly viruses, and psychologic factors may play an important role (Komaroff and Buchwald, 1991; Buchwald and Komaroff, 1991).

Historical review suggests that chronic fatigue may be an old clinical problem. Epidemic and sporadic cases of unexplained fatigue associated with a variety of symptoms including fever, pharyngitis, headache, myalgias, and depression have been reported for the last 100 years. In particular, epidemics of an illness remarkably similar to CFS, variously labeled myalgic encephalomyelitis, Icelandic disease, or Royal Free disease, have been described for half a century (Medical Staff, 1957; Acheson, 1959). A post-infectious syndrome of debilitating fatigue following exposure to a variety of agents also has been reported. Finally, fibromyalgia, a chronic condition characterized by myalgias, fatigue and tender points on examination, is similar, if not identical, to CFS (Buchwald et al., 1987a; Goldenberg et al., 1990).

Note: Supported in part by grants RO1AI26788 and RO1AG08240 from the National Institute of Allergy and Infectious Diseases and the National Institute of Mental Health, and by a Young Investigator Award from the National Alliance for Research on Schizophrenia and Depression to Dr. Buchwald.

Address all correspondence to: Dr. Dedra Buchwald, Harborview Medical Center, 325 9th Avenue, ZA-60, Seattle, WA 98104.

Heterogeneous and loosely drawn case definitions have probably contributed to an apparent increase in CFS. In 1988, a consensus panel convened by the Centers for Disease Control (CDC) attempted to identify a discrete and homogeneous syndrome and proposed formal diagnostic criteria for CFS (Holmes, et al, 1988). This case definition is intended for use by providers as a guide in evaluating patients with chronic fatigue of unknown cause. Although the CDC case definition may facilitate research on CFS, it probably excludes most patients who report symptoms of chronic fatigue.

To meet criteria, patients must have both of 2 major and 8 of 11 minor symptoms or 6 of 11 minor symptoms and 2 of 3 physical findings (Table 1). These criteria are controversial because they may not identify a discrete group of patients and because the presence of psychiatric dysfunction is both a minor symptom criterion and potential grounds for exclusion. Moreover, because these criteria were developed for adult populations, they may not be appropriate for children and adolescents in whom the average number of symptoms is lower and CFS appears to be less common.

Pathophysiology

Many theories on the etiology of CFS have been postulated; however, strong evidence is lacking to support any of them. Most investigators believe that no single agent will prove to be the cause in most, or all, cases. A plausible hypothesis is that CFS occurs in a vulnerable individual in whom depression, atopy, viral infections, or other factors have resulted in some degree of immune compromise that may not be measurable by routine testing. Subsequently, latent viruses are reactivated and contribute directly to the morbidity by causing symptoms and indirectly by eliciting an ongoing immune response. More specifically, the production of cytokines may be responsible for the symptoms of CFS. Cytokines are known to cause the flu-like symptoms associated with many acute infections and have been shown to induce a CFS-like syndrome when administered for therapeutic purposes (McDonald et al., 1987; Denicoff et al., 1987). As in any chronic illness, symptoms lead to disability which is both due to, and results in, psychologic dysfunction.

Clinical Manifestations

Symptoms. We have recently reviewed the signs and symptoms common to CFS (Komaroff and Buchwald, 1991). One compelling piece of evidence that CFS is an "organic" illness is the sudden onset in up to 85% of patients. The majority of patients state that their illness began on a particular day with a flu-like syndrome characterized by fever, pharyngitis, adenopathy, myalgias and related symptoms. However, they never fully recovered as with the usual flu.

Thereafter, the most prominent symptom is fatigue. Approximately 25% of CFS patients describe themselves as regularly bedridden or shut-in and unable to work, and one third can work only part-time. Patients were typically very physically active; however, since they became ill, even modest physical exertion results in an exacerbation of many symptoms.

TABLE 1. Centers for Disease Control Criteria for CFS

MAJOR CRITERIA

1. New onset of debilitating fatigue that does not resolve with bedrest and that impairs average daily activity below 50% of the patient's premorbid level for at least six months.
2. Exclusion of clinical condition that may mimic CFS, including:
   a. Malignancy
   b. Autoimmune disease
   c. Infection
      Localized (e.g., occult abscess)
      Chronic/subacute bacterial (e.g., Lyme disease)
      Fungal (e.g., histoplasmosis)
      Parasitic (e.g., toxoplasmosis)
      Human immunodeficiency virus infection
   d. Psychiatric disease
   e. Chronic use, abuse or side effects of prescription or illicit drugs (e.g., antidepressant, alcohol)
   f. Exposure to toxic agents (e.g., solvents, heavy metal)
   g. Chronic inflammatory disease (e.g., sarcoidosis)
   h. Neuromuscular disease (e.g., multiple sclerosis)
   i. Endocrine disease (e.g., hypothyroidism, diabetes mellitus)
   j. Other chronic pulmonary, cardiac, gastrointestinal, hepatic, renal, or hematologic disorder

MINOR CRITERIA

Symptom Criteria
To fulfill a symptom criterion, a symptom must have begun at or after the time of onset of increased fatigability, and must have persisted or recurred over a period of at least six months.

1. Mild fever—oral temperature between 37.5°C and 38.6°C—or chills
2. Sore throat
3. Painful axillary, anterior, or posterior cervical nodes
4. Unexplained generalized muscle weakness
5. Muscle discomfort or myalgia
6. Prolonged (24 hours or greater) generalized fatigue after levels of exercise that would have been easily tolerated in the patient's premorbid state
7. New or different generalized headaches
8. Migratory arthralgia without joint swelling or redness
9. Neuropsychologic complaints (photophobia, transient visual scotomata, forgetfulness, excessive irritability, confusion, difficulty thinking, inability to concentrate, depression)
10. Sleep disturbance (hypersomnia or insomnia)
11. Description of the main symptom complex as initially developing over a few hours to a few days

PHYSICAL EXAMINATION CRITERIA

Physical criteria must be documented by a physician on at least two occasions, at least one month apart.

1. Low-grade fever—oral temperature between 37.6° and 38.6°C, or rectal temperature between 37.8°C and 38.8°C
2. Nonexudative pharyngitis
3. Palpable or tender anterior or posterior cervical or axillary lymph nodes

Although the exertion itself may be tolerated reasonably well, the majority of patients report a marked worsening of their fatigue, cognitive function, adenopathy, pharyngitis, and fevers 6-24 hours later.

Other chronic symptoms, such as fever, myalgias, sleep disorders, headaches, and pharyngitis, frequently accompany the fatigue (Table 2). These symptoms are present virtually all the time, and many are of sufficient severity to impair normal functioning. Furthermore, in response to explicit questioning, the majority of patients state that these symptoms were not a problem before their illness but became common after the illness began.

TABLE 2. Frequency of Symptoms Reported by CFS Patients

SYMPTOM	FREQUENCY (%)*
Fatigue	100
Low-grade fever	60-95
Myalgias	20-95
Sleep disorder	15-90
Impaired cognition	50-85
Depression	70-85
Headaches	35-85
Pharyngitis	50-75
Anxiety	50-70
Muscle weakness	40-70
Post-exertional malaise	50-60
Pre-menstrual worsening	50-60
Stiffness/"gelling"	50-60
Visual blurring	50-60
Nocturia	50-60
Nausea	50-60
Dizziness	30-50
Arthralgias	40-50
Tachycardia	40-50
Paresthesias	30-50
Dry eyes	30-40
Dry mouth	30-40
Diarrhea	30-40
Anorexia	30-40
Cough	30-40
Finger swelling	30-40
Night sweats	30-40
Painful lymph nodes	30-40
Rash	30-40

*Frequencies are estimated from our experience and from studies reported in the literature and reviewed by us (Komaroff and Buchwald, 1991). (Note: This table is as it appears in that review article.)

Finally, neuropsychiatric symptoms, such as word groping, decreased concentration ability, and impaired short-term memory, are common, and a third of patients report at least one episode of complete confusion or disorientation. Some investigators have noted neurocognitive deficits in more than half the patients with CFS. Although these neurocognitive symptoms are suggestive of organicity, they may also be frequent symptoms in major depression. No controlled studies of neurocognitive testing, comparing CFS and depressed patients, have been performed.

Physical Findings. With the exception of non-tender, shotty anterior or posterior cervical adenopathy in 40% of patients, abnormalities in the physical examination are unusual (Table 3). Fevers, low basal body temperatures, posterior and anterior cervical adenopathy and "tender points" have all been described (Goldenberg et al., 1990; Komaroff and Buchwald, 1991).

TABLE 3. Frequency of Physical Findings in CFS Patients

FINDING	FREQUENCY (%)*
Tender points	33-75
Inflamed pharynx	40-60
Posterior cervical	20-40
Impaired tandem gait	15-25
Macular rash	10-20
Fever (T > 99.6°F)	10-20
Low body temperature (T < 97.0°F)	20-30
Hepatomegaly	5-20
Splenomegaly	5-20
Axillary adenopathy	5-15

* Frequencies are from our experience and from studies reported in the literature and reviewed by us (Komaroff and Buchwald, 1991).

Laboratory Studies. Although a diversity of laboratory abnormalities has been reported mostly in uncontrolled studies, no consistent findings have been associated with CFS (Buchwald and Komaroff, 1991). Leukocytosis and leukopenia are each seen in up to a quarter of the patients, and lymphocytosis has been described in up to 71% of patients (Tobi et al., 1982; Dubois et al, 1984; Salit, 1985; Komaroff[1]). Atypical lymphocytes have been reported in up to 30% and in about half the patients in serial studies[2]. Approximately 20% of patients have modestly elevated transaminases (Tobi et al., 1982; Dubois et al., 1984; Straus et al., 1985; Buchwald et al., 1992).

Unusual and often conflicting findings have also been noted in many immunologic studies. Of the various circulating autoantibodies reported, the most common are antinuclear antibodies, found in up to a third of patients (Salit, 1985; Straus et al., 1985; Gold et al., 1990; Buchwald et al., 1992). Immunoglobulin levels have yielded inconsistent findings: both decreased and, less commonly, increased immunoglobulins of the IgA, IgD, IgG, or IgM class have been reported (Dubois et al.,1984; Salit, 1985; Straus et al., 1985; Tosato et al., 1985;

[1] Komaroff, personal communication.
[2] Komaroff, personal communication.

Lloyd et al., 1989; Buchwald et al., 1992). Up to 73% of patients have low levels of circulating immune complexes (Straus et al., 1985; Behan et al.,1985), but only a minority of patients show depressed complement, and none have clinical manifestations of immune-complex-mediated disease (Dubois et al., 1984; Salit, 1985; Behan et al., 1985).

Several groups have noted abnormalities in the lymphokine and interleukin responses in CFS. Leukocyte activity of the interferon-induced enzyme, 2'5'-oligoadenylate synthetase was elevated in a small number of CFS patients (Morag et al., 1982; Straus et al., 1985). Circulating interferon has been detected infrequently (Jones et al., 1985; Straus et al., 1985; Ho-Yen et al., 1988). Although several studies found reduced synthesis of gamma interferon and interleukin-2 by mitogen-stimulated lymphocytes (Kibler et al., 1985; Klimas et al., 1990), others have found elevated interleukin-2 levels in CFS patients (Cheney et al., 1989).

Abnormalities of lymphocyte number and function have also been noted. Both elevated and diminished numbers of B cells and CD4 and CD8 cells and increased and decreased CD4/CD8 ratios have been found (Dubois et al., 1984; Hamblin et al., 1983; Straus et al., 1985; Behan et al., 1985; Lloyd et al., 1989; Klimas et al., 1990; Buchwald et al., 1992; Komaroff[3]). Lymphocyte function may also be abnormal. In some studies, one-half the patients are anergic, and significantly delayed hypersensitivity to multiple antigens has been demonstrated (Murdoch, 1988; Lloyd et al., 1989). Decreased responsiveness in vitro to standard mitogen stimulation assays also has been reported (Behan et al., 1985; Lloyd et al., 1989; Klimas et al., 1990; Buchwald et al., 1992). A recent controlled study found a significant reduction in the display of CD3, a common T-cell membrane structure (Subira et al., 1989). Because CD3 interacts with antigen in the transduction of T-cell activation signal, this finding suggests a mechanism for the T-cell dysfunction observed in CFS. Finally, decreased numbers of natural killer cells and reductions in the normally dominant (NKH1 positive-CD3 negative) subset have been reported. Natural killer cell function has been found to be increased (Gold et al., 1990) and decreased (Aoki, 1987; Caligiuri et al., 1987; Klimas et al., 1990) as measured by cytolytic activity against several different target cell lines. These abnormalities are interesting, given the central role played by these cells in viral infections.

CFS has been reported as a post-infectious syndrome in patients with several different infectious agents (Salit, 1985). Although early studies linked Epstein-Barr virus and CFS, further investigations have indicated that it is unlikely an etiologic agent in most cases (Lamy et al., 1982; Buchwald et al., 1987b; Hellinger et al., 1988). Many studies have found elevated IgG to viral capsid antigen and early antigen, generally in the absence of viral capsid antigen-IgM; however, there is extensive overlap between patients and healthy control subjects (Salit, 1985; Straus et al., 1985; Tosato et al., 1985; Buchwald et al., 1987b; Kroenke et al., 1988). A low or absent antibody to Epstein-Barr nuclear antigen, thought to be an unusual finding in healthy seropositive individuals, has been found in some CFS patients (Jones et al., 1985; Straus et al., 1985; Tosato et al., 1985; Buchwald and Komaroff, 1991). Several other latent

---

[3] Komaroff, personal communication.

herpes viruses—herpes simplex, cytomegalovirus, measles virus, and human herpes virus-6—may be reactivated in patients with CFS (Holmes et al., 1987; Wakefield et al., 1988; Buchwald et al., 1992). If this is true, it is unclear whether this reactivation contributes to the morbidity of CFS, or whether it is an epiphenomenon.

Enteroviruses also have been implicated as possible etiologic agents in CFS. Two intriguing studies report finding circulating enteroviral antigen and IgM complexes in the majority of patients, and isolation of the virus in 22% (Yousef et al., 1988; Halpin and Wessely, 1989). Moreover, enteroviral nucleic acid has been found in muscle cells more often in patients with CFS than in control subjects (Archard et al., 1988).

Finally, the frequently reported allergic and muscle symptoms of CFS have been evaluated extensively. Increased cutaneous reactivity to allergens, increased levels of circulating IgE and IgE-bearing T and B cells, and greater lymphocyte responsiveness to allergens have all been found in adults (Olson et al., 1986a and 1986b; Straus et al., 1988a). Muscle studies have not demonstrated abnormal muscle enzymes or metabolic or enzymatic evidence of a defect in the intermediary energy pathway (Behan et al., 1985; Byrne and Trounce, 1987). Although muscle biopsies have shown structural abnormalities (Behan et al., 1985; Warner et al., 1989) and 75% of patients tested had abnormal single fiber electromyography (Jamal and Hansen, 1985), various techniques have documented normal muscle strength, endurance, and recovery (Lloyd et al., 1988b; Stokes et al., 1988). However, on graded exercise testing, CFS patients have a markedly limited exercise capacity characterized by an inability to achieve their target heart rate, a lower exercise heart rate, and an abbreviated exercise duration (Montague et al., 1989).

In summary, the laboratory abnormalities observed in adults with CFS are diverse, sometimes conflicting, and frequently modest in degree. Although evidence of diffuse immunologic dysfunction exists, reflecting both deficient function and hyperactivity, these findings have not been shown to explain the symptomatology of CFS or correlate with changes in symptoms over time. Most likely, when more systematic blinded studies using healthy control subjects are performed, some of the abnormalities currently thought to be characteristic of CFS will disappear.

Psychologic Dysfunction. Recent investigations have found a high prevalence of psychiatric disorders, primarily major depression, in CFS patients (Manu et al., 1988 and 1988b; Kruesi et al., 1989; Wessely and Powell, 1989; Katon et al., 1991). Those studies, which included control groups, demonstrate a higher prevalence of psychiatric disorder in CFS patients than in control subjects (Wessely and Powell, 1989; Katon et al., 1991). In one specialty fatigue clinic, 95% of patients did not meet CDC criteria for CFS, primarily because a psychiatric disorder was thought to be responsible for the fatigue (Manu et al., 1988a). Although patients' recall may be inaccurate, studies that examined the temporal sequence of CFS and psychiatric disorder found the latter often preceded the onset of fatigue (Kruesi et al., 1989; Katon et al.,

1991). At the present time, controversy remains regarding a primary or secondary role for psychologic dysfunction in CFS.

Diagnosis

A detailed history, physical examination, and selected laboratory tests are indicated in individuals with chronic fatigue that significantly affects daily activity. Equal emphasis must be placed on the psychosocial evaluation, searching, in particular, for anxiety and depression. Although no laboratory tests are diagnostic, a reasonable laboratory test battery to evaluate patients for CFS consists of a complete blood count with manually performed differential white blood cell count; erythrocyte sedimentation rate; chemistry panel; thyroid stimulating hormone; antinuclear antibodies; and in some cases, circulating immune complexes and immunoglobulin levels. These tests may be helpful in supporting a diagnosis of CFS, or in ruling out other diseases that can produce chronic fatigue. Extensive serologic testing for viruses is not usually helpful in identifying a causal link between fatigue and persistent or reactivated viruses.

Treatment

There is no specific treatment for CFS. Treatment should be directed at relief of specific symptoms and reintegration into the individual's social network. Symptomatic therapy may include antipyretic and nonsteroidal anti-inflammatory agents. Low doses of antidepressants, such as fluoxetine in the morning and amitriptyline or doxepin taken at bedtime, may be beneficial. Anecdotal reports of successful treatment include antiviral and immunomodulating drugs, vitamins, holistic remedies, diet modifications, and rest. At present, only two agents, acyclovir (Straus et al., 1988b) and a liver extract, B12-folate preparation (Kaslow et al., 1989), have undergone vigorous evaluation, and neither was more effective than placebo.

A supportive approach aimed at enhancing self-efficacy and reconditioning is effective in many cases. Physical activities previously enjoyed should be gradually increased in a structured manner because complete inactivity appears to promote fatigue. Finally, continuity of care coupled with education, instruction in coping skills, and counseling are often neglected but important aspects of treatment for CFS. Periodic reassessment for organic and psychiatric disorder should be performed at regular intervals; however extensive laboratory testing without additional reasons is not indicated.

Course of Illness and Prognosis

CFS is usually not progressive. Symptoms are usually most severe during the first 6 months of illness, plateau relatively early, and recur with varying degrees of severity. Although longitudinal studies are lacking, most patients report gradual recovery, usually punctuated with relapses precipitated by overexertion, stress, or infection.

STUDIES AT THE UNIVERSITY OF WASHINGTON

Collaborators from various subspecialties have evaluated patients at the University of Washington Chronic Fatigue Clinic for evidence of immunologic, virologic, and psychologic dysfunction. Subjects for the following studies were a group of adult patients who sought consultation and underwent thorough evaluation at this university-based, specialty clinic devoted to evaluation and treatment of chronic fatigue. Clinic patients were not required to meet the CDC criteria to be seen. All patients underwent standardized, comprehensive review of their current and past medical history, physical examination, laboratory studies, and psychiatric examination. Patients with suggestive responses on a standardized sleep questionnaire were referred for polysomnography. Selected patients were queried about symptoms of multiple chemical sensitivities.

Routine laboratory studies included complete blood count, 12-factor automated chemical analysis, thyroid stimulating hormone, liver-function tests, and erythrocyte sedimentation rate. Virologic and spirochettsial studies included antibodies to adenovirus (complement fixation), human herpesvirus 6 (ELISA) cytomegalovirus (Western blot), Coxsackie B types 1-6 (neutralization), Borrelia Burgdorffei (ELISA), IgM and IgG antibodies to Epstein-Barr virus viral capsid antigen, and IgG to the early antigens (IFA) and nuclear antigens (anticomplement immunofluorescence). Antinuclear antibodies were examined with indirect immunofluorescence at screening dilutions of 1:40 on rat liver and 1:100 on Hep 2 cells, and immunoglobulins quantified by radial immunodiffusion or immunonephelometry. Evaluation of T-, B-, and natural killer cell populations was performed on a flow cytometer using a whole blood lysis technique for preparing the cells. Antibodies against CD2, CD19, CD4, CD8, and CD56 were used. Immune complexes were detected using C1q binding in a solid-phase, enzyme-linked-immunosorbent assay as well as a fluid-phase radioimmunoassay.

Patients were examined using an abridged version of the National Institute of Mental Health Diagnostic Interview Schedule (DIS) Version III-A (Robins and Helzer, 1985), a highly structured diagnostic interview that assigns current and lifetime diagnoses based on DSM -III-R criteria (American Psychiatric Association, 1987). Patients were considered to have a current diagnosis if criteria were met within the previous month. To avoid the assumption that psychiatric illness caused the fatigue, major depression was scored without the symptom of fatigue and somatization disorder without the seven symptoms common to both CFS and somatization.

Sleep laboratory data were obtained during standardized monitoring of the electro-oculogram, electroencephalogram, electromyogram, electrocardiogram, and nasal air flow. In addition, a multiple sleep latency test was performed on all patients undergoing formal sleep evaluation.

The mean age of our patients was 40 years, ranging from 18 to 77 years (Table 4). Seventy-five percent were female, and 51% were married. Patients were generally well-educated with an average of 15 years of school. The average duration of illness was more than 5 years, and 17% of patients had been bedridden. Almost a third had a close contact with CFS. When patients with a current psychiatric diagnoses were excluded, 32% met the CDC criteria for CFS.

TABLE 4. Characteristics of University of Washington CFS Patients (N = 300)

CHARACTERISTICS		PATIENTS
Age	(mean years)	40
	(range)	18-77
Female		75%
Education	(mean years)	15
	(range)	11 - 26
Married		51%
Duration of illness (mean years)		5.5
Bedridden or shut-in		17%
Close contact with CFS		31%
Met CDC criteria		32%

As expected, symptoms in the CDC criteria were common in our patients (Table 5). Most common were disturbances of sleep in 95% of patients. Other prominent symptoms include myalgias, neurocognitive complaints, and post-exertional fatigue.

TABLE 5. Prevalence of Symptoms in the CFS Criteria Among University of Washington Patients

SYMPTC 1	% OF PATIENTS
Sleep disturbance	95
Myalgias	92
Neurocognitive symptoms	89
Post-exertional fatigue	80
Arthralgias	79
Muscle weakness	76
Headaches	71
Onset with flu-like illness	68
Lymphadenopathy	65
Pharyngitis	60
Fever	60

Clinical findings in our patients varied in frequency (Table 6). The physical findings identified in the CDC case definition were found in 21-45% of patients. The most common laboratory finding was the presence of antibody levels ≥ 1:128 to Coxsackie B virus in 40% of patients. Antinuclear antibodies, usually in low titers in the absence of anti-double-stranded DNA or other evidence of lupus erythematosus, were noted in 35% of patients. This frequency is higher than previously reported in CFS and may be the result of using two cell lines for determining antinuclear antibodies in our patients. Abnormalities of the absolute B4 lymphocyte count, CD4/CD8 ratio, and percent of natural killer cells were all observed in more than 20% of patients. Other findings, in decreasing order of frequency, included atypical lymphocytosis (18%), an abnormal absolute CD4 lymphocyte count (16%), elevated transaminases (12%) and an abnormal white blood cell count (11%).

TABLE 6. Clinical Findings in University of Washington CFS Patients

FINDING	% OF PATIENTS
Physical Examination	
Lymphadenopathy	45
Fever	33
Pharyngitis	21
Laboratory Studies	
Coxsackie B 1-6 antibodies ≥ 1:128	40
Antinuclear antibodies present	35
Abnormal absolute B4 count	25
Abnormal CD4/CD8 ratio	23
Elevated natural killer cell number	21
Atypical lymphocytes present	18
Abnormal absolute T4 count	16
Elevated transaminases	12
Abnormal white blood cell count	11

Fifty-seven patients underwent comprehensive evaluation including 37 immunologic and virologic tests and the DIS. The number of abnormalities ranged from 0 (4 patients) to 11 (1 patient). Patients had a median of 3.0 and a mean of 3.54 abnormalities. When patients were grouped according to the absolute number of abnormalities observed, no significant differences were found between patients with and without psychiatric illness (Table 7). For example, among patients with any current disorder, 37% had 0-2 abnormalities vs. 42% of those without any disorder, and 63% had 3-11 abnormalities compared with 58% of those without any disorder. This lack of an association between immunologic and psychiatric dysfunction also was observed for any lifetime diagnosis, and for patients meeting and not meeting the CDC case definition. Among patients not meeting the criteria, 39% had 0-2 abnormalities vs. 44% of those who did meet the criteria, and 61% had 3-11 abnormalities compared with 56% of those who fulfilled the case definition.

This observation also held true when mean numbers of diagnoses were examined. Among patients without current psychiatric diagnoses, the mean number of abnormalities was 3.47 compared with 3.68 for patients with one or more diagnoses. Examination of the CDC criteria yielded similar findings. In comparing patients who did and did not meet the CDC case definition, no significant differences were found in the average number of abnormalities on tests of immune function (3.67 versus 3.28).

Using a similar strategy, we compared the prevalence of physical findings among patients with and without psychiatric diagnoses (Table 8). The majority of patients in both groups had no findings. Nor were differences detected in the prevalence of fever or one or more physical findings as described in the CDC criteria. This held true for both current and lifetime diagnoses and the mean number of abnormalities on examination.

TABLE 7. Immunologic Abnormalities in University of Washington CFS Patients, by CDC Criteria and Psychiatric Diagnoses

	PERCENT OF PATIENTS WITH GIVEN NUMBER OF ABNORMALITIES		
	0 - 2	3 - 9	p
Met CDC Criteria for CFS			
No	38	62	ns
Yes	44	56	
Current DSM III-R Diagnoses			
None	42	58	ns
One or more	37	63	
Lifetime DSM III-R Diagnoses			
None	44	56	ns
One or more	39	61	

ns = not significant

TABLE 8. Physical Findings and Current Psychiatric Diagnosis in University of Washington CFS Patients

	CURRENT PSYCHIATRIC DIAGNOSIS		
Physical Findings	Present (n=40) (% patients)	Absent (N=60) (% patients)	p
Fever	10	12	ns
None	62	59	ns
1 finding	33	31	ns
2 findings	5	10	ns
Mean number of findings	.38	.41	ns

ns = not significant

As noted above, 95% of patients in our clinic reported at least one symptom suggestive of disturbed sleep (Table 9). These symptoms included usually needing to nap (41%), awakening unrefreshed (35%), having difficulty staying (33%) or falling (32%) asleep, sleeping at least ten hours at night (31%), regular early awakenings (27%), and frequent snoring (17%). As a result of their answers to a ten-item screening questionnaire about sleep behaviors, 32 patients underwent polysomnography. Preliminary results indicate a total of 39 sleep disorders were present with at least one disorder in 91% of patients. In decreasing order of frequency, the sleep pathologies noted were obstructive sleep apnea in 47%, other disorders of excessive sleepiness in 41%, disorders of initiating and maintaining sleep in 19%, and movement disorders in 16% of patients. The multiple sleep latency test was abnormal in 43% of patients. Although 52% had current psychiatric diagnoses, the type of sleep disorders observed were not those seen in association with psychiatric disorders.

TABLE 9. Results of Polysomnography in University of Washington CFS Patients

Number of patients	32
Total number of sleep disorders present	39
	% OF PATIENTS
Sleep disorder present	91
Current psychiatric diagnosis	52
Lifetime psychiatric diagnosis	48
Abnormal Multiple Sleep Latency Test	43
Type of Sleep Disorder	
Obstructive sleep apnea	47
Other disorders of excessive sleepiness	41
Disorders of initiating and maintaining sleep	19
Movement disorders	16

Finally, we administered questionnaires to 21 patients with multiple chemical sensitivities (MCS), 30 with fibromyalgia (FM), and 24 with CFS to determine the nature and extent of similarities between these illnesses. All three groups comprised primarily females, ranging from 58% of CFS patients to 93% of those with FM (Table 10). The groups were similar in mean age (39-44) and years of education (15) but differed substantially in percentage of married individuals (24% MCS, 40% FM, 67% CFS). The MCS patients had been ill for an average of 8.5 years compared with 5.0 for FM and 4.1 for CFS. Patients in all three groups had high rates of health care utilization ranging from 21 visits to a health care provider during the preceding year for CFS patients to 37 visits for patients with FM. Over half the MCS and FM patients believed themselves to be in control of their health, although 38% MCS and 53% FM patients were disabled. Among patients with CFS, 63% believed themselves in control and only 29% were disabled.

TABLE 10. Characteristics of MCS, FM, and CFS Patients in University of Washington Study

CHARACTERISTIC	DIAGNOSIS		
	MCS	FM	CFS
Number of patients	21	30	24
Mean age (years)	39	44	42
Female	81%	93%	58%
Married	24%	40%	67%
Mean education (years)	15	15	15
Mean illness duration (years)	8.5	5.0	4.1
Mean # visits per year	24	37	21
In control of health	52%	50%	63%
Disabled	38%	53%	29%

MCS = multiple chemical sensitivity
FM = fibromyalgia
CFS = chronic fatigue syndrome

TABLE 11. Symptoms and Environmental Sensitivities among MCS and CFS Patients in University of Washington Study

	PERCENTAGE OF PATIENTS WITH DIAGNOSIS		
	MCS	FM	CFS
SYMPTOM			
Fatigue for six months with 50% reduction activity	76	77	96
Post-exertional fatigue	50	86	78
Confusion	81	43	46
Memory loss	71	43	42
Sleep difficulties	48	50	46
Myalgias	38	83	46
Headaches	38	43	50
Sore throat	19	37	21
Lymphadenopathy	19	43	17
Fever	10	13	17
SENSITIVITIES			
Pollution	86	27	25
Cigarette smoke	48	33	29
Fumes	81	43	21
Perfumes	62	23	17

MCS = multiple chemical sensitivity
FM = fibromyalgia
CFS = chronic fatigue syndrome

Symptoms included in the CDC case definition of CFS were surprisingly common among both FM and MCS patients (Table 11). Fatigue of six-months duration with a 50% reduction in activity was present in two-thirds of patients with MCS and FM. As expected, environmental sensitivities were most commonly reported by MCS patients, with sensitivities ranging from 86% for pollution to 48% for cigarette smoke.

REFERENCES

ACHESON, E.D. (1959). The clinical syndrome variously called benign myalgic encephalomyelitis, Iceland disease, and epidemic neuromyasthenia. Am. J. Med. 26:569-595.

AMERICAN PSYCHIATRIC ASSOCIATION DIAGNOSTIC AND STATISTICAL MANUAL OF MENTAL DISORDERS. (1987). (3rd ed. revised) DSM-III-R. Washington, DC.

AOKI, T., USUDA, Y., MIYAKOSHI, H., TAMURA, K., and HERBERMAN, R.B. (1987). Low natural killer syndrome: Clinical and immunologic features. Nat. Immun. Cell Growth Regul. 6:116-128.

ARCHARD, L.C., BOWLES, N.E., BEHAN, P.O., BELL, E.J., and DOYLE, D.(1988). Postviral fatigue syndrome: Persistence of enterovirus RNA in muscle and elevated creatine kinase. J. Royal Soc. Med. 81:326-329.

BEHAN, P.O., BEHAN, WM.H., and BELL, E.J. (1985). The postviral fatigue syndrome-An analysis of the findings in 50 cases. J.Infect. 10:211-222.

BUCHWALD, D., GOLDENBERG, D.L., SULLIVAN, J.L., and KOMAROFF, A.K. (1987a). The 'chronic, active Epstein-Barr virus infection' syndrome and fibromyalgia. Arthritis Rheum. 30:1132-1136.

BUCHWALD, D., SULLIVAN, J.L., and KOMAROFF, A.L. (1987b). Frequency of 'chronic active Epstein-Barr virus infection' in a general medical practice. J.A.M.A. 257:2303-2307.

BUCHWALD, D. and KOMAROFF, A.K. (1991). Review of laboratory findings for patients with chronic fatigue syndrome. Rev. Infect. Dis. 13:S12-S18.

BUCHWALD, D., CHENEY, P.R., PETERSON, D.L., HENRY, B., WORMSLEY, S.B., GEIGER, A., ABLASHI, D.V., SALAHUDDIN, S.Z., SAXINGER, C., BIDDLE, R., KIKINIS, R., JOLESZ, F.A., FOLKS, T., BALACHANDRAN, N., GALLO, R.C., and KOMAROFF, A.L. (1992). A chronic illness characterized by fatigue, neurologic and immunologic disorders: Association with human herpesvirus-6. Ann. Intern. Med. (in press).

BYRNE, E. and TROUNCE, I. (1987). Chronic fatigue and myalgia syndrome: Mitochrondrial and glycolytic studies in skeletal muscle. J. Neurol. Neurosurg. Psychiatry 50:743-746.

CALIGIURI, M., MURRAY, C., BUCHWALD, D., LEVINE, H., CHENEY, P., PETERSON, D., KOMAROFF, A.L., and RITZ, J. (1987). Phenotypic and functional deficiency of natural killer cells in patients with chronic fatigue syndrome. J. Immunol. 139:3306-3313.

CHENEY, P.R., DORMAN, S.E., and BELL, D.S. (1989). Interleukin-2 and the chronic fatigue syndrome. Ann. Intern. Med. 110:321.

DENICOFF, K., RUBINOW, D.R., PAP, M.Z. et al. (1987). The neuropsychiatric effects of treatment with interleukin-2 and lymphokine-activated killer cells. Ann. Intern. Med. 107:293-300.

DUBOIS, R.E., SEELEY, J.K., BRUS, I., SAKAMOTO, K., BALLOW, M., HARADA, S., BECHTOLD, T.A., PEARSON, G., and PURTILO, D.T. (1984). Chronic mononucleosis syndrome. South. Med. J. 77:1376-1382.

GOLD, D., BOWDEN, R., SIXBEY, J., RIGGS, R., KATON, W.J., ASHLEY, R., OBRIGEWITCH, R., and COREY, L. (1990). Chronic fatigue: A prospective clinical and virologic study. J.A.M.A. 264:48-53.

GOLDENBERG, D.L., SIMMS, R.W., GEIGER, A., and KOMAROFF, A.K. (1990). High frequency of fibromyalgia in patients with chronic fatigue seen in a primary care practice. Arthritis Rheum. 33:381-387.

HALPIN, D. and WESSELY, S. (1989). VP-1 antigen in chronic postviral fatigue syndrome. Lancet 1:1028-1029.

HAMBLIN, T.J., HUSSAIN, J., AKBAR, A.N., TANG, Y.C., SMITH, J.L., and JONES, D.B. (1983). Immunological reason for chronic ill health after infectious mononucleosis. Br. Med. J. 287:85-82.
HELLINGER, W.C., SMITH, T.F., VAN SCOY, R.E., SPITZER, P.G., FORGACS, P., and EDSON, R.S. (1988). Chronic fatigue syndrome and the diagnostic utility of antibody to Epstein-Barr virus early antigen. J.A.M.A. 260:971-973.
HO-YEN, D.O., CARRINGTON, D., and ARMSTRONG, A.A. (1988). Myalgic encephalitis and alpha-interferon. Lancet 1:125.
HOLMES, G.P., KAPLAN, J.E., STEWART, J.A., HUNT, B., PINSKY, P.F., and SCHONBERGER, L.B. (1987). A cluster of patients with a chronic mononucleosis-like syndrome: Is Epstein-Barr virus the cause? J.A.M.A. 257:2297-2302.
HOLMES, G.P., KAPLAN, J.E., GANTZ, N.M. et al. (1988). Chronic fatigue syndrome: A working definition. Ann Intern Med. 108:387-389.
JAMAL, G.A. and HANSEN, S. (1985). Electrophysiological studies in the post-viral fatigue syndrome. J. Neurol. Neurosurg. Psychiatry 48:691-694.
JONES, J.F., RAY, G., MINNICH, L.L., HICKS, M.J., KIBLER, R., and LOCAS, D.O. (1985). Evidence for active Epstein-Barr virus infection in patients with persistent, unexplained illnesses: Elevated anti-early antigen antibodies. Ann. Intern. Med. 102:1-7.
JONES, J.F. and STRAUS, S.E. (1987). Chronic Epstein-Barr virus infection. Ann. Rev. Med. 38:195-209.
KASLOW, J.E., RUCKER, L., and ONISHI, R. (1989). Liver extract-folic acid-cyanocobalamin vs. placebo for chronic fatigue syndrome. Arch Intern. Med. 149:2501.
KATON, W.J., BUCHWALD, D.S., SIMON, G.E., RUSSO, J.E., and MEASE, P.J. (1991). Psychiatric illness in patients with chronic fatigue and rheumatoid arthritis. J. Gen. Intern. Med. 6:277-285.
KIBLER, R., LUCAS, D.O., HICKS, M.J., POULOS, B.T., and JONES, J.F. (1985). Immune function in chronic active Epstein-Barr virus infection. J. Clin. Immunol. 5:46-54.
KLIMAS, N.G., SALVATO, F.R., MORGAN, R., and FLETCHER, M.A. (1990). Immunologic abnormalities in chronic fatigue syndrome. J. Clin. Microbiol. 28:1403-1410.
KOMAROFF, A.K. and BUCHWALD, D. (1991). Symptoms and signs of chronic fatigue syndrome. Rev. Infect. Dis. 13:S8-S11.
KROENKE, K., WOOD, D.R., MANGELSDORFF, A.D., NEIER, N.J., and POWELL, J.B. (1988). Chronic fatigue in primary care: Prevalence, patient characteristics, and outcome. J.A.M.A. 260:929-934.
KRUESI, M.J.P., DALE, J., and STRAUS, S.E. (1989). Psychiatric diagnoses in patients who have chronic fatigue syndrome. J. Clin. Psychiatry 50:53-56.
LAMY, M.E., FAVART, A.M., CORNU, C., MENDEZ, M., SEGAS, M., and BURTONBOY, G. (1982). Study of Epstein Barr virus (EBV) antibodies: IgG and IgM anti-VCA, IgG anti-EA and Ig anti-EBNA obtained with an original microtiter technique: -Serological criterions of primary and recurrent EBV infections and follow-up of infectious mononucleosis; -Seroepidemiology of EBV in Belgium based on 5178 sera from patients. Acta. Clin. Belgica 37:281-298.
LINDE, A., HAMMARSTROM, L., and SMITH, C.I.E. (1988). IgG subclass deficiency and chronic fatigue syndrome. Lancet 1:885-886.
LLOYD, A.R., HALES, J.P., and GANDEVIA, S.C. (1988b). Muscle strength, endurance and recovery in the post-infection fatigue syndrome. J. Neurol. Neurosurg. Psychiatry 51:1316-1322.
LLOYD, A., ABI, H.D., and WAKEFIELD D. (1988). Interferon and myalgic encephalomyelitis. Lancet 1:471.
LLOYD, A.R., WAKEFIELD, D., BOUGHTON, C.R., and DWYER, J.M. (1989). Immunological abnormalities in the chronic fatigue syndrome. Med. J. Aust. 151:122-148.
MANU, P., LANE, T.J., and MATTHEWS, D.A. (1988). The frequency of the chronic fatigue syndrome in patients with symptoms of persistent fatigue. Ann. Intern. Med. 109:554-556.
MANU, P., MATTHEWS, D.A., and LANE, T.J. (1988b). The mental health of patients with a chief complaint of chronic fatigue: A prospective evaluation and follow-up. Arch. Intern. Med. 148:2213-2217.

McDONALD, E.M., MANU, A.H., and THOMAS, H.C. (1987). Interferons as mediators of psychiatric morbidity. Lancet 2:1175-1177.

MEDICAL STAFF OF THE ROYAL FREE HOSPITALS (1957). An outbreak of encephalomyelitis in the Royal Free Hospital Group, London, in 1955. Br. Med. J. 2:895-904.

MONTAGUE, T.J., MARRIE, T.J., KLASSEN, G.A., BERWICK, D.J., and HORACEK, B.M. (1989). Cardiac function at rest and with exercise in the chronic fatigue syndrome. Chest 95:779-784.

MORAG, A., TOBI, M., RAVID, Z., REVEL, M., and SCHATTNER, A. (1982). Increased (2'-5')-oligo-a synthetase activity in patients with prolonged illness associated with serological evidence of persistent Epstein-Barr virus infection. Lancet 1:744.

MURDOCH, J.C. (1988). Cell-mediated immunity in patients with myalgic encephalomyelitis syndrome. NZ. Med. J. 101:511-512.

OLSON, G.B., KANAAN, M.N., GERSUK, G.M., KELLEY, L.M., and JONES, J.F. (1986). Correlation between allergy and persistent Epstein-Barr virus infections in chronic-active Epstein-Barr virus infected patients. J. Allergy. Clin. Immunol. 78:308-314.

OLSON, G.B., KANAAN, M.N., KELLEY, L.M., and JONES, J.F. (1986b). Specific allergen-induced Epstein-Barr nuclear antigen-positive B cells from patients with chronic-active Epstein-Barr virus infections. J. Allergy Clin. Immunol. 78:315-320.

ROBINS, L.N. and HELZER, J.E. (1985). Diagnostic Interview Schedule (DIS): Version III-A. St. Louis, Missouri: Dept. of Psychiatry, Washington Univ. School of Med.; May.

SALIT, I.E. (1985). Sporadic postinfectious neuromyasthenia. Can. Med. Assoc. J. 4385;133:659-663.

STOKES, M.J., COOPER, R.G., and EDWARDS, R.H.T. (1988). Normal muscle strength and fatigability in patients with effort syndromes. Br. Med. J. 297:1014-1017.

STRAUS, S.E., TOSATO, G., ARMSTRONG, G., LAWLEY, T., PREBLE, O.T., HENLE, W., DAVEY, R., PEARSON, G., EPSTEIN, J., BRUS, I., and BLAESE, R.M. (1985). Persisting illness and fatigue in adults with evidence of Epstein-Barr virus infection. Ann. Int. Med. 102:7-16.

STRAUS, S.E., DALE, J.K., WRIGHT, R., and METCALFE, D.D. (1988a). Allergy and the chronic fatigue syndrome. J. Allergy Clin. Immunol. 81:791-795.

STRAUS, S.E., DALE, J.K., TOBI, M. et al. (1988b). Acyclovir treatment of the chronic fatigue syndrome: Lack of efficacy in a controlled placebo trial. N. Engl. J. Med. 319:1692-1698.

SUBIRA, M., CASTILLA, A., CIREIRA, M., and PRIETO, J. (1989). Deficient display of CD3 on lymphocytes of patients with chronic fatigue syndrome. J. Infect. Dis. 160:165-166.

TOBI, M., MORAG, A., RAVID, Z., CHOWERS, I., WEISS, V.F., MICHAELI, Y., BEN-CHETRIT, E., SHALIT, M., and KNOBLER, H. (1982). Prolonged atypical illness associated with serological evidence of persistent Epstein-Barr virus infection. Lancet 1:61-64.

TOSATO, G., STRAUS, S., HENLE, W., PIKE, S.E., and BLAESE, R.M (1985). Characteristic T cell dysfunction in patients with chronic active Epstein-Barr virus infection (chronic infectious mononucleosis). J. Immunol. 134:3082-3088.

WAKEFIELD, D., LLOYD, A., and DWYER, J. (1988). Human herpes virus 6 and myalgic encephalomyelitis. Lancet 1:1059.

WARNER, C.L., COOKFAIR, D., MEFFUER, R., BELL, D., LEY, D., and JACOBS, L. (1989). Neurologic abnormalities in the chronic fatigue syndrome. Neurology 39 (suppl):420.

WESSELY, S. and POWELL, R. (1989). Fatigue syndromes: A comparison of chronic "postviral" fatigue with neuromuscular and affective disorders. J. Neurol. Neurosurg. Psychiatry 52:940-548.

YOUSEF, G.E., BELL, E.J., MANN, G.F., MURUGESAN, V., SMITH, D.G., McCARTNEY, R.A., and MOWBRAY, J.F. (1988). Chronic enterovirus infection in patients with postviral fatigue syndrome. Lancet 1:146-147.

RESEARCH STRATEGIES FOR INVESTIGATING MULTIPLE CHEMICAL SENSITIVITY

DON L. JEWETT, MD, DPhil
Professor Emeritus, Department of Orthopaedic Surgery
University of California, San Francisco
Director of Research, Abratech Corporation, Mill Valley, California

As the number and amount of chemicals in the human environment increase, some of these chemicals can be found in the human body in varying (though sometimes minute) amounts. If the amount of chemicals increases over time in a large group of humans, standard physiologic and pharmacologic principles dictate that inevitably mild symptoms will appear in a minority of the population. What will these symptoms be? Clearly, the challenge is to determine the causes of such chemically related symptoms and to distinguish them from other causes of the same mild symptoms. It is certainly unclear that symptoms reported by patients evincing "multiple chemical sensitivities" have been precipitated by exposure to chemicals. Although in hyperosmia, odor and symptom are coincident, even this does not establish the underlying cause of the symptom. Thus, the title of this workshop seems to pre-judge the cause. Non-prejudicial terminology, such as "hypersensitivity syndrome," is much to be preferred until the causes of these symptoms are determined scientifically. Hence, this neutral syndrome name will be used in this paper.

The symptoms of hypersensitivity syndrome characteristically involve multiple-organ systems and are commonly being reported in medical practice in our society. Jewett (1992) provided a list of such symptoms. The history of these symptoms begins in the 19th century or before. One of the earliest references to symptoms that may be part of a hypersensitivity syndrome thought to be environmentally caused is found in a letter written in 1608 by the composer Claudio Monteverdi after a long period of being overworked and underpaid (Schonberg, 1981):

> if I do not take a complete rest . . . my span of life will be shortened; for in consequence of my tremendous over-exertions in the past I have developed headaches and a severe, maddening rash on my body which neither the cauterizing I have undergone, nor the purgatives I have taken, nor even the blood-letting and other measures to which I have submitted, have succeeded in curing more than partially. The Signor Padrone ascribes the cause of the headaches to strenuous study, but the maddening rash to the air. . .which does not suit me, and he even goes so far as to fear that the air will be the cause of my death not long hence.

Even in 1992, we may not do any better with either diagnosis or treatment!

Address all correspondence to: Dr. Don L. Jewett, Abratech Corporation, 150 Shoreline Hwy., Bldg. E, Mill Valley, CA 94941-6610.

The multiple-organ symptoms of hypersensitivity syndrome have a long history in medical practice. Thus, these symptoms must be placed in an historical perspective, as Layzer (1991) has done with the symptoms of chronic fatigue syndrome (CFS). All the CFS symptoms Layzer described are included in the symptoms of hypersensitivity syndrome described by Jewett (1992). The medical diagnoses applied over the years to explain the CFS symptoms include hypochondriasis, neurasthenia, Da Costa's syndrome, soldier's heart, effort syndrome, neurocirculatory asthenia, vasoregulatory asthenia, functional hyperadrenalism, functional hyperinsulinism, subclinical hypothyroidism, iron deficiency, vitamin deficiency, food allergy, spasmophilia, subclinical magnesium deficiency, chronic brucellosis, depression, anxiety neurosis, somatoform disorder, generalized fibrositis, mitral valve prolapse, environmental allergy, systemic candidiasis, chronic Epstein-Barr virus infection, and chronic mononucleosis syndrome. Clearly, "multiple chemical sensitivity" could be added to this list, and despite considerable study and a variety of treatments, we have yet to establish a basis for these symptoms.

How can a set of symptoms have so many explanations? As Layzer (1991) says:

> ... a detailed perusal of old and recent hypotheses about the medical causes of [these symptoms] provides a fascinating insight into the pitfalls and fallacies of medical research. Time and again one encounters examples of physicians obsessed by a pet idea, of research conducted without appropriate controls, of hypotheses confirmed by the subjective responses of patients to treatment, and of a failure to define the normal range of clinical measurements.

Under such circumstances, carefully controlled, double-blind studies are absolutely essential if explanations of the causes and treatment of hypersensitivity syndrome symptoms are to be trusted. Neither therapeutic success nor clinical experience are acceptable as adequate proof of cause; such evidence from medical practice should only be used to form hypotheses that must then be tested in a rigorous, scientific manner. Regrettably, enthusiastic advocates can generate hypotheses in greater numbers than time or money is available to appropriately study them, and some of these advocates tacitly assume that someone else is responsible for testing their hypotheses. Yet, the proponent of a given medical hypothesis should be responsible for either providing initial data from a well-controlled, double-blind study, or seeking out and cooperating in such a study. Indeed, only the first experimenter can morally withhold a treatment from the control group (an essential element of a good double-blind study). Once a treatment has been publicized as successful (even if the evidence is poor), patients do not reasonably wish to be part of a control group. Proper testing must precede publicity if patients, physicians, and third-party payers are not to be misled by the vagaries of clinical experience, which in turn drives medical fads.

In this social context, I wish to discuss research strategies for studying the hypersensitivity symptoms of "multiple chemical sensitivity". Our society holds ambivalent views of the expert. The expert of the first kind (who is viewed positively) has desirable, valuable, valid knowledge. The expert of the second kind (who is viewed negatively) offers opinions biased by conflicts of interest and supports a viewpoint with unsubstantiated statements designed to confuse issues and mislead the unwary. In terms of media-driven public opinion, the advocates of an environmental cause for "multiple chemical sensitivity" are experts of the first kind. By contrast, professional societies and governmental spokespersons are experts of the second kind. Any organization interested in establishing a credible position on "multiple chemical sensitivity" or the hypersensitivity syndrome must firmly take the following positions.

1. Only scientifically valid studies will be seriously considered.
2. The organization will put a good-faith effort into determining the causes of hypersensitivity symptoms rather than evaluating or promoting a specific, special-interest hypothesis.

Otherwise, the organization will have difficulty gaining acceptance for the results of negative studies.

Good science is practiced by forming hypotheses to explain interesting observations and then trying to disprove all reasonable hypotheses. To accomplish this, an overall strategy that considers the technical and financial aspects of a scientific project as well as the political and social concerns related to the research is needed. At least three strategies might be applied to the hypersensitivity syndrome. The first strategy could be to determine the effects of low-level chemicals on human symptoms. Trying to determine the effects of the minute amounts of chemicals that can be detected by modern instrumentation will clearly involve an immense expenditure of time and money. Such a project would satisfy growing political and social concerns about the dangers from chemical pollution, but such research will not likely quell concerns about "multiple chemical sensitivity" because experts of the first kind will doubt negative studies whether the basis for doubt is adequate or not. Also, the public will have no way of determining the quality of the research reported.

A second strategy is to study the methods used by advocates to establish the initial claims about the "multiple chemical sensitivity" syndrome. Because a study by a non-advocate will be criticized for a lack of experience and knowledge on the part of the investigator, this strategy requires the cooperation of advocates with the investigator, as in the study by Jewett, Fein, and Greenberg (1990). Even though both proponents and opponents of clinical ecology had reached preliminary agreement on the validity of the study, when the results did not support the clinical ecologists, other advocates offered different opinions about the implications. This lead to the presentation of markedly different expert opinions to the public, even though the study design was so tight that it could not be refuted without presenting a hypersensitive patient tested under the same protocol. (The study methodology would permit advocates to readily produce just a single patient demonstrating hypersensitivity under double-blind conditions, using only the testing methods in use in their practices. The absence of evidence for that single patient is essentially a passive admission that the provocation method claimed to demonstrate hypersensitivity to environmental agents is fatally flawed and is believed to be so by the advocates.) Also, it is difficult to gain further cooperation of the advocates after a negative study. Indeed, many advocates will not cooperate at any time. Finally, a negative study does not disprove any new variations on the methods, so this strategy leads to an apparently endless series of increasingly difficult tests of alternative hypotheses, even though some of them are contrary to existing knowledge (Langmuir, 1989).

A third and final strategy is to determine the causes of hypersensitivity symptoms without pre-conceived notions about these causes. This is intellectually direct and may provide a more useful outcome. This also has the advantage of tying in with other stress-related problems in occupational health, rather than considering the symptoms as resulting only from chemical exposure. A disadvantage is that the medical constituency for these multi-organ symptoms is so large that no single specialty is likely to take on the problem. A problem this broad and pervasive should really be taken on by some organization with a larger view or by a group of organizations (perhaps motivated by political/social concerns). Even the structure of NIH is fragmented regarding the cross-specialty and cross-organ-system nature of hypersensitivity syndrome.

The interest of occupational health professionals in hypersensitivity symptoms provides an important opportunity to review the issues broadly and consider alternative hypotheses. The reviews might be on the following topics.

1. The physiologic and pathophysiologic consequences of stress.
2. The influence of the nervous system on other organ systems, including but not limited to the endocrine and immune systems.
3. The physiologic effects elicited by Pavlovian conditioning in the context of patients' autonomic responses to environmental stimuli.
4. Evaluation of various classifications of "somatization" disorder and the hypothesized causes of that disorder.
5. The influence of non-specific causes (such as stress) on allergy symptoms and what accentuates or modulates them.
6. Compilation and description of unexplained symptoms from all medical specialties, including _____itis and _____algia and chronic _______ syndrome, especially if they involve multiple-organ-system complaints.
7. Compilation and description of all symptoms from placebo responses in well-documented, placebo-controlled studies.
8. Compilation and description of all symptoms resulting from chemical exposures, including the possibility that early symptoms are aggravated or modulated by stress.
9. Research on hyperosmia, including its relationship to adrenal insufficiency or changes in adrenal function (hyperosmia is a common complaint in hypersensitivity syndrome).
10. Review of newer methods for measuring adrenal function and insufficiency.

11. Review of the mechanisms of premenstrual syndrome because many hypersensitivity symptoms are aggravated by premenstrual swelling and may be swelling-related.
12. Review of the psychiatric diagnoses of hypersensitivity syndrome patients and the diagnostic criteria.
13. Compilation and review of the symptoms of indoor air pollution.
14. The physiologic effects of hopelessness and helplessness (learned or not).
15. Review of the symptoms, methods, and conclusions for other disorders with related symptoms (e.g., chronic fatigue syndrome).
16. Review of modern methods of measuring autonomic nervous system hyperactivity.

Trying to study hypersensitivity symptoms should recall the story of the blind men attempting tactile analysis of an elephant. For this problem of elephantine proportions, we need a broader view that incorporates both the history of these symptoms and the multidisciplinary nature of the problem. And we need a broad base of support in both finances and interested scientific investigators. Research in this area might reduce the current costs of many fruitless diagnostic studies and therapeutic attempts, even if it these symptoms were shown not to be chemically induced. Other cost savings could result if a rational basis for evaluating chemicals in the human environment were established.

REFERENCES

JEWETT, D.L. (1992). Diagnosis and treatment of hypersensitivity syndrome. Toxicol. Ind. Health (this issue).

JEWETT, D.L., FEIN, G., and GREENBERG, M.H. (1990). A double-blind study of symptom provocation to determine food sensitivity. N. Engl. J. Med. 323:429-433.

LANGMUIR, I. (1989). Pathological science. Physics Today 42:36-48.

LAYZER, R.B. (1991). Clinical symptoms in the Chronic Fatigue Syndrome. In: Chronic Fatigue Syndrome (G. Serratrice and J.L. Vilde, eds.). Springer-Verlag, Paris. pp. 5-9.

SCHONBERG, H.C. (1981). The Lives of the Great Composers. Norton, New York. p.23.

POSSIBLE MODELS FOR MULTIPLE CHEMICAL SENSITIVITY: CONCEPTUAL ISSUES AND ROLE OF THE LIMBIC SYSTEM

CLAUDIA S. MILLER, MD, MS
Assistant Professor, Allergy and Immunology
University of Texas Health Science Center San Antonio, Texas

INTRODUCTION

It was six men of Hindostan,
To learning much inclined,
Who went to see the elephant,
(Though all of them were blind);
That each by observation,
Might satisfy his mind.

J. G. Saxe

There is an accumulating number of clinical observations concerning patients who report multiple chemical sensitivities (MCS), and investigators who might propose a mechanism or model to explain this syndrome should be aware of these observations. Some observations have come from researchers, some from clinicians, and some from MCS patients. Few of these observations have been tested in a scientific manner. Nevertheless, taken as a whole, they may provide us with important clues about the MCS "elephant." The purpose for reviewing these observations is 1) to provide a common body of data, anecdotal or otherwise, to facilitate discussion of the problem and thereby 2) to enhance the probability of generating fruitful mechanistic hypotheses. Consider this a kind of brainstorming session—an attempt to write all our ideas on the blackboard while suspending judgment about their correctness or utility. Later we can edit them as we learn more.

Remarkably, despite the lack of an accepted clinical case definition for MCS, most physicians who are aware of this problem almost immediately recognize these patients. This process has a gestalt-like quality. We must find ways to dissect these clinical intuitions, frame them as hypotheses, and test them in a scientific manner.

Persons who have worked with MCS patients can readily attest that getting involved with their medical problems and lifestyle difficulties can be both time-consuming and exceedingly frustrating, for a variety of reasons. Yet MCS may represent an unprecedented opportunity to further our knowledge of environmental health, psychology and psychiatry, or all of these

Address all correspondence to: Dr. Claudia S. Miller, Department of Pediatrics, University of Texas Health Science Center, 7703 Floyd Curl Drive, San Antonio, TX 78284.

(depending upon our personal views of the etiology of this problem). Throughout the history of science, the diligent pursuit of rare, serendipitous events has produced enormous yields. Wilson's cloud chamber allowed physicists to capture rare atomic events, providing new information concerning protons, electrons, and other sub-atomic particles. In medicine, rare genetic events have provided valuable clues about the human genome. The diligent study of MCS patients may also lead us toward unanticipated discoveries.

It is hoped that the following observations (and they are *only* observations) will lead to clearer hypotheses and experiments to test those hypotheses. Science is not about "belief." It is about "guess and test," that is, formulating hypotheses that can then be subjected to scientific scrutiny. Many, if not all, of us have strong personal beliefs about chemical sensitivity. Despite our differing views, we should nevertheless be able to agree on which hypotheses and mechanisms are most plausible and how to test them in an objective manner.

OBSERVATIONS ON MCS

Induction (Priming or Sensitization)

Onset Following a Major Chemical Exposure. In many cases, MCS appears to follow a major exposure to any of a wide range of environmental chemicals. The exposure may occur briefly at high levels or repeatedly or continuously at lower levels. Commonly reported inducing (priming or sensitizing) exposures include pesticides, solvents, or complex mixtures, such as combustion products or indoor air pollutants. However, onset of illness has been attributed to various other events, for example, surgery with anesthesia, pregnancy, extreme stress, or courses of various drugs. In one survey of 6800 persons reporting chemical sensitivities, almost 50% blamed pesticides (National Foundation for the Chemically Hypersensitive, 1989). Between 1984 and 1990, the National Pesticide Telecommunications Network received more than 1000 calls on chemical sensitivity attributed to pesticide exposures (NPTN, 1991). A disproportionate number (relative to exposure opportunity) involved organophosphate pesticides.

Demographic Diversity. MCS-like illnesses have been reported by industrial workers, individuals exposed in contaminated communities, tight-building occupants, and individuals with personal and unique exposure histories, for example to pesticides in their homes (Table 1). These groups differ greatly in terms of age, sex, social group, and the kinds of medical specialists they consult. What they share in common is a history of a major chemical exposure event and similar kinds of complaints following that event, particularly complaints about symptoms occurring after exposure to common environmental substances that were formerly tolerated. The demographic diversity of these groups suggests that a real problem may be occurring.

Temporal Cohesiveness. The temporal cohesiveness of MCS-like illnesses occurring within a group of individuals sharing a recognized, major chemical exposure (for example, near-simultaneous development of symptoms in several family members, co-workers, or community members exposed to the same chemical or mix of chemicals) helps point to a problem that is real in those circumstances.

TABLE 1. Chemically Sensitive Groups

Group	Nature of Exposure	Demographics
Industrial workers	Acute and chronic exposure to industrial chemicals	Primarily males; blue collar; 20 to 65 years old
Tight-building occupants	Off-gassing from construction materials, office equipment or supplies; tobacco smoke; inadequate ventilation	Females more than males; white collar office workers and professionals; 20 to 65 years old; school children
Contaminated communities	Toxic waste sites; aerial pesticide spraying; ground water contamination; air contamination by nearby industry and other community exposures	All ages, male and female; children or infants may be affected first or most; pregnant women with possible effects on fetuses; middle to lower class
Individuals	Heterogeneous; indoor air (domestic), consumer products, drugs, and pesticides	70-80% females; 30 to 59 years old (Johnson and Rea, 1989); white middle to upper middle class and professionals

Source: Ashford and Miller, 1991.

Total Load. An apparent threshold effect, referred to by some as the patient's "total load," has been noted and cited by some practitioners to help explain why an individual develops this syndrome at a particular time. Illness is said to occur when the total load of biologic, chemical, physical, and psychologic stressors exceeds some threshold for the patient. This concept has emerged from clinical observations. No direct experiments have been done to test its validity in humans. However, Hans Selye's work demonstrating the General Adaptation Syndrome in animals exposed to various stressors lends support to this idea (Selye, 1946).

Triggering

Triggering by Extremely Low Levels of Chemicals. Following induction of the illness, patients report subsequent triggering by extremely low levels of exposure to common chemicals, levels that are tolerated by most of the population.

Spreading. Spreading of sensitivities to other, often chemically dissimilar, substances is commonly reported, with each substance triggering a different but reproducible constellation of symptoms.

Characteristic Constellation. For a given patient with a given exposure, a characteristic constellation of symptoms results. Indeed, even when no odor is present, MCS patients may be able to identify the nature of the exposures based solely upon the pattern of symptoms they experience.

Dose dependence. The severity of symptoms is proportional to the "dose" (duration and magnitude) of the exposure in the sensitized individual.

Pronounced Stimulatory and Withdrawal Symptoms. MCS patients sometimes report that when they are exposed to environmental incitants, either foods or chemicals, they may initially experience a stimulated feeling that is either pleasant (euphoria) or unpleasant (jitteriness); this is followed by withdrawal symptoms (headache, irritability, depression, fatigue) when the exposure is reduced or terminated (Ashford and Miller, 1991). Very sensitive individuals report having experienced fewer and shorter-lived stimulatory effects and more pronounced withdrawal symptoms as their illness progressed. In an extreme case, onset of exposure might be associated with manic behavior and withdrawal from the same exposure might be linked to depression.

Inter-Individual Variability. Different MCS patients exposed to the same substance(s) may experience different symptoms and symptom severity.

Intra-Individual Reproducibility. Individual MCS patients report similar symptoms occurring each time a specific exposure is encountered.

Multiple Exposure Routes Trigger Symptoms. Patients' symptoms may be triggered by inhalation, ingestion, mucosal contact, or injection. Eyedrops, dental materials, suppositories, IVs, and other exposures have been reported to elicit symptoms. Exposure to a given substance in a given patient but via different routes has been reported to result in similar symptoms.

Subsensory Triggering. Onset of symptoms may occur with exposure to chemicals at concentrations below the olfactory threshold. For example, patients may report entering a room, experiencing a "reaction," and learning only later that pesticides had been applied recently. Anosmia and chemical sensitivity are not mutually exclusive; a certain number of MCS patients have no sense of smell, but nevertheless report symptoms with exposure.

Rapid Onset of Symptoms. MCS patients frequently report onset of symptoms within a few seconds of being exposed. In addition, the symptoms have an on/off quality. As quickly as symptoms may begin, an almost immediate cessation of symptoms has been reported following neutralization or use of alkali salts. Such responses might suggest rapid shifts of ions across cell membranes, a neurologic or immunologic mechanism, or possibly conditioning.

Variable Threshold for Triggering. Following development of MCS, major exposures, for example, an airplane trip or exposure to pesticides at work, may result in a general decrease in tolerance for chemical substances, even substances for which tolerance may have been regained. Following such exposures, gradual improvement usually occurs, but it may take days, weeks, or even months, depending on the patient and the severity of the exposure. Patients may say that following a particular exposure, they "lost their foods," meaning they once again lost tolerance for foods they had been able to reintroduce into their diets. This variability in responsiveness in some ways resembles the increased bronchial reactivity that occurs in asthma and increased basophil releasibility in atopic dermatitis following major exposures to incriminated inhalants or foods, respectively.

Decrease in Symptoms with Oral Versus Nasal Inhalation. Some MCS patients have noticed that, faced with an unavoidable exposure (for example, traffic exhaust or perfume in an elevator), inhaling via their mouths instead of their noses mitigates symptoms.

Limited Voluntary Control. Certain patients report that biofeedback, self-hypnosis, or variations on these techniques improve their ability to cope during exposures and lessen symptom severity. Such techniques seem to be less helpful for more severe exposures and symptoms.

Adaptation. Adaptation occurs with frequent or continuous exposure, and acute symptoms may become chronic and no longer appear related to particular exposures. This will be discussed later in more detail. Potentially, adaptation may have enormous significance for future research in this area.

Other Intolerances

Food and Alcohol Intolerances. Food and alcohol intolerances are estimated to occur in a sizeable percentage, perhaps up to 80% (Rea, 1988) of individuals with MCS. Foods are complex mixtures of many chemicals, some of which can also be air contaminants. For example, pinene is a constituent of both bathroom air fresheners and oranges. Many MCS patients first become aware of their intolerance for ingestants when they consume an alcoholic beverage, such as red wine or beer, and become ill afterward.

Drug Intolerances. MCS patients often respond poorly to usual doses of medications and may even report difficulties with excipients, binders, coloring, and flavoring agents. Several physicians have reported that their patients are intolerant of the anticholinergic side-effects of various drugs (Schottenfeld, 1987).

Hyperresponsiveness to Physical Stimuli. Some MCS patients may perceive not only chemical but also physical stimuli, including touch, sound, and light, as greatly magnified. These individuals describe their nerves as being easily irritated, and they experience pain or extreme discomfort when someone bumps their bed or they hear conversational-level noises.

Symptoms/Signs

Multisystem Symptoms. Commonly, this patient group experiences symptoms involving virtually any system in the body and often several systems simultaneously. These do not coincide with other known multisystem disorders, such as an endocrinopathy.

Frequent Central Nervous System (CNS) Complaints. Mood, memory, and concentration difficulties are among the CNS symptoms these patients report most frequently.

Rhinitis. The majority of MCS patients appear to have nasal inflammation or rhinitis, suggesting that inflammatory processes are involved in MCS. Rhino-laryngoscopy, nasal biopsies, and assays of mediators in nasal lavage fluid may provide useful data in the future. Doty (1988) reported greater nasal airway resistance among MCS patients than normal controls, both before and after unblinded exposure to phenyl ethyl alcohol and methyl ethyl ketone. Despite increased nasal airway resistance, Doty's subjects had olfactory thresholds for phenyl ethyl alcohol and methyl ethyl ketone that did not differ significantly from the olfactory thresholds of controls.

Personal/Predisposing Characteristics

Apparent Female Preponderance. Among certain demographic groups in which chemical sensitivities have been reported, a greater number of females than males appear to be affected. Although exposure to domestic chemicals in poorly ventilated homes cannot be discounted as a factor, numerous medical conditions and diseases, such as most connective tissue diseases, show a disproportionate number of females affected. Because the maternal immune system is challenged with paternal antigens during pregnancy, its immunologic repertoire and responses may somehow make it more vulnerable to chemical exposure. In a Soviet experiment, females were shown to be more susceptible to the effects of low-level organophosphate exposure than their male counterparts (Krasovskii et al.). The incidence of cutaneous drug reactions in females is 35% greater than among males (Bigby et al., 1986).

Food Cravings and Addiction. Patients often report having carbohydrate or other food cravings, being "chocoholics" or addicted to certain foods, such as baked goods, peanuts, popcorn, sugar (especially corn sugar), potatoes, a particular soft drink, etc. Some report sipping coffee, tea, or soda all day long, carrying it with them wherever they go. Intense food cravings following chemical exposures may be reported, for example, an uncontrollable urge to stop and buy candy after driving in heavy traffic.

Pre-Morbid History of Ill-defined Medical Conditions. Diagnoses such as hypothyroidism, chronic fatigue, temporomandibular joint dysfunction, and fibrositis seem to be more prevalent among MCS patients. Simon et al. (1990) reported an average of 6.2 unexplained physical symptoms before workplace exposure among workers who developed "environmental illness," versus 2.9 such symptoms for controls.

Pre-Morbid History of Depression. Simon et al. (1990) also found that 54% of workers who developed chemical sensitivity reported anxiety or depression prior to their workplace exposure, versus 4% of controls. Others corroborated these impressions, although the sources of these patients' depression and anxiety, whether endogenous or chemically-related, have yet to be demonstrated.

Target Organs Vary Over Time. Patients may report lifelong medical problems, but these affect different organ systems over time. For example, they may have experienced colic as babies, asthma in grade school, stomach problems in high school, headaches in college, and finally chronic fatigue and memory and mood difficulties following a major exposure event.

Family History of Chemical Intolerance. Many patients report a family history of chemical intolerances or addiction. The fact that similarly-exposed co-workers or household members of patients often do not develop MCS suggests that MCS patients have a possible genetic, nutritional, psychologic, or other predisposition, or an accumulation of effects from prior exposures.

Laboratory Findings

Lack of a Consistent Immunologic Abnormality. A variety of immunologic tests have been conducted in this patient group including measures of both cellular immunity (especially T and B lymphocyte counts and ratios) and humoral immunity (especially immunoglobulin levels). Some of the most striking data are those of Levin and Byers (1987), which show an apparent decrease in helper/suppressor ratios among various exposed populations compared with controls. By contrast, other investigators claim that immune parameters for these patients generally fall within expected normal ranges (Terr, 1986). Further studies of T and B lymphocytes in these patients, using healthy age- and sex-matched controls, are needed before conclusions can be drawn. In any event, because helper and suppressor T-cell numbers and ratios of patients and controls overlap greatly, one-time measurements of these parameters are of little value for clinical diagnosis.

Increased Frequency of Autoimmune Antibodies. A few investigators have reported an increased frequency of low titers of various autoimmune antibodies (ANA, anti-smooth antibody, thyroid antibodies, and others) among individuals who report becoming ill after an identifiable exposure to pesticides, solvents, and, more recently, pentachlorophenol (Broughton and Thrasher, 1990; McConnachie and Zahalsky, 1991). One group of researchers has reported that 35 (91%) of 39 individuals with ongoing exposure to trichloroethylene in drinking water had one or more positive results in a panel of autoantibody tests, versus 6 (21%) of 28 controls. Two years after removal from exposure, 41% of the exposure group continued to have one or more positive autoantibody tests, suggesting that avoidance of exposure led to some improvement (Broughton and Thrasher, 1990). The physiologic significance of low titers of autoantibodies is unknown and, although of considerable research interest, their diagnostic utility is compromised because 20% of healthy controls had similar findings. Such data could suggest an autoimmune response to host proteins altered by reactive environmental chemicals. In future work, careful attention should be directed to employing matched controls and blinded specimens and to assessing inter- and intra-laboratory reproducibility and the day-to-day variation in individual patients' results.

Increased Frequency of Activated T-Lymphocytes. The same laboratories that report an increased frequency of autoimmune antibodies in these patients also report an increased frequency of activated T-lymphocytes (Ta1 or CD26). If borne out, such a finding might suggest an inappropriate or persistent immune response that fails to clear chemicals from the tissues. Activated T cells have been reported in association with multiple sclerosis, infections, and various other inflammatory states. The significance of activated T-cells in this setting is unknown but merits further study.

EEG Abnormalities. A recent study compared EEGs from 58 "universal reactors," 55 healthy controls, and 89 patients from a psychology practice (Staudenmayer and Selner, 1990). EEGs of universal reactors more closely resembled EEGs of psychologic patients than those of controls. The authors, however, dismissed the environmental hypothesis that psychologic and psychiatric disturbances could be consequences, rather than causes, of MCS (Davidoff et al., 1991). Similar EEG changes with increased beta and decreased alpha activity have been reported among organophosphate-exposed workers (Duffy et al., 1979)

ADAPTATION[1]

Adaptation is also called "acclimation" or "acclimatization," "habituation," or "developing tolerance." "Acclimatization" often refers to the process of workers gradually becoming accustomed to exposures on the job, for example, heat stress. Understanding the possible role of adaptation in MCS is important for two reasons. (1) Adaptation may interfere with discovering effects on the body from a particular exposure, and (2) chemical exposures may adversely impact adaptative mechanisms and thus lead to illness.

That human beings adapt after chronic exposure to environmental challenges is widely recognized for a variety of substances. For example, the first cigarette an individual smokes might result in eye and throat irritation, but, as they continue to smoke, most individuals adapt over time and experience primarily the pleasurable effects of nicotine on the brain. After months or years, more cigarettes may be required for the same amount of lift, and, eventually, smokers may exhibit addictive behavior, seeking cigarettes more frequently. Subsequent quitting may lead to withdrawal symptoms including irritability, drowsiness, fatigue, moodiness, and headache. The reformed smoker may become extremely intolerant of the smoke of others, even in tiny amounts, and suddenly recall the irritation and unpleasant

[1] Modified in part from Ashford, N.A. and Miller, C.S. (1991). Chemical Exposures: Low Levels and High Stakes. Van Nostrand Reinhold, New York.

feelings associated with the first cigarette ever smoked. Over time, the individual had "adapted" to those effects.

Adaptation, which on the surface seems a good thing for the organism, may in fact be a two-edged sword. Tolerance for the noxious properties of exposure may make the individual more comfortable in the exposure and allow other harmful consequences of exposure to continue. Thus, heavy smokers who are "adapted" to tobacco smoke are also at increased risk for developing emphysema, lung cancer, or cardiovascular disease. Parallel analogies are valid for caffeine, alcohol, and other drugs, and frequent exposure may result in adaptation (irritation and other warning signals may disappear), continued exposure may lead to addiction, and reduction or cessation of exposure may precipitate withdrawal symptoms.

Some physicians have observed food and chemical adaptation and addiction in their chemically sensitive patients (Randolph, 1962, 1965). For MCS patients, multiple incitants, not only tobacco smoke, may be involved and, conceivably, all incitants may need be avoided simultaneously for improvement to occur.

Confusion could arise if patients seek physicians' help for withdrawal symptoms when they are no longer exposed (or less exposed) to the offending agents. For example, irritability may occur when individuals smoke fewer cigarettes than usual, or headaches may develop when they drink less caffeine. Such unpleasant withdrawal symptoms may be forestalled by smoking another cigarette or taking another drink of coffee, thus perpetuating addiction. Because smoking a cigarette or drinking a cup of coffee in the morning (after about 8 hours abstinence) relieves irritability or headache and makes patients feel better, they may not suspect that cigarettes or coffee might also be the source of their difficulties.

Occupational health presents other examples of acclimatization, inurement, or tolerance development, for example, after repeated exposure to ozone, nitroglycerin, or solvents (Ashford and Miller, 1991). The incitants mentioned thus far differ considerably from one another; some are ingestants, others inhalants; and their physical states and chemical composition vary greatly. The human body appears able to adapt to an enormous range of substances.

Ozone has been the focus of considerable research on adaptation. Intrigued by how little respiratory illness and death occurred in very polluted cities that had high levels of ozone and suspecting that adaptation might play a protective role, Hackney and associates (1977a) compared the responses of four Canadians (not adapted) and four Californians (adapted) to ozone challenges. Although individual reactivity varied greatly, the Californians were only minimally reactive to levels that caused coughing, substernal discomfort and airway irritation, pulmonary-function-test decrements, and increased red blood cell fragility for the Canadians.

In another experiment, six volunteers with respiratory hyperreactivity were placed in an environmental chamber with ozone at 0.5 ppm, typical of high ambient levels, for four days (Hackney et al., 1977b). Five of six showed decreased pulmonary function during days 1 to 3, but gradually improved almost to baseline by day 4, suggesting adaptation had occurred. The authors note that not all adverse effects of ozone may be prevented by adaptation; for example, increased red blood cell fragility may persist. Thus, adaptation or masking of some (usually acute or irritant) symptoms may occur while other physiologic alterations continue.

Individuals' abilities to adapt to ozone appear to depend on their initial sensitivity to it. More sensitive persons adapt more slowly and cannot maintain adaptation as long, usually less than seven days after exposure ceases (Horvath et al., 1981). Although adaptation to ozone may be related to what occurs in MCS, chemically sensitive patients frequently blame solvents for their symptoms (Terr, 1989; Cone et al., 1987) and adaptation to solvents has been well-documented. Vapors from various solvents are the most prevalent of indoor air contaminants (Molhave, 1982), and indoor air contaminants are of major concern to MCS patients. The volatile organic compounds associated with sick building syndrome are in large part solvent vapors.

Persons who have painted or used solvents extensively are well aware of the olfactory fatigue (nasal adaptation) that occurs and may have experienced the stimulatory and depressive (withdrawal) effects of solvents. Alcoholic beverages contain the solvent ethanol, which has related and familiar stimulatory and withdrawal effects.

Studies of xylene, one of the most prevalent solvents in indoor air, demonstrate that its effects are attenuated as exposure continues, presumably due to adaptation (Riihimaki and Savolainen, 1980). Riihimaki and Savolainen exposed healthy male volunteers to constant (100 or 200 ppm) and varying (200 or 400 ppm hourly peak) concentrations of xylene, adjusting baseline concentrations in the latter case so that a mean concentration of 100 or 200 ppm was maintained. Exposures occurred over a six-hour period (with a one-hour break at noon) for five days, followed by a two-day weekend without exposure, and one to three more days of active exposure to xylene. The investigators measured a variety of psychophysiologic parameters, including reaction time, body balance, manual dexterity, and nystagmus.

Most of the adverse effects of xylene on normal subjects "tended to disappear after a few succeeding days of exposure." However, "after the weekend away from exposure, the effects were again discernible." Riihimaki and Savolainen conclude: "This phenomenon suggests that tolerance had developed over a few days with regard to psychophysiologic effects by xylene."

We have mentioned several exposures that may involve adaptation. Clearly, individuals with or without multiple chemical sensitivities undergo adaptation to a wide variety of substances in their environment. However, the specific role adaptation plays in the dramatic responses patients report with MCS is not clear.

Patients may experience the effects of adaptation and de-adaptation without realizing what has occurred (Ashford and Miller, 1991). For example, a person with asthma may feel well after spending a week on a Caribbean island, breathing relatively uncontaminated air, and eating a diet devoid of usual foods. Then, exhaust emissions from the engine of a boat the individual takes to go home may bring on a severe, life-threatening asthmatic response. Back home in a metropolis, the person readapts, acclimatizes to auto emissions, combustion products and other air pollutants in the area, and experiences only chronic wheezing. Thus, following deadaptation (removal from incitants), the individual exhibits a more acute and convincing reaction upon re-exposure. This is what is reportedly observed when patients are tested in an environmental unit following deadaptation. Some of these reactions are so acute and convincing when they occur in daily life, that patients themselves may erroneously (at least in the eyes of some) surmise they must have an "allergy" to the incitant. Physicians may be skeptical and elect to challenge a patient with a supposed incitant. If, however, the patient is not deadapted or unmasked when tested, no reaction may occur, and the physician will be convinced the "allergy" was all in the patient's mind.

The deadaptation of patients may be critical to the study and the diagnosis of MCS for several reasons.

1. People are often exposed to dozens of potential incitants simultaneously, such as volatile organic compounds in a tight home or building, and literally hundreds over the course of a single day, so that health effects of these exposures may overlap, making it difficult to discern cause-and-effect relationships (Figure 1).

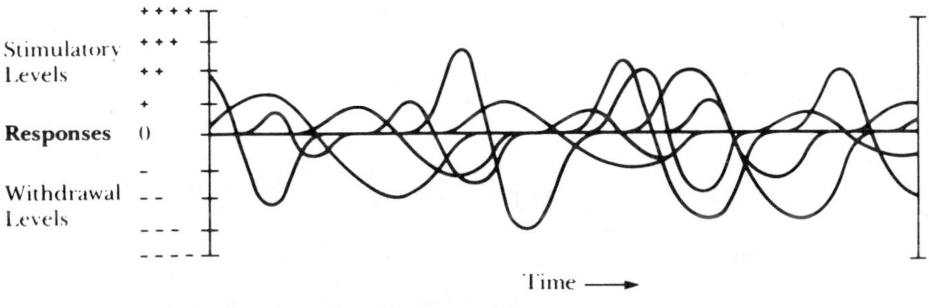

FIGURE 1. Overlapping of responses to food and chemical incitants in an individual with multiple exposures and multiple chemical sensitivities. *Source:* Ashford and Miller, 1991. Reprinted with permission of the publisher.

2. With continuous or frequent exposure to the same substance or chemically-related substances (such as xanthines in coffee, tea, chocolate, and colas), individuals adapt or develop tolerance to certain aspects of those exposures. Acute symptoms may gradually give way to chronic symptoms that may bear no apparent relationship to exposure. Exposures may never stop long enough for the patient to reach a baseline from which responses to challenges can be accurately gauged.

Graphical Representation of an Individual's Symptoms Before and After Entering an Environmental Unit

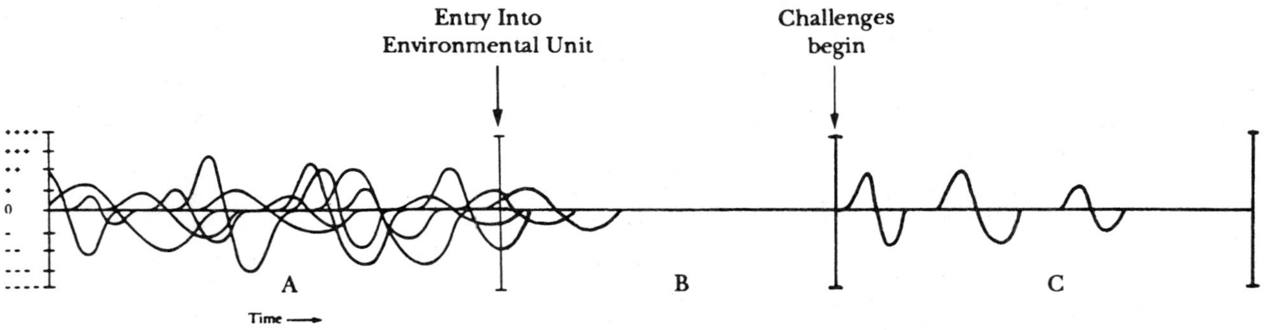

FIGURE 2. In time period A an individual is responding to multiple incitants encountered in normal daily living (chemicals and/or foods), with stimulatory and withdrawal effects that overlap in time. At any particular time, how the person feels is determined not only by ongoing exposures, but by previous exposures whose effects may still be waning. In time period B, the individual enters an environmentally controlled facility, fasting. With cessation of contributory exposures, withdrawal effects occur, for example, headache, fatigue, and myalgias. Symptoms continue for some time (typically 4-7 days) until the individual reaches "0" baseline. In some time period C, single challenges to suspected incitants are administered. Symptoms, often robust, develop soon after challenges, allowing patient and physician to begin to observe the relationship between exposures and symptoms for that individual. *Source:* Ashford and Miller, 1991. Reprinted with permission of the publisher.

Environmental units have been used to help overcome the masking effect of adaptation and the problems of overlapping exposures that result in overlapping responses to multiple agents. Such environmental units may in fact allow backing up or reversal of adaptation. Figure 2 graphically depicts the changes in symptoms that might occur in a patient who enters an environmental unit. Advantages of such an approach over conventional methods for determining toxicity may include facilitating detection of subclinical, prepathologic effects of chemicals and providing more than just a snapshot of an individual's response to substances. Removing a person from interacting, time-dependent stimuli in this way could allow the unraveling of multiple causes. Environmental units are an essential research tool. Many carefully conducted studies of chemical effects with negative or equivocal outcomes may have been flawed by failure to take adaptive mechanisms into account. The potential consequences of such an oversight are major.

Important questions that must be addressed in future studies of chemical sensitivity include:

1) Are subjects in a deadapted state before challenges so that extraneous exposures during and before a challenge (up to several days before) do not interfere with testing?

2) Are open challenges performed first to confirm that the placebo (clean air or a masking odor) is in fact a placebo and that the "active" challenge is something to which the patient has had a demonstrable response?

3) What is the recency and latency of the patient's exposure to the substance being tested? In other words, has enough time elapsed (about a week or so) so that the person is no longer adapted or responding to the last exposure but not too much time so that sensitivity has waned? Time elapsed since last exposure is recognized as a crucial variable in conducting challenges of patients with, for example, occupational asthma.

Carefully designed studies of deadapted patients in an environmental unit, using double-blind, placebo-controlled challenges, are an essential first step for helping resolve current professional antagonisms and placing this subject on a scientific footing.

POSSIBLE MECHANISMS FOR MCS: THE LIMBIC SYSTEM[2]

The most frequently cited physiologic theories to explain chemical sensitivity involve the nervous system, the immune system, or an interaction between them, because these two systems most clearly link the external environment and the internal milieu. Both can respond to exquisitely low concentrations of chemicals. The rapid responsiveness of these systems also makes them attractive candidates because symptoms of food or chemical sensitivity have been

[2] Modified in part from Ashford, N.A. and Miller, C.S. (1991). Chemical Exposures: Low Levels and High Stakes. Van Nostrand Reinhold, New York.

reported to develop within seconds of exposure. Many chemicals, such as polybrominated biphenyls (PBBs) and trichloroethylene, affect both the nervous system and the immune system. Until 1980, the idea that direct communication between the nervous and immune systems might occur was widely debated. Subsequently, the existence of a neuroimmunoendocrine axis has been increasingly realized. Several discoveries have helped to confirm the presence of two-way communication between the nervous and immune systems (Payan et al., 1986).

Kilburn (1989) proposes that the human nervous system, because it is so highly evolved, may be most susceptible to environmental agents:

> "Sensitivity may be its undoing. The intuitive hypothesis is advanced that the nervous system is the most liable of the body's systems to damage from environmental toxins. Appreciation of damage may be masked because subtle dysfunction is concealed by the nervous system's remarkable redundancy and substitution of functions, or it is overlooked in clinical evaluations which are usually only qualitative."

The hypothalamus, which is in close association with the limbic system, has attracted considerable attention because it is the focal point in the brain where the immune, nervous, and endocrine systems interact (Bell, 1982). Bell notes that assuming a direct cause-and-effect relationship would be premature, but that the hypothalamus could mediate food and chemical addictions in patients with multiple chemical sensitivities. The olfactory system has known links to the hypothalamus and other parts of the limbic system, which has led Bell to speculate that "the olfactory system, hypothalamus and limbic system pathways would provide the neural circuitry by which adverse food and chemical reactions could trigger certain neural, psychologic and psychiatric abnormalities." Patients with chemical sensitivities have reported food cravings, binges, violence, or hypersexual activity following chemical exposures. A model involving the hypothalamus could help explain such behavioral changes.

Some authors have alleged that psychologic conditioning to odors is responsible for patients' reactions to chemicals. Of course, odor conditioning could be involved in selected cases. However, physiologic mechanisms involving the limbic system may also occur. A direct pathway from the oropharynx to the brain and the hypothalamic and limbic region has been demonstrated in rats (Kare, 1968; Maller et al., 1967). Substances placed in the oropharynx migrated to the brain in minutes via a pathway other than the blood stream and in higher concentrations than if administered via the gastrointestinal tract, suggesting a direct route from mouth to brain. Similarly, Shipley (1985) showed that inhaled substances that contact the nasal epithelium may cross into the brain and be distributed widely via transneuronal transport. Thus, molecules that are inhaled and contact the olfactory apparatus could conceivably influence functions in other parts of the brain.

Ryan and associates (1988) studied 17 workers who attributed changes in thought processes, particularly memory and concentration difficulties, or changes in mood, to their exposure to solvents. Workers with "cacosmia" (a heightened sensitivity to odors resembling that reported by chemically sensitive individuals) performed most poorly on neurobehavioral tests requiring verbal learning or visual memory. The authors believed their findings supported a hypothesis that chronic solvent exposure may affect the "rhinencephalic structures" (primitive "smell" brain), the evolutionary precursor of the limbic system.

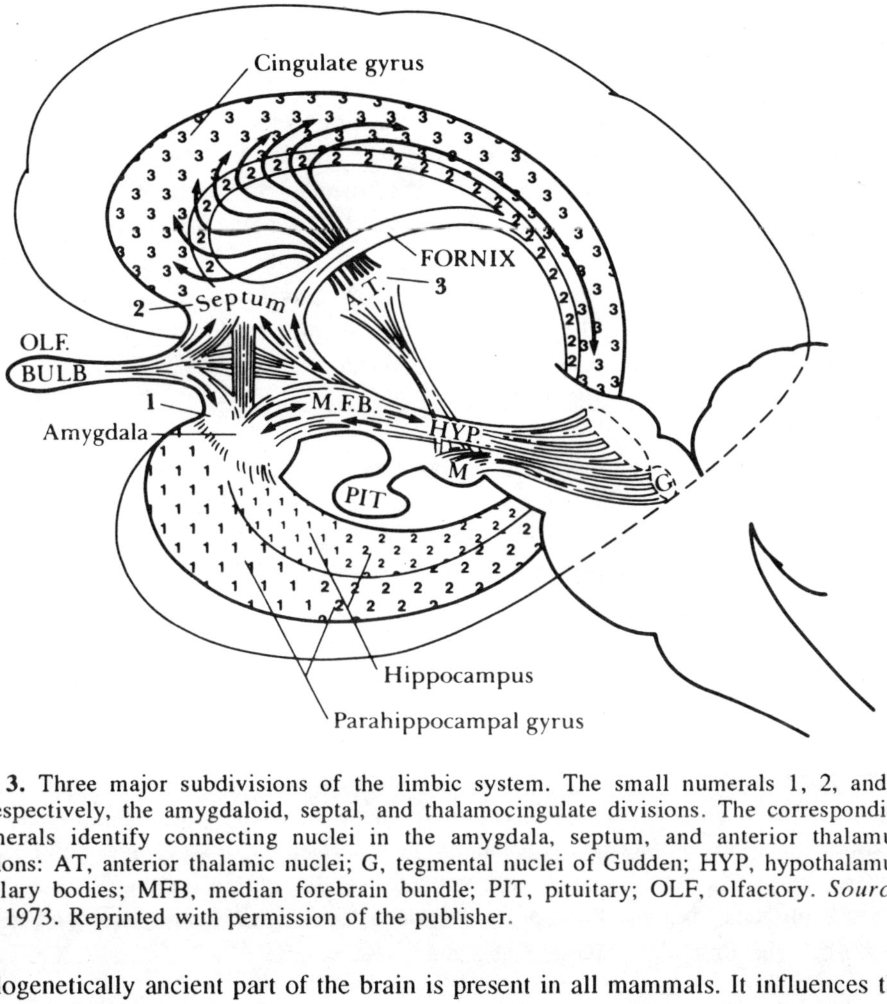

FIGURE 3. Three major subdivisions of the limbic system. The small numerals 1, 2, and 3 overlie, respectively, the amygdaloid, septal, and thalamocingulate divisions. The corresponding large numerals identify connecting nuclei in the amygdala, septum, and anterior thalamus. Abbreviations: AT, anterior thalamic nuclei; G, tegmental nuclei of Gudden; HYP, hypothalamus; M, mamillary bodies; MFB, median forebrain bundle; PIT, pituitary; OLF, olfactory. *Source*: MacLean, 1973. Reprinted with permission of the publisher.

This phylogenetically ancient part of the brain is present in all mammals. It influences the organism's interaction with its environment in many subtle ways essential for preservation of the individual, its offspring, and the species. Limbus (Latin for "margin" or "rim") refers to its rim-like appearance around the inner edge of the cerebral hemispheres. Figure 3 shows its component parts. Note the close anatomical relationship to the olfactory bulb. The olfactory nerves are the brain's most direct contact with the external environment. No protective blood-brain barrier exists here. Each olfactory nerve consists of bipolar neurons with one end in the upper part of the nose and the other in the brain (olfactory bulb). Odors can elicit electrical activity in the amygdala and hippocampal areas of the limbic system via this pathway

(Monroe, 1986). Subsensory exposure to chemicals may cause protracted, if not permanent, alterations in the electrical activity of the limbic region, beginning first with the most sensitive structures, particularly that portion of the amygdala that analyzes odors (Bokina et al., 1976). All parts of the limbic system are intimately interconnected. Interestingly, ablation of the olfactory bulb in laboratory animals serves as a model for depression that investigators have used to test the efficacy of various antidepressants (Jesberger and Richardson, 1988).

The amygdala, the most sensitive portion of the brain to chemical stimuli, is involved in feelings and activities related to self-preservation, such as searching for food, feeding, fighting, and self-protection (MacLean, 1986). It has been dubbed "emotion central," signifying its importance in mood states. The cingulate gyrus appears to influence maternal care and nursing, separation cries between mother and offspring, and playful behavior, including wit and humor (MacLean, 1986), qualities that many MCS patients feel have been impaired. The septum involves feeling and expression relating to procreation. Lesions in the septal area may cause hyperresponsiveness to physical stimuli (such as touching, sounds, or temperature changes), hyperemotionality, loss of motivation, excessive sugar and water intake, and fear of unfamiliar situations (Isaacson, 1982), phenomena reported by some chemically sensitive individuals.

Another limbic region, the hippocampus, appears important for laying down new memories and is thus essential for learning (Gilman and Winans, 1982). Hippocampal lesions may cause difficulty in retaining recent memories (Isaacson, 1982). The hippocampus, at the intersection of numerous neural pathways and in a critical position to affect the transfer of information from one brain region to another, acts as an information switching center. Learning and memory decrements are a frequent consequence of exposure to toxic substances, and some researchers view the hippocampus as a prime target for such toxins (Office of Technology Assessment, 1990; Walsh and Emerich, 1988). Damage to the hippocampus itself, or to nerves leading to or from it, may adversely affect the synthesis, storage, release, or inactivation of the excitatory and inhibitory amino acids that serve as neurotransmitters in this region of the brain. Toxins may disrupt the delicate balance of these amino acids, perhaps leading to the release of a flood of excitatory neurotransmitters that damage neighboring cells, a phenomenon that has been called "excitotoxicity" (Office of Technology Assessment, 1990). Remarkably small perturbations of hippocampal function may have large and long-lasting effects on behavior and cognition (Walsh and Emerich, 1988).

The hypothalamus, which interacts closely with the limbic region, governs (1) body temperature via vasoconstriction, shivering, vasodilation, sweating, fever, and behaviors such as moving to a cooler or warmer environment or putting on or taking off clothing; (2) reproductive physiology and behavior; (3) feeding, drinking, digestive, and metabolic activities, including water balance, addictive eating, and complete refusal of food and water leading to death; (4) aggressive behavior and physical manifestations of emotion such as increased heart rate, elevated blood pressure, dry mouth, and gastrointestinal responses (Gilman and Winans, 1982).

The hypothalamus is also the locus at which sympathetic and parasympathetic nervous systems converge. Many symptoms that patients experience with food and chemical sensitivities could be related to the autonomic nervous system. For example, altered smooth muscle tone produces Raynaud's phenomenon, diarrhea, constipation, and other symptoms reported by these individuals. The hypothalamus also appears to influence anaphylaxis and other aspects of immunity (Stein et al., 1981). Conversely, antigens may affect electrical activity in the hypothalamus (Besedovsky et al., 1977).

It is important also to recognize that thoughts arising in the cerebral cortex that have strong emotional overtones may trigger hypothalamic responses and recreate the physical effects associated with intense anger, fear, and other feelings. To implement its effects, the hypothalamus not only has a direct electrical output to the nervous system but also produces its own hormones, many of which stimulate or inhibit the pituitary's production of hormones (Gilman and Winans, 1982).

Most of the neural input to the hypothalamus comes from the nearby limbic and olfactory areas (Isaacson, 1982). Lesions in the limbic region may be associated with irrational fears, feelings of strangeness or unreality, wishing to be alone, and sadness (MacLean, 1967). A feeling of being out of touch with or out of control of one's feelings and thoughts, not unlike that described by many patients with chemical sensitivity, may be perceived. Some patients report feeling "spacey" or that "the camera isn't on" unless they make an enormous effort to focus their attention.

Doane and Livingston (1986) describe potential difficulties for patients with limbic dysfunction. "Activity controlled by the limbic system may seem largely irrational and often is not perceived within one's self in ways that are easily understood or communicated in verbal language."

The dynamic involvement of the hypothalamus and limbic system in virtually every aspect of human physiology and behavior makes injury to these structures an intriguing hypothesis to explain the development of chemical sensitivity with its diverse manifestations. Rich neural connections lie between the olfactory system and the limbic and temporal regions of the brain. Surgical or epileptic patients with damage to the limbic or medial temporal portions of the brain may experience persistent alterations in odor perception (for example, an unusual smell that characteristically precedes seizure activity), as well as learning and memory difficulties (Ryan et al., 1988).

Bell (1990) hypothesizes that chemically sensitive patients may have olfactory-limbic-temporal pathways that are more easily "kindled." In other words, a small signal or insult would more readily trigger firing of nerve cells in brain regions where kindling was present. Kindling might be enhanced by genetic endowment, prior environmental exposures, psychologic stress, hormonal variations, nutritional or other factors. Unlike surgical ablation, which destroys a brain area, kindling is a kind of stimulatory lesion (Girgis, 1986). Kindling has previously been described in the context of seizures. The amygdala, for example, which is particularly susceptible to electrical discharge following either electrical (Girgis, 1986) or chemical provocation (Bokina et al., 1976), is subject to long-lasting alteration when strong or repetitive stimuli are administered. Very potent or repeated stimuli, whether electrical or chemical, may permanently augment the tendency for neurons to fire in the presence of future stimuli, even when challenged with much lower levels than those originally involved.

Girgis (1986) reports a decrease in acetylcholinesterase (AChE), an enzyme that breaks down the neurotransmitter acetylcholine in junctions between nerve cells, that parallels the increase in supersensitivity to stimuli. The limbic system is especially rich in AChE, which is strongly bound to the nerve cell membranes and is very stable. The AChE may play a protective role by enzymatically maintaining acetylcholine concentrations at nerve junctions within safe bounds and protecting susceptible cells in the limbic system from developing "bizarre sensitivity" (Girgis, 1986). Interestingly, physicians who see patients with chemical sensitivities have noted that some of the most severe and debilitating exposures for these patients have involved organophosphate pesticides, which inhibit AChE (Ashford and Miller, 1991; Rosenthal and Cameron, 1991) and have reported that these patients may be especially sensitive to the anticholinergic properties of drugs (Schottenfeld, 1987).

Animals primed by high or repeated lower concentrations of various chemicals, such as formaldehyde or ozone, and subsequently re-exposed to low concentrations of the same chemicals showed an increased tendency toward paroxysmal electrical discharge in the amygdala (Bokina et al., 1976). Bokina observed that although the chemicals he used to sensitize the animals were different in terms of their structure and physical and chemical properties, their effects on the limbic system were remarkably similar, paralleling observations in MCS.

Kindling could help explain the apparent loss of adaptive capacity and spreading of sensitivities to chemically unrelated substances reported in multiple chemical sensitivity. Formerly well-tolerated, low-level exposures to, for example, tobacco smoke or perfume might trigger symptoms in individuals whose limbic areas had been kindled by a prior pesticide or solvent exposure.

One intriguing aspect of the limbic system as a possible mechanism for multiple chemical sensitivities is the system's responsiveness to both chemical and cortical stimuli. Therefore, conscious thought processes and emotional states may influence limbic activity just as chemical or physical stimuli can. The former may be under more or less conscious control of the individual, whereas the latter are almost entirely unconscious and automatic. However, conscious efforts that play into the delicate circuitry of the limbic system may be able to alter or suppress concurrent electrical activity evoked by environmental agents. Some patients with chemical sensitivities report being able to "will" their way out of a mild reaction to a food or chemical, and attempt to control their symptoms in this manner. Most patients, however, say such efforts do not work for their most problematic incitants. In fact, the ability to exercise any conscious effort, even that of simply getting away from the exposure, reportedly may be lost during a reaction.

Monroe (1986) described a man for whom exposure to the odor of stale beer caused greatly increased electrical activity in the limbic area (amygdala and hippocampal areas). Various memories, some associated with beer, also increased electrical activity in the same region. However, simple arithmetic computations would immediately stop such activity. Therefore, conscious thought processes could alter some electrical activity in the limbic system.

Detection of chemical stimuli in the nose is not limited to the olfactory nerve but involves the trigeminal nerve and its afferens, which may also play a role in this condition. Trigeminal free nerve endings in the nose and mouth detect noxious chemicals and reflexively initiate protective responses including cessation of breathing, constriction or dilatation of the airways, reduction in heart rate and cardiac output, constriction of most blood vessels (except capillaries in the head), increased epinephrine release, changes in blood pressure and efforts to withdraw (Silver and Maruniak, 1981). This powerful reflex serves an obvious protective role. Most familiar is the trigeminal reflex response to smelling salts. Apparently trigeminal responses can occur with non-irritating stimuli. Thus far, no odor has been found that stimulates the trigeminal nerve alone or the olfactory nerve alone (Silver and Maruniak, 1981) making study of either system in isolation difficult. Potentially both play a role in chemical sensitivity.

SUMMARY

Conceivably, chemicals contacting olfactory nerve projections in the nose could either be transported into or relay electrical signals to the limbic region, leading to a vast array of symptoms. Likewise, thought processes and mood states may trigger or interrupt pre-existing limbic activity. At present, however, no evidence suggests that limbic activity triggered by environmental exposures can be entirely overcome by psychologic interventions. One important ramification of a limbic hypothesis, if true, is that no convenient biologic marker for multiple chemical sensitivity may exist at the present time.

Ten years from now, we may finally confirm the existence of multiple chemical sensitivities (by careful, blinded challenges) but still have no single mechanism to explain it; that is, after all avenues of biochemical and immunologic inquiry have been exhausted, no single cause or marker for this disorder may be apparent. The theory that adaptation plays a role in MCS is based on the observed responses of patients in a deadapted state who have been housed in an environmental unit. Although adaptation is only an observation at this time, not a mechanism, biologic limits might regulate how much an organism can adapt. Such limits could be highly individual and vary by orders of magnitude. Certainly adaptation occurs at all levels of biologic systems, from enzyme systems to cells, tissues, organs, and even behavior

(Fregly, 1969). Theoretically, a major insult or the accumulation of lower-level injuries within these systems could lead to a kind of "overload" or "saturation" effect with respect to adaptive capacity. This might cause an individual to have environmental responses, which, instead of being flexible and fluid, would become fragile and overly responsive. Many MCS patients report that years, and in some cases decades, after the onset of their problems, they have recovered only a portion of their former energies and tolerance for their environment. Their descriptions seem to suggest the loss of an intangible capacity to adapt, parts of which may be temporary and recoverable and other parts of which may not. Perhaps our patients have been telling us the diagnosis.

And so these men of Hindostan
Disputed loud and long,
Each in his own opinion
Exceeding stiff and strong,
Though each was partly in the right,
And each was partly wrong.

J.G. Saxe

REFERENCES

ASHFORD, N.A. and MILLER, C.S. (1991). Chemical Exposures: Low Levels and High Stakes. Van Nostrand Reinhold, New York.

BELL, I.R. (1982). Clinical Ecology. Common Knowledge Press, Bolinas, CA.

BELL, I.R. (1990). The biopersonality of allergies and environmental illness. Paper presented at the Eighth Annual International Symposium on Man and His Environment in Health and Disease. Dallas, TX.

BESEDOVSKY, H., SORKIN, E., FELIX, D., and HAAS, H. (1977). Hypothalamic changes during the immune response. Eur. J. Immunol. 7:323-325.

BIGBY, M., JICK, S., JICK, H., and ARNDT, K. (1986). Drug-induced cutaneous reactions. J.A.M.A. 256(24):3358-3363.

BOKINA, A.I., EKSLER, N.D., SEMENENKO, A.D., and MERKURYEVA, R.V. (1976). Investigation of the mechanism of action of atmospheric pollutants on the central nervous system and comparative evaluation of methods of study. Environ. Health Perspect. 13:37-42.

BROUGHTON, A. and THRASHER, J.D. (1990). Personal communication. Cited in Ashford, N. A. and Miller, C. S. (1991). Chemical Exposures: Low Levels and High Stakes. Van Nostrand Reinhold, New York.

CONE, J.E., HARRISON, R., and REITER, R. (1987). Patients with multiple chemical sensitivities: Clinical diagnostic subsets among an occupational health clinic population. In: Workers with Multiple Chemical Sensitivities. (M.R. Cullen, ed.) Occup. Med: State Art Rev. 2(4):721-738.

DAVIDOFF, L.L., CALLENDER , T.J., MORROW, L.A., and ZIEM, G. (1991). Letter to the Editor, J. Psychosom. Res. 35(4):1-3.

DOANE, B.K. and LIVINGSTON, K.E., eds. (1986). Clinical psychiatry and the physiodynamics of the limbic system. In: The Limbic System: Functional Organization and Clinical Disorders. Raven Press, New York. pp. 285-315.

DOTY, R. et al. (1988). Olfactory sensitivity, nasal resistance, and autonomic function in patients with multiple chemical sensitivities. Arch. Otolaryngol. - Head and Neck Surg. 114:1422-1427.

DUFFY, F.H., BURCHFIELD, J.L., BARTELS, P.H., GAON, M., and SIM, V.M. (1979). Long term effects of an organophosphate upon the human electroencephalogram. Toxicol. Appl. Pharmacol. 47:161-176.

FREGLY, M.J. (1969). Comments on cross-adaptation. Environ. Res. 2:435-441.

GILMAN, S. and WINANS, S.W. (1982). Manter and Gatz's Essentials of Clinical Neuroanatomy and Neurophysiology. F. A. Davis Company, Philadelphia, PA.

GIRGIS, M. (1986). Biochemical patterns in limbic system circuitry: Biochemical-electrophysiological interactions displayed by chemitrode techniques. In: The Limbic System: Functional Organization and Clinical Disorders. Raven Press, New York. pp. 55-65.

HACKNEY, J.D., KARUZA S.K., and LINN, W.S. (1977a). Effects of ozone exposure in Canadians and southern Californians, evidence for adaptation? Arch. Environ. Health. 32:110-116.

HACKNEY, J.D., LINN, W.S., MOHLER, J.G., and COLLIER, C.R. (1977b). Adaptation to short-term respiratory effects of ozone in men exposed repeatedly. J. Appl. Psychol. 43:82-85.

HORVATH, S.M., GLINER, J.A., and FOLINSBEE, L.J. (1981). Adaptation to ozone: Duration of effect. Am. Rev. Respir. Dis. 123:496-499.

ISAACSON, R.L. (1982). The Limbic System. Plenum Press, New York.

JESBERGER, J.A. and RICHARDSON, J.S. (1988). Brain output dysregulation induced by olfactory bulbectomy: An approximation for the rat of major depressive disorders in humans. Int. J. Neurosci. 38:241-265.

JOHNSON, A. and REA, W.J. (1989). Review of 200 Cases in the Environmental Control Unit, Dallas. Presented at the Seventh International Symposium on Man and His Environment in Health and Disease. February 25-26, Dallas, TX.

KARE, M. (1968). Direct pathways to the brain. Science. 163:952-953.

KILBURN, K.H. (1989). Is the human nervous system most sensitive to environmental toxins? Arch. Environ. Health. 44(6):343-344.

KRASOVSKII et al. cited in Calabrese, E.J. (1985). Toxic Susceptibility: Male/Female Differences. John Wiley and Sons, New York.

LEVIN, A. and BYERS, V. (1987). Environmental illness: A disorder of immune regulation. In: Workers With Multiple Chemical Sensitivities. (M. R. Cullen, ed). Occup. Med.: State Art Rev. 2(4):669-682.

MACLEAN, P.D. (1967). The brain in relation to empathy and medical education. J. Nerv. Ment. Dis. 144(5):374-382.

MACLEAN, P.D. (1973). "A Triune Concept of the Brain and Behavior." In: The Hincks Memorial Lectures (T. Boag and D. Campbell, eds.). University of Toronto Press, Toronto, Ontario.

MACLEAN, P.D. (1986). Culminating developments in the evolution of the limbic system: The thalamocingulate division. In: Organization and Clinical Disorders. (B.K. Doane and K.E. Livingston, eds). Raven Press, New York. pp. 1-28.

MALLER, O., KARE, M.R., WELT, M., and BEHRMAN, H. (1967). Movement of glucose and sodium chloride from the oropharyngeal cavity to the brain. Nature 213(2):713-714.

McCONNACHIE, P.R. and ZAHALSKY, A.C. (1991). Immunological consequences of exposure to pentachlorophenol. Arch. Environ. Health. 46(4):249-253.

MOLHAVE, L. (1982). Indoor air pollution due to organic gases and vapors of solvents in building materials. Environ. Internat. 8:117-127.

MONROE, R.R. (1986). Episodic behavioral disorders and limbic ictus. In: The Limbic System: Functional Organization and Clinical Disorders. (B.K. Doane and K.E. Livingston, eds). Raven Press, New York. pp. 251-266.

NATIONAL FOUNDATION FOR THE CHEMICALLY HYPERSENSITIVE. (1989). Cheers. 1:6.

NATIONAL PESTICIDE TELECOMMUNICATIONS NETWORK (NPTN). (1991). Active Ingredients from Possible Chemical Sensitivities. Texas Tech University Health Science Center, Lubbock, TX.

OFFICE OF TECHNOLOGY ASSESSMENT, CONGRESS OF THE UNITED STATES. (1990). Identifying and Controlling Poisons of the Nervous System. United States Government Printing Office, Washington, DC.

PAYAN, D.G., MCGILLIS, J.P., and GOETZL, E.J. (1986). Neuroimmunology. Adv. Immunol. 39:299-323.

RANDOLPH, T.G. (1962). Human Ecology and Susceptibility to the Chemical Environment. Charles C. Thomas, Springfield, IL.

RANDOLPH, T.G. (1965). Ecologic orientation in medicine: Comprehensive environmental control in diagnosis and therapy. Ann. Allergy 23:7-22.

REA, W.J. (1988). Chemical hypersensitivity and the allergic response. Ear, Nose and Throat J. 67(1):50-56.

RIIHIMAKI, V. and SAVOLAINEN, H. (1980). Human exposure to m-xylene. Kinetics and acute effects on the central nervous system. Ann. Occup. Hyg. 23:411-432.

ROSENTHAL, N.E. and CAMERON, C.L. (1991). Exaggerated sensitivity to an organophosphate pesticide (letter). Am. J. Psychiatry 148(2): 270.

RYAN, C.M., MORROW, L.A., and HODGSON, M. (1988). Cacosmia and neurobehavioral dysfunction associated with occupational exposure to mixtures of organic solvents. Am. J. Psychiatry 145(11):1442-1445.

SCHOTTENFELD, R. (1987). Workers with multiple chemical sensitivities: A psychiatric approach to diagnosis and treatment. In: Workers With Multiple Chemical Sensitivities. (M.R. Cullen, ed.). Occup. Med.: State Art Rev. 2(4) 739-754.

SELYE, H. (1946). The general adaptation syndrome and the diseases of adaptation. J. Allergy. 17:231-247, 289-323, 358-398.

SHIPLEY, M., (1985). Transport of molecules from nose to brain: Transneuronal antegrade and retrograde labeling in the rat olfactory system by wheat germ agglutinin-horseradish peroxidase applied to nasal epithelium. Brain Res. Bull. 15:129-142.

SILVER, W.L. and MARUNIAK, J.A. (1981). Trigeminal chemoreception in the nasal and oral cavities. Chemical Senses 6(4):295-305.

SIMON, G.E., KATON, W.J., and SPARKS, P.J. (1990). Allergic to life: Psychological factors in environmental illness. Am. J. Psychiatry 147:901-906.

STAUDENMAYER, H. and SELNER, J.C. (1990). Neuropsychology during relaxation in generalized, universal 'allergic' reactivity to the environment: A comparison study. J. Psychosom. Res. 34(3):259-270.

STEIN, M., KELLER, S., and SCHLEIFER, S. (1981). The hypothalamus and the immune response. In: Brain, Behavior and Bodily Disease. (H. Weiner, M. A. Hofer, and A.J. Stunkard, eds). Raven Press, New York. pp. 45-65.

TERR, A.I. (1986). Environmental illness: A clinical review of 50 cases. Arch. Intern. Med. 146:145-149.

TERR, A.I. (1989). Clinical ecology in the workplace. J. Occup. Med. 31(3):257-261.

WALSH, T.J. and EMERICH, D.F. (1988). The hippocampus as a common target of neurotoxic agents. Toxicol. 49:137-140.

MULTIPLE CHEMICAL SENSITIVITIES AND THE IMMUNE SYSTEM

WILLIAM J. MEGGS, MD, PhD
Division of Clinical Toxicology
Department of Emergency Medicine
East Carolina University, School of Medicine
Greenville, North Carolina

INTRODUCTION

Patients claiming sensitivity to multiple chemical inhalants in the environment are a challenge to both the clinician and researcher. Many aspects of the syndrome are reminiscent of allergy, with specific odorous chemicals triggering symptoms in an acquired manner after an initial encounter. A spreading phenomenon similar to the 'environmental-adjuvant' effect exists in which one environmental exposure can potentiate the acquisition of immune sensitivity to a second agent. Just as avoidance of allergen, whenever possible, is the treatment of choice for allergic diseases, avoidance of incriminating inhalants is said to provide relief to the chemical-sensitive patient.

Despite similarities to immune-mediated allergic diseases, multiple chemical sensitivities (MCS) present distinct differences from traditional allergy. Patients with allergy have well-defined, clinical reactions to allergens, and symptoms of rhinitis, asthma, urticaria, or gastrointestinal symptoms occur shortly after exposure. Patients with MCS describe a host of symptoms, some mental (e.g., depression, psychosis, and mania) and some physical (involving the respiratory system, gastrointestinal tract, skin, cardiovascular system, or other organ systems).

The basic step in developing allergic disease is the production of specific IgE directed against environmental proteins. By contrast, the mechanism of MCS is still unknown, but hypotheses range from a somatization disorder to a cell-mediated immune reaction. Establishing the mechanism of this disorder is essential if we are to arrive at a consensus necessary to help these patients.

DESCRIPTIONS OF MCS WITH ATTENTION TO MECHANISMS

Randolph (1962) described patients with chemical sensitivities who developed chronic debilitating illnesses, as summarized in Table 1. He described a chronic, debilitating illness in which patients developed a variety of mental and physical symptoms related to exposures to

Address all correspondence to: Dr. William J. Meggs, Division of Clinical Toxicology, Department of Emergency Medicine, East Carolina University, Greenville, NC 27858.

environmental chemicals. From his private practice as an allergist, Randolph described a trial-and-error procedure for eliminating chemical inhalants from the patients' environment and then reexposing them in open, unblinded challenges. Foods and chemical inhalants were tested. Treatment consisted of avoiding substances shown to cause illness, and, in extreme cases, maintaining patients in porcelain trailers on a diet of rare or exotic foods on strict rotation.

TABLE 1. Randolph's Description of Chemical Sensitivity Disease

Acquired, often after chronic insidious or acute exposure to a petrochemical.

Chemical exposures can trigger physical (arthritis, asthma, colitis, etc.) and mental (depression, difficulty with concentration, mania, psychosis, etc.) symptoms, depending on individual susceptibility.

Specific Adaptation Syndrome. Adaptation to specific chemicals with chronic exposure is followed by maladaptation and chronic illness, withdrawal symptoms when removed from the chemical environment, and "shock reactions" on reexposure.

Spreading phenomena. As an individual becomes maladapted to the chemical environment, intolerance to increasing numbers of environmental chemicals develops.

Avoidance. By avoiding the chemical environment, chronic illnesses may resolve.

Randolph introduced a special hospital unit, the Environmental Control Unit, to isolate patients from the chemical environment for testing. Patients fasted in this environment for about five days before being reexposed to foods and then inhalants, one-by-one. By 1982, 20,000 patients had been treated with this routine (Randolph, 1982), but no scientific data on this population have yet been presented. Randolph's patients had various severe illnesses, such as rheumatoid arthritis, steroid-dependent asthma, multiple sclerosis, and psychiatric illnesses, such as depression, bipolar affective disorders, and schizophrenia.

Chemicals triggering these illnesses are given in Table 2. Studies have verified that the greatest exposure to organic inhalants comes from indoor air contaminants rather than industrial or other sources (Wallace et al., 1985, Wallace et al., 1987). Current scientific interest in MCS is motivated partly by work done on exposures during the 1980s.

Ashford and Miller (1991) gave an operational definition of MCS: "The patient with multiple chemical sensitivities can be discovered by removal from the suspected offending agents and by rechallenge, after an appropriate interval, under strictly controlled environmental conditions. Causality is inferred by the clearing of symptoms with removal from the offending environment and recurrence of symptoms with specific challenge." This approach circumvents the problems of the multitude of environmental exposures common in our society.

TABLE 2. Chemicals Alleged to Cause Reactions in Chemically Sensitive Individuals (Randolph, 1962)

Indoor air contaminants	utility gas coal smoke fumes from fresh paint, turpentine, mineral spirits
Detergents	fragrances from toiletries and perfumes cleaning products such as bleach, ammonia, disinfectants insect sprays and repellents odorous synthetic carpets, pads, adhesives, and building materials
Outdoor air contaminants	automobile and diesel exhaust industrial air pollutants paint manufacturing and sulfur-processing fumes fumes from roof tar and roads
Chemical additives and contaminants of food and water	insecticides and fumigant residues some chemical preservatives sulfur residues chemical flavoring and sweetening agents plastic containers and lined tins chloride in water
Synthetic drugs, cosmetics, and miscellaneous chemicals	medications, including aspirin, sulfonamides, synthetic vitamins allergic extracts or other biologic materials that contain phenol cosmetics synthetic textiles bed linens washed with detergents, dried in gas driers, or impregnated with synthetic starch or sizing

The publication of Cullen's case definition of the illness, based on experiences he and others had at occupational medicine clinics around the country, gave a boost to the scientific study of MCS (Cullen, 1987) (Table 3). Although this definition may be modified as the illness is better defined scientifically, it gives researchers a standard for defining study populations. A comparison of Tables 1 and 3 leads to the conclusion that Randolph and Cullen are most likely discussing the same syndrome, although Randolph's description includes patients at different stages of "adaptation" to chemicals.

To help understand the different stages of the illness, we introduce the "Hypothetical Chemical Stress Syndrome," a concept that draws heavily on the work of others (Selye, 1946; Randolph, 1962). "Hypothetical" connotes the current lack of sufficient data to establish a consensus about this syndrome's role in human disease, and "chemical" highlights our focus on exposure to environmental chemicals, although emotional and other stresses may cross-react with chemical exposures.

TABLE 3. Cullen's Case Definition of the Multiple Chemical Sensitivity Syndrome (1987)

The illness is acquired as a result of some environmental exposure, insult, or illness that can be documented.

Symptoms involve more than one organ.

Symptoms recur and abate in response to some predictable stimuli.

Symptoms are elicited by exposures to chemicals of diverse structural classes and toxicological modes of action.

Symptoms are elicited by demonstrable exposures.

Exposures that elicit symptoms must be very low—many standard deviations below the "average" exposures known to cause adverse human responses.

No single widely available test of organ system function can explain the symptoms.

The stages of the Hypothetical Chemical Stress Syndrome (HCSS) are given in Table 4. Individuals in Stage 0 tolerate chemical exposures without difficulty. Progression from Stage 0 to Stage 1 occurs when an individual is chemically stressed by either an acute, high-dose chemical exposure or a chronic, insidious exposure. Individuals in Stage 1 have symptoms from chemical exposure, but no physical findings on examination. Hence, Stage 1 is the "-algia" (Greek for pain) stage.

TABLE 4. The Hypothetical Chemical Stress Syndrome

Stage Zero	Tolerance of the chemical environment
Stage One	Multiple complaints following chemical exposures, such as headache, nausea, pruritis, flushing, etc., but few findings on examination
Stage Two	Inflammation initially isolated to one organ system (arthritis, vasculitis, dermatitis, etc.)
Stage Three	Fibrosis, with damage to tissues

In Stage 2, an individual develops foci of inflammation from chemical exposures. Findings appear on physical examination, and a medical diagnosis can be given, such as rheumatoid arthritis, multiple sclerosis, some forms of dermatitis and vasculitis, non-allergic asthma, and rhinitis. This is the "-itis" (Greek for inflammation) stage. Progression from Stage 1 to Stage 2 may be triggered by chemical exposures, and, if tissues are not damaged, the inflammation can be reversed by removing the chemical stimuli. Thus, progression from Stage 1 to Stage 2 is a reversible process if chemicals are avoided. Medications, such as corticosteroids and the non-steroidal anti-inflammatory agents, can reduce but not cure the inflammation of Stage 2. If chemical stimuli are not removed and anti-inflammatory medications are discontinued, a relapse of inflammation occurs. Medications do not prevent progression from Stage 2 to Stage 3.

Damage to tissue occurs in Stage 3. This may be called the "-osis" (Greek for change in living tissue) stage. Once tissue is damaged, reversal is unlikely, and organ function is lost.

Progression into Stages 2 and 3 usually occurs in one organ system, such as the skin, joints, or lungs, and later may involve a second organ system. Despite the irreversible tissue damage of Stage 3, an individual should be removed from the chemical environment to prevent inflammation in other tissues.

Stage 1 of the Hypothetical Chemical Stress Syndrome is what many clinicians call the Multiple Chemical Sensitivity Syndrome in which patients have many complaints they attribute to chemical exposures but few, if any, findings on physical examination. Hence, these patients may be regarded as having a somatization disorder with no real medical illness. Studies of adverse reactions to environmental chemicals should include patients with diagnosable medical conditions (Stage 2 and 3 patients), and hypothesized progression between these stages should be studied. Longitudinal studies will be very important.

The concept of a stress syndrome was originally introduced as the "Generalized Adaptation Syndrome" to explain observations in which animal models exposed to stresses progress through stages of adaptation similar to the stages described above (Selye, 1946). Although the syndrome was initially discovered while exposing animals to chemical stresses, other stresses (e.g., confinement), were observed to lead to similar pathology, so the word "generalized" was introduced. "Specific Adaptation Syndrome" has been used when one specific stress, such as a single chemical, causes maladaptation (Randolph, 1962). Although these stress syndromes may not be equivalent, they are very similar.

TYPES OF IMMUNE REACTIONS AND THEIR POSSIBLE RELEVANCE TO MCS

The immune system has several branches or arms, each with a particular function (Table 5). These branches interact in a common immunoregulation, and activating one may shut down the others in certain situations (Moore et al., 1990). In addition to providing host defense, these branches of the immune system can lead to disease and tissue damage. The IgE system is involved in allergic diseases such as anaphylaxis, asthma, rhinitis, and urticaria. The IgG system can lead to serum sickness, and specific IgG for autoantibodies is found in various autoimmune diseases, including myasthenia gravis, Wegener's granulomatosis, and systemic lupus erythematosus. Cellular immunity plays a role in contact dermatitis, in which the skin receives an initial sensitizing contact with an organic chemical and then becomes inflamed after reexposure. Cellular infiltrates are seen in the skin.

TABLE 5. Branches of the Immune System and Their Function

BRANCH	FUNCTION
IgE-Mast Cell-Eosinophil System	Destruction of tissue parasites
IgG-IgM-Neutrophil/Macrophage System	Destruction of encapsulated bacteria
Lymphocyte/Cellular Immune System	Destruction of fungi and viruses, immunoregulation
IgA-Secretory System	Destruction of pathogens at mucosal surfaces

MCS is clearly distinct from IgE-mediated allergy, as Randolph recognized in introducing new terminology for this illness. Of the known types of immune reactivity, cellular immunity is most likely to play a role in chemical sensitivities. The delay of inflammatory symptoms described in these patients is similar to the delay seen in cellular inflammation. Two interleukins (IL-1 and IL-2) are known to affect the central nervous system (Obal et al., 1990 and De Sarro et al., 1990). Several authors (Rea et al., 1986; Levin and Byer, 1989; Broughton and Thrasher, 1988) have claimed abnormalities of cellular immunity in these patients, such as alterations of helper and suppressor ratios and the presence of activated lymphocytes in peripheral blood.

The concept of an environmental adjuvant has been introduced to describe substances that induce an immune response to other substances. In immunology, an adjuvant potentiates an immune response to other substances. For example, an animal injected with the adjuvant alum mixed with a protein will begin producing IgE specific for that protein. Inhaled environmental adjuvants in experimental models include sulfur dioxide, which can induce allergic asthma to co-administered aeroallergens in the guinea pig (Riedel et al., 1988), and ozone, which potentiates platinum sensitivity in monkeys (Biagini et al., 1986). Animal models for respiratory exposure to environmental adjuvants are given in Table 6.

The mechanism of environmental adjuvancy may simply be that inflamed tissue processes antigen. Substances like sulfur dioxide and ozone produce tissue irritation, which then leads to inflammation and the recruitment of macrophages to the area. The macrophages then present antigens to lymphocytes which in turn produce an immune response to the antigen. Subsequent exposure to either the adjuvant or the antigen could produce further inflammation and the processing of other antigens. This process may be initiated by the interaction of chemical irritants with sensory nerves. The repertoire of triggers for inflammation would grow, with the increased risk of more substances becoming triggers. Although such an immunologic spreading phenomena has not been demonstrated to my knowledge, it is theoretically possible, and this mechanism may be operative in the spreading phenomenon of MCS, if MCS is mediated by cellular immunity in the airway.

TABLE 6. Examples of Environmental Adjuvants in Animal Models

SPECIES	ENVIRONMENTAL ADJUVANT	ALLERGEN	REFERENCE
guinea pig	sulfur dioxide, ozone, and nitrogen dioxide	ovalbumin	Matsumura et al., 1970
guinea pig	sulfur dioxide	ovalbumin	Riedel et al., 1988
monkey	ozone	platinum	Biagini et al., 1986
mouse	diesel exhaust particles	Japanese cedar pollen	Muranaka et al., 1986

REVIEW OF DATA ON IMMUNE PARAMETERS IN MCS

Studies have attempted to establish an immunologic basis for patients with MCS (Table 7). These studies depict increases in T4/T8 ratios in chemically sensitive patients, decreases in this ratio, and no abnormalities in this ratio. However, patient populations differ in these studies. One study described three groups of chemically sensitive patients, those with vascular dysfunction (vasculitis), asthma, and rheumatoid arthritis (Rea et al., 1986). Chemical sensitivity was verified by chemical challenges. The patients would be classified in Stages 2 and 3 in the HCSS. Statistically significant decreases in T8 (suppressor) cells and increases in the T4/T8 (helper:suppressor ratio) were found in patients with vasculitis and rheumatoid arthritis but not in those with asthma.

Levin and Byer (1989) have studied patients with documented environmental exposures to such chemicals as trichlorethylene, PCBs, or a mixture of industrial dyes, solvents, and pesticides. These patients, who had decreased T4:T8 ratios, were said to be chemically sensitive following these exposures and may represent Stage 1 in the HCSS.

Terr (1986) found no consistent pattern of immunologic abnormality in the medical records of 50 patients referred by the California Worker's Compensation Board of Appeals because of difficulties with workplace chemicals. In contrast to the patients Rea and his collaborators studied, these patients had various medical diagnoses but were not challenged with chemical exposures to determine any chemical sensitivities. Instead, because another medical diagnosis could be given, it was concluded that they had a somatization disorder rather than chemical sensitivities. Terr performed no laboratory measurements but picked data from the medical records of a diverse group of patients.

These studies, with their diverse patient populations and conflicting results in T4:T8 ratios, illustrate the importance of using well-designed clinical research protocols and well-defined patient populations to assess chemically sensitive patients. The expression, multiple chemical sensitivity, might be better used as an etiology than as a medical diagnosis. Several inflammatory conditions can arise from the chronic stress of exposure to environmental chemicals, and the various conditions developing from chemical exposures may have different immunologic profiles reflecting the medical diagnosis rather than the chemical etiology.

TABLE 7. Selected Published Data on Immune Abnormalities in Chemically Exposed Patients

REFERENCE	PATIENT GROUP	NUMBER	ABNORMALITY
Rea et al., 1986	Vasculitis	70	T8 decreased[a], T4/T8 increased[a]
	Asthma	27	T4% decreased[b]
	Rheumatoid arthritis	7	T8 decreased[a], T4/T8 increased[c]
Terr, 1986	Worker comp appeals	50	No immunological abnormality
Broughton and Thrasher, 1988	Formaldehyde exposure	67	Antibodies to formaldehyde-albumin increased CD26 cells
Levin and Byers, 1989	TCE exposure	25	T4/T8 decreased[d]
	Industrial chemical exposure	10	T4/T8 decreased[e]
	PCB exposure	21	T4/T8 decreased[e]
	Electronic workers	78	T4/T8 decreased[e]
Patterson et al., 1989	Chemically sensitive		No antibodies to formaldehyde-albumin
	Dialysis patients		Antibodies to formaldehyde-albumin
McConnachie and Zahalsky, 1991	Pentachlorophenol	38	Activated T cells Decreased mixed lymphocyte response

[a] $p < 0.0001$ [b] $p < 0.05$
[c] $p < 0.001$ [d] $p < 10^{-8}$ [e] $p < 10^{-6}$

Although Rea divided patients into groups according to medical diagnosis, Terr combined patients with various medical diagnoses that were allegedly exacerbated by chemical exposures. Terr claimed his patients had somatization disorders not chemical sensitivities even though some of their conditions—asthma, rhinitis, and dermatitis—are known to be exacerbated by environmental chemicals and no challenge studies were performed to evaluate chemical sensitivity. Rea found normal T4:T8 ratios in the patients with asthma who are chemically sensitive. Of course, patients with chemical sensitivities are a heterogeneous group, and those with illnesses primarily involving the respiratory system will be very different from those with psychiatric or autoimmune manifestations. Research protocols should allow for these differences.

One group of investigators has proposed that antibodies to formaldehyde coupled to human serum album (f-HSA) may be a good marker for chemical sensitivity (Broughton and Thrasher, 1988; Thrasher et al., 1988; Thrasher et al., 1989). These investigators found low level IgG, IgE, and IgM titers to f-HSA (in the 1:4 to 1:16 range) in chemically sensitive patients. Well-established associations between diseases and antibody titers, such as systemic lupus erythematosus and anti-DNA titers, are characterized by extremely high titers (1:1,000 or higher). Non-specific binding at the high serum concentrations found in low-titer wells make low-titer associations with disease of dubious clinical significance. Another group found no association between formaldehyde conjugates and chemical sensitivity but showed that titers may be a possible marker of exposure to formaldehyde in renal dialysis patients (Patterson et al., 1989).

An increased frequency of the CD26 lymphocyte marker in chemically sensitive individuals has been reported (Madison et al., 1991). This marker has also been reported with greater frequency in pentachlorophenol-exposed individuals (McConnachie and Zahalsky, 1991), although this study did not address the issue of chemical sensitivity. Because CD26 is a lymphocyte-activation marker, the notion of chronic activation of cellular immunity by chemical exposures is an attractive one that deserves further study.

A recent retrospective review of ten patients who met the Cullen case definition for MCS showed rhinitis and abnormal rhinolaryngoscopy results (Meggs et al., 1992a). Limited data from biopsies of the nasal mucosa in MCS patients have shown chronic lymphocytic infiltrates in the submucosal tissue, basement membrane thickening, and desquamation of the nasal epithelium (Meggs, 1992b). If these results are sustained in a blinded, controlled study now being planned, the importance of cellular immunity in MCS will become clear.

Another important finding concerns effects in the upper airway of exposures to volatile organic chemicals (VOCs). A group at the U. S. Environmental Protection Agency's Human Experimentation Research Laboratory has performed nasal lavage on normal subjects and neutrophil counts on the nasal lavage fluid (Koren et al., 1990). Subjects were exposed for 2 or 4 hours to a 25 mcg/m^3 VOC mixture typical of that found in the indoor air. Nasal lavages were performed before and after exposure and at 18 hours post exposure. A cross-over design was used, with air challenges used as controls. A statistically significant increase in polymorphonuclear cells in the lavage fluid occurred immediately after and 18 hours after exposure to VOCs; no increase was seen after exposure to air.

THE NEUROIMMUNOLOGIC HYPOTHESIS

It may not be sufficient to look at either the immune or the nervous system to understand the mechanism of MCS. Perhaps we must investigate the interaction between these two systems before we understand this illness. Aspects of MCS are clearly reminiscent of immunity, such as the specificity of triggering agents, the acquired nature following an initial exposure, and the spreading phenomena. Furthermore, the nervous system is clearly involved with the cognitive and psychiatric aspects. The autonomic nervous system may also play a role in some peripheral manifestations through vasomotor phenomena. To reconcile this duality, we have proposed the neuroimmunologic hypothesis of MCS.

This hypothesis proposes that inhaled VOCs stimulate irritant receptors in the upper airway, which then lead to neurogenic inflammation. The inflamed tissue processes antigen, leading to antigen-specific lymphocytes in the upper airway. Subsequent exposures trigger these lymphocytes to release mediators including interleukins that have a known effect on the central nervous system (Obal et al., 1990; De Sarro et al., 1990). In addition, a "hyperesthesia" of irritant receptors is postulated to occur, with stimulation of these receptors feeding into acquired pathways that can cause vasomotor response at distant sites. Primitive stress pathways may be activated, leading to limbic-system dysfunction and the psychiatric manifestations of MCS.

RESEARCH AGENDA TO STUDY IMMUNOLOGIC MECHANISMS OF MCS

MCS is a very special disorder, and the proper study of this illness will be an exercise in walking through a mine field without getting blown up. So much controversy surrounds this illness, and eminent clinicians as well as prestigious professional organizations have taken the public position that no physical basis for this disorder exists. Any research protocols that attempt to go beyond the psychiatric hypothesis must be impeccable in design and execution. Problems include performing double-blinded challenges to odorous materials and dealing with patient populations that report psychiatric reactions, such as depression and hallucinations, from chemical exposures and that often have subjective symptoms in great excess of objective findings. Immune parameters may constantly be fluctuating up and down with exposure to the chemical environment, infectious agents, and other stresses. Furthermore, the ubiquitous nature of chemicals in the environment will make it difficult to define normal values, particularly given the documentation of an immune aberration in normal subjects exposed to a VOC mixture (Koren et al., 1990).

Nonetheless, the full power of the scientific method can be brought to bear on this problem. Patient selection is of utmost importance. Studies of hysterical patients who have no physical findings or objective parameters will be less rewarding than studies of patients with definable, quantifiable, inflammatory illnesses such as rheumatoid arthritis, asthma, and multiple sclerosis. To isolate patients and controls from the chemical environment, a specially constructed residential clinical research unit will be necessary. Changes in immune parameters must be studied as subjects are removed from the chemical environment and then reexposed to chemicals in a blinded fashion. Concomitantly, clinical signs and symptoms must be followed and measured.

CONCLUSIONS

No consensus about MCS will be reached until the pathophysiology of this disorder is understood. Investigators designing studies of this illness should incorporate tests of proposed mechanisms into their protocols. MCS is not an allergic disease in the traditional sense, but the inflammation that characterizes subgroups of patients with chemical sensitivities implies that the immune system is involved. The nervous system is also involved, and this author believes that the interaction of the nervous and the immune systems will be crucial in understanding this disorder.

The following conclusions can be drawn from published studies of immune abnormalities in MCS patients. Antibody titers to formaldehyde-human serum albumin conjugates are not useful as a biomarker for MCS. Activated lymphocytes in the peripheral blood may be a sensitive but non-specific marker of this disorder, but further study is needed. Ratios of helper to suppressor lymphocytes in peripheral blood are valuable as a sensitive but non-specific marker of specific diagnoses involving sensitivity to environmental chemicals, but these ratios do not correlate with MCS per se.

Well-designed challenge studies performed in an environmental control unit will be needed to elucidate this illness. Verification of adaptation syndromes, such as the generalized adaptation syndrome, the specific adaptation syndrome, and the hypothetical chemical-stress syndrome, would unite areas of medicine currently thought to be unrelated and would establish environmental chemicals as the etiology of those autoimmune and psychiatric disorders currently thought to be idiopathic. The impact on clinical medicine could be enormous

REFERENCES

ASHFORD, N.A. and MILLER, C.S. (1990). Multiple Chemical Sensitivities. Report to the New Jersey State Department of Health.

ASHFORD, N.A. and MILLER, C.S. (1991). Chemical Exposures: Low Levels and High Stakes. Van Nostrand Reinhold, New York.

BIAGINI, R.E., MOORMAN, W.J., LEWIS, T.R., and BERNSTEIN, I.L. (1986). Ozone enhancement of platinum asthma in a primate model. Am. Rev. Respir. Dis. 134:719-725.

BROUGHTON, A. and THRASHER, J.D. (1988). Antibodies and altered cell mediated immunity in formaldehyde exposed humans. Clin. Toxicol. (2):155-174.

COCHRAN, C.G. (1988). Immune complex diseases. In: Cecil Textbook of Medicine. (J.B. Wyngaarden and L.H. Smith, Jr., eds.). 18th ed. W.B. Saunders, Philadelphia, PA. p. 1960.

CULLEN, M. (1987). The worker with multiple chemical sensitivities: An overview. In: Workers with Multiple Chemical Sensitivities (M.R. Cullen, ed.). Occup. Med.: State Art Rev. 2(4):655-662.

DE SARRO, G.B., MASUDA, Y., ACIOTI, C., AUDINO, M.G., and NISTICO, G. (1990). Behavioral and ECoG spectrum changes induced by intracerebral infusion of interferons and interleukin 2 in rats are antagonized by naloxone. Neuropharmacol 29:167-179.

KOREN, H.S., DEVLIN, R.B., HOURSE, D., STEINGOLD, S., and GRAHAM, D.E. (1990). The inflammatory response to the human upper airways to volatile organic compounds.

Proceedings of the 5th International Conference on Indoor Air Quality and Climate. Ottawa, Canada. p. 325.

LEVIN, A.S. and BYERS, V.S. (1989). Environmental illness: A disorder of immune regulation. In: Workers with Multiple Chemical Sensitivities. (M.R. Cullen, ed.). Occup. Med.: State Art Rev. 2(4)669-681.

MADISON, R.E., BROUGHTON, A., and THRASHER, J.D. (1991). Immunologic biomarkers associated with an acute exposure to exothermic byproducts of a ureaformaldehyde spill. Environ. Health. Perspect. 94:219-223.

MATSUMA, Y. (1970). The effects of ozone, nitrogen dioxide, and sulfur dioxide on the experimentally induced allergic respiratory disorder in guinea pigs. I. The effects of sensitization with albumin through the airways. Amer. Rev. Resp. Dis. 102:430-437.

McCONNACHIE, P.R. and ZAHALSKY A.C. (1991) Immunological consequences of exposure to pentachlorophenol. Arch. Environ. Health 46(4):249-253.

MEGGS, W.J. and CLEVELAND, C.H. (1992a). Rhinolaryngoscopy findings in patients with the multiple chemical sensitivity syndrome. Arch. Environ. Health In press.

MEGGS, W.J., CLEVELAND, C.H., JR., METZGER, W.J., LARKIN, E.L., and ALBERNAZ, M. (1992b). Reactive upper airways dysfunction syndrome (RUDS): A form of chemical irritant rhinitis induced by a chemical exposure. J. Allergy Clin. Immunol. 89:170a.

MOORE K.W., VIEIRA, P., FIORENTINO, D.F., TROUNSTINE, M.L., KHAN, T.A., and MOSMANN, T.R. (1990). Homology of cytokine synthesis inhibitory factor (IL-10) to the Epstein-Barr virus gene BCRFI. Science 248:1230-1234.

MURANAKA, M., SUZUKI, S., KOIZUMI, K., TAKAFUJI, S. et al. (1986). Adjuvant activity of diesel-exhaust particles for the production of IgE antibody in mice. J. Allergy Clin. Immunol. 77:616-623.

OBAL, F. Jr., OPP, M., CADY, A.B., JOHANNSEN, L., POSTLETHWAITE, A.E., POPPLETON, H.M., SEYER, J.M., and KRUEGER, J.M. (1990). Interleukin 1 alpha and an interleukin 1 beta fragment are somnogenic. Am. J. Physiol. 259:R439-46.

PATTERSON, R., DYKEWICZ, M.S., EVANS, R., GRAMMER, L.C., GREENBERGER, P.A., HARRIS, K.E., LAWRENCE, I.D., PRUZANSKY, J.J., ROBERTS, M. SHAUGHNESSY, M.A. et al. (1989). IgG antibodies against formaldehyde human serum proteins: A comparison of other IgG antibodies against inhalant proteins and reactive chemicals. J. Allergy Clin. Immunol. 84:359-366.

RANDOLPH, T. G. (1962). Human Ecology and Susceptibility to the Chemical Environment. Charles C. Thomas, Springfield, IL.

RANDOLPH, T.G. and MOSS, R. (1980). An Alternative Approach to Allergies. Lippincott & Cromwell, New York.

REA, W.J., JOHNSON, A.R., YOUDIM, S. et al. (1986). T&B lymphocyte parameters measured in chemically sensitive patients and controls. Clin Ecol (4):11.

RIEDEL, F., KRAMER, M., SCHEIBENBOGEN, C., and RIEGER, C.H. (1988). Effects of SO2 exposure on allergic sensitization in the guinea pig. J. Allergy Clin. Immunol. 82:527-534.

SELYE, H. (1950). Stress. Acta, Inc. Montreal, Canada.

TERR, A.I. (1986). Environmental illness: A clinical review of 50 cases. Arch. Intern. Med. 146:145.

THRASHER, J.D., BROUGHTON, A., and MICEVICH, P. (1988). Antibodies and immune profiles of individuals occupationally exposed to formaldehyde: 6 case reports. Amer. J. Indust. Med. 14:479-488.

WALLACE, L. et al. (1985). Organic chemicals in indoor air: A review of human exposure studies and indoor air quality studies. In: Indoor Air and Human Health. (R. Gammage and S. Kaye, eds.). Lewis Publishers, Chelsea, MI. pp. 361-378.

WALLACE, L., PELLIZZARI, E.D., HARTWELL, T.D., SPARACINO, C., WHITMORE, R., SHELDON, L., ZELON, H., and PERRITT, R. (1987). The TEAM study: Personal exposures to toxic substances in air, drinking water, and breath of 400 residents of New Jersey, North Carolina, and North Dakota. Environ. Res. 43:290-307.

IS THERE EVIDENCE OF AN IMMUNOLOGIC BASIS FOR MULTIPLE CHEMICAL SENSITIVITY?

JOSEPH F. ALBRIGHT, PhD
ROBERT A. GOLDSTEIN, MD, PhD
Division of Allergy, Immunology and Transplantation
National Institute of Allergy and Infectious Diseases
National Institutes of Health
Bethesda, Maryland

INTRODUCTION

The increasing prevalence in recent years of toxicants (xenobiotics) in the environments where people live, work, and play has apparently resulted in significant numbers of persons who profess to be suffering from a diffuse group of complaints. These include migraine, irritable bowel syndrome, neurasthenia, cardiac arrhythmias, depression and cognitive disorders, seizures, and flu-like malaise. Collectively, these complaints have been lumped together as symptoms of a disorder called multiple chemical sensitivity (MCS) (also called environmental illness, total allergy syndrome, and twentieth century disease). For some, MCS has been viewed as a disorder resulting from dysregulation of the immune system. At present, aberrations of the immune system are included with neuro-psychologic and endocrinologic aberrations as possible leading players in MCS. Many believe that, in time, MCS will be shown to result from psycho-neuro-immuno-endocrinologic disturbances.

The purpose of this essay is to a) urge persons interested in this topic to initiate programs of research on MCS that embrace the scientific method so that results of their efforts can be accepted by all biomedical scientists; and b) offer suggestions about potentially promising directions of research if MCS does, in fact, exist and if it has an immunologic component.

PRESENT STATUS OF EVIDENCE

At the present time, MCS does not appear to be a typical allergic condition because few if any patients purported to suffer from the condition show clinical evidence of allergies associated with well-established, IgE-mediated reactions. Rather, an unusual type of allergic reaction associated with chemicals (food additives, in particular) has been postulated, called total allergy (American College of Physicians, 1989), allergic tension fatigue syndrome (Speer, 1954), or cerebral allergy (Miller, 1972). The most compelling evidence for the existence of such a syndrome appears to be anecdotal, that is, claims of dramatic improvement in patients who

Address all correspondence to: Dr. Joseph F. Albright, Solar Building, Room 4A22, Division of Allergy, Immunology and Transplantation, National Institute of Allergy and Infectious Diseases, National Institutes of Health, Bethesda, MD 20892.

have been separated from offensive chemicals. Experimental evidence to support the idea of an immunologic basis for MCS is meager. In fact, a recent, well-controlled, double-blind study of multiple food sensitivities, following protocols agreed on by clinical ecologists, failed to provide any evidence of multiple sensitivities in a group of subjects previously judged by subjective criteria to be sensitive (Jewett et al., 1990).

Other components of the immune system, in addition to those involved in allergic reactions, have been invoked to explain the MCS syndrome. For example, there have been claims of deficiencies in one or more complement components in patients evincing environmental illness (McGovern et al., 1983; Rea, 1976, 1977, 1978a). These claims have not been clearly substantiated (Terr, 1987). Circulating levels of C4 have been reported to be either low (McGovern et al, 1983) or high (Rea, 1978a). In general, reports have suggested that overall complement activity was lower, C3 levels were lower, and C4 levels higher than normal in patients diagnosed as being chemically sensitive (Rea, 1976, 1977; Terr, 1987). However, many of the reported values were within normal ranges (Terr, 1987). Furthermore, the significance, if any, of modest variations in complement levels is unclear. Whether variations in complement levels, if shown to be significant by future studies, could be associated with the presence of immune complexes remains to be determined.

Attempts to detect and measure immune complexes in MCS patients appear to be lacking or meager. If small organic molecules do elicit antibodies presumed to be immunopathologic in MCS patients, they would be expected to react with the patient's proteins to form hapten-carrier complexes. This has been amply demonstrated in conventional allergic reactions to small molecules, such as trimellitic anhydride (Zeiss et al., 1977). It may also occur in the postulated sensitivity of certain individuals to substances such as formaldehyde (Pan et al., 1988), although the reported generation of sensitizing antibodies to formaldehyde (Thrasher et al., 1990) has not been confirmed (Patterson et al., 1987).

Currently, the component of the immune system considered likely to be responsible for MCS is the T-cell population. In particular, it is argued that a change in the ratio of $CD4^+$ to $CD8^+$ cells occurs in individuals who have been exposed to environmental chemicals that are toxic for T cells (Levin and Byers, 1987; Rea et al., 1987). Consequently, the ability of T cells to regulate immune responses is impaired. For example, increased CD4/CD8 ratios would suggest selective effects of toxic chemicals on $CD8^+$ T suppressor cells accompanied by nonspecific expansion of B lymphocytes (polyclonal B-cell activation). The possible consequences of such a change and how such observations, if corroborated, might lead to an interesting research program will be discussed below. However, it should be noted that many of the reported changes in CD4/CD8 ratios fall within normal range, which can vary widely for different individuals and increases significantly with age (Lifson et al., 1986). Furthermore, few, if any, examples of disorders can be related simply to altered CD4/CD8 ratios in the absence of detectable changes in other objectively measured, clinical parameters.

HYPOTHESES TO STIMULATE RESEARCH

If some of the reported changes in CD4/CD8 ratios in individuals exposed to certain chemicals do have a degree of statistical validity, what research program could be developed around this observation? The first step would be to formulate a testable hypothesis. One hypothesis might be that substance X, an environmental pollutant suspected of involvement in MCS, is selectively toxic for $CD8^+$ suppressor T cells. This could be readily tested in laboratory animals without any regard to whether the animals succumb to MCS. It could also be tested on leukocytes collected from MCS patients and maintained in vitro.

Another hypothesis could be formulated if one were confident in the reliability of the observation that the CD4/CD8 ratio is substantially altered in persons exposed to chemical X. This hypothesis might be that the consequence of an abnormally low number/proportion of suppressor T cells could be a mild-to-moderate polyclonal expansion of B cells. Thus, indirectly, chemical X would be a nonspecific stimulant of B cells. This could be tested both in laboratory animals and in humans known to have been exposed to chemical X. It would be essential to include all appropriate controls in the study and, in the case of humans, to perform the study in a double-blind fashion.

If, indeed, the evidence for altered CD4/CD8 ratios in persons exposed to chemical X was compelling, a third hypothesis might be that the toxicity of chemical X for $CD8^+$ suppressor cells results in expansion of $CD4^+$ cells that elaborate cytokines. Thus, exposure to chemical X results in an imbalance in the production of cytokines. A differential effect of the chemical on the T_H1 and T_H2 subsets of $CD4^+$ cells (now known to exist in humans) could easily account for an imbalance in the normal levels of cytokines. It would be a simple matter to determine whether or not the T cells of a given patient, exposed to one or more chemicals that resulted in symptoms consistent with MCS, produced an abnormal pattern of cytokines or abnormal amounts of certain cytokines. Such an experiment would have to be designed carefully, however, to include all appropriate controls.

The preceding are only three possible hypotheses that could be developed from a simple observation. These are straightforward and lend themselves readily to the design of simple, quantifiable experiments. The formulation and testing of realistic hypotheses is certain to be more rewarding than is the continued accumulation of subjective impressions and anecdotes. Too frequently impressions and anecdotes lead to confusing concepts such as chemical overload and spreading phenomenon—all concepts that are not consistent with known immunologic phenomena.

Chemical overload, for example, is intended to convey the idea that the immune system has a finite capacity to deal with antigens, at least in some patients (Rea et al., 1978b). When that capacity is exceeded, symptoms associated with MCS appear. This concept is meaningless to an immunologist. If capacity means the ability of the immune system to deal with a spectrum of different antigens (chemicals), the concept is without merit because the specificity repertoire of the immune system is known to be in the range of 1 x 10^7. If capacity means the quantitative potential to respond to a battery of antigens, the concept still lacks merit because it has been known for decades that the immune system makes the same quantity of antibody to a given antigen whether that antigen is given alone or together with 20 or more other antigens.

Furthermore, the reserve capacity of the immune system to respond to a test antigen is at least ten times that normally engaged in even a prolonged response to the antigen. Studies on the effects of chemical or physical insults on the immune system (e.g., ionizing radiation) have shown that a major portion (as much as 90%) of the system must be damaged before the effect is detectable in otherwise normal adult animals. Similarly, adolescent and elderly individuals, in whom the reserve immune potential is much less than in the young adult, cope perfectly well with ordinary exposures to infectious agents. Only when exposure to an insult or stress has lowered their limited immune potential do they succumb to infection. Thus, it appears that the concept of chemical overload, viewed from any perspective, is unacceptable to an immunologist.

CONCLUSIONS AND RECOMMENDATIONS

At present, there is a lack of convincing evidence that MCS results from functional, biologic impairment of the immune system. Nor is there evidence of any recognized type of immunologic hyper-reactivity. If certain individuals are exquisitely sensitive to a variety of chemicals, they must suffer from a disorder substantially different from known allergic phenomena or other known immunologic aberrations.

Traditional, acceptable approaches to the study of immune mechanisms—approaches that have served so well in discovering and explaining immunologic diseases—have apparently not been fully or properly employed in exploring chemical sensitivity disorders. Therefore, if there is to be real progress toward determining whether such disorders involve the immune system in any way, studies must be conducted that embrace the scientific method, that is, the formulation and careful testing of precise hypotheses. That the National Institute of Allergy and Infectious Diseases is prepared to support such research is evidenced by the recent publication of a Program Announcement on immunotoxicology, which mentions interest in research dealing with chemical sensitivities (National Institutes of Health, 1991).

REFERENCES

AMERICAN COLLEGE OF PHYSICIANS. (1989). Position paper: Clinical ecology. Ann. Intern. Med. 111:168-178.

JEWETT, D.L., FEIN, G., and GREENBERG, M.H. (1990). A double-blind study of symptom provocation to determine food sensitivity. N. Eng. J. Med. 323:429-437.

LEVIN, A.S. and BYERS, V.S. (1987). Environmental illness: A disorder of immune regulation. Occup. Med.: State Art Rev. 2:699-681.

LIFSON, J.O., FINCH, S.L., SASAKI, D.T., and ENGLEMAN, E.G. (1986). Variables affecting T-lymphocyte subsets in a volunteer blood donor population. Clin. Immunol. Immunopathol. 36:151-160.

McGOVERN, J.J., LAZARONI, J.A., SAIFER, P., LEVIN, A.S., RAPP, D.J., and GARDNER, R.W. (1983). Clinical evaluation of the major plasma and cellular measures of immunity. Orthomol. Psychiatry 12:60-71.

MILLER, J.B. (1972). Food Allergy: Provocative Testing and Injection Therapy. Charles C. Thomas, Springfield, IL.

NATIONAL INSTITUTES OF HEALTH. (1991). NIH Guide for Grants and Contracts. (June 14, 1991). Studies on the Interactions between Environmental Toxicants and the Immune System. 20(23).

PAN, Y., REA, W.J., JOHNSON, A.R., and FENYVES, E.J. (1988). Formaldehyde sensitivity. Clin. Ecol. VI:79-83.

PATTERSON, R., DYKEWICZ, M.S., GRAMMER, L.C., PRUZANSKY, J.J., ZEISS, C.R., and HARRIS, K.E. (1987). Formaldehyde reactions and the burden of proof (editorial). J. Allergy Clin. Immunol. 79:705-706.

REA, W.J. (1976). Environmentally-triggered thrombophlebitis. Ann. Allergy 37:101-109.

REA, W.J. (1977). Environmentally-triggered small vessel vasculitis. Ann. Allergy 38:245-251.

REA, W.J. (1978a). Environmentally-triggered cardiac disease. Ann. Allergy 40:243-251.

REA, W.J., BELL, I.R., SUITS, C.W., and SMILEY, R.E. (1978b). Food and chemical susceptibility after environmental chemical overexposure: Case histories. Ann. Allergy 41:101-109.

REA, W.J., PAN, Y., JOHNSON, A.R., and FENYVES, E.J. (1987). T and B lymphocytes in chemically sensitive patients with toxic volatile organic hydrocarbons in their blood. Clin. Ecol. V:171-179.

SPEER, F. (1954). The allergic tension-fatigue syndrome. Pediatr. Clin. North Am. 1:1029-1038.

TERR, A.J. (1987). 'Multiple chemical hypersensitivity': Immunologic critique of clinical ecology theories and practice. Occup. Med.: State Art Rev. 2:683-694.

THRASHER, J.D., BROUGHTON, A., and MADISON, R. (1990). Immune activation and autoantibodies in humans with long-term inhalation exposure to formaldehyde. Arch. Environ. Health 45:217-223.

ZEISS, C.R., PATTERSON, R., PRUZANSKY, J.J., MILLER, M.M., ROSENBERG, M., and LEVITZ, D. (1977). Trimellitic anhydride (TMA)-induced airway syndromes: Clinical and immunologic studies. J. Allergy Clin. Immunol. 60:96-103

MULTIPLE CHEMICAL SENSITIVITY: A RESPIRATORY DISORDER?

REBECCA BASCOM, MD, MPH
Associate Professor of Medicine
Director, Environmental Research Facility
Division of Pulmonary and Critical Care Medicine
Department of Medicine
University of Maryland School of Medicine, Baltimore

INTRODUCTION

This manuscript will evaluate the evidence that multiple chemical sensitivity (MCS) is related to altered function of the respiratory mucosa. The respiratory mucosa are defined as beginning at the nose and extending to the lower airways. Although no direct evidence exists for this broad hypothesis, indirect evidence indicates that it may be a fruitful area for investigation.

Features of multiple chemical sensitivity point to the respiratory mucosa as both a possible target and an effector organ. Exposures associated with the onset of the condition are frequently airborne, as are exposures that subsequently trigger symptoms. "Multiple chemicals" are, therefore, likely to have encountered the respiratory mucosa. In this discussion, the term "exposures" refers to a patient's perceived contact with a chemical, without an implied quantitative exposure level (e.g., the one-tenth of the threshold limit value used by industrial hygienists). A patient's perception of exposure may occur either with the detection of an odor or with a sensation of respiratory mucosal irritation. Exposures that trigger symptoms involve diverse chemical structures, making a specific immune response an unlikely mechanism for augmented responsiveness. However, most of these chemicals can probably stimulate the c-fiber neurons, a non-specific response pathway (Nielsen, 1991).

The basic hypothesis of this paper is that multiple chemical sensitivity represents an amplification of the non-specific immune response to low-level irritants. This alteration may occur through altered function of the c-fiber neurons, an altered function of the respiratory epithelium, or an altered neuro-epithelial interaction.

Studies over the last decade have demonstrated that the acute inflammatory response to inhaled irritants may involve chemosensitive neurons. C-fiber neurons are unmyelinated neurons that ramify extensively in the mucosa, with fibers ending near blood vessels, glands, and in intraepithelial spaces. These fibers contain neuropeptides such as vasoactive intestinal peptide,

Note: The author is funded by NIH HL 40945.
Address all correspondence to: Dr. Rebecca Bascom, MSTF - 800, 10 S. Pine Street, Baltimore, MD 21201.

substance P, and calcitonin-gene-related peptide (Barnes, 1987). Species differences exist, and little is known about how the distribution and function of these nerves may be altered in disease states in humans. Neonatal administration to rodents of capsaicin, the active ingredient in red pepper, will result in removal of the c-fiber neurons. Challenge studies in capsaicin-pretreated animals can then be used to determine the role of these neurons in a specific response. Stimulation of these nerves in adult animals results in a local release of neuropeptides, with effects ranging from vasodilation, vascular extravasation, and bronchoconstriction. In the adult animal, capsaicin is used acutely to stimulate the c-fiber neurons and is used repetitively to induce local denervation.

In the rat, exposure to tobacco smoke causes vascular extravasation in the nose and trachea. Filtration of the particulate and nicotine portions of the smoke does not block this response, indicating that the active component of smoke is in the organic vapor phase (Lundberg et al., 1984). Administration of local anesthesia or neonatal capsaicin (selectively ablating the neurons) will block the vascular extravasation. Substance P, a biologically active neuropeptide contained in the neurons, will cause inflammation, and administration of a substance-P antagonist will block the inflammatory response to smoke (Lundberg et al., 1984). Increased vascular permeability of the respiratory tract with other chemical irritants, including ether and formalin, also acted via activation of capsaicin-sensitive neurons (Lundberg and Saria, 1983). These data indicate that the local response to chemical irritants in rodents occurs via stimulation of c-fiber neurons.

Nasal irritation can activate systemic reflexes. For example, increased blood pressure and tachycardia were induced by local application of capsaicin or nicotine to the nasal mucosa of guinea pigs. These effects could be entirely blocked by local anesthesia or neonatal capsaicin and partially blocked by phentolamine and propranolol (Lundblad et al., 1984). Allison et al., (1971) demonstrated a cigarette smoke-induced vagal activation followed by bradycardia and then an increase in peripheral vascular resistance in guinea pigs. Whether these reflexes may be altered in humans is unknown, although nasal stimulation with irritants can cause bronchoconstriction.

The airway epithelium and the airway surface fluid it produces and regulates are the first line of defense against the multiple, "non-self" constituents in the 10-20,000 liters of air inhaled each day. The airway epithelium is not an inert barrier, as was once thought, but contains a great capacity for xenobiotic metabolism (Dahl and Hadley, 1991; Brittebo et al., 1986; Bond, 1986; Harkema, 1990). The metabolic capacity of nasal cavity tissues was extensively reviewed recently (Dahl and Hadley, 1991). The reviewer noted that human nasal cytochrome P-450 activities and other human nasal enzyme activities differ sharply from those of similar hepatic enzymes. In contrast to activities in the liver, nasal cytochrome P-450 activities are relatively resistant to induction but are easily inhibited (Dahl and Hadley, 1991).

The ability of airway epithelium to release pro-inflammatory arachidonic acid products (Churchill et al., 1989) and immunoregulatory molecules, such as cytokines, was recently recognized (Mattoli et al., 1990; Cox et al., 1990). The epithelium is stimulated by environmentally relevant concentrations of irritants such as acrolein and ozone (Doupnik and Leikauf, 1990; Leikauf et al., 1988) and can participate in initiating and perpetuating inflammation at the airway surface (Cox et al., 1990). A study by Mattoli et al., (1990) showed that epithelial cells, proliferating after *in vitro* exposure to toluene diisocyanate, demonstrated increased release of the cytokines interleukin 1 alpha, interleukin 1 beta, and interleukin 6.

The observation that the airway epithelium can release cytokines, such as interleukins 1 and 6, is particularly interesting when considering MCS. Anecdotally, patients with MCS describe flu-like symptoms, low-grade fever, and fatigue that follow an exposure. These symptoms usually begin several hours after the exposure and may last up to 48 hours. These symptoms may be a type of acute phase response. Cytokines are now recognized as responsible for many of the manifestations of the acute-phase response (Kelley, 1990). Nasal lavage studies, following controlled exposure to ozone and the Molhave VOC mixture, are showing increased numbers of neutrophils (Koren et al., 1992; Graham and Koren, 1990), while ozone exposure is associated with increases in interleukin 8.[1] Increased neutrophils were also demonstrated in bronchoalveolar lavage samples obtained following ozone exposure (Graham and Koren, 1990). A key point about cytokines is that they normally act locally, and when these effects are widespread, such as in septic shock, the effects on tissues can be devastating. At present, it is unknown whether local activation of the airway epithelium can in any circumstance cause a systemic reaction. Another airway cell that has the potential to produce large quantities of acute-phase-related cytokines is the alveolar macrophage. Whether this cell could be activated following exposures that trigger symptoms in MCS patients is unknown. Because ample evidence exists for local activation and phenotypic alteration of monocytes-macrophages, every effort should be made to study airway cells or tissue.

Animal studies have produced evidence that exposure to an irritant can alter the subsequent response of the airway to neuropeptides. Dusser et al., (1989) administered substance P to guinea pigs and saw no change in airway tone. Following inhalation of the smoke from one cigarette, however, the response to substance P was markedly amplified. Pretreatment with superoxide dismutase, an antioxidant enzyme, prevented the amplification. These same authors measured neutral endopeptidase in the airway epithelium. This enzyme inactivates substance P, thereby reducing the effect of chemosensitive nerve stimulation. The administration of smoke reduced the quantity of neutral endopeptidase in the airway mucosa. The authors concluded that cigarette smoke induced an amplification of substance-P-induced bronchoconstriction through an oxidant injury to the airway epithelium. This study suggests that MCS patients demonstrate VOC intolerance because they have a reduced quantity or activity of

[1] J. Koenig, personal communication.

airway epithelial neutral endopeptidase. This reduction could result in an amplified response to subsequent c-fiber stimulation by other inhaled irritants.

The capacity of airway mucosa to be substantially inflamed even in the presence of normal pulmonary function, as occurs in asymptomatic asthmatics, has recently been recognized (Laitinen et al., 1985). Presumably, airway inflammation is central to the pathogenesis of asthma. Many patients with asthma report an intolerance to low-level VOC mixtures such as perfumes and cigarette smoke (Shim and Williams, 1986; Stankus et al., 1988). This VOC intolerance often decreases with the administration of anti-inflammatory agents, or when the patient is away from allergen exposure. One hypothesis is that MCS patients who describe lower respiratory symptoms have airways inflammation despite the usual presence of normal pulmonary function tests. To my knowledge, no systematic biopsy studies have been conducted of individuals meeting case criteria for MCS, nor have studies been done of mucosal function following controlled challenge.

In a population-based, cross-sectional survey, a large group of American office workers were asked "During the last year, how often do you believe you have experienced eye, nose, throat, or respiratory irritation at your workstation from: tobacco smoke; fumes from a photocopying machine; fumes from printing processing; fumes from other chemicals such as adhesives, glues,cleaners, white out, rubber cement; fumes from pesticides; fumes from new carpeting; fumes from new drapes, curtains, or furniture; fumes from paint; fumes from the cleaning of carpets, drapes, or other furnishings?" The next question on the survey was "Do you consider yourself especially sensitive to any of the items in the previous question?" In that survey, 29-31% of the workers answered that they were especially sensitive to one or more of the above-named substances. Furthermore, this perceived chemical sensitivity was positively associated with mucosal symptoms and negatively associated with comfort in the work environment.[2] This survey suggests a relationship between the presence of mucosal symptoms and self-reported chemical sensitivity.

Specific areas of investigation suggested by the above discussion include the careful study of the respiratory mucosa of MCS patients compared with appropriate controls. Attention should be focused on the distribution and contents of peptidergic neurons, the morphology of the epithelium, and the presence and type of mucosal inflammation. Epithelial neutral endopeptidase should be measured and the nose, trachea, and smaller airways should all be studied. These basic, descriptive studies could then be followed with *in vitro* challenges of tissue, such as respiratory epithelium or macrophages derived from MCS patients, to determine whether they demonstrate increased release of inflammatory mediators or cytokines upon direct stimulation with an incitant or with neuropeptides. Clinical challenge studies could begin with a determination of the systemic, reflex response to nasal irritation.

---

[2] C.J. Nelson, personal communication.

At the University of Maryland Environmental Research Facility, a focus of research has been respiratory sensitivity to environmental tobacco smoke (ETS). The reasons for this interest are several. Tobacco smoke is a complex irritant mixture that contains oxidants and aldehydes. Animal studies (cited above) indicate that the organic vapor phase of smoke stimulates chemosensitive neurons. ETS is a common indoor air pollutant because nearly 30% of the American population still smokes cigarettes (CDC, 1989). Most adults have been exposed to ETS and are aware of the nature of their responses. A range of responses can be elicited by individuals who describe themselves as healthy; however, intolerance to cigarette smoke is an almost universal complaint among patients with MCS (a group we have not studied). We have hypothesized that differential responsiveness to tobacco smoke reflects differential function of c-fiber neurons in humans. The focus of our work has been to characterize the upper respiratory response. Other investigators have studied the lower respiratory response to tobacco smoke (Stankus et al., 1988).

More than 20 years ago, a survey of smoke-related symptoms showed that 29% of nonallergic adults reported nasal symptoms with ETS exposure (Speer, 1968). On a questionnaire administered to 77 healthy young adults to ask about their response to ETS exposure, 34% reported a history of one or more symptoms of rhinitis and were classified as "ETS-sensitive" (Bascom et al., 1991). We then challenged historically ETS-sensitive and ETS-nonsensitive subjects to brief, high levels of tobacco smoke (45 ppm CO, 15 minutes) on two separate days (Bascom et al., 1991). Symptoms showed significant concordance on the two days and were greater in the historically sensitive subjects. Nasal congestion was significantly greater in the ETS-sensitive subjects as demonstrated by posterior rhinomanometry. This congestive response was not due to IgE allergy (or mast cell activation via a non-IgE mechanism) because no histamine was detected in nasal lavage fluid. Interestingly, no vascular transudation nor glandular stimulation was detected, either. These data provided objective evidence of differential responsiveness to tobacco smoke and suggested that vascular congestion was the mechanism of increased nasal resistance.

To address the possibility that the increased response to tobacco smoke was due to an increased dose or altered metabolism of nicotine, the physiologic response was compared with the urinary cotinine:creatinine ratio. No significant correlation was found.[3] Prolonged exposure to a lower concentration of smoke (15 ppm CO, 2 hours) also induced nasal congestion in both historically ETS-sensitive and ETS-nonsensitive subjects. A spectrum of congestive responses occurred but was unrelated to historical sensitivity. We have determined the response to filtered tobacco smoke in a small group of subjects. No response to the filtered smoke occurred in the historically nonsensitive subjects, but significant congestion occurred in the historically sensitive subjects (Willes et al., 1991). These data suggest that people may be responding to different components of the smoke.

---

[3] R. Bascom, unpublished data.

development of the disease for many chronic conditions. He recommends determining the genetics of specific components of the response to an environmental stressor and suggests that common allelic variations may be found that could be important in determining disease phenotypes. The role of acute responsiveness to ETS in the susceptibility to the chronic effects of tobacco smoke is unknown.

We have recently initiated studies to address the hypothesis that interstrain differences exist in the acute-airways response to tobacco smoke and in the induction of hyperreactivity by smoke. Highly inbred strains of mice are being used in these studies. Because animals within a given strain are genetically identical, interstrain differences in the response to smoke can be attributed to genetic differences; intrastrain differences will be due to variability in the exposure protocol. This approach was used by Levitt et al. (1988, 1989) to demonstrate that responsiveness to acetylcholine and serotonin in mice are inherited in an autosomal recessive pattern, each at a separate locus. Studies by other investigators have demonstrated interstrain differences in the RD_{50} (the concentration of an irritant causing a 50% inhibition of the respiratory rate) to irritants such as sulfur dioxide (Alarie et al., 1980). To avoid differential exposure to irritants, animals are anesthetized, tracheostomized, and ventilated during the exposure period. Our studies to date have demonstrated interstrain differences in the effect of smoke on serotonin-induced changes in lung compliance (Bascom et al., 1992).

These avenues of research can contribute to an understanding of multiple chemical sensitivity by helping to characterize the acute response to a common irritant in humans. They can also increase understanding of the basis for biologic variability in the acute response and help identify factors responsible for susceptibility to a specific effect of that irritant.

Multiple chemical sensitivity is a syndrome, that is, a cluster of symptoms that occur in concert. The formulation of the specific, testable hypotheses listed above does not exclude the need to study other divergent hypotheses. However, the development of appropriate experimental approaches and models to answer specific research questions will be a critical component for further understanding this syndrome.

REFERENCES

ALARIE, Y., KANE, L., and BARROW, C. (1980). Sensory irritation: The use of an animal model to establish acceptable exposure to airborne chemical irritants. In: Toxicology: Principles and Practices, Vol. 1 (A.L. Reeves, ed.). John Wiley, New York. pp. 48-91.

ALLISON, D.J. and POWIS, D.A. (1971). Adrenal catecholamine secretion during stimulation of the nasal mucous membrane in the rabbit. J. Physiol. 217:327-339.

BARNES, P.J. (1987). Neuropeptides in the lung: Localization, function and pathophysiologic implication. J. Allergy Clin. Immunol. 79:285-295.

BASCOM, R., KULLE, T., KAGEY-SOBOTKA, A., and PROUD, D. (1991). Upper respiratory tract environmental tobacco smoke sensitivity. Am. Rev. Respir. Dis. 143:1304-1311.

BASCOM, R., CHIN, C-Y., LEVITT, R.C., and KLEEBERGER, S.R. (1992). Interstrain differences in the acute response to tobacco smoke. Am. Rev. Respir. Dis. (in press).

BARNES, P.J. (1987). Neuropeptides in the lung: Localization, function and pathophysiologic implication. J. Allergy Clin. Immunol. 79:285-295.
BASCOM, R., KULLE, T., KAGEY-SOBOTKA, A., and PROUD, D. (1991). Upper respiratory tract environmental tobacco smoke sensitivity. Am. Rev. Respir. Dis. 143:1304-1311.
BASCOM, R., CHIN, C-Y., LEVITT, R.C., and KLEEBERGER, S.R. (1992). Interstrain differences in the acute response to tobacco smoke. Am. Rev. Respir. Dis. (in press).
BOND, J.A. (1986). Bioactivation and biotransformation of xenobiotics in rat nasal tissue. In: Toxicology of the Nasal Passages (C.S. Barrow, ed.). Hemisphere Publishing Corporation, Washington. pp. 249-261.
BRITTEBO, E.B., CASTONGUAY, A., RAFTER, J.J., KOWALSKI, B., AHLMAN, M., and BRANDT, I. (1986). Metabolism of xenobiotics and steroid hormones in the nasal mucosa. In: Toxicology of the Nasal Passages (C.S. Barrow, ed.). Hemisphere Publishing Corporation, Washington. pp. 211-234.
CENTERS FOR DISEASE CONTROL (CDC). (1989). Reducing the Health Consequences of Smoking: 25 Years of Progress. A Report of the Surgeon General. U.S. Department of Health and Human Services, Public Health Service, Centers for Disease Control, Center for Chronic Disease Prevention and Health Promotion, Office on Smoking and Health, Atlanta.
CHURCHILL, L., CHILTON, F.H., RESAU, J.H., BASCOM, R., HUBBARD, W.C., and PROUD, D. (1989). Cyclooxygenase metabolism of endogenous arachidonic acid by cultured human tracheal epithelial cells. Am. Rev. Respir. Dis. 140:449-459.
COX, G., OHTOSHI, T., GAULDIE, J., DENBURG, J., and JORDANA, M. (1990). Human bronchial epithelial cells cause differentiation of HL-60 cells and prolong the in vitro survival of eosinophils. Am. Rev. Respir. Dis. 141:A108.
DAHL, A.R. and HADLEY, W.M. (1991). Nasal cavity enzymes involved in xenobiotic metabolism: Effects on the toxicity of inhalants. Crit. Rev. Toxicol. 21:345-372.
DOUPNIK, C.A. and LEIKAUF, G.D. (1990). Acrolein stimulates eicosanoid release from bovine airway epithelial cells. Am. J. Physiol. 259:L222-229.
DUSSER, D.J., DJOKIC, T.D., BORSON, D.B., and NADEL, J.A. (1989). Cigarette smoke induces bronchoconstrictor hyperresponsiveness to substance P and inactivates airway neutral endopeptidase in the guinea pig. Possible role of free radicals. J. Clin. Invest. 84:900-906.
GRAHAM, D.E. and KOREN, H.S. (1990). Biomarkers of inflammation in ozone-exposed humans. Comparison of the nasal and bronchoalveolar lavage. Am. Rev. Respir. Dis. 142:152-156.
HARKEMA, J.R. (1990). Comparative pathology of the nasal mucosa in laboratory animals exposed to inhaled irritants. Environ. Health Perspect. 85:231-238.
KAZAZIAN Jr., H.H. (1976). A geneticist's view of lung disease. Am. Rev. Respir. Dis. 113:261-266.
KELLEY, J. (1990). State of the art: Cytokines of the lung. Am. Rev. Respir. Dis. 141:765-788.
KOREN, H.S., GRAHAM, D.E., and DEVLIN, R.B. (1992). Exposure of humans to volatile organic compounds (VOC) results in inflammation in the nasal passages. Arch. Environ. Health (in press).
LAITINEN, L.A., HEINO, M., LAITINEN, A., and HAAHTELA, T. (1985). Damage of the airway epithelium and bronchial reactivity in patients with asthma. Am. Rev. Respir. Dis. 131:599-606.
LEIKAUF, G.D., DRISCOLL, K.E. and WEY, H.E. (1988). Ozone-induced augmentation of eicosanoid metabolism in epithelial cells from bovine trachea. Am. Rev. Respir. Dis. 137:435-442.
LEVITT, R.C. and MITZNER, W. (1988). Expression of airway hyperreactivity to acetylcholine as a simple autosomal recessive trait in mice. FASEB J. 2:2605-2608.
LEVITT, R.C. and MITZNER, W. (1989). Autosomal recessive inheritance of airway hyperreactivity to 5-hydroxytryptamine. J. Appl. Physiol. 67:1, 125.
LUNDBERG, J.M. and SARIA, A. (1983). Capsaicin-induced desensitization of airway mucosa to cigarette smoke, mechanical and chemical irritants. Nature 302:251-253.
LUNDBERG, J.M., LUNDBLAD, L., SARIA, A., and ANGGARD, A. (1984). Inhibition of cigarette smoke induced oedema in the nasal mucosa by capsaicin pretreatment and a substance P antagonist. Naunyn-Schmiedeberg's Arch. Pharmacol. 326:181-185.

LUNDBLAD, L., HUA, X-Y., and LUNDBERG, J.M. (1984). Mechanisms for reflexive hypertension induced by local application of capsaicin and nicotine to the nasal mucosa. Acta Physiol. Scand. 121:277-282.

MATTOLI, S., MIANTE, S., CALABRO, F., MEZZETTI, M., FASOLI, A., and ALLEGRA, L. (1990). Bronchial epithelial cells exposed to isocyanates potentiate activation and proliferation of T-cells. Am. J. Physiol. 259:L320-L327.

NIELSEN, G.D. (1991). Mechanisms of activation of the sensory irritant receptor by airborne chemicals. CRC Crit. Rev. Toxicol. 21:183-208.

SHIM, C. and WILLIAMS, M.H. (1986). Effect of odors in asthma. Amer. J. Med. 80:18-22.

SPEER, F. (1968). Tobacco and the non-smoker. A study of subjective symptoms. Arch. Environ. Health 16:443-446.

STANKUS, R.P., MENON, P.K., RANDO, R.J., GLINDMEYER, H., SALVAGGIO, J.E., and LEHRER, S.B. (1988). Cigarette smoke-sensitive asthma: Challenge studies. J. Allergy Clin. Immunol. 82:331-338.

WILLES, S.J., FITZGERALD, T.K., PERMUTT, T., SAUDER, L., and BASCOM, R. (1991). Respiratory effects of prolonged sidestream tobacco smoke exposure and effect of filtration. Am. Rev. Respir. Dis. 143:A90.

MODELS OF MULTIPLE CHEMICAL SENSITIVITIES (MCS) SYNDROME: USING EMPIRICAL DATA (ESPECIALLY INTERVIEW DATA) TO FOCUS INVESTIGATIONS

LINDA L. DAVIDOFF, PhD
Division of Occupational Health
School of Hygiene and Public Health
Johns Hopkins University
Baltimore, Maryland

INTRODUCTION

In this paper, three popular general models or hypotheses of multiple chemical sensitivities (MCS) syndrome, also called environmental illness, will be examined. Each model has been advanced to explain both biologic and psychosocial-psychiatric aspects of MCS. The discussion will focus on how well research findings and the author's interview data support each model. Questions about these models that arise because of empirical findings will be identified; they should be addressed in future investigations. Interview and other self-report data are assumed to be useful in formulating hypotheses that can be tested in more rigorous ways. Descriptive self-report data (i.e., based on the subject's simple self-observations) are assumed to be more informative than self-report data that rely on the subject's inferences.

INTERVIEW DATA

The interview findings come primarily from an ongoing pilot study of MCS syndrome at Johns Hopkins University, which uses four study groups—one MCS group and three comparison groups. The first study group consists of 60 patients reporting MCS; the group has three equal subgroups, each reporting different origins of their illness (sick building syndrome, organic solvents, and organophosphate pesticides). The criteria for selecting MCS patients were as follows. First, patients must report that to feel good each day they had to avoid three or more different environmental chemical compounds or mixtures common in everyday life (e.g., tobacco smoke, hardware stores, perfume, exhaust fumes, gasoline, chlorinated compounds, pesticides, detergents). Second, they must describe a clear change from non-MCS to MCS status within the past five years and an exposure in one of the targeted categories that seemed to precipitate the change. The target age range was 30-55 years, but the actual range was 28-65 years. MCS subjects came from various referral sources, including occupational physicians, support groups for injured workers, registries for pesticide poisonings, and word of mouth. Subjects may not be representative of the population of MCS patients but may be at the severe end of a continuum.

Address all correspondence to: Dr. Linda L. Davidoff, Division of Occupational Health, School of Hygiene and Public Health, Johns Hopkins University, 615 North Wolfe Street, Baltimore, MD 21205.

This study includes three comparison groups, two of which comprise persons who are thought to experience environmental hypersusceptibilities but not be disabled by them. The first comparison group consists of 20 patients deemed to be atopic by one of two cooperating physicians on the bases of challenge testing and clinical history. The second comparison group consists of 20 office workers who describe a history of at least four months of feeling ill at work in a particular workplace at least once a week, with their symptoms confined largely to that workplace (SBS without generalization). The third comparison group consists of 60 members from the general population; names of potential volunteers were selected at random from Baltimore telephone directories. The atopic and general population samples are still being interviewed, so comparisons in this paper will be confined to the MCS group and the SBS group, which, unfortunately, may be two ends of the same continuum.

Among MCS subjects, approximately 70% of the building and pesticide MCS subgroups were female, but only 40% of the solvent MCS subgroup was female. The mean ages of the MCS subgroups ranged from 42.6 to 45.1 years. Years of education and training of MCS participants ranged from a mean of 18.1 years for building MCS subjects to a mean of 13.5 years for solvent MCS subjects. SBS and MCS groups were similar in gender, age, and number of years of education or training. MCS subjects reported becoming ill an average of 3-4 years previously.

The telephone interview took 30-60 minutes to complete. Most questions channeled responses into categories (yes, no, don't know, often, a little, moderately, a lot). Although three interviewers are currently involved, the forced-response option makes high reliability likely.

Three MCS subgroups were interviewed because MCS may be a heterogeneous collection of conditions. However, very significant differences were found among MCS participant subgroups on the basis of symptoms reported, sensitivity indices, and other variables of interest. The demographics of the solvent and building MCS subgroups did differ markedly in education, job category, and gender distribution. Although perceived originating exposures differed somewhat, all MCS subjects were probably exposed to organic solvents. Moreover, very few differences were found between subjects who had or had not seen a clinical ecologist. Therefore, the MCS subgroups were not differentiated, and aggregated data are described on the entire MCS sample.

Characteristically, both SBS and MCS subjects in this study reported reactions to even brief exposures of multiple, unrelated environmental chemicals at levels that did not bother others. MCS participants reported more incitants as a problem, said they got sicker, and experienced more symptoms than did SBS subjects. SBS subjects reported experiencing nine symptoms a week, and MCS subjects reported experiencing 20-21 symptoms per week.

The data confirm accounts of others and suggest that MCS patients report symptoms involving multiple systems: nonspecific manifestations such as overwhelming fatigue and low energy (93%) and headaches (75%); CNS symptoms such as memory impairment (85%), sudden or unexplained emotional responses (depression, rage, tension) (72%), sudden sleepiness (68%), dizziness (58%), unusual clumsiness (58%), and numbness and tingling (54%); muscle and joint symptoms such as muscle pain or spasm (80%) and joint pain or discomfort (70%); respiratory symptoms such as chest pain and shortness of breath or wheezing (both 63%); visual disturbances such as periodic blurring or floaters (63%); gastrointestinal symptoms especially bloating or stomach upset (77%); and eye-ear-nose-throat complaints especially rhinitis (82%), eye irritation (83%), ringing or itching or aching ears (72%), and sinusitis (68%).

Individuals reporting MCS described 3-4 changes in tolerance and subsequent intolerances for odors, allergens, medications, alcohol, and food, beginning in the period when the hypersensitivity developed. Reports of changed tolerances were much less common among persons with SBS who had an average of 1.3 lifetime changes, not necessarily associated with anything specifiable. Of MCS groups, 85-100% reported changed odor awareness (SBS, 40%), and 70-80% reported increased problems with allergens (SBS, 40%). Contrasts between MCS and SBS subjects were greater for other changed tolerances: food intolerance (60-90% in MCS vs. 25% in SBS), alcohol intolerance (50-55% vs. 5%), and medication intolerance (45-75% vs. 10%).

MCS subjects reported not being especially sick before the age of 18 (1.2-3.5 conditions in MCS vs. 2.4 conditions in SBS). Fewer MCS participants (31%) reported having allergies as children compared with SBS participants (50%). On average, MCS subjects said they did not respond adversely to any environmental chemical until the mean age of 27-34 (SBS, 28). MCS subjects date the onset of MCS considerably later than their first response, usually during their early 40s after an exposure lasting more than three weeks (85%); in 46% of cases, exposures lasted 1-3 years or longer.

MODEL I: PSYCHIATRIC DISORDER MODEL (FOCUS: SOMATIZATION DISORDER)

Authors who propose psychiatric diagnoses for MCS patients tend to assume the MCS population is heterogeneous psychiatrically as well as medically (Terr, 1986, 1989; Simon et al., 1990; Stewart and Raskin, 1985; Black et al., 1990). Among specific categories of psychiatric conditions, depression, anxiety, and somatoform disorders are repeatedly mentioned in conjunction with MCS. Persons espousing the psychiatric model of MCS seem to agree that the only type of psychopathology invariably present in MCS patients is unexplained somatic concerns. Among somatotype disorders, the somatization disorder fits the MCS profile best.

According to the DSM-III-R, the current diagnostic manual of the American Psychiatric Association, the defining features of a somatization disorder include 1) recurrent and multiple physical symptoms, 2) no demonstrable organic findings or known physiologic mechanisms, 3) psychologic or psychiatric symptoms, 4) positive evidence, or strong presumption, that physical symptoms are linked etiologically to psychologic factors or conflicts, and 5) a history of many physical complaints or a belief the patient is sickly beginning before age 30 and persisting several years.

Empirical Support for the Somatization Disorder Model

The interview data are somewhat consistent with two descriptive criteria of the somatization disorder: recurrent and multiple physical symptoms and psychologic or psychiatric symptoms. Other criteria are more problematic.

DSM-III-R Criterion 1: Recurrent and Multiple Physical Symptoms

MCS subjects reported multiple and recurrent physical symptoms. They reported feeling at least somewhat sick 60-67 hours per week (SBS, 23 hours per week). MCS subjects also reported long mean recovery periods after exposure. The modal MCS subject (40%) required 1-3 days to recuperate. The modal SBS subject (35%) required 1-3 hours to recuperate, with 25% requiring 1-3 days. Because of their health, 82% of MCS participants said they were now unable to do certain kinds or amounts of work, housework, or schoolwork on a regular basis (15% of SBS subjects reported comparable disability).

An essential question is whether specific physical symptoms reported in MCS match those considered particularly characteristic of somatization disorder. Seven cardinal physical symptoms are often used to screen for somatization disorder: painful menstruation, burning sensation in sexual organs or rectum, amnesia, difficulty swallowing, shortness of breath without exertion, pain in extremities, and vomiting (DSM-III-R, 1987). Although MCS subjects reported an average of 20-21 symptoms weekly, they commonly reported only one cardinal symptom: 63% of the MCS sample reported shortness of breath during certain exposures vs. 45% of the SBS sample. No specific questions about certain cardinal symptoms were asked but some were mentioned in general terms, and subjects were asked to report additional symptoms not mentioned specifically.

DSM-III-R Criterion 3: Psychologic and Psychiatric Symptoms

All but one MCS subject (98%) reported symptoms referable to the central nervous system (CNS), described previously. These symptoms are also consonant with a chronic illness and conditioning model. Other data collected are more relevant to current psychologic health. In the course of the interview, the Positive Affect Negative Affect Scales (PANAS), modified for clarity, were administered (Watson et al., 1988). This standardized test presents 20 adjectives (ten positive and ten negative) and asks to what extent participants felt that way during the past few weeks. This test avoids reading psychiatric meaning into medical symptoms, a serious confounder in some psychiatric MCS investigations. Both SBS and MCS groups reported experiencing various positive feelings (interested in things, excited in a positive way, emotionally strong, mentally active, and determined) anywhere from moderately to a lot during the past few weeks. Usually, MCS subjects reported feeling alert and attentive less frequently than SBS subjects. Greater differences between MCS and SBS groups were found with negative emotions. MCS subjects reported experiencing more distress, fear, and irritable-jittery-nervous feelings than did SBS subjects. Within the MCS group, the solvent-exposed, predominantly male subjects reported feeling more negative and less positive than did the building and pesticide subjects.

Depression, according to DSM-III-R, entails either loss of interest or pleasure or a depressed mood most of the day, nearly every day, for a period of time. A substantial proportion of MCS subjects did not report a pattern consistent with depression: 65% of pesticide MCS subjects and 45% of both building and solvent MCS subjects had positive emotion scores that exceeded negative emotion scores by ten or more points. Most remaining MCS subjects had experienced a lot of both positive and negative emotions. Only 5-15% of MCS subjects had negative emotion scores that exceeded positive emotion scores by ten or more points, and they sounded depressed. These pilot data suggest that a substantial proportion of MCS patients do not fit traditional definitions of depression when medical symptoms are deleted from and positive affectivity is added to the equation for assigning depressed status.

In another strategy to pinpoint possible psychopathology, subjects were asked, "Has a doctor ever treated you for any kind of condition that she or he considered psychologic or psychiatric?" ("Yes" responses were followed with, "What do you think caused the condition?") The percentage of college-educated people in the general population who have sought counseling at some point in their lives is unknown. MCS and SBS subjects differed significantly on prior treatment for psychologic or psychiatric conditions; 28% of MCS subjects but no SBS subjects said the cause of past psychologic or psychiatric treatment was related to environmental exposures or an illness due to such exposures. Building MCS and SBS subjects were similar (about 30%) in reporting psychologic or psychiatric treatment for conditions not related to exposure; both differed from pesticide and organic solvent MCS subjects (5-10%).

Two additional observations are relevant. First, reported psychiatric treatment tended to relate to specific situations, such as marital problems or deaths of family members, and rarely to longstanding, chronic anxiety or mood disorders. Second, contrary to popular belief, 45% of MCS subjects in this study had seen mental health professionals for problems related to MCS. Although no information about the perceived efficacy of psychologically oriented therapies was collected, some MCS patients volunteered comments. Most found counseling helpful when oriented toward improved coping, handling of life problems, and support, but thought it did not alter their hypersensitivity.

Only rarely did MCS patients report histories of sexual abuse. No specific questions about sexual abuse were included, but when asked generic questions about stress before the age of 18, and its duration and intensity, 15% of SBS participants said they had endured prolonged or intense stress as children compared with 30% of solvent MCS subjects and 20% of pesticide MCS subjects. Differences between these three groups were not significant. By contrast, 50%

of the building MCS group reported intense or prolonged stress in childhood; that is significantly different from that reported by the SBS group. This finding will be discussed later.

DSM-III-R Criterion 5: Medical History

The hypothesis of somatization disorder requires a history of many physical complaints or a belief that the patient is sickly, beginning before age 30 and persisting for several years. In this study, MCS subjects said their overall health had changed and that MCS appeared during their 40s. Solvent MCS subjects recalled one physical symptom before the age of 18; building MCS subjects recalled two. Only pesticide MCS subjects recalled three to four symptomatic conditions in childhood. As in other studies, many MCS subjects spontaneously reported how healthy they had been before MCS onset.

DSM-III-R Criterion 2: Absence of Organic Factors or Known Physiologic Mechanisms

The hypothesis of somatization disorder postulates that no evidence of organic pathology and no pathophysiologic mechanism could account for the physical symptoms. Several observations challenge this requirement. First, an unknown percentage of MCS patients receive Social Security Disability and other forms of compensation for documented medical symptoms. Second, medical records on two cohorts of MCS patients examined by the author and Dr. G. Ziem contained observations supporting some of the symptoms reported, such as upper airway abnormalities consonant with sinusitis and rhinitis, and spirometric patterns and methacholine challenge results consistent with obstructive dysfunction and hyperreactive airway disorders. Third, although negative studies are more well known (Terr, 1986; 1989), rigorous studies have shown evidence of aberrant physiologic function in at least part of an MCS sample (Doty et al., 1988; Staudenmeyer and Selner, 1990; McConnachie and Zahalsky, 1991a,b[1]; Morrow et al., 1991; Broughton et al., 1988a,b, 1990a,b; Thrasher et al., 1987, 1988, 1989, 1990; Galland, 1987; Bardana and Montanaro, 1989; and McGovern et al., 1983). Currently, evidence implicating mechanisms that could account for the condition MCS patients report is missing. This would include the original change of status from non-MCS to MCS, the spreading responses to incitants, and the changed tolerances following the original induction.

DSM-III-R Criterion 4: Link Between Psychologic or Psychiatric Factors and Physical Illness

The most important criterion of somatization disorder is positive evidence, or strong presumption, that illness is linked to psychologic or psychiatric factors. Unfortunately, no gold standards for making that judgment exist. In the context of MCS, this criterion is questionable for several reasons.

First, psychologic or psychiatric factors seen in MCS can be explained in another way. Although MCS patients in this study and in clinic settings commonly report distress, depression, anxiety, obsessiveness, tension, and nervousness, they attribute these complaints either

---

[1] McConnachie, P.R. and Zahalsky, A.C. (1991b). Immune alternations following exposure to chlordane/heptachlor. (Submitted for publication).

to certain types of exposures or to the experience of being chronically ill. One patient, described by her physician as conscientious and hardworking, said:

> I did not feel depressed despite my overwhelming fatigue and arthritis. I did realize that I had lost my spontaneity, my mental quickness and memory for detail, my enthusiasm and my high level of energy, and most of all, my special creativity. I had also lost the joy from life that I have, and have always had, throughout my life before the _____ experience. These things were gone or buried/subdued by my overwhelming fatigue. I replaced them with step by step thought processes. I became methodical and slow. I simply began to operate in survival mode.
>
> Details of everyday life that, at one time, I would have noticed only in passing, now became central and of great concern. For example, ... I began to focus and gripe about the details of office business and personalities. In my home life, I focused a negative eye on small details and characteristics of family members and ceased to consider the larger picture. I became a negative, complaining person with nothing good to say about anyone.
>
> As I regained my overall health, which took place over the year after I was assigned to alternate workspace, I noticed that from time to time my old joy in life would return, even though only for short periods of time. I now realize that the feeling of well-being and joy are directly tied to the overwhelming fatigue and the low grade fever that often is present with my overwhelming fatigue/flulike symptoms. I now maintain my feeling of well-being and joy by assiduously avoiding sources of chemicals and products that I have found bring about suppression of my feeling of well-being.

A second reason for doubting the link between psychologic and psychiatric factors and medical illness in MCS is the absence of credible supportive evidence. To date, no studies have shown that anxiety, depression, distress, and the like precede and contribute to the onset of MCS, although this remains a popular belief. Existing psychiatric studies that appear to support this impression have numerous methodologic weaknesses.

1) *Psychiatric investigations often lack information about how common depression, anxiety, and distress are in well-documented chronic medical conditions.* When psychologists study persons with certain identifiable medical conditions—in which the pathophysiology is well understood—and compare them with patients being treated for psychiatric conditions (such as affective or anxiety disorders) by administering psychologic or psychiatric tests, they cannot tell the patient groups apart on test performance alone (Cripe, 1988; Enders, 1985). Well persons from the general population, as used by Black et al. (1990) for example, are inappropriate controls. Medical controls for MCS subjects should be drawn from persons with chronic, disabling, stigmatizing, and isolating illnesses.

2) *Findings from psychiatric or psychologic tests are often interpreted without reference to known limitations of their validity.* Some authors do not discuss the many limitations on the interpretation of psychologic tests and psychiatric interviews (Simon et al., 1990; Black et al., 1990). Interpretations of test data must be guided by general knowledge about tests and test-taking and by knowledge about the reliability, validity, and norms of specific tests (how particular reference subjects have responded on the test). Most psychologic tests or psychiatric interviews were designed for physically healthy individuals; many ask questions like "Are you fatigued a lot?" "Do you have difficulty sleeping?" "Have you been experiencing headaches?" Persons with medical problems are likely to answer affirmatively. Yet, test scales are often formulated so that affirmative answers are taken as evidence of psychiatric problems because the test-takers were assumed to be healthy medically. This assumption is not justified with a group whose physical complaints are central and poorly understood.

3) *The mere presence of distress or neuropsychiatric problems is sometimes assumed to mean that presenting symptoms have a psychologic origin (e.g., conflict, another stressor, or an adjustment or personality disorder).* Psychologic or neuropsychiatric symptoms, however, are nonspecific in etiology and may have many different causes. For example, major depression is currently thought to involve a functional deficiency in monoamines, such as serotonin, norepinephrine, or both, and attendant subjective states. The term depression describes but does not explain. The same negative affects (hopelessness, fatigue), cognitions (pessimism), passivity, and anhedonia that may arise because of psychologic trauma, conflict, or other type of psychologic stress may also originate because of structural brain lesions (stroke, brain tumor), metabolic or endocrine dysfunction (e.g., hypothyroidism, hypoadrenocorticalism, B12 deficiency), medication (e.g., steroids, sedatives, estrogens, cardiovascular medications, analgesics, antihypertensives), toxic exposure (e.g., lead, organic solvents), or a disruption of circadian rhythms, as seen in jet lag. Regardless of origin, depression presents in the same way, and no known test or clinician can distinguish origin when presented with a symptom complex alone. Only systematic investigation and elimination of medical possibilities—which is often cursory and careless (Extein and Gold, 1986; Hall et al., 1978; McIntyre and Romano, 1977, as cited by Watson, 1989; Strickland, 1979)—can establish the plausibility of a psychosocial origin.

4) *Some medical practitioners base a diagnosis of somatoform disorder on the occurrence of several associated conditions: the appearance of secondary gain, improvement due to suggestion, and selective focusing. But these conditions are common in medical illnesses with a clear pathophysiologic origin.*
Secondary gain. Secondary gain is sometimes defined as the conscious or unconscious use of illness symptoms for other perceived benefits (e.g., avoiding noxious activities; gaining support not otherwise forthcoming). For several reasons, secondary gain is difficult to assess and distinguish from adaptive coping whether the condition be medical or psychosomatic, First, the sick role (e.g., staying home from work, staying in bed) has benefits

for symptomatic people. Second, there is no objective way to measure motivation and, hence, no way to know whether the coping action is being used for anything other than trying to get better. Third, the assumption that seeking a benefit is somehow descriptive only of psychosomatic conditions is questionable because most psychologically healthy people also make the best of what they have to do and emphasize the positive. Indeed, a positive strategy is so common that psychologists call it rationalization or the sweet-lemon strategy.

Improvement due to suggestion. Some clinicians erroneously assume that if suggestion or support lead to even short-term improvement, the condition improved must have been largely psychosomatic. However, medical literature is filled with information about placebo effects in persons with documented medical conditions. Medical patients who believe they have benefitted substantially from supportive counseling are easy to find.

Presentation with multiple organ symptoms. Some clinicians may assume that a report of multiple-organ symptoms in the absence of positive medical findings supports a psychiatric origin for a condition. Specialization in modern medicine encourages physicians to ask about symptoms relating to just one organ system, and if patients answer only those questions, other problems will be missed. Specialization may also bias the specialists' psychiatric judgments. Specialists tend to encounter patients after complaints appear; they have little or no direct knowledge about the patient's premorbid personality or adjustment, and hence, make mental health judgments without considering premorbid factors. Furthermore, many practitioners are unfamiliar with research data on health complaints in healthy populations. When introductory psychology students were asked to recall symptoms over the past month, using a 55-item checklist to refresh their memories, they reported an average of 17 different physical symptoms per month (Pennebacker, 1983). These data suggest that multiple-organ complaints may be common.

Selective focusing. Well persons probably minimize health complaints because our culture tends to stigmatize reports of illness as unseemly, self-indulgent, and weak. However, persons who seek help for a medical condition probably focus on and emphasize the symptoms and ill health. If MCS is a real medical condition that is poorly understood, then persons who have it will likely be eager to convince potential caregivers of the reality of their condition. Some will do so by obsessive reporting of substantiating encounters. Such behavior would be expected from a medically ill but untreated person who is seeking help. Some authors have used psychopathology as an explanation for the patient's preoccupation with illness (Rosenberg et al., 1990), but this assumes the patient is not medically ill, and such an assumption cannot legitimately be made at present.

Difficulty in confirming or disproving medical illness is the most important problem with a somatization diagnosis. In bold letters, DSM-III-R states, "It is necessary to rule out physical disorders that present with vague, multiple, and confusing somatic symptoms, e.g., hyperparathyroidism, porphyria, multiple sclerosis, and systemic lupus erythematosus. The onset of multiple physical symptoms late in life is almost always due to physical disease (p. 263)." It is simply premature to say that MCS has no physiologic or environmental basis when serious investigation has barely begun.

Needed Research on the Psychiatric Model

To support a somatization disorder or other psychiatric model of MCS, investigators must first rule out organic factors. Psychiatric diagnoses are supposed to be diagnoses of last resort, because they cannot be tested directly; they are simply inferred after testable explanations have proven inadequate. In addition, the case for psychiatric factors must be strengthened. Existing data are currently very weak and rely generally on:

1) Retrospective designs, which cannot differentiate neuropsychiatric and psychosocial causes from neuropsychiatric and psychosocial consequences;
2) Self-report data, including psychologic tests and psychiatric interviews that are biased by the participant's current perceptions and memories;
3) Uncontrolled observations, often on a single occasion; many physicians who make such observations are specialists who see the patient for the first and only time during a period of high distress;
4) Inappropriate measurement instruments, commonly methods of appraisal that were standardized and validated on medically healthy subjects and that confound medical symptoms and distress with evidence for psychogenic origin;
5) Small samples;
6) Inadequate controls; at the very least, psychiatric studies must control for the effects of medical symptoms that are baffling, isolating, and hard to control.

If MCS is to be considered psychiatric in origin, it will be important to demonstrate that:

1) Affected individuals show signs of psychopathology that exceed those expected for a comparable medical condition;
2) Affected individuals show signs of psychopathology that preceded the entire exposure period;
3) Psychopathology appears consistently in different subgroups such as young and old, chronically ill and newly ill, males and females;
4) Modifications of anxiety, depression, and other forms of psychopathology reduce or eliminate symptoms and signs of MCS, including chemical hypersensitivity and other intolerances.

At present, no credible data support any of these observations.

MODEL II: CONDITIONING MODEL

Several investigators have advanced and popularized the classical (respondent, Pavlovian) conditioning model of MCS (Schusterman et al., 1988; Dager et al., 1987; Bolla-Wilson et al., 1988). This model is straightforward. It assumes MCS begins with an originating exposure to one or more toxic compounds, the "unconditioned stimulus," which then produces a distinctive symptom complex or "unconditioned response." Ivan Pavlov originally used meat powder to produce salivation in dogs. The third component in the conditioning sequence is the "neutral stimulus," defined as any stimulus that does not initially evoke the unconditioned response. Pavlov used a tone as a neutral stimulus in some of his original studies. After the neutral stimulus is paired with the unconditioned stimulus, often repeatedly, the neutral stimulus will evoke a similar response, known as the "conditioned response." The conditioned response for Pavlov's dogs was salivation. In MCS, odors are considered neutral stimuli because they do not initially evoke symptoms, but it is hypothesized that odors come to evoke symptoms in MCS. Pavlov discovered that conditioned responses tended to generalize to similar stimuli. Dogs that salivated to one tone often salivated to other tones, especially similar ones. Presumably, the symptoms of individuals affected with MCS are evoked by odors that resemble the odor of the original exposure.

The conditioning theory assumes that conditioning is physiologic; that is, the animal is somehow reprogrammed to respond reflexively to stimuli that did not initially evoke such a response. Proponents of the conditioning model for MCS can borrow from another model to explain the distress and neuropsychiatric symptoms, or they may assume that those, too, are conditioned because they were experienced during the original exposure.

Empirical Support for the Conditioning Model

Testable predictions can be derived from the classical conditioning model. Major predictions include: 1) Odor will be salient during the originating exposure and will be an important characteristic of incitants. 2) Incitants will have somewhat similar odors. 3) Responses will be stereotyped and reflex-like and will occur predictably whenever certain odors are encountered. 4) The originating exposure will be distinctive and well defined. Animals do not appear to be indiscriminately and continually forming new classically conditioned responses to every new stimulus. 5) Prior psychopathology is not required because emotional responses can be reflexively induced. 6) Extinction should occur unless the conditioned stimuli are re-paired with the unconditioned stimulus (presumably, a toxic exposure).

Case reports presented in support of the classical conditioning model for MCS clearly show that MCS patients report symptomatic responses to numerous environmental chemicals to which they were not initially overexposed and which are not chemically related to those in the originating exposure. Also, odor awareness is often salient in MCS. Of the MCS subjects interviewed, 93% described being more aware of odors than are other persons they know (76% for SBS). The absence of a psychiatric history predating the condition and the presence of emotional responses during exposure to incitants, reported in MCS, are also consistent with the conditioning model.

The present interview data are inconsistent with the conditioning model in the following ways:

1) *The conditioning model predicts that odor perceptions are salient during the original exposure.* Some interviewees say that odor perceptions were prominent during the initiating exposure, but others say they were not. Some solvent workers fitting the criteria for MCS reported being anosmic at the time of the exposure and, in some cases, for months following the incident. In a previous pilot study, 4 of 11 individuals fitting the criteria for MCS were unaware of an identifiable odor in the sensitizing incident.

2) *The conditioning model predicts that odor perceptions invariably precede and initiate illness from incitants.* In previous pilot study, 6 of 11 individuals reporting MCS syndrome were infrequently aware of the odors of presumed incitants before becoming ill. Commonly, affected individuals described becoming ill after non-odorous incitants, such as pesticide residues on food and dyes in clothing or food.

3) *The conditioning model predicts that an odor gradient can explain the link between dissimilar incitants in MCS, that is, all incitants are presumed to be at least somewhat similar in odor to the initiating chemical exposure.* Contrary to this prediction, the present interview data suggest that the chemical incitants for individuals with MCS show multidimensional odor, for example, the odors of solvents, perfumes, tobacco smoke, natural gas, and chlorine. Pleasant odors (that were formerly liked) may also acquire symptom-eliciting properties, odors such as pine trees, burning leaves or logs, perfumes, fragrant flowers like peonies, lilacs, or roses, and cooking food. MCS patients can often differentiate between similar smelling items; for example, many of the 60 MCS subjects reported that only certain perfumes made them ill. Moreover, an odor gradient cannot explain reports of changed tolerances for allergens, foods, medications, or alcohol.

4) *The conditioning model predicts a reflex-like, all-or-nothing response to odor. In other words, individuals will respond with symptoms whenever they encounter incitants that resemble the original stimulus in odor.* In this interview study, all 60 MCS subjects mentioned at least one influence on their response to incitants; and the vast majority mentioned more than one. Many of these are conditions that influence dose, such as ventilation, length of exposure, and concentration of incitant. Participants also described general factors, such as diet and other sources of stress, that modulate whether they get sick and how sick they get when exposed to environmental chemicals. Some MCS patients tolerate brief odorous exposures under certain conditions.

5) *Conditioning theory, as usually elaborated, assumes a unitary response to an incitant.* MCS patients, however, report responding in multiple ways, depending on the incitant. Roughly 65% of MCS subjects and 50% of SBS subjects said they had several, distinctive response patterns to incitants; that is, for some incitants, one response (e.g., headaches or upper respiratory symptoms) and for other incitants, another response (e.g., lower respiratory symptoms or burning eyes). In some cases, responses resembled irritant responses that either did not occur previously or were not previously noticed.

6) *Conditioning theory does not address the changes MCS patients report in olfactory quality.* In MCS, SBS, and atopy, subjects report being more aware of odors than are other persons they know. In this interview study, however, only 40% of SBS subjects said their awareness of odors had changed, but 90% of MCS subjects said their awareness of odors had changed. Reports about how awareness changed are especially noteworthy. Although formal data are lacking, SBS subjects appear likely to say "I notice odors more now because I don't want to get sick." In contrast, MCS subjects are likely to say that awareness of odors has become vivid to the point of being annoying; odors now capture their consciousness; and that, despite their best efforts, odors have taken on a persisting, pervasive character and have become impossible to ignore. One person likened the change in olfactory quality to feeling like a human bloodhound. In other words, individuals reporting MCS describe qualitative changes in odor awareness. Some say odor awareness eventually reverts to normal; others say it does not.

7) *According to conditioning theory, responses that have been conditioned tend to extinguish if they are not periodically paired again with an unconditioned stimulus.* The present interview data and clinical reports suggest that MCS subjects commonly say they improve over time. Many report that certain types of symptoms are less frequent than at the peak-severity phase of the condition, that their responses are less debilitating, and that they have learned to cope better. In general, however, the vast majority interviewed continue to report aversive responses to numerous low-level stimuli, as many as five or more years after MCS onset. Conditioning theory does not explain why symptomatic responses persist even though subjects seldom report subsequent toxic re-exposures.

Needed Research on the Conditioning Model

Experimental and observational studies are needed to substantiate the alleged role(s) of conditioning in MCS. The issues above should be considered in designing such research. Counter-conditioning treatments are a potentially promising avenue of inquiry, but, although Bolla-Wilson et al. (1988) (two patients) and Staudenmayer and Selner (1990) (unspecified number of patients) have claimed successful conditioning treatments, neither has published systematic outcome and follow-up data on counter-conditioning therapy.

A major difficulty in testing the conditioning model is that chemicals and chemical odors are present together. Hence, investigators must differentiate whether individuals are responding to the odors (as predicted by conditioning theory) or to the chemicals themselves. In either case, one would predict physiologic responses. Isolating the effect of odor will be very challenging. Obvious solutions, like using topical anaesthetics or masking odorants, as the author and others have discussed, introduce both new chemicals and new odors and do not solve the problem. Sub-olfactory-threshold chemical challenges are one way to test the conditioning hypothesis. As yet, no credible findings on such challenges have been published.

MODEL III: CHRONIC ILLNESS MODEL

Chronic illness models suggest that MCS has a pathophysiologic origin like other medical conditions. Most current chronic-illness models assume that MCS syndrome is induced by a sensitizing exposure to environmental chemicals or drugs—chronic or acute—and that immune, metabolic, neural, and/or endocrinologic functions and the systems that depend on these functions are disrupted. Then, according to most such models, exposures to relatively low levels of environmental chemicals provoke or worsen the illness. If depression, malaise, tension, and unexplained nonspecific medical symptoms predate the diagnosis of MCS, these symptoms are often attributed to preclinical manifestations of MCS or to unidentified hypersusceptibilities that predated the onset of MCS and predisposed the individual to develop MCS. The common psychosocial or psychiatric profile (e.g., depression, anxiety) that follows MCS syndrome is attributed to various sources.

1) Social-psychologic consequences of having a stigmatized, untreatable, debilitating, unpredictable, frustrating, expensive, baffling, isolating, and chronic medical condition.
2) Mourning over the loss of former health and freedom of movement;
3) Neuropsychiatric or neurotoxic consequences or correlates of responses to exposures; and
4) Loss of esteem due to loss of valued social and economic roles.

The avoidant and phobic-like responses of MCS patients are interpreted as therapeutic and adaptive because they avoid illness. Obsessive actions are attributed to the patient's attempts to make sense of and to gain control over otherwise unpredictable flares of illness. Unusual distress is sometimes attributed to a selection effect, with the most distressed patients seeking out medical care. Unusual distress may also occur at an early stage when patients try to understand their experiences and to convince derisive or skeptical doctors and doubting family members, friends, or employers of the credibility of their problem.

Empirical Support for the Chronic Illness Model

Present data do not provide persuasive support for a physiologic or environmental mechanism for 1) the original change in status from non-MCS to MCS, 2) ongoing responses to incitants, and 3) changed tolerances for food and other substances. Nonetheless, the chronic illness model should be evaluated first for three principal reasons.

1) Ethics require testing the chronic illness model for any syndrome with prominent medical symptoms first. If MCS has an environmental or physiologic origin, then diagnosis, treatment, and prevention depend on better understanding the mechanism(s). Without that knowledge, we may unfairly stigmatize people and prolong their suffering unnecessarily.

2) Physiologic and environmental hypotheses of MCS must be eliminated before choosing psychiatric hypotheses, which are exclusionary diagnoses and currently impossible to test

directly. Moreover, physiologic and environmental hypotheses may be the most easily tested of the MCS hypotheses considered.

3) Environmental and physiologic explanations are not implausible. Case studies of MCS patients—although lacking adequate controls, for the most part—suggest that certain subsets of the MCS population may have documentable physiologic abnormalities that could contribute to the syndrome (Ashford and Miller, 1991; Doty et al., 1988; Staudenmeyer and Selner, 1990; McConnachie and Zahalsky, 1991a,b[2] Morrow et al., 1991; Thrasher and Broughton, et., al., 1987, 1988, 1989, 1990; Galland, 1987; Bardana and Montanaro, 1989; McGovern et al., 1983).

The present interview data are consistent with the chronic-illness model of MCS in several respects.

1) The psychosocial profile of MCS patients resembles that found in chronic medical conditions. Studies of social and psychologic adjustments following chronic dialysis (Kaplan-DeNour, 1982), major heart surgery (Zyzanski et al., 1981), and other conditions (Derogatis, 1986) indicate that, in many medical conditions, psychosocial disruption can be major. An intensive interview study of patients chronically ill with documentable chronic conditions, such as lupus erythematosus, multiple sclerosis, cancer, cardiovascular disease, and diabetes, identified four sources of suffering commonly seen in chronic illnesses, which sound very much like those reported by MCS patients: leading restricted lives and focusing on their illness; wearing out family and friends and experiencing social isolation as a result; feeling unattractive and inadequate and being discredited by others; and finally, burdening others as they became more dependent (Charmaz, 1983).

2) The uniformity of self-reports about MCS, even from naive patients, give credibility to a physiologic and environmental basis for the condition. In some cases, affected individuals have never heard of other sufferers, let alone support groups or clinical ecologists. For example, the author is studying a cohort of workers—102 men, mostly unskilled laborers—who were exposed to gasoline fumes while working underground on a subway system. Because these men were contacted after they became sick and subway construction was halted, it is not known whether they were sensitive before illness; but that possibility seems unlikely. At least some of the tunnel workers breathed gasoline fumes at levels estimated between 150 and 500 ppm for up to three months. Some exposure data, as well as liver-function data, spirometry, and symptom reports shortly after the men left the tunnel, are available. They are now being screened for MCS. At least a few of the men now appear to be reporting hypersusceptibilities and chronic symptom patterns resembling those seen in MCS syndrome. No members of the cohort are in support groups; currently, none has seen a clinical ecologist.

---

[2] McConnachie, P.R. and Zahalsky, A.C. (1992b). Immune alterations following exposure to chlordane/heptachlor. (Submitted for publication).

The author and Dr. G. Ziem are also examining a cohort of relatively naive individuals who were exposed to chlorine dioxide while working in a factory that processes potatoes in rural Wisconsin. Thirteen of fifteen affected workers report conditions that look very much like MCS syndrome; objective data (e.g., positive methacholine challenges, spirometric abnormalities, examinations of nasal mucosa, and sinus x-rays) substantiate some of their symptomatic reports.

3) The diversity of sources for persons reporting hypersusceptibility to environmental chemical incitants is consistent with a physiologic and environmental basis for the hypersusceptibility. Perceived intolerance to low levels of environmental chemicals was reported by 19% of a sample of 114 California patients diagnosed by physicians as having occupational disease attributable to organophosphates (Tabershaw and Cooper, 1966); by 13% of a sample of 160 solvent-exposed workers consecutively admitted for clinical services (Gyntelberg et al., 1986); by 95% of 60 asthmatics (30 consecutively admitted for hospitalization) (Shim and Williams, 1986); and by 80% of 105 consecutively admitted allergy clinic patients.[3] A Honeywell Technalysis survey using a random national sample suggested that 24% of 600 office workers studied reported problematic air quality in their workplaces; 20% believed their performance at work was hampered, and 7-11% experienced significant building-related symptoms (Woods et al., 1987).

Although some SBS complaints may be due simply to poor ventilation or to biologic agents such as mold, a substantial number of complainants perceive themselves as becoming hypersusceptible to environmental chemicals. In a study of office workers at EPA headquarters, roughly 30% of 3955 respondents (81% of an estimated 4900 EPA employees) described being unusually sensitive to environmental chemicals in the workplace in terms of eye, nose, throat, and respiratory irritative symptoms (EPA, 1989). In the present interview study, subjects for the SBS category were difficult to find because many potential SBS subjects reported consciously avoiding three or more different types of chemicals to maintain their health, even though they did not classify themselves as having MCS.

4) The apparent psychologic health of subsets of MCS patients (before and after MCS) gives a physiologic and environmental basis for MCS plausibility. In some cases, patients evaluated by mental health professionals have been pronounced sound. Many times, affected individuals appear to have been well adjusted at work and in the community and lacked premorbid psychiatric histories suggesting chronic maladjustment. Some affected individuals report concealing symptoms from others and denying symptoms to themselves for years to avoid admitting a stigmatizing condition; some report struggling for years to hold onto jobs before leaving the workplace; many continue working. Moreover, in a pilot study of MMPI-2 performance in five subjects reporting MCS, the author found

---

3 Meggs, W. (1991). Personal correspondence.

no evidence of psychopathology when medical complaints were segregated from test items. A larger sample is scheduled for future study.

Needed Research on the Chronic Illness Model

Interview data suggest directions for needed research on the chronic illness model of MCS. Regardless of which physicians MCS patients have seen and what their exposure histories and life circumstances are, they are likely, during interviews, to report developing numerous multisystem symptoms that fall predominantly into certain categories (overwhelming fatigue and low energy, memory impairment and other CNS symptoms, upper respiratory symptoms, muscle and joint symptoms, respiratory symptoms, and visual disturbances). They also report changed tolerances for food, alcohol, medication, and odors of many types. Premorbid histories do not appear to be characterized by these complaints. In the present interview study, MCS subjects were also likely to describe autoimmune diagnoses (35% vs. 5% for SBS), thyroid dysfunction (42% vs. 20%), and other endocrine dysfunctions (19% vs. 0%).

Physiologically-based research is needed to examine mechanisms that could underlie such complaints. Clinical reports of immune, metabolic, and neural disturbances in MCS populations suggest several promising directions: glucose metabolism in CNS and electrophysiologic indices of CNS function (Davidoff et al., 1991; Morrow et al., 1991); immune abnormalities (Thrasher et al., 1987, 1988, 1989, 1990; McConnachie and Zahalsky, 1991a, b[4]; McGovern et al., 1983; Levin and Byers, 1987); amino acid and fatty acid imbalances and other biochemical abnormalities (Galland, 1987; Davies et al., 1969; Truss, 1984); and elevated uptake of serotonin in platelets (Beving et al., 1984). Because of its multisystem effects, the trigeminal reflex is also of interest (Alarie, 1973). Rigorously controlled studies of well defined groups of index and comparison subjects are needed during both a) baseline conditions when subjects report feeling well and b) challenges that produce subjective reports of illness. Prospective studies are also needed to evaluate risk factors in the onset of MCS, including the role of psychiatric factors, such as chronic stress during childhood (reported by subjects developing MCS in the present interview study).

REFERENCES

ALARIE, Y. (1973). Sensory irritation by airborne chemicals. CRC Crit. Rev. in Tox. 2:299-363.

ASHFORD, N.A. and MILLER, C.S. (1991). Chemical Exposures: Low Levels and High Stakes. Van Nostrand Reinhold, New York.

BARDANA, E.J. and MONTANARO, A. (1989). 'Chemically sensitive' patients: Avoiding the pitfalls. J. Respir. Dis. 10:32-45.

BEVING, H., KRISTENSSON, J., MALMGREN, R., OLSSON, P., and UNGE G. (1984). Effects on the uptake kinetics of serotonin (5-hydroxytryptamine) in platelets from workers with long-term exposure to organic solvents: A pilot study. Scand. J. Work Environ. Health 10:229-234.

BLACK, D.W., RATHE, A., and GOLDSTEIN, R.B. (1990). Environmental illness: A controlled study of 26 subjects with '20th century disease.' J.A.M.A. 64:3166-3170.

---

4 McConnachie, P.R. and Zahalsky, A.C. (1991b). Immune alterations following exposure to chlordane/heptachlor. (Submitted for publication).

BOLLA-WILSON, K., WILSON, R., and BLEECKER, M.L. (1988). Conditioning of physical symptoms after neurotoxic exposure. J. Occup. Med. 30:684-686
BROUGHTON, A., THRASHER, J.D., and MADISON, R. (1990a). Chronic health effects and immunological alterations associated with exposures to pesticides. Comments on Toxicol. 4(1):59-71.
BROUGHTON, A., THRASHER, J.D., and MADISON, R. (1990b). Biological monitoring of indoor air pollution: A novel approach. In: Indoor Air 90: Proceedings of the 5th International Conference on Indoor Air Quality and Climate. Toronto. pp. 145-150.
BROUGHTON, A., THRASHER, J.D., and GARD, Z. (1988a). Immunological evaluation of four arc welders exposed to fumes from ignited polyurethane (isocyanate) foam: Antibodies and immune profiles. Am. J. Ind. Med. 13:463-472.
BROUGHTON, A. and THRASHER, J.D. (1988b). Antibodies and altered cell mediated immunity in formaldehyde exposed humans. Comments on Toxicol. 2:155-174.
CHARMAZ, K. (1983). Loss of self: A fundamental form of suffering in the chronically ill. Sociol. Health Ill. 5:168-195.
CRIPE, L.I. (1988). The clinical use of the MMPI with neurologic patients: A new perspective. Presented at the Army Medical Department Psychology Conference, Seattle, WA.
DAGER, S.J., HOLLAND, J.P., COWLEY, D.S., et al. (1987). Panic disorder precipitated by exposure to organic solvents in the work place. Am. J. Psychiatry 144:1056-1058.
DAVIDOFF, L.L., CALLENDER, T.J., MORROW, L.A., and ZIEM, G. (1991). Letter to the editor. J. Psychosom. Res. 35:621-623.
DAVIES, J.E., MANN, J.B., and TOCCI, P.M. (1969). Renal tubular dysfunction and amino acid disturbances under conditions of pesticide exposure. Ann. N.Y. Acad. Sci. 160 (Art 1):323-333.
DEROGATIS, L.R. (1986). The psychosocial adjustment to illness scale (PAIS). J. Psychosom. Res. 30:77-91.
DOTY, R.L., DEEMS, D.A., FRYE, R.E., PELBERG, R., and SHAPIRO, A. (1988). Olfactory sensitivity, nasal resistance, and autonomic function in patients with Multiple Chemical Sensitivities. Arch. Otolaryngol. Head Neck Surg. 114:1422-1427.
DSM-III-R. Diagnostic and Statistical Manual of Mental Disorders. (1987). 3rd ed.-revised ed. American Psychiatric Association, Washington, D.C.
ENDERS, P.L. (1985). The value of psychological testing with chronic pain patients. In: Evaluation and Treatment of Chronic Pain (G.M. Aronoff, ed.). Urban & Schwartzenberg, Baltimore, MD. pp. 589-601.
EXTEIN, I. and GOLD, M.S. (1986). Medical Mimics of Psychiatric Disorders. Progress in Psychiatry Series. American Psychiatric Association, Washington, DC.
GALLAND, L. (1987). Biochemical abnormalities in patients with multiple chemical sensitivities. In: Workers with Multiple Chemical Sensitivities. (M.R. Cullen, ed.) Occup. Med: State Art Rev. 2(4):713-720.
GYNTELBERG, F., VESTERHAUGE, S., FOG, P., et al. (1986). Acquired intolerance to organic solvents and results of vestibular testing. Am. J. Ind. Med. 9:363-370.
HALL, R.C., POPKIN, M.K., DEVAUL, R.A., FAILLACE, L.A., and STICKNEY, S.K. (1978). Physical illness presenting as psychiatric disease. Arch. Gen. Psychiatr. 35:1315-1320.
KAPLAN-DENOUR, A. (1982). Social adjustment of chronic dialysis patients. Am. J. Psychiatr. 139:97-100.
LEVIN, A.S. and BYERS, V.S. (1987). Environmental illness: A disorder of immune regulation. In: Workers with Multiple Chemical Sensitivities. (M.R. Cullen, ed.) Occup. Med: State Art Rev. 2(4):669-681.
MADISON, R.A., BROUGHTON, A., and THRASHER, J.D. (1991). Immunologic biomarkers associated with an acute exposure to urea-formaldehyde resin. Environ. Health Perspect. (in press).
McCONNACHIE, P.R. and ZAHALSKY, A.C. (1991a). Immunological consequences of exposure to pentachlorophenol. Arch. Environ. Health 46:249-253.

McGOVERN, J.J., JR., LAZARONI, J.A., HICKS, M.F., ADLER, J.C., CLEARY, P. (1983). Food and chemical sensitivity: Clinical and immunologic correlates. Arch. Otolaryngol. 109:292-297.

MORROW, L.A., STEINHAUER, S.R., ROBIN, N. HODGSON, M.J., TORTORA, S., and BABER, S. (1991). Neurophysiological and neuropsychological impairment following chemical exposure. Presented to the International Neuropsychological Society, San Antonio.

PENNEBAKER, J.W. (1983). The Psychology of Physical Symptoms. Springer-Verlag, New York.

ROSENBERG, S.J., FREEDMAN, M.R., SCHMALING, K.B., and ROSE, C. (1990). Personality styles of patients asserting environmental illness. J. Occup. Med. 32:678-681.

SCHUSTERMAN, D., BALMES, J., and CONE, J. (1988). Behavioral sensitization to irritants/odorants after acute overexposures. J. Occup. Med. 30:565-67.

SHIM, C. and WILLIAMS, M.H. (1986). Effects of odors in asthma. Am. J. Med. 80:18-22.

SIMON, G.E., KATON, W.J., and SPARKS, P.J. (1990). Allergic to life: Psychological factors in environmental illness. Am. J. Psychiatry 147:901-906.

STAUDENMAYER, H.S. and SELNER, J.C. (1990). Neuropsychophysiology during relaxation in generalized universal 'allergic' reactivity to the environment: A comparison study. J. Psychosomat. Res. 34:259-270.

STEWART, D.E. and RASKIN, J. (1985). Psychiatric assessment of patients with '20th century disease (total allergy syndrome'). Can. Med. Assoc. J. 133:1001-1006.

STRICKLAND, B.R. (1979). Depression as a concomitant of physical disorders: Difficulty of differentiation. Presented as part of a symposium, "Neglected aspects in the treatment of depression and schizophrenia." American Psychological Association.

TABERSHAW, I.R. and COOPER, W.C. (1966). Sequelae of acute organic phosphate poisoning. J. Occup. Med. 8:5-20.

TERR, A.I. (1986). Environmental illness: A clinical review of 50 cases. Arch. Intern. Med. 146:145-149.

TERR, A.I. (1989). Clinical ecology in the workplace. J. Occup. Med. 31:257-261.

THRASHER, J.D., WOJDANI, A., HEUSER, G., and CHEUNG, G. (1987). Evidence for formaldehyde antibodies and altered cellular immunity in subjects exposed to formaldehyde in mobile homes. Arch. Environ. Health 42:347-350.

THRASHER, J.D., BROUGHTON, A., and MICEVICH, P. (1988). Antibodies and immune profiles of individuals occupationally exposed to formaldehyde: Six case reports. Am. J. Ind. Med. 14:479-488.

THRASHER, J.D., MADISON, R., BROUGHTON, A., and GARD, Z. (1989). Building-related illness and antibodies to albumin conjugates of formaldehyde, toluene diisocyanate and trimellitic anhydride. Am. J. Ind. Med. 15:187-195.

THRASHER, J.D., BROUGHTON, A., and MADISON, R. (1990). Immune activation and autoantibodies in long-term inhalation to formaldehyde. Arch. Environ. Health 45(4): 217-223.

TRUSS, C.O. (1984). Metabolic abnormalities in patients with chronic Candidiasis. J. Orthomol. Psychiatry 13:66-93.

U.S.ENVIRONMENTAL PROTECTION AGENCY (EPA). (1989). Indoor Air Quality and Work Environment Study of EPA Headquarter Buildings. Volume I: Employee Survey, 19K-1003.

WATSON, D., CLARK, L.A., and TELLEGEN, A. (1988). Development and validation of brief measures of positive and negative affect. The PANAS scales. J. Pers. Soc. Psychol. 54:1063-1070.

WATSON, V.M. (1989). Mental symptoms can mask physical ailments. APA Monitor 10:24-25.

WOODS, J.E., DRAWNY, G.M., and MOREY, P.R. (1987). Office worker perceptions of indoor air quality effects on discomfort and performance. In: Indoor Air '87, Proceedings of the 4th International Conference on Indoor Air Quality and Climate. (B. Seifert, H. Esdorn, M. Fischer, H. Ruden, and J. Wegner, eds.). Institute for Water, Soil, and Air Hygiene, Berlin.

ZYGANSKI, S.J., STANTON, B.A., JENKINS, C.D., and KLEIN, M.D. (1981). Medical and social outcomes in survivors of major heart surgery. J. Psychosom. Res. 23:213-221.

AEOC MULTIPLE CHEMICAL SENSITIVITY WORKSHOP PLANNING COMMITTEE

Nicholas Ashford, PhD, JD
Center for Technology, Policy and Industrial Development
Massachusetts Institute of Technology
E40-239
Cambridge, MA 02139

Mark Cullen, MD, MPH
Yale Occupational Medicine Program
School of Medicine
333 Cedar Street
New Haven, CT 06510

Robert Harrison, MD, MPH
Occupational and Environmental Medicine Clinic
University of California at San Francisco
350 Parnassus, Suite 609
San Francisco, CA 94117-0924

Claudia Miller, MD, MS
Dept of Pediatrics
Division of Allergy and Immunology
University of Texas Health Science Center
7703 Floyd Curl Drive, Room 613L
San Antonio, TX 78284-7805

Patricia Price, DO
Division of Health Education
ATSDR
1600 Clifton Road, NE, Mail Stop E-33
Atlanta, GA 30333

Linda Rosenstock, MD, MPH
Occupational Medicine Program
University of Washington
Harborview Medical Center
325 9th Avenue, ZA66
Seattle, WA 98104

Workshop Coordinators
Judie Guerriero, RN, MPH
Occupational Health Clinical Program
San Francisco General Hospital
Bldg 9, Room 109
1001 Potrero Avenue
San Francisco, CA 94110

Kathleen Rest, PhD, MPA
Occupational Health Program
Dept of Family and Community Medicine
University of Massachusetts Medical Center
55 Lake Avenue North
Worcester, MA 01655

Staff
Edmund Kelly, Executive Director
AEOC
1010 Vermont Avenue, NW, Suite 513
Washington, DC 20005

Katherine Kirkland
AEOC
1010 Vermont Avenue, NW, Suite 513
Washington, DC 20005

AOEC MULTIPLE CHEMICAL SENSITIVITY WORKSHOP LIST OF PARTICIPANTS

Note:
[1] Presenter; [2] Responder; [3] Session Presider; [4] Small Group Leader

Joseph F. Albright, PhD[1]
National Institute of Allergy and Infectious Disease
Westwood 757
Bethesda, MD 20892

Nicholas Ashford, PhD, JD[3]
Center for Technology, Policy and Industrial Development
MIT, E40-239
Cambridge, MA 02139

Rebecca Bascom, MD[1,2]
Dept of Medicine
University of Maryland
10 South Pine Street, MSTF 800
Baltimore, MD 21201

Charles Becker, MD
Occupational Health Clinic
San Francisco General Hospital
Bldg 9, Room 109
1001 Potrero Ave
San Francisco, CA 94110

Gershon Bergeisen, MD, MPH
184 Gardenia Way
Gaithersburg, MD 20879

Eddy Bresnitz, MD, MS[2]
Division of Occupational and Environmental Health
Medical College of Pennsylvania
3300 Henry Street
Philadelphia, PA 19129

Dedra Buchwald, MD[1]
Harborview Medical Center
University of Washington
325 9th Avenue, ZA 60
Seattle, WA 98104

Thomas Callender, MD
108 Harolyn Park Drive
Lafayette, LA 70503

Peter G. Casten, MD
Ochsner Clinic
1514 Jefferson Highway
New Orleans, LA 70121

David Christiani, MD
Occupational Health Service
Massachusetts Respiratory Hospital
2001 Washington Street
South Braintree, MA 02184

Crawford Cleveland, MD
East Carolina School of Medicine
Greenville, NC 27858-4354

Linda Cochiarelli, MD
MacNeal Occupational Medicine Clinic
Bridgeview Medical Center
7217 West 84th Street
Bridgeview, IL 60455

James E. Cone, MD[1]
224 Channing Way
Berkeley, CA 94704

Brenda Cuccerini, PhD, CRNP
Division of Occupational and Environmental Medicine
School of Medicine
George Washington University
2300 K Street, NW
Washington, DC 20037

Mark Cullen, MD[3,4]
Occupational Medicine Program
Yale University School of Medicine
333 Cedar Street
New Haven, CT 06510

William Daniell, MD, MPH[1]
Occupational Medicine Progam
University of Washington Medical Center
325 9th Avenue, ZA 66
Seattle, WA 98104

Dennis J. Darcey, MD, MPH
Duke University Medical Center
Division of Occupational and Environmental Medicine
Box 2914
Durham, NC 27710

Linda Davidoff, PhD[1,2]
Division of Occupational Health
School of Hygiene and Public Health
Johns Hopkins University
615 North Wolfe Street
Baltimore, MD 21205

Raymond Demers, MD, MPH[3,4]
Division of Occupational Health
Dept of Family Medicine
Wayne State University
4201 St. Antoine
Detroit, MI 48201

Christopher DeRosa, PhD
ATSDR
1600 Clifton Road, NE , Mail Stop E-28
Atlanta, GA 30333

Peter Dillard, MD, MPH
26 Pasho Street
Andover, MA 01810

Talla Eslami, MD
11279 Knights Griffin Road
Thonotosassa, FL 33592

Dennis Estep, DO
University Occupational Health Services
Oklahoma Memorial Hospital
800 NE 13th St, Room 5E 109
Oklahoma City, OK 73104

Kathleen Fagan, MD
Occupational and Environmental Health Clinic
Dept of Family Medicine
Metrohealth Medical Center
2500 Metrohealth Drive
Cleveland, OH 44109-1998

Elissa Favata, MD
West Jersey Medical Plaza
120 Carnie Boulevard, Suite 7
Voorhees, NJ 08043

Nancy Fiedler, PhD
Occupational and Environmental Medicine Clinic
Dept of Environmental and Community Medicine
UMDNJ-RW Johnson Medical Center
675 Hoes Lane
Piscataway, NJ 08854

Alfred Franzblau, MD
Occupational Health Program
Dept of Occupational Health
School of Public Health
University of Michigan
1420 Washington Heights
Ann Arbor, MI 48109-2029

Howard Frumkin, MD
Division of Occupational and Environmental Health
Emory University
School of Public Health
1599 Clifton Road, NE
Atlanta, GA 30329

Janie Gordon, ScM
Division of General Internal Medicine
Occupational Health Project
School of Medicine
University of Maryland
405 Redwood Street
Baltimore, MD 21202

Gary Greenberg, MD, MPH
Duke University Medical Center
Division of Occupational and Environmental Medicine
Box 2914
Durham, NC 27710

Judie Guerriero, RN, MPH
Occupational Health Clinic
San Francisco General Hospital
Bldg 9, Room 109
1001 Potrero Avenue
San Francisco, CA 94110

Robert Harrison, MD, Director[3,4]
Occupational and Environmental Medicine Clinic
University of California at San Francisco
400 Parnassus Avenue, A585
San Francisco, CA 94143

David Hartman, PhD
Occupational Medicine Clinic
Cook County Hospital
720 South Wolcott
Chicago, IL 60612

Stephen Hessl, MD
Occupational Medicine Clinic
Cook County Hospital
720 South Wolcott
Chicago, IL 60612

Michael Hodgson, MD
Occupational Medicine Program
University of Connecticut
263 Farmington Avenue
Farmington, CT 06030

Don Jewett, MD, DPhil[1]
Abratech Corporation
Bldg E, 2nd Floor
Mill Valley, CA 94941-6610

Donald Johnson, MD
Columbia Park Medical Group
6401 University Avenue, NE #200
Fridley, MN 55432

Michael Kamrin, PhD
Institute for Environmental Toxicology
C231 Holden Hall
Michigan State University
East Lansing, MI 48824-1206

Karl Kelsey, MD
Occupational Health Service
Massachusetts Respiratory Hospital
2001 Washington Street
South Braintree, MA 02184

David G. Kern, MD, MPH
Occupational Health Service
Brown University Program in Occupational Medicine
111 Brewster Street
Pawtucket, RI 02860

David King, PhD[1]
Dept of Psychiatry
University of California at Los Angeles
760 Westwood Plaza
Los Angeles, CA 90024

Howard Kipen, MD[1]
Dept of Environmental and Community Medicine
UMDNJ-RW Johnson Medical Center
675 Hoes Lane
Piscataway, NJ 08854

Steven Kirkhorn, MD
Occupational Health Services
Ramsey Clinic
640 Jackson Street
St. Paul, MN 55101-2595

Rosemary Klein, MSN
Central NY Occupational Health Clinic Center
550 Harrison Center, Suite 300
Syracuse, NY 13202

Michael B. Lax, MD, MPH
Central NY Occupational Health Clinic Center
550 Harison Center, Suite 300
Syracuse, NY 13202

Virginia Lee, MD
Center for Prevention Services
1600 Tullie Circle
Atlanta, GA 30329

Michael Lenoir, MD[1]
401 29th, Suite 201
Oakland, CA 94609

Allen S. Levin, MD[1]
Immunology, Inc.
500 Sutter Street, #511
San Francisco, CA 94102-1114

Max Lum, EdD, MPA
Director, Division of Health Education
ATSDR
1600 Clifton Rd, NE, Mail Stop E-28
Atlanta, GA 30333

Robert McLellan, MD
Center for Occupational and Environmental Medicine
60 Washington Avenue
Hamden, CT 06518

William Meggs, MD[1,2]
Division of Clinical Toxicology
Dept of Emergency Medicine
East Carolina School of Medicine
Greenville, NC 27858-4354

Susan Metcalf, MD
Division of Health Studies
ATSDR
1600 Clifton Rd, NE
Atlanta, GA 30333

Claudia Miller, MD, MS[1]
Dept of Pediatrics
Division of Allergy and Immunology
University of Texas
Health Science Center
7703 Floyd Curl Drive
San Antonio, TX 78284-7805

James Nethercottt, MD
Occupational Medicine Program
Johns Hopkins University
615 North Wolfe Street
Room 7509
Baltimore, MD 21205

Frank Mitchell, DO
Office of Assistant Administrator
ATSDR
1600 Clifton Road, NE, Mail Stop E-28
Atlanta, GA 30333

Stephen Mooser, MPH
Irving J. Selikoff Occupational Health
Clinic Center
Mount Sinai School of Medicine
1 Gustave Levy Place - P.O. Box 1058
New York, NY 10029

Sean Muldoon, MD, MPH
Occupational and Environmental Medicine
Program
University of Pittsburg
School of Medicine
149 Lothrop Hall
Pittsburg, PA 15261

Karen Mulloy, DO
Division of Occupational and Environmental
Health
Dept of Family and Community Medicine
Marshal University School of Medicine
Huntington, WV 25755

Tara O'Toole, MD
Occupational Health Project
School of Medicine
Division of General Internal Medicine
University of Maryland
405 Redwood Street
Baltimore, MD 21202

L. Christine Oliver, MD
Pulmonary Critical Care Unit - Bulfinch 1
Massachusetts General Hospital
Boston, MA 02114

David L. Orgel, MD, MPH
Eastern NY Occupational Health Program
1201 Troy Schnenectady Road
Latham, NY 12110

Donna Orti, MS
Division of Health Education
ATSDR
1600 Clifton Road, NE, Mail Stop E-33
Atlanta, GA 30333

John E. Parker, MD
NIOSH
944 Chesnut Ridge Road, Mail Stop 122
Morgantown, WV 26505

Sue Perlin, ScD
Office of Health Research
Environmental Protection Agency
401 M Street, SW, Mail Drop RD683
Washington, DC 20460

Patricia Price, DO
Division of Health Education
ATSDR
1600 Clifton Road, NE, Mail Stop E-33
Atlanta, GA 30333

Patricia Quinlin, MPH, CIH
Occupational Health Clinic
San Francisco General Hospital
Bldg 9, Room 10
1001 Potrero Avenue
San Francisco, CA 94110

Kathleen Rest, PhD, MPA[1]
Occupational Health Program
Dept of Family and Community Medicine
University of Massachusetts Medical Center
55 Lake Avenue North
Worcester, MA 01605

Carrie Redlich, MD[1]
Occupational Medicine Program
Yale University of Medicine
333 Cedar Street
New Haven, CT 06510

Marie-Claude Rigaud, MD
Occupational Medicine Clinic
Cook County Hospital
720 South Wolcott
Chicago, IL 60612

Linda Rosenstock, MD, MPH[3,4]
Occupational Medicine Program
Haborview Medical Center
University of Washington
325 9th Avenue, ZA 66
Seattle, WA 98104

Gerald Ross, MD[1,2]
Environmental Health Center
8345 Walnut Hill Lane, Suite 205
Dallas, TX 75231

Anne Sassaman, PhD
Divisional Director, Extramural Programs
DERT, NIEHS
North Campus Bldg 3, Room 301A
Research Triangle Park, NC 27709

Barbara Scolnick, MD
Occupational and Environmental Health Center
Cambridge Hospital
1493 Cambridge Street
Cambridge, MA 02139

John C. Selner, MD[1]
Allergy Respiratory Institute of Colorado
5800 East Evans
Denver, CO 80222

Greg Simon, MD[1]
Center for Health Studies
Dept of Psychiatry and Behavioral Sciences
University of Washington
Seattle, WA 98101

Wil A. Spaul, MD
College of Public Health, MHH 104
11279 Knights Griffin Road
Thonotasassa, FL 33592

Herman Staudenmayer, PhD
Allergy Respiratory Institute of Colorado
5800 East Evans
Denver, CO 80222

Susan Stock, MD
175 Duke Street
Hamilton, Ontario
L8P 1X9 CANADA

Virginia Sublet, PhD
Division of Health Education
ATSDR
1600 Clifton Road, NE, Mail Stop E-33
Atlanta, GA 30333

Charles P. Sweet, MD, MPH
Occupational Health Program
Dept of Family and Community Medicine
University of Massachusetts Medical Center
55 Lake Avenue North
Worcester, MA 01605

Doug Trout, MD
Division of Occupational and Environmental Health
Medical College of Pennsylvania
3300 Henry Street
Philadelphia, PA 19129

Mark Upfal, MD
Division of Occupational Health
Dept of Family Medicine
Wayne State University
4201 St. Antoine, Suite 4-J
Detroit, MI 48201

Robert Vogt, PhD
Center for Environmental Health and Injury Control
Centers for Disease Control
1600 Clifton Road, NE
Chamblee Bldg 17, Room 1520
Mail Stop F-19
Atlanta, GA 30333

Bailus Walker, Jr., MD
Dean, College of Public Health
University of Oklahoma Health Sciences Center
P.O. Box 26901
Oklahoma City, OK 73190

Andrea Wargo, PhD
ATSDR
Hubert H. Humphrey Bldg, Room 727E
200 Independence Avenue, SW
Washington, DC 20201

Laura Welch, MD[1]
Division of Occupational and Environmental Medicine
George Washington University
2300 K Street, NW
Washington, DC 20037

John D. White, MD, MPH
Georgetown University Hospital
3800 Reservoir Road, NW
Washington, DC 20007

Thomas Wilcox, MD
NIOSH
4676 Columbia Parkway
Cincinnati, OH 45226

Grace E. Ziem, MD[1]
Occupational and Environmental Health Research Office
1722 Linden Avenue
Baltimore, MD 21217

Staff

Edmund Kelly, Executive Director
AOEC
1010 Vermont Avenue, NW, Suite 513
Washington, DC 20005

Katherine Kirkland
AOEC
1010 Vermont Avenue, NW, Suite 513
Washington, DC 20005

Toxicology and Industrial Health

AN INTERNATIONAL JOURNAL

PROCEEDINGS OF THE CONFERENCE ON LOW-LEVEL EXPOSURE TO CHEMICALS AND NEUROBIOLOGIC SENSITIVITY

6–7 April, 1994
Baltimore, Maryland

Editor
Frank L. Mitchell, D.O., M.P.H.
Agency for Toxic Substances and Disease Registry

Co-Editor
Patricia Price, D.O.
Agency for Toxic Substances and Disease Registry

Sponsored by:
U.S. Department of Health and Human Services
Public Health Services
Agency for Toxic Substances and Disease Registry

Published by:
Princeton Scientific Publishing Co., Inc.
Princeton, New Jersey

Toxicology and Industrial Health

TABLE OF CONTENTS
VOLUME 10, NUMBER 4/5

PROCEEDINGS OF THE CONFERENCE ON LOW-LEVEL EXPOSURE TO CHEMICALS AND NEUROBIOLOGIC SENSITIVITY

PREFACE

The development in some individuals of sensitivity to a wide variety of substances presents difficult and complex medical and scientific issues. This increased sensitivity, often called Multiple Chemical Sensitivity (MCS), has been the subject of a great deal of interest and controversy over the past several years. For example, a 1992 National Academy of Science report entitled "Multiple Chemical Sensitivities" noted that "The diagnostic label of multiple chemical sensitivity is being applied increasingly, although definition of the phenomenon is elusive and its pathogenesis as a distinct entity is not confirmed. Multiple chemical sensitivity has become more widely known and increasingly controversial as more patients have received the label."

Given the need for additional scientific research on mechanisms of actions and clinical experience with chemical sensitivity, ATSDR has provided support for two major national conferences on the subject of MCS. The first was a National Academy of Sciences meeting held in March, 1991. The second was a meeting sponsored by the Association of Occupational and Environmental Clinics (AOEC) in September, 1991. Both meetings made valuable contributions toward identifying key scientific, clinical, and public health issues concerning MCS.

In fiscal year 1993, ATSDR received a congressional mandate that directed funds to be used ". . . for chemical sensitivity/low-level chemical and environmental exposure workshops." To address the mandate, a panel of experts was convened in April, 1993. The purpose was to provide guidance on what initiatives the Agency could undertake with the available resources that would offer potential for advancing the knowledge base of low-level chemical sensitivity. High on the list of suggestions developed by the expert panel was the sponsorship of a national meeting to consider the neurobiological aspects of chemical sensitivity. The meeting was held on April 6–7, 1994, in Baltimore, Maryland.

With the assistance of an outstanding planning group, a program was produced that offered the opportunity to explore several issues relating to the neurobiological effects of chemical exposures. A large number of papers were presented that explored what role the nervous system might play in MCS.

Prior to the meeting, the invited speakers were asked to supply copies of the papers they intended to present at the conference. All did so, and some chose to expand on the remarks they intended to present verbally. Their submissions were also subjected to a peer review process following the meeting. Therefore, the papers that appear in this issue may differ from the exact presentations given during the meeting.

Toxicology and Industrial Health, Vol. 10, No. 4/5, pp. i–ii

ISSN: 0748-2337

The two white papers, prepared by Drs. Claudia Miller and Iris Bell, present an overview of the history and current knowledge of chemical sensitivity.

It was the aim of ATSDR to maintain the greatest possible number of opportunities for dialogue throughout the conference. During several response panels and question and answer sessions, there was a very lively and beneficial exchange of points of view and information among speakers, panel members, and conference attendees. In order to capture this exchange to the extent possible, selected comments from those sessions have been included in edited form.

For those interested in reviewing the conference in its entirety, videotapes and audiotapes of the conference are commercially available.

ATSDR believes that this conference served its intended purpose of bringing together scientists and clinicians to discuss, from viewpoints perhaps not previously considered, the issues of sensitivity to low levels of chemicals in the environment. The Agency hopes that this dialogue will stimulate further research, thereby increasing knowledge about the ways humans react to low levels of chemicals in the environment.

MULTIPLE CHEMICAL SENSITIVITY WORKSHOP

MEETING ADMINISTRATION

Chairman: Frank Mitchell, D.O.
Chief Medical Officer, ATSDR
Atlanta, GA

Co-Chairman: Patricia Price, D.O.
Medical Officer, Division of Health Education
ATSDR
Atlanta, GA

PLANNING COMMITTEE

Iris Bell, M.D., Ph.D.
University of Arizona
Tucson, Arizona

Peter Kalivas, Ph.D.
Department of Veterinary and Comparative Anatomy, Pharmacology and Physiology
College of Veterinary Medicine
Washington State University
Pullman, Washington

Howard Kipen, M.D.
University of Medicine and Dentistry
Robert Wood Johnson Medical Center
Piscataway, New Jersey

Mary Lamielle
National Center for Environmental Health Strategies
Voorhees, New Jersey

Richard Letz, Ph.D.
Emory University
Atlanta, Georgia

Max Lum, Ed.D.
Director, Division of Health Education
ATSDR
Atlanta, Georgia

Claudia Miller, M.D.
University of Texas
San Antonio, Texas

Kathy Rest, Ph.D.
University of Massachusetts Medical Center
Worcester, Massachusetts

Norman Rosenthal, M.D.
Chief, Environmental Psychiatry Section
Clinical Psychobiologic Branch
National Institute of Mental Health
Bethesda, Maryland

Cynthia Wilson
Chemical Injury Information Network
White Sulphur Springs, Montana

WHITE PAPER: CHEMICAL SENSITIVITY: HISTORY AND PHENOMENOLOGY

CLAUDIA S. MILLER

Department of Family Practice
The University of Texas Health Science Center
San Antonio, Texas

Nearly everyone has heard something about chemical sensitivity, either from personal experience with someone who has the condition or from the media. The television series Northern Exposure recently featured a chemically sensitive attorney who lived in a geodesic dome in Alaska, and L.A. Law depicted the struggles of a Persian Gulf veteran with chemical sensitivities who lost his case against the Veterans Administration, but may appeal later in the season. Television news programs and the printed media have showcased patients living spartan existences in remote areas or in aluminum foil-lined rooms. Our views of the illness no doubt are colored by our own personal experiences of it. While some discount or make jokes about chemical sensitivity or these patients, physicians who have seen a number of them are discovering that many appear to be credible individuals with prior good work records who say they became ill following an identifiable exposure to chemicals.

HISTORICAL OVERVIEW

Chemical sensitivity was first described about 40 years ago by an allergist named Theron Randolph who practiced in Chicago. His first patient was a physician's wife and cosmetic saleswoman with rhinitis, asthma, headache, fatigue, irritability, depression, weight swings, and intermittent loss of consciousness, who reported that each time she drove from her home in southern Michigan to his office in Chicago she became ill while passing through the heavily industrialized areas of northwest Indiana and South Chicago (Randolph, 1987). There appeared to be a common thread to her complaints — she reported becoming ill when exposed to combustion products and other derivatives of gas, oil, or coal. Similar patients with the "petrochemical problem," as Randolph called it, followed. Patients were advised to avoid a wide range of everyday chemical exposures and foods to see whether they improved, and, if so, to reintroduce single substances one at a time while observing the effect of each.

1. Address all correspondence to: Claudia S. Miller, M.D., M.S., Environmental and Occupational Medicine, Department of Family Practice, The University of Texas Health Science Center at San Antonio, San Antonio, TX 78284. Tel.:(210)567-4557. Fax:(210)567-4579.

Toxicology and Industrial Health, Vol. 10, No. 4/5, pp. 253–276

ISSN: 0748-2337

Randolph's approach toward these patients was, and continues to be, vehemently opposed by other physicians, particularly some allergists, who have been critical of the anecdotal nature of his work, his reliance upon patients' self-reporting of symptoms, and certain diagnostic practices and treatments adopted by his followers who became known as "clinical ecologists." These practices include sauna therapy, vitamin and mineral supplementation, and sublingual or intradermal administration of chemicals to diagnose and treat the condition. Position papers by the California Medical Association (1986), the American College of Physicians (1989), the American Academy of Allergy and Immunology (1986), and other physician groups have criticized the ecologists for a lack of critical thinking and use of unproven practices. Clinical ecology has been labelled as "junk science," a "medical subculture," and its patients as "true believers" (Brodsky, 1987; Staudenmayer and Selner, 1987; Huber, 1991). In contrast, others regard chemical sensitivity as a potentially real and growing problem and, to some degree, view such attacks on the clinical ecologists as "killing the messenger."

An acrimonious debate between allergists and clinical ecologists concerning this condition and how to treat it has been ongoing for nearly a decade in professional meetings, medical journals, and courtrooms. Other voices recently have joined in — e.g., those of occupational medical physicians and environmental health researchers. Over the past five years, more and more patients reporting chemical sensitivity have sought the help of academically-based occupational medicine physicians.

Two recent books and two reports written by academicians for the states of Maryland and New Jersey on chemical sensitivity have brought focus to the subject (Cullen, 1987; Ashford and Miller, 1989, 1991; Bascom, 1989). The New Jersey Report on chemical sensitivity suggested that legitimate professional concerns over unorthodox diagnostic and treatment approaches employed by the clinical ecologists be separated carefully from the question "Does chemical sensitivity exist as a clinical entity?" In the spring of 1991, in response to growing public and professional interest in chemical sensitivity, the National Research Council (NRC) convened a workshop to develop research recommendations for multiple chemical sensitivity or "MCS," as it has come to be called. Clinicians, toxicologists, immunologists, epidemiologists, psychiatrists, psychologists, and others with relevant skills or interests were invited to attend. The Environmental Protection Agency (EPA) and the National Institute of Environmental Health Sciences (NIEHS) shared sponsorship and the Agency for Toxic Substances and Disease Registry (ATSDR) provided ancillary support for the meeting. Participants offered diverse perspectives and achieved consensus as to future research directions, despite general concern that "definition of the phenomenon was elusive and its existence as a distinct clinical entity had not been confirmed" (National Research Council, 1992).

In the fall of 1991, ATSDR sponsored the second national meeting devoted exclusively to MCS in conjunction with the Association of Occupational and Environmental Clinics (AOEC), inviting occupational medicine physicians from across the country. Participants agreed that there were many unanswered questions about the illness and that further research on MCS was needed. In the past three years, a handful of academic researchers have applied for federal funding for MCS-related projects and a few, small pilot studies have been funded.

One problem that mitigates funding agencies' willingness to support studies on MCS is the fact that there is no reliable estimate of the number of patients affected by it. In part, this is due to lack of an agreed-upon case definition. What is known about the size of the problem is that approximately twenty MCS patient newsletters are published, including some with national circulations and several thousand subscribers apiece. Recently, the Social Security Administration and Housing and Urban Development (HUD) have recognized MCS as a disability, despite the fact that there is no generally accepted case definition, no identified mechanism for the disorder, and no laboratory marker to confirm its presence. The Americans with Disabilities Act (ADA) also affords some protections, on a case by case basis, to those disabled by MCS. Some fear that the illness is rapidly becoming politically defined before there is adequate science to support its existence.

FEDERAL AGENCY INTEREST

With growing public concern and litigation over MCS, several federal agencies now find themselves facing important policy questions related to the condition. One of these agencies is the Environmental Protection Agency (EPA). The EPA's mission includes preventing adverse human health effects from pesticides and indoor air pollution. Many MCS patients point to pesticides or sick buildings as the original causes of their illness. Ironically, several years ago the EPA itself installed 27,000 square yards of new carpeting, painted and remodeled space in its Waterside Mall headquarters in Washington, D.C., and had the unwelcome opportunity to study MCS firsthand (Ashford and Miller, 1991). Out of 200 or so agency employees who developed symptoms associated with sick building syndrome, several dozen reported developing MCS. These employees complained of being unable to tolerate tobacco smoke, perfume, engine exhaust, and other low-level exposures that they say had not been a problem for them before the remodeling took place. Some left the agency claiming they could no longer work. Some went to other jobs or now work at home. Some moved into specially-furnished offices the EPA provided which had no carpeting, disinfectants, perfume, etc., and where occupants could open windows. Litigation currently is in process. Nationwide, indoor air pollution is estimated to cost tens of billions of dollars annually (Environmental Protection Agency, 1989). MCS cases are part of this costly burden, exacting an enormous financial toll on patients, building owners, and product manufacturers. In addition to helping sponsor the NAS meeting on MCS, the EPA recently initiated its own in-house study to characterize the condition.

About a year ago, the U.S. Congress asked the Agency for Toxic Substances and Disease Registry (ATSDR) to direct $250,000 from its budget toward chemical sensitivity and low-level environmental exposure workshops. In the spring of 1993, ATSDR convened a panel of physicians, scientists, and MCS patients who recommended that a conference be held to explore the extent to which the central nervous system might be involved in the disorder.

Superfund monies fund ATSDR to investigate and provide information regarding health effects related to toxic wastes. Many citizens who live near Superfund hazardous waste sites report being ill, yet exposures frequently are "low-level," that is, well below recognized safety limits. Consequently, ATSDR is interested in a wide range of possible health effects from low-level chemical exposures, including MCS. Thus far, twelve hundred toxic waste dump sites have been placed on a national priority list for remediation out of an estimated 400,000 sites throughout the United States. Nevertheless, there is a paucity of data concerning health effects associated with most of the exposures involved. Billions of dollars have been spent for clean-up of Superfund hazardous waste sites over the past decade, and results of research on MCS could affect future policy and expenditures in this area in important ways.

In the past year, the Department of Veterans Affairs (DVA) and Department of Defense also have been drawn into the MCS debate. Many Persian Gulf veterans returned from the War complaining of multi-system health problems, such as fatigue, depression, irritability, memory and concentration difficulties, muscle aches, shortness of breath, diarrhea, and a host of other problems which they attribute to exposures in the Gulf. Exposures included combustion products from oil well fires, paints, fuels, pesticides, solvents, and others. Some congressmen and veterans have raised the specter of possible chemical or biological war agent use. A number of veterans have expressed dissatisfaction with the DVA's inability to link their illnesses with their wartime exposures and have sought help from clinical ecologists in private practice who in turn have diagnosed them as having MCS. Some veterans who have seen clinical ecologists subsequently have tried to obtain medical benefits and compensation from the DVA for war-related injuries only to be told that MCS is not a recognized medical condition. Angry, frustrated and sick, more and more veterans have turned to their legislators for assistance. In early 1994, the DVA issued a request for proposals to establish Environmental Hazards Research Centers in up to three VA hospitals. Researchers in these centers would focus on the role of environmental exposures in the Gulf veterans' health problems, including chemical sensitivity.

Thus, a growing number of federal agencies involved in occupational and environmental health issues, including the EPA, ATSDR, DOD, and the DVA, but also NIEHS, OSHA (Occupational Safety and Health Administration), and NIOSH (National Institute for Occupational Safety and Health), have encountered MCS, and several of these agencies have become interested in advancing scientific knowledge concerning this condition. In the past five years, the controversies surrounding MCS have exploded far beyond the narrow confines of a professional dispute between allergists and ecologists into a national debate with far-reaching policy and regulatory implications.

OPPOSING VIEWS OF THE ILLNESS

A wide variety of names have been applied to MCS and to those affected by it. Many of the names themselves seem to invite controversy.

Terms for the condition:

Multiple Chemical Sensitivity (MCS)
Chemical Sensitivity
Environmental Illness (EI)
Cerebral Allergy
Twentieth Century Disease
Chemically-induced Immune Dysregulation
Total Allergy Syndrome
Ecologic Illness
Chemical Hypersensitivity Syndrome
Environmental Maladaptation Syndrome
Universal Allergy
Chemical AIDS

Terms patients use to describe themselves:

universal reactors
canaries
chemies

Those who feel MCS is the consequence of chemical exposures point to the exponential rise in the use of synthetic organic chemicals and pesticides in homes and workplaces since World War II; to the construction of "tight," energy-efficient housing, offices, and commercial buildings since the oil embargo of the 1970s, with consequent reduction in fresh air indoors and higher volatile organic chemical (VOC) levels inside buildings; and to the fact that the average American now spends 90% or more of the day indoors. Indeed, EPA studies have demonstrated that indoor concentrations of certain VOCs may be orders of magnitude above outdoor levels (Wallace, 1985). Besides outgassing from interior finishes and furnishings, VOCs and other pollutants might emanate from toxic waste sites, percolate up from the soil via foundation cracks and crevices (e.g., unsealed pipe runs) into structures, and accumulate indoors, analogous to radon. MCS proponents point out that Americans living today are the first of their species to live indoors with relatively high levels of a peculiar mix of VOCs.

MCS patients often attribute onset of their illness to a specific exposure, for example, a chemical spill, repeated exposure to a solvent, application of a pesticide, a sick building or combustion products from a fire. Subsequently, patients generally report that they experience their greatest difficulties indoors where perfume, air fresheners, cleaners, etc., are used and interior finishes or furnishings such as carpet or particle-board "outgas," releasing VOCs.

MCS patients express frustration with many physicians, whom they perceive as not believing them and not understanding that their symptoms are caused by chemical exposures. A number of patients are teachers, lawyers, health care providers, and other professionals who appear to be credible historians and who say they experience reproducible symptoms with specific exposures, e.g., to tobacco smoke, a certain perfume, etc. They express anger toward physicians who prescribe antidepressants or who refer them to psychiatrists or psychologists. From the patients' perspectives, they have lost their health, their livelihoods, and many of their pleasures in life, while skeptical physicians treat them as psychological patients or malingerers, and cast doubt among the patients' families, friends, and employers concerning the reality of their condition.

Clinicians who question the existence of an organic basis for MCS point to the ongoing medical debate and lack of a generally accepted case definition, although several definitions have been proposed (Table 1). Some skeptics feel that well-established diagnoses, such as somatoform disorder, depression, asthma, migraine, post–traumatic stress disorder, and other conditions, account for symptoms in the majority of cases. Some practitioners believe that MCS patients hold an inappropriate belief that chemicals are causing illness. Some feel patients can be deprogrammed from their beliefs (Selner, 1988). Skeptics point to other problems with the illness, including the absence of a clinical or laboratory marker and the lack of an identified mechanism for the condition. MCS patients, who are notorious for the panoply of distressing symptoms they report, may not always realize how physicians perceive their litany of complaints — in medical school, young doctors-to-be often are taught that the more symptoms a patient reports, the less likely there is anything to them, i.e., the diagnosis is probably a psychological one.

Clinicians who view MCS as a possible new diagnosis point out that before multiple sclerosis was known as "multiple sclerosis" and before lupus was known as "lupus," there was an interim phase during which clinicians simply observed some patients with novel presentations who seemed to share certain features in common. A number of physicians and scientists believe that MCS currently may be in such an early observational stage.

CLINICAL OBSERVATIONS AND PHENOMENOLOGY

Of necessity, clinical observations concerning an illness generally precede case definitions, the discovery of markers, and elucidation of mechanisms. It has been observed that many cases of MCS appear to involve a two-step process (Figure 1) (Ashford and Miller, 1991):

1. *Sensitization*, also referred to as "priming" or "induction." In many MCS patients, symptoms appear to develop following a major exposure to any of a wide range of environmental chemicals. The "sensitizing event" may be either an acute high-level exposure, such as a chemical spill, or it may be a chronic (repeated or continuous) exposure, occurring at much lower levels, for example, a sick building. The nature of the events MCS patients say led to their illness is extraordinarily diverse and includes exposures to pesticides, solvents, combustion products, indoor air pollutants, drugs, anesthetics, and, in a few instances, extreme stress without any obvious chemical exposure.

TABLE 1. Proposed Case Definitions for Multiple Chemical Sensitivity

Ashford and Miller (1989):

The patient with multiple chemical sensitivities can be discovered by removal from the suspected offending agents and by rechallenge, after an appropriate interval, under strictly controlled environmental conditions. Causality is inferred by the clearing of symptoms with removal from the offending environment and recurrence of symptoms with specific challenge.

Association of Environmental and Occupational Clinics 1992 Workshop on Multiple Chemical Sensitivity, Working Group on Characterizing Patients:

- A change in health status identified by the patient
- Symptoms triggered regularly by multiple stimuli
- Symptoms experienced for at least six months
- A defined set of symptoms reported by patients
- Symptoms that occur in three or more organ systems
- Exclusion of patients with other medical conditions (psychiatric conditions are not considered exclusionary)

Clinical Ecologists (definition appearing in each issue of the journal *Clinical Ecology*):

Ecologic illness is a chronic multi-system disorder, usually polysymptomatic, caused by adverse reactions to environmental incitants, modified by individual susceptibility and specific adaptation. The incitants are present in air, water, food, drugs, and our habitat.

Cullen (1987):

Multiple chemical sensitivities (MCS) is an acquired disorder characterized by recurrent symptoms, referable to multiple organ systems, occurring in response to demonstrable exposure to many chemically unrelated compounds at doses far below those established in the general population to cause harmful effects. No single widely accepted test of physiologic function can be shown to correlate with symptoms.

Nethercott et al. (1993):

1. The symptoms are reproducible with exposure.
2. The condition is chronic.
3. Low-level exposure results in manifestations of syndrome.
4. Symptoms improve or resolve when incitants are removed.
5. Responses occur to multiple, chemically unrelated substances.

National Research Council (1992), Workshop on Multiple Chemical Sensitivities, Working Group on Research Protocol for Clinical Evaluation:

1. Sensitivity to chemicals. By sensitivity we mean symptoms or signs related to chemical exposures at levels tolerated by the population at large that is distinct from such well recognized hypersensitivity phenomena as IgE-mediated immediate hypersensitivity reactions, contact dermatitis, and hypersensitivity pneumonitis.
2. Sensitivity may be expressed as symptoms and signs in one or more organ systems.
3. Symptoms and signs wax and wane with exposures.

It is not necessary to identify a chemical exposure associated with the onset of the condition. Preexistent or concurrent conditions, e.g., asthma, arthritis, somatization disorder, or depression, should not exclude patients from consideration.

2. *Triggering*. Following sensitization, patients report that extremely low levels of common chemicals tolerated by the majority of the population, for example, tobacco smoke, perfume, and traffic exhaust, trigger severe symptoms. Commonly, they report that, in addition to the chemicals involved in the original exposure event, over time more and more *chemically unrelated* substances trigger symptoms. The latter observation is referred to by patients as the "spreading phenomenon."

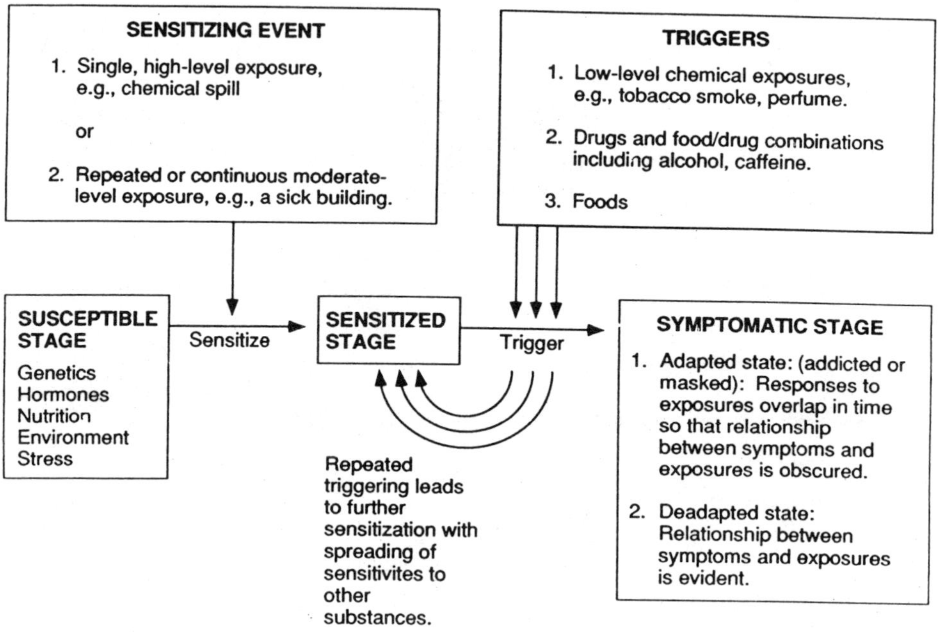

FIGURE 1. Phenomenology of Multiple Chemical Sensitivity.

This two-step process, sensitization and triggering, is reminiscent of allergic sensitization. Indeed, these patients often describe themselves as being allergic. Notably, when von Pirquet first coined the word "allergy" in 1906, he defined it as "altered reactivity" of whatever origin. However, in the 1920s, following the discovery of antibodies, allergy was redefined in immunological terms over the protests of some allergists who cautioned that certain nonimmunologic forms of hypersensitivity might be excluded. The discovery of IgE in 1967 further solidified the immunologic view of allergy.[1]

[1]Early in the development of their specialty, allergists had been accused by their colleagues of practicing witchcraft or "voodoo" medicine when they treated their patients by injecting them with tiny amounts of the same substances to which they reacted. With the discovery of IgE, allergists at last had a scientific basis for their practices.

MCS differs from classical allergies in at least one important respect: IgE formation is exquisitely specific for particular substances, e.g., ragweed or bee venom. In contrast, MCS patients report that their sensitivities spread to chemically unrelated substances. This discrepancy further enhances many allergists' doubts concerning MCS.

The limited data available at this time suggest that any mechanism or model that would purport to explain MCS would need to address the following clinical observations associated with this illness (Ashford and Miller, 1991):

1. Symptoms involving virtually any system in the body or several systems simultaneously, but most frequently the central nervous system (fatigue, mood changes, memory and concentration difficulties).
2. Different symptoms and severity in different individuals, even among those experiencing the same exposure.
3. *Induction* or *sensitization* by a wide range of environmental agents, including pesticides, solvents, and combustion products.
4. Subsequent *triggering* by lower levels of exposure than those involved in initial induction of the illness.
5. *Spreading* of sensitivity to other, often chemically dissimilar substances. Each substance may trigger a different but reproducible constellation of symptoms.
6. Concomitant food, alcohol, and medication intolerances, estimated to occur in a sizeable percentage of MCS patients.

Multiple chemical sensitivity has been reported among several distinct demographic groups (Table 2). To these groups might be added a fifth group — Persian Gulf veterans. When clinicians see patients one at a time (as in group 4 of Table 2), they are unlikely to attribute an individual patient's symptoms to an environmental exposure, even if the patient happens to mention the exposure (unless of course the hazard is one already well known to the physician, e.g., lead or benzene). It is especially easy to overlook environmental causes if symptoms are subjective and nonspecific, such as headache, fatigue, depression, or difficulty concentrating. Many of the first cases of MCS described involved individual, upper-middle class women. Such cases often were viewed as depression or "hysterical housewife syndrome" and referred accordingly. Notably, the first cases of sick building syndrome occurring in offices and schools were attributed to "mass psychogenic illness," rather than poor indoor air quality, before sick buildings were recognized. Physicians who happen to see a series of cases, all of whom share an identifiable exposure, are more likely to view the illness as "real" and investigate its origins. Thus, the *temporal cohesiveness* of symptoms occurring in a group of individuals sharing a recognizable exposure event, for example, several family members, co-workers, or community residents exposed to the same chemicals, helps physicians recognize the possibility of an environmentally caused illness in those circumstances. The outbreak of MCS among technical staff at the EPA headquarters who had been present during remodeling and carpet installation facilitated recognition of a possible environmentally related illness in that circumstance.

TABLE 2. Chemically Sensitive Groups (Ashford and Miller, 1991)

Group	Nature of Exposure	Demographics
Industrial workers	Acute and chronic exposure to industrial chemicals	Primarily males; blue collar; 20 to 65 years old
Sick building occupants	Off-gassing from construction materials, office equipment or supplies; tobacco smoke; inadequate ventilation	Females more than males; white-collar office workers and professionals; 20 to 65 years old; school children
Contaminated communities	Toxic waste sites, aerial pesticide spraying, groundwater and air contamination by nearby industry and other community exposures	All ages, male and female; children or infants may be affected first or most; pregnant women with possible effects on fetuses; middle to lower class
Individuals	Heterogeneous; indoor air (domestic), consumer products, drugs and pesticides	70–80% females; 50% 30 to 50 years old (Johnson and Rea, 1989); white, middle to upper middle class and professionals

Notably, the four groups represented in Table 2 vary greatly in terms of their age, sex, social group, and the kinds of medical specialists they consult. Nevertheless, all report onset of MCS-like symptoms following an identifiable exposure event. The *demographic diversity* of the groups reporting MCS (Gulf veterans, school children, office workers, industrial workers, etc.) also suggests the possibility that a real problem may be occurring. The complaints voiced by each of these groups appear unusual:

1. Odor intolerances. Patients frequently avoid tobacco smoke, gasoline, hairspray, cleaning agents, and many other substances because they say they feel ill around them;
2. Adverse reactions to medications or medical or dental materials, for example, anesthetics, radiographic contrast dye, antibiotics, decongestants, eye drops, suppositories;
3. Alcohol and/or caffeine intolerance; and
4. Food intolerances.

MCS patients report symptoms following inhalation, ingestion, mucosal contact, or injection of an enormous variety of substances. Sometimes they describe reacting to chemicals at concentrations below the olfactory threshold. Some have no sense of smell (anosmia), yet report reactions to chemicals. Symptoms are said to begin as soon as a few seconds after the exposure. Some note that breathing through their mouths instead of their noses slows the onset of symptoms, and they report using this mouth-breathing technique to their advantage, for example, when entering an elevator where passengers are wearing perfume. Some patients report not only being overly sensitive to chemicals, but also to physical stimuli, such as bright light, noise, and being touched, not unlike some post–traumatic stress disorder (PTSD) patients. They may complain of extreme discomfort just from someone bumping their bed or from hearing conversational-level noises. Some describe feeling as though the amplitude or gain in their nervous systems were turned up too high. Except for military and industrial populations that are primarily male, most samples of MCS patients have been predominantly female (in the majority of studies, approximately three-quarters of the patients have been women).

MCS patients often, but not always, have a life-long history of medical problems. Premorbidly, one group of MCS patients reported an average of 6.2 unexplained physical symptoms *prior to* their workplace exposure versus 2.9 for controls (Simon et al., 1990). Likewise, 54% of the same MCS patient group reported anxiety or depression prior to their workplace exposure versus only 4% of controls. In contrast, Fiedler and colleagues did not find that premorbid psychiatric conditions accounted for MCS in a group of eleven patients they studied (Fiedler et al., 1992). MCS proponents argue that even if some MCS patients were depressed prior to florid onset of the illness, the question still remains whether MCS is caused by depression, whether depressed people are more susceptible to MCS, or whether the prior depression was in fact the result of earlier, undiagnosed chemical or food sensitivities.

PARALLELS WITH ADDICTION

Patients' descriptions of MCS share striking parallels with alcohol and drug addiction (Randolph, 1980; Ashford and Miller, 1991). An interesting feature of MCS is the carbohydrate or other food cravings described by many patients. Some refer to themselves as "chocoholics" or report addiction to certain foods, such as baked goods, popcorn, sweets, colas, etc. Some report a past history of having sipped coffee, tea, cola or other caffeinated beverages throughout the day, carrying one of these with them wherever they went. Following a chemical exposure, for example, after driving in heavy traffic, some MCS patients report experiencing intense food cravings. MCS reportedly may interact with other appetitive behaviors: Most patients describe a loss of interest in sex; however, a few report hypersexuality especially following chemical exposures. Some MCS patients report extreme thirst in conjunction with their chemical or food reactions.

Randolph was impressed by his patients' descriptions of food cravings and compulsive eating behaviors. In addition, he observed that symptoms following chemical exposures or foods seemed to begin with "stimulatory" symptoms and terminate in "withdrawal symptoms" (Figure 2). He viewed food and chemical addiction as the base of a pyramid of addiction, with alcohol and street drugs being at the apex (Figure 3).

Notably, from our own observations and those of other clinicians familiar with this problem, MCS patients tend to shun alcohol, xanthines (including tea, cola, coffee, and chocolate), and drugs in general because their responses to these, they say, are so unpleasant. The nature of the symptoms they report seems consistent with what one might expect from a normal person who ingested large quantities of such substances. For example, MCS patients may report excessive irritability, agitation, and headaches following one cup of coffee or a chocolate bar.

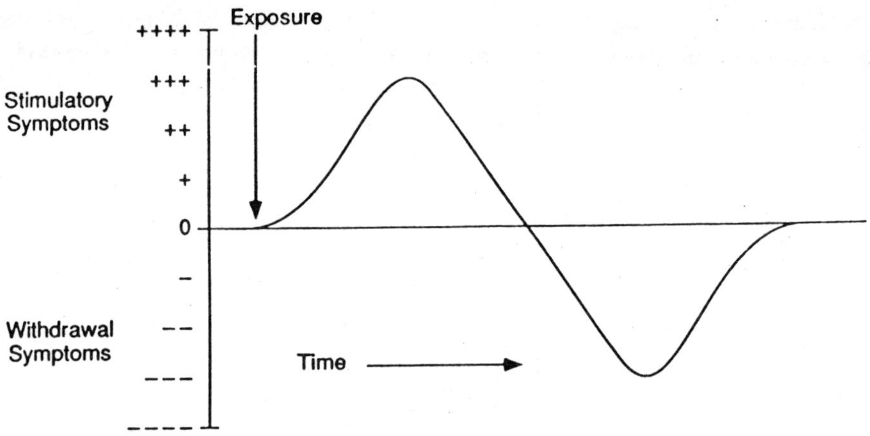

FIGURE 2. Graphical representation of symptom progression in a sensitive individual following exposure to a single substance (from O'Banion, Ecological and Nutritional Treatment of Health Disorders, 1981; courtesty of Charles C. Thomas, Publisher, Springfield, Illinois). The sine wave is a graphical representation of symptoms experienced by a sensitive individual following exposure to a *single* chemical, food, or drug. Initial symptoms, associated with the *onset* of exposure have been described as *stimulatory* in nature, for example, feeling "hyper," jittery, talkative, overly enthusiastic, anxious, or panicky. Symptoms associated with the *offset* of exposure have been described as *withdrawal* symptoms, that is, lethargy, sleepiness, depressed feelings, headache, concentration difficulties, etc. A normal individual who is not sensitive to the substance would experience no symptoms and the sine wave would be flat, with zero amplitude.

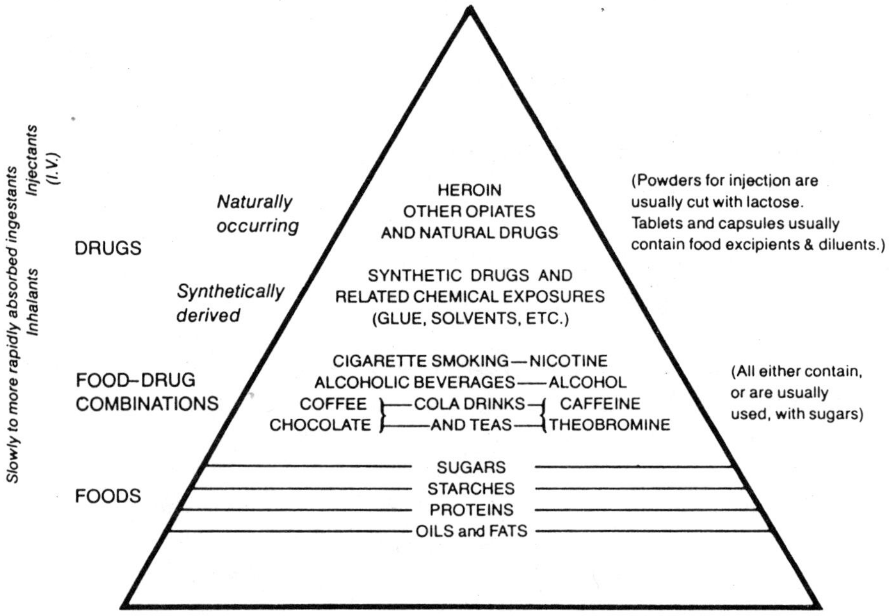

FIGURE 3. Addiction pyramid, after Randolph (1980).

Yet before they eliminated caffeine from their diets, MCS patients often say they consumed it frequently or addictively (e.g., half a pound of chocolate or 20 cups of coffee a day) and felt *chronically* ill. In this respect, MCS patients appear analogous to certain reformed smokers or alcoholics who after quitting tobacco or alcohol become exquisitely sensitive to even minute amounts of tobacco smoke or alcohol. Many MCS patients claim that they formerly were addicted to foods or chemicals, but that in the beginning the addiction was unwitting and unrecognized by them (for example, an addiction to corn or to pollutants in a sick building). Initially, when they first attempted to avoid problem chemicals and foods, MCS patients say they experienced a kind of "detox" (also called "unmasking" or "de-adaptation") accompanied by several days of intense "withdrawal" symptoms. Indeed, "detox" and "withdrawal" are terms MCS patients use to describe this experience. One patient characterized MCS as being "like drug abuse without any of the fun." The medical teaching, "Listen to the patient — he is telling you the diagnosis," would suggest that MCS patients' choice of drug addiction terminology is a clue that invites further scrutiny.

Following avoidance or "withdrawal," patients often report pronounced symptoms when they are *reexposed* to substances they appeared to tolerate before. To decipher which exposures affected which patients, Randolph developed the concept of an environmental control unit, a hospital ward that excludes disinfectants, deodorizers, strong cleaners, pesticides, perfumes, etc., and in which patients fast for several days. If patients improved in the controlled environment (and many say they did), they were then reexposed to single chemicals and foods, one at a time, to see which, if any, provoked symptoms. Tolerated foods remained in the diet, but were not to be eaten more than once every four days to prevent addiction to these from also developing. According to Randolph, "problem" foods tended to be those that were eaten frequently or addictively, such as corn, wheat, milk, or eggs. For example, he reported that some patients would drink sodas and eat candy containing corn sugar on a frequent basis and become addicted to corn without realizing it. MCS patients (once they are diagnosed as such) frequently shun alcohol, tobacco, and caffeine, and often say they are prone to addiction to foods, sugar, etc. Alcoholic beverages, such as beer or red wine, frequently are the first ingestants patients identify as causing problems. Many MCS patients say they become "hooked on" substances far less addicting than alcohol, nicotine, or caffeine and considered to have little, if any, stimulatory effect. On the other hand, certain foods contain or are metabolized to neuroactive peptides which could play a role in this process (Bell et al., 1992). In effect, MCS patients appear to be "hyper-holic," not to be confused with "hypergolic" — the property that certain rocket propellants have to ignite spontaneously upon contact between the components. The similarity between the terms, however, may not be entirely inapt.

SYMPTOMATOLOGY

Patients with this illness report multi-system health complaints. Their most frequent complaint is fatigue, which is one of the most frequent presenting symptoms of Americans who see primary care physicians (Cathebras et al., 1992). Other disabling symptoms reported by MCS patients include changes in their mood and cognitive abilities. MCS patients with professional careers are likely to view their cognitive difficulties as the most disabling feature of their illness. Miller and Mitzel (1994) surveycd 75 MCS patients who reported onset of their illness following remodeling in a building and 37 who reported onset following exposure to a cholinesterase-inhibiting pesticide. The most frequently reported symptoms in each group were quite similar; the majority involved the central nervous system (Table 3). The most common gastrointestinal complaint was "problems digesting food," and the most common respiratory complaint was "shortness of breath or being unable to get enough air."

TABLE 3. Top 20 Symptoms (of 119 Symptoms) Reported by MCS Patients Attributing Their Illness to Pesticides (N = 37) Versus Remodeling (N = 75) (Miller and Mitzel, 1994)

	Ranking		Mean symptom severity**	
Symptom	Pesticide	Remodel	Pesticide	Remodel
*Tired or lethargic	1	1	2.49	2.44
*Fatigue > 6 months	2	3	2.43	2.10
*Memory difficulties	3	4	2.32	2.09
*Difficulty concentrating	4	2	2.32	2.17
*Dizziness, lightheadedness	5	6	2.19	1.85
*Depressed feelings	6	8	2.19	1.83
*Spacey	7	12	2.19	1.74
*Groggy	8	5	2.14	1.96
*Loss of motivation	9	7	2.11	1.84
*Tense, nervous	10	15	2.11	1.64
*Short of breath	11	18	2.11	1.61
*Irritable	12	10	2.03	1.79
Problem focusing eyes	13	43	2.03	1.27
Chest pain	14	52	2.00	1.19
*Muscle aches	15	11	2.00	1.79
Problems digesting food	16	33	1.97	1.35
*Joint pain	17	9	1.95	1.83
Tingling fingers/toes	18	59	1.95	1.12
*Headache	19	14	1.92	1.67
*Head fullness or pressure	20	19	1.92	1.60
Difficulty making decisions	21	13	1.89	1.69
Eye irritation	22	16	1.89	1.64
Slowed responses	34	17	1.72	1.63
Nausea	36	20	1.65	1.56

* = Among top 20 symptoms in both pesticide and remodeling patients.
** = Symptoms scored on 0 to 3 scale: 0 = not a problem; 1 = mild; 2 = moderate; 3 = severe.

Many patients report that exposure to a particular chemical or mixture of chemicals produces a characteristic constellation of symptoms. For example, they may say they feel spacey and have an upset stomach with diesel exhaust; become irritable when walking down the detergent aisle of a grocery store; or experience confusion around a particular perfume. Some say these symptoms are so specific that they can identify the exposure source in the absence of any detectable odor. Individuals who shared the same initial exposure event (e.g., the EPA workers who became ill) often report markedly different symptoms. For example, one individual may report more problems with concentration, another with breathing, and still a third with digestion. This reported variability in presentation coupled with the lack of a case definition, has thwarted attempts to conduct epidemiological investigations of MCS. Following onset of their illness, many MCS patients report that their sensitivities rapidly spread to more and more chemicals, foods and drugs. If they are able to reduce their overall exposure to chemicals, some patients say they gradually regain some tolerance over a period of months and years, but that this is quickly lost if they are not continuously vigilant about minimizing their exposures.

A poorly understood, but potentially crucial, variable that may affect symptom expression and intensity is adaptation, or the development of tolerance. With repeated exposures or continuous exposure, humans adapt to many substances; acute symptoms tend to become chronic in nature and may no longer appear related to particular exposures. Adaptation has been described for substances as diverse as solvents (Riihimaki and Savolainen, 1980); ozone (Hackney et al., 1977a,b); nitroglycerin (Daum, 1983); tobacco smoke; caffeine; and many drugs (tachyphylaxis). MCS patients refer to adaptation as "masking." Many report that their illness began with "flu-like" symptoms, similar to Chronic Fatigue Syndrome. Indeed, a large number of MCS patients carry this diagnosis as well. Patients often report they were unaware of having any sensitivities to chemicals or foods in this early, flu-like stage of their illness. It was not until they avoided exposures (unintentionally, or intentionally upon someone's recommendation) that they noticed their symptoms improved. Then when they reexposed themselves (unintentionally or intentionally) to a particular environment or chemical and their symptoms recurred, they began to suspect environmental causes. This process of avoidance, whether intentional or not, has been termed "unmasking" or "deadaptation." MCS patients who travel to a large city often report that they "remask" or "adapt" and feel like they have the flu again. As long as they are "masked," they say they do not experience acute, robust symptoms with exposure to perfume or diesel exhaust — they just feel bad all of the time. Figure 4 illustrates this concept.

SUMMARY OF RESEARCH AND RESEARCH RECOMMENDATIONS

Pivotal medical, compensation, litigation, regulatory and policy questions rest upon a full understanding of MCS. Notwithstanding, remarkably little funding has been directed toward researching this illness, for a variety of reasons. Because of limited funding, the few studies that have been done have had "shoestring" budgets. Consequently, data on MCS are meager. Scientists involved in other rapidly expanding fields may find this paucity of research on an apparently vital topic surprising. Economic stakes are high. Insurers, agencies such as the DVA and DOD that provide medical care and compensation, the chemical industry, manufacturers of consumer products including carpets, building materials, fragrances and other goods could be affected greatly by the outcome of research on MCS. Some progress has occurred, however. In the past two years, two national meetings, both of which ATSDR helped sponsor, have brought together professionals with divergent views on this subject who have made recommendations for research (Table 4).

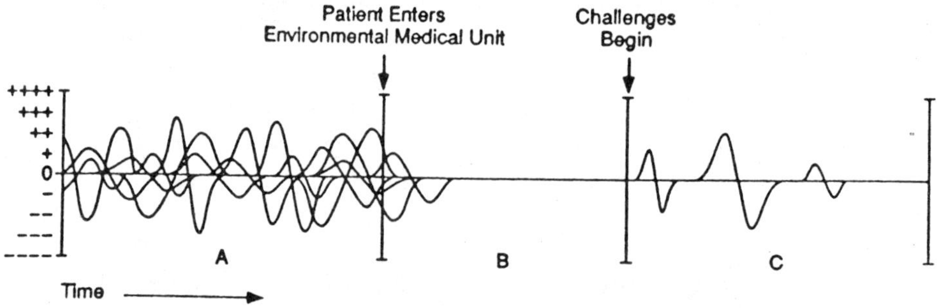

FIGURE 4. Hypothetical graphical representation of an individual's symptoms before and after entering an environmental unit (Ashford and Miller, 1991). In Time Period A, prior to entering an environmental medical unit (EMU), a chemically sensitive individual is responding to multiple exposures (chemicals and/or foods), with stimulatory and withdrawal effects that overlap in time. At any particular time, how the person feels is determined not only by ongoing exposures, but also by previous exposures whose effects still may be waning. The effect of any single exposure is not discernible. Some MCS patients refer to this as being "masked." In Time Period B, the individual enters an environmental medical unit. With cessation of contributory exposures, withdrawal effects occur, for example, headache, fatigue, and muscle aches. Symptoms continue for some time (typically 4–7 days) until the individual reaches "0" (baseline). Some MCS patients refer to themselves as being "unmasked" when they are in this state, i.e., when they are avoiding all of their problem exposures. In Time Period C, single challenges to suspected chemicals or foods are administered. Symptoms, if they occur, develop soon after challenges, allowing patient and physician to observe the relationship between exposures and symptoms for that individual.

TABLE 4. Summary of Research Recommendations from Federally Sponsored Meetings on Multiple Chemical Sensitivity (MCS)

I. National Research Council meeting on Multiple Chemical Sensitivities, March, 1991 (NRC, 1992).

   A. Sponsors: EPA, ATSDR, NIEHS
   B. Participants: Invited clinicians, immunologists, toxicologists, epidemiologists, psychiatrists, psychologists, and others involved in research or clinical activity relevant to MCS
   C. Recommendations (3 groups):

      1. Clinical Evaluation Group: Proposed a case definition for research (see Table 1 Proposed Case Definitions). Also suggested:
         a. Development of a uniform patient database
         b. Hypothesis-driven specialized evaluations
         c. Development of an environmental control unit for study of adaptation/deadaptation hypothesis, control of exposures, and challenging subjects
         d. Prospective studies of exposure events
      2. Exposures and Mechanisms Group

a. Double-blind controlled exposure challenges, examining the possible role of "adaptation" and "deadaptation"
b. Evaluation of MCS patients in their usual environment, as symptoms and exposures vary over time
c. Development of animal models that mimic the human syndrome
d. Evaluation of biopsy or necropsy tissue for pathologic changes
e. Development of database of chemicals, foods, drugs, and associated symptoms and signs

3. Epidemiology Working Group
   a. Improvement of case definition
   b. Multi-center clinical case-comparison studies using agreed-upon set of criteria and tests
   c. Use of information from case-comparison study to construct a population-based study to determine the prevalence of MCS
   d. Follow-up of a defined population subjected to a discrete and sudden exposure to assess the initiation of hypersensitivity and its natural history
4. Consensus was reached among all workshop participants that challenging subjects in a well-defined environment should have the highest priority for future research

II. Association of Occupational and Environmental Clinics (AOEC) Meeting on Chemical Sensitivity, September, 1991 (AOEC, 1992).

A. Sponsor: ATSDR
B. Participants: Invited speakers representing divergent views on MCS, members of the AOEC which includes occupational medicine physicians from academia and private practice
C. Recommendations (4 groups):
   1. Group on Characterizing Patients: Proposed a case definition for research (see Table 1, Proposed Case Definitions)
   2. Group on Characterizing Events
      a. Assessment of incidence and prevalence of MCS
      b. Surveys of specific occupational cohorts and cross-cultural studies of "naive" populations, such as pesticide-exposed agricultural workers in the Third World
      c. Longitudinal studies of populations exposed in "natural" experiments such as a sick building
      d. Case registries for descriptive and future serologic studies of panels of MCS patients
      e. Double-blind, placebo-controlled challenge studies
      f. Studies to determine whether chemical exposures truly can be blinded
   3. Group on Treatment Methods
      a. A study of the effects of early intervention in an exposed population, such as critical incident counseling
      b. Randomized, controlled trials of therapies that have some reasonable theoretical basis
   4. Group on Mechanisms
      a. Challenge studies, including but not limited to chamber studies (the latter should address the issue of adaptation)
      b. Studies of olfactory function and the nasal-olfactory-limbic pathway
      c. Neuro-imaging studies including the use of pharmacologic probes
      d. Prospective studies of cohorts of persons sensitive to chemicals
      e. Studies of families of MCS patients, both medical and psychological

Based upon the many clinical observations and few studies available, a number of mechanistic hypotheses have been advanced to explain MCS. Among these hypotheses are:

1. Immune dysfunction or sensitization
2. Neurological damage or sensitization
3. Impaired detoxification pathways
4. Inflammation
5. Vasoconstriction/vasculitis
6. Psychiatric or psychological disorders, such as:
    a. An inappropriate belief that chemicals are causing illness
    b. Post–traumatic stress disorder
    c. Conditioned behavior (odor conditioning)
    d. Somatoform disorder
    e. Depression
7. Combinations of the above mechanisms

These proposed mechanisms are discussed in detail elsewhere (Cullen, 1987; Ashford and Miller, 1991; Bell et al., 1992). The first two have enjoyed the most attention by proponents of the illness. Up to now, most clinical studies of MCS patients have focused on markers of immunological, neurological, inflammatory and psychological responses (Table 5).

Clinical ecologists, a few other physicians in the private sector, and some commercial laboratories have reported alterations in a number of parameters in these patients, including T and B lymphocyte counts; helper/suppressor T cell ratios; immunoglobulin levels; autoimmune antibodies (including antinuclear, antismooth muscle, antithyroid, antiparietal cell and other autoantibodies); activated T lymphocytes (TA1 or CD26); quantitative EEGs; evoked potentials; SPECT and other brain scans; levels of various vitamins, minerals, amino acids, and detoxification enzymes; and blood or tissue levels of pesticides, solvents and other "pollutants." Flaws in these studies are many and varied, but include: Failure to define the study population (no case definition used); failure to compare cases with age- and sex-matched controls; failure to blind specimens so that those performing the analyses are unaware of whether specimens came from subjects or controls; and failure to assess the accuracy and reproducibility of the test method. For these reasons, results of studies performed by clinical ecologists or commercial laboratories have been viewed with considerable skepticism by regulatory agencies and academic researchers. Some MCS proponents claim that different immunological abnormalities occur in different patients. However, if enough tests are done, statistically a certain number will be abnormal (e.g., one in twenty). This is not always taken into account. With regard to claims of immunological dysfunction, to date no single, consistently abnormal immunological parameter has been demonstrated in these patients. This could be because there are no changes in immune system function, or because relevant cells or cytokines have not been examined, or because changes are spurious, for example, if neurological disruption led to a spectrum of immunological sequelae.

TABLE 5. Multiple Chemical Sensitivity Clinical Studies Measuring Nonneuropsychiatric Endpoints (See Table 2 in accompanying paper by I.R. Bell for results of neuropsychiatric studies)

Author(s) Date	Subject selection criteria	No. of subjects (M:F) Source	Controls? No. of subjects (M:F) Type	Chemical challenges performed? Type	Parameters measured	Major findings
Doty et al. 1988	Self-reported responses to chemicals + high score on Randolph chemical exposure questionnaire	18 (6:12) Ads in patient newsletter, physician referrals	Yes. 18 (6:12) Matched for gender, age, ethnicity, smoking + no evidence of MCS	Olfactory threshold determinations for phenylethyl alcohol, followed by methyl ethyl ketone	Physiological: Nasal airflow resistance, odor detection threshold, heart rate, BP, respiratory rate	Physiological: Subjects had significantly higher nasal resistances pre- & post-challenge than controls, higher average respiratory rate. No differences between subjects and controls for olfactory thresholds, heart rate, blood pressure
Fiedler et al. 1992	Cullen MCS criteria + good health prior to exposure + no psychiatric history	11 (3:8) Patients seen in university occupational medicine clinic	No	No	Immunological: IgG, IgA, IgM, IgE, CH50, C3, C4. Allergy skin tests (10 antigens); Delayed-type hypersensitivity skin tests (4 antigens); Delayed-type patch tests (8 antigens)	Immunological: Scattered tests out of normal lab ranges, but no significant or consistent abnormalities
Leznoff 1993	Self-reported MCS with Type A or B symptoms: A: Breathlessness & lightheadedness B: throat-related symptoms	15 (3:12) (in another 5 subjects sex not specified); Workers' Comp, disability & other referrals for allergy consultation	No	Yes. Single, unblinded challenge tailored to patient (e.g., perfume, cigarette smoke, hairspray, detergent)	Type A (15 patients): Pulmonary function, pCO_2, pO_2. Type B (5 patients): Laryngoscopy, phonograms	Type A: In 10/15 subjects symptoms were reproduced; no changes in pulmonary functions; pCO_2 fell & pO_2 rose post-challenge. Author's interpretation: hyperventilation and chemophobia. 5/15 had neither symptoms nor changes in pCO_2 or pO_2. Type B: 5/5 subjects had no symptoms or laryngoscopic changes

TABLE 5. Multiple Chemical Sensitivity Clinical Studies Measuring Nonneuropsychiatric Endpoints (Cont'd.)

Author(s) Date	Subject selection criteria	No. of subjects (M:F) Source	Controls? No. of subjects (M:F) Type	Chemical challenges performed? Type	Parameters measured	Major findings
Meggs and Cleveland 1993	Cullen MCS criteria; current smokers (2/10) included	10 (6:4) University allergy clinic referrals; self-referred	No	No	Fiberoptic rhinolaryngoscopy in 10/10; allergy skin tests in 7/10; Pulmonary function tests in 7/10; methacholine challenge in 2/10	Abnormal rhinolaryngoscopic findings in 10/10 patients including edema, excess mucus, cobblestoning, mucosal injection, blanching around vessels; + skin tests in 5/7; normal PFTs in 7/7; normal methacholine challenges where performed
Simon et al., 1993	Recorded diagnosis of MCS + ill for 3 months or longer + symptoms in CNS & at least 2 other systems + self-reported sensitivity to 4 or more of 14 common exposures	41 (6:35) Community allergy practice billing records (56% of eligibles enrolled)	34 (6:28) Musculoskeletal/back injury patients from university clinics (41% of eligibles enrolled)	No	Immunological: Blinded samples for T cells (total, CD4, CD8); B cells; CD25 (IL-2R+); CD26 (TA1, activated T cells); IL-1 generation by monocytes; autoantibodies against parietal cells, mitochondria, smooth muscle, brush border, nuclear components	Immunological: No significant differences between patients and controls; lower IL-1 generation among cases interpreted as probably due to laboratory methods
Terr 1986	Prior diagnosis by clinical ecologist of environmentally induced illness	50 (11:39) 43 = evaluations for Workers' Comp; 3 = evaluations for litigation; 4 = self or family referral	No; laboratory normal ranges used for comparison	No	T cells (total CD4, CD8); B cells; IgG, IgA, IgM, IgE, C3, C4	T cells in normal ranges; B cells elevated in one-third; high B cells & high IgA in 4 consistent with history of infections; other tests within expected ranges

A major limitation of studies of MCS up to the present time is the fact that all but one have been performed on patients under non-exposure conditions. In order to maximize the opportunity for detecting an abnormality, it may be important to compare markers in patients before, during, and after a salient exposure. Physicians who evaluate individuals with suspected occupational asthma often have patients keep a record of their peak flow readings before, during, and after exposures at work. Some physicians perform a provocative inhalation challenge with the suspected substance. At baseline or random points in time, patients with occupational asthma may exhibit normal pulmonary function. In parallel fashion, provocative challenges may be key to detecting and diagnosing MCS.

Because adaptation could affect patients' responses, exposure challenges may need to be performed after patients have been removed from their usual background of everyday exposures, including the challenge substance itself, for a sufficient period of time that any tolerance they may have developed does not interfere with responses during testing. Again, in the case of occupational asthma it is recognized that inhalation challenges should not be conducted either too soon or too long after removal from the workplace; in the former case, tolerance may have developed, and, in the latter, sensitivity may be waning. Thus, in order to observe the most robust effect of a particular exposure, patients may need to be tested within a narrow window of time, perhaps seven to ten days after the last exposure, and in the absence of background exposures that may trigger extraneous symptoms. For this purpose, it has been proposed that patients be housed in an environmentally controlled hospital unit prior to challenges (Figure 4) (Ashford and Miller, 1991; Miller, 1992). At the conclusion of the NRC workshop on MCS, participants unanimously endorsed human challenge studies using a controlled environment, assigning this approach their highest priority for research on MCS (National Research Council, 1992).

Future research on MCS depends upon the development of a case definition for the condition. The six case definitions that have been proposed thus far differ greatly in terms of the minimum number of organ systems that must be affected (one to three); whether patients with other definable clinical or psychological conditions should be excluded; whether blinded, provocative challenges are required; and whether the illness has to have been acquired following a documented exposure (Table 6). Half of the proposed case definitions assert that symptoms in one organ system are sufficient for diagnosing the condition (Ashford/Miller, Nethercott et al., and NRC). All but two (AOEC and Cullen) agree that other definable clinical conditions, such as asthma, arthritis, vasospasm, and seizure disorder should not be excluded; the majority also agree that chemical sensitivity could be an etiology for these diagnoses, which themselves are simply descriptive clinical labels. None of the case definitions excludes psychological conditions, such as somatization disorder or depression. The AOEC and Cullen definitions exclude other clinical diagnoses, but not psychological ones; such an approach might tend to bias study populations toward those with psychological problems.

TABLE 6. Features of Proposed Research Case Definitions for MCS[1]

	Ashford/ Miller	AOEC	Clinical Ecology	Cullen	Nethercott et al.	NRC
Minimum number of organ systems that must be affected	1	3	2	2	1	1
Excludes other definable clinical conditions such as asthma, arthritis, vasospasm, seizure disorder	No	Yes	No	Yes	No	No
Excludes psychological conditions such as somatization disorder, depression	No	No	No	No	No	No
Provocative challenge required to document	Yes	No	No	No	No	No
Must be acquired in relation to a documentable environmental exposure	No	No	No	Yes	No	No

[1]Sources: Ashford and Miller, 1991; Association of Occupational and Environmental Clinics (AOEC), 1992: Clinical Ecology Journal (definition appears in each issue); Cullen, 1987; Nethercott et al., 1993; National Research Council, 1992.

The case definition proposed by Dr. Ashford and this author requires blinded, provocative challenge in a controlled environment to document chemical sensitivity. It has been our opinion that such an approach is required to define the etiology of MCS, as well as the etiology of other clinical conditions in which environmental triggers have been alleged by some, such as chronic fatigue, headaches, depression, and asthma. In our view, other case definitions prematurely exclude potential cases from study. For example, an unknown but perhaps sizeable number of patients with asthma might have bronchoconstriction and inflammation on the basis of low-level chemical exposures. We have urged that such patients not be excluded from study, and that a broader perspective be adopted, i.e., that chemical sensitivity may not be a single illness, but perhaps an etiology for multiple disorders, just as infectious agents are an etiology for meningitis, syphilis, and pneumonia which differ greatly in their clinical presentations. In other words, we have viewed chemical sensitivity as a possible new mechanism for a variety of chronic illnesses. Others argue that challenges in a controlled environment would be costly and that not everyone could be evaluated in such a specialized unit, especially given the fact that no such research facility currently is available. Finally, only one case definition requires that the condition be acquired in relation to a documentable environmental exposure (Cullen). The other definitions acknowledge that some patients report life-long illness or becoming ill following a series of less well-defined exposures over several years and recognize that such individuals may be chemically sensitive as well.

CONCLUSION

Understanding MCS is pivotal to establishing sound environmental policy. If there is a subset of the population that is especially sensitive to low-level chemical exposures, a strategy for protecting this subset must be found. If it were to be determined that certain chemical exposures can lead to MCS, then perhaps these (sensitizing) exposures could be avoided. Perhaps by preventing chemical accidents, forbidding occupancy of buildings prior to finish-out or completion, avoiding use of cholinesterase-inhibiting pesticides indoors, etc., society could protect more vulnerable individuals from becoming sensitized in the first place. It would make little sense to regulate chemicals at the parts per billion level or lower if what was required was to keep people from becoming sensitized in the first place. Indeed, by understanding the true nature of MCS and who is at risk, we may prevent unnecessary and costly overregulation of environmental exposures in the years to come.

Chemical sensitivity could be a new paradigm that has the potential to explain many chronic and costly illnesses, including fatigue, depression, headaches, and asthma, or it could be nothing at all. Not understanding MCS, we take an immense gamble. But knowledge will not come cheaply. Future studies on chemical sensitivity that involve blinded challenges in a controlled environment, that utilize brain imaging, state-of-the-art immunological testing or other sophisticated tests, and that compare adequate numbers of patients and controls, will be costly. Funding agencies will need to invest adequate sums to acquire answers in this area as they have for other diseases, such as breast cancer and AIDS. Until sufficient research funds become available, chemical sensitivity no doubt will continue to pit physician against physician, perplex policy makers, and impoverish patients and corporations alike.

REFERENCES

AMERICAN ACADEMY OF ALLERGY AND IMMUNOLOGY. EXECUTIVE COMMITTEE OF THE AMERICAN ACADEMY OF ALLERGY AND IMMUNOLOGY (1986). "Position Statements — Clinical Ecology." J. Allergy Clin. Immunol. 78(8):269–271.

AMERICAN COLLEGE OF PHYSICIANS (1989). "Position paper: Clinical Ecology." Ann. Intern. Med. 111(2):168–178.

ASHFORD, N.A. and MILLER, C.S. (1989). Chemical Sensitivity. A Report to the New Jersey State Department of Health.

ASHFORD, N.A. and MILLER, C.S. (1991). Chemical Exposures: Low Levels and High Stakes. Van Nostrand Reinhold, New York.

ASSOCIATION OF OCCUPATIONAL AND ENVIRONMENTAL CLINICS (1992). Advancing the understanding of multiple chemical sensitivity. Toxicol. and Ind. Health. 8(4):1–257.

BASCOM, R. (1989). Chemical Hypersensitivity Syndrome Study: Options for Action, a Literature Review, and a Needs Assessment. A Report to the State of Maryland Department of Environment.

BELL, I.R., MILLER, C.S., and SCHWARTZ, G.E. (1992). "An olfactory-limbic model of multiple chemical sensitivity syndrome: Possible relationships to kindling and affective spectrum disorders." Biol. Psychiatry 32:218–242.

BRODSKY, C.M. (1987). "Multiple chemical sensitivities and other 'environmental illness': A psychiatrist's view." Occup. Med.: State Art Rev. 2(4):695–704.

CALIFORNIA MEDICAL ASSOCIATION TASK FORCE ON CLINICAL ECOLOGY (1986). "Clinical ecology: a critical appraisal." West. J. Med. 144(2):239–245.

CATHEBRAS, P.J., ROBBINS, J.M., KIRMAYER, L.J., and HAYTON, B.C. (1992). "Fatigue in primary care: Prevalence, psychiatric comorbidity, illness behavior, and outcome." J. Gen. Intern. Med. 7:276–286.

CULLEN, M.R., ed. (1987). "Workers with multiple chemical sensitivities." Occup. Med.: State Art Rev. 2(4):655–806.

DAUM, S. (1983). "Nitroglycerin and alkyl nitrates." In: Environmental and Occupational Medicine (W. Rom, ed.). Little, Brown and Co., Boston. pp. 639–648.
DOTY, R.L., DEEMS, D.A., FRYE, R.E., PELBERG, R., and SHAPIRO, A. (1988). "Olfactory sensitivity, nasal resistance, and autonomic function in patients with multiple chemical sensitivities." Arch. Otolaryngol. — Head Neck Surg. 114:1422–1427.
ENVIRONMENTAL PROTECTION AGENCY (1989). Report to Congress on Indoor Air Quality, Volume II. Assessment and Control of Indoor Air Pollution.
FIEDLER, N., MACClA, C., and KIPEN, H. (1992). "Evaluation of chemically sensitive patients." J. Occup. Med. 34(5):529–538.
HACKNEY, J.D., KARUZA S.K., and LINN, W.S. (1977a). "Effects of ozone exposure in Canadians and southern Californians, evidence for adaptation?" Arch. Environ. Health 32:110–116.
HACKNEY, J.D., LINN, W.S., MOHLER, J.G., and COLLIER, C.R. (1977b). "Adaptation to short-term respiratory effects of ozone in men exposed repeatedly." J. Appl. Psychol. 43:82–85.
HUBER, P.W. (1991). Galileo's Revenge: Junk Science in the Courtroom. Basic Books, New York.
JOHNSON, A. and REA, W.J. (1989). Review of 200 cases in the Environmental Control Unit, Dallas, presented at the Seventh International Symposium on Man and His Environment in Health and Disease, February 25–26, Dallas, TX.
LEZNOFF, A. 1993. "Multiple chemical sensitivity: Myth or reality." Practical Allergy Immunol. 8(2):48–52.
MEGGS, W.J. and CLEVELAND, C.H. (1993). "Rhinolaryngoscopic examination of patients with the multiple chemical sensitivity syndrome." Arch. Environ. Health 48(1):14–18.
MILLER, C.S. (1992). "Possible models for multiple chemical sensitivity: Conceptual issues and role of the limbic system. Advancing the understanding of multiple chemical sensitivity." Association of Occupational and Environmental Clinics. Toxicol. Ind. Health 8(4):181–202.
MILLER, C.S. and MITZEL, H.C. (1994). "Chemical sensitivity attributed to pesticide exposure versus remodeling." Arch. Environ. Health. In press.
NATIONAL RESEARCH COUNCIL (NRC) (1992). Multiple Chemical Sensitivities: Addendum to Biologic Markers in Immunotoxicology. National Academy Press, Washington, D.C.
NETHERCOTT, J.R., DAVIDOFF, L.L., CURB, W.B., and ABBEY, H. (1993). "Multiple chemical sensitivities syndrome: Toward a working case definition." Arch. Environ. Health 48(1):19–26.
O'BANION, D.R. (1981). Ecological and Nutritional Treatment of Health Disorders. Charles C. Thomas, Springfield.
RANDOLPH, T.G. (1987). Environmental Medicine — Beginnings and Bibliographies of Clinical Ecology. Clinical Ecology Publications, Inc., Fort Collins, CO.
RANDOLPH, T.G. and MOSS, R.W. (1980). An Alternative Approach to Allergies. Harpeı and Row, New York.
RIIHIMAKI, V. and SAVOLAINEN, H. (1980). "Human exposure to m-xylene. Kinectics and acute effects on the central nervous system." Ann. Occup. Hyg. 23:411–432.
SELNER, J.C. (1988). "Chemical sensitivity." In: Current Therapy in Allergy, Immunology and Rheumatology (L.M. Lichtenstein and A.S. Fauci, ed.). B.C. Decker, Inc., Philadelphia, PA. pp. 48–52.
SIMON, G.E., DANIELL, W., STOCKBRIDGE, H., CLAYPOOLE, K., and ROSENSTOCK, L. (1993). "Immunologic, psychological, and neuropsychological factors in multiple chemical sensitivity." Ann. Int. Med. 19(2):97–103.
SIMON, G.E., KATON, W.J., and SPARKS, P.J. (1990). "Allergic to life: Psychological factors in environmental illness." Am. J. Psychiatry 147:901–906.
STAUDENMAYER, H. and SELNER, J.C. (1987). "Post–traumatic stress syndrome (PTSS): Escape into the environment." J. Clin. Psychology 43(1):156–157.
TERR, A.I. (1986). "Environmental illness: A clinical review of 50 cases." Arch. Intern. Med. 146:145–149.
WALLACE, L., PELLIZZARI, E.D., and GORDON, S.M. (1985). "Organic chemicals in indoor air: A review of human exposure studies and indoor air quality studies." In: Indoor Air and Human Health (R. Gammage and S. Kaye, ed.). Lewis Publishers, Chelsea, MI. pp. 361–378.

WHITE PAPER: NEUROPSYCHIATRIC ASPECTS OF SENSITIVITY TO LOW-LEVEL CHEMICALS: A NEURAL SENSITIZATION MODEL

IRIS R. BELL

Department of Psychiatry
University of Arizona Health Sciences Center
and
Tucson Veterans Affairs Medical Center
Tucson, Arizona

The present paper summarizes the proposed time-dependent sensitization (TDS) and partial limbic kindling model for illness from low-level chemicals; reviews and critiques prior studies on CNS aspects of multiple chemical sensitivity (MCS); and outlines possible experimental approaches to future studies. TDS is the progressive and persistent amplification of behavioral, neurochemical, endocrine, and/or immunological responses to repeated intermittent stimuli over time. Partial limbic kindling is a progressive and persistent lowering of the threshold for eliciting electrical afterdischarges, but not motor seizures, in certain brain structures such as amygdala and hippocampus; behavioral consequences include increased avoidant behaviors.

The focus of the paper is the controversial claim of altered sense of smell and illness from low *levels of environmental chemicals (i. e., "cacosmia"), levels that should not have any biologically harmful effects by the rules of classical neurotoxicology. A major perspective of this paper is that the phenomenology of MCS is similar to that of time-dependent sensitization (reverse tolerance) and tolerance as studied in the substance abuse literature.*

The TDS model for MCS proposes that neurobiological amplification underlies the symptoms and phenomenology of these patients, including their behavioral features of heightened affective and somatic distress. It is hypothesized that MCS patients, who are mostly women, may be individuals who sensitize to substances rapidly and to the extreme, to the point of aversive symptomatology with less complete capacity for development of tolerance. Possible parallels between MCS and TDS include: (a) initiation by single or multiple intermittent

1. Address all correspondence to: Dr. Iris R. Bell, Tucson VA Medical Center, 3601 S. 6th Avenue, MS 116A, Tucson, AZ 85723. Tel.:(602)792-1450x6392. Fax:(602)629-4632.

Toxicology and Industrial Health, Vol. 10, No. 4/5, pp. 277–312

ISSN: 0748-2337

stimuli; (b) lasting changes in subsequent reactivity to low levels of chemically unrelated substances; (c) cross-sensitization between the stressors and pharmacological agents; (d) greater vulnerability of individuals who are female, who have certain genetic characteristics, and/or who may be hyperreactive to novelty (cf. trait shyness); (e) lack of obvious differences between sensitized and unsensitized individuals at baseline without eliciting exposures; (f) bidirectionality (bipolarity) of sensitized responses; (g) both context-dependent (conditioned) and context-independent (unconditioned) amplification of responses.

To minimize variability between studies, research in this area needs (a) consensus on a working case definition of MCS or at least of cacosmia as a specific symptom in a subset of well-defined medical and psychiatric disorders; and (b) proper design of chemical challenge studies in MCS, controlling for individual differences in sensitizability and for the properties of sensitization (e.g., repeated intermittent exposure tests) and tolerance (e.g., removal from customary ambient air exposures prior to testing).

Several studies have shown that psychiatric histories in MCS include increased rates of major depression, panic and other anxiety disorders, and somatization disorder. Skeptics of MCS and cacosmia suggest that the problem is one of psychogenic symptom amplification and misattribution of psychiatric symptoms to chemicals; they dismiss MCS as the twentieth century version of "hysteria (i.e., somatization disorder)." However, no previous investigations of psychopathology in MCS patients have reported systematic assessments, including chemical exposure challenge studies, to rule out organic factors in etiology.

A number of investigators have previously proposed that TDS and/or partial limbic kindling offer a neurobehavioral model for the same psychiatric conditions found at increased rates in MCS. Furthermore, recent data suggest that psychological variables are a correlate of but not a full explanation for cacosmia. Medical histories reported by cacosmics include increased rates of physician-diagnosed irritable bowel, migraine headache, nasal allergies, hypothyroidism, and breast and ovarian cysts. Such medical histories suggest possible directions in the search for objective markers, such as functional studies of the hypothalamic-pituitary-adrenal system, endogenous opioid peptides, the thyroid axis, and breast and ovarian ultrasound combined with reproductive hormone patterns. "Hysteria" has no known causes or treatments in modern psychiatry. Systematic research on MCS may provide new insights into old questions to facilitate advances in medicine, neuropsychiatry, and clinical neuroscience.

OVERVIEW

As was well summarized by Dr. Miller (this volume), the problem of illness from low-level chemicals has received little scientific attention, but it has generated much emotional debate over the fifty years since it was first raised. Although the original controversy focused on individuals who reported illness in various settings from multiple chemicals encountered in the home or occupational environment (multiple chemical sensitivity, MCS), additional cohorts have recently emerged (Ashford and Miller, 1991). These are (a) white collar professionals who report setting-specific illness, sometimes from low-level chemicals encountered in the modern office building, i.e., the so-called "sick building syndrome" (Kreiss, 1989; Kjaergaard et al., 1991; Cone and Sult, 1992; Middaugh et al., 1992); (b) residents of contaminated communities; and (c) veterans from the Persian Gulf War. The lack of a case definition has also significantly hindered research on the problem (Cullen, 1987; National Research Council, 1992; Rest, 1992; Nethercott et al., 1993). Nonetheless, as indicated by Dr. Miller, certain themes emerge from clinical evaluations of diverse patient samples. These themes provide a starting point for consideration of possible mechanisms. This paper will focus on possible central nervous system (CNS) aspects of the most controversial form of illness from low-level chemicals, i.e., MCS, with primary emphasis on a neuropharmacological model for chemical sensitivities involving time-dependent sensitization and/or partial limbic kindling (Bell et al., 1992a, 1993a,b,c,d,e).

PHENOMENOLOGY TO EXPLAIN

To review the key features to be explained, from the perspective of individual differences, the following may be essential: (a) patients report symptoms especially in the central nervous system (e.g., difficulty concentrating, poor memory, affective symptoms, clumsiness/impaired balance, facial and/or extremity numbness and tingling), as well as in many other somatic systems; (b) on average, the prevalence is higher in women than men; (c) on average, clinical illness emerges during middle age (30–50) years old); (d) only certain individuals of a given similarly exposed group develop problems of illness from low-level chemicals; (e) the range of incitants varies between individuals, and may include not only environmental chemicals, but also alcohol and drugs, foods, and natural inhalants; (f) a given individual may experience the same symptoms from exposure to chemically different agents; (g) MCS patients often acknowledge hyperreactivity in various other sensory modalities, including noise, light, and touch. The phenomenology of MCS appears to involve a two-step process, i.e., initiation and subsequent elicitation (Ashford and Miller, 1991) If such patients are not simply misattributing the cause of their symptoms to chemicals, then clinical data suggest that any explanation of MCS must also address the following: (a) initiation of clinical-level illness may derive from a specific exposure to one or more chemical agents either at acutely high levels within a toxic range (i.e., peak exposure) or in an environment in which subtoxic levels are relatively low but fluctuate in time over repeated intermittent exposures; (b) either one or a mixture of chemical agents may initiate illness from low-level chemicals; (c) not only the inducing chemical(s), but also numerous other substances with little or no structural or pharmacological relationship to the original agent(s) subsequently elicit symptoms; and (d) low levels of chemicals that may have previously had no effect or minor irritant effects now trigger debilitating degrees of illness, often to the point of disability (Ross, 1992); (e) the same chemical may trigger different symptoms in different chemically sensitive individuals; (f) any given adverse reaction may be bipolar in nature (Randolph, 1978), with both stimulatory and withdrawal characteristics; (g) illness remits partially or fully during avoidance of chemical exposures; (h) development of tolerance (called "masking" in the MCS literature) during chronic exposures may obscure evidence of established sensitivity to a given agent. In essence, MCS patients appear to be amplifiers of exogenous and endogenous stimuli; they report that environmental chemicals have initiated and continue to elicit such amplification.

POSSIBLE MODELS FOR ILLNESS FROM LOW-LEVEL CHEMICALS

On the one hand, such observations make it difficult to account for illness from low-level chemicals within any traditionally recognized mechanisms such as atopic allergy, pharmacological actions, or dose-response toxicology (Terr, 1989; Simon et al., 1993). On the other hand, systematic review of the ideas that have been proposed to explain low-level reactivity reveals a finite set of alternative hypotheses, grouped by category: (a) psychogenic factors (Schottenfeld and Cullen, 1985; Bolla-Wilson et al., 1988; Black et al., 1990; Simon et al., 1990, 1993; Staudenmayer and Selner, 1990; Staudenmayer et al., 1993), i.e., misattribution, mass hysteria, classical conditioning, psychogenic amplification resulting from depression, panic disorder, somatization disorder, childhood abuse, or post–traumatic stress disorder, delusional thinking; (b) neurogenic factors (Bell et al., 1992a, 1993c), i.e., direct neurotoxic damage (Kaplan et al., 1993), partial limbic kindling, time-dependent sensitization, peripheral neurogenic inflammation (Meggs, 1993) and/or vasculitis (Heuser et al., 1993), (c) inborn or acquired impairments of detoxification pathways (Rosenthal and Cameron, 1991; Overstreet 1993; Williams, 1993); and (d) nonatopic immune dysfunction (Thrasher et al., 1989). Leading skeptics of MCS have largely rejected certain possible explanations, especially immunological ones (Simon et al., 1993). Many skeptics appear to assume that low levels of chemical are inherently nontoxic to all individuals and thus incapable of causing biologically based illness, and/or to overlook the fact that patients give histories of symptom flares during exposures, not at rest. Systematic definitive studies to test each of the above possibilities, including misattribution, have not been published. The most common flaw in existing studies is reliance on assessments at rest, without properly designed blinded chemical challenges. It should be emphasized that the present paper focuses on the question of illness from low-level chemicals; the additional controversies surrounding nonmainstream treatment methods for MCS are a separate, though equally complex, issue.

Although patients who report illness from low-level chemicals often indicate multiple symptoms in multiple somatic systems, the central nervous system is the most commonly identified as dysfunctional in terms of cognitive, affective, and/or sensory symptoms. To date, skeptics of the possibility of low-dose reactivity have produced strong arguments against the two most obvious possible mechanisms; i.e., (a) classical toxicology and neurotoxicology require a dose-response relationship such that it is higher, not lower, doses that induce more toxic responses; and (b) immunology-allergy requires identification of consistently activated or impaired elements of the immune system (e.g., specific immunoglobulin E to specific allergens). The ability of structurally and immunologically diverse agents purportedly to trigger symptoms in sensitive patients and the lack of reliable findings of immune system involvement in such persons cast doubt on a primary role for the immune system in chemical sensitivity. Moreover, even if classical toxicological or immunological mechanisms were found, it is not obvious how these could lead to the polysymptomatic conditions with prominent central nervous system features, but without consistently identified abnormal laboratory tests.

This paper addresses a rational hypothesis that could accommodate much of the existing literature on MCS patients as well as the clinical claims summarized above. The central perspective is that illness from low-level chemicals involves individual differences in the well-documented pharmacological processes of sensitization and tolerance to exogenous substances (Bell et al., 1993a,b,c,d,e, 1994a,b,c). Much of the research that pertains to MCS derives from the substance abuse literatures on stimulants, opiates, alcohol, and stress (Antelman and Eichler, 1979; McComb and Goldstein, 1979; Antelman et al., 1980, 1988, 1991, 1992a,b; Kalivas et al., 1986, 1993; Robinson and Becker, 1986; Antelman, 1988; Cabib and Puglisi-Allegra, 1991; Newlin and Thomson, 1991; Snyder-Keller, 1991; Sorg and Kalivas, 1991; Hooks et al., 1992; Hunt and Lands, 1992; Khanna et al., 1992; Post, 1992; Karler et al., 1993; Kokka et al., 1993). The purpose of this paper is to outline the rationale and possible experimental tests for the hypotheses that (a) sensitization of the nervous system and perhaps peripheral organ systems underlies reactivity to low-level chemicals (Bell et al., 1992a, 1993a,b,c,d,e, 1994a,b,c); and (b) tolerance to chronic exposures confuses or obscures evidence of sensitization (Ashford and Miller, 1991; Newlin and Thomson, 1991). Toward this objective, the paper will summarize the neural sensitization model, critique previous research studies relevant to psychogenic and neurogenic hypotheses, review some of the author's preliminary data in support of the model, and indicate future research considerations.

THE CNS/LIMBIC SYSTEM SENSITIZATION MODEL

For purposes of discussion, time-dependent sensitization (TDS) is defined as the progressive increase in behavioral, neurochemical, endocrine, and/or immunological responses to a foreign substance or stressor by the passage of time between successive exposures (Antelman, 1988). Partial limbic kindling is defined as a special type of TDS, initiated by repeated intermittent low-level electrical or perhaps chemical stimuli, involving development of a persistent ability to elicit afterdischarges without motor seizures in limbic system pathways; electrophysiological and behavioral endpoints provide measures of partial kindling (Adamec, 1990, 1993). Table 1 lists a number of the possible overlapping phenomena between MCS and TDS (Bell et al., 1993a,b,c,d,e, 1994a,b,c). The leading caveat in considering these overlaps is that MCS is a controversial clinical phenomenon that some skeptics doubt exists as a discrete syndrome, whereas TDS is a well-replicated preclinical process documented mainly in animal studies. Despite the obvious risks of extrapolating between human and animal phenomenology, let alone of confronting the unresolved controversies surrounding MCS, the apparent similarities between MCS and TDS are sufficiently compelling to support a TDS model for illness from low-level chemicals. In postulating a long-needed theoretical framework for MCS, we can use the TDS model to generate specific, testable hypotheses for both clinical and animal research studies. The model also leads to specific guidelines for proper experimental design in testing for chemical sensitivity; as such, this conceptualization provides us with a more systematic approach to resolving controversies surrounding MCS than has been previously available.

TABLE 1. Possible Overlaps between MCS and Cacosmia and Time-Dependent Sensitization Phenomenology

Process	
	two-step process: initiation, then elicitation
	initiation by single event/exposure
	initiation by repeated intermittent events/exposures
	response amplification grows over time without exposures
	response amplification persists over time
	multiple target organs/systems involved, especially CNS
	sensitized and nonsensitized individuals differ overtly only during eliciting exposures
	amplified responses may be increases or decreases compared with unsensitized
	context-dependent (conditioned) and context-independent (unconditioned) elicitation
Agents	
	pharmacological and nonpharmacological (stress) stimuli can initiate and elicit responses
	low doses can elicit amplified response
	agents chemically unrelated to initiator can elicit amplified response
	pesticides reportedly initiate or elicit amplified responses
	ethanol reportedly initiates or elicits amplified responses
Individual Differences	
	females more susceptible
	genetic differences in susceptibility
	hyperreactivity to novelty associated with susceptibility

MCS reviews: Randolph, 1978; Ashford and Miller, 1991; Ross, 1992. TDS reviews: Antelman, 1988; Robinson and Becker, 1986; Kalivas et al., 1993

The list in Table 1 includes aspects of the sensitization process, properties of agents initiating or eliciting amplified responses in TDS, and individual differences in sensitizability. For example, both MCS and TDS involve initiation and subsequent elicitation steps. MCS patients usually report fairly normal or only slightly impaired tolerance of chemicals, alcohol, drugs, and foods for most of their lives before the initiation of their chronic poor health by some type of chemical exposure and/or stressful life event (Black et al., 1990). Antelman (1988) has repeatedly observed the ability of multiple chemically unrelated agents (e.g., cocaine, amphetamine, amitriptyline, haloperidol, diazepam, opiates, ethanol, clonidine) and nonpharmacological stressors (e.g., needle jab, tail pinch, extremely novel cage) to initiate TDS. As a result, he has proposed that it may be the foreignness and/or novelty of a stimulus combined with the passage of time after the initiating event, not necessarily the pharmacological properties, that is the necessary factor for sensitization (Antelman, 1988; Antelman et al., 1992a,b). Such a process would answer the concern of classical toxicologists that the purported triggers for MCS reactions are "nontoxic." To the nonsensitized individual, low doses are indeed nontoxic; but, to the sensitized person, the response amplification might make seemingly "low" doses quite toxic.

The relative interchangeability of drugs and stress in TDS (Antelman et al., 1980; Kalivas et al., 1986; Snyder-Keller, 1991) may allow us to accomodate observations of increased life stress in MCS patients without needing to conclude that finding evidence of stress rules out chemicals as bona fide candidates for eliciting adverse reactions. Rather, a TDS model predicts that finding either a chemical exposure event or a stressful life event at the initiation point of illness in an MCS patient increases, not decreases, the likelihood of future amplified reactivity to both chemicals and stress (Bell et al., 1994a,b,c). The gender difference that is marked in MCS (Ashford and Miller, 1991) is also mirrored in the greater susceptibility of female animals to TDS (Robinson et al., 1982; Peris et al., 1991).

In environmental health research, a history of a peak exposure, i.e., a single episode of a toxic exposure requiring medical treatment, was a predictor of persistent cognitive deficits after 16 months for solvent-exposed workers (Morrow et al., 1991a) and after two years in organophosphate exposed workers (Rosenstock et al., 1991). The peak exposure history was more relevant than total cumulative exposure. Taken together, these studies indicate that CNS function, especially of regions involved in attention and memory, might undergo lasting changes following one episode of high-level exposure. While preliminary, this research offers guidance for the development of TDS-based animal models for MCS. In addition, the animal literature indicates that multiple episodes of repeated lower level (Hooks et al., 1992) or subtoxic level (Burchfiel et al., 1976; Burchfiel and Duffy, 1982) exposures to specific substances may initiate TDS. Morrow (1993) also has reported preliminary evidence of increased autonomic reactivity to a challenging cognitive task on six-month follow-up in workers with chronic, subtoxic occupational solvent exposures compared with nonexposed controls. In other studies, she and colleagues have observed that up to 60% of solvent-exposed workers also reported cacosmia; and cacosmia was the primary variable that predicted performance on neurophsychological tests of learning and memory (Ryan et al., 1988; Morrow et al., 1990). However, only some MCS patients can identify a specific toxic exposure that initiated their illness to low-level chemicals. Skeptics of MCS also note that solvent-exposed workers have more chronic exposures at higher doses and more fixed cognitive deficits than do many MCS patients.

Certain environmental chemicals, such as various pesticides, can induce not only TDS, but also facilitate full limbic kindling by electrical stimuli (Joy, 1982; Gilbert, 1992). Kindling is the ability of an initially subthreshold stimulus without convulsant properties to elicit a tonic-clonic seizure on subsequent reexposure following repeated intermittent exposures to the stimulus (e.g., daily for 10–14 days.) Kindling appears to be permanent, or at least long-term, although animal testing has assessed kindled animals only up to one year. Limbic structures, such as the olfactory bulb and amygdala, are especially vulnerable to kindling. Although not directly demonstrated in humans to date, electrical kindling is a mechanism invoked as a possible explanation of development of additional irritative foci in temporal lobe epileptics over time. Chemical kindling by cocaine has been invoked as a possible explanation of newly acquired panic disorder in cocaine abusers without genetic risk factors or premorbid anxiety disorder. Dager et al. (1987) originally proposed limbic kindling as a possible explanation for newly acquired panic disorder in workers with low-level solvent exposures.

Agents that can initiate or facilitate kindling include stimulant drugs such as cocaine and amphetamine, endogenous opioid peptides such as beta-endorphin or enkephalin (Cain and Corcoran, 1985), local anesthetics such as lidocaine (Post et al., 1984), cholinergic agonists such as physotigmine (Girgis, 1986) (cf. pyridostigmine used in American troops to prepare for the effects of possible organophosphate chemical warfare agents in the Gulf War), GABA antagonists (Adamec, 1993), one-time exposure to corticotropin releasing hormone (Weiss et al., 1986), and pesticides such as dieldrin and lindane (Joy, 1982; Gilbert, 1992). As pointed out above, removal of the olfactory bulbs in animals additionally facilitates kindling of the amygdala. Once kindling has been induced in an animal by any means, it may be susceptible to seizures from stimuli such as otherwise nonconvulsant doses of the opiate drug morphine (opioids) that have no effect on unkindled animals. Physical stressors, such as handling, that have no inherent convulsant effect may also elicit seizures in previously kindled animals (Cain and Corcoran, 1985). Kindling of the right centromedial amygdala in rats increases restraint-stress–induced stomach ulcers (Henke and Sullivan, 1985). If cacosmia were to reflect amygdala dysfunction in part (i.e., in view of the amygdala's role in processing olfactory information), then cacosmics and/or MCS patients could have increased prevalence of peptic ulcers, among other disorders. We have found increased lifetime self-reported prevalence of ulcers in a recent college student survey of nonpatient cacosmics vs. noncacosmics (8% vs. 2%, $p = 0.03$) (Bell et al., 1994a).

STUDIES OF SENSITIZATION TO ALCOHOL AND PESTICIDES

The only direct laboratory study of initiation of xenobiotic sensitization in humans was performed by Newlin and Thomson (1991). They gave nonalcoholic sons of alcoholics a moderate dose of alcohol during three identical sessions occurring over a two-week period. They found progressive increases over time (i.e., sensitization) between sessions in finger pulse amplitude and lack of development of chronic tolerance in skin conductance and finger temperature in the sons of alcoholics, in contrast with the autonomic changes in sons of nonalcoholics. This study demonstrated individual differences in sensitizability of autonomic responses to a common solvent and drug, ethanol, in persons with genetic vulnerability to substance abuse. However, no investigations have reported the results of systematic repeated exposures to low-level chemicals over time in the same laboratory environment, comparing MCS patients and controls on progressive changes in responses of CNS, autonomic, endocrine, immune, or other systems.

Is there evidence for chemically induced sensitization phenomena in the CNS of animals? Yes. In 1976, Bokina et al. reported that a variety of common air pollutants such as acetone, benzene, ammonia, formaldehyde, and ozone had adverse effects on the olfactory bulb and amygdala. They stated, "a study of sensitization during repeated exposure reveals that, when animals have been subjected to an initial exposure to high concentrations of chemical substances (e.g., ozone at 4–6 mg/m^3), the use of those same substances in low concentrations (e.g., ozone at 0.03 mg/m^3), combined with a functional load (rhythmic light), provokes paroxysmal activity in the olfactory analyzer structures (olfactory bulb and corticomedial amygdala) . . . The results also indicate that the effect on functions of the CNS of the investigated compounds, though they differ in chemical structure and physical and chemical properties, is alike in principle. . ." They found that 20-minute exposures were optimal for inducing the EEG changes they observed.

In a different series of animal studies, Burchfiel and colleagues (Burchfiel et al., 1976; Burchfiel and Duffy, 1982) examined the effects of one toxic dose (producing motor seizures) or multiple (ten) subtoxic (producing few or no signs of poisoning) weekly doses of an organophosphate (sarin) and a chlorinated hydrocarbon pesticide (dieldrin) on rhesus monkey spectral EEG. Of note, sarin is the "nerve gas" agent that some have claimed was detected in low levels during the Persian Gulf War. For the single large dose of sarin and dieldrin, they found greater EEG beta activity while awake in darkness in the chemically treated animals vs. the diluent-treated controls at 24 hours and one year later at occipital-temporal sites. Similarly, for the repeated subtoxic dose regimen with sarin, they observed increased relative beta activity while awake in darkness and while drowsy at both occipital-temporal and frontal-central sites.

Thus, the literature on TDS and kindling makes it plausible that MCS might involve neural sensitization, but direct tests of the model are needed. In preparation for further research, it will be useful to review the clinical studies of MCS patients and nonpatient cacosmics for information to guide the design of sensitization studies.

REVIEW AND CRITIQUE OF PREVIOUS NEUROPSYCHIATRIC STUDIES

Doty et al.

Table 2 summarizes the past studies involving the CNS in which heightened sensitivity to environmental chemicals is a feature of subject selection (Doty et al., 1988; Black et al., 1990; Simon et al., 1990,1993; Staudenmayer and Selner, 1990). Although previous clinical observers had published case series on MCS purporting to demonstrate or disprove the existence of illness from low-level chemicals over the years (Randolph, 1978), Doty et al. (1988) published the first laboratory study of MCS patients. They recruited 18 MCS patients

TABLE 2. Controlled Neuropsychiatric Studies in MCS

First author (year)	Chem test used	Tolerance removed	Controls	Major findings
Doty (1988)	yes	no	normals	normal olfactory thresholds; higher Beck depression inventory; increased nasal resistance and respiratory rate; no reported major changes after chemicals
Black (1990)	no	no	psychiatric normals	increased mood, anxiety, somatization disorders in MCS
Simon (1990)	no	no	workers without MCS	increased past, not current, anxiety or depression dx in MCS; higher SCL-90-R somatization, depression; trend anxiety; Barsky Amplification
Simon (1993)	no	no	chronic musculoskeletal patients	increased current, not past, anxiety or depression dx in MCS; increased current and past somatic symptoms in MCS; higher SCL-90-R somatization, depression, anxiety; no major neuropsychological deficits in MCS
Staudenmayer (1990)	data not given	no	mixed psych. patients and normals	increased parietal EEG beta in MCS; increased scalp EMG activity
Staudenmayer (1993)	data not given	no	mixed psych. patients	increased childhood abuse histories in MCS
Bell (1994c)	no	no	healthy cacosmics and normals	increased lifetime anxiety and depression dx in MCS; higher scores on SCL-90–somatization, obsessiveness, depression, anxiety, phobic anxiety; Barsky Amplification in MCS; increased rhinitis, bronchitis, migraine, irritable bowel, arthritis, CFS, low adrenal, candidiasis, ovarian cysts, menstrual dysfunctions in MCS; trend toward more hypothyroidism, osteoporosis, fibrocystic breast disease in MCS; no excess family histories of psychiatric dx in MCS; trend toward more family histories of epilepsy in MCS
T. Simon (1992; abstr/SPECT)	data not given	data not given	depressives and normals	increased diffuse cortical defects
Heuser (1993; abstr/SPECT)	data not given	data not given	depressives and CFS	increased diffuse cortical defects especially dorsofrontal, temporal, parietal, R worse than L (vs. orb–fr and temporal in depressives)

from patient newsletters and physician referrals, as well as 18 normal controls; in addition to clinical history of long-standing health problems and cacosmia, the MCS group scored higher than controls on the Randolph Environmental Questionnaire, a self-report measure of adverse somatic reactivity to multiple chemicals. Potential subjects who gave a history suggestive of psychiatric or other medical problems that might account for their clinical presentation were excluded from this study. Most MCS patients in this study did report symptoms in the CNS (89%), respiratory system (67%), and gastrointestinal system (67%). As in clinical series and later studies, the MCS cases were well educated, and mostly women (67%). Cases and controls were compared for olfactory detection thresholds with phenyl ethyl alcohol (PEA) and methyl ethyl ketone (MEK); nasal resistance by anterior rhinomanometry; blood pressure, heart rate, and respiratory rate; and Beck Depression Inventory. They found no differences between MCS and controls for olfactory thresholds or blood pressures, but MCS patients had higher total nasal resistances on inhalation and exhalation before and after the olfactory stimuli than controls. Over trials, MCS women showed a decrease in exhalation nasal resistance to PEA vs. controls, a difference not seen in men. MCS patients had higher respiratory rates than did controls, but groups did not differ in changes of respiratory rate after olfactory stimuli. They found no average heart rate differences between groups, but did observe a trend toward a decrease in heart rates of the control, but not the MCS, women, after exposure to MEK. Finally, the Doty et al. study noted significantly higher Beck Depression scores in MCS patients than in the controls, although the means for both groups fell within the range of normal mood (means of 7 vs. 2).

This study showed that MCS patients have objectively documented increases in nasal resistance and respiratory rate, but no differences in olfactory threshold, as well as greater, though subclinical, depression. The analyses did not include comparisons of the olfactory thresholds after controlling for the nasal and respiratory findings. If nasal patency had been equivalent between the MCS and control subjects (cf. Meggs and Cleveland, 1993), it is possible that the MCS patients might then have shown a lower olfactory detection threshold. The study did not report any blinding of exposures, but did not find dramatic increases in respiratory rate or other autonomic symptoms that might suggest acute anxiety or panic reactions to the exposures. While the Doty et al. study had many rigorous features, it did not evaluate or control for chronic tolerance to ambient chemical exposures and did not perform the testing over multiple sessions to test for sensitization processes. Nevertheless, the subtle trend toward lack of heart rate habituation in the MCS women over MEK trials may also be some evidence of either a mildly sensitized response or a failure of developing tolerance (cf. Newlin and Thomson, 1991). The Doty data showing trial effects might be consistent in part with previously established TDS in the MCS patients.

Another observation that distinguishes MCS from simple, predictable placebo effect or classical conditioning is that patients may undergo bidirectional changes in symptoms during a given reaction. Thus, the same person might show marked arousal, e.g., anxiety or irritability, or marked sleepiness to the same stimulus at different times after the beginning of a given reaction or on different days during separate reactions to the same stimulus

(Randolph, 1978). Similarly, Doty et al. (1988) observed decreased rather than the expected increased nasal resistance to PEA in the MCS women. In TDS research, Antelman et al. (1991) have shown that time-sensitized reactions can occur in either direction (bipolar, increase or decrease) dependent on prior history of sensitizing exposures. While the bipolarity concept itself in preclinical sensitization research requires additional replication and will require much more study, it might explain some of the variability in clinical studies of specific reactions in MCS patients. Future research on MCS may need to allow for the possibility of bidirectional outcomes in dependent measures or risk Type II error.

Black et al.

Black et al. (1990) performed the first controlled psychiatric study of MCS patients. They recruited 26 patients who had received a diagnosis of MCS (Environmental Illness, EI) from a physician specializing in such disorders. Subjects were recruited from MCS support groups and health food stores as well as university psychiatry and occupational medicine clinics. The 33 controls were recruited as normals in a family study of 13 obsessive-compulsive disorder patients. They found increased rates of major depression (30% vs. 7%), various anxiety disorders (especially panic and simple phobia) (43% vs. 17%), and somatization disorder (17% vs. 0%) in the MCS group. Overall, 65% of the MCS patients had lifetime psychiatric diagnoses vs. 28% of the controls. The MCS patients were middle-aged married women (mean age 49, SD 13.88%) with some college education, in white collar and teaching occupations. The MCS patients claimed initiation of their illness by "fumes" at work or home (50%), pesticides/insecticides (27%), oral contraceptives/pregnancy/hysterectomy (19%), psychological stress (15%), and antibiotics (12%). Black et al. commented that many patients had histories of multiple somatic complaints but did not meet criteria for somatization disorder because of age of onset after age 30 or insufficient numbers of symptoms. The leading conclusion was that "commonly recognized psychiatric disorders could explain some or all of their symptoms."

The Black study was limited by small sample size and by potential recruitment bias issues for both patients and controls. Once a clinical diagnosis is suspected, DSM-III-R diagnostic guidelines include the expectation to rule out possible organic etiologies. The Black et al. paper did not present any clinical data to rule out nor did it evaluate the possibility that chemicals or other medical pathology (e.g., hypothyroidism, nutritional deficiencies, medications) might have been organic factors in the psychiatric findings, especially given the relatively late age of onset in many patients. Later age of onset is often a clinical indicator of organic factors in etiology of psychiatric disorders. The authors also offered no explanation for etiology in the 35% of MCS subjects without evidence of psychopathology.

Simon et al.

Simon et al. (1990, 1993) have published two well known, but controversial psychiatric studies on MCS patients. In the first (1990), they compared aerospace industry plastics workers who did (n = 13) and did not (n = 23) develop MCS symptoms following introduction of new manufacturing materials. They found no group differences in current psychiatric diagnoses determined by Diagnostic Interview Schedule methods, but did find higher rates of past major depression and panic disorder diagnoses in MCS. They also reported no cases of somatization disorder, but did find increased scores on the Symptom Checklist 90 (revised) somatization subscale, the Barsky Amplification Scale of distress from physical symptoms, and other markers of past subclinical somatization-like histories in the MCS patients.

They acknowledged that their small sample and potential selection bias of using workers involved in disability and litigation claims could have influenced the results. Nonetheless, they concluded that conditioning mechanisms and psychological distress were sufficient to explain MCS from low-level chemicals. They did not discuss possible biological mechanisms for MCS other than to state that the subtoxic levels of exposures ruled out classical neurotoxicological mechanisms. They did not integrate their findings into the large body of literature of research on biological mechanisms of depression and panic that might indicate a biological basis for heightened effects of low-level chemicals in patients with depression or panic histories. For example, animal models for depression that respond to antidepressant medications include (a) supersensitivity to cholinesterase blockers such as organophosphates (i.e., pesticides, war nerve gases such as sarin in animals tested for such neurochemical properties (Overstreet, 1993); and (b) facilitated amygdala (limbic) kindling in animals given olfactory bulb lesions (Girgis, 1986). Though still not fully established, accumulating evidence in panic disorder indicates heightened instability of autonomic respiratory regulation in response to specific triggering agents such as lactate infusions or inhaled carbon dioxide, but not simply to increased airway resistance (Papp et al., 1993). In other words, the data from Simon et al. offer potentially useful observations about a subset of MCS patients in litigation, but the findings do not test for or rule out neurobiologic mediation.

In their second study, Simon et al. (1993) recruited patients (n = 41) from the practice of a local allergist. Inclusion criteria included chronic illness (> 3 months), symptoms in at least three organ systems including the CNS; and reported sensitivity to at least 4 out of 14 common environmental chemicals. Controls (n = 34) were recruited from university-based clinics for patients with musculoskeletal injuries. No normal control group was assessed. They indicated that the control group with pain was an attempt to address criticisms of previous studies in which controls without chronic medical problems had been used to assess rates of psychiatric pathology. They did not report any effort to screen or measure their control group for presence or absence of reported sensitivity to the same 14 chemicals. As in earlier studies, they found that the MCS patients were mostly well educated women. From the CNS perspective, they found higher rates of current panic disorder (24% vs. 3%), with trends toward more major depression (29% vs. 12%) and generalized anxiety disorder (10% vs. 0%). Cases also had higher scores on the SCL-90-R subscales for depression, anxiety, and somatization. In contrast with their previous study, these case and control groups did not differ for past psychiatric diagnoses. The chemically sensitive group did report significantly more past medically unexplained somatic symptoms than did the controls (8 vs. 2). Cases reported lower rates of any current alcohol use (24% vs. 56%). On neuropsychological testing, they found somewhat poorer performance on immediate verbal recall on the Wechsler Memory Scale and a trend toward poorer performance on the Rey Auditory Verbal learning Test, but not on other measures. They reported that covarying for psychological distress eliminated the differences between cases and controls.

In their discussion, Simon et al. acknowledged that 56% of the "MCS" patients in their sample did not have current psychiatric diagnoses, and 25% had no findings of "clinically significant psychologic distress." They suggested that the psychiatric data could reflect either causes or effects of chemical sensitivity. They also noted that current dysphoria might explain the inflated rates of reporting of past somatic symptoms. Simon et al. concluded, without citing supporting evidence, that "no studies have found plausible evidence of any such effect" of ("biological injury from continued low-level chemical exposure"). As in other studies of MCS, it is uncertain whether or not they actually selected representatively ill MCS patients; for example, the "MCS" cases in the second Simon et al. study used caffeine (61%) at unusually high rates for individuals who claim marked illness from environmental exposures. Either these patients were not especially ill, a possibility suggested by the low cutoff of 4/14 chemicals in the inclusion criterion (cf. Bell et al., 1994a,b,c); and/or they may have been under the influence of extensive cross-tolerance effects of caffeine and other substances that could obscure any environmental chemical factors (Ashford and Miller, 1991). The control subjects, who were apparently not screened for the symptom of cacosmia, may or may not have differed in degree of cacosmia from the "MCS" patients (cf. Bell et al., 1993a,b,c, 1994b).

As in several others, this study did not test for the effects of chemical exposures on the outcome measures. As every proposed definition of MCS includes a requirement for symptom aggravation during exposures and relative improvement on avoidance (Cullen, 1987; Ashford and Miller, 1991; National Research Council, 1992; Rest, 1992; Nethercott et al., 1993), such studies without exposure challenge conditions cannot address the question of whether or not low levels of chemicals cause or aggravate symptoms in certain individuals. In addition to the baseline nonexposure conditions of the neuropsychological testing, this study did not employ especially sensitive measures for cognitive dysfunction in this presumably above-average intelligence sample with superior educational attainment. In contrast, Fiedler et al. (1992) used the more sensitive California Verbal Learning Test in a group of MCS patients with current but not premorbid psychiatric diagnoses and found markedly below-expected performance in 6 of 11 patients. Other standard tests of cognitive function, including several in the second Simon et al. study, did not reveal comparable rates of deficits. By analogy, Sandman et al. (1993) examined memory in patients with chronic fatigue syndrome (CFS), another controversial diagnosis that overlaps MCS. The CFS patients had normal standard neuropsychological profiles, but they nonetheless demonstrated deficits in more challenging memory tasks, e.g., tests of proactive interference, that were far worse than those of either depressed or normal control subjects.

Staudenmayer and Selner

The remaining controlled study of MCS patients was a resting baseline investigation of spectral electroencephalography (EEG) by Staudenmayer and Selner (1990). They compared 58 MCS patients; 88 mixed psychiatric patients with depression, anxiety, bipolar disorder, phobias, panic disorder, and insomnia; and 55 controls without psychological disorders but with asthma, atopic allergy, gastrointestinal disorders, dermatologic conditions, tension and migraine headaches, hypertension, and menstrual pain. Staudenmayer and Selner report that none of the "MCS" patients exhibited "objective chemical sensitivity" under controlled laboratory chemical challenges, but they did not present their methods or results. They do state that the tests were blinded and included use of "olfaction maskers" (possibly peppermint). The use of presumptively "inert" substances for placebo controls without specific pretests in the individuals with "universal" reactivity under study is a potentially unreliable procedure. An alternative methodology that circumvents the use of masking odors would have included both subolfactory threshold and supraolfactory threshold levels of each test substance, as well as confirmation of presence or absence of self-reported detection by each subject on each trial.

Without specification of outcome measures, it is also impossible to evaluate the sensitivity of their dependent measure to detect any changes in the patients; that is, computerized EEG, PET scans, and certain cognitive tests would provide a much more sensitive outcome measure than would subjective symptom reports. For example, even in normal young adults, Lorig et al. (Lorig and Schwartz, 1988; Lorig et al., 1988, 1990, 1991; Lorig, 1989) have demonstrated that consciously undetected (subolfactory threshold, without masking odors) levels of certain chemical substances, including perfume derivatives, will induce changes in computerized EEG frequencies and cognitive performance not seen during placebo, sham exposures. Moreover, it would be reasonable to assume that at least a subset of Staudenmayer and Selner's MCS patients had panic disorder, if their sample were similar to those of Black and Simon. If so, then chemical exposures, blinded or unblinded, should have produced different shifts in EEG spectral frequency patterns of certain MCS patients with panic disorder than of the controls (Locatelli et al., 1993). In any event, the Staudenmayer and Selner published data address only resting EEG status. They found greater proportions of MCS patients and mixed psychiatric patients with high right parietal beta power compared with the controls. The MCS sample also had markedly higher scalp electromyographic (EMG) activity compared with either the psychiatric or the control group. They concluded that as the MCS and psychiatric groups had similar resting EEG patterns, the MCS patients were simply psychiatric patients who misattribute their symptoms to chemicals instead of psychogenic factors.

The data from the Staudenmayer and Selner study may be useful, though their methodology and conclusions have significant limitations. First, they did not provide replicable criteria for selection of their MCS or comparison group subjects. Second, although benzodiazepines and other drugs can increase EEG beta activity, they did not specify drug histories in their subjects. Third, anxiety reportedly can increase EEG beta activity, but they did not use a standardized mood scale that might have permitted covarying for state anxiety level at the time of the EEG recording to determine whether or not anxiety was in fact the deciding factor in the beta finding. Fourth, many researchers in EEG are reluctant to assume that high beta is a true reflection of brain electrical activity because of the ability of EMG to introduce high-frequency artifacts; the exceedingly high scalp EMG in the MCS group would particularly raise this possibility. Finally, previous studies of monkeys have shown that a single toxic dose or ten repeated intermittent doses of either organophosphate or chlorinated hydrocarbon pesticides will induce a lasting increase in EEG beta activity one year after the end of the exposures (Burchfiel et al., 1976; Burchfiel and Duffy, 1982); Staudenmayer and Selner did not address the pesticide exposure histories of their subjects in any of the groups. Logically, they have shown the possibility that some MCS patients have increased parietal beta activity and/or scalp muscle tension, but not that MCS patients have functional psychiatric disorders. They have also not provided any data to indicate that chemicals or other organic factors were ruled out as possible causes of the psychiatric clinical pictures they claim to be present in the MCS or the mixed psychiatric comparison group. Doty et al. (1988), for example, reported that 22% of their MCS sample had a history of thyroid problems (see also Bell et al., 1994b); and subclinical hypothyroidism is associated with increased vulnerability to lifetime diagnoses of major depression (Haggerty et al., 1993). Furthermore, generalized resistance to thyroid hormone has a possible association with elevated rates of attention-deficit-hyperactivity disorder (ADHD) (Hauser et al., 1993); a potentially relevant finding in view of reported increased ADHD among some young adult cacosmics (Bell et al., 1994a) and MCS patients (Ashford and Miller, 1991).

Functional Neuroimaging Studies

Finally, two different groups have reported single photon emission tomography (SPECT) brain scan studies of MCS patients and controls (Simon et al., 1992; Heuser et al., 1993). Both reports appear only in abstract form, limiting ability to evaluate methodology and results. Nevertheless, the findings reported by the two studies seem similar. That is, they found diffuse cortical defects, especially in frontal (dorsofrontal), temporal, and parietal regions (see also Callender et al., 1993). The abnormalities appear bilaterally, but more over the right hemisphere. Simon et al. (1992) concluded that the patterns resembled those of substance abusers of stimulants and solvents with possible small infarctions, whereas Heuser et al. (1993) concluded that the patterns resembled those of vasculitis. Much work remains to refine subject sampling and imaging procedures in future attempts at replication and extension of these findings. If replicated, these data would indicate that MCS patients could differ in functional CNS activity from major depressives, thereby supporting the possibility of organic factors in the etiology of the neuropsychiatric symptoms. If functional lesions are present on SPECT, the nature of the underlying pathophysiology requires much more study. No investigators have reported positron emission tomography (PET) or SPECT brain scans of MCS patients during acute chemical challenges or sensitization protocols in the laboratory. One PET study has shown multiple areas of low glucose metabolism (e.g., left parietal, temporal, superior frontal regions; left amygdala and hippocampus; right putamen; right thalamus) in a worker with chronic neuropsychiatric symptoms (headaches, parosmia [cf.

cacosmia – Ryan et al., 1988], memory loss, depression, irritability) following a single peak exposure to the solvent tetrabromoethane (Morrow et al., 1991b).

Cacosmia as a Possible Feature of DSM-III-R Diagnoses

It might seem tempting to dismiss MCS as a belief system of a subset of psychiatric patients. Many skeptics do so (Brodsky, 1983; Staudenmayer et al., 1993). However, even if there are only some objectively chemically sensitive individuals within the population of MCS patients, then studying them might lead to new insights into old problems with no current solutions, e.g., the unknown cause(s) and treatment(s) of certain patients with "somatization disorder" (Smith et al., 1986; Smith, 1990). Investigating the biology of cacosmia in the subset of patients with apparent "somatization disorder" might expand our understanding of the mechanisms of that chronic illness (Bell, 1994). No studies have directly tested the hypothesis that MCS patients or cacosmics sensitize to environmental chemicals. Table 3 highlights the various neuropsychiatric diagnoses such as major depression, panic disorder, and PTSD in which a number of clinical neuroscientists have implicated possible CNS sensitization (Post and Kopanda, 1976; Post et al., 1984; Dager et al., 1987; Antelman, 1988; Friedman, 1991; Post, 1992; Kokka et al., 1993; Yehuda and Antelman, 1993). If MCS patients with true chemical sensitivity represent even a portion of individuals with such psychiatric disorders, then it is reasonable to consider TDS as a mechanism of their illnesses as well. Then the key empiric question becomes whether or not environmental chemicals in low doses can elicit symptoms in MCS or cacosmic patients who also meet clinical criteria for the psychiatric diagnoses. The preclinical data on interchangeability of stress and pharmacological agents in TDS (Antelman et al., 1980; Kalivas et al., 1986) suggest that such patients should exist.

PHENOMENOLOGICAL DIFFERENCES BETWEEN MCS AND NON–MCS PSYCHIATRIC DISORDERS

Taken together, several studies show that MCS patients may have increased psychiatric morbidity, either before or after onset of chemical sensitivity. All of the studies of MCS patients point to the central nervous system as a major site of involvement. At the same time, these studies do not provide any data to assess whether or not chemical sensitivity is a cause, an effect, or a correlate of the psychiatric problems. In many of these individuals, the data on depression, anxiety, and somatization suggest limbic system dysfunction. However, even at the clinical level, the pattern of MCS affective symptoms (Ross, 1992) is often atypical for classical psychiatric disorders. For example, MCS patients state that they become irritable, depressed, or confused for a few minutes or hours, only during days involving chemical exposures, but return to euthymia and normal cognition thereafter. They express interest in their usual activities except when "reacting" to chemicals. Many seek alternative hobbies and activities if prevented by chemical exposures from pursuing their customary interests. In contrast, most depressed patients report depressed mood most of the day, nearly every day, as

TABLE 3. Conditions in Which Time-Dependent Sensitization and/or Kindling Have Been Hypothesized

Condition	CNS sensitization hypothesized?	Increased prevalence claimed in MCS?
Temporal Lobe Epilepsy	yes	not tested
Bipolar Disorder	yes	proposed but not tested
Major depression	yes	yes
Panic Disorder	yes	yes
Post–Traumatic Stress Disorder	yes	yes
Somatization Disorder	yes	yes
Childhood abuse	yes	yes
Cocaine abuse	yes	no
Alcohol abuse	yes	no
Chemical, drug, food intolerance	yes	yes

well as a generalized loss of interests without compensatory effort. MCS symptoms include poor balance and/or clumsiness (Bell et al., 1993e), a rare presenting complaint in depression. The prior studies also demonstrate that a sizeable proportion of MCS patients do not have any psychiatric pathology in their histories as a potential cause or effect (e.g., Simon al., 1990, 1993); this implies the need for alternative, probably nonpsychogenic, explanations at least in this subset.

Age of onset may be important as well. A substantial subset of MCS patients in the various case series report onset of their multiple system symptoms after age 30 (in contrast with the diagnostic criteria for somatization disorder), making organic factors more likely (Bowler et al., 1991). While the outcome of middle-aged depressives with cognitive complaints is unstudied, convergent evidence in older depressives with reversible dementias now indicates that many such patients are biologically similar to Alzheimer's patients and deteriorate over time into irreversible dementias at an increased rate (Alexopoulos et al., 1993). Finally, many patients with psychiatric disorders have increased rates of comorbid substance abuse disorders and smoking; in contrast, chemically sensitive patients usually report marked intolerance of small amounts of beverage alcohol, recreational and therapeutic drugs, and tobacco smoke. Thus, even within the subset of MCS patients who meet criteria for psychiatric disorders, alcohol and drug sensitivity rather than tolerance or abuse distinguishes them from other psychiatric patients.

The TDS model for low-level chemical sensitivities may have additional clinical implications for neuropsychiatric dysfunction. That is, premorbid histories of an affective, anxiety, or somatization (irritable bowel) diagnosis more than double the relative risk for another neurodegenerative disorder, Parkinson's disease (PD) (Rajput et al., 1987); premorbidly, PD patients are also nonsmokers and often nondrinkers (Poewe et al., 1983). Other epidemiological evidence, though itself still controversial, indicates that PD patients may have had increased premorbid environmental toxicant exposures via rural living (Koller et al., 1990). Solvents have a particular toxicity toward dopaminergic pathways in the brain (Mutti et al., 1984; Mutti and Franchini, 1987; Pezzoli et al., 1990), a finding that may provide one starting point for studying the intersection of illness from low-level chemicals and the neuropharmacology of TDS (Kalivas et al., 1993). Changes in dopaminergic function have been associated with development of TDS in some animal studies (Robinson and Becker, 1986; Kalivas et al., 1993). While no data directly link MCS with PD at the present time (Bell et al., 1994d), even this, albeit speculative, possibility merits consideration.

BIOLOGICAL CORRELATES OF CACOSMIA IN PSYCHIATRIC PATIENTS

Our preliminary pilot data on biological correlates of cacosmia in psychiatric patients are limited, but may provide useful clues for subsequent research. We have operationally defined cacosmia as self-reported illness from various chemical odors (e.g., pesticide, paint, car exhaust, perfume, and new carpet), each item rated on a five-point Likert scale. For example, in a recent study of community elderly cacosmics and noncacosmics with and without depression, we have found more rather than less postdexamethasone (0.5 mg test) suppression of 8 A.M. cortisol levels (cacosmics: 2.8 μg/dl; noncacosmics 5.7 μg/dl, $p = 0.03$) (Bell and Amend, 1994), a finding that may be compatible with observations from studies of patients with post–traumatic stress disorder (Yehuda et al., 1993), chronic fatigue syndrome (Demitrack et al., 1991), or fibromyalgia (Griep et al., 1993), rather than of classical major depression.

In two other small series of psychiatric patients (one of geriatric depressed outpatients and one of mixed age, mixed diagnosis psychiatric inpatients), Bell et al. (in preparation) have found correlations between greater cacosmia scores and lower serum B12 ($r = -0.54$, $n = 12$, $p = 0.04$; $r = -0.78$, $n = 6$, $p = 0.03$) as well as trends towards lower serum folate ($r = -0.46$, $n = 11$, $p = 0.08$; $r = -0.51$, $n = 6$, $p = 0.15$). These observations might suggest specific metabolic differences between cacosmic and noncacosmic psychiatric patients, e.g., in methylation and related pathways involved in neurotransmitter synthesis, myelin synthesis, and xenobiotic detoxification (Bell et al., 1992b; Bottiglieri et al., 1992; Selhub et al., 1993). The cacosmia had a unique relationship to the vitamin levels per se, as cacosmia score in those studies did not correlate with overall nutritional status, as measured by serum total protein ($r = 0.04$, $n = 9$, $p = 0.5$) and serum albumin ($r = 0.2$, $n = 9$, $p = 0.3$). Table 4 summarizes the weak positive correlation in our studies between cacosmia ratings and olfactory identification ability. Notably, Kurtz et al. (1993) have shown that MCS patients and normal controls do not differ in olfactory identification ability (Doty et al., 1984), but the MCS patients rate the test odors as more intense and less pleasant. Taken together, the data suggest that the cacosmics and the MCS population deserve specific study from a biological psychiatry perspective (Bell, 1994; Bell et al., 1994b).

TABLE 4. Correlations between Cacosmia Ratings and Cain Olfactory Identification Test Scores

Sample	r	n	p
Elderly			
Community-living depressives (study 1)	0.27	29	0.08
Community-living depressives (study 2)	0.14	12	0.3
Community-living normals (health fair)	0.28	28	0.07
Young adults			
College student cacosmics & noncacosmics	0.24	65	0.03

SURVEYS OF CHEMICAL SENSITIVITY/CACOSMIA IN NONPATIENT POPULATIONS

One of the most common doubts about the existence of MCS derives in part from the extensive patient education on the purported nature of MCS done by doctors who specialize in treating such patients. Skeptics suggest that the patients are ill, but that they have simply learned from their doctors and the popular press to attribute their illness to chemicals rather than to accept a psychogenic etiology (Staudenmayer et al., 1993). While this may be a factor in the clinical populations studied to date, it may be possible to find persons who report episodic illness from chemicals in the general population but who are not patients. The psychological and medical profiles of such cacosmic nonpatients compared with normal controls should be instructive. Some of Simon and colleagues' (1990, 1993) findings on increased premorbid numbers of medically unexplained symptoms in their MCS samples suggest that it might be feasible to find cacosmic persons in the community.

Consequently, Bell et al. (1993a,b,c, 1994a,b,c) completed six surveys on over 2,000 individuals (e.g., Table 5), including five on nonpatient populations and one on MCS patients. We have found that more than half of young adults and active retired older adults in southern Arizona report at least one chemical odor that makes them feel ill. However, roughly 15% identify at least 3 to 5 chemicals (pesticide, drying paint, perfume, car exhaust, new carpet) as causing illness frequently. In contrast, 30% of the samples endorse as true the single question, "Do you consider yourself to be especially sensitive to certain chemicals?" By comparison, Wallace et al. (1991) also found that 30% of office workers in the Environmental Protection Agency answered the latter phrasing affirmatively. In four of our five surveys of nonpatients, women outnumbered men in the highly cacosmic groups compared with the noncacosmic groups. This gender difference was similar to that reported in the MCS patient studies and in the sick building syndrome literature; in TDS, female animals sensitize to drugs and stress more readily than do males. Cacosmics had generally elevated scores on most subscales of the SCL-90-R and the Barsky Amplification Scale, but did not differ from noncacosmics on scales of repressive defensiveness or sense of mastery over external events. Both MCS

TABLE 5. Comparison of Reported Physician Diagnoses and Chemical Sensitivity Ratings across Bell et al. Surveys

	Bell et al., 1994c		Bell et al., 1994b		Bell et al., 1994a	
Item	MCS patients (n = 28)	Normals (n = 20)	Cacosmics (n = 46)	Noncacosmics (n = 51)	Cacosmics (n = 131)	Noncacosmics (n = 117)
Demographics						
Age	53	52	70	77	19	19
% women	100	100	83	61	71	44
Cacosmia Index*	22.0 ± 4.5	6.9 ± 1.9	18.2 ± 3.0	5 ± 0	14.2 ± 3.2	5 ± 0
Simon Survey**	3.6 ± 0.7	0.1 ± 0.2	0.7 ± 1.2	0.1 ± 0.3	0.4 ± 0.8	0.13 ± 0.4
% with lifetime physician diagnoses						
chem. sensitivity	100	0	9	0	0	0
food sensitivity	93	31	26	4	2	0
nasal allergies	71	25	48	18	34	21
migraine headache	39	10	17	4	12	6
irritable bowel	43	10	22	4	6	3
hypothyroidism	32	5	41	10	1	0
breast cysts	39	15	47	19	8	0
ovarian cysts	32	5	16	16	6	2
depression	54	15	13	3	8	4
anxiety	39	5	13	3	6	2

* Cacosmia Index range = 5–25, based on five-point Likert scale ratings of frequency of illness from five chemicals: pesticide, paint, perfume, car exhaust, and new carpet.

** Simon Survey range = 0–4, based on true/false answers to four questions on changes in lifestyle (diet, home furnishings, clothing, shopping/restaurant use) because of chemical sensitivity.

patients and cacosmics rated the earlier decades of their lives as more stressful than did the controls. Notably, in all of the surveys, including one on MCS patients vs. controls (see below), covarying for depression, anxiety, somatization, amplification, and other psychological measures did not eliminate the difference in cacosmia scores, alcohol intolerance scores, or life stress ratings between cacosmics and controls. That is, psychological variables did not explain the symptom of cacosmia, even though cacosmia was clearly associated with psychological distress. Less than 1% of the young adult sample and only 9% of older adult cacosmics (4% of the overall sample) reported a physician diagnosis of MCS.

In young adults, base rates of psychiatric and medical diagnoses are low. Nonetheless, the data on young cacosmics indicated increased prevalence or trends toward more lifetime diagnoses of nasal allergies, sinusitis, hives, food intolerances, irritable bowel, migraine headache, ulcers, chronic pain, childhood hyperactivity in men, breast cysts and premenstrual syndrome in women, anxiety disorders, and chronic fatigue syndrome. In geriatric samples, cacosmics reported increased prevalence of nasal allergies, breast cysts in women, hypothyroidism, sinusitis, food intolerances, irritable bowel, and migraine headache. Our recent survey of MCS female patients (Bell et al., 1994c) recruited by patient newsletters and newspaper advertisements (mean number of 4.6 out of 5 chemicals causing illness on a self-report index of cacosmia and a mean of 3.6 out of 4 on the Simon lifestyle change Inventory) revealed increased rates of nasal allergies, chronic bronchitis, migraine headache, irritable bowel, arthritis, chronic fatigue syndrome, low adrenocortical function, candidiasis, ovarian cysts, irregular menstrual cycles, heavy menstrual bleeding, painful menses, chronic pelvic pain, depression, anxiety, and panic disorder, with trends toward more hypothyroidism, osteoporosis, and fibrocystic breast disease. Sixty percent of the MCS patients vs. none of the controls considered themselves occupationally disabled. Comparison groups were cacosmics and noncacosmics who rated themselves in good health. Notably, the family histories of the MCS patients included increased rates of nasal allergies and diabetes mellitus, with trends toward more candidiasis, nasal polyps, and epilepsy, but not depression, anxiety, panic disorder, or substance abuse. While requiring replication with more systematic structured interview protocols and examination of medical records, the data raise the possibility of specific overlapping medical and psychiatric patterns in ill cacosmic and MCS patients. The family history data, if replicated, support the possibility of genetic loading for organic (e.g., epilepsy-related) rather than functional psychiatric disorders.

Interestingly, in most of the above studies, the cacosmic groups were more shy on standardized scales than were the comparison groups. This finding is significant from a genetic and a neurobiological point of view. Kagan et al. (1988) have shown that approximately 15% of children are extremely inhibited (shy) from infancy. Shy children reportedly have increased rates of nasal allergies, colic, constipation, and insomnia. Stably shy children develop clinical anxiety disorders at increased rates; and panic disorder patients with major depression have the highest rates of offspring who are behaviorally inhibited. The essential nature of shyness is hyperreactivity to novelty (Kagan et al., 1988). It is striking that

(a) partial limbic kindling is Adamec's animal model for shyness (Adamec, 1990); and (b) a predictor variable for sensitizability to drugs in the TDS literature is hyperreactivity to novel physical environments (Hooks et al., 1992). Given the possibly heightened genetic and/or familial vulnerability to panic disorder in shy individuals (Rosenbaum et al., 1992), it is possible that the shyness constitutes a neurobiological risk factor in chemical sensitivity. Conversely, chemical exposures might initiate amygdala kindling, as various researchers have shown for pesticides in animals (Joy 1982; Gilbert, 1992), thereby inducing an acquired tendency toward increased shy or avoidant/defensive behaviors. The available data do not permit differentiation between shyness as a cause, a result, or a noncausal correlate of chemical sensitivity. Nevertheless, in conjunction with the animal data on hyperreactivity to novelty as a predictor of drug sensitizability, they support the potential value of studies incorporating novel stressors and habituation protocols into the experimental design of MCS studies.

The limitations of surveys are numerous; they include the subjective nature of the information and the lack of objective verification of histories. Data may be forgotten or distorted. The key findings must be viewed with skepticism until more systematic, prospective studies with objective measures can be performed. Nevertheless, the similarities of the psychiatric pictures of large samples of cacosmics recruited without clinical or litigation biases to those of smaller samples of MCS patients recruited with such biases (see above) supports the possibility of neurobehavioral dysfunction in a subset of chemically sensitive persons. The data further implicate limbic involvement in at least some individuals with cacosmia and/or MCS. However, the fact that psychological measures, including scales of somatization and psychogenic symptom amplification, do not account in full for the variance of measures of cacosmia suggests that psychiatric disorders are not sufficient to explain MCS. The fact that it is possible to find so many cacosmics in the nondisabled populations with clinical histories and psychological profiles that overlap those of MCS, but who have not received MCS diagnoses or made lifestyle changes because of chemical sensitivity, indicates that subjective illness from chemicals is unlikely to be only a belief system taught to patients by specific doctors. The data do not rule out excess disability in certain MCS patients as a consequence of iatrogenic factors, but they suggest the need in future studies to distinguish questions on the existence of cacosmia *per se* from those on (mis)attribution of poor health to chemicals (Bell et al., 1994a,b).

ADDITIONAL EVIDENCE FOR LIMBIC INVOLVEMENT

Convergent lines of evidence suggest that sensitization of the central nervous system might be the primary phenomenon underlying cacosmia and MCS. In the first place, the psychiatric diagnoses that have been found at increased rates in MCS patients are ones in which biological psychiatrists have previously postulated sensitization mechanisms, especially for the long term course (Table 3). Another psychiatric disorder that overlaps MCS (Schottenfeld and Cullen, 1985; Morrow et al., 1989), for which several clinical neuroscientists have previously proposed time-dependent sensitization and kindling mechanisms (Antelman, 1988;

Friedman, 1991; Charney et al., 1993; Pitman et al., 1993; Yehuda and Antelman, 1993), is post–traumatic stress disorder. For example, Morrow et al. (1989) have demonstrated that the Minnesota Multiphasic Personality Inventory (MMPI) psychological profiles of workers with organic solvent exposure are more similar to those of war veterans with post–traumatic stress disorder than to those of patients with chronic medical illnesses. The PTSD-like psychological profile occurred not only in workers who had recognizable, chemical accident traumatic events, but also in those who had no such history. The latter observation raises the possibility of organic, solvent- induced PTSD initiated by a primarily chemical rather than a psychogenic factor. In their cohort of MCS patients without premorbid psychiatric disorders, Fiedler et al. (1992) commented that the MMPI profiles resembled those of the solvent-exposed workers in the Morrow study. Furthermore, Levy (1988) showed that Vietnam combat veterans with medical evidence (i.e., chloracne) of Agent Orange (dioxin) exposure had higher rates of post–traumatic stress disorder and more severe symptoms than did a control group with comparable combat exposure but no medical evidence of Agent Orange. In preclinical neuroscience research and in epileptology, it is common to stimulate neural pathways with electrical or chemical stimuli in order to activate a variety of behaviors and/or "psychological" experiences that might otherwise occur naturally triggered by psychogenic input. It is a testable hypothesis that environmental chemicals might be able to induce directly or to act synergistically with life stressors and produce some of the biological features of PTSD and associated neurobehavioral dysfunctions (cf. Carlson and Rosellini, 1987).

Although Schottenfeld and Cullen (1985) proposed MCS to be an atypical, somatoform version of PTSD, none of the studies of MCS patients has found clinical evidence for criteria symptoms of post–traumatic stress disorder per se. Staudenmayer et al. (1993) have recently claimed to find symptoms of PTSD in some self-reported MCS patients who had histories of childhood abuse. However, they presented no systematic data using structured PTSD–oriented interviews or standardized scales on PTSD symptomatology on both patients and controls. Given the evidence for high rates of childhood abuse in patients with psychiatric disorders, especially somatization disorder and irritable bowel (Bell, 1994), it is not surprising that they found more frequent histories of childhood abuse in some MCS patients with multiple, medically unexplained somatic symptoms than in controls (Williams et al., 1987). Such individuals may overlap with a subset of MCS patients previously noted by Black et al. (1990) and Simon et al. (1990, 1993) to have somatization disorder.

In themselves, these data do not rule out neurobiological mediation of the pathology in such patients. For example, previous research has shown that children with histories of abuse have more electrophysiological abnormalities over frontal, temporal, and/or anterior regions than do those without abuse (Ito et al., 1993). Teicher et al. (1993) have demonstrated that adult psychiatric patients with childhood abuse histories have higher scores on a questionnaire based on tenporal lobe epilepsy symptomatology than do nonabused psychiatric patients. If abused individuals start with an unstable baseline function in limbic regions, then it would be plausible to hypothesize altered reactivity to stimuli that activate limbic regions, i.e., emotionally charged events and/or olfactory stimuli such as environmental chemicals. As stated above, some patients could have initiated their limbic instability with genetic factors, others with life stressors, others with chemicals, and still others with various combinations of genetics, stressors, and chemicals. Recent studies of temporal lobe symptomatology in psychiatric patients have utilized symptoms such as easy startle or feeling unreal (depersonalization, derealization) as clinical indicators of limbic dysfunction (Locatelli et al., 1993). In cacosmic young adults, we have found additional evidence for limbic involvement in their symptom picture; e.g., we showed higher ratings for fearfulness and startle, with a trend for more irritability, compared with noncacosmic controls, even after covarying for SCL-90-R symptom checklist scores, symptom amplification, and shyness scores (Bell et al., 1994a). In an earlier study (Bell et al., 1993e), we also found the symptom of "spaciness" in 46% of polysymptomatic patients with food intolerances (vs. 13% of normal controls), half of whom also reported chemical sensitivities. Studies in TDS would predict that a range of stressors and chemicals might later elicit symptoms, regardless of whether or not they were originally involved in the initiation process. Systematic manipulation of variables would permit clarification of these possibilities. If so, useful dependent measures for studies to test this hypothesis would include functional neuroimaging procedures such as spectral EEG, evoked potentials, and PET/SPECT scans comparing baseline after removal of tolerance with status after chemical challenges on a sensitizing schedule.

Some of the medical history data such as the ovarian cysts in MCS also support limbic involvement (Bell et al., 1994c). That is, Herzog et al. (1984) have observed that women with temporal lobe epilepsy (TLE) have an inordinately increased prevalence of comorbid menstrual cycle irregularities, dysfunctional menstrual bleeding, and polycystic ovary disease. The amygdala, which is the site of the TLE seizure focus for experiential phenomena in up to 30% of such seizures (Mesulam, 1985), has direct input into the areas of the hypothalamus that control gonadal hormone regulation. In addition, polycystic ovary disease is associated with anovulatory cycles, which in turn lead to unopposed estrogen (failure of luteal phase progesterone release). Anovulatory cycles and/or unopposed estrogen lower seizure threshold in TLE patients; in such individuals, progesterone replacement reduces the frequency of seizures (Herzog, 1986). Dysfunction of the amygdala could induce the menstrual anomalies and cyst formation in TLE, which could feed back excess estrogen and thereby favor worsening of the seizure disorder (Herzog et al., 1984). What about less severe limbic dysfunction? Orenstein (1989) reported that patients with Briquet's syndrome, the predecessor of the DSM-III-R somatization disorder, had lifetime histories of ovarian cysts at a rate of 37% and of irritable bowel at a rate of 48%. He did not report on breast cysts or other medical conditions. However, anovulatory cycles and unopposed estrogen also favor formation of breast cysts (Schurz et al., 1991), such as reported in the cacosmic and MCS women from three of our surveys (Bell et al., 1994a,b,c). Such data imply that at least some cases of somatization disorder reflect serious, though nonmotor, limbic system pathology distinct from simply subjective, psychogenic amplification of somatic symptomatology (cf. Orenstein et al., 1986). As the olfactory system has direct neuroanatomical and neurophysiological input to the amygdala and eventually hippocampus, it is conceivable that chemical stimuli at low levels could trigger limbic dysfunction in MCS patients who happen to meet descriptive criteria for somatization disorder (Bell, 1994). The key question is how low levels might end up producing toxiclike effects in the brain; again, neural sensitization could be a plausible answer (Bell et al., 1992a, 1993c).

Interactions between xenobiotics and the neuroanatomy and neurophysiology of the olfactory and limbic systems could underlie some of the findings. Inhaled and injected chemicals such as certain organic solvents or herbicides become concentrated in the olfactory mucosa, which has its own cytochrome P-450 enzyme system (Ghantous et al., 1990). Substances placed on the mucosa can find their way via axonal or transneuronal transport into the olfactory bulbs and other structures of the limbic system (Shipley, 1985). The olfactory pathways, for example, send efferents directly to the amygdala, a limbic structure involved in emotional tone and regulation of hypothalamic activity. In turn, disruption of hypothalamic activity could produce abnormalities in eating, drinking, reproductive function, immune system responses, and autonomic tone. Of interest, olfactory bulbectomy is an animal model for depression which responds to antidepressant medications (Jesberger and Richardson, 1988; van Riezen and Leonard, 1990). Olfactory bulbectomy facilitates kindling of the amygdala (Girgis, 1986) and increases hyperreactivity to novel stimuli (van Riezen and Leonard, 1990). Thus, stimuli that can alter olfactory bulb function might produce marked changes in activity of other limbic regions.

On the other hand, few MCS patients present clinically with documentable motor seizures or even temporal lobe epilepsy; MCS is probably not undiagnosed TLE. Rather, the data imply that the mechanisms of MCS might fall squarely between the phenomenology of interest to behavioral neurology (TLE) and that of interest to psychiatry (somatization disorder with irritable bowel and ovarian cysts). In parallel, partial rather than full limbic kindling might be a more reasonable hypothesis to account for some of the clinical phenomena, including the cyst formation in cacosmic and MCS women. If limbic structures such as amygdala or hippocampus were partially kindled in chemical sensitivity, then one might expect certain clinical symptoms such as lower thresholds for emotional and/or memory dysfunction generated by CNS events. The CNS site of the dysfunction could mean that peripheral tissues could show no objective evidence of abnormality, e.g., dizziness, unreality, extremity numbness and tingling (without hyperventilation). Moreover, as in temporal lobe epilepsy, chemically induced limbic-generated abnormalities would be deep in the brain and difficult to detect with surface electroencephalographic (EEG) electrodes.

Mesulam (1985) points out that the behavioral manifestations of temporolimbic seizures include sensory (e.g., "headache, focal pain, discomfort, tingling or numbness . . . dizziness or vertiginous sensations. . ."), motor (e.g., "complex automatisms, abnormal eye movements or nystagmus, starting, twitching, or jerking of upper or lower extremities . . . transient weakness or motor paresis [with] intact muscle tone or reflexes. . ."), autonomic (e.g., "flushing or a 'hot sensation' that may be bilateral or unilateral, sometimes restricted only to the head or upper torso . . . shortness of breath [without] pulmonary function abnormalities, apneic attacks, chest pain or a feeling that the heart has stopped, is racing, or is pounding hard . . . nausea or epigastric distress . . . experienced as a risking or sinking feeling 'like an elevator ride'"), hallucinations or illusions in any sensory modality, experiential (e.g., "memory flashbacks, illusions of familiarity or unfamiliarity"), emotional (e.g., "sudden and unpredictably without apparent precipitant . . . and generally ceases as abruptly as it began . . . [including] embarrassment, sadness, depression or sudden crying, explosive laughter, peacefulness, . . . or perhaps most commonly, fear . . . sudden anger or increased irritability . . .)." We need to remember that even patients diagnosed with "hysterical" or pseudoseizures at certain times, many of whom have a clinical picture of somatization disorder, also have bona fide seizures at other times. While we might prefer to reduce the debate over MCS to a dualistic set of possibilities, a CNS sensitization model suggests that a multifactorial, biopsychosocial approach may be more appropriate. The proportionate contribution of psychosocial stressors versus that of environmental chemicals to illness in a given patient (0–100%) would need comprehensive study of such factors both in the initiation of sensitization and in the elicitation of symptoms.

IMPLICATIONS OF THE MODEL FOR EXPERIMENTAL APPROACHES

The model of CNS/limbic sensitization for cacosmia and MCS begins to offer a testable series of hypotheses to accommodate existing observations and guide design of future studies. The body of research on MCS and related controversial conditions suggests a need for particular care in experimental design. A number of methodological issues emerge from the existing studies. Only Doty et al. (1988) presented any data on actual chemical exposures, but no CNS outcome measures other than olfactory threshold were used. No studies focused on time factors in outcome measures, especially multiple sessions separated by days to weeks. None addressed the issue of concomitant tolerance or novelty of the laboratory environment as potential factors in obscuring possible sensitization. For the latter considerations, it may be useful to review the differences in methodology in studies of single challenge studies with alcohol in substance abuse research. In their work on alcohol sensitization in sons of alcoholics, Newlin and Thomson (1991) have concisely summarized many such issues also of likely relevance to studies on environmental chemical challenges in MCS. These would include individual differences in current sensitivity to the test agent, acute tolerance during a session, chronic tolerance, and interactions between the response potential and the effects of novelty in the laboratory environment. Newlin's group (Newlin and Pretorius, 1991) had previously observed that responses to alcohol were greater in human subjects given alcohol after undergoing familiarization over three sessions with the laboratory setting and procedures (without prior alcohol) than in controls given alcohol on their first session (i.e., context dependence).

The research priorities (Bell et al., 1992a, 1993c) include (a) adaptation of TDS protocols to develop an animal model for initiation and perpetuation of chemical sensitivity; (b) formalization of a consensus based case definition for MCS; (c) development of clinically feasible approaches, including objective markers (e.g., endocrine challenge tests for adrenal, thyroid, and gonadal hormone axes) for identifying sensitizable individuals before and after onset of illness; (d) development of rational medical and behavioral therapies based on underlying mechanisms. Meanwhile, other studies should assess patients with well-defined psychiatric (e.g., major depression, panic disorder) and neurological disorders (e.g., TLE), who either do or do not also report cacosmia as an adjunctive symptom; chemical challenges given on a sensitizing schedule (e.g., repeated exposures, widely separated in time by days to weeks) in such samples may be the most productive while we await a usable case definition for MCS per se. Another subset of the MCS population for separate study needs to be those without lifetime or current psychopathology.

Previous studies indicate that the experimental design of animal and human studies testing for TDS must include time factors, both within a given session and over multiple sessions, spanning at least 7–14 days. Promising dependent measures for chemical challenge studies suggested by prior research include quantitative EEG (Lorig and Schwartz, 1988; Lorig et al., 1988, 1990, 1991; Lorig, 1989); visual and auditory evoked potentials (Lorig et al., 1991; Morrow et al., 1992a) as well as olfactory evoked potentials (Kobal and Hummel, 1988); high-resolution SPECT (Simon et al., 1992; Heuser et al., 1993) and PET brain scans; sensitive neuropsychological tests (Ryan et al., 1988; Morrow et al., 1992b; Sandman et al., 1993); and autonomic measures such as respiratory rate, finger pulse volume, skin conductance, and skin temperature (Doty et al., 1988; Newlin and Thomson, 1991). Standardized mood scales in human subjects and behavioral measures in animals such as responses to the plus maze anxiety paradigm, motor activity in a novel cage, and acoustic startle reactions may provide important psychological data. Lorig and colleagues' (1988, 1989, 1990, 1991) work on olfactory stimuli in normals suggests that a rational approach to blinding chemical exposures without confounds of "masking odors" might be use of consciously undetectable, subolfactory threshold doses in addition to suprathreshold doses. Olfactometer equipment permits carefully titrated exposures to give test chemicals and sham exposures (Kobal and Hummel, 1988). Environmentally controlled units are the optimal approach to allow removal of chronic tolerance to ambient chemicals in the everyday environment (Ashford and Miller, 1991). As a less desirable but necessary compromise, it may be feasible to study context-dependent sensitization in ordinary room air by mimicking the Newlin and Thomson (1991) design of inducing and testing for TDS in the same laboratory room over a period of weeks. Finally, animal studies may be able to compare those strains already recognized to be high vs. low in susceptibility to sensitization to drugs in order to test for sensitization to inhaled low-level chemicals.

Behavioral medicine research on psychogenic modifiers of clinical status and outcome is an important potential component of future studies in MCS, as it has been in many chronic illnesses such as hypertension, heart disease, cancer, diabetes, and arthritis. Especially in situations in which subtoxic doses of substances may be involved, psychological factors such as expectation, inescapable stress (Carlson and Rosellini, 1987), and/or familiarity of the environment (Newlin and Pretorius, 1991) could well play a role in modulating responses to a particular substance. Previous studies on exogenous triggers and psychological factors in panic disorder might provide options for effective experimental designs. However, before limiting our considerations solely to behavioral therapies, as several skeptics of MCS have already done (Staudenmayer et al., 1993), it is important to test the TDS model for MCS and attempt to come to a greater understanding of the full range of possible factors involved.

CONCLUSIONS

With the limited data available, the perspective presented here is necessarily speculative and synthetic. It offers a feasible starting point for organizing thinking about the complex clinical problem of illness from low-level chemicals. Extrapolating from animal studies to human clinical research is always a difficult endeavor. Nevertheless, the many apparent overlaps between animal studies on sensitization and clinical studies of MCS offer the possibility of a rational, testable model for human illness from low-level chemicals. The model provides an integration of existing data and various proposed mechanisms for MCS to address the most contested aspect of the problem, i.e., the claims of "toxicity" from low, "nontoxic" levels of xenobiotics. If we were dealing with recognized higher levels of substances, then most researchers would accept that chemicals could cause psychiatric and neuropsychiatric dysfunctions (Morrow et al., 1991a, 1992b, 1993).

Even if sensitization is a major mechanism for MCS, it is unlikely to be the sole and complete explanation of the problem; other mechanisms are undoubtedly involved (e.g., Meggs, 1993). Some skeptics have proposed dismissing MCS as merely the twentieth century version of hysteria. In the end, we may do so. For those who would appear to meet diagnostic criteria for somatization disorder (hysteria, Briquet's syndrome), we do not know the causes or cures of "hysteria" in mainstream psychiatry or medicine (Bell, 1994). Like MCS, somatization disorder has high rates of lifetime comorbid depression and panic disorders, both of which have underlying neurobiological mediation (Bell, 1994). At present, we have no data to prove or disprove a hypothesis that xenobiotics might contribute to causing certain cases of apparent "somatization" disorder (cf. Hudson et al., 1992). Chronic or recurrent debilitating symptoms in multiple systems constitute a serious and costly health care problem at both the individual and societal levels (Smith et al., 1986). Preliminary data on MCS and cacosmia suggest that the phenomenology and biology of sensitivity to low-level chemicals may differ from that of noncacosmic psychiatric disorders in interesting ways. We must still also explain the subset of MCS patients who have no psychopathology. Consequently, research on MCS and the symptom of cacosmia may open some valuable new insights into key neuropsychiatric issues with broad applicability within medicine and clinical neuroscience.

REFERENCES

ADAMEC, R.E. (1990). "Does kindling model anything clinically relevant?" Biol. Psychiat. 27:249–279.

ADAMEC, R.E. (1993). "Partial limbic kindling — brain, behavior, and the benzodiazepine receptor." Physiol. Behav. 54:531–545.

ALEXOPOULOS, G.S., MEYERS, B.S., YOUNG, R.C., MATTIS, S., and KAKUMA, T. (1993). "The course of geriatric depression with 'reversible dementia': a controlled study." Am. J. Psychiat. 150:1693–1699.

ANTELMAN, S.M. (1988). "Time-dependent sensitization as the cornerstone for a new approach to pharmacotherapy: drugs as foreign/stressful stimuli." Drug Devel. Res. 14:1–30.

ANTELMAN, S.M. and EICHLER, A.J. (1979). "Persistent effects of stress on dopamine-related behaviors: clinical implications." In: Catecholamines: Basic and Clinical Frontiers (E. Usdin, I.J. Kopin, and J. Barchas, ed.). Pergamon Press, New York. pp. 1759–1761.

ANTELMAN, S.M., EICHLER, A.J., BLACK, C.A., and KOCAN, D. (1980). "Interchangeability of stress and amphetamine in sensitization." Science 207:329–331.

ANTELMAN, S.M., KNOPF, S., KOCAN, D., EDWARDS, D.J., RITCHIE, J.C., and NEMEROFF, C.B. (1988). "One stressful event blocks multiple actions of diazepam for up to at least a month." Brain Res. 445:380–385.

ANTELMAN, S.M., CAGGIULA, A.R., KOCAN, D., KNOPF, S., MEYER, D., EDWARDS, D.J., and BARRY, H. (1991). "One experience with 'lower' or 'higher' intensity stressors, respectively, enhances or diminishes responsiveness to haloperidol weeks later: implications for understanding drug variability." Brain Res. 566:276–283.

ANTELMAN, S.M., CAGGIULA, A.R., KNOPF, S., KOCAN, D.J., and EDWARDS, D.J. (1992a). "Amphetamine or haloperidol two weeks earlier antagonized the plasma corticosterone response to amphetamine: evidence for the stressful/foreign nature of drugs." Psychopharm. 107:331–336.

ANTELMAN, S.M., KOCAN, D., KNOPF, S., EDWARDS, D.J., and CAGGIULA, A.R. (1992b). "One brief exposure to a psychological stressor induces long-lasting, time-dependent sensitization of both the cataleptic and neurochemical responses to haloperidol." Life Sci. 51:261–266.

ASHFORD, N.A. and MILLER, C.S. (1991). Chemical Exposures. Low Levels and High Stakes. Van Nostrand Reinhold, New York.

BELL, I.R. (1994). "Somatization disorder: health care costs in the decade of the brain." Biol. Psychiat. 35:81–83.

BELL, I.R. and AMEND, D. (1994) "More cortisol suppression in depressed cacosmics." Presentation at the 147th Annual meeting of the American Psychiatric Association. Philadelphia, PA, 21–26 May.

BELL, I.R., MILLER, C.S., and SCHWARTZ, G.E. (1992a). "An olfactory-limbic model of multiple chemical sensitivity syndrome: possible relationship to kindling and affective spectrum disorders." Biol. Psychiat. 32:218–242.

BELL, I.R., EDMAN, I.S., SELHUB, I., MORROW, F.D., MARBY, D.W., KAYNE, H.L., and COLE, J.O. (1992b). "Plasma homocysteine in vascular disease and nonvascular dementia of depressed elderly people." Acta Psychiatrica Scand. 86:386–390.

BELL, I.R., SCHWARTZ, G.E., PETERSON, J.M., and AMEND, D. (1993a). "Self-reported illness from chemical odors in young adults without clinical syndromes or occupational exposures." Arch. Environ. Health 48:6–13.

BELL, I.R., SCHWARTZ, G.E., PETERSON, J.M., and AMEND, D. (1993b). "Symptom and personality profiles of young adults from a college student population with self-reported illness from foods and chemicals." J. Am. Coll. Nutr. 12:693–702.

BELL, I.R., SCHWARTZ, G.E., PETERSON, J.M., AMEND, D., and STINI, W.A. (1993c). Possible time-dependent sensitization to xenobiotics: self-reported illness from chemical odors, foods, and opiate drugs in an older adult population." Arch. Environ. Health 48:315–327.

BELL, I.R., AMEND, D., KASZNIAK, A.W., and SCHWARTZ, G.E. (1993d). "Memory deficits, sensory impairment, and depression in the elderly." Lancet 341:62.

BELL, I.R., MARKLEY, E.I, KING, D.S., ASHER, S., MARBY, D., KAYNE, H., GREENWALD, M., OGAR, D.A, and MARGEN, S. (1993e). "Polysymptomatic syndromes and autonomic reactivity to nonfood stressors in individuals with self-reported adverse food reactions." J. Am. Coll. Nutr. 12:227–238.

BELL, I.R., MILLER, C.S. SCHWARTZ, G.E., PETERSON, J.M., and AMEND, D. (1994a). "Neurobehavioral and somatic characteristics of young adults with and without self-reported chemical odor intolerance and chemical sensitivity." Submitted.

BELL, I.R., SCHWARTZ, G.E., AMEND, D., PETERSON, J.M., and STINI, W.A. (1994b). "Sensitization to early life stress and response to chemical odors in older adults." Biol. Psychiat. 35:857–863.

BELL, I.R., PETERSON, J.M., and SCHWARTZ, G.E. (1994c). "Medical histories and psychological profiles of middle-aged women with and without self-reported illness from environmental chemicals." J. Clin. Psychiatry. In press.

BELL, I.R., SCHWARTZ, G.E., AMEND, D., PETERSON, J.M., KASZNIAK, A.W., and MILLER, C.S. (1994d). "Psychological characteristics and subjective intolerance for xenobiotic agents of normal young adults with trait shyness and defensiveness: a 'Parkinsonian-like' personality type?" J. Nerv. Ment. Dis. 182:367–374.

BLACK, D.W., RATHE, A., and GOLDSTEIN, R.B. (1990). "Environmental illness. A controlled study of 26 subjects with '20th Century disease'." J. Am. Med. Assoc. 264:3166–3170.

BOKINA A.I., EKSLER, N.D., SEMENENKO, A.D., and MERKURYEVA R.V. (1976). "Investigation of the mechanism of action of atmospheric pollutants on the central nervous system and comparative evaluation of methods of study." Environ. Health Perspect. 13:37–42.

BOLLA-WILSON, K., WILSON, R., and BLEECKER, M.L. (1988). "Conditioning of physical symptoms after neurotoxic exposure." J. Occup. Med. 30:684–686.

BOTTIGLIERI, T., HYLAND, K, LAUNDY, M., GODFREY, P., CARNEY, M.W.P., TOONE, B.K., and REYNOLDS, E.H. (1992). "Folate, homocysteine, and monoamine metabolism in depression." Pteridines 3:151–152.

BOWLER, R.M., MERGLER, D., RAUCH, S.S., HARRISON, R., and CONE, J. (1991). "Affective and personality disturbances among female former microelectronics workers." J. Clin. Psych. 47:41–52.

BRODSKY, C.M. (1983). "'Allergic to everything': a medical subculture." Psychosom. 24:731–742.

BURCHFIEL, J.L. and DUFFY, F.H. (1982). "Organophosphate neurotoxicity: chronic effects of sarin on the electroencephalogram of monkey and man." Neurobehav. Toxicol. Teratol. 4:767–778.

BURCHFIEL, J.L., DUFFY, F.J., and SIM, V.M. (1976). "Persistent effects of sarin and dieldrin upon the primate electroencephalogram." Toxicol. Appl. Pharm. 35:365–379.

CABIB, S. and PUGLISI-ALLEGRA, S. (1991). "Genotype-dependent effects of chronic stress on apomorphine-induced alterations of striatal and mesolimbic dopamine metabolism." Brain Res. 542:91–96.

CAIN, D.P., and CORCORAN, M.E. (1985). "Epileptiform effects of met-enkephalin, B-endorphin, and morphine: kindling of generalized seizures and potentiation of epileptiform effects by handling." Brain Res. 338:327–336.

CALLENDER, T.J., MORROW, L., SUBRAMANIAN, K., DUHON, D., and RISTOVV, M. (1993). "Three-dimensional brain metabolic imaging in patients with toxic encephalopathy." Environ. Res. 60:295–319.

CARLSON, J.N., and ROSELLINI, R.A (1987). "Exposure to low doses of the environmental chemical dieldrin causes behavioral deficits in animals prevented from coping with stress." Psychopharm. 91:122–126.

CHARNEY, D.S., DEUTCH, A.Y., KRYSTAL, J.H., SOUTHWICK, S.M., and DAVIS, M. (1993). "Psychobiologic mechanisms of post–traumatic stress disorder." Arch. Gen. Psychiat. 50:294–305.

CONE, J.E., and SULT, T.A. (1992). "Acquired intolerance to solvents following pesticide/solvent exposures in a building: a new group of workers at risk for multiple chemical sensitivity." Toxicol. Ind. Health 8:29–39.

CULLEN, M.R. (1987). "The worker with multiple chemical sensitivities: an overview." In: Occupational Medicine: State of the Art Reviews (M.R. Cullen, ed.). Hanley and Belfus, Philadelphia. pp. 655–662.

DAGER, S.R., HOLLAND, J.P. COWLEY, D.S., and DUNNER, D.L. (1987). "Panic disorder precipitated by exposure to organic solvents in the work place." Am. J. Psychiat. 144:1056–1058.

DEMITRACK, M.A, DALE, J.K, STRAUS, S.E., LAUE, L., LISTWAK, S.J., KRUESI, M.J.P., CHROUSOS, G.P., and GOLD, P.W. (1991). "Evidence for impaired activation of the hypothalamic-pituitary-adrenal axis in patients with chronic fatigue syndrome." J. Clin. Endocrin. Metab. 73:1224–1234.

DOTY, R.L., SHAMAN, P., KIMMELMAN, C.P., DANN, M.S. (1984). "University of Pennsylvania smell identification test: a rapid quantitative olfactory function test for the clinic." Laryngoscope 94:176–178.

DOTY, R.L., DEEMS, D.A, FRYE, R.E., PELBERG, R., and SHAPIRO, A. (1988). "Olfactory sensitivity, nasal resistance, and autonomic function in patients with multiple chemical sensitivities." Arch. Otolaryngol Head Neck Surg. 114:1422–1427.

FIEDLER, N., MACCIA, C., and KIPEN, H. (1992). "Evaluation of chemically sensitive patients." J. Occup. Med. 34:529–538.

FRIEDMAN, M.J. (1991) "Biological approaches to the diagnosis and treatment of post–traumatic stress disorder." J. Traum. Stress 4:67–91.

GHANTOUS, H., DENCKER, L., GABRIELSSON, J., DANIELSSON, B.R.G., and BERGMAN, K. (1990). "Accumulation and turnover of metabolites of toluene and xylene in nasal mucosa and olfactory bulb in the mouse." Pharm. Toxicol. 66:87–92.

GILBERT, M.E. (1992). "Neurotoxicants and limbic kindling." In: The Vulnerable Brain and Environmental Risks. Vol.1: Malnutrition and Hazard Assessment. (R.L. Isaacson and K.F. Jensen, ed.). Plenum Press, New York. pp. 173–192.

GIRGIS, M. (1986). "Biochemical patterns in limbic system circuitry: biochemical-electrophysiological interactions displayed by chemitrode techniques." In: The Limbic System: Functional Organization and Clinical Disorders (B.K. Doane and K.E. Livingston, ed.). Raven Press, NewYork. pp. 55–65.

GRIEP, W.N., BOERSMA, J.W., and DE KLOET, E.R. (1993). "Altered reactivity of the hypothalamic-pituitary-adrenal axis in the primary fibromyalgia syndrome." J. Rheum. 20:469–474.

HAGGERTY, J.J., STERN, R.A., MASON, G.A., and BECKWITH, T., (1993). "Subclinical hypothyroidism: a modifiable risk factor for depression?" Am. J. Psychiat. 150:508–510

HAUSER, P., ZAMETKIN, A.M., and MARTINEZ, P., (1993). "Attention deficit-hyperactivity disorder in people with generalized resistance to thyroid hormone." N. Engl. J. Med. 328:997–1001.

HENKE, P.G. and SULLIVAN, R.M. (1985). "Kindling in the amygdala and susceptibility to stress ulcers." Brain Res. Bull. 14:5–8.

HERZOG, A.G. (1986). "Intermittent progesterone therapy and frequency of complex partial seizures in women with menstrual disorders." Neurol. 36:1607–1610.

HERZOG, A.G., SEIBEL, M.M., SCHOMER, D., VAITUKAITIS, J., and GESCHWIND, N. (1984). "Temporal lobe epilepsy: an extrahypolthalamic pathogenesis for polycystic ovarian syndrome?" Neurol. 34:1389–1393.

HEUSER, G., MENA, I., GOLDSTEIN, J., THOMAS, C., and ALAMOS, F. (1993). "NeuroSPECT findings in patients exposed to neurotoxic chemicals." Clin. Nucl. Med. 18:923.

HOOKS, M.S., JONES, G.H., NEUILL, D.B., and JUSTICE, J.B. (1992). "Individual differences in amphetamine sensitization: dose-dependent effects." Pharm. Biochem. Behav. 41:203–210.

HUDSON, J.L., GOLDENBERG, D.L., POPE, H.G., KECK, P.E., and SCHLESINGER, L. (1992). "Comorbidity of fibromyalgia with medical and psychiatric disorders." Am. J. Med. 92:363–367.

HUNT, W.A. and LANDS, W.E.M. (1992). "A role for behavioral sensitization in uncontrolled ethanol intake." Alcohol 9:327–328.

ITO, Y., TEICHER, M.H., GLOD, C.A., HARPER, D., MAGNUS, E., and GELBARD, H.A. (1993). "Increased prevalence of electrophysiological abnormalities in children with psychological, physical, and sexual abuse." J. Neuropsychiat. Clin. Neurosci. 5:401–408.

JESBERGER, J.A. and RICHARDSON, J.S. (1988). "Brain output dysregulation induced by olfactory bulbectomy: an approximation in the rat of major depressive disorder in humans." Int. J. Neurosci. 38:241–265.

JOY, R.M. (1982). "Mode of action of lindane, dieldrin, and related insecticides in the central nervous system." Neurobehav. Toxicol. Teratol. 4:813–823.

KAGAN, J., REZNICK, J.S., and SNIDMAN, N. (1988). "Biological bases of childhood shyness." Science 240:167–171.

KALIVAS, P.W., RICHARDSON-CARLSON, R., and VAN ORDEN, G. (1986). "Cross-sensitization between foot shock stress and enkephalin-induced motor activity." Biol. Psychiat. 21:939–950.

KALIVAS, P.W., SORG, B.A., and HOOKS, M.S. (1993). "The pharmacology and neural circuitry of sensitization to psychostimulants." Behav. Pharm. 4:315–334.

KAPLAN, J.G., KESSLER, J., ROSENBERG, N., PACK, D., and SCHAUMBERG, H.H. (1993). "Sensory neuropathy associated with Dursban (chlorpyrifos) exposure." Neurol. 43:2193–2196.

KARLER, R., FINNEGAN, K.T., and CALDER, L.D. (1993). "Blockade of behavioral sensitization to cocaine and amphetamine by inhibitors of protein synthesis." Brain Res. 603:19–24.

KHANNA, J.M., KALANT, H., WEINER, I., and SHAH, G. (1992). "Rapid tolerance and cross-tolerance as predictors of chronic tolerance and cross-tolerance." Pharm. Biochem. Behav. 41:355–360.

KJIAERGAARD, S.K, MOLHAVE, L., and PETERSEN, O.F. (1991). "Human reactions to a mixture of indoor air volatile organic compounds." Atm. Environ. 25A:1417–1426.

KOBAL, G. and HUMMEL, C. (1988). "Cerebral chemosensory evoked potentials elicited by chemical stimulation of the human olfactory and respiratory nasal mucosa." EEC Clin. Neurophys. 71:241–250.

KOKKA, N. SAPP, D.W., TAYLOR, A.M., and OLSEN, R.W. (1993). "The kindling model of alcohol dependence: similar persistent reduction in seizure threshold to pentylenetetrazol in animals receiving chronic ethanol or chronic pentylenetetrazol." Alc. Clin. Exp. Res. 17:525–531.

KOLLER, W., VETERE-OVERFIELD, B., GRAY, C., ALEXANDER, C., CHIN, T., DOLEZAL, J., HASSANEIN, R., and TANNER, C. (1990). "Environmental risk factors in Parkinson's disease." Neurol. 40:1218–1221.

KREISS, K (1989). "The epidemiology of building-related complaints and illness." Occup. Med.: State Art Rev. 4:575–592.

KURTZ, D., WHITE, T., and BELKNAP, E. (1993). "Perceived odor intensity and hedonics in subjects with multiple chemical sensitivity." Chem. Senses 18:584.

LEVY, C.J. (1988). "Agent Orange exposure and post–traumatic stress disorder." J. Nerv. Ment. Dis. 176:242–245.

LOCATELLI, M., BELLODI, L., PERNA, G., and SCARONE, S. (1993). "EEG power modifications in panic disorder during a temporolimbic activation task: relationships with temporal lobe clinical symptomatology." J. Neuropsychiat. Clin. Neurosci. 5:409–414.

LORIG, T.S. (1989). "Human EEG and odor response." Prog. Neurobiol. 3:387–398.

LORIG, T.S. and SCHWARTZ, G.E. (1988). "Brain and odor: I. Alteration of human EEG by odor administration." Psychobiology 16:281–284.

LORIG, T.S., SCHWARTZ, G.E., HERMAN, K.B., and LANE, R.D. (1988). "Brain and odor: II. EEG activity during nose and mouth breathing." Psychobiol. 16:285–287.

LORIG, T.S., HERMAN, K.B., SCHWARTZ, G.E., and CAIN, W.S. (1990)." EEG activity during administration of low-concentration odors." Bull. Psychonom. Soc. 28:405–408.

LORIG, T.S., HUFFMAN, E., DEMARTINO, A., and DEMARCO, J. (1991). "The effects of low-concentration odors on EEG activity and behavior." J. Psychophysiol. 5:69–77.

MCCOMB, J.A. and GOLDSTEIN, D.B. (1979). "Additive physical dependence: evidence for a common mechanism in alcohol dependence." J. Pharm. Exp. Ther. 210:87–90.

MEGGS, W.J. (1993). "Neurogenic inflammation and sensitivity to environmental chemicals." Environ. Health Perspect. 101:234–238.

MEGGS, W.J. and CLEVELAND, C.H. (1993). "Rhinolaryngocopic examination of patients with the multiple chemical sensitivity syndrome." Arch. Environ. Health 48:14–18.

MESULAM, M.M. (1985). Principles of Behavioral Neurology. F.A. Davis, Philadelphia.

MIDDAUGH, D.A, PINNEY, S.M., and LINZ, D.H. (1992). "Sick building syndrome. Medical evaluation of two work forces." J. Occup. Med. 34:1197–1203.

MORROW, L.A (1993). "Exposure to organic solvents: cognitive, psychiatric, and psychophysiological correlates and consequences." Presented at the 101st Annual Meeting of the American Psychological Association, Symposium on Neuropsychotoxicology, Toronto, Ontario, Canada.

MORROW, L.A, RYAN, C.M., GOLDSTEIN, G., and HODGSON, M.J. (1989). "A distinct pattern of personality disturbance following exposure to mixtures of organic solvents." J. Occup. Med. 31:743–746.

MORROW, L.A, RYAN, C.M., HODGSON, M.J., and ROBIN, N. (1990). "Alterations in cognitive and psychological functioning after organic solvent exposure." J. Occup. Med. 32:444–450.

MORROW, L.A., RYAN, C.M., HODGSON, M.J., and ROBIN, N. (1991a). "Risk factors associated with persistence of neuropsychological deficits in persons with organic solvent exposure." J. Nerv. Ment. Dis. 179:540–545.

MORROW, L.A, CALLENDER, T.J., LOTTENBERG, S., BUCHSBAUM, M.S., HODGSON, M.J., and ROBIN, N. (1991b). "PET and neurobehavioral evidence of tetrabomoethane encephalopathy." J. Neuropsychiat. Clin. Neurosci. 2:431–435.

MORROW, L.A., STEINHAUER, S.R., and HODGSON, M.J. (1992a). "Delay in P300 latency in patients with organic solvent exposure." Arch. Neurol. 49:315–320.

MORROW, L.A, ROBIN, N., HODGSON, M.J., and KAMIS, H. (1992b). "Assessment of attention and memory efficiency in persons wtih solvent neurotoxicity." Neuropsychologia 30:911–922.

MORROW, L.A, KAMIS, H., and HODGSON, M.J. (1993). "Psychiatric symptomatology in persons with organic solvent exposure." J. Consult. Clin. Psych. 61:171–174.

MUTTI, A. and FRANCHINI, I. (1987). "Toxicity of metabolites to dopaminergic systems and the behavioral effects of organic solvents." Br. J. Ind. Med. 44:721–723.

MUTTI, A., VESCOVI, P.P., FALZOI, M., ARFINI, F., VALENTI, G., and FRANCHINI, I. (1984). "Neuroendocrine effects of styrene on occupationally exposed workers." Scand. J. Work Environ. Health 10:225–228.

NATIONAL RESEARCH COUNCIL (1992). Multiple Chemical Sensitivities. Addendum to Biologic Markers in Immunotoxicology. National Academy Press, Washington, D.C.

NETHERCOTT, J.R., DAVIDOFF, L.L., CURBOW, B., and ABBEY, H. (1993). "Multiple chemical sensitivities syndrome: toward a working case definition." Arch. Environ. Health 48:19–26.

NEWLIN, D.B. and PRETORIUS, M.B. (1991). "Prior exposures to the laboratory enhance the effect of alcohol." J. Stud. Alc. 52:470–473.

NEWLIN, D.B. and THOMSON, J.B. (1991). "Chronic tolerance and sensitization to alcohol in sons of alcoholics." Alc. Clin. Exp. Res. 15:399–405.

ORENSTEIN, H. (1989). "Briquet's syndrome in association with depression and panic: a reconceptualization of Briquet's syndrome." Am. J. Psychiat. 146:334–338.

ORENSTEIN, H., RASKIND, M.A, WYLLIE, D., RASKIND, W.H., and SOULES, M.R. (1986). "Polysymptomatic complaints and Briquet's syndrome in polycystic ovary disease." Am. J. Psychiat. 143:768–771.

OVERSTREET, D.H. (1993). "The Flinders sensitive line rats: a genetic animal model of depression." Neurosci. Biobehav. Rev. 17:51–68.

PAPP, L.A., KLEIN, D.F., MARTINEZ, J., SCHNEIER, F., COLE, R., LEIBOWITZ, M.R., HOLLANDER, E., FYER, A.J., JORDAN, F., and GORMAN, J.M. (1993). "Diagnostic and substance specificity of carbon dioxide–induced panic." Am. J. Psychiat. 150:250–257.

PERIS, J., DECAMBRE, N., COLEMAN-HARDEE, M.L., and SIMKINS, J.W. (1991). "Estradiol enhances behavioral sensitization to cocaine and amphetamine-stimulated striatal [3H] dopamine release." Brain Res. 566:255–264.

PEZZOLI, G., RICCIARDI, S., MASOTTO, C., MARIANI, C.B., and CARENZI, A. (1990). "n-Hexane induces Parkinsonism in rodents." Brain Res. 531:355–357.

PITMAN, R.K., ORR, S.P., and SHALEV, A.Y. (1993). "Once bitten, twice shy: beyond the conditioning model of PTSD." Biol. Psychiat. 33:145–146.

POEWE, W., GERSTENBRAND, F., RANSMAYR, G., and PLORER, S. (1983). "Premorbid personality of Parkinson patients." J. Neural Trans. 19(Suppl):215–224.

POST, R.M. (1992). "Transduction of psychosocial stress into the neurobiology of recurrent affective disorder." Am. J. Psychiat. 149:999–1010.

POST, R.M. and KOPANDA, R.T. (1976). "Cocaine, kindling, and psychosis." Am. J. Psychiat. 133:627–634.

POST, R.M., RUBINOW, D.R., and BALLENGER, J.C. (1984). "Conditioning, sensitization, and kindling: implications for the course of affective illness." In: Neurobiology of Mood Disorders (R.M. Post and J.C. Ballenger, ed.). Williams and Wilkins, Baltimore. pp. 432–466.

RAJPUT, A.H., OFFORD, K.P., BEARD, B., and KURLAND, L.T. (1987). "A case-control study of smoking habits, dementia, and other illnesses in idiopathic Parkinson's disease." Neurol. 37:226–232.

RANDOLPH, T.G. (1978). "Specific adaptation." Ann. Allergy 40:333–345.

REST, K.M. (1992). "Advancing understanding of multiple chemical sensitivity (MCS): overview and recommendations from an AOEC workshop." Toxicol. Ind. Health 8:1–13.

ROBINSON, T.E. and BECKER, J.B. (1986). "Enduring changes in brain and behavior produced by chronic amphetamine administration: a review and evaluation of animal models of amphetamine psychosis." Brain Res. Rev. 11:157–198.

ROBINSON, T.E., BECKER, J.B., and PRESTY, S.K (1982). "Long-term facilitation of amphetamine-induced rotational behavior and striatal dopamine release produced by a single exposure to amphetamine: sex differences." Brain Res. 253:231–241.

ROSENBAUM, J.F., BIEDERMAN, J., BOLDUC, E.A., HIRSHFELD, D.R., FARAONE, S.V., and KAGAN, J. (1992). "Comorbidity of parental anxiety disorders as risk for childhood-onset anxiety in inhibited children." Am J. Psychiat. 149:475–481.

ROSENSTOCK, L., KEIFER, M., DANIELL, W.E., McCONNELL, R., and CLAYPOOLE, K. (1991). "Chronic central nervous system effects of acute organophosphate pesticide intoxication." Lancet 338:223–227.

ROSENTHAL, N.E. and CAMERON, C.L. (1991). "Exaggerated sensitivity to an organophosphate pesticide." Am. J. Psychiat. 148:270.

ROSS, G.H. (1992). "History and clinical presentation of the chemically sensitive patient." Toxicol. J. Ind. Health 8:21–28.

RYAN, C.M., MORROW, L.A, and HODGSON, M. (1988). "Cacosmia and neurobehavioral dysfunction associated with occupational exposure to mixtures of organic solvents." Am. J. Psychiat. 145:1442–1445.

SANDMAN, C.A, BARRON, J.L., NACKOUL, K., GOLDSTEIN, J., and FIDLER, F. (1993). "Memory deficits associated with chronic fatigue immune dysfunction syndrome." Biol. Psychiat. 33:618–623.

SCHOTTENFELD, R.S. and CULLEN, M.R. (1985). "Occupation-induced post–traumatic stress disorders." Am. J. Psychiat. 142:198–202.

SCHURZ, B., SCHON, H.J., WENZL, R., KUBISTA, E., SPONA, J., HUBER, J., and WEINDLMAYRGOETTEL, M. (1991). "Breast cyst fluid concentrations of beta-endorphin, steroids, and gonadotrophins in premenopausal women with gross cystic disease." Maturitas 13:123–128.

SELHUB, J., JACQUES, P.F., WILSON, P.W.F., RUSH, D., and ROSENBERG, I.H. (1993). "Vitamin status and intake as primary determinants of homocysteinemia in an elderly population." J. Am. Med. Assoc. 270:2693–2698.

SHIPLEY, M.T. (1985). "Transport of molecules from nose to brain: transneuronal anterograde and retrograde labeling in the rat olfactory system by wheat germ agglutin-horseradish peroxidase applied to the nasal epithelium." Brain Res. Bull. 15:129–142.

SIMON, G.E., KATON, W.J., and SPARKS, P.J. (1990). "Allergic to life: psychological factors in environmental illness." Am. J. Psychiat. 147:901–906.

SIMON, G.E., DANIELL, W., STOCKBRIDGE, H., CLAYPOOLE, K., and ROSENSTOCK, L. (1993). "Immunologic, psychological, and neuropsychological factors in multiple chemical sensitivity. A controlled study." Ann. Int. Med. 19:97–103.

SIMON, T.R., HICKEY, D.C., REA, W.J., JOHNSON, A.R., and ROSS, G. (1992). "Breast implants and organic solvent exposure can be associated with abnormal cerebral SPECT studies in clinical impaired patients." Radiology 185(P):234.

SMITH, G.R. (1990). "Somatization Disorder in the Medical Setting." U.S. Dept. of Health and Human Services/National Institute of Mental Health, Rockville, MD.

SMITH, G.R., MONSON, R.A., and RAY, D.C. (1986). "Patients with multiple unexplained symptoms. Their characteristics, functional health, and health care utilization." Arch. Int. Med. 146:69–72.

SNYDER-KELLER, A.M. (1991). "Striatal c-fos induction by drugs and stress in neonatally dopamine-depleted rats given nigral transplants: importance of NMDA activation and relevance to sensitization phenomena." Exp. Neurol. 113:155–165.

SORG, B.A. and KALIVAS, P.W. (1991). "Effects of cocaine and footshock stress on extracellular dopamine levels in the ventral striatum." Brain Res. 559:29–36.

STAUDENMAYER, H. and SELNER, J.C. (1990). "Neuropsychophysiology during relaxation in generalized, universal 'allergic' reactivity to the environment: a comparison study." J. Psychosom. Res. 34:259–270.

STAUDENMAYER, H., SELNER, M.E., and SELNER, J.C. (1993). "Adult sequelae of childhood abuse presenting as environmental illness." Ann. Allergy 71:538–546.

TEICHER, M.H., GLOD, C.A., SURREY, J., and SWETT, C. (1993). "Early childhood abuse and limbic system ratings in adult psychiatric outpatients." J. Neuropsychiat. Clin. Neurosci. 5:301–306.

TERR, A.I. (1989). "Clinical ecology in the workplace." J. Occup. Med. 31:257–261.

THRASHER, J.D., MADISON, R., BROUGHTON, A., and GARD, Z. (1989). "Building-related illness and antibodies to albumin conjugates of formaldehyde, toluene diisocyanate, and trimellitic anhydride." Am. J. Ind. Med 15:187–195.

VAN RIEZEN H. and LEONARD, B.E. (1990). "Effects of psychotropic drugs on the behavior and neurochemistry of olfactory bulbectomized rats." Pharm. Ther. 47:21–34.

WALLACE, L., NELSON, C.J., KOLLANDER, M., LEADERER, B., BASCOM, R. and DUNTEMAN, G. (1991). "Indoor air quality and work environment study. Multivariate statistical analysis of health, comfort, and odor perceptions as related to personal and workplace characteristics." U.S. Environmental Protection Agency vol. 4, EPA Headquarters Buildings. Atmospheric Research and Exposure Asessment Laboratory.

WEISS, S.R.B., POST, R.M., GOLD, P.W., CHROUSOS, G., SULLIVAN., T.L., WALKER, D., and PERT, A. (1986). "CRF-induced seizures and behavior: interaction with amygdala kindling." Brain Res. 372:345–351.

WILLIAMS, A (1993). "Hypersusceptibility to chemicals: risk factors for neurological disease?" J. Neurol. Neurosurg. Psychiat. 56:943–946.

WILLIAMS, C.L., PETERSON, J.M., VILLAR, R.G., and BURKS, T.F. (1987). "Corticotropin-releasing factor directly mediates colonic responses to stress." Am. J. Physiol. 253:G582–586.

YEHUDA, R. and ANTELMAN, S.M. (1993). "Criteria for rationally evaluating animal models of post–traumatic stress disorder." Biol. Psychiat. 33:479–486.

YEHUDA, R., SOUTHWICK S.M., KRYSTAL, J.H., BREMNER, D., CHARNEY, D.S., and MASON, J.W. (1993). "Enhanced suppression of cortisol following dexamethasone administration in post–traumatic stress disorder." Am. J. Psychiat. 150:83–86.

PATIENT STATEMENT: CHEMICALLY SENSITIVE

HELEN KEPLINGER

Bethesda, Maryland

There are three areas of my experience with chemical sensitivity that may interest you. The first is the onset of this condition and what happened to me over time as I became chemically sensitive in my work place. The second is my experience with the "medical community." The third is the effect chemical sensitivity has had on my career.

ONSET

I am a lawyer with the United States Environmental Protection Agency. I have worked at EPA since I graduated from law school in 1979. About 1985, chemical exposures at the main EPA building in Washington D.C. began to have a serious impact on my professional life and on my personal life, but I get ahead of my story by telling you that at this point in my talk. However, I believe that to be a true statement, but I speak today from perfect hindsight. In 1985, my office moved into the EPA Headquarters building, Waterside Mall, after having been located elsewhere in Washington.

When I moved into Waterside Mall in 1985, my office had been freshly painted, with new drapes and new carpet. Almost from the first day there, my mouth always had a bad taste in it. A metallic taste. I kept rolls of lifesavers, and chewing gum in my desk and ate them constantly because of that taste in my mouth. I also had a headache nearly every day, and thus I began to keep giant-sized bottles of aspirin and Tylenol in my desk. Most of the secretaries also had frequent headaches, and came to my office for an aspirin.

I often flew to Chicago on business, perhaps as frequently as once a month. I began to have problems with my ears when I flew. An ear, nose, and throat specialist prescribed a decongestant that I would take a day or two before I flew and during the trip. The problem was that after I would fly into to Chicago my ears would be so stopped up that I had trouble hearing, not a good situation for a lawyer to be in during negotiations.

I was often congested during that time, all of the time. I had my hearing tested, but it was normal. Some mornings when I woke up, I could hear a whooshing sound in my ears, as if I

1. Address all correspondence to: Helen Keplinger, 5001 Nahant St., Bethesda, MD 20816.

Toxicology and Industrial Health, Vol. 10, No. 4/5, pp. 313-317

ISSN: 0748-2337

were driving up a mountain in a car and my ears were adjusting to the altitude. I also began to find on Mondays, after a week end away from the office, I was prone to losing my voice. I would go in the office on Monday morning, and my voice would fade away. I never understood what was going on, but I would sometimes be barely able to talk by mid-afternoon. Believe it or not, I rationalized that the problem was that I spent so much time on the phone, and this talking caused my voice to become hoarse, as if to say that I didn't use my voice at home over the weekend.

In 1987, my office was recarpeted. It had been newly carpeted only two years before. The first week after the new carpet was installed, glued to cement, I had such terrible headaches at work that I could barely function. I remember in the middle of that first week I had to fly to Chicago and was out of the office a day or two. The next week the headaches seemed to be less severe, but I had numerous headaches at work all the time.

After 1987, I also began to be so tired in the evening that by the time I got home and began my after-work jog, I could barely jog up the small hills in my neighborhood. I had been a regular jogger for at least 12 years, and it was depressing to me to find myself dragging through the motions of exercise, something I'd previously enjoyed. I was tired all the time. I also began to experience dizziness, especially when I would lie on the floor in the evening to do back exercises before jogging. Eventually, I discussed the tiredness and dizziness with my family doctor. He really had no suggestions, nor offered any remedy except an iron supplement.

Between 1986 and 1990 I worked as a supervisor at EPA. This meant supervising at least six other lawyers and handling my own casework and policy work as well. Bear in mind, I also had family responsibilities. My two sons were teenagers, and my spouse was traveling out of the country on business as much as a week out of every month. It was a very stressful period of my life in every respect. I was tired all the time. I also found myself forgetting things that were previously routine, like the names of people I was interviewing for a job, or the names of guests at a party.

Although I see it now with the same perfect hindsight I mentioned earlier, I saw nothing at the time that implicated my building in the problems I was having. The fatigue continued. I was depressed often. No matter how hard I worked, I seemed to be behind, whether work related or family related. I couldn't keep track of work in progress. I would read and reread documents that I needed to digest for work.

In addition, I realize now that I was so inefficient at work that everything took me much longer to do than it should have. I just couldn't remember things. I forgot commitments, I forgot things I had just read, I forgot where I parked my car, what people's phone numbers were, how to get to places where I was driving. These are things a lawyer can't even admit to herself.

Beginning in the fall of 1988, several things happened in my life which caused me further significant chemical exposures. At work, after a week at the beach in August 1988, I returned to my office to find that it had been painted again. All my books and files had been moved from the tops of file cabinets and stacked on my desk. The office was a wreck. I spent the entire day digging out my files. The next day when I awoke, my eyes were so red and swollen that I could barely see. I went to a dermatologist the next day. He assumed I'd had an allergic reaction. I had no history of allergy, and he did not suggest tests. He prescribed cortisone cream. He asked me about cosmetic use, and that sort of thing. He didn't have much interest in hearing about my office. After two weeks of using the cortisone several times a day the puffiness and redness left, but over time they returned off and on during the rest of the time I worked in Waterside Mall.

On the personal side, in October 1989, my family moved to a different house. This was something that both my spouse and I were excited about because it meant being closer to work and having more room to entertain. It also meant we would have to do renovations on the new place and paint the old place. In addition, at Waterside Mall, they were reroofing for the second time, painting, and wallpapering, all during work hours.

In September 1989, one month prior to my family's move, at Waterside Mall one morning a solvent smell was so extreme that the entire mall area of the building was evacuated at about 9:00. I'd already been at work an hour when we were told to leave the building. The following day, half of the mall employees were again told to stay out of the building. After that incident, I noticed that some evenings after 5:30 when the ventilation system was shut off, the building was so stuffy that I would have shortness of breath and dizziness.

During this same time my family moved to the other house, and we were trying to spend weekends painting inside. Within about two months, I found that I could not tolerate paint at all and we stopped all renovation and agreed that the following summer we would hire someone to finish the work. In the meantime, I found that I became nauseated driving behind a bus, or riding in one. Perfume on people near me seemed so disagreeable to me that I'd move away from them. I found myself having recurring dreams of smothering. I started having nose bleeds, and a twitching in my left eye.

By now you may wonder how I could have just stood by and watched my life deteriorate like this. How about a reality check? By this time in my life, my two sons had graduated from high school or college. I had finally gotten myself into a management position in order to be promoted. My spouse was increasingly successful in his job, but I was struggling, struggling, struggling just to keep my head above water, at a time when I should have been able to enjoy life and take it easier. I seemed to be losing ground on all fronts. I couldn't find my car when I shopped alone. I would leave my office at work and have to tell myself several times where I was going. I was depressed all the time. I lost my temper at trivial things. There seemed to be more jobs to do than people to do them, and as a female manager in a male-dominated profession, I felt that I had to do it all. If I couldn't delegate work, I'd do it myself. But because my memory and concentration were so impaired, I was less efficient all the time. I worked longer hours day in and day out. I often ate lunch at my desk.

In March, 1990, I finally asked my supervisor if I could work at home for a few days. I had been sick with a recurrent respiratory infection for six months. I had taken three different antibiotics, and I was no better. I could no longer stand to pump gas in my car, walk through the detergent aisle at the super market, or read the newspaper. Fabric stores made me sick, malls made me sick, there was almost no place where I found any relief, even my home.

My family physician wrote my supervisor a letter, advising him that it appeared that I was suffering from something related to my building. On the strength of that letter, I continue to work at home. That summer, we finally hired painters to finish painting in the house in July of 1990, when I had a week's litigation training in Colorado. When I returned home, I could not tolerate the paint in the house. I slept on the screened porch from July to October of that year, until it got so cold that the dog and I moved back inside. I consulted an Occupational Health specialist that December, and began to learn additional ways to isolate myself from the things that caused me to have the most problems. Of course isolation is just that, isolation. Not particularly compatible with modern life and career.

Now, it's been more than four years since March 1990, when my supervisor agreed I could work at home for a few days. I still have my job. Certainly I have not achieved any career growth in four years. Many of my professional colleagues think something has happened to me, but they don't understand what, and they don't ask. They understand that I work at home, but I don't think even the people in my immediate office understand what life is like for someone who is chemically sensitive.

I can't start every conversation with someone new by saying "You understand that I'm chemically sensitive, don't you?" So instead, I go to lunch with the people in my office, and some of them are wearing perfume. I sit as far away from the worst offenders as possible, and I feel sick and congested for the next few hours or the next few days. Mood swings and fatigue associated with chemical exposure are real, but that's another story.

THE MEDICAL COMMUNITY

Let me shift gears, and talk about the medical community's response to chemical sensitivity. In April of 1990, after I had worked at home for a month, I consulted with a physician who had previously been the Health Unit Physician at EPA. Although I did not know it at the time, he was fired from his job at EPA after he wrote a letter to the Administrator of EPA advising him to evacuate Waterside Mall because of the serious indoor air problem. This doctor promised to help me learn how to make my house healthier, and advised me to stay out of Waterside Mall. That was the only time I ever saw him, and eventually I learned that he had retired because of failing eyesight. He never followed up with me by sending the articles and documents he promised or providing information to my management. It was a dead end for me. In December of that year I consulted with another doctor whose specialty is occupational medicine. This doctor wrote a note about my condition which I could use at work because my management was preparing to order me to return to the building.

The primary thing this doctor did was to tell me where to find information about chemical sensitivity. Three years later, in 1993, this doctor further advised me to have a blood test, an Activated Lymphocyte Profile. The test showed immune changes consistent with chemical injury and chemical sensitivity. Otherwise, my efforts to find a physician that can "treat" my injury, or with whom I can consult regularly, have been fruitless. Practicing avoidance has isolated me from almost all activities that would otherwise be part of my personal and professional life. I am angry and alone with this problem. I am, however, still employed.

MCS AND MY CAREER

The third issue, the effect of chemical sensitivity on my career, is best addressed by what I don't say. You can be the judge. I am a woman who went to law school after her children were born, several years out of undergraduate school. I was highly motivated to go back to school and enter the professional world. I worked like hell to successfully complete law school and find meaningful work. During the first 10 years of my legal career, I became gradually debilitated by low levels of chemicals in my workplace. And please remember that my workplace is the United States Environmental Protection Agency, and that there are more than 60 employees there currently who cannot work in an ordinary office building any longer, and probably several hundred who suffer ill effects in that building. Recently a small number of injured employees have been successful in suing the building owner for damage to their health. That case is being appealed by both sides.

Today, when I travel for business, I fear that I will be ineffective because of mental confusion and short-term memory problems. My last business trip was in 1991, although I've recently agreed to work on a case in Philadelphia this summer and I'll have to see how train trips to Philadelphia work out. I haven't had a vacation since 1990 partly because of my difficulty in staying in hotels and traveling where people smoke and wear perfume. It is fair to say, I don't think my life will ever be the same. I see nothing on the horizon that encourages me to think otherwise. The medical establishment can't even agree on a definition of chemical sensitivity, and many practitioners still feel it necessary to point out that this condition may be a mental condition. One of my friends at EPA who is also chemically sensitive said to me when I mentioned addressing this group today, "Helen, I'm glad you're doing it, because at least people won't think you're crazy." Again, you be the judge.

PATIENT STATEMENT: CHEMICAL SENSITIVITY — ONE VICTIM'S PERSPECTIVE

CYNTHIA WILSON

White Sulphur Springs, Montana

Not a day goes by that I don't miss my old life and the old me. To illustrate how my life has changed, I have two brief stories. In 1982, I developed a lending procedure in conjunction with Banker's Life Insurance Company that enabled commercial real estate developers to secure permanent financing for property that had not yet been developed — in essence using a permanent loan in place of a construction loan. It fixed the interest rate, at a time when new construction rates were bankrupting many projects and it allowed the developer to invest the excess funds to offset interest expenses. I received national recognition for this loan.

In 1989, the police found me wandering around in 15 inches of snow, in below zero weather with no shoes or coat. The officer took me to the hospital because I was obviously disoriented. I didn't even know my name or where I lived. These stories show the disparity between my life as a successful, independent business woman and my life as someone who is chemically disabled.

In November of 1986, my chemical sensitivity forced me to walk away from two businesses I had started and a life I loved. I didn't walk away willingly — or easily. In fact, I didn't walk away at all until the simple act of walking to my car one morning ended with my waking up in an emergency room being treated for shock.

It was another year before I was diagnosed with formaldehyde poisoning. As a commercial real estate developer and a general contractor, how I had gotten exposed to formaldehyde was no mystery, but I had no idea that formaldehyde posed such a health threat. Once I learned what was behind my health problems, I started to read everything I could find on the health effects of formaldehyde, and most everything I was experiencing was described in the materials I read. Only for me these symptoms were happening at exposure levels that were believed to be safe according to the government and scientific literature.

Since moving from an urban setting to a rural area, my overall health has unproved. For the most part, this improvement has allowed me to isolate my chemical exposures to specific

1. Address all correspondence to: Cynthia Wilson, Chemical Injury Information Network, P.O. Box 301, White Sulphur Springs, MT 59645.

Toxicology and Industrial Health, Vol. 10, No. 4/5, pp. 319–321

ISSN: 0748-2337

reactions. Some days I can almost forget how ill I am, but if I let down my guard, then I pay with my health.

I get hemolytic hematūria from exposures to freshly cut wood, new carpeting, and a variety of other products that contain chemicals known to cause destruction of red blood cells. The first time it happened, I called the doctor who had originally diagnosed my formaldehyde poisoning. He explained that hemolytic hematuria had only been known to happen with industrial exposures, and that I would have to isolate the exposure to be sure that what I thought had triggered the bleeding was in fact the cause. A month later I purposely exposed myself to fresh cut wood and the hemolytic hematuria returned.

Bronchial pneumonia has become part of my winters. The pneumonia will last anywhere from five to seven months, depending on when my neighbors start and stop using their wood stoves. Handling too much printed material will cause my hands and feet to swell and bruise. Exposure to phenol-based products sends me into cardiac distress, and is the principle reason I am not attending this conference in person. I suffer from horrible headaches that have sometimes been described as explosive by people who have seen their rather violent onset. The headache starts with a pain so intense that it feels like I've been physically struck on the side of my head. They can be triggered by a multitude of different exposures. The headaches distort my vision and hearing and make me extremely light sensitive, and for the next 7 to 36 hours, my world consists of nothing but darkness and the pain. Several times I've been taken to an emergency room with these headaches, not because anything has been found that has effectively dealt with the pain, but because my blood pressure has dropped below 60/40. During many of my chemical reactions I have trouble maintaining my blood pressure, so much so that I bought a blood pressure monitor in order to make certain that I could get medical attention before it dropped to a dangerously low level.

Fatigue is also a major factor in my life. Sometimes these bouts of fatigue come on so fast and are so absolute that I've actually fallen off chairs that didn't have arm rests or crumpled to the floor where I stood. Yet, of all the health problems my chemical sensitivities have created for me, coping with the neurological damage has been the most difficult.

My short-term memory is so bad that, more often than not, I won't remember the name of the person I'm talking with by the end of the conversation. To remember things that are important, I rely on Post-it notes. The more important it is for me to remember, the more Post-it notes I write and place around the house where I'm most likely to see them again.

When I realized how far my math skills had deteriorated, I studied for three years to regain some of what I had lost. I was finally able to relearn basic math. I can add and subtract in my head, but I still need a calculator if I want to multiply or divide. Oddly enough, I was never able to relearn how to count. I can't count to 20 without making a mistake and most of the time I don't get past 10. By contrast, in my freshman year of college I took a math placement test, and based on my score, I was given 24 credits of math without ever entering a classroom.

I can't say the alphabet correctly anymore either. I once spent two hours trying to figure out which letter was lemon. I actually thought lemon was a letter by itself and I'd simply forgotten what it looked like. My husband patiently explained that lemon was most probably the letters l, m, and n. When I wrote my book, *Chemical Exposure and Human Health,* my husband had to write out the alphabet for me and I still needed help with alphabetizing the index. I forget the names of common items such as a dishwasher or a refrigerator.

Learning was once easy. If I could read it, it was mine. I once over heard a male employee explaining to a less-than-understanding subcontractor why he worked for a woman. He said, "She's smart. Give her a book and 24 hours and there isn't any problem she can't solve."

Before I became ill, the various reading comprehension tests I'd taken over my school years had always shown that I had 98 to 100 percent retention. Now, I read materials a minimum of three times to get past my short-term memory problems and then to integrate what I've learned takes hours of talking about it with someone. Discussing what I've learned seems to help me get past my problems with logic sequencing.

I suppose the most dramatic representation of what has happened to my intellect is reflected in my IQ scores. In high school and college, my IQ scores averaged out at 187. Since becoming ill, I have had three IQ tests. During the first test, I experienced minor reactions to something in the doctor's office I was never able to identify, and my IQ came out at 117. During another test, I reacted to the new leather furniture in the office. My eyes were burning, my vision and hearing became distorted, my speech started to break up, and my IQ was recorded at 93. The best IQ score I've been able to achieve was 143 during the last test I had. Particular attention was paid to finding me the safest possible place in the doctor's office to take the test. Unfortunately, part of the test was administered in the doctor's medical library and I started to react to the printed materials. My IQ has never been tested in a totally safe environment, so I don't know if my scores represent permanent brain damage or if those scores are reflecting the inconsistent IQ profile characteristic of formaldehyde problems.

In addition, I've had a challenge electroencephalogram assessing my cognitive evoked potentials. My preexposure testing showed that all my brain waves were within the normal range. After being exposed to dried hair spray on a cotton ball, the latency of the P300 component doubled and the pattern of my brain waves became extremely abnormal.

This illness has made my life a waking nightmare. I don't believe there is a simple answer for the problem of chemical sensitivity, but I do believe there is an answer. However, finding that answer will require major funding for credible, unbiased research.

REFERENCE

WILSON, C. (1993). Chemical Exposure and Human Health: A Reference to 314 Chemicals with a Guide to Symptoms and a Directory of Organizations. McFarland Co., Jefferson, NC.

PATIENT STATEMENT: A CANARY'S TALE

JACOB BERKSON

Hagerstown, Maryland

I am Jacob B. Berkson, a 68-year-old resident of Hagerstown, Maryland. I was a trial lawyer for some 40 years. I am now retired and writing a book on Environmental Pollution and Environmental Illness, titled A Canary's Tale. I was invited to speak to you as a patient — one who was poisoned by an organophosphate pesticide and who subsequently developed Multiple Chemical Sensitivity (MCS, or sometimes referred to as Environmental Illness, EI).

A BRIEF BIOGRAPHICAL SKETCH

I am a graduate of the University of Virginia College of Arts and Sciences and School of Law; the Federal Executive Institute, Charlottesville, Virginia, Residential Program In Executive Education; the U.S. Naval Reserve Midshipmen's School, Columbia University; various naval training schools, including oriental languages and naval justice; and a participant in federal, state, and county seminars.

I have been admitted to practice law in Maryland, Virginia, and California, and before the Supreme Court of the United States. I have held public office, both elected and appointed, and served as a member of the Maryland State Legislature and as Deputy General Counsel, United States General Services Administration. I am a veteran of World War II and the Korean War and served overseas and aboard ships.

THE CHEMICAL EXPOSURE

The first chemical exposure (after which I became ill) occurred at my home and office in Hagerstown, Maryland, on March 23, 1988. Licensed exterminators or pest control operators (PCOs) used a chemical formulation to exterminate an infestation of insects which they identified as subterranean termites. A noxious odor followed their departure and I became ill that very night. When the odor did not dissipate after days of ventilation, they inspected their work again. They explained to me they had used Dursban TC, which they called a safe termiticide approved by the EPA. They said there was nothing to worry about. Just ventilate

1. Address all correspondence to: Jacob Berkson, 1419 Potomac Ave., Hagerstown, MD 21742-3315. Tel.:(301)739-4793.

Toxicology and Industrial Health, Vol. 10, No. 4/5, pp. 323–326

ISSN: 0748-2337

the house. The noxious odor persisted for a long time. Air sampling taken later indicated a "low level" of Dursban TC and xylene.

I was forced to leave the house. I lived in various places waiting for the air in my house to "clear out" — including a tent in my backyard, an RV, an apartment, motels, homes of relatives and friends.

THE ADVERSE HEALTH EFFECTS

Since my exposure of March 23, 1988, I have experienced the following adverse health effects (including, but are not limited to): acute and chronic exposure to Dursban TC with xylene with development of: nasal irritation and burning, sinusitis requiring sinus endoscopy, excessive salivation, lacrimation, eyes burning, frequent and urgent urination, diarrhea, frequent defecation, nausea, abdominal cramping, difficulty breathing, agitation, nervousness, hyperactivity, depression, confusion, disorientation, acute weight loss, cholinesterase intoxication, development of submandibular gland enlargement requiring surgery, and onset of multiple chemical sensitivity, severe cacosmia.

Subsequently, exposure to tobacco smoke, wood smoke, paint, pesticides, household cleaners, perfumes, auto exhaust, newsprint, carpeting, etc., trigger: nasal burning; nervous system dysfunction — agitation, fatigue, lethargy, nervousness, explosiveness, depression, confusion, disorientation, memory loss; G.I. distress — nausea, abdominal cramps; urinary frequency and urgency, and eventual prostate surgery.

INSTITUTIONAL RESPONSES

I was examined and treated by licensed medical physicians and surgeons of the following specialties: general practice; internal medicine; otolaryngology; ophthalmology; urology; environmental medicine; gastroenterology; psychiatry; allergy and immunology; toxicology; taste and smell; ear, nose and throat; head and neck surgery; dermatology; and radiology.

All of these doctors were competent, capable, interested, and devoted and furnished me the best medical advice and service of which they were capable. Most of them are recognized as top men in their profession, including those at the Johns Hopkins Hospital, Baltimore, Maryland; the Veterans Administration Hospital, Martinsburg, West Virginia; the Washington County Hospital in Hagerstown; the Center for Environmental Medicine, North Charleston, South Carolina; and the Johns Hopkins Asthma and Allergy Center, Baltimore, Maryland. I am still under the care of specialists in several of the specialties mentioned above and have follow-up visits scheduled as needed.

SOCIOLOGICAL OUTCOME

My life is ruled and controlled by Air Quality (AQ). I can no longer lead a normal, unrestricted life as I did prior to March 23, 1988. This is generally due to pollutants in the air, both indoors and outdoors, especially from petrochemicals and petrochemical derivatives.

Because of my sensitivity to chemicals that are ubiquitous in the environment in which we live, I can no longer participate in many of my former regular activities. Or if I do participate, I must assume the risk and suffer the consequences of recurrent adverse health effects. I have had to reduce, restrict, curtail, or give up entirely many routine activities such as: auto and airline travel, bowling, attending baseball and football games, bar association activities, conventions, dining out, boat rides, intimacy, lawn mowing, legal research, shopping and walking in malls, certain office visits, political meetings, public speaking, painting, reunions, railroad trips, shopping and walking in inner cities where the air is heavily polluted, swimming in chlorine-treated water, visiting the homes of friends, relatives and even children, and many other activities too numerous to list. There is no safe place for a patient with MCS.

To my knowledge, there is no cure for MCS but there are treatments to relieve symptoms for some period of time unless and until the patient returns to an offending environment. Avoidance is the best method for a patient to cope with this illness, but avoidance of chemicals in a chemical society is difficult. That is why it is most important to prevent pollution and prevent the exposure in the first place, rather than accept pollution and attempt to clean it up after the patient has developed MCS or other illness. Prevention is the key, and education of the public and all professions is the key to prevention.

The financial expenses incurred to date for medical and nonmedical items exceed $100,000. Space does not permit a listing here of all the expenses incurred in fixing my house in order to make it tolerable to live in. No amount of money can compensate me for what has occurred.

CONCLUSION

I have spent the past six years, from 1988 to the present, on an odyssey attempting to get well. In the process I have studied MCS, chemical exposure and disease, the link between Environmental Pollution and Environmental Illness, and environmental matters ranging from A to Z.

Over the past six years, I have traveled to many parts of this country in search of clean air. I have personally met and interviewed many victims of pesticide poisoning and of environmental pollution. Their lives have been devastated by their illness and inability to obtain effective medical treatment. Thousands of victims have been poisoned to the extent that they are dysfunctional and unable to speak for themselves.

Physical problems affecting persons exposed to chemical, industrial, and environmental pollution include nearly every system in the body, especially the central nervous system, the immune system, and the brain.

SELECTED BIBLIOGRAPHY

ASHFORD, N.A. and MILLER, C.S. (1989). Chemical Sensitivity: A Report to the New Jersey State Department of Health. Dec.

ASHFORD, N.A. and MILLER, C.S. (1991). Chemical Exposures: Low Levels and High Stakes. Van Nostrand Reinhold, New York.

BERKSON, J. A Canary's Tale. Environmental Pollution and Environmental Illness. Submitted.

BRIGGS, S., ed. (1992). Basic Guide to Pesticides: Their Characteristics and Hazards. Rachel Carson Council. Washington, D.C.

DAVIDOFF, L.L. (1991). "Multiple Chemical Sensitivities: Research on Psychiatric/Psychosocial Issues." Paper presented at the symposium Multiple Chemical Sensitivity and the Environment II: Diagnosis and Therapy. American Public Health Association, Atlanta, Georgia, Nov. 13.

HEUSER, G. (1992). "Diagnostic markers in clinical immunotoxicology and neurotoxicology." J. Occup. Med. Toxicol. 1(4):v–x.

HILLEMAN, B. and WASHINGTON, O. (1991). "Multiple chemical sensitivity." Chem. Engineering News, 22 July. pp. 26–42.

NATIONAL RESEARCH COUNCIL(1992). Multiple Chemical Sensitivities: Addendum to Biologic Markers in Immunotoxicology. National Academy Press, Washington, D.C.

PULLMAN, C. and SZYMANSKI, S. (1993). Multiple Chemical Sensitivities at Work: A Training Workbook for Working People. The Labor Institute, New York.

RANDOLPH, T.G. (1987). Environmental Medicine — Beginnings and Bibliographies of Clinical Ecology. Clinical Ecology Publications, Inc., Fort Collins, CO.

REINHOLD, R. (1990). "When life is toxic: They suffer agonizing reactions from contact with almost anything chemical, forcing them into protective cocoons. They call it a disease, but is it? Not all doctors are convinced." New York Times Mag., 16 Sept. pp. 50–51, 65–70.

ROSENTHAL, N. and CAMERON, C.L. (1991). "Letter to the editor: Exaggerated sensitivity to an organophosphate pesticide." Am. J. Psychiatry 148(2):270.

ROUECHÉ, B. (1988). "Annals of medicine: The fumigation chamber." New Yorker, 4 Jan. pp. 60–65.

SHERMAN, J.D. (1988). Chemical Exposure and Disease: Diagnostic and Investigative Techniques. Van Nostrand Reinhold, New York.

THRASHER, J.D. (1993). "Immunologic abnormalities in humans exposed to chlorpyrifos: Preliminary observations." Arch. Environ. Health 48(2):89–93.

WEISSKOPF, M. (1990). "Hypersensitivity to chemicals called rising health problem. Some cannot adapt to low doses of toxics, study says." Washington Post, 10 Feb. p. A2.

ZIEM, G. (1993). "Diagnosing and treating chemically injured people." Pesticides and You Summer/Fall:8–17.

RESPONSE PANEL #1

HOWARD KIPEN, MARY LAMIELLE, AND CHARLES HINSHAW

Dr. Kipen: I think the previous speakers did a very creditable job of presenting to you the phenomenology which is the reason that we are all here today.

I would reiterate, from my own ten years of experience with patients who were experiencing sensitivities such as the ones described, that it's a very real problem. In environmental and occupational practice, it is quite common for people to present with such complaints.

These are not people who have been taught these belief systems by practitioners or others. That's not to say that we have established a real link between an exposure and a chemical, but this is not some kind of teaching on behalf of certain practitioners out there in the field.

Beyond that, we know very little about its causes and whether the symptoms people are describing were really caused by the remodeling, pesticide exposure, or other event. Second, we also know very little about whether the exacerbations of symptoms, subsequent to the onset of the syndrome, are in fact truly triggered by the perceived chemical exposures. That is undefined. In addition, we know almost nothing, in public health terms, about the prevalence of this phenomenon, its possible other causes, or other determinants.

Many of the speakers emphasized something that I think is taking on increasing importance, which is the role of fatigue as a symptom, and the possibility of a relationship between chemical sensitivities and the idea, or the syndrome of, chronic fatigue syndrome. Certainly it has occurred to many of us that there may be a relationship there, and I think you will hear some papers presented which deal with that relationship.

All of this strongly suggests the need to explore the psychology and biology of these symptoms, and their relationship to the chemicals in the world around us.

Mrs. Lamielle: I'm president of the National Center for Environmental Health Strategies, a nonprofit tax-exempt organization based in New Jersey. The center has been there for nine years, working with those struggling to live with chemical sensitivity, and those struggling to respond to patient needs. I, and many others, have worked to provide information and services to those injured by chemical and environmental exposures, to achieve recognition of chemical sensitivity as a public health problem, and to facilitate and catalyze programs and policies that protect public health, and assist patient populations.

Toxicology and Industrial Health, Vol. 10, No. 4/5, pp. 327–332

ISSN: 0748-2337

In many ways, we've been the pathfinders, working to initiate the New Jersey study of chemical sensitivity by Drs. Ashford and Miller, testifying before Congress on the Indoor Air Quality Act on three occasions, securing HUD recognition of chemical sensitivity as a physical disability, entitling those so disabled to protection under the Fair Housing Act and Rehabilitation Act of 1973, and insuring that guidelines under the Americans With Disabilities Act address this disability.

I've served on numerous local, state and federal committees and review panels, including appointment to an EPA lawn care and pesticide advisory committee, and a more recent appointment to the President's committee on employment of persons with disabilities. I serve on the ATSDR expert panel, whose recommendations bring us to Baltimore today, and was responsible for securing the directed Congressional funds, which sponsored this two-day conference.

Our center receives up to 1000 requests for information each month. Calls and correspondence come from across the United States, Canada, New Zealand, Australia, Germany, and other countries. Many calls to us are referrals from federal agencies, including the National Pesticide Telecommunications Network, the EPA Interwire Hotline, the Job-Accommodations Network, from physicians and environmental clinics, from environmental and disability organizations, and from the media.

It's not unusual for a first time caller, recently ill, to be perplexed by their declining health, and the circuitous route to information and diagnosis. These individuals seek understanding and validation, explanation for their medical situation; a diagnosis, treatment, and of course, a cure. Calls and letters from those who are already hypersensitive focus less on the cure, and more on the ways to deal with the cascading problems from chronic illness and disability, and the burdens peculiar to chemical sensitivity.

These individuals may be unable to work, to find suitable housing, to use everyday consumer products, to be able to access stores and buildings without severe health problems. They may continue to have a difficult time finding accessible, affordable, and appropriate medical care, insurance reimbursement and compensation.

The calls and information include more questions than there are answers, more problems than there are solutions. Research is direly needed to fill in the gap, the chasm that exists between the phenomenon and its health and social implications.

I would estimate that we've heard from well over 10,000 people with chemical sensitivity, and that I've spoken with about half that number. These contacts have provided a glimpse at exposure and disability patterns. Those who would dismiss chemical sensitivity are quick to label the illness as psychiatric, neither a toxicological nor a traditional allergic response.

I would offer that those with chemical sensitivities develop predictable symptom patterns, that evolve over time, that may vary from person to person. We're also aware of the changing face of the chemically sensitive. In the past, pesticides were implicated as a primary culprit. Clearly, they remain a very serious problem, particularly because of the intense neurologic involvement. And there are repeat performers among pesticides, such as chlorpyrofos (Dursban), identified as the cause of hypersensitivity in the majority of those who reported pesticide injury and poisoning to our center.

However, we've been hearing from several new or increasingly visible populations. Injured workers, particularly office workers, are reporting chemically induced illnesses with greater frequency, and seeking assistance, accommodations, compensation, and action. A number of Gulf War veterans appear to exhibit the illness patterns of chemical sensitivity. When we first started to receive calls and correspondence from Gulf War veterans, I was struck by the remarkable similarity of patterns of illness that these individuals reported to those reporting hypersensitivity to chemicals in the civilian population. Pesticides and diesel fuel are among those exposures identified by veterans as those causing their health problems.

In Germany, over 10,000 people have reported hypersensitivity following exposure to a particular wood stain and paint contaminated with pentachlorophenol and lindane. These patients reportedly experienced severe central nervous system symptoms, respiratory involvement, and aggressive and antisocial behavior.

"Does chemical sensitivity exist?" and "How many people are affected?" are the two most frequently asked questions. The inability to respond, due to the lack of research and analysis, is frequently cited as a validation of the position of those who would oppose the possibility of this health problem.

The directed congressional funds which we sought for research on chemical sensitivity, were initially intended to answer these questions, and to fund efforts toward an environmental control unit, or an environmental medical unit, and to fund a data base to collect information on the patient experience. When issues were prioritized by the ATSDR expert panel last spring, the environmental unit did receive the most votes, however, funding is inadequate for this project.

As we explore newer science issues over the next two days, it is important to keep in mind that the unit remains an important research and diagnostic tool, to establish, once and for all, "Does the illness exist?"

And, the patient experience may hold the key to explain and explore this new illness paradigm.

Today's meeting, for me, is a culmination of nine years of hard work in this area, and, I hope, a new direction toward a resolution of this problem. For you and your colleagues, the work is just beginning. Research in the area of chemical sensitivities, and low-level chemical exposures, has been inadequate. It has failed to respond to the urgency of this growing medical problem.

It is no longer acceptable from either a moral or a scientific perspective, to dismiss this segment of our society. It is time for those of you in the medical, scientific and research communities, to set the pace for advancement of valid scientific inquiry, and we'll be there to assist, to facilitate, to advocate, to seek funding for your work.

It is in part because we suspect that chemicals (consumer products) cause potentially debilitating illness, that we have avoided facing it head-on, and describing and analyzing what we see. It's time to end the fractious debates within the industry and the medical community. It's time to address the needs of those who are sick, and those who may join their ranks.

I know that there is great fear of disturbing the status quo. There are powerful forces fighting against research, though few would admit to this in a public forum. It is helpful to know that the only constant is change, and that it is time to change. There will be a significant price to pay for inaction.

It is becoming more and more difficult to avoid the swell of grass-roots activities from those who are chemically sensitive, who are demanding that society open its doors, and its windows, to accommodate them. Research holds the key. Not to whether these doors will open, but how wide. And whether the welcome will be hostile or inviting.

These two days in Baltimore can have major medical, economic, and societal ramifications. It's time to do what's necessary, and it's time to do what's right.

Dr. Hinshaw: In the 1940s and 1950s, Dr. Theron Randolph was recognizing phenomena which he described as chemical sensitivity. Interestingly, at about the same time, Dr. Hans Selye was working at McGill University in Montreal, was working on what he called the "general adaptation syndrome." Their courses were remarkably parallel, and they arrived at similar, though not identical conclusions.

These conclusions have formed the basis for the practice of environmental medicine. In this type of medicine, we like to think of the environment as furnishing an innumerable number of insults that act upon the individual, the biologic organism. In spite of the many insults, we only have a limited number of ways of reacting. For a time, we can adapt to these insults. But, at some point, each of us is susceptible to losing our adaptations, and developing symptoms of illness. Prior to the loss of adaptation, we may be able to mask or disguise the symptoms of illness, even though the process is proceeding.

Coming from Wichita, Kansas, I am very familiar with the aircraft industry. They often speak of redundancy, or back-up systems. The same things prevail in all biologic organisms. We generally have more than one, often several ways, of coping with various insults. But in the end, when we use up all our methods of adaptation, we become ill. Environmental medicine is thus very interested in locating, or identifying, the cause of an illness. The feeling is, if we can remove the cause, or somehow deal with the cause, natural healing can occur.

That preface leads me to what I consider the basic tenets of environmental medicine. First, we speak of total load. This means the number of environmental insults that may impinge upon us. A second principal in environmental medicine is the idea of adaptation. In adaptation, the body will compensate for the stressors.

Another phenomena is known as bipolarity. Initially, a response is often stimulative, and this is well described by Dr. Randolph. At the end of the response, there may be a very depressive, fatigue-type reaction.

Similar work was published, about the same time that Dr. Randolph did his work, by Dr. Leland Spears in Kansas City. He was talking about food allergy, and he called it the "tension fatigue syndrome."

Another finding in environmental medicine is the concept of the "spreading phenomenon." Oftentimes, a person will begin to react to a single chemical, or stressor, but later will find himself or herself reacting to many different stressors. There is also a phenomenon known as the "switch" phenomenon, in which symptoms initially may involve the central nervous system, but later involve the gastrointestinal tract, the skin, the respiratory tract, and so forth. Finally, and very importantly, is the concept of biochemical susceptibility or individuality. All of us have certain characteristics which are partially determined by heredity, but also are a result of our environmental exposures. All of us, except for identical twins living in the same environment, have different susceptibilities.

Finally, I would like to propose, for scientific interest, a concept that I think ties in very closely with neuroscience. That is the question of whether we might be dealing with a problem related to organic chemistry. Carbon compounds seem to be the most common cause of the problems in multiple chemical sensitivities. More particularly, the aromatic amines, the benzene ring hydrocarbons, are most often the problem. It may be that the organism initially deals with the problem at the barrier level. This may be the skin surface, the blood-brain barrier, the gastrointestinal mucosa, or the lining of the respiratory tract. When the compensatory mechanisms for dealing with those hydrocarbons are exceeded, they gain entrance to the circulation or the brain, and more systemic symptoms occur. This also leads us into the problem, recognized by environmental medicine physicians, that the liver's detoxification pathways may be exceeded and prolong the event.

The aromatic hydrocarbons have two properties which bear on this problem. One, they tend to preserve the life of vasoactive amides and catecholamines, acting sometimes as antioxidants. Secondly, they affect membrane barriers, so that the vasoactive amides and catecholamine can more readily access the cells. This can more readily lead to the production of such factors as cerebral edema and electrolyte imbalance within the cells, resulting in various central nervous symptoms.

I leave you with the thought that the aromatic hydrocarbons, and the organic chemistry of carbon atoms, may be a major problem in MCS, perhaps accounting for many of the neuroscience phenomena that we have observed.

QUESTIONS AND ANSWERS #1

IRIS BELL, HOWARD KIPEN, AND CLAUDIA MILLER

Dr. Rosenthal, NIMH: What about all those people with multiple chemical sensitivities who may not have such high IQs, may not be so articulate as the patients who spoke today, and may come across as "flaky"? The question for Dr. Bell is: You have pointed out that a lot of your folks with MCS also have depression or other psychiatric illnesses. Do you think that MCS may actually be a more specific syndrome than some of those?

Dr. Bell: I think there is a distinct possibility that we may be looking at a subtype of these particular syndromes that will help us in psychiatry become more precise about the description of patients. As I indicated in some of my data, we do have some indication that you can pick apart people that claim depression from people who have depression and chemical sensitivity. The nature of the illness they have is in fact different. The nature of their response to chemicals that I have seen in laboratory challenges I have performed, appears to be different. In Seasonal Affective Disorder, we have a subtype of depression whose characteristics are not identical to what we would diagnose as generalized major depression. That is what we call an atypical picture and I think it may be simply one of many subtypes of depression that we haven't characterized yet.

Dr. Art Sehalski, Memorial Medical Center: Dr. Bell, Dr. Kipen: Where is the metabolic evidence or evidence based on metabolic loading experiments that will support this phenomenology? It seems to me, from what I have read, including your work especially Dr. Bell, that much of the treatment has been based on phenomenological observation. Much of that treatment, as I understand it, either lacks or requires additional understanding. For example, the University of Arizona has a PET scan. I don't know whether you have performed any positron emission transmission tomography scans of your patients to determine whether there are metabolic defects. Dr. Kipen, you are working on a chamber. Do you intend to test patients in a blinded fashion to determine whether these chemical sensitivities are real, and in what manner they present as biological markers or metabolic evidence that can be used both diagnostically and as a rational basis for treatment?

Dr. Bell: I think you have a very good point about a methodology that may be available to look at this question systematically. There will be talks tomorrow on SPECT scanning that has been done, and there will be a discussion with Dr. Mayberg, who is an expert in PET scanning, who can address some of the technical issues. You are absolutely correct that we need objective measures to try to identify what the problems are in these patients in a way that will help us get through some of the subjective concerns. There are ways, which I will be discussing tomorrow, of using subolfactory threshold levels of chemicals that are blinded,

Toxicology and Industrial Health, Vol. 10, No. 4/5, pp. 331–334

ISSN: 0748-2337

without having to get into the problem of adding masking orders and arguing about whether a masking odor is benign. There are some methodological approaches available that have not been done, in part, that is because there is no funding in this field right now. We are very hopeful that this conference and other work related to it, will lead to some interest in providing funding for this needed research.

Dr. Kipen: I'm not aware of any metabolic evidence that would support the phenomenology, or any credible studies that have attempted to address that issue. I don't believe that metabolic studies ought to be the first of our priorities. Nancy Fiedler and I are planning to do double-blinded challenge studies in the controlled exposure facility at Robert Wood Johnson Medical School. We are also planning to do some studies involving SPECT scanning if we get pending funding. We do not have a PET scanner at our institution. Finally, I can't really comment on the treatment you asked about, since at our institute we do not prescribe much in the way of pharmacological treatment.

From the Audience: The organophosphates are excitatory neurotoxins. Given the apparent correlation between hyperactivity, Attention Deficit Disorder, and intelligence, perhaps the intelligent people are the sensitive subpopulation. Somebody with an IQ of 180 may end up with brain damage due to organophosphate because it's a learning mechanism.

Dr. Miller: That's an interesting hypothesis. The reason insects, in general, are susceptible to organophosphates and people are not is that we have covered nerve endings. However, there are areas of the body, for example the olfactory nerves, which are not covered nerves so there may be some exceptions. In addition, I've heard that the limbic system is particularly rich in acetylcholinesterases and some people have hypothesized that if acetylcholine levels get out of certain neural ranges at nerve synapses, then the cells in that area show what Guergis has called bizarre sensitivities. In other words, you have erratic electric discharges in the area of the brain that effects things like mood and ability to retain short-term information. So, there are reasons to think that organophosphate may be among the more potent initiators of this process, simply because the patients with organophosphate exposures as the initiating event report higher symptom severity and more chemicals and more foods as triggering their symptoms, as do the remodeling people.

TIME-DEPENDENT SENSITIZATION IN ANIMALS: A POSSIBLE MODEL OF MULTIPLE CHEMICAL SENSITIVITY IN HUMANS

SEYMOUR M. ANTELMAN

Department of Psychiatry
Western Psychiatric Institute and Clinic
University of Pittsburgh
School of Medicine
Pittsburgh, Pennsylvania

It often happens in science that clues to the nature of a problem under study come from a completely different, seemingly unrelated, line of investigation. This may be the case with MCS and Time-Dependent Sensitization (TDS), a phenomenon we discovered in rats in the late 1970s and later named. TDS refers to the ability of mild stressors — whether pharmacological or environmental — to induce physiological and behavioral effects which then progress, i.e., get stronger, entirely as a function of the passage of time since stressor presentation. This strengthening is revealed when the organism is later exposed to either the original or another stressor. The characteristics of TDS bear a remarkable resemblance to the features of MCS and that similarity is the subject of this manuscript.

In recent years the general public has seen numerous reports on television and in print media of illnesses presumably resulting from low-level exposure to chemicals. These have ranged from problems associated with "tight" buildings and installation of new carpets to service in the Gulf War. Such syndromes have come to be referred to as multiple chemical sensitivity (MCS). While many in the medical and scientific communities have ridiculed MCS and simply dismissed it as due to preexisting psychological disturbance and/or misattribution of symptoms to chemicals (e.g., Terr, 1993), others have adopted a more systematic approach.

Due principally to the efforts of the latter investigators (e.g., Ashford and Miller, 1991; Bell et al., 1992), would-be modellers of MCS now have a potential framework against which to measure their findings. According to Ashford and Miller (1991) those features most closely associated with MCS are:

1. Address all correspondence to: Seymour M. Antelman, Department of Psychiatry, Western Psychiatric Institute and Clinic, University of Pittsburgh, School of Medicine, 3811 O'Hara Street, Pittsburgh, PA 15213. Tel.: (412)624-4523. Fax:(412)624-1772.

2. Abbreviations: MCS, Multiple Chemical Sensitivity; TDS, Time-Dependent Sensitization.

3. Key Words: Multiple Chemical Sensitivity, Stress, Time-Dependent Sensitization.

Toxicology and Industrial Health, Vol. 10, No. 4/5, pp. 335–342

ISSN: 0748-2337

1. Symptoms involving any system in the body or several systems simultaneously

2. Varied symptoms and severity in different individuals, even those with the same exposure

3. Induction (that is, sensitization) by a wide range of environmental agents

4. Subsequent "triggering" by lower levels of exposure than those involved in initial induction of the illness

5. Concomitant food intolerances, estimated to occur in a sizable percentage of those with chemical sensitivities

6. "Spreading" of sensitivity to other, often chemically dissimilar substances; each substance may trigger a different constellation of symptoms

7. Adaptation (masking), that is, acclimatization to environmental incitants, both chemical and food, with continued exposure; loss of this tolerance with removal from the incitant(s); and augmented response with reexposure after an appropriate interval (for example, four to seven days)

8. An apparent threshold effect referred to by some as the patients **total load**. Total load is a theoretical construct that has been invoked by clinical ecologists to help explain why an individual develops this syndrome at a particular time. Illness is said to occur when the total load of biological, chemical, physical, and psychological stressors exceeds some threshold for the patient.

It often happens in science that clues to the nature of a problem under study come from a completely different, seemingly unrelated, line of investigation. This may be the case with MCS and Time-Dependent Sensitization (TDS), a phenomenon we discovered in rats in the late 1970s and later named. TDS refers to the ability of mild stressors — whether pharmacological or environmental — to induce physiological and behavioral effects which then progress, i.e., get stronger, entirely as a function of the passage of time since stressor presentation. This strengthening is revealed when the organism is later exposed to either the original or another stressor. The characteristics of TDS bear a remarkable resemblance to the features of MCS outlined above. We briefly noted a correspondence between these phenomena in a 1988 article on TDS (Antelman, 1988a). However, it was Iris Bell who realized the remarkable fidelity of this correspondence and who introduced TDS to the community of MCS investigators (Bell et al., 1992). In this article we point up some of the apparent similarities between TDS and MCS.

SIMILARITIES BETWEEN MCS AND TDS: EVIDENCE FROM THE NEUROSCIENCE LITERATURE

1. Just as MCS can apparently involve any bodily system or multiple systems simultaneously, the same appears to be true of TDS. TDS has been demonstrated in animals:

a) using indices of at least five different neurotransmitter systems: dopamine (DA) (Antelman and Eichler, 1979), norepinephrine (Lace and Antelman, 1983; Martin et al., 1985), serotonin (Antelman et al., 1983), GABA (Antelman et al., 1989a) and aspartate (Caggiula et al., 1994);
b) in various hormonal systems, e.g., corticosterone (Caggiula et al., 1989), ACTH and beta-endorphin (Denicoff et al., 1989; this study was done in humans);
c) in the immune system (Antelman et al., 1990);
d) in the cardiovascular system (Antelman et al., 1981, 1989b);
e) on carbohydrate metabolism (Cunnick et al., unpublished observations);
f) on vitellogenesis (the activation of genes encoding major yolk proteins; Deeley et al., 1977; Baker and Shapiro, 1977; Burch and Evans, 1986). A number of these reports found simultaneous TDS in more than one system (e.g., Caggiula et al., 1989).

2. As is apparently the case with MCS (Cullen, 1987), TDS can be induced by a wide array of agents, often chemically quite dissimilar. For example, among drugs it has been demonstrated with:

a) antidepressants (imipramine, Chiodo and Antelman, 1980a); desmethylimipramine, Lace and Antelman, 1983; amitriptyline, Antelman et al., 1981, 1983; bupropion, Antelman, 1988b; phenelzine, Antelman et al., unpublished observations);
b) stimulants (amphetamine, see Robinson and Becker, 1986 and Antelman, 1988a for review; cocaine, Kalivas and Stewart, 1991);
c) neuroleptics (haloperidol and fluphenazine hydrochloride, Antelman et al., 1986; clozapine, Antelman et al., unpublished observations);
d) an anxiolytic (diazepam, Antelman et al., 1989a);
e) an anxiogenic (pentylenetetrazol, Antelman et al., 1988);
f) ethanol (Antelman et al., 1991a);
g) a beta-adrenergic agonist (clenbuterol, Martin et al., 1985);
h) a benzodiazepine antagonist (RO15-1788, Antelman et al., 1989a);
i) a glucose antimetabolite (2-deoxy-D-glucose, Antelman et al., 1991a);
j) hormones (estrogen, Deeley et al., 1977; Baker and Shapiro, 1977; Burch and Evans, 1986; corticosterone, Cunnick et al., unpublished observations);
k) a cytokine (interleukin-2, Denicoff et al., 1989);
l) nicotine (Antelman and Caggiula, unpublished observations).

Among nondrug stressors, TDS has been induced by a single experience with:

a) a 1–4 hr period of immobilization by wrapping in muslin (e.g., Antelman et al., 1989b; 1991a);

b) 0.7 sec of electroconvulsive shock (Chiodo and Antelman, 1980b);

c) insertion and immediate withdrawal of an empty, standard 26 gauge 0.5 in. syringe needle (Antelman et al., 1988; 1991a);

d) an injection of isotonic saline (Antelman et al., 1989a);

e) the psychological stimulus of 10 minutes in a strange environment (Antelman et al., 1992).

In addition to the ubiquitous ability of drugs and stressors to act as inducers of TDS, they can also serve to reveal TDS once it has been induced. In MCS terminology, they also act as triggers of TDS. This is noteworthy, since MCS patients apparently frequently display considerably magnified responses both to standard doses of medications as well as physical stimuli (Miller, 1992).

3. The spreading or generalization of sensitivity to stimuli other than and often chemically different from the inducing agent, seen in MCS, is also prominently evident in TDS, where it is referred to as cross-sensitization. Bidirectional cross-sensitization has frequently been reported between nonpharmacological stressors and drugs, as well as between a given drug and one which is chemically dissimilar (e.g., Antelman et al., 1980; Antelman and Chiodo, 1983; Robinson et al., 1987; Antelman, 1988a; Robinson, 1988; Kalivas et al., 1988; Kalivas and Stewart, 1991).

4. The food and alcohol intolerances which frequently accompany MCS (Miller, 1992) may also have a parallel in TDS. We have shown, for instance, that in rats a single exposure to a stressor weeks earlier altered the effect of ethanol on plasma ACTH and glucose (Antelman et al., 1991b). The alterations in glucose — which we have observed on several occasions — may be related to either or both the food intolerances or carbohydrate cravings often reported in MCS patients (Miller, 1992).

5. Just as the manifestation of MCS requires a provocative stimulus, so too does TDS. Neither is obvious between provocations. Thus, MCS apparently remits partially or completely in the absence of chemical exposures (e.g., Bell, this volume) and symptoms of hyperresponsiveness are similarly not obvious in sensitized animals without the presence of stressful stimuli.

6. Both MCS and TDS can be induced by a single exposure to an inciting stimulus and each can persist for extraordinary periods of time. MCS can apparently persist in some individuals for decades. In rats, we have shown that TDS can last for at least two months following a single exposure to a stressor (Antelman et al., 1986), this being the longest period we tested. While two months is obviously not the same order of time magnitude as decades, it should be noted that is a significant percentage of the rat's approximately two-year life span.

7. Both TDS and MCS have the capacity to show a bipolarity of response. However the extent to which these bipolarities are similar remains to be determined. In the case of MCS, bipolarity has been described as a withdrawal type of reaction following a stimulatory response to a provocative agent. TDS, on the other hand, displays sensitization of stimulatory responses following exposure to "lower-intensity" stressors and sensitization of inhibition after exposure to "higher-intensity" stressors (Antelman et al., 1991a). However, these seeming differences may be more apparent than real. Thus, according to Miller (1991), stimulatory effects decrease while withdrawal symptoms increase, in very sensitive individuals, a state which may be analogous to our exposing animals with "normal sensitivity" to "higher-intensity" stressors.

8. TDS (Antelman et al., 1992) and MCS each appear to require some threshold to be reached before induction can take place. As noted above, "total load" is the term used to describe this in MCS. In animal experiments using graded levels of a psychological stressor — a brief exposure to a novel enclosure — we found that a minimum level of stress (as indexed by plasma corticosterone levels) was necessary for the induction of TDS (Antelman et al., 1992).

9. Females appear to be more vulnerable to both MCS (Miller, 1992) and TDS (Robinson, 1988).

10. Multiple exposure routes are able to induce and/or trigger symptoms both in MCS and TDS. For example, TDS has been induced both by injection of a drug and putting it into the animal's drinking water (Antelman et al., 1986) as well as by interoceptive (e.g., food deprivation, Antelman and Chiodo, 1983), exteroceptive (e.g., immobilization) and psychological (novel environment) stressors (Antelman, 1988a; Antelman et al., 1992).

DISCUSSION

The foregoing similarities suggest that TDS in animals has face validity as a model of MCS in humans. Only research deliberately focused on further exploring this apparent relationship can determine whether such promise will be realized or turn out to be merely superficial. Should the former be the case, an animal model of MCS based on TDS in addition to its more obvious virtues, could aid in dispelling the argument that MCS patients are psychologically disturbed, rather than chemically sensitive.

Animal models of human disorders such as MCS are desirable for several reasons. First, they offer the possibility of simulating a human condition under controlled circumstances, with large numbers of subjects, in a simpler more readily understandable system. Second, in contrast to the human disorders being modeled, which can be studied only after they have already developed, animal models are observable as they evolve, permitting study of symptoms and processes as they are developing. Third, they allow the testing of pharmacological and other prospective "treatments" that might be difficult in humans.

Animal models, of course, also have their limitations. Depending on the species being used, their behavioral repertoires may be limited compared to humans. Also, as already alluded to above, the life span of species most likely to be studied in the laboratory is only a fraction of that of human beings and it is therefore not possible to study an animal model of MCS for decades, even though the disorder may persist that long. Despite these limitations of animal models in general, it should be noted that TDS has already proven its clinical value (in a context unrelated to MCS) by predicting that the therapeutic effects of drugs as well as nonpharmacological treatments such as electroconvulsive shock would grow with the passage of time after only acute administration, a prediction that has now been borne out in several clinical studies (e. g., Pollock et al., 1989; Denicoff et al., 1989; Jagadeesh et al., 1992).

REFERENCES

ANTELMAN, S.M. (1988a). "Time-dependent sensitization as the cornerstone for a new approach to pharmacotherapy: drugs as foreign/stressful stimuli." Drug Develop. Res. 14:1–30.

ANTELMAN, S.M. (1988b). "Is it possible that drugs have been given the wrong way for centuries?" In: Receptors and Ligands in Psychiatry and Neurology (A.K. Sen and T. Lee, eds.). Cambridge University Press, Cambridge, MA. pp. 484–503.

ANTELMAN, S.M. and CHIODO, L.A. (1983). "Amphetamine as a stressor." In: Stimulants: Neurochemical, Behavioral and Clinical Perspectives (I. Creese, ed.). Raven Press, New York, NY. pp. 269–299.

ANTELMAN, S.M. and EICHLER, A.J. (1979). "Persistent effects of stress on dopamine-related behaviors: clinical implications." In: Catecholamines: Basic and Clinical Frontiers (E. Usdin, I.J. Kopin, and J. Barchas, eds.). Pergamon Press, New York, NY. pp. 1759–1761.

ANTELMAN, S.M., EICHLER, A.J., BLACK, C.A., and KOCAN, D. (1980). "Interchangeability of stress and amphetamine in sensitization." Science 207:329–331.

ANTELMAN, S.M., KOCAN, D., ROWLAND, N.E., DEGIOVANNI, L.A., and CHIODO, L.A. (1981). "Amitriptyline provides long-lasting immunization against sudden cardiac death from cocaine." Eur. J. Pharmacol. 69:119–120.

ANTELMAN, S.M., DEGIOVANNI, L.A., KOCAN, D., PEREL, J.M., and CHIODO, L.A. (1983). "Amitriptyline sensitization of a serotonin-mediated behavior depends on the passage of time and not repeated treatment." Life Sci. 33:1727–1730.

ANTELMAN, S.M., KOCAN, D., EDWARDS, D.J., KNOPF, S., PEREL, J.M., and STILLER, R. (1986). "Behavioral effects of a single neuroleptic treatment grow with the passage of time." Brain Res. 385:58–67.

ANTELMAN, S.M., KNOPF, S., KOCAN, D., EDWARDS, D.J., RITCHIE, J.C., and NEMEROFF, C.B. (1988). "One stressful event blocks multiple actions of diazepam for up to at least a month." Brain Res. 445:380–385.

ANTELMAN, S.M., KOCAN, D., EDWARDS, D.J., and KNOPF, S. (1989a). "Anticonvulsant and other effects of diazepam grow with time after a single treatment." Pharmacol. Biochem. Behav. 33:31–39.

ANTELMAN, S.M., DEGIOVANNI, L.A., and KOCAN, D. (1989b). "A single exposure to cocaine or immobilization stress provides extremely long-lasting, selective protection against sudden cardiac death from tetracaine." Life Sci. 44:201–207.

ANTELMAN, S.M., CUNNICK, J.E., LYSLE, D.T., CAGGIULA, A.R., KNOPF, S., KOCAN, D.J., RABIN, B.S., and EDWARDS, D.J. (1990). "Immobilization 12 days (but not one hour) earlier enhanced 2-deoxy-D-glucose-induced immunosuppression: evidence for stressor-induced time-dependent sensitization of the immune system." Progr. Neuropsychopharmacol. Biol. Psychiatry 14:579–590.

ANTELMAN, S.M., CAGGIULA, A.R., KOCAN, D., KNOPF, S., MEYER, D., EDWARDS, D.J., and BARRY III, H. (1991a). "One experience with "lower" or "higher" intensity stressors, respectively enhances or diminishes responsiveness to haloperidol weeks later: implications for understanding drug variability." Brain Res. 566:276–283.

ANTELMAN, S.M., CAGGIULA, A.R., KNOPF, S., EDWARDS, D.J., and BARRY III, H. (1991b). "Exposing rats to a single brief stressor two weeks earlier modifies the response of plasma ACTH and glucose to ethanol (ETOH)." In: Society for Neuroscience Abstracts, Vol. 17. Society for Neuroscience, Bethesda, MD. p. 1250.

ANTELMAN, S.M., KOCAN, D., KNOPF, S., EDWARDS, D.J., and CAGGIULA, A.R. (1992). "One brief exposure to a psychological stressor induces long-lasting, time-dependent sensitization of both the cataleptic and neurochemical responses to haloperidol." Life Sci. 51:261–266.

ASHFORD, N.A. and MILLER, C.S. (1991). Chemical Exposures. Low Levels and High Stakes. Van Nostrand Reinhold, New York. pp. 87–88.

BAKER, H.J. and SHAPIRO, D.J. (1977). "Kinetics of estrogen induction of Xenopus laevis vitellogenin messenger RNA as measured by hybridization to complementary DNA." J. Biol. Chem. 252:8428–8434.

BELL, I.R., MILLER, C.S., and SCHWARTZ, G.E. (1992). "An olfactory-limbic model of Multiple Chemical Sensitivity Syndrome: possible relationships to kindling and affective spectrum disorders." Biol. Psychiatry 32:218–242.

BURCH, J.B.E. and EVANS, M.I. (1986). "Chromatin structural transitions and the phenomenon of vitellogenin gene memory in chickens." Mol. Cell. Biol. 6:1886–1893.

CAGGIULA, A.R., ANTELMAN, S.M., AUL, E., KNOPF, S., and EDWARDS, D.J. (1989). "Prior stress attenuates the analgesic response but sensitizes the corticosterone and cortical dopamine responses to stress 10 days later." Psychopharmacol. 99:233–237.

CAGGIULA, A.R., ANTELMAN, S.M., PALMER, A.M., KISS, S., EDWARDS, D.J., and KOCAN, D. (1994). "The effects of ethanol on striatal dopamine and frontal cortical D-[^{3}H]Aspartate efflux oscillate with repeated treatment: relevance to individual differences in drug responsiveness." Submitted.

CHIODO, L.A. and ANTELMAN, S.M. (1980a). "Repeated tricyclics induce a progressive dopamine autoreceptor subsensitivity independent of daily drug treatment." Nature 287:451–454.

CHIODO, L.A. and ANTELMAN, S.M. (1980b). "Electroconvulsive shock: progressive dopamine autoreceptor subsensitivity independent of repeated treatment." Science 210:799–801.

CULLEN, M.R. (1987). "The worker with multiple chemical sensitivities: An overview." In: Workers with Multiple Chemical Sensitivities (M.R. Cullen, ed.). Occup. Med.: State Art Rev. 2:655–662.

DEELEY, R.G., UDELL, D.S., BURNS, A.T.H., GORDON, J.I., and GOLDBERGER, R.F. (1977). "Kinetics of avian vitellogenin messenger RNA induction." J. Biol. Chem. 252:7913–7915.

DENICOFF, K.D., DURKIN, T.M., LOTZE, M.T., QUINLAN, P.E., DAVIS, C.L., LISTWAK, S.J., ROSENBERG, S.A., and RUBINOW, D.R. (1989). "The neuroendocrine effects of interleukin-2 treatment." J. Clin. Endocrinol. Metab. 69:402–410.

JAGADEESH, H.N., GANGADHAR, B.N., JANAKIRAMAIAH, D.K., SUBBAKRISHNA, D.K., and JAIN, S. (1992). "Time dependent therapeutic effects of single electroconvulsive therapy (ECT) in endogenous depression." J. Affective Disorders 24:291–296.

KALIVAS, P.W. and STEWART, J. (1991). "Dopamine transmission in the initiation and expression of drug- and stress-induced sensitization of motor activity." Brain Res. Rev. 16:223–244.

KALIVAS, P.W., DUFFY, P., ABHOLD, R., and DILTS, R.P. (1988). "Sensitization of mesolimbic dopamine neurons by neuropeptides and stress." In: Sensitization in the Nervous System (P.W. Kalivas and C.D. Barnes, eds.). The Telford Press, N.J. pp. 119–143.

LACE, J.W. and ANTELMAN, S.M. (1983). "Cortical beta-adrenergic subsensitivity after desmethylimipramine may depend on the passage of time rather than daily treatment." Brain Res. 278:359–361.

MARTIN, P., SOUBRIE, P., and SIMON, P. (1985). "Does a single priming injection of clenbuterol alter behavioral response to beta-adrenoceptor agonists and antagonists in mice through a time-dependent process?" Eur. J. Pharmacol. 115:91–96.

MILLER, C.S. (1991). "Possible models for multiple chemical sensitivity: conceptual issues and role of the limbic system." Conference on Multiple Chemical Sensitivity, Association of Occupational and Environmental Clinics, Washington, D.C., September 20–21.

MILLER, C.S. (1992). "Possible models for multiple chemical sensitivity: conceptual issues and role of the limbic system." Toxicol. Ind. Health 8:181–202.

POLLOCK, B.G., PEREL, J.M., NATHAN, S.R., and KUPFER, D.J. (1989). "Acute antidepressant effect following pulse loading with intravenous and oral clomipramine." Arch. General Psychiatry 46:29–35.

ROBINSON, T.E. (1988). "Stimulant drugs and stress: factors influencing individual differences in the susceptibility to sensitization." In: Sensitization in the Nervous System (P.W. Kalivas and C.D. Barnes, eds.). The Telford Press, N.J. pp. 145–173.

ROBINSON, T.E. and BECKER, J.B. (1986). "Enduring changes in brain and behavior produced by chronic amphetamine administration: a review and evaluation of animal models of amphetamine psychosis." Brain Res. Rev. 11:157–198.

ROBINSON, T.E., BECKER, J.B., YOUNG, E.A., AKIL, H., and CASTANEDA, E. (1987). "The effects of footshock stress on regional brain dopamine metabolism and pituitary beta-endorphin release in rats previously sensitized to amphetamine." Neuropharmacology 26:679–691.

TERR, A.I. (1993). "Multiple chemical sensitivities." Ann. Internal Med. 119:163–164.

THE PHENOMENOLOGY OF LIMBIC KINDLING

M.E. GILBERT

ManTech Environmental Technology, Inc.
Research Triangle Park, North Carolina
and
University of North Carolina
Department of Psychology
Chapel Hill, North Carolina

Kindling is a model of epilepsy whereby repeated administration of brief low-intensity trains of electrical stimulation come to elicit electrographic and behavioral manifestations of seizure. In the absence of overt tissue damage, an animal that has been kindled is rendered in a permanent state of increased susceptibility to seizures. A number of persistent biochemical and physiological alterations in function accompany kindling, some of which may impact upon behavior of the organism for a long period of time despite the absence of further seizure activation. The sensitivity of limbic structures to kindling may contribute to the behavioral categories of cognition and affect that are particularly impacted by the kindling process. The increased proclivity for seizure disorders that characterizes kindling is not restricted to the initial kindling stimulus, but generalizes to other agents with convulsive properties. This paper provides an overview of the phenomenology of kindling, describes some of the conditions necessary for its induction, and some of the functional alterations that accompany its development and endure when overt convulsive behavior has subsided. Finally, a series of studies in our laboratory is presented which provides evidence of chemically induced kindling by repeated low-level exposure to some pesticides, namely those of the chlorinated hydrocarbon class.

Epilepsy is a clinical disorder manifested by recurrent self-sustaining paroxysmal bursts of abnormal electrical brain activity. Kindling refers to the gradual development of

1. Address all correspondence to: M. E. Gilbert, Ph.D., Neurotoxicology Division (MD-8), Health Effects Research Laboratory, U.S. Environmental Protection Agency, Research Triangle Park, NC 27711. Tel.:(919)541-4394. Fax:(919)541-4849. Email:gilbert@herl45.herl.epa.gov

2. The research described in this article has been supported by the U.S. Environmental Protection Agency, through contract #68-02-4450 to Mantech Environmental Technology Services Inc. The manuscript has been reviewed by the Health Effects Research Laboratory, U.S. EPA, and approved for publication. Approval does not signify that the contents necessarily reflect the views and policies of the Agency nor does mention of trade names or commercial products constitute endorsement or recommendation for use.

3. Key Words: amygdala, epilepsy, kindling, limbic, pesticides, seizures.

Toxicology and Industrial Health, Vol. 10, No. 4/5, pp. 343–358

ISSN: 0748-2337

electrographic and behavioral seizures in response to brief, low-intensity trains of electrical stimulation. The phenomenon was coined kindling by Goddard et al. (1969) to imply a plastic neuronal alteration whereby an initially subthreshold stimulus with repetition eventually reaches threshold and leads to the development of generalized seizures. Kindling serves well as a model of clinical epilepsy because of its chronicity and its overt manifestations, electrographically and behaviorally, closely mimic partial complex seizures in humans. Unlike most other experimental epilepsy models, kindling is progressive and persistent, and eventually it displays spontaneity. There is good pharmacological evidence that correspondence exists between anticonvulsants that are effective in the control of human epilepsy and those that protect against kindled seizures. Its strength as a model of "activity-dependent" plasticity includes the creation of a permanent state of enhanced seizure susceptibility in the absence of major destruction of tissue.

The purpose of this paper is to review the phenomenology of limbic kindling. I will provide an overview of electrical kindling and describe both ictal and interictal manifestations of kindling-induced alterations in brain function. Electrographic and behavioral phenomena that accompany kindling during the seizure (ictus) and persist between seizure episodes (interictal) will be described. Finally, a summary of studies performed in our laboratory on chemical kindling with low doses of pesticide exposure will follow.

ICTAL PHENOMENOLOGY

Electrographic Seizure Development

The electrographic properties of the kindled seizure can be seen in Figure 1. A period of baseline quiescence precedes the delivery of a brief train of electrical stimulation to the amygdala. An abrupt increase in amplitude and synchrony of the EEG recorded from that site in direct response to the electrical stimulus immediately follows. Initially brief, with repetition this ictal phase in response to an identical stimulus train increases in duration, frequency, and complexity. Eventually, the seizure spreads to regions beyond the focal site and recruits other anatomically related regions into the discharge, in this case the contralateral amygdala. Following the ictal period, the baseline EEG is suppressed, corresponding to a period of postictal depression observed in epileptics immediately postseizure. Postictal spikes are frequent immediately following the seizure, and may be accompanied by behavioral seizure components.

Behavioral Seizure Development

No behavioral alteration is evident with subthreshold stimulation. Initial stimulation that leads to very brief afterdischarges (ADs) is often not accompanied by any behavioral manifestations of seizure, but as the kindling process develops, subtle behavioral symptoms emerge and accompany the electrographic seizure indices. Behaviorally, these seizures are focal in nature, representing activation of the stimulation site, and then become generalized

Electrical Kindling

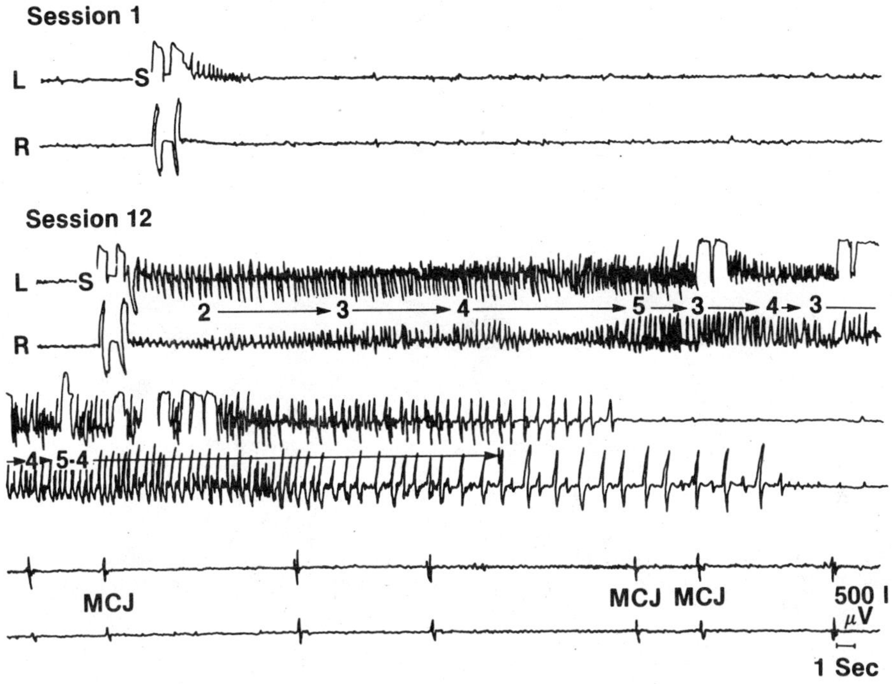

FIGURE 1. Recording from electrodes placed in the right (R) and left (L) amygdala of a rat. A one-second train of 60 Hz pulses was delivered at an intensity of 200 μA at the arrow. The response evoked to this stimulus on the first and twelfth trial is shown. Electrographic seizure discharge (afterdischarge, AD) is brief upon initial stimulation, is not accompanied by behavioral symptoms, and is limited to the focal stimulation site. Once daily stimulation for 12 consecutive days yields a more prolonged AD with complex spike and wave discharge, and a generalized convulsion. Behavioral seizure stages (see Table 1) are indicated on the record. The AD is evident in the amygdala contralateral to the stimulated site. Interictal spikes are evident after termination of the behavioral and electrographic seizure and are accompanied on occasion with myoclonic jerks (MCJ).

and involve activation of motor components similar to secondary generalized seizures in human patients. In the amygdala, the most common kindling site, immobility, eye closure and chewing motions are the most common focal seizure signs (see Table 1).

Subsequent stimulation produces these behaviors and stronger chewing motions and vertical nodding of the head. The focal seizure has now propagated to recruit a larger limbic area, including the contralateral amygdala, hippocampus, septum, and the pyriform cortex. Significant activation of the pyriform cortex appears to precede the transition from primarily focal to generalized seizure states, when generalized motor seizures involve clonic movements of the forelimbs initially, and then the forelimbs and hindlimbs (McIntyre and Kelly, 1993).

TABLE 1. Stages of Seizure Development*

1. Motor Arrest, Immobility, Staring, Eyeblink, Chewing
2. Headnodding
3. Unilateral Forelimb Clonus
4. Rearing with Bilateral Forelimb Clonus
5. Rearing with Bilateral Clonus and Loss of Postural Control
6. Multiple Bouts of Rearing and Falling
7. Running Fits Precede Rearing and Falling
8. Spontaneous Seizures

*Racine, 1972 and Pinel and Rovner, 1978.

PHASES OF KINDLING

Kindling is not monolithic, but can be divided into at least four phases, each of which may model something distinct about the epileptic process, and each of which may depend on a different mechanism.

Prekindling

In the *prekindling* phase, repeated stimulation that is not epileptiform leads to the lowering of AD thresholds. We know very little about the underlying mechanisms of AD reduction with repeated subthreshold activation. It is possible that this type of stimulation results in a potentiation of synaptic responses as has been demonstrated in plasticity models such as long-term potentiation (LTP).

Kindling Process

Although stimulation below the AD threshold will lower the threshold for evoking an AD, unless an AD is induced, kindling will never occur (Racine, 1972). The *kindling process* refers to the progressive development of behavioral and electrographic seizure manifestations that gradually emerge with repeated application of the stimulus. Although the most common AD-eliciting stimulus is application of electrical current to the amygdala, the phenomenon is not limited to an electrical eliciting stimulus. Intermittent injections of subconvulsive dosages of a variety of chemicals can also lead to development of an AD and behavioral seizures (Post, 1980; Cain, 1986). Thus the sufficient and necessary condition for *initiation* of the kindling process is the elicitation of an AD, regardless of its source. The necessary condition for the *progression* of the kindling process to the kindled state entails repetition of the initial AD-inducing events. If a treatment promotes the activation of an AD, and that AD can be repeatedly elicited with further treatment, that treatment will likely produce kindling (see Racine and Burnham, 1984; Cain, 1986).

Kindled State

The *kindled state* represents a period of prolonged maintenance of susceptibility to seizures. In the absence of further stimulation, the animal remains in a state of seizure proneness. A generalized seizure response will be triggered with subsequent stimulation, despite a prolonged intervening period of no stimulation. The permanency of the predisposition to seizures that is induced by kindling is another attractive feature of this plasticity model.

Spontaneity

Unlike most epilepsy models, kindling offers the added feature of chronicity. If stimulation is continued, *spontaneous seizures* will develop in the absence of the initiating stimulus. Many stimulations are required in the rat, and the seizures appear to be nonfocal, a generalized convulsion erupting suddenly in response to handling. The focal site of stimulation that produced the state of spontaneity does not necessarily lead in terms of seizure discharge (Pinel and Rovner, 1978). These seizures resemble true generalized seizures in human patients.

TRANSFER PHENOMENON IN KINDLING

The transfer phenomenon in kindling was first described by Goddard et al. (1969) who reported that electrical kindling of the amygdala facilitates subsequent kindling of the contralateral amygdala or ipsilateral septum. As the electrical stimulus is repeatedly applied to the primary kindling focus and the seizure propagates to areas beyond the stimulated site, the regions transynaptic to the focal stimulation site begin to display discharge independent of the primary focus (Figure 1). If the amygdala is kindled to stage 5 seizures, fewer stimulations will be required in subsequent kindling of a number of other sites (e.g., dorsal or ventral hippocampus, septum, contralateral amygdala), indicating a significant facilitation in transmission among forebrain structures during the development of the kindling process (Racine, 1972; Burnham, 1975). That positive transfer occurs even upon removal of the primary kindled focus implies that this facilitation occurs beyond the primary kindled site (Racine, 1978). Transfer, however, is not immediate, but requires several stimulations of the secondary transfer site to reestablish maximal behavioral responsiveness. It appears that the kindled circuitry supporting behavioral convulsive responses is gradually built with each evoked AD. This circuit is in a permanent state of readiness with heightened propensity for seizure activation. Positive transfer occurs when another site is activated, begins to build its circuitry up with repetition of the AD-inducing stimulus, and eventually accesses the existing generalized seizure circuitry produced by primary site kindling.

Increased sensitivity after initial kindling with one agent to other agents capable of inducing epileptiform discharge is another aspect of the transfer phenomenon (see review by Cain, 1986). For example, rats kindled by traditional focal electrical stimulation of the amygdala are rendered permanently more susceptible to convulsions induced by alcohol withdrawal (Pinel and Van Oot, 1975) or pentylenetetrazol injection (Cain, 1981, 1982). Conversely, rats kindled with repeated subthreshold administration of convulsant drugs such as pentylenetetrazol, lidocaine, carbacol, bicucculine, or picrotoxin require many fewer stimulations to reach generalized seizures when kindled electrically (e.g., Post, 1980; Cain, 1981, 1983, 1987; Corda et al., 1990; Uemura and Kimura, 1990). The "savings" that occurs between chemical and electrical kindling implies that the same seizure circuitry is enhanced with repeated chemical exposure, as occurs with repeated electrically-induced ADs. Therefore, strong evidence of chemical kindling is provided when facilitation of subsequent electrical kindling can be demonstrated.

INTERICTAL PHENOMENA

Spontaneous Interictal Events

Interictal spikes (IIS) are a common marker of epileptic brain in a nonictal state. In the clinic they are used as an indication of focal activity and a means to identify the epileptogenic tissue. It is believed that spontaneous IIS represent normal transients present in a nonepileptic brain that become augmented in size and occur with higher probability following kindling (Racine et al., 1986). Amygdala kindling results in the appearance of IIS following about the fourth AD. There is a progressive increase in frequency of IIS as kindling develops, and they are most prevalent immediately following generalized seizures in kindled animals. In some areas, i.e. the hippocampus, IIS increase initially, then decline over the course of several days (Leung, 1990). In other regions, however, there is an initial decline in frequency, but the IIS persist and remain stable for periods of 6 weeks to 12 months (Gotman, 1984; Kairiss et al., 1984; Racine et al., 1988). The incidence and increase in amplitude of IIS following kindling is greatest in the pyriform region regardless of the primary kindling site (Kairiss et al., 1984; Racine et al., 1988).

Kindling-Induced Potentiation

Long-term potentiation (LTP) refers to a long-lasting augmentation in synaptic responses that is produced by driving those pathways to high levels of activity by the application of high frequency, *nonepileptiform* producing trains of electrical stimulation. It is most prominent in the hippocampus, where it has been studied extensively as a model of learning and memory. One of the important features of LTP is restriction of the potentiation to the stimulated pathways. A more prolonged and higher magnitude potentiation of synaptic responsivity also accompanies kindling stimulation and is referred to as kindling-induced potentiation (Racine et al., 1983). There has been considerable debate as to the contribution of LTP-mechanisms to the kindling process. Regardless of whether potentiation mechanisms *contribute* to kindling development, it is clear that a long-lasting increase in synaptic responsivity nevertheless *accompanies* kindling. The source of the high activity stimulus in kindling-induced potentiation is the AD as successive augmentation of synaptic signals occur as the seizures propagate from focal to extrafocal sites. Racine et al. (1983) have demonstrated potentiation of evoked field responses throughout the forebrain in kindled animals. The synaptic potentiation of extrafocal circuits that accompanies nonconvulsive states of kindling may well have long-lasting effects on behavior in the interictal period (see Adamec and Stark-Adamec, 1983a,b; Leung and Shen, 1991).

We typically think of potentiation as representing increases in the excitability of neuronal circuits. In the monosynaptic connections between the perforant path — dentate gyrus synapse and the olfactory bulb — pyriform cortex synapse, where an interpretable field potential can be reliably recorded in a chronically implanted animal, clear evidence of synaptic potentiation of excitatory neurotransmission has been well documented (e.g., Russell and Stripling, 1985; deJonge and Racine, 1986; Gilbert and Mack, 1990; Racine et al., 1991; Robinson et al., 1991). However, it is also clear that *inhibitory* circuitry is also potentiated in these regions by the kindling process (Russell and Stripling, 1985; deJonge and Racine, 1986; Stripling et al., 1988; Gilbert, 1991; Patneau and Stripling, 1992).

Interictal Behavioral Sequelae of Kindling

It is well established that in humans, nonhuman primates, and other mammals, damage to medial temporal structures can disrupt normal function in two broad behavioral categories, memory and emotion (Zola-Morgan et al., 1991). Both the amygdala and hippocampus have been implicated in these behavioral categories, in temporal lobe epilepsy, in partial complex seizures, and in kindling. Several animal studies have demonstrated cognitive impairment when the learning or memory demand is placed upon the animal during the presentation of the relevant stimuli, or a kindled seizure is given in the immediate posttraining period. Deficits occur if the seizure occurs during acquisition of a task or in a well-trained animal stimulated just prior to performance of the task (see Table 2). This is not surprising if the seizure is thought of as creating a functional lesion in areas important for learning and memory. Information is neither processed or appropriately stored in these regions when the mechanisms of information transfer, i.e., spatial temporal patterning of neuronal firing, are inaccessible. These impairments may be best thought of as acute and transient disruptions of behavior directly tied to the ictal event.

TABLE 2. Effects of Kindling and Kindled Seizures on Cognitive Function

Learning task	Reference
Seizures Transiently Disrupt Learning and Memory	
Passive Avoidance	Peele and Gilbert, 1992
Conditioned Avoidance	McIntyre, 1979; Majkowski et al., 1989
Flavor Aversion	LePaine and Phillips, 1978; Mikulka and Freeman, 1984; Peele and Gilbert, 1992
Holeboard Test	Beldhuis et al., 1992
Radial Arm Maze	Knowlton et al., 1989, Leung et al., 1990; Feasey-Truger et al., 1993
Water Maze	McNamara et al., 1992
History of Seizures Leads to Persistent Cognitive Deficits	
Passive Avoidance	Boast and McIntyre, 1977; Stone and Gold, 1988; Peele and Gilbert, 1992
Conditioned Emotional Response	McIntyre, 1979
Eye Blink Conditioning Reversal	Robinson, 1992
Radial Arm Maze	Lopes de Silva et al., 1986; Leung et al., 1990, 1991

More importantly, persistent memory disturbances have also been reported in kindled animals in the absence of ictal discharge. Either fully kindled or partially kindled animals are deficient in a variety of tasks including passive avoidance learning, radial arm maze performance, conditioned emotional responding, and reversal learning in an eyeblink conditioning task (see examples in Table 2). These deficits can occur in animals pretrained on the learning task prior to kindling, or when initial learning takes place in well kindled subjects. Deficits have been reported to occur from 7 to more than 21 days after the last seizure (see, e.g., Leung et al., 1990; Leung and Shen, 1991; Peele and Gilbert, 1992).

Kindling and Psychopathology

A high percentage of epileptics experience serious psychological problems that often take the form of depression and anxiety (see Adamec, 1989, 1990a for review). Fear auras implicate activation of the amygdala in the seizure response of epileptics (Gloor et al., 1982). It has been proposed that patients with temporal lobe foci and patients with fear auras are more likely than "other epileptics" and other "nonepileptic patients with psychiatric disorders" to have interictal psychological disturbances (Weiser, 1983). The clinical data suggest that these patients have a persistent seizure-induced alteration in limbic function that manifests itself in the form of depression and anxiety (see Adamec, 1989, 1990a).

Some data are available on the emotional disposition of kindled animals that may reflect the clinical picture of psychopathology in some epileptics. Increases in emotionality as indexed by irritability and aggression in handling tests, resistance to capture, and agitation and aggression in response to tactile stimuli have been reported in kindled rats. Recently, Adamec (1990b) has shown kindling-induced differences in activity in an elevated plus maze, a standard test of anxiety in rodents. However, probably the best studied animal model of human affective disorders is defensive behavior in cats. Defensive behavior is taken as a measure of animal anxiety and is produced by and requires repeated focal limbic seizures that need not generalize to a convulsive response. Partial kindling produces interictal mood lability in which animals display explosive defensive reactions to mild provocation (Adamec and Stark-Adamec, 1983a). The augmentation in defensive reaction persists for long periods of time in the absence of continued limbic seizure activation. Explosive defense reactions in kindled cats are associated with a long lasting kindling-induced potentiation of responses evoked in the hypothalamus from single pulse stimulation of the amygdala, circuitry known to be involved in the expression of emotional behavior (see Adamec and Stark-Adamec, 1983b). So in kindling, as in epilepsy associated with limbic activation, there may exist an enhanced propensity for pathological responses to normal stressors.

CHEMICAL KINDLING

There are many toxicants that produce convulsions in humans and animals in response to acute high-dose exposure (Morgan, 1989; Hayes and Laws, 1991). We and others had previously demonstrated the capacity for a variety of drugs and pesticides to facilitate the development of electrical kindling (see Gilbert, 1992a for review). In this type of study, the drug of interest is administered prior to the delivery of the electrical stimulus. Some chemicals augment the electrically induced AD and facilitate seizure development, others suppress the process by either elevating seizure threshold or limiting the spread of seizure propagation. A more interesting question to us, however, was whether or not repeated administration of the pesticide alone at low levels could promote kindling.

Environmental chemicals with neuroactive properties may induce seizures at high dosages by directly stimulating excitatory pathways in the brain, or by suppressing inhibitory neurotransmission. However, seizures are also a symptom of global brain dysfunction, arising from a variety of other nonspecific etiologies such as encephalopathy, ionic imbalances, metabolic disturbances, and periods of anoxia. Therefore, although many convulsants do induce kindling when administered repeatedly at subthreshold levels (PTZ and cocaine are common kindling agents), seizures arising from acute high-dose administration of chemicals that secondarily and only indirectly disrupt central nervous system (CNS) function may not produce kindling-like changes with repeated low-level exposure.

It has also become clear that interpretations of chemical kindling can be confounded by several methodological difficulties that lead to apparent but spurious kindling (see Nutt et al., 1982). For example, increased seizure severity may result from an accumulation of the chemical in the body, aging may lead to increased sensitivity to convulsant agents, increases in body weight over the course of the dosing period may result in higher absolute target doses with a constant delivered dose, or alterations in pharmacokinetics of the administered dose may lead to increased plasma and brain levels and account for an augmented behavioral response. Therefore, a number of criteria should be met in a chemical kindling experiment prior to concluding that repeated exposure to a chemical produces kindling. The chemical kindling process should begin with an initially subconvulsive dose of the chemical of interest in order to monitor the gradual emergence of a behavioral response with repeated dosing. There must also be a maintained behavioral/electrographic responsiveness to the initially subconvulsive dose, following a washout period of sufficient duration to permit elimination of the chemical from the body. Finally, if kindling is produced by repeated exposure to the chemical, there should be positive transfer to electrical kindling. Just as transfer occurs between kindling sites, if chemical kindling has occurred, a significant savings in electrical kindling development must follow in the absence of further exposure.

DO PESTICIDES CHEMICALLY KINDLE?

We have currently studied two pesticides, endosulfan and lindane, for their kindling potential. The mechanism of action of these chlorinated hydrocarbon pesticides is not known, but both have been shown to decrease GABA function in *in vitro* binding assays of rodent brain (e.g., Lawrence and Casida, 1984; Abalis et al., 1985). Additionally, lindane decreased GABAergic synaptic transmission in hippocampus *in vivo* (Joy and Albertson, 1987). Repeated oral dosing of either pesticide of this class at subconvulsive dosages led to the emergence of behavioral signs of seizure, myoclonic jerks, and clonic seizures (see Figure 2). Both of these chemicals have relatively long elimination half-lives, on the order of 27 hours for lindane (Tusell et al., 1987). A washout period instituted to test a cumulative toxicity vs. a kindling mechanism, revealed a maintained behavioral responsiveness upon challenge up to one month following the last dose.

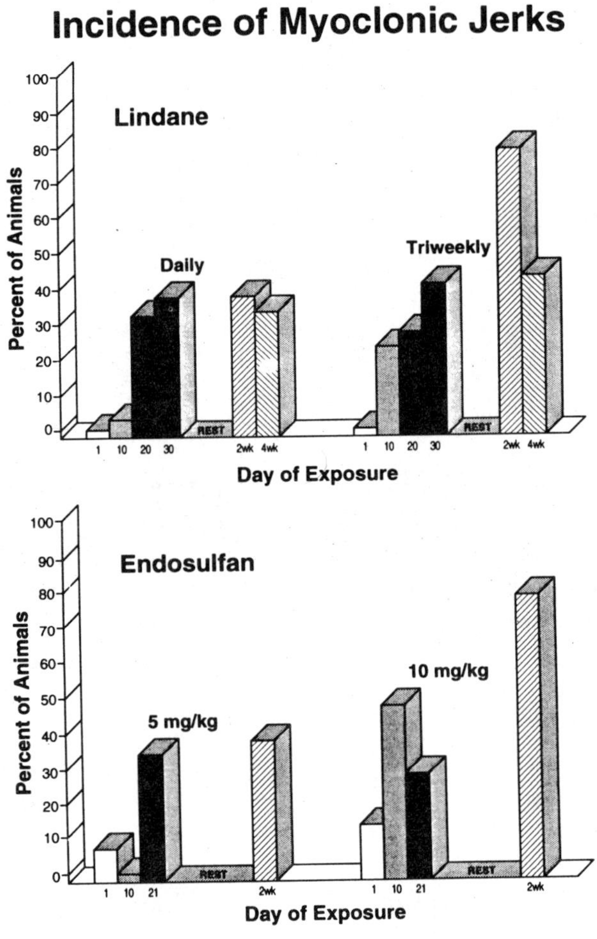

FIGURE 2. Incidence of myoclonic jerks (MCJ) in animals treated daily with 5 or 10 mg/kg of endosulfan (A), or daily or triweekly with 10 mg/kg lindane (B). Behavioral observations were performed for 1 hour, 30 minutes after dosing on days 1, 10, 20–21, and 30. The endosulfan-treated animals and one group of animals treated with lindane were challenged 14 days after the final dose (2wk). A second group of lindane-treated animals was challenged 30 days after the final dose. Control animals in the 30-day challenge group were also administered 10 mg/kg of lindane; the 14-day challenge control group received corn oil. Data from Gilbert, 1992b, 1994.

Electrographic seizure indices also revealed a heightened sensitivity to lindane in animals showing minimal behavioral seizure responsiveness. Overt clonic convulsions accompanied by discrete bouts of AD were observed in a portion of animals upon challenge. However, the electrographic record of the population of subjects demonstrating minimal behavioral responsivity is of particular interest. Figure 3 is a sample of an EEG recording from the amygdala of a control animal previously treated with corn oil for 30 days and challenged with lindane for the first time. A few spikes are evident over the course of 1.5 hours of recording. In contrast, an animal responding with only minimal behavioral signs (i.e., MCJs) shows many interictal spike events throughout the recording period. We counted IIS during a 30-minute baseline period before dosing animals with 10 mg/kg lindane on the one-month challenge, excluding subjects that responded with overt behavioral convulsions. Naive

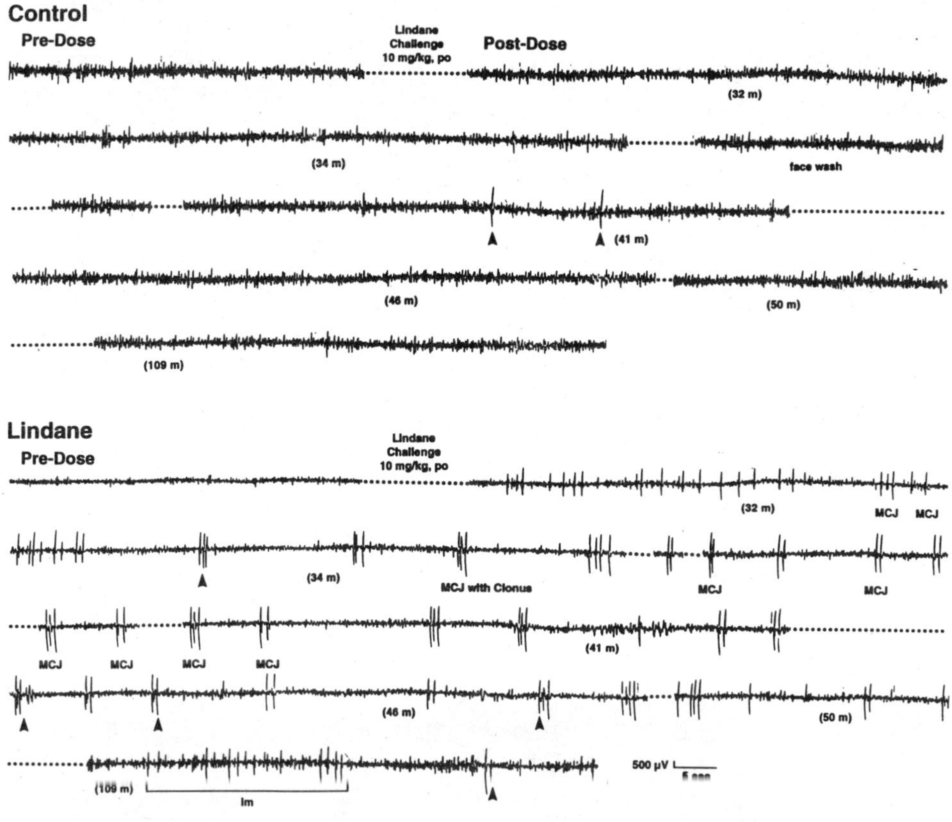

FIGURE 3. Electroencephalogram (EEG) recorded from the amygdala in a control (A) and a lindane-pretreated (B) animal, 30 days after the last dose. A sample taken from the 30-minute period of baseline EEG preceding administration of a 10 mg/kg challenge dose of lindane is followed by selected segments of the 1.5-hr recording session that followed. No behavioral response to lindane was observed in the control animal. Two spike-and-wave interictal discharges of higher amplitude than any observed over the baseline recording session are indicated by the arrowheads in A. No convulsion was observed in the treated animal in record B, but myoclonic jerks (MCJ) were prevalent, and always associated with high amplitude spike and wave discharges occurring singly or in clusters. A 20-sec period of AD-like activity was recorded 109 minutes after dosing but no ictal behavior patterns were evoked. Data from Gilbert, 1994.

animals had fewer spikes during the baseline period than treated animals, and both groups increased the frequency of IIS following dosing. There was not a significant increase in control animals, but previously treated animals had a much higher incidence of IIS relative to their predose baselines. These data indicate that despite the absence of overt behavioral seizures, the amygdala is much more responsive to the excitatory effects of lindane for periods of up to one month.

Beginning two weeks after the challenge dose, animals were then kindled electrically with once daily stimulation applied to the amygdala. Both lindane and endosulfan pretreated animals responded with behavioral convulsions with fewer stimulations than corn oil-treated controls (Figure 4). A history of *repeated* exposure to the pesticides was essential to facilitating subsequent electrical kindling — i.e., a single convulsive experience produced by

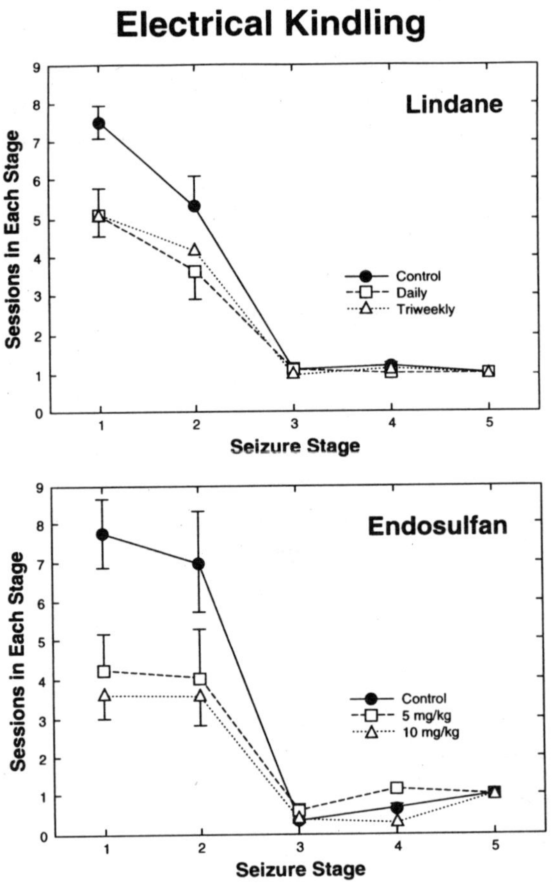

FIGURE 4. Preexposure to endosulfan (A) and lindane (B) facilitated the development of electrical kindling that began 4–6 weeks after the last dose. The figure shows the mean number of electrical stimulation sessions in each stage of motor seizure. Data from Gilbert, 1992b, 1994.

a high dose of lindane or endosulfan did not change kindling rate relative to corn oil-treated controls. Positive transfer to electrical kindling provides further evidence of long-lasting augmentation in limbic excitability.

Animals chemically kindled with agents such as pentylenetetrazol and picrotoxin demonstrate bidirectional transfer, i.e. a heightened sensitivity to the convulsive properties of these drugs in electrically kindled animals, in addition to positive savings from chemical to electrical stimulation paradigms (see Cain, 1986). To ascertain the bidirectionality of our findings with pesticides, we performed two additional experiments. In well kindled but pesticide-naive animals, we determined the effects of low dosages of endosulfan on AD thresholds and found a significant increase in the number of animals responding with generalized convulsions at subthreshold stimulations. Secondly, we compared the behavioral seizure responsiveness in well kindled vs. nonkindled animals to lindane and endosulfan. Both pesticides led to a heightened seizure response at lower doses in kindled than nonkindled subjects. A history of kindling increased the mortality rate in animals administered a high dose of endosulfan. Electrographic responsiveness was also markedly enhanced at low dosages of both pesticides in animals failing to respond with overt convulsions to the chemical challenge (Gilbert and Mack, 1994).

Collectively, these data indicate that kindling increases limbic system excitability and lowers the threshold for convulsions and lethality to pesticides. Furthermore, they suggest that kindling is produced by repeated low-level exposure to pesticides and creates a long-lasting augmentation in CNS excitability, even in the absence of overt behavioral convulsions. The bidirectional transfer between electrical and chemical kindling suggests that similar neural mechanisms are invoked by repeated delivery of either chemical or electrical stimuli.

In summary, kindling occurs when repeated subthreshold electrical or chemical stimuli are applied to trigger abnormal bursts of paroxysmal activity in susceptible CNS structures. The physiological responses provoked by these bursts promote a persistent alteration in nervous system function, one manifestation of which is the development of convulsive behavior and a permanent predisposition to seizures from a variety of other sources. The similarity in the gradual and progressive nature of behavioral seizure development and the transfer between kindling induced by electrical and chemical means suggests that they rely upon the same mechanisms. The evocation of electrographic and behavioral seizures and IIS by repeated low-level exposure to some pesticides indicates that persistent neurological sequelae may also accompany prolonged exposure to some environmental pollutants. If pesticides produce changes in CNS function similar to electrical kindling, exposures so severe as to lead to convulsive behavior are not necessary to produce long-term alterations in behavior. Partial kindling induced by repeated induction of focal ADs to a nonconvulsive state disrupts cognitive processes and emotionality (e.g., Leung et al., 1991; Adamec and Stark-Adamec, 1983a). Therefore, it is not unreasonable to propose that neurological sequelae known to accompany electrical kindling may also occur with chemical kindling induced by pesticides.

The possibility of persistent neurological sequelae to *nonafterdischarge*-inducing exposures to electrical stimulation or pesticides, however, has not been established.

ACKNOWLEDGMENTS

The expert technical assistance of Cina Mack is gratefully acknowledged. The author would also like to thank Drs. W.K. Boyes and Bill Wilson for their comments on an earlier version of this manuscript.

REFERENCES

ABALIS, I.M., ELDEFRAWI, M.E. and ELDEFRAWI, A.T. (1985). "High-affinity stereospecific binding of cyclodiene insecticides and gamma-hexachlorocyclohexane to gamma-aminobutyric acid receptors of rat brain." Pesticide Biochem. Physiol. 24:95–102.

ADAMEC, R.E. (1989). "Kindling, anxiety and personality." In: The Clinical Relevance of Kindling (T.G. Bolwig and M.R. Trimble, ed.). John Wiley and Sons, New York, NY. pp. 117–263.

ADAMEC, R.E. (1990a). "Does kindling model anything clinically relavent?" Biol. Psychiatry 27:249–279.

ADAMEC, R.E. (1990b). "Amygdala kindling and anxiety in the rat." NeuroReport. 1:255–258.

ADAMEC, R.E. and STARK-ADAMEC, C. (1983a). "Partial kindling and emotional bias in the cat: Lasting aftereffects of partial kindling of the ventral hippocampus: Physiological changes." Behav. Neural. Biol. 38:205–222.

ADAMEC, R.E. and STARK-ADAMEC, C. (1983b). "Partial kindling and emotional bias in the cat: Lasting aftereffects of partial kindling of the ventral hippocmapus: Behavioral changes." Behav. Neural. Biol. 38:205–222.

BELDHUIS, H.J.A., EVERTS, H.G.J., VAN DER ZEE, E.A. LUITEN, P.G.M. and BOHUS, B. (1992). "Amygdala kindling-induced seizures selectively impair spatial memory. 1. Behavioral characteristics and effects on hippocampal neuronal protein kinase C isoforms." Hippocampus 2:397–410.

BOAST, C. and McINTYRE, D.C. (1977). "Bilateral kindled amygdala foci and inhibitory avoidance behavior in rats: A functional lesion effect." Physiol. Behav. 18:25–28.

BURNHAM, W.M. (1975). "Primary and transfer seizure development in the kindled rat." Can. J. Neurol. Sc. 2:417–428.

CAIN, D.P. (1981). "Transfer of pentylenetetrazol sensitization to amygdaloid kindling." Pharmacol. Biochem. Behav. 15:533–536.

CAIN, D.P. (1982). "Bidirectional transfer of intracerebrally administered pentylenetetrazol and electrical kindling." Pharmacol. Biochem. Behav. 17:1111–1113.

CAIN, D.P. (1983). "Bidirectional transfer of electrical kindling and carbachol kindling." Brain Res. 260:135–138.

CAIN, D.P. (1986). "The transfer phenomenon in kindling." In: Kindling 3 (J.A. Wada, ed.). Raven Press, New York, NY. pp. 231–248.

CAIN, D.P. (1987). "Kindling by repeated intraperitoneal or intracerebral injection of picrotoxin transfers to electrical kindling." Exper. Neurol. 97:243–254.

CORDA, M.G., GIORGI, O., ORLANDI, M., LONGONI, B., and BIGGIO, G. (1990). "Chronic administration of negative modulators produces chemical kindling and $GABA_A$ receptor down-regulation." In: GABA and Benzodiazepine Receptor Subtypes (G. Biggio and E. Costa, ed.). Raven Press, New York, NY. pp. 1–25.

DEJONGE, M. and RACINE, R.J. (1986). "The development and decay of kindling-induced increases in paired pulse depression in the dentate gyrus." Brain Res. 412:318–328.

FEASEY-TRUGER, K.J., KARGL, L., and TEN BRUGGENCATE, G. (1993). "Differential effects of dentate kindling on working and reference spatial memory in the rat." Neurosci. Lett. 151:25–28.

GILBERT, M.E. (1991). "Potentiation of inhibition with perforant path kindling: An NMDA-dependent process." Brain Res. 564:109–116.
GILBERT, M.E. (1992a). "Neurotoxicants and limbic kindling." In: The Vulnerable Brain and Environmental Risks (R.L. Isaacson and K.F. Jensen, ed.). vol. 1: Malnutrition and Hazard Assessment. Plenum Press, New York, NY. pp. 173–192.
GILBERT, M.E. (1992b). "A characterization of chemical kindling with the pesticide endosulfan." Neurotoxicol. Teratol. 14:151–158.
GILBERT, M.E. (1994). "Repeated low-level exposure to lindane leads to behavioral sensitization and facilitates electrical kindling." Neurotoxicol. Teratol. In press.
GILBERT, M.E. and MACK, C.M. (1990). "The NMDA-receptor antagonist, MK-801, suppressed long-term potentiation, kindling, and kindling-induced potentiation in the perforant path of the unanesthetized rat." Brain Res. 519:89–96.
GILBERT, M.E. and MACK, C.M. (1994) "Seizure thresholds in kindled animals are reduced by the pesticides lindane and endosulfan." Neurotoxicol. Teratol. In Press.
GLOOR, P., OLIVIER, O., QUESNEY, L.F., ANDERMANN, F. and HOROWITZ, S. (1982). "The role of the limbic system in experiential phenomena of temporal lobe epilepsy." Ann. Neurol. 12:129–144.
GODDARD, G.V., McINTYRE, D.C. and LEECH, C. (1969). "A permanent change in brain function resulting from daily electrical stimulation." Exper. Neurol. 25:295–330.
GOTMAN, J. (1984). "Relationships between triggered seizures, spontaneous seizures, and interictal spiking in the kindling model of epilepsy." Exper. Neurol. 84:259–273.
HAYES, W.J. and LAWS, E.R. (1991). Handbook of Toxicology. Academic Press, New York, NY.
JOY, R.M. and ALBERTSON, T.E. (1987). "Interactions of lindane with synaptically mediated inhibition and facilitation in the dentate gyrus." Neurotoxicology 8:529–542.
KAIRISS, E.W., RACINE, R.J. and SMITH, G. (1984). "The development of the interictal spike during kindling in the rat." Brain Res. 322:101–110.
KNOWLTON, B.J., SHAPIRO, M.L., and OLTON, D.S. (1989). "Hippocampal seizures disrupt working memory performance but not reference memory acquisition." Behav. Neurosci. 103:1144–1147.
LAWRENCE, L.J. and CASIDA, J.E. (1984). "Interaction of lindane, toxephine cyclodienes with brain specific t-butylbicyclophophorothionate receptor." Life Sci. 35:171–178.
LEPAINE, F.G. and PHILLIPS, A.G. (1978). "Differential effects of electrical stimulation of amygdala, caudate-putamen, or substantia nigra pars compacata on taste aversion and passive avoidance in rats." Physiol. Behav. 21:979–985.
LEUNG, L. S. (1990). "Spontaneous hippocampal interictal spikes following local kindling: Time-course of change and relations to behavioral seizures." Brain Res. 513:308–314.
LEUNG, L.S., BOON, K., KAIBARA, T., and INNIS, N.K. (1990). "Radial maze performance following hippocampal kindling." Behav. Brain Res. 40:119–129.
LEUNG, L.S. and SHEN, B. (1991). "Hippocampal CA1 evoked response and radial 8-arm maze performance after hippocampal kindling." Brain Res. 555:353–357.
LOPES DA SILVA, F.H., GORTER, J.A., and WADMAN, W.J. (1986). "Kindling of the hippocampus induces spatial memory deficits in the rat." Neurosci. Lett. 63:115–120.
MAJKOWSKI, J., SOBIESZEK, A., and DLAWICHOWSKA, E. (1989). "EEG afterdischarge patterns and performance of the avoidance response in hippocampally kindled cats." Acta Neurobiol. Exper. 49:13–22.
McINTYRE, D.C. (1979). "Effects of focal vs generalized kindled convulsions from anterior neocortex or amygdlal on CER acquisition in rats." Physiol. Behav. 23:855–859.
McINTYRE, D.C. and KELLY, M.E. (1993). "Are differences in dorsal hippocampal kindling related to amygdala-piriform area excitability?" Epilepsy Res. 14:49–61.
McNAMARA, R.K., KIRKBY, R.D., DEPAPE, G.E., and CORCORAN, M.E. (1992). "Limbic seizures, but not kindling, reversibly impair place learning in the Morris water maze." Behav. Brain Res. 50:167–175.
MILKULKA, P.J. and FREEMAN, F.G. (1984). "The effect of amygdala-kindled seizures on the acquisition of taste and odor aversions." Physiol. Behav. 32:967–972.

MORGAN, D.P. (1989). Recognition and Management of Pesticide Poisonings, 4th Edition. US EPA, Washington, D.C.

NUTT, D.J., COWEN, P.J., BATTS, C.C., GRAHAME-SMITH, D.G. and GREEN, A.R. (1982). "Repeated administration of subconvulsant doses of GABA antagonist drugs. I. Effects on seizure threshold (Kindling)." Psychopharmacology 76:84–87.

PATNEAU, D. and STRIPLING, J.S. (1992). "Functional correlates of selective long-term potentiation in the olfactory cortex and olfactory bulb." Brain Res. 585:219–228.

PEELE, D.B. and GILBERT, M.E. (1992). "Functional dissociation of acute and persistent cognitive deficits accompanying amygdala-kindled seizures." Behav. Brain Res. 48:65–76.

PINEL, J.P.J. and ROVNER, L.I. (1978). "Experimental epileptogenesis: Kindling-induced epilepsy in rats." Exper. Neurol. 58:190–202.

PINEL, J.P.J. and VAN OOT, P.H. (1975). "Intensification of the alcohol withdrawal syndrome by repeated brain stimulation." Nature 254:510–511.

POST, R.M. (1980). "Minireview: Intermittent versus continious stimulations: Effects of time interval on the development of sensitization and tolerance." Life Sci. 26:1275–1282.

RACINE R.J. (1972). "Modification of seizure activity by electrical stimulation. II. Motor seizures." Electroenceph. Clin. Neurophysiol. 32:281–294.

RACINE, R.J. (1978). "Kindling, the first decade." Neurosurgery 3:234–252.

RACINE, R.J. and BURNHAM, W.M. (1984). "The Kindling Model." In: Electrophysiology of Epilepsy (P. Schwartzkroin and H. Wheal, ed.). Academic Press, London. pp.153–171.

RACINE, R.J., BURNHAM, W.C., GILBERT, M.E., and KAIRISS, E.W. (1986). "Kindling mechanisms: I. Electrophysiological studties." In: Kindling 3 (J.A. Wada, ed.). Raven Press, New York, NY. pp. 263–282.

RACINE, R.J., MILGRAM, N.W., and HAFNER, S. (1983). "Long-term potentiation phenomena in the rat limbic forebrain." Brain Res. 260:217–231.

RACINE, R.J., MOSHER, M., and KAIRISS, E.W. (1988). "The role of the pyriform cortex in the generation of interictal spikes in the kindled preparation." Brain Res. 454:251–263.

RACINE, R.J., MOORE, K.-A., and EVANS, C. (1991). "Kindling-induced potentiation in the piriform cortex." Brain Res. 556:218–225.

ROBINSON, G.B. (1992). "Reversal learning of the rabbit nictitating membrane response following kindling-induced potentiation within the hippocampal dentate gyrus." Behav, Brain Res. 50:185–192.

ROBINSON, G.B., SCLABASSI, R.J., and BERGER, T.W. (1991). "Kindling-induced potentiation of excitatory and inhibitory inputs to hippocampal dentate granule cells. I. Effects on linear and nonlinear response characteristics." Brain Res. 562:17–25.

RUSSELL, R.D. and STRIPLING, J.S. (1985). "Effect of olfactory bulb kindling on evoked potentials in the pyriform cortex." Brain Res. 361:61–69.

STONE, W.S. and GOLD, P.E. (1988). "Amygdala kindling effects on sleep and memory in rats." Brain Res. 449:135–140.

STRIPLING, J.S., PATNEAU, D., and GRAMLICH, C.A. (1988). "Selective long-term potentiation in the pyriform cortex." Brain Res. 441:281–291.

TUSELL, J.M., SUNOL, C., GELPI, E., and RODRIGUEZ-FARRE, E. (1987). "Relationship between lindane concentration in blood and brain and convulsant response in rats after oral or intraperitoneal administration." Arch. Toxicol. 60:432–437.

UEMURA, S. and KIMURA, H. (1990). "Common epileptic pathway in amygdaloid bicuculline and electrical kindling demonstrated by transferability." Brain Res. 537:315–317.

WIESER, H.G. (1983). "Depth recorded limbic seizures and psychopathology." Neurosci. Behav. Rev. 7:427–440.

ZOLA-MORGAN, S., SQUIRE, L., ALVAREZ-ROYO, P., and CLOWER, R.P. (1991). "Independence of memory functions and emotional behaviour: Contributions of the hippocampal formation and the amygdala." Hippocampus 1:207–220.

OLFACTION AND MULTIPLE CHEMICAL SENSITIVITY

RICHARD L. DOTY

Smell and Taste Center
University of Pennsylvania Medical Center
Philadelphia, Pennsylvania

In this paper, a description of olfactory anatomy is presented, followed by a brief review of modern procedures for testing olfactory function. Information from the sole study which has quantitatively examined olfactory function in patients with apparent multiple chemical sensitivity (MCS) is presented. In essence, this study suggests that MCS is associated with increased nasal airflow resistance, respiration rate, heart rate, and scores on the Beck Depression Inventory, but not with significant changes in odor detection threshold sensitivity to phenyl ethyl alcohol and methyl ethyl ketone, the two target stimuli evaluated. Whether MCS patients evidence hypersensitivity to other chemicals is unknown.

INTRODUCTION

The human sense of smell determines to a large degree the flavor of foods and beverages, and is responsible for a wide variety of aesthetic experiences. Losses or distortions of this sense significantly alter the quality of life, including the appreciation of fine wine, perfumes, and fresh spring air. Importantly, abnormalities in the ability to smell can lead to death or severe injury, as when persons with limited mobility, such as the aged, are unable to detect the smell of fire, leaking natural gas, fireplace effluence, automobile exhaust fumes, or smoke. In a recent study of 750 consecutive patients presenting to the University of Pennsylvania Smell and Taste Center with complaints of disordered chemosensation, 68% viewed their dysfunction as effecting their quality of life, 56% indicated that the problem altered their daily living and/or psychological well being, and 46% reported that the problem changed either their body weight or appetite (Deems et al., 1991).

Decreased ability to smell is quite common. For example, approximately half of individuals between the ages of 65 and 80 years of age have major difficulty in smelling odors (Doty et al., 1984b). In contrast, increased ability to smell is rare, and its prevalence in the general

1. Address all correspondence to: Richard L. Doty, Smell and Taste Center, Hospital of the University of Pennsylvania, 3400 Spruce Street, Philadelphia, PA 19104.

2. Key Words: Chemical Hypersensitivity, Odor Threshold, Olfaction, UPSIT.

Toxicology and Industrial Health, Vol. 10, No. 4/5, pp. 359–368

ISSN: 0748-2337

population is unknown. Despite the fact that there are reports of heightened olfactory sensitivity in patients with adrenal cortical insufficiency (Henken and Bartter, 1966) and cystic fibrosis (Henken and Powell, 1962), such hypersensitivity has not been confirmed by others. Indeed, in the case of cystic fibrosis, decreased, not increased, sensitivity has been reported. Thus, Hertz et al. (1975) found that cystic fibrosis patients exhibit decreased sensitivity to the odor of n-butanol. A similar decrease in threshold sensitivity to the odorant phenylethylmethylethylcarbinol, as well as in odor identification ability, was noted by Weiffenbach and McCarthy (1984) in such patients. Importantly, only a small subgroup of the 20 patients examined in this study evidenced marked olfactory dysfunction, with most being either normal or only minimally different from controls.

To our knowledge, only one study has examined the olfactory sensitivity and nasal resistance of individuals reporting symptoms consistent with the diagnosis of multiple chemical hypersensitivity (MCS)(Doty et al., 1988b). Although this study, which is described in detail later in this chapter, found no evidence of MCS-related heightened olfactory sensitivity to the two odorants studied, phenyl ethyl alcohol and methyl ethyl ketone, it did find that MCS patients had more resistance to nasal airflow than did matched normal controls.

The purpose of this chapter is to briefly describe (a) the anatomy of the olfactory system, (b) basic psychophysical procedures for assessing its function, and (c) the olfactory function of patients with presumed multiple chemical hypersensitivity. The reader is referred elsewhere for more detailed reviews of the first two of these topics (e.g., Getchell et al., 1991; Serby and Chober, 1992; Doty, 1995).

BRIEF DESCRIPTION OF THE ANATOMY OF THE OLFACTORY SYSTEM

Volatile chemicals enter the nose where they may, depending upon a number of physicochemical factors, activate cells of the olfactory nerve (CN I) and/or free nerve endings from the ophthalmic and maxillary divisions of the trigeminal nerve (CN V). The latter nerve filaments are distributed throughout the nasal mucosa, including the olfactory neuroepithelium (Figure 1). Some inhaled chemicals also stimulate free nerve endings within the oral pharynx and cavity, such as those from the glossopharyngeal (CN IX) and vagus (CN X) nerves. However, qualitative sensations commonly termed odor are mediated solely by the olfactory nerves (CN I); the sensations mediated by CN V, IX, and X are largely somatosensory in nature, such as warmth/coolness, irritation, and pungency (Silver, 1987). Although olfactory receptor cells have the propensity to replenish themselves at regular intervals, replacement does not always occur, particularly when marked damage to the olfactory membrane has taken place (e.g., Hurtt et al., 1988). Furthermore, the regenerative process is influenced by a number of factors, including the quality of the air, metabolic factors associated with age or health, and hormonal status (see Matulionis, 1982; Hinds et al., 1984; Breipohl et al. 1986; Rehn et al., 1986).

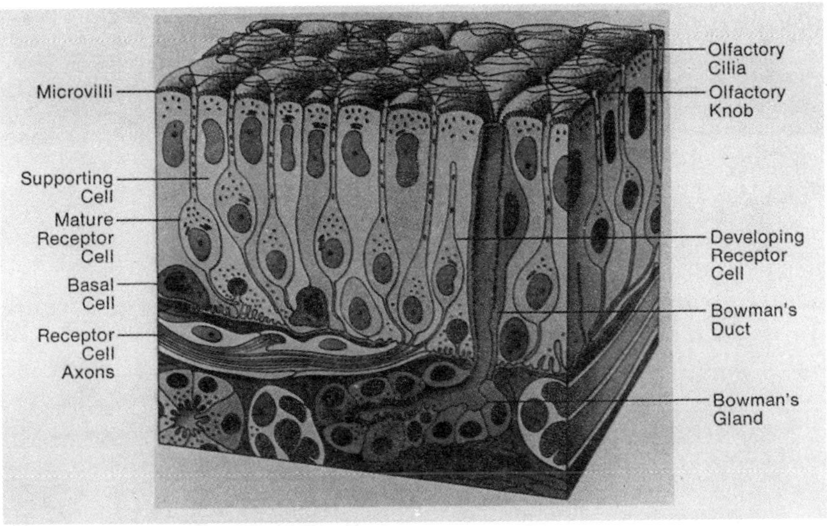

FIGURE 1. The cellular organization of the vertebrate olfactory epithelium. From Warwick and Williams, 1973, used with permission.

The unmyelinated axons of the olfactory receptor cells collect into bundles and course through the foramina of the cribriform plate into the olfactory bulb. Synapses with second order neurons are then made in the bulb within rotund structures termed glomeruli, which receive many more incoming axons than outgoing dendrites from the second order cells. The major second order neurons (mitral and tufted cells) project to higher brain centers within the primary olfactory cortex, which includes the anterior olfactory nucleus, the olfactory tubercle, the piriform cortex, the lateral entorhinal cortex, and the periamygdaloid cortex. Interbulbar communication occurs via the anterior commissure, and complex intrabulbar connections are present between interglomular cells, periglomerular cells, granule cells, and numerous colateral projections from the mitral and tufted cells. The olfactory bulb receives centrifugal fibers from various parts of the brain, including most of those to which it projects. Many of these fibers arrive via the ipsilateral lateral olfactory tract and typically synapse within the external plexiform and granule cell layers of the bulb. However, terminations are found in all layers of the bulb, with the exception of the olfactory nerve fiber layer (Powell and Cowan, 1963; Powell et al., 1965). Centrifugal fibers also project from the contralateral side, but are small in number and rarely extend beyond the mitral cell layer (Powell et al., 1965; for recent reviews of olfactory anatomy, see Scott and Harrison, 1986; Price, 1987; Greer, 1991; Moran et al., 1991; Shipley and Reyes, 1991).

The entorhinal cortex, in addition to receiving direct projections from the bulb, also receives projections from the prepiriform and periamygdaloid cortices (Powell et al., 1965). Although there is no direct pathway between the olfactory bulb and the hippocampus, connections are present between the entorhinal cortex and the hippocampus, reportedly giving the olfactory system the most direct access of all sensory systems to the latter structure (Shipley and Reyes,

1991). The hypothalamus receives projections from both the olfactory bulbs and the amygdala (Powell et al., 1965). Importantly, the orbitofrontal cortex receives direct projections from the primary olfactory cortex, as well as indirect projections from the entorhinal cortex via a relay in the dorsomedial nucleus of the thalamus (Powell et al., 1965).

MODERN STIMULUS PRESENTATION AND PSYCHOPHYSICAL PROCEDURES FOR ASSESSING OLFACTORY FUNCTION

A wide variety of procedures have been developed for presenting odorants to subjects (see Doty and Kobal, 1995, for review). With the exception of the measurement of event-related electrical responses (where precise timing of stimulus presentation is required), meaningful assessment of olfactory function can be made using rather simple stimulus presentation equipment. Thus, for most purposes, it is not necessary to know the number of molecules which enter the nose in order for an olfactory test to be valid. The key issues are whether a test procedure is sensitive and reliable, and whether normative data are available to establish whether a subject's performance on the test is normal or not. Among the most popular means of presenting odors to subjects are (i) glass sniff bottles (e.g., Cheesman and Townsend, 1956; Doty et al., 1986), (ii) glass rods, wooden sticks, or strips of blotter paper dipped into odorants (e.g., Toyota et al., 1978), (iii) plastic squeeze bottles (e.g., Amoore and Ollman, 1983; Cain et al., 1988) and (iv) microencapsulated "scratch and sniff" odorized strips (Doty et al., 1984a). Because tests based upon electrical responses are presently (i) largely experimental, (ii) not widely available, (iii) applicable to only some subjects, and (iv) comparatively expensive, they are not discussed here. The reader is referred elsewhere for such a discussion (Doty and Kobal, 1995).

A common psychophysical means for assessing chemosensory function is to establish, operationally, the lowest concentration of a stimulus that can be discerned, i.e., the absolute threshold. Since the olfactory system is dynamic and olfactory sensitivity fluctuates from moment to moment, such a measure is only an estimate of basal sensitivity. At very low odorant concentrations qualitative odor sensations (e.g., "banana-like") are rarely perceived, as only a faint presence of something is noted. In most modern olfactory detection threshold tests, a subject is asked to indicate, on a given trial, which of two or more stimuli (e.g., an odorant and one or more blanks) smells strongest, rather than to report whether an odor is perceived or not. Such "forced-choice" tests are less susceptible than non-forced choice tests to contamination by response biases (i.e., the conservatism or liberalism in reporting the presence of an odor under uncertain conditions) and are more reliable and sensitive (Blackwell, 1953; Doty et al., 1994a).

Two types of threshold procedures which have received wide use in the clinic are the *ascending method of limits* and the *single staircase* procedures. In the ascending method of limits procedure, odorants are presented sequentially from low to high concentrations and the point of transition between detection and no detection is estimated. In the single staircase method (a variant of the method of limits technique; see Cornsweet, 1962; Deems and Doty,

1987), the concentration of the stimulus is increased following trials on which a subject fails to detect the stimulus, and decreased following trials where correct detection occurs.

An alternative and equally sensitive approach for assessing olfactory function derives from test measurement theory and focuses on the comparative ability of subjects to identify a number of odorants at the suprathreshold level. The most popular test using this approach — the University of Pennsylvania Smell Identification Test (UPSIT; commercially available as the Smell Identification Test,™ Sensonics, Inc., Haddon Heights, NJ) — has been administered in over 2500 clinics in North America. The number of correct items out of 40 (presented in a forced-choice multiple alternative format) serves as the test measure. This value is compared to normative data and a percentile rank is determined, depending upon the age and gender of the patient (Doty, 1983). The UPSIT has several unique features, including amenability to self administration, a means for detecting malingering, and lack of dependence upon complex equipment or cumbersome sniff bottles. Importantly, this test is very reliable (test-retest reliability $r > 0.90$) and sensitive to even subtle olfactory deficits present in a number of diseases, including Alzheimer's disease, amyotrophic lateral sclerosis, Huntington's disease, Kallmann's syndrome, Korsakoff's psychosis, Parkinson's disease, and schizophrenia (e.g., Deems et al., 1991; Doty, 1991, 1992; Doty et al., 1988).

It should be noted that test scores from most nominally distinct psychophysical tests of olfactory function (e.g., tests of odor identification, detection, discrimination, and memory) are intercorrelated and likely measure, at least in normal subjects spanning a wide age range, a common source of variance (Doty et al., 1994b). Since, however, different tests have different degrees of reliability, they can be differentially sensitive to olfactory deficits, a problem which is particularly critical when small groups of subjects are tested. Thus, when an olfactory deficit is found by one type of test but not by another, the discrepant findings must be cautiously interpreted, as the test which does not detect a deficit may simply be less sensitive or reliable than the one that does. The lack of independence among nominally distinct tests can also lead to problems in the interpretation of test results. For example, while a patient may do poorly on a test of "odor memory," such performance could reflect an inability to detect or recognize odorants, rather than a deficit in odor memory, per se.

OLFACTORY FUNCTION IN PATIENTS WITH APPARENT MULTIPLE CHEMICAL SENSITIVITY

A common complaint of persons with symptoms of multiple chemical sensitivity (MCS) is heightened ability to smell (Randolph and Moss, 1980). Therefore, it was of interest to us whether olfactory thresholds are altered in persons complaining of somatic hypersensitivity to environmental chemicals (Doty et al., 1988b). We hypothesized that odor detection thresholds for phenyl ethyl alcohol (a major component of rose oil) and methyl ethyl ketone (a common solvent) would be lower in MCS patients than in normal controls. Since no common underlying physiologic basis for MCS has been identified and there is lack of consensus regarding the specific factors necessary for a primary diagnosis of MCS, we operationally defined MCS on the basis of the following: (i) historical evidence of adverse effects to environmental chemicals; (ii) the achievement of a high score on a quantified version of Randolph's Environmental Questionnaire (REQ), an index of somatic reactions to a variety of chemicals and household products (Randolph and Moss, 1980); and (iii) a careful evaluation of the patient's medical history to determine whether factors other than exposure to chemicals might explain the symptoms. The section of the REQ used to help establish the MCS classification consisted of 91 items to which subjects made the responses of "like," "neutral," "dislike," or "made sick from." These categories were assigned the respective values of 0, 1, 2, and 4.

Control subjects were matched to the 18 MCS subjects on the basis of age, gender, ethnic background, and smoking habits, and received equivalent testing, as described below. To ensure that the control group did not include persons with MCS, a prerequisite for inclusion in the study was an REQ item average score of less than 1.86 (a value that fell 2.5 SDs below the mean of an independent group of 72 persons who were asked to complete the questionnaire in a pilot study). Compared to the mean (± SD) MCS patient item score of 2.38 (± 0.54), the mean item score of the controls was 1.38 (± 0.24). All experimental and control subjects completed the Beck Depression Inventory before participation (Beck, 1967).

The design of the study was straightforward. Each subject participated in two test sessions separated from one another by 15 to 30 minutes. In the first session, blood pressure, heart rate, respiration rate, and nasal airflow resistance in each nasal chamber (as determined by anterior rhinomanometry) were determined, followed by bilateral detection threshold sensitivity to phenyl ethyl alcohol (PEA) (Doty et al., 1984a, 1986). Following this testing, measures of blood pressure, heart rate, respiration rate, and nasal airflow resistance were again obtained. In the second session, the same sequence of testing was performed, with the exception that unilateral detection thresholds for ethyl methyl ketone, rather than bilateral thresholds for phenyl ethyl alcohol, were measured.

The PEA detection threshold was established using a single-staircase procedure. On a given trial, two glass sniff bottles were presented to the subject in rapid succession. Each bottle was immediately opened and held over the nose in a standardized manner (see Doty et al., 1978). One bottle contained 20 ml of a given concentration of odorant dissolved in propylene glycol, whereas the other contained propylene glycol alone. The task of the subject was to report which of the two bottles produced the strongest sensation. If no sensation was discernable from the bottles or if the sensations appeared equivalent, the subject was required to guess which bottle contained the strongest stimulus (i.e., the test was forced-choice). No feedback as to the correctness of a subject's responses was provided. The initial trial was begun at the -6.00 log concentration step of a half-log step dilution series extending from -7.50 to -1.00 log concentrations. If an incorrect response occurred on this trial, a concentration one log step higher was presented. If an incorrect response occurred on this trial, a stimulus one log step higher was again presented. When four consecutive correct trials occurred at a given concentration, the subsequent stimulus was presented one half-log concentration step lower.

From this point on, sampling of the concentrations occurred at half-log steps, with one or two trials occurring at each step (i.e., if the first trial was missed, the second one was not given, and the next highest concentration was presented; if both were responded to correctly, the next lowest concentration was presented). The geometric mean of the last four of seven reversal points served as the threshold value.

The MEK thresholds were established in a similar manner, although separate threshold values were determined for each side of the nose and the stimuli were presented using an air-dilution olfactometer. Ten stimulus concentrations (0.6, 1.0, 1.8, 3.2, 5.6, 10.0, 17.8, 31.6, 56.2, and 100 ppm) were shunted, along with blank air, to 1.5-L stimulus delivery flasks arranged around the circumference of a rotating stimulus sampling table. The contents of each flask were sampled from small glass nozzles. The side of the nose to be tested was determined randomly from trial to trial. To eliminate knowledge as to which stimulus was presented, the subject was required to wear opaque goggles during testing. Odorant concentrations were determined empirically using a photoionization detector (Analytical Instruments, Avondale, PA, model 910). The olfactometer was calibrated before each session using a 250-ppm butadiene span gas and calibration factors provided by the manufacturer.

PEA was tested prior to MEK in all cases, since MEK has a greater propensity to stimulate intranasal trigeminal afferents and thereby alter mucus secretion and nasal tubinate engorgement. All tests were performed in a 290 m^3 stainless-steel environmental chamber specifically designed for olfactory testing (Environmental Growth Chambers, Chagrin Falls, OH). The ambient air temperature within this 3.25 m W x 3.35 m D structure was 23 ± 0.3 C. The chamber's air was reconditioned through the plenum every 45 seconds and was completely replenished with filtered outside air every eight minutes.

No evidence was found to support the view that patients with MCS were more sensitive to the two target stimuli than were matched controls. Thus, for the MCS group, the mean (SD) male and female PEA threshold values, respectively, were -4.56 (1.67) and -4.96 (1.67), compared to analogous control group values of -4.27 (1.50) and -4.46 (1.05) (all values log vol/vol). The mean (SD) male and female MEK threshold values (in ppm) were 5.69 (5.40) and 7.63 (8.41) respectively, compared to analogous control values of 8.25 (4.46) and 8.10 (13.37). A group by gender analysis of variance demonstrated no significant group, gender, or interaction effects for these data. Also, within the MCS group no correlation was present between the degree of self-reported hypersensitivity and the PEA or MEK olfactory thresholds (respective Pearson r values = -0.01 and -0.13, ps > 0.20).

In the case of nasal airflow measurement, the MCS group evidenced higher nasal resistance on both inhalation and exhalation than did the controls both before and after the two test sessions (see Doty et al., 1988, for specific values). In addition, the MCS group had a higher average respiratory rate than that of the controls in both test sessions. While Beck Depression Inventory scores were significantly higher in the MCS than in the control subjects, heart rate and blood pressure did not differ between the two groups.

In conclusion, these data suggest that, on average, patients with complaints of MCS do not evidence higher olfactory threshold sensitivity than do matched normal controls. Our finding of increased nasal resistance in MCS individuals is in accord with previous suggestions that such patients have increased nasal congestion (Cullen, 1987; Meggs and Cleveland, 1993). This observation, along with the finding of increased respiration and heart rates in such patients, suggests that MCS patients may have labored breathing, which might explain some of their somatic complaints. It is well documented that the nasal airway represents the single largest component of total airway resistance and can influence tidal volume, respiratory function, and expiratory time (Swift and Proctor, 1977; Cole, 1982). Thus, understanding the basis of such problems may shed considerable light on the MCS syndrome.

ACKNOWLEDGMENTS

Supported by Grant P01 DC 00161 from the National Institute on Deafness and Other Communication Disorders.

REFERENCES

AMOORE, J.E. and OLLMAN, B.G. (1983). "Practical test kits for quantitatively evaluating the sense of smell." Rhinology 21:49–54.

BLACKWELL, H.R. (1953). "Psychological thresholds: Experimental studies of methods of measurement." Bulletin of the Engineering Research Institute No. 36, University of Michigan Press, Ann Arbor.

BECK, A.T. (1967). Depression: Clinical, Experimental, and Theoretical Aspects. Staples Press, London.

BREIPOHL, W., MACKAY-SIM, A., GRANDT, D., REHN, B., and DARRELMANN, C. (1986). "Neurogenesis in the vertebrate main olfactory epithelium." In: Ontogeny of Olfaction (W. Breipohl, ed.). Springer-Verlag, Berlin. pp. 21–33.

CAIN, W.S., GENT, J.P., GOODSPEED, R.B., and LEONARD, G. (1988). "Evaluation of olfactory dysfunction in the Connecticut Chemosensory Clinical Research Center." Laryngoscope 98:83–88.

CHEESMAN, G.H. and TOWNSEND, M.J. (1956). "Further experiments on the olfactory thresholds of pure chemical substances, using the 'sniff-bottle' method." Q. J. Exp. Psychol. 8:8–14.

COLE, P. (1982). "Upper respiratory airflow." In: The Nose: Upper Airway Physiology and the Atmospheric Environment. (D. Proctor and I. Andersen, eds.). Elsevier Biomedical Press, Amsterdam. pp. 163–189.

CORNSWEET, T.N. (1962). "The staircase-method in psychophysics." Am. J. Psychol. 75:485–491.

CULLEN, M.R., ed. (1987). "Workers with multiple chemical hypersensitivities." State Art Rev. Occup. Med. 2:655–806.

DEEMS, D.A. and DOTY, R.L. (1987). "The nature of age-related olfactory detection threshold changes in man." Trans. Penn. Acad. Opthalmol. Otolargyngol. 39:646–650.

DEEMS, D.A., DOTY, R.L., SETTLE, R.G., MOORE-GILLION, V., SHAMAN, P., MESTER, A.F., KIMMELMAN, C.P., BRIGHTMAN, V.J., and SNOW, J.B., JR. (1991). "Smell and taste disorders: A study of 750 patients from the University of Pennsylvania Smell and Taste Center." Arch. Otolaryngol. Head Neck Surg. 117:519–528.

DOTY, R.L. (1983). The Smell Identification Test™ Administration Manual. Sensonics, Inc., Haddon Hts., NJ.

DOTY, R.L. (1991). "Olfactory dysfunction in neurodegenerative disorders." In: Smell and Taste in Health and Disease (T.V. Getchell, R.L. Doty, L.M. Bartoshuk, and J.B. Snow, eds.). Raven Press, N.Y. pp. 735–751.

DOTY, R.L. (1992). "Diagnostic tests and assessment." J. Head Trauma Rehab. 7:45–64.

DOTY, R.L., ed. (1995). Handbook of Olfaction and Gustation. Marcel Dekker, N.Y.

DOTY, R.L. and KOBAL, G. (1995). "Current trends in the measurement of olfactory function." In: Handbook of Olfaction and Gustation (R.L. Doty, ed.). Marcel Dekker, N.Y. pp. 191–225.

DOTY, R.L., BRUGGER, W.E., JURS, P.C., ORNDORFF, M.A., SNYDER, P.J., and LOWRY, L.D. (1978). "Intranasal trigeminal stimulation from odorous volatiles: Psychometric responses from anosmic and normal humans." Physiol. Behav. 20:175–185.

DOTY, R.L., SHAMAN, P., and DANN, M. (1984a). "Development of the University of Pennsylvania Smell Identification Test: A standardized microencapsulated test of olfactory function." Physiol. Behav. (Monograph) 32:489–502.

DOTY, R.L., SHAMAN, P., APPLEBAUM, S.L., GIBERSON, R., SIKORSKI, L., and ROSENBERG, L. (1984b). "Smell identification ability: Changes with age." Science 226:1441–1443.

DOTY, R.L., GREGOR, T., and SETTLE, R.G. (1986). "Influences of interval and sniff bottle volume on the phenyl ethyl alcohol olfactory detection threshold." Chemical Senses 11:259–264.

DOTY, R.L., DEEMS, D.A., and STELLAR, S. (1988a). "Olfactory dysfunction in Parkinson's Disease: A general deficit unrelated to neurologic signs, disease stage, or disease duration." Neurology 38:1237–1244.

DOTY, R.L., DEEMS, D.A., FRYE, R., PELBERG, R., and SHAPIRO, A. (1988b). "Olfactory sensitivity, nasal resistance, and autonomic function in multiple chemical sensitivities (MCS) syndrome." Arch. Otolaryngol.: Head Neck Surg. 114:1422–1427.

DOTY, R.L., McKEOWN, D.A., and SHAMAN, P. (1994a). "Test-retest reliability of 10 olfactory tests." Submitted.

DOTY, R.L., SMITH R., McKEOWN, D., and RAJ, J. (1994b). "Tests of human olfactory function: Principal components analysis suggests that most measure a common source of variance." Percept. Psychophys. In press.

GETCHELL, T.V., DOTY, R.L., BARTOSHUK, L.M., and SNOW, J.B, JR., Eds. (1991). Smell and Taste in Health and Disease. Raven Press, N.Y.

GREER, C.A. (1991). "Structural organization of the olfactory system. In: Smell and Taste in Health and Disease (T.V. Getchell, R.L. Doty, L.M. Bartoshuk, and J.B. Snow, Jr., eds.). Raven Press, N.Y. pp. 65–81.

HENKIN, R.I. and BARTTER, F.C. (1966). "Studies of olfactory thresholds in normal man and in patients with adrenal cortical insufficiency: the role of adrenal cortical steroids and of serum concentration." J. Clin. Invest. 45:1631–1639.

HENKIN, R.I. and POWELL, G.F. (1962). "Increased sensitivity of taste and smell in cystic fibrosis." Science 138:1107.

HERTZ, J., CAIN, W.S., BARTOSHUK, L.M., and DOLAN, T.F., JR. (1975). "Olfactory and taste sensitivity in children with cystic fibrosis." Physiol. Behav. 14:89–95.

HINDS, J.W., HINDS, P.L., and McNELLY, N.A. (1984). "An autoradiographic study of the mouse olfactory epithelium: evidence of long lived receptors." Anat. Rec. 210:375–383.

HURTT, M.E., THOMAS, D.A., WORKING, P.K., MONTICELLO, T.M., and MORGAN, K.T. (1988). "Degeneration and regeneration of the olfactory epithelium following inhalation exposure to methyl bromide: Pathology, cell kinetics, and olfactory function." Toxicol. Appl. Pharmacol. 94:311–328.

MATULIONIS, D.H. (1982). "Effects of the aging process on olfactory neuron plasticity." In: Olfaction and Endocrine Regulation (W. Breipohl, ed.). IRL Press, London. pp. 299–308.

MEGGS, W.J. and CLEVELAND, C.H. (1993). "Rhinolaryngoscopic examination of patients with MCS syndrome." Arch. Environ. Health 48:14–18.

MORAN, D.T., ROWLEY, C., III, and JAFEK, B. (1991). "Anatomy of the human olfactory epithelium." In: The Human Sense of Smell (D.G. Laing, R.L. Doty, and W. Breipohl, eds.). Springer-Verlag, Amsterdam. pp. 3–28.

POWELL, T.P.S. and COWAN, W.M. (1963). "Centrifugal fibers in the lateral olfactory tract." Nature 199:1296–1297.

POWELL, T.P.S., COWAN, W.M., and RAISMAN, G. (1965). "The central olfactory connections." J. Anat. 99:791–813.

PRICE, J.L. (1987). "The central and accessory olfactory systems." In: Neurobiology of Taste and Smell (T.E. Finger and W.S. Silver, eds.). John Wiley, New York. pp. 179–204.

RANDOLPH, T.G. and MOSS, R.W. (1980). An Alternative Approach to Allergies. Lippincott and Crowell, New York.

REHN, B., BREIPOHL, W., MENDOZA, A.S., and APFELBACH, R. (1986). "Changes in granule cells of the ferret olfactory bulb associated with imprinting on prey odours." Brain Res. 373:114–125.

SCOTT, J.W. and HARRISON, T.A. (1986). "The olfactory bulb: anatomy and physiology." In: Neurobiology of Taste and Smell (T.E. Finger and W.L. Silver, eds.). Wiley and Sons, New York. pp. 151–178.

SERBY, M.J. and CHOBER, K.L., eds. (1992). Science and Olfaction. Springer-Verlag, New York.

SHIPLEY, M.T. and REYES, P. (1991). "Anatomy of the human olfactory bulb and central olfactory pathways." In: The Human Sense of Smell (D.G. Laing, R.L. Doty, and W. Breipohl, eds.). Springer-Verlag, Amsterdam. pp. 29–60.

SILVER, W.S. (1987). "The common chemical sense." In: Neurobiology of Taste and Smell (T.E. Finger and W.L. Silver, eds.). Wiley and Sons, New York. pp. 65–87.

SWIFT, D.L. and PROCTOR, D.F. (1977). "Access of air to the respiratory tract." In: Respiratory Defense Mechanisms (J.D. Brian, D.F. Proctor, and L.M. Reid, eds.). Marcel Dekker, Inc., New York. pp. 63–93.

TOYOTA, B., KITAMURA, T., and TAKAGI, S.F. (1978). Olfactory Disorders, Olfactometry and Therapy. Igaku-Shoin, Ltd., Tokyo.

WARWICK, R. and WILLIAMS, P.L. (1973). Gray's Anatomy. Saunders, Philadelphia.

WEIFFENBACH, J.M. and McCARTHY, V.P. (1984). "Olfactory deficits in cystic fibrosis: Distribution and severity." Chem. Senses 9:193–199.

NEUROANATOMY AND NEUROCHEMICAL MECHANISMS OF TIME-DEPENDENT SENSITIZATION

BARBARA A. SORG, M. STACY HOOKS, AND PETER W. KALIVAS

Program in Neuroscience
Department of Veterinary and Comparative Anatomy,
Pharmacology and Physiology
Washington State University
Pullman, Washington

Time-dependent sensitization (TDS) is a phenomenon described in rodents as an enhancement in the behavioral and neurochemical responses to intermittent exposure to psychostimulant drugs. Time-dependent sensitization also occurs after repeated encounters with environmental stress. Several features of TDS parallel those of multiple chemical sensitivity (MCS) in humans, and these similarities have led to the hypothesis that MCS may be explained in part by a similar sensitization process that occurs in rodents. In the studies presented here, we discuss some of the critical features of TDS following repeated exposure to cocaine and environmental stress, including the anatomical and neurochemical pathways utilized in expressing TDS. In addition, we discuss the possible neurochemical basis for individual differences in responsiveness to stimuli, including novelty and cocaine. The striking similarities between TDS and MCS suggest it may be possible to develop an animal model of MCS, using TDS in rodents as its basis.

INTRODUCTION

In rodents, time-dependent sensitization (TDS) is characterized by the progressive and enduring enhancement of a variety of behavioral responses after repeated, intermittent exposure to environmental or pharmacological stressors (Antelman, 1988). In humans, TDS can be induced by repeated use of psychostimulant drugs such as cocaine and amphetamine, or by repeated environmental stress, and is manifest as paranoid schizophrenia, panic

1. Address all correspondence to: Barbara A. Sorg, Ph.D., Dept. of VCAPP, Washington State University, Pullman, WA 99164-6520. Tel.:(509)335-4709. Fax:(509)335-4650.

2. Abbreviations: CNS, central nervous system; MCS, multiple chemical sensitivity, mPFC, medial prefrontal cortex; PTSD, post–traumatic stress disorder; TDS, time-dependent sensitization; VP, ventral pallidum; VTA, ventral tegmental area.

3. Key words: medial prefrontal cortex, nucleus accumbens, ventral pallidum, ventral tegmental area, sensitization, stress.

Toxicology and Industrial Health, Vol. 10, No. 4/5, pp. 369–386

ISSN: 0748-2337

disorder, and post–traumatic stress disorder (PTSD) (Angrist, 1983; Antelman, 1988; Post and Weiss, 1988). The onset of symptoms associated with multiple chemical sensitivity (MCS) largely parallels that of TDS. The following observations are common to the onset and progression of MCS and TDS: 1) Both can be induced by few exposures to large doses or repeated, intermittent exposure to lower doses of the stimulus, and the response to the same dose becomes progressively greater (Post and Weiss, 1988; Bell, 1994); 2) After exposure to only one type of stimulus, spreading of sensitivity to other stimuli can occur (cross-sensitization) (Antelman et al., 1980; Bell et al., 1992); 3) No changes in baseline activity in animals or clinical symptoms in humans are apparent if exposure to the stimulus is avoided (Ashford and Miller, 1991; Kalivas and Duffy, 1993); 4) Tolerance of a stimulus may develop with continuous exposure, while sensitization appears after withdrawal and acute reexposure to the stimulus (Ashford and Miller, 1991; Kalivas and Duffy, 1993); 5) Wide individual variability occurs in the behavioral response to stimuli (Hooks et al., 1992); and 6) Neurobiological changes occurring in TDS and MCS in many cases appear to be irreversible, lasting for months in rodents and for years in humans (Robinson and Becker, 1986; Ashford and Miller, 1991). The parallels between TDS and MCS outlined above have been discussed in greater detail by Bell (1994). The similarities between TDS and MCS indicate that the symptoms may be initiated and reinstated by similar neuronal mechanisms. Consequently, the study of TDS in rodents may provide an appropriate animal model system for studying MCS.

The broad range of symptoms reported by MCS patients involves almost every system in the body, including the central nervous system (CNS) (Ashford and Miller, 1991). Bell (1982) has speculated that the hypothalamus and other limbic structures such as the amygdala, hippocampus, and olfactory pathways are involved in producing MCS. Limbic pathways are also involved in the production of TDS in animals. While some of the pathways important for TDS overlap with those proposed to be involved in MCS, such as the amygdala and hippocampus (Kalivas et al., 1993), additional limbic pathways have been identified as playing a primary role in TDS in animals. Most prominent are the mesolimbic and mesocortical dopamine projections, which originate in the ventral tegmental area (VTA), and extend to the nucleus accumbens and medial prefrontal cortex (mPFC) (Fallon and Moore, 1978; Swanson, 1982). We propose that it is this dopamine circuitry that modulates the behavioral response to incoming stimuli via limbic structures such as the amygdala and hippocampus (see Kalivas et al., 1993). Figure 1 depicts this circuitry as a gain mechanism by which the relative intensity of stimuli is increased or decreased to produce sensitization or tolerance, respectively. The circuitry involving these brain regions appears critical for the translation of environmentally relevant stimuli into motor responses. Thus, environmental cues are filtered through this circuitry before adaptive behavior can take place. In the case of an organism's repeated exposure to more extreme pharmacologically or nonpharmacologically stressful circumstances, the gain is increased, producing maladaptive behaviors such as repetitive behavior (stereotypy) in laboratory rodents or psychopathologies in humans. It should be emphasized that this is a simplistic model of TDS in animals with which to extend our thinking to MCS for the purpose of designing MCS studies in animals. This paper focuses on studies conducted in this laboratory, with an emphasis on the basic phenomena of sensitization by psychostimulants, cross-sensitization between cocaine and environmental stress, and individual variability among animals in their response to pharmacological and environmental stimuli. We have highlighted the experiments that investigate the role of dopamine-related circuitry, important in modulating behavioral output, including the mesocorticolimbic pathway and some of its major afferents and efferents to the VTA, nucleus accumbens and mPFC. The first section will concentrate on anatomical and neurochemical substrates involved in psychostimulant and stress-induced TDS.

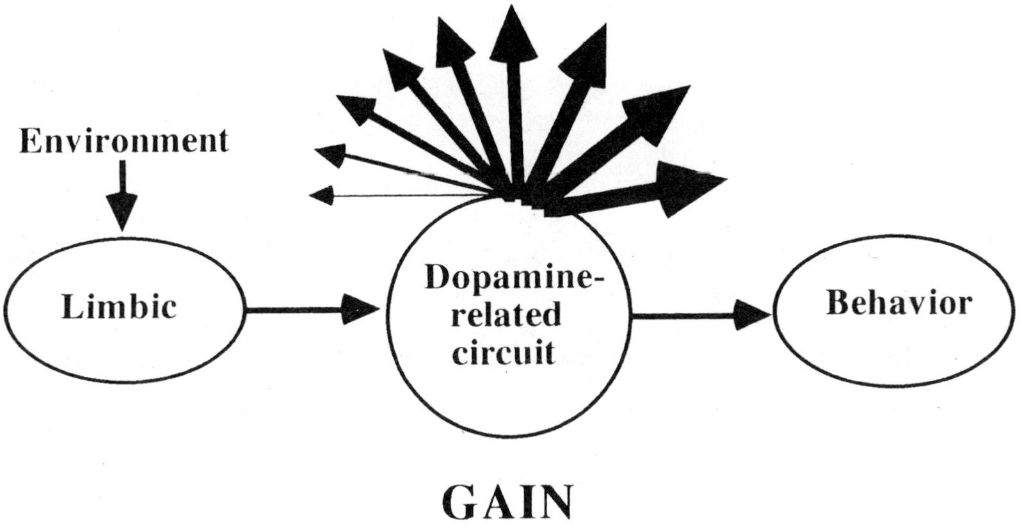

FIGURE 1. Regulation of environmental stimuli by limbic pathways and dopamine-related circuit in the translation of environmental cues to behavior. Increasing thickness of arrows represents increased gain or responsivity of dopamine-related circuitry to the environment (see Kalivas et al., 1993).

While the study of TDS in animals has been widely used as an animal model for human psychoses, animal models for MCS remain unexplored. If vulnerability to stress and psychostimulant drugs does indeed provide an indicator of the sensitivity of the dopamine-related circuitry to stimuli in gènéral, we would predict that higher vulnerability to these stimuli would be produced by other noxious stimuli, including a wide range of organic chemicals. The ability to predict which animals might be most prone to develop MCS is useful to the development of an animal model for MCS. In the second portion of the paper we discuss the neurochemical differences between animals with high and low vulnerability to stimuli resulting in TDS. The results presented in this paper represent a compilation of previously published work that demonstrates some of the critical aspects of TDS (see Kalivas et al., 1993; Kalivas and Duffy, 1993; Sorg and Kalivas, 1993a; Hooks and Kalivas, 1994, for details).

METHODS

Animals

Male Sprague-Dawley rats (Simonsen, Gilroy, CA) were single-housed (Experiments 1 and 2) or group-housed (Experiment 3) following surgery, with water and food made available *ad libitum.* Animals were maintained on a 12 hr light/dark cycle, with lights on at 07:00 hr. Rats were anaesthetized with Equithesin and placed in a stereotaxic apparatus. The coordinates used for placement of chronic guide cannulae for dialysis are described by Kalivas and Duffy (1993) and Sorg and Kalivas (1993a). Coordinates for microinjection of cannulae into the nucleus accumbens, VTA, and ventral pallidum (VP) are described by Hooks and Kalivas (1994). The rats were allowed a minimum five-day recovery period.

Treatment Schedule and Behavior

Three experiments are described. The first two utilized the *in vivo* microdialysis technique in the nucleus accumbens and mPFC and the third experiment utilized microinjection into the nucleus accumbens, VP, and VTA. For all experiments, locomotor behavior was monitored in photocell cages (Omnitech Electronics, Columbus, OH). Locomotion was analyzed by obtaining horizontal photocell counts (consecutive breaking of adjacent photocell beams) or total distance traveled (in cm). The apparatus used for analyzing locomotion and delivery of footshock are described in detail elsewhere (Kalivas and Duffy, 1993; Sorg and Kalivas, 1993a; Hooks and Kalivas, 1994).

Table 1 describes each experimental protocol. For Experiments 1 and 2, animals were implanted with a dialysis probe in the nucleus accumbens or mPFC the evening before the experiment. The next day, dialysis probes were perfused at a rate of 2 μl/min with artificial cerebral spinal fluid as described by Sorg and Kalivas (1993a) for a minimum of 2 hr to stabilize basal extracellular dopamine levels. Dialysates were collected every 20 min for 60 to 80 min., and saline (1 ml/kg, i.p.) or cocaine (15 mg/kg, i.p.) was injected. Sample collection continued for 120–180 min.

Experiment 2 was similar to Experiment 1, except that repeated footshock stress was given instead of daily cocaine (see Table 1). Each daily session consisted of administering footshock (0.45 mA/200 msec/sec) for 20 min. This level of footshock is considered mild and does not produce any overt escape behavior such as jumping or vocalization.

For Experiment 3 (see Table 1), the rats' locomotor response to novelty was assessed by placing animals in the photocell cages and measuring horizontal photocell counts accumulated over a 1 hr period. The median number of photocell counts was determined, and all animals above the median were considered high responders (HR), while those animals below the median were low responders (LR). After this screening, rats were given a sham microinjection and i.p. saline injection. On the day of microinjection of drugs, rats were given either saline, various doses of dopamine (Sigma Chemical Co., St. Louis, MO), an excitatory amino acid agonist, AMPA (Research Biochemicals, Natick, MA), or a mu-opioid agonist,

DAMGO (Bachem, Torrance, CA) (Note: only one dose is shown in the present study; for full dose-response see Hooks and Kalivas, 1994). All microinjections were made bilaterally 1 mm below the chronic guide cannulae. Drugs were infused at a rate of 0.5 µl/min, and locomotor activity was monitored for 120 min postinjection.

TABLE 1. Experimental Protocols

Experiment 1A (Figure 2)

Nucleus Accumbens Dialysis Day 1	Days 2–6	Day 7	10–14 days	Nucleus Accumbens Dialysis Challenge
Saline	Saline	Saline	Home cage[1]	Cocaine 15
Cocaine 15[2]	Cocaine 30[3]	Cocaine 15	Home cage	Cocaine 15

Experiments 1B and 2 (Figures 3–5)

	Days 1–5	7 days	Dialysis Challenge
Exp. 1B	Saline	Home cage	Cocaine 15 (mPFC)
	Cocaine	Home cage	Cocaine 15 (mPFC)
Exp. 2	Sham Shock[4]	Home cage	Cocaine 15 (mPFC or accumbens)
	Footshock[5]	Home cage	Cocaine 15 (mPFC or accumbens)

Experiment 3 (Figures 6–8)

Screen for Novelty,[6] Day 1	Days 2–3	Microinjection Day 4
High responder	Home cage	Saline, dopamine, AMPA, DAMGO
Low responder	Home cage	Saline, dopamine, AMPA, DAMGO

[1] Home cage = no treatment during withdrawal time.
[2] Cocaine 15 = cocaine (15 mg/kg, i.p.).
[3] Cocaine 30 = cocaine (30 mg/kg, i.p.).
[4] Sham Shock = animals placed in shock apparatus without delivery of shock.
[5] Footshock = 0.45 mA/200 msec/sec for 20 min.
[6] Screen included a sham microinjection into the brain after a one-hour adaptation period in photocell cages followed by an additional one-hour adaptation to the cages.

In Vivo *Dialysis Probes and Dopamine Measurement*

Dialysis probes were designed according to Robinson and Whishaw (1988) and contained a membrane region 0.25 mm in diameter and 2 mm (nucleus accumbens) or 3 mm (mPFC) in length. Dialysis samples were collected as described previously (Sorg and Kalivas, 1993a). The content of extracellular dopamine was measured using high performance liquid chromatography with coulometric electrochemical detection.

Histology and Data Analysis

All animals were given an overdose of pentobarbital and perfused as described (Sorg and Kalivas, 1993a). Placement of dialysis and microinjection probes were verified in coronal slices of brain stained with cresyl violet using light microscopy. Behavioral and neurochemical data were analyzed using a one-way ANOVA.

RESULTS

Experiments 1 and 2: Cocaine and Stress-Induced TDS

Figure 2A demonstrates the phenomenon of TDS in rats administered repeated cocaine injections. The locomotor response on the last day of cocaine was significantly augmented compared to the first day of treatment. Similarly, extracellular dopamine levels in the nucleus accumbens were significantly elevated compared to the first day of cocaine administration (Figure 2B).

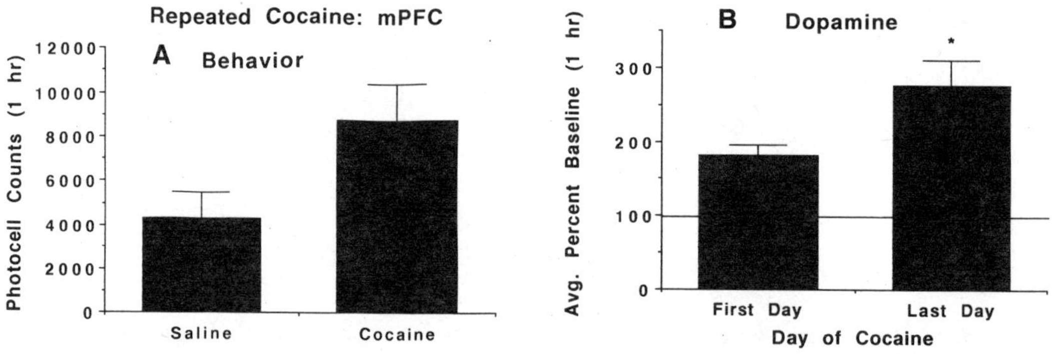

FIGURE 2. Effect of acute and repeated cocaine on locomotor activity and extracellular dopamine levels in the nucleus accumbens. Data are shown as mean ± SEM. A. Total horizontal activity accumulated over a one-hour period after cocaine injection on the first or last (challenge) day. First day N = 22, last day N = 9. B. Average percent of baseline levels of extracellular dopamine in the nucleus accumbens over a one-hour postinjection period from animals in (A). First day N = 22, last day N = 9. *$p < 0.05$ compared to first day of cocaine, using a one-way ANOVA. (Note: from the 22 rats used for first day measurements, 13 were used for analysis on a variety of test days other than those shown here; see Kalivas and Duffy, 1993 for details.)

Repeated daily footshock stress produced a response similar to an acute cocaine challenge (Figures 3A and B). Although the behavioral response was not as robust as that observed in daily cocaine-pretreated animals, the neurochemical profile was virtually identical to cocaine-pretreated rats; that is, repeated stress augmented cocaine-induced extracellular dopamine levels in the nucleus accumbens.

Figures 4 and 5 demonstrate behavioral and neurochemical responses to an acute cocaine injection in animals given repeated daily cocaine or footshock stress and implanted with a dialysis probe in the mPFC. In both cases the behavior showed a trend toward a sensitized locomotor response in cocaine- or shock-pretreated rats compared to their controls. In general, locomotor activity was significantly augmented at 20 min, but not at 40 or 60 min, after the cocaine challenge (data not shown). In contrast to the nucleus accumbens, the mPFC demonstrated a decrease in the ability of a cocaine challenge to elevate extracellular dopamine levels. Thus, Figures 4B and 5B demonstrate that repeated cocaine or stress produce a tolerance rather than a sensitization of the dopamine response to a subsequent stimulus in the mPFC.

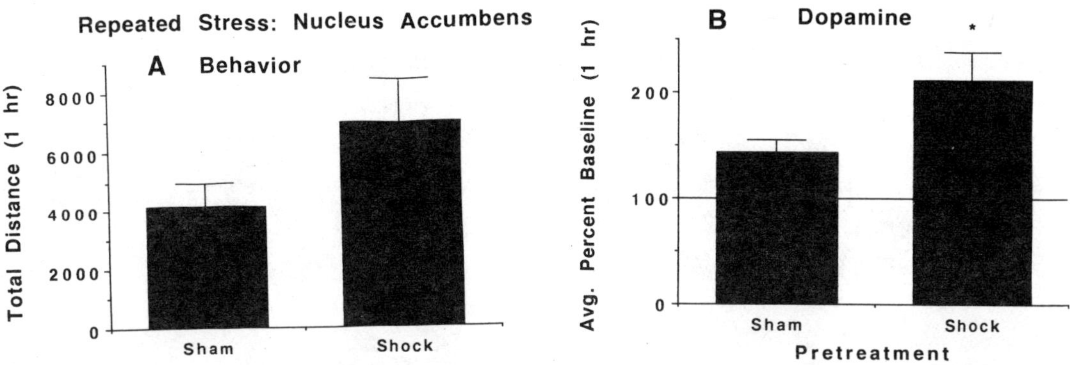

FIGURE 3. Effect of pretreatment with daily sham or footshock stress on acute cocaine-induced changes in behavior and extracellular dopamine levels in the nucleus accumbens. Data are shown as mean ± SEM. A. Total distance (in cm) accumulated over a one-hour period after cocaine challenge. Sham N = 17, shock N = 14. B. Average percent of baseline levels of extracellular dopamine in the nucleus accumbens over a one-hour post–cocaine injection period from animals in (A). Sham N = 7, shock N = 8. *p < 0.05, compared to sham controls, using a one-way ANOVA (see Sorg and Kalivas, 1991, for details).

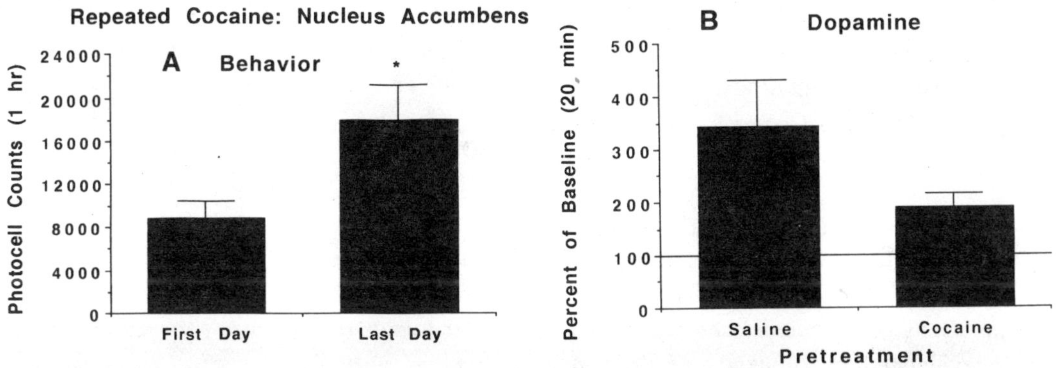

FIGURE 4. Effects of daily saline and cocaine pretreatment on locomotor activity and extracellular dopamine levels in the mPFC. Data are shown as mean ± SEM. A. Total horizontal activity accumulated over a one-hour period after cocaine challenge. Saline N = 9, cocaine N = 11. B. Change in percent of baseline levels of extracellular dopamine levels in the mPFC over a 20 min post–cocaine injection period from animals in (A). Saline N = 11, cocaine N = 12 (from Sorg and Kalivas, 1993b).

Experiment 3: Individual Differences

Figures 6–8 show the locomotor response of rats classified as HR and LR according to their motor response to a novel environment. Figure 6 demonstrates that microinjection of dopamine and AMPA into the nucleus accumbens produced greater locomotor activity in HR than in LR, while DAMGO did not elicit any differences between these groups. Similarly, Figure 7 indicates that dopamine and AMPA, but not DAMGO, microinjection into the VP induced a greater locomotor response in HR than in LR. In the VTA, AMPA microinjection

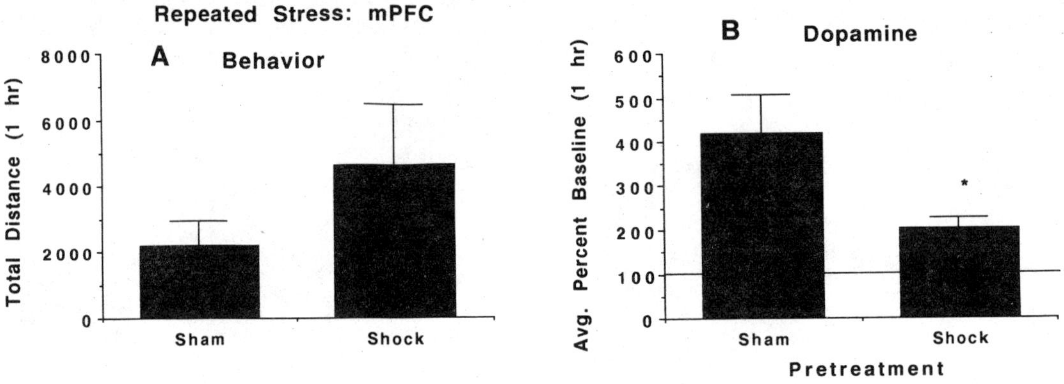

FIGURE 5. Effects of daily sham and footshock stress pretreatment on locomotor activity and extracellular dopamine levels in the mPFC. Data are shown as mean ± SEM. A. Total distance (in cm) accumulated over a one-hour period after cocaine challenge. Sham N = 9, shock N = 9. B. Average percent of baseline levels of extracellular dopamine levels in the mPFC over a one-hour post–cocaine challenge from animals in (A). Sham N = 9, shock N = 9. *$p < 0.05$, compared to sham controls using a one-way ANOVA (see Sorg and Kalivas, 1993a for details).

resulted in a greater locomotor response in HR than in LR, while DAMGO again did not produce any differences (Figure 8). These data suggest that dopaminergic and excitatory amino acid, but not mu-opioid–mediated mechanisms in the VTA-accumbens-pallidal circuitry are important in modulating individual differences in the response to novelty stress and psychostimulant drugs.

DISCUSSION

Mesocorticolimbic Dopamine and TDS

The mesolimbic dopamine pathway has been widely implicated in the locomotor activating effects of psychostimulant drugs (Wise and Bozarth, 1987; DiChiara and Imperato, 1988). Involvement of nucleus accumbens dopamine in the locomotor response to psychostimulants is evidenced by blockade of drug-induced hyperactivity by dopamine antagonists or destruction of dopamine terminals in this region (Kelly and Iverson, 1976; Scheel-Kruger et al., 1977). Both cocaine and amphetamine increase dopamine levels by blocking and/or promoting release of dopamine from axon terminals (Koe, 1976; Moore, 1978). In more recent years, the *in vivo* microdialysis technique has been used in attempts to estimate the levels of dopamine released into the synapse (Ungerstedt, 1984). We and others have reported an enhancement in extracellular dopamine levels in the nucleus accumbens following repeated cocaine exposure using the *in vivo* microdialysis technique (Robinson et al., 1988; Kalivas and Duffy, 1990; 1993; Parsons and Justice, 1993; see Figures 2A and B). Furthermore, the augmented dopamine levels in response to a cocaine challenge are correlated with the degree of locomotor activity in cocaine-sensitized animals (Kalivas and Duffy, 1990). Increases in nucleus accumbens dopamine levels induced by sex-related olfactory cues

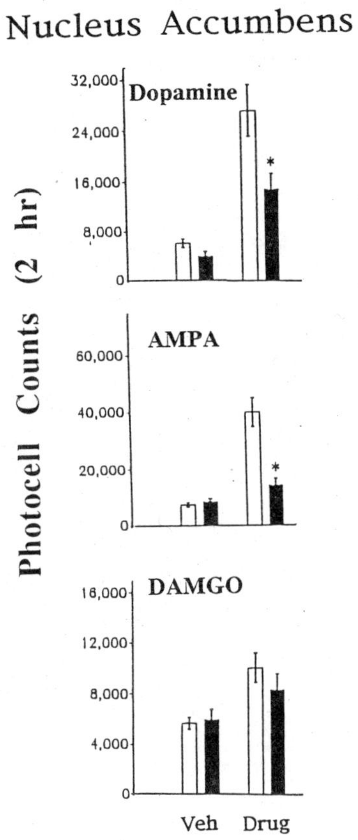

FIGURE 6. Effect of intranucleus accumbens microinjections of dopamine, AMPA and DAMGO on locomotor activity in HR and LR. Data represent mean ± SEM of horizontal photocell counts accumulated over a two-hour postinjection period. HR are indicated by open bars, LR are shown in black bars. Veh = saline vehicle in which drug was dissolved. The number of determinations and dose of each drug are given. For dopamine: 30 μg/side, HR N = 9, LR N = 10; AMPA: 0.03 μg/side, HR N = 11, LR N = 11; DAMGO: 0.03 nmol/side, HR N = 7, LR N = 7. *$p < 0.05$, comparing HR to LR using a two-way ANOVA with repeated measures over dose (only one dose is shown) and comparison of simple main effects (see Hooks and Kalivas, 1994 for details for Figures 6–8).

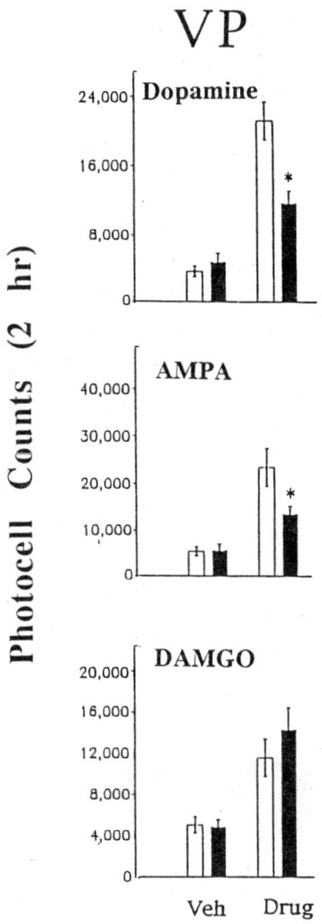

FIGURE 7. Effect of intra-VP microinjections of dopamine, AMPA and DAMGO on locomotor activity in HR and LR. Data represent mean ± SEM of horizontal photocell counts accumulated over a two-hour period postinjection. HR are shown by open bars. Veh = saline vehicle in which drug was dissolved. The number of determinations and dose of each drug are given. For dopamine: 30 μg/side, N = 11 for HR and LR; AMPA: 0.03 μg/side, N = 11 for HR and LR: DAMGO: 0.03 nmol/side, N = 7 for HR and LR. *$p < 0.05$, comparing HR to LR as described in Figure 6.

also are enhanced after repeated stimulation, suggesting that sensitization of mesolimbic circuitry can occur through olfactory pathways (Mitchell and Gratton, 1992).

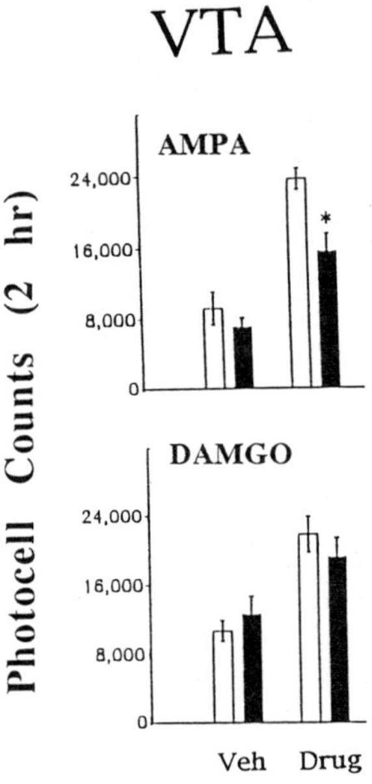

FIGURE 8. Effect of intra-VTA microinjections of AMPA and DAMGO on locomotor activity in HR and LR. Data represent mean ± SEM of horizontal photocell counts accumulated over a two-hour period postinjection. HR are shown in open bars. Veh = saline vehicle in which drug was dissolved. The number of determinations and dose of each drug are given. For AMPA: 0.01 µg/side, HR N = 10, LR N = 11; DAMGO: 0.01 nmol/side, N = 7 for HR and LR. *$p < 0.05$, comparing HR to LR as described in Figure 6.

Nonpharmacological stressors, such as footshock, restraint, tail pinch, handling, and tailshock produce increases in extracellular dopamine levels in the mPFC and/or nucleus accumbens (Abercrombie et al., 1989; Sorg and Kalivas, 1991, 1993a; Cenci et al., 1992; Doherty and Gratton, 1992). Behavioral cross-sensitization of psychostimulants with stress has also been well documented. Repeated stress enhances the behavioral response to a later drug challenge (Antelman et al., 1980; Herman et al., 1984; Robinson et al., 1985; Sorg and Kalivas, 1991). Our present results indicate that a picture neurochemically similar to cocaine-induced TDS emerges when repeated footshock stress is administered followed by an acute cocaine challenge. As shown in Figures 3A and B, daily footshock stress produced not only an increased locomotor response to an acute cocaine challenge but also an augmentation in extracellular dopamine levels in the nucleus accumbens.

In mPFC, the response to repeated cocaine and footshock stress was opposite that observed in the nucleus accumbens. Both repeated daily cocaine and repeated stress produced a decrease in extracellular dopamine levels in response to an acute cocaine injection given one week later. These data support several findings that the mPFC modulates dopamine release in the nucleus accumbens by its inhibitory action on glutamate neurons projecting from the mPFC to the nucleus accumbens (Carter, 1980; Carter and Pycock, 1980; Louilot et al., 1989). Thus, increased dopamine in the mPFC decreases dopamine turnover in the nucleus accumbens (Louilot et al., 1989), and depletions of dopamine in the mPFC increase dopamine metabolism in the nucleus accumbens (Carter and Pycock, 1980; Deutch et al., 1990). The opposite dopamine responses observed between the nucleus accumbens and mPFC in both cocaine and footshock stress-sensitized rats suggest that a decrease in mPFC dopamine levels in response to cocaine may promote the augmentation in nucleus accumbens dopamine levels and thus, behavioral sensitization. To summarize, by sensitizing animals to one type of stimulus, such as stress, one can produce a sensitized response to a pharmacological stimulus, such as cocaine. This process resembles the spreading phenomenon that occurs in MCS, in which individuals initially exposed to one type of chemical become sensitive to the effects of multiple chemicals (Ashford and Miller, 1991).

Role of Corticosteroids in Sensitization

The striking behavioral and neurochemical similarities between cocaine and stress-induced sensitization support the hypothesis of Antelman et al. (1980) that drugs may be acting as stressors to the organism. However, the factors common to mediation of TDS by cocaine and stress are unknown. A growing body of literature suggests one link: both drugs and stress activate the hypothalamic-pituitary-adrenal (HPA) axis (Knich and Eisenberg, 1979; Munck and Guyre, 1986; Rivier and Vale, 1987). In the rat, corticosterone is the naturally occurring glucocorticoid released from the adrenal cortex in response to stress. Recently, investigators have found that behaviors such as the locomotor response to a novel environment, behavioral sensitization produced by repeated stress or amphetamine, and the acquisition of amphetamine self-administration are dependent on corticosterone secretion (Rivet et al., 1989; Piazza et al., 1991; Deroche et al., 1992; 1993). Furthermore, repeated doses of corticosterone that resemble levels induced by environmental stress produce a behaviorally sensitized response to amphetamine (Deroche et al., 1992). Animals that demonstrate an enhanced response to novelty and predisposition to self-administer cocaine have a longer duration of increased plasma corticosterone levels following exposure to novelty stress (Piazza et al., 1990; 1991). Although the central sites at which corticosterone may mediate its action in sensitization are unknown, several candidates exist. The two receptors for corticosteroids, type I and type II glucocorticoid receptors, are present throughout many brain regions, including the hippocampus, amygdala, olfactory nucleus, paraventricular nucleus, and regions of the cortex, striatum, and thalamus (de Kloet et al., 1990). Also, the hippocampus and mPFC can regulate circulating corticosterone levels (de Kloet, 1991; Diorio et al., 1993). Thus, corticosterone levels can be modulated by specific brain regions and, in turn, corticosterone can alter the neuronal responsiveness of those neurons that contain type I and/or type II glucocorticoid receptors. Several findings support the regulation by corticosterone on dopaminergic systems via changes in dopamine levels and their receptors and vice versa (Imperato et al., 1989; Biron et al., 1992; Casolini et al., 1993; Piazza et al., 1993).

Individual Differences

Multiple chemical sensitivity develops in only a subpopulation of humans. Thus, a prediction of which individuals may be most susceptible to developing sensitivity to chemicals would be useful. In the TDS animal model, individual differences among animals in the response to stress and psychostimulants have only recently been considered. How might these differences be mediated by dopaminergic systems relevant to the development of TDS? Piazza et al. (1990) first classified differences in a rat's likelihood of developing amphetamine self-administration based on its locomotor response to a novel environment. Animals were divided into high and low responders (HR and LR, respectively), and HR were found to readily acquire amphetamine self-administration while LR did not. Subsequently, Piazza and other investigators demonstrated increased locomotor activity and dopamine transmission after exposure to acute psychostimulants in HR compared to LR (see Hooks and Kalivas, 1993, for review). Thus, the behavioral and neurochemical responses of HR are similar to rats that have already been sensitized to stress or psychostimulants, while LR are similar to naive rats.

Individual differences in the response to dopamine, AMPA and DAMGO were analyzed in three brain nuclei intimately involved with the initiation of locomotor activity. In addition to the importance of the VTA-accumbens pathway in initiating motor behavior as discussed above, the VP is a critical component of the circuitry involved in locomotor activation initiated by increased dopamine transmission in the nucleus accumbens (Mogenson et al., 1993). Excitatory amino acid (EAA), dopamine, and mu-opioid stimulation increase locomotor activity when microinjected into all three brain regions (Hooks and Kalivas, 1994). However, only EAA and dopaminergic systems responded differentially between HR and LR, indicating that mu receptors do not play a role in novel stimulus-induced activity.

Motive Circuit and TDS

Sensitization of neuronal responses after repeated stressful experiences may provide a means by which an organism can readily adapt to the environment. For example, a heightened response to aversive stimuli would result in rapid escape behavior or avoidance of high levels of noxious stimuli, thus increasing the chances of survival. The dopamine-related circuitry is ultimately impacted by these stimuli, in turn altering the response to subsequent stimuli. Studies examining TDS have provided ample evidence that portions of this dopamine-related circuit, including the VTA, nucleus accumbens, and mPFC, are important in the production of a hyperlocomotor response to psychostimulant drugs. Figure 9 illustrates that several other brain nuclei are also important in the circuitry involved in translation of information to the motivation of action, a series of pathways referred to as the motive circuit. These regions are believed to participate in the neurochemical flow of information important for initiating behavioral responses to various environmental stimuli (see Kalivas et al., 1993). Although all the structures shown in Figure 9 have not been discussed in detail, these major brain nuclei are involved in initiation of adaptive locomotor responses. Several reciprocal interactions between these nuclei suggest that a continual fine-tuning of the circuitry occurs to maintain homeostasis (see Mogenson et al., 1993). Among the several transmitter systems that participate in maintaining this balance, perturbation of dopaminergic and EAA transmission can link communication between limbic structures to increased or decreased motivation to respond to the environment. Dopamine projections from the VTA to several of these sites provides a powerful modulatory influence on structures in the motive circuit. The mPFC appears to play a role in assessing the importance of environmental and interoceptive cues and modulating appropriate behavioral responses to those cues. The oft-reported psychological component of symptoms associated with MCS implies that perception of a stressful event can be critical and can likely amplify the responses to chemical and environmental stimuli that trigger a stress response. In turn, the stress response involves activation of the HPA axis and increased levels of circulating glucocorticoid, which may further amplify dopaminergic responses to subsequent stimuli.

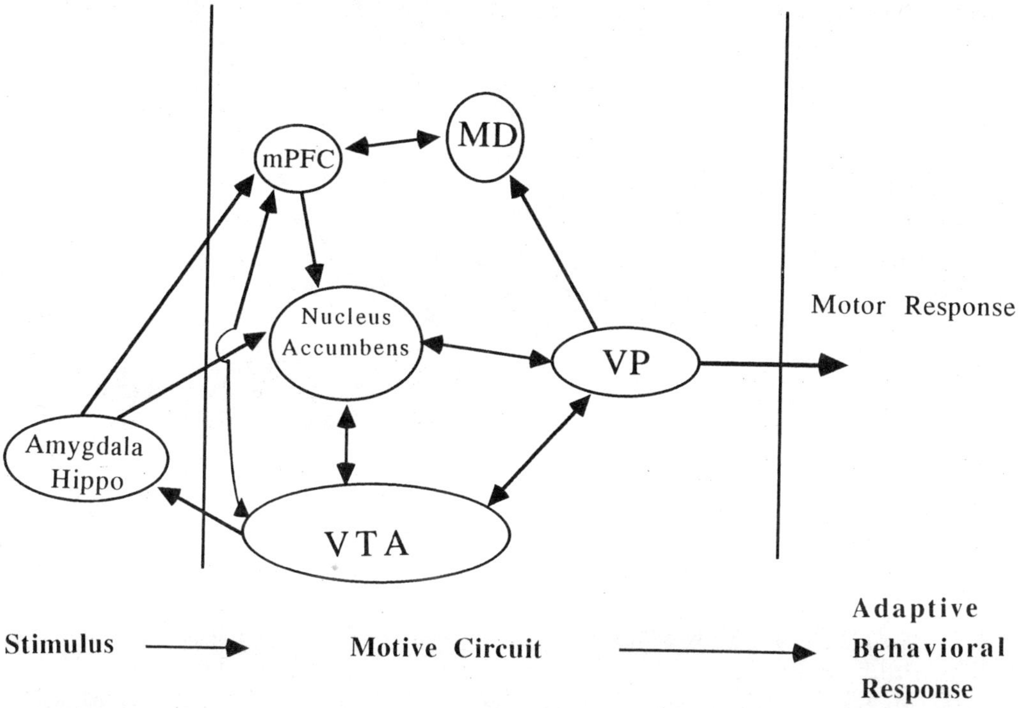

FIGURE 9. Motive circuit illustrating the translation of information from environmental stimuli to adaptive behavioral responses. Reciprocal connections are indicated by bidirectional arrows. Abbreviations: Hippo, hippocampus; mPFC, medial prefrontal cortex; MD, mediodorsal nucleus of thalamus; VP, ventral pallidum; VTA, ventral tegmental area [see Kalivas et al. (1993) for discussion].

Relationship Between TDS and MCS

What is the link between the vulnerability to develop neurochemical and behavioral sensitization to stress/psychostimulants and the propensity to develop MCS? Throughout the literature describing TDS in animals, psychostimulant-induced locomotion is one of the most well-studied phenomena. The link with locomotor activity in rodents and MCS in humans may lie in alterations of the same limbic circuitry that appears to be involved in precipitating psychiatric symptoms such as panic attacks (Post and Weiss, 1988), paranoid psychosis (Robinson and Becker, 1986), and PTSD (Lieberman et al., 1990; Post, 1992). Psychomotor stimulant-induced locomotion in animals and the symptoms described above in humans are believed to be analogous behaviors, since they may occur via sensitization of the same limbic pathways (see Figure 9). Psychiatric symptoms reported by individuals with MCS include depression, anxiety and irritability, which are associated with the symptoms mentioned above. Schottenfeld and Cullen (1985) have presented the possibility that MCS is an atypical form of PTSD. Although there is no direct evidence to support the hypothesis that TDS and MCS are mediated by similar mechanisms in the CNS, some animal studies have reported sensitization and kindling, a form of sensitization, in olfactory pathways and amygdala after repeated exposure to chemicals (Bokina et al., 1976; Burchfiel and Duffy, 1982; Joy, 1982; Gilbert and Mack, 1989). In addition, the time-dependent nature of the onset of symptoms and amplification of those symptoms after withdrawal and reexposure to stimuli provides ample evidence to begin a search within limbic circuitry known to modulate both locomotor activity in animals and the psychiatric symptoms discussed above.

The conflicting results from MCS studies undoubtedly stem from the complexity of genetic differences and environmental influences found in the human population. One example of a difficulty one encounters in the study of MCS is consideration of the length of time spent away from the problem chemical before subsequent reexposure (provocation testing) and appearance of subjective symptoms (Bell, 1994). For example, Ashford and Miller (1991) describe the difficulty in identifying a sensitized response to chemical or food reexposure in humans due to adaptation and deadaptation (analogous to tolerance and withdrawal, respectively, in the TDS literature). Compounding this problem are the overlapping effects of responses to the many chemicals that are encountered in the environment. In studies on TDS in animals, the withdrawal aspect of the sensitization process alone has lead to confusion even when using a single compound within a single strain of animals. It is therefore not surprising that there is disagreement among MCS studies in which withdrawal time has not been controlled. Regardless of whether the origin of MCS is psychogenic, as reasoned by Simon et al. (1990) and Black et al. (1990), the proposed sensitization of CNS components may result from synergistic effects of chemicals with the existing neurochemical state of the individual, as suggested by Bell et al. (1992). The diversity of events posited to contribute to the onset and elicitation of MCS symptoms in humans points to the strong need for establishing an animal model system in which many variables can be easily controlled.

SUMMARY

TDS can be induced in experimental animals by repeated exposure to psychostimulant drugs or environmental stressors, and these agents can also cross-react with each other to produce TDS. Major neurochemical changes associated with TDS include an increase in dopamine transmission in the nucleus accumbens and a decrease in dopamine transmission in the mPFC in response to a subsequent stimulus. Individual differences in the initial responsiveness to novelty stress or psychostimulants are at least partially dependent on differences in dopaminergic and EAA transmission in the VTA, nucleus accumbens and VP. Several features are characteristic of both TDS and MCS, including the onset of symptoms after repetitive stimuli and withdrawal, the progressive nature of sensitivity to low doses of the compound (or stressor) and the possible permanence of neurobiological changes. Individuals who are more prone to develop MCS may sensitize pathways involved in the motive circuit more easily after repeated exposure to stimuli. As with HR rats, individuals who develop MCS may, by virtue of differences in genetic and environmental history, possess differences in their response to perturbations in dopaminergic or excitatory amino acid transmission in discrete brain regions of the motive circuit. Although modification of the response from only a few brain areas illustrates a simplified view of differences between individuals prone to develop MCS and those who do not, the data presented here provide information upon which a testable animal model system for MCS could be based.

ACKNOWLEDGMENTS

This work was supported by U.S. Public Health Service Grants DA-07827 and DA-08212 (BAS) and DA-03906, MH-40817, and a Research Career Development Award DA 00158 (PWK).

REFERENCES

ABERCROMBIE, E.D., KEEFE, K.A., DIFRISCHIA, D.S. and ZIGMOND, M.J. (1989). "Differential effect of stress on *in vivo* dopamine release in striatum, nucleus accumbens, and medial frontal cortex." J. Neurochem. 52:1655–1658.

ANGRIST, B. (1983). "Psychoses induced by central nervous system stimulants and related drugs." In: Stimulants: Neurochemical, Behavioral, and Clinical Perspectives (I. Creese, ed.). Raven Press, New York. pp. 1–30.

ANTELMAN, S.M. (1988). "Stressor-induced sensitization to subsequent stress: implications for the development and treatment of clinical disorders." In: Sensitization in the Nervous System (P.W. Kalivas and C.D. Barnes, ed.). TelfordPress, New Jersey. pp. 227–256.

ANTELMAN, S.M., EICHLER, A.J., BLACK, C.A., and KOCAN, D. (1980). "Interchangeability of stress and amphetamine in sensitization." Science 207:329–331.

ASHFORD, N.A. and MILLER, C.S. (1991). Chemical Exposures: Low Levels and High Stakes. Van Nostrand Reinhold, New York, NY.

BELL, I. (1982). Clinical Ecology. Common Knowledge Press, Bolinas, CA.

BELL, I. (1994). "Neuropsychiatric aspects of sensitivity to low-level chemicals: a neural sensitization model." Toxicol. Ind. Health. 10:277–312.

BELL, I., MILLER, C.S., and SCHWARTZ, G.E. (1992). "An olfactory-limbic model of multiple chemical sensitivity syndrome: possible relationships to kindling and affective spectrum disorders." Biol. Psychiatr. 32:218–242.

BIRON, D., DAUPHIN, C., and DI PAOLO, T. (1992). "Effects of adrenalectomy and glucocorticoids on rat brain dopamine receptors." Neuroendocrinol. 55:468–476.

BLACK, D.W., RATHE, A., and GOLDSTEIN, R.B. (1990). "Enviornmental illness. A controlled study of 26 subjects with '20th century disease'." J. Am. Med. Assoc. 264:3166–3170.

BOKINA, A.I., EKSLER, N.D., SEMENENKO, A.D., and MERKURYEVA, R.V. (1976). "Investigation of the mechanism of action of atmospheric pollutants on the central nervous system and comparative evaluation of methods of study." Environ. Health Perspect. 13:37–42.

BURCHFIEL, J.L. and DUFFY, F.H. (1982). "Organophosphate neurotoxicity: chronic effects of sarin on the electroencephalogram of monkey and man." Neurobehav. Toxicol. Teratol. 4:767–778.

CARTER, C.J. (1980). "Glutamatergic pathways from the medial prefrontal cortex to the anterior striatum, nucleus accumbens and substantia nigra." Brit. J. Pharmacol. 70:50P–51P.

CARTER, C.J. and PYCOCK, C.J. (1980). "Behavioral and biochemical effects of dopamine and noradrenaline depletion within the medial prefrontal cortex of the rat." Brain Res. 192:163–176.

CASOLINI, P., PIAZZA, P.V., KABBAJ, M., LEPRAT, F., ANGELUCCI, L., SIMON, H., LE MOAL, M., and MACCARI, S. (1993). "The mesolimbic dopaminergic system exerts an inhibitory influence on brain corticosteroid receptor affinities." Neurosci. 55:429–434.

CENCI, M.A., KALEN, P., MANDEL, R.J., and BJORKLUND, A. (1992). "Regional differences in the regulation of dopamine and noradrenaline release in medial frontal cortex, nucleus accumbens and caudate-putamen: a microdialysis study in the rat." Brain Res. 581:217–228.

DE KLOET, E.R. (1991). "Brain corticosteroid receptor balance and homeostatic control." Front. Neuroendocrinol. 12:95–164.

DE KLOET, E.R., REUL, J.M.H.M., and SUTANTO, W. (1990). "Corticosteroids and the brain." J. Steroid Biochem. Molec. Biol. 37:387–394.

DEROCHE, V., PIAZZA, P.V., CASOLINI, P., LE MOAL, M., and SIMON, H. (1993). "Sensitization to the psychomotor effects of amphetamine and morphine induced by food restriction depends on corticosterone secretion." Brain Res. 611:352–356.

DEROCHE, V., PIAZZA, P.V., CASOLINI, P., MACCARI, S., LE MOAL, M., and SIMON, G. (1992). "Stress-induced sensitization to amphetamine and morphine psychomotor effects depend on stress-induced corticosterone secretion." Brain Res. 598:343–348.

DEUTCH, A.Y., CLARK, W.A., and ROTH, R.H. (1990). "Prefrontal cortical dopamine depletion enhances the responsiveness of mesolimbic dopamine neurons to stress." Brain Res. 521:311–315.

DICHIARA, G. and IMPERATO, A. (1988). "Drugs abused by humans preferentially increase synaptic dopamine concentrations in the mesolimbic system of freely moving rats." Proc. Natl. Acad. Sci. 85:5274–5278.

DIORIO, D., VIAU, V., and MEANEY, M.J. (1993). "The role of the medial prefrontal cortex (cingulate gyrus) in the regulation of hypothalamic-pituitary-adrenal responses to stress." J. Neurosci. 13:3839–3847.

DOHERTY, M.D. and GRATTON, A. (1992). "High-speed chronoamperometric measurements of mesolimbic and nigrostriatal dopamine release associated with repeated daily stress." Brain Res. 586:295–302.

FALLON, J.H. and MOORE, R.Y. (1978). "Catecholamine innervation of basal forebrain. IV. Topography of the dopamine projection to the basal forebrain and striatum." J. Comp. Neurol. 180:545–580.

GILBERT, M.E. and MACK, C.M. (1989). "Enhanced susceptibility to kindling by chlordimeform may be mediated by a local anesthetic action." Psychopharmacol. 99:163–167.

HERMAN, J.-P., STINUS, L., and LE MOAL, M. (1984). "Repeated stress increases locomotor response to amphetamine." Psychopharmacol. 84:431–435.

HOOKS, M.S. and KALIVAS, P.W. (1993). "Methods to analyze individual vulnerability to drugs of abuse." Pharmacol. Tech. 1:125–142.

HOOKS, M.S. and KALIVAS, P.W. (1994). "Involvement of dopamine and excitatory amino acid transmission in novelty-induced motor activity." J. Pharmacol. Exp. Ther. 269:976–988.

HOOKS, M.S., JONES, G.H., NEUILL, D.B., and JUSTICE, J.B., JR. (1992). "Individual differences in amphetamine sensitization: dose-dependent effects." Pharmacol. Biochem. Behav. 41:203–210.

IMPERATO, A., PUGLISI-ALLEGRA, P., CASOLINI, P., ZOCCHI, A., and ANGELUCCI, L. (1989). "Stress-induced enhancement of dopamine and acetylcholine release in limbic structures: role of corticosterone." Eur. J. Pharmacol. 165:337–338.

JOY, R.M. (1982). "Mode of action of lindane, dieldrin, and related insecticides in the central nervous system." Neurobehav. Toxicol. Teratol. 4:813–823.

KALIVAS, P.W. and DUFFY, P. (1990). "Effect of acute and daily cocaine treatment on extracellular dopamine in the nucleus accumbens." Synapse 5:48–58.

KALIVAS, P.W. and DUFFY, P. (1993). "Time course of extracellular dopamine and behavioral sensitization to cocaine. I. Dopamine axon terminals." J. Neurosci. 13:266–275.

KALIVAS, P.W., CHURCHILL, L., and KLITENICK, M.A. (1993). "The circuitry mediating the translation of motivational stimuli into adaptive motor responses." In: Limbic Motor Circuits and Neuropsychiatry (P.W. Kalivas and C.D. Barnes, eds.). CRC Press, Ann Arbor, MI. pp. 237–288.

KELLY, P. and IVERSEN, S.D. (1976). "Selective 6-OHDA-induced destruction of mesolimbic dopamine neurons: abolition of psychostimulant-induced locomotor activities in rats." Eur. J. Pharmacol. 40:45–56.

KNICH, E.T. and EISENBERG, R.M. (1979). "Effect of amphetamine on plasma corticosterone in the conscious rat." Neuroendocrinol. 29:110–118.

KOE, B.K. (1976). "Molecular geometry of inhibitors of the uptake of catecholamines and serotonin in synaptosomal preparations of rat brain." J. Pharmacol. Exp. Ther. 199:649–661.

LIEBERMAN, J.A., KINON, B.J., and LOEBEL, A.D. (1990). "Dopaminergic mechanisms in idiopathic and drug-induced psychoses." Schiz. Bull. 16:97–110.

LOUILOT, A., LE MOAL, M., and SIMON, H. (1989). "Opposite influences of dopaminergic pathways to the prefrontal cortex or the septum on the dopaminergic transmission in the nucleus accumbens. An *in vivo* voltammetric study." Neurosci. 29:45–56.

MITCHELL, J.B. and GRATTON, A. (1992). "Partial dopamine depletion of the prefrodental cortex leads to enhanced mesolimbic dopamine release elicited by repeated exposure to naturally reinforcing stimuli." J. Neurosci. 12:3609–3618.

MOGENSON, G.J., BRUDZYNSKI, S.M., WU, M., YANG, C.R., and YIM, C.C.Y. (1993). "From motivation to action: a review of dopaminergic regulation of limbic-nucleus accumbens-ventral pallidum-pedunculopontine nucleus circuitries involved in limbic-motor integration." In: Limbic Motor Circuits and Neuropsychiatry (P.W. Kalivas and C.D. Barnes, eds.). CRC Press, Ann Arbor, MI. pp. 193–236.

MOORE, K.E. (1978). "Amphetamines: biochemical and behavioral actions in animals." In: Handbook of Psychopharmacology (L.L. Iversen, S.D. Iversen, and S.H. Snyder, eds.). Vol. 11. Plenum Press, New York. pp. 41–98.

MUNCK, A. and GUYRE, P.M. (1986). "Glucocorticoid physiology, pharmacology and stress." Adv. Exp. Med. Biol. 196:81–96.

PARSONS, L.H. and JUSTICE, J.B., JR. (1993). "Serotonin and dopamine sensitization in the nucleus accumbens, ventral tegmental area and dorsal raphe nucleus following repeated cocaine administration." J. Neurochem. 61:1611–1619.

PIAZZA, P.V., DEMINIERE, J.M., MACCARI, S., LE MOAL, M., MORMEDE, P., and SIMON, H. (1990). "Individual reactivity to novelty predicts vulnerability to amphetamine self-administration." Behav. Pharmacol. 1:339–345.

PIAZZA, P.V., MACCARI, S., DEMINIERE, J.M., LE MOAL, M., MORMEDE, P., and SIMON, H. (1991). "Corticosterone levels determine individual vulnerability to amphetamine self-administration." Proc. Natl. Acad. Sci. 88:2088–2092.

PIAZZA, P.V., ROUGE-PONT, F., DEROCHE, V., KHAROUBY, M., LE MOAL, M., and SIMON, H. (1993). "Corticosterone and sensitivity to drugs of abuse: role of dopamine release." Soc. Neurosci. Abstr. 19:1860.

POST, R.M. (1992). "Transduction of psychosocial stress into the neurobiology of recurrent affective disorder." Am. J. Psychiatr. 149:999–1010.

POST, R.M. and WEISS, S.R.B. (1988). "Sensitization and kindling: implications for the evolution of psychiatric symptomatology." In: Sensitization in the Nervous System (P.W. Kalivas and C.D. Barnes, eds.). Telford Press, NJ. pp. 257–292.

RIVET, J.-M., STINUS, L., LE MOAL, M., and MORMEDE, P. (1989). "Behavioral sensitization to amphetamine is dependent on corticosteroid receptor activation." Brain Res. 498:149–153.

RIVIER, C. and VALE, W. (1987). "Cocaine stimulates adrenocorticotropin (ACTH) secretion through a corticotropin-releasing factor (CRF)-mediated mechanism." Brain Res. 422:403–406.

ROBINSON, T.E. and BECKER, J.B. (1986). "Enduring changes in brain and behavior produced by chronic amphetamine administration: A review and evaluation of animal models of amphetamine psychosis." Brain Res. 11:157–198.

ROBINSON, T.E. and WHISHAW, I.W. (1988). "Normalization of extracellular dopamine in striatum following recovery from a partial unilateral 6-OHDA lesion of the substantia nigra: a microdialysis study in freely moving rats." Brain Res. 450:209–224.

ROBINSON, T.E., ANGUS, A.L., and BECKER, J.B. (1985). "Sensitization to stress: the enduring effects of prior stress on amphetamine-induced rotational behavior." Life Sci. 37:1039–1042.

ROBINSON, T.E., JURSON, P.A., BENNETT, J.A., and BENTGEN, K.M. (1988). "Persistent sensitization of dopamine neurotransmission in ventral striatum (nucleus accumbens) produced by prior experience with (+)-amphetamine: a microdialysis study in freely moving rats." Brain Res. 462:211–222.

SCHEEL-KRUGER, J., BRAESTRUP, C., MIELSON, M., GOLEMBIOWSKA, K., and MOGLNICKA, E. (1977). "Cocaine: discussion on the role of dopamine in the biochemical mechanism of action." In: Cocaine and Other Stimulants, Advances in Behavioral Biology (E.H. Ellinwood, Jr. and M.M. Kilbey, eds.). Plenum Press, New York. pp. 373–407.

SCHOTTENFELD, R.S. and CULLEN, M.R. (1985). "Occupation-induced post–traumatic stress disorders." Am. J. Psychiatr. 142:198–202.

SIMON, G.E., KATON, W.J., and SPARKS, P.J. (1990). "Allergic to life: psychological factors in environmental illness." Am. J. Psychiatr. 147:901–906.

SORG, B.A. and KALIVAS, P.W. (1991). "Effects of cocaine and footshock stress on extracellular dopamine levels in the ventral striatum." Brain Res. 559:695–703.

SORG, B.A. and KALIVAS, P.W. (1993a). "Effects of cocaine and footshock stress on extracellular dopamine levels in the medial prefrontal cortex." Neurosci. 53:695–703.

SORG, B.A. and KALIVAS, P.W. (1993b). "Effects of repeated cocaine on subsequent cocaine- and amphetamine-induced changes in extracellular dopamine levels in the medial prefrontal cortex." Soc. Neurosci. Abstr. 19:1814.

SWANSON, L.W. (1982). "The projections of the ventral tegmental area and adjacent regions: A combined fluorescent retrograde tracer and immunofluorescence study in the rat." Brain Res. Bull. 9:321–353.

UNGERSTEDT, U. (1984). "Measurement of neurotransmitter release by intracranial dialysis." In: Measurements of Neurotransmitter Release In Vivo (C.A. Marsden, ed.). John Wiley & Sons, Ltd., London. pp. 81–105.

WISE, R.A. and BOZARTH, M.A. (1987). "A psychomotor stimulant theory of addiction." Psychol. Rev. 94:469–492.

QUESTIONS AND ANSWERS #2

BARBARA SORG, ROBERT DOTY, SEYMOUR ANTELMAN,
AND CLAUDIA MILLER

Dr. Bill Meggs, Carolina University and New York University: We've seen very convincing evidence that time-dependent sensitization exists, but nobody has talked about the mechanism. There seems to be an element of memory and specificity reminiscent of the immune response. Is there any clue about the mechanism of this?

Dr. Sorg: What we know about the mechanism of sensitization is a lot of what you've seen today, in terms of measuring the changes in neurochemistry in specific brain areas that seem to be responsible for bringing about the sensitization. The ventral integumental area in particular seems to be very important, and it seems to be necessary for the whole process of sensitization. In terms of the mechanism, I think that it is dependent on changes or alterations in gene expression.

We also know that the hippocampus is very important for the induction of sensitization. You can lesion certain sections of the hippocampus and prevent behavioral sensitization, so there may be a mechanism where memory is very important for the animal to remember the stimulus he has received. It has been shown, although it's not published yet, that animals can be anesthetized and will not develop sensitization to cocaine. It's as if they need to be preconceiving this daily stimulus, so that could very well be extremely important in the sensitization process.

Jamie Tessler, occupational and safety health educator in Brooklyn, MCS in the Workplace Task Force: If the blood/brain barrier is compromised in the olfactory bulb, which leads to a lot of this kindling, do other routes of transmission kick off the kindling in the same way?

Dr. Doty: That is a very good question. In the late 1930s, there were some otolaryngologists who tried to use zinc sulfate to damage the receptors so that the polio virus couldn't make its way from the nasal cavity to the brain. There was a very large series of studies in primates showing that if you damaged that region with various chemicals, there were fewer animals that got polio after intranasal inoculation of the polio virus. Subsequent to that, they did human clinical trials in Montreal, as well as in Alabama, and in these studies, when there was a period of a polio epidemic, around 3,000 school children were administered zinc sulfate into the nose, and it was not shown to be successful. So, if there are agents that make their way into the brain through that route, it is conceivable that blockage of that area can be done. In the one trial I am aware of in humans, with the polio virus, it didn't seem to work. Now part

Toxicology and Industrial Health, Vol. 10, No. 4/5, pp. 387–389

ISSN: 0748-2337

of that may be that they did not spray the areas as much as they did the primates, but these are issues that are related to dose, and how it was done.

Howard Kipen, Robert Wood Johnson Medical School: Dr. Doty: Inhalation challenge studies, which have been done in human beings, I'm thinking mostly of Scandinavia but also by our EPA, have shown adaptation of the olfactory response to VOCs, but not of the irritative symptoms. I'm wondering how that may be integrated into the concept of time-dependent sensitization as it was presented today?

Dr. Doty: There is evidence that there are some chemicals which are trigeminal stimulants which may induce irritation, over time. As a result, over time, irritation will occur at even lower doses than originally presented, so that's probably time-dependent or time-related. The levels being given are somewhat higher than, or considerably higher than the olfactory thresholds, and I can conceive of a situation where the olfactory system would adapt, and trigeminal stimulation would continue based on whatever the temporal sequence of the irritating chemicals was.

Ken Lindeman, EBIS: I have been under the impression that one of the difficulties of evaluating these patients was the inconsistency in their response. If this is in fact true, how does this hold up to the time-dependent sensitization explanation?

Dr. Antelman: Different systems show different sensitization at different rates. As the two slides of Diazapam showed, when we were looking at convulsions as an end point, the effect was no longer there after a four-week interval. In the same animals, when we looked at dopamine as an end point, it was there. So you could see that if things sensitize at different rates, you could end up with very different symptoms.

Lindeman: As a clarification, I'm talking about differences within the same individual, not between individuals. In terms of responsiveness to exposures, do you see that phenomenon?

Dr. Miller: Patients do report differences in the intensity of their responses on different days with the same exposure. They tend to say there is a constellation of symptoms they report with diesel exhaust and a certain constellation with a certain perfume, and a constellation with certain other things that they say is reproducible for them, but the intensity of that response may vary from day to day and they feel often that this is related to how much they have been exposed to prior to that date. In other words, if they are in New York City around all the pollution, they don't notice their responses to diesel exhaust or even to a particular food as much as when they are in a cleaner environment, away from all that background noise. The other piece of this is that we think adaptation may play a role. The use of an environmental unit, which has been proposed, allows you to remove people from their background of cross-sensitizing exposures, and test them at a clean baseline. In that way, one would expect you could get reproducible symptoms and reproducible intensities of responses. If you have a lot of background noise and adaptation going on, you will see variability and I do think this is what patients report. One other comment, patients do say that breathing through their mouth does mitigate their responses in many cases, and they'll use that when they get on an elevator where someone is wearing perfume. They find that they can tolerate that elevator ride a little better if they breathe through their mouths, which may be a very interesting observation. Eventually, that breaks through, and they will start having problems, but it will decrease the acuteness of their response.

Richard Conrad, Biochemist: I'd like to add that there is another side to the variability in the person. In addition to adaptation it can go the other way. The other day I had just sniffed an air filter which I knew had aired out really well. A friend came over who had just had some cough drops a few hours before. I reacted severely to the smell from his lungs. I was showing him this air filter and said, "see it has no smell" and I sniffed it and I nearly fell over it was so strong. Two days later I smelled it again and it had no odor to me. So sometimes when you're exposed to only one thing depending on the similarity or dissimilarity of the molecules of what you are about to smell, the other thing can be very enhanced in its effect, not just of the smell, but of the brain fog it produced.

Dr. Doty: Do you ever experience a very different response, sometimes an opposite response from the same stimulation on different occasions?

Conrad: I wouldn't say opposite. Sometimes it just causes nasal pain, sometimes the nasal pain and brain fog. There are certain things that which I can't smell which just cause brain fog, at too low a level to smell it. You don't have to smell it and it doesn't even have to be very odoriferous. There are some things with very strong odors which have no effect, so it's all over the board.

From the Audience: You don't ever get different responses to the same stimulus all the time?

Dr. Conrad: Some things give me bronchitis and they continue to do so. In reference to kindling, you have the induction and then retriggering. What I want to mention is in terms of the odorant getting through to the brain, or is it stimulating the nerve endings in the nose, and do we have a problem with no longer having the negative feedback to keep the brain down to cause adaptation. I don't mean to make it complex, but my point is that some things can get through to the brain, but that takes time. Our reactions often take half a second, not just to smell it, but to have brain fog. When we can't smell it, let's say one to two seconds to have brain fog. I don't think those molecules can get through the cribiform plate in two seconds. I think there has to be a direct stimulation of the olfactory nerve receptor in the mucosa.

Dr. Doty : I think that is an important point. I think that things that go into the brain take a matter of a few hours, perhaps, or longer through the transport mechanisms of the olfactory fibers, so those can't be direct triggering agents, but they might sensitize things.

MODELLING ANXIETY DISORDERS FOLLOWING CHEMICAL EXPOSURES

ROBERT ADAMEC

Department of Psychology
Basic Medical Science
Memorial University of Newfoundland
Newfoundland, Canada

The effects of kindling and inverse benzodiazepine receptor agonist β-carbolines on animal models of anxiety are briefly reviewed in relation to affective disorder associated with chemical exposure. Recent experimental results are described. In the present study, cats were given the inverse benzodiazepine receptor agonist, FG-7142, a powerful anxiogenic compound in humans and animals. Neural transmission in pathways involved in defensive behavior in the cat was monitored using evoked potential techniques. Change in these pathways was related to behavioral changes induced by the drug. It was found that a single dose of FG-7142 lastingly increased defensive response to rodents for at least 40 days after drug administration. Behavioral change was specific to defensive response, since approach-attack behavior remained unchanged, replicating previous studies. The benzodiazepine receptor antagonist, Flumazenil, reversed the increase in defensiveness in a drug-dependent manner, replicating previous findings. Increased defensiveness was paralleled by a delayed onset potentiation of neural transmission between the amygdala and the medial hypothalamus of the left hemisphere. Potentiation in the left hemisphere was transient, decaying between 6 and 12 days after the drug. There was a longer lasting potentiation (LTP) of activity evoked in the left and right amygdalo-periacqueductal gray pathways and in the right amygdalo-medial hypothalamic pathway. Potentiation in these pathways appeared at the time of behavioral change. Potentiation of the right amygdalo-periacqueductal gray and right amygdalo-medial hypothalamic pathways persisted until the end of the experiment. In contrast, potentiation of

1. Address all correspondence to: Dr. Robert Adamec, Dept. of Psychology, Memorial University of Newfoundland, Newfoundland, CANADA, A1B 3X9. Tel.:(709)737-7671. Fax:(709)737- 4000.

2. Key Words: Anxiety, Cat, Chemical Exposure, Defense, FG-7142, Limbic Physiology.

3. Abbreviations: AP7, 2-amino-7-phosphonoheptanoic acid; BM, medial basolateral amygdala; EEG, electroencephalogram; LTP, long-term potentiation; MCS, multiple chemical sensitivity; NMDA, N-methyl-D-aspartate; PAG, periacqueductal gray; VAF, ventral amygdalofugal pathway; VMH, ventromedial hypothalamus.

4. This work was supported by a grant from the Medical Research Council of Canada (MRC MT-7022). The technical assistance of Tanya Shallow and John Budgell is gratefully acknowledged.

Toxicology and Industrial Health, Vol. 10, No. 4/5, pp. 391–420

ISSN: 0748-2337

the left amygdalo-periacqueductal gray pathway faded by 40 days after the drug. Flumazenil decreased potentiation only in the right amygdalo-periacqueductal gray pathway. These data strongly suggest that lasting affective change is mediated by lasting changes in particular efferents of the amygdala of the right hemisphere. Behavioral and physiological effects of FG-7142 were blocked by the N-methyl-D-Aspartate (NMDA) receptor blocker, AP7. The data suggest that failure of neural inhibition induced by FG-7142 engages NMDA receptor processes to produce lasting potentiation of transmission in neural circuits that mediate defensive response with behavioral consequences. Since FG-7142 interferes with GABA mediated neural inhibition and is proconvulsant, its action might mimic the action of other environmental chemicals with similar properties, such as chlorinated hydrocarbon insecticides. The relationship of the present data to the literature on the neural and behavioral effects of insecticide exposure is discussed. The significance of these findings for multiple chemical sensitivity disorder is also briefly discussed.

INTRODUCTION

Multiple Chemical Sensitivity, Anxiety, and Acute Chemical Exposure

Multiple chemical sensitivity (MCS) is a putative syndrome of multiple somatic, cognitive, and affective symptoms that supposedly arise in vulnerable individuals as a result of acute exposure to common chemicals (Ashford and Miller, 1991). This syndrome is not universally accepted, and the literature is divided between proponents of MCS and those who question the validity of the concept. Proponents argue that acute high-level exposure to common environmental chemicals can initiate chemical sensitivity, and chronic lower level exposures, including exposures to foods, perpetuate the syndrome (Randolf 1970, 1974; Rea 1977, 1978; Boyles, 1985).

Concern about the validity of the syndrome comes from several sources. It is troublesome that the symptoms arise without any consistently abnormal findings in routine laboratory testing (Bell 1975, 1982, 1987; Ashford and Miller 1991). In addition, there are other explanations for the somatic and affective symptoms. Many have suggested that the syndrome is really a manifestation of psychiatric disorders, especially somatization disorders, anxiety, and depression, combined with a misattribution of symptoms to chemical etiology (Brodsky, 1983; Stewart and Taskin, 1985; Terr, 1986; Staudenmayer and Camazine, 1989; Black et al., 1990; Rosenberg et al., 1990; Staudenmeyer and Selner, 1990).

Bell et al. (1992) recently critically reviewed both sides of the issue. They argue that there is insufficient evidence to date to rule out a combined effect of chemical exposure, vulnerability, psychosocial stresses, and neurological problems in precipitation of the symptoms of MCS. Further, they propose a possible mechanism for the symptoms of MCS. They suggest that chemical exposure, and the activation of the olfactory-limbic system, may result in a limbic kindling–like change in limbic function that mediates many of the symptoms of MCS.

Therefore, kindling has been invoked as a possible explanatory concept for MCS, and as a justification for the validity of the syndrome.

In this paper, my purpose is not to resolve the controversy. My purpose is to review aspects of long-lasting changes in affective behavior produced by chemical exposure in cats, which may be relevant to anxiety associated with particular kinds of chemical exposure in humans.

Environmental Pollutants and Lasting Affective Change

Among common chemicals used commercially that produce neurological sequelae are insecticides. Two major classes are organophosphates and chlorinated hydrocarbon compounds. Both classes have potent effects on the nervous systems of organisms from invertebrates to humans (Burchfiel et al., 1976). Single high-dose exposures of primates (*Macaca mulatta*) to insecticides such as sarin or dieldrin produced convulsions acutely, and lasting changes in electroencephalograph (EEG) without convulsions, for up to one year after exposure (Burchfiel et al., 1976). These changes were restricted to the high beta band of the EEG in the temporal lobe EEG and lasted for at least one year after the exposure. Multiple low subconvulsant doses had similar effects on the EEG.

These data are of interest because there is a literature that suggests exposure to insecticides may lead to lasting changes in affect, visuomotor, attention, and language functioning in humans (Savage et al., 1988; Rosenstock et al., 1991). Lasting increases in anxiety have been reported following acute and chronic organophosphate insecticide poisoning (Dile and Smith, 1964; Midtling et al., 1985; Savage et al., 1988; see also Burchfiel et al., 1976 for a review of other studies). In contrst, Rosenstock et al. (1991) found disturbances of visuomotor and attention functioning following acute pesticide exposure without changes in anxiety. This discrepancy suggests that cognitive sequelae are not dependent on affective changes. Moreover, affective changes may not be inevitable, but are dependent on nonpesticide vulnerability factors. One possible factor may be stress.

While high doses of insecticides are convulsant, lower doses are not. Yet lower doses do have lasting neurological effects. In this regard, it is of interest that one chlorinated hydrocarbon insecticide (endosulfan) has been shown experimentally to be proconvulsant in the amygdala kindling model (Gilbert, 1992a). It is appropriate, therefore, to briefly consider kindling and its relationship to anxiety.

Kindling — Definition and Properties

Goddard et al. (1969) were the first to investigate systematically the effects of applying repeated brief high-frequency trains of electrical stimulation to the brains of animals. When such trains were applied daily, the response to the stimulus changes from no response, through local seizure discharges recordable only on the EEG, to full motor convulsions outlasting the stimulus train. Convulsive sensitization to the stimulus was very long lasting, persisting for as long as one year after brain stimulation had stopped (Wada et al., 1974). This convulsive sensitization was called kindling by Goddard et al. (1969).

As kindling proceeds, seizure activity spreads to areas distal to the brain area stimulated (Racine, 1978; Adamec, 1991d). Kindling is produced in many species, including reptiles, rodents, cats, primates (Racine, 1978), and possibly humans (Adamec, 1990a).

A second type of kindling is partial kindling. Partial kindling involves repeated stimulation that lowers the electrical threshold to produce EEG seizures without producing motor convulsion (Adamec, 1990a). Both kindling and partial kindling have been associated with lasting changes in affective behavior in animals.

Kindling and Anxiety in Animals
Full kindling of the amygdala in rodents increases anxiety-like behavior, measured in the elevated plus maze for at least one to two weeks after the completion of kindling (Adamec, 1990c; Nieminen et al., 1992; Adamec and Morgan, 1994). A much longer-lasting increase in defensive response to threat is observed in felines after partial kindling of the amygdala by direct stimulation of the amygdala, or as a result of spread of seizure activity from the entorhinal cortex-ventral hippocampal system to the amygdala (Adamec, 1978, 1991b).

It has been argued that the changes in feline defensive response can be a model for the fear component of human anxiety. Defensive response in cats does meet the criteria of animal analogues of anxiety proposed by Skolnick (1983).

Insecticides and Kindling
Joy et al. (1980) showed that the chlorinated hydrocarbon pesticide, dieldrin, was proconvulsant. Repeated doses led to seizures of increasing severity until death resulted. No tests of effect on kindling were done. Another chlorinated hydrocarbon pesticide, endosulfan, has been shown to be proconvulsant. That is, repeated subconvulsant dosing leads to increased seizure response to such doses. This effect has been called chemical kindling. Endosulfan also reduces the number of stimulations to full kindled convulsions by amygdala stimulation. Proconvulsant effects and the facilitation of kindling by endosulfan were observed after the poison had cleared the system of the animals (Gilbert, 1992b). This is an important observation, since many of the pesticides are stored in fat. If stored in fat, persisting effects could then be due to slow release into the bloodstream. This is apparently not the case with endosulfan, which does not persist in fat or other tissues after consumption (Maier-Bode, 1967). Moreover, endosulfan is one of the few chlorinated hydrocarbon insecticides which is still used extensively.

Clearly, insecticides produce or facilitate kindling. Moreover, kindling has been associated with lasting changes in affect (anxiety-like behavior). These findings suggest some commonality of mechanism may exist between either the convulsant or proconvulsant properties of these chemicals and their lasting effects on anxiety.

Chemical Sensitization, FG-7142, and Anxiety in Animals
Systemically administered proconvulsant chemicals have been shown to have kindling-like effects on seizure proneness and behavior. Post et al. (1981) found that repeated systemic injections of lidocaine produced convulsions to initially subconvulsant doses. Effects were long lasting. Lidocaine-treated animals electrically kindled more rapidly than vehicle controls, and became more defensive. They also became more defensively aggressive in response to *icv* administration of the stress related neuropeptide, CRF. Moreover, lidocaine activated limbic system structures (Post et al., 1984).

Another class of proconvulsant compounds of interest is the inverse benzodiazepine agonists. Beta carboline inverse agonists work in a manner opposite to the benzodiazepine anxiolytics, such as vallium. They produce a behavioral and endocrine syndrome characterized as "anxiety" and "fear" in both primates (Crawley et al., 1985) and humans (Dorow et al., 1983). This syndrome appears to involve the benzodiazepine receptor (Crawley et al., 1985). Of particular interest is the β-carboline, FG-7142, since it produces a benzodiazepine receptor-dependent anxiety in humans (Dorow et al., 1983), rodents (File and Pellow, 1986), and primates (Crawley et al., 1985).

There have been several reports of lasting aftereffects of FG-7142 on animal anxiety in rodent models (Drugan et al., 1985; Jeevanjee et al., 1985). Even longer-lasting effects on behavior have been reported in felines. A single systemic injection of FG-7142 increases fearfulness in cats for months (Adamec, 1991a, 1993a). The lasting increase in defensive response produced by FG-7142 is identical to that produced by partial kindling, even though the drug does not produce seizures (Adamec, 1990b). The drug also increases neural communication between the amygdala and ventromedial hypothalamus (VMH) in the left hemisphere transiently (for six days after drug) (Adamec, 1990b, 1993a). This is in contrast to the lasting increase in left amygdala–VMH transmission seen following partial kindling. These data suggest that pathways other than the amygdalo–VMH pathway are involved in lasting increases in defensiveness. Exploring what those other pathways might be is one of the purposes of the present report.

FG-7142 transiently blocks inhibition in limbic structures (Glushankov et al., 1985). The lasting changes in defensive behavior can be blocked with the benzodiazepine receptor blocker, flumazenil (unpublished observations). Flumazenil prevents FG-7142 from binding to the benzodiazepine receptor (Hunkeler et al., 1981). In addition, one can prevent the behavioral changes by increasing neural inhibition with injection of GABA transaminase inhibitors (gamma vinyl gaba) prior to injection of FG-7142 (unpublished observations). These data suggest that the failure of GABA mediated inhibition produced by the binding of FG-7142 to the benzodiazepine receptor is somehow responsible for lasting changes in defensive response.

A candidate neural mechanism for increased fearfulness suggested by these data is NMDA (N-methyl-D-Aspartate) receptor-mediated long-term potentiation (LTP). The reasoning is as follows. FG-7142–induced failure of inhibition might engage NMDA mechanisms of LTP that lastingly potentiate neural communication within the circuitry that mediates feline defensive response. Inhibition failure is known to potentiate LTP (Wigstrom and Gustaffson, 1985). Moreover, NMDA receptor-dependent processes have been identified within limbic circuits that produce and modulate feline defensiveness. The limbic circuits in question all involve the amygdala and its outputs to the VMH and periacqueductal gray (PAG).

The available data suggest NMDA receptor-dependent processes facilitate defensive behavior organized by activation of the PAG. The evidence is as follows. Defensive behavior in cats can be produced by stimulation of the VMH and the dorsolateral PAG (Maeda, 1976; Zhang et al., 1990; Hiyoshi et al., 1991; Lu et al., 1992). It is of interest that intense anxiety is produced in humans when the PAG is electrically stimulated (Nashold et al., 1969, 1974). The basal amygdala facilitates VMH and PAG elicited defense (Shaikh et al., 1993a), and is involved in defensive response to visual threats (Maeda and Hirata, 1978). Moreover, the VMH facilitates PAG-elicited defense. Finally, the facilitation of PAG-elicited defense by the amygdala and VMH is blocked by the NMDA receptor blocker, AP7 (2-amino-7-phosphonoheptanoic acid), administered systemically, or locally into the PAG (Lu et al., 1992; Shaikh et al., 1993b). Anatomically, there are direct projections from the medial hypothalamus and amygdala to PAG sites from which defense can be elicited. Many of the projecting neurons carry aspartate (Schubert et al., 1993). NMDA-dependent processes also exist within the amygdala. There are NMDA-dependent excitatory inputs from the lateral amygdala onto both inhibitory neurons and other neurons in the basolateral amygdala (Rainnie et al., 1991a,b). The NMDA input appears to exert a tonic influence on amygdala cell excitability (Rainnie et al., 1991a,b).

FG-7142 could block the action of local inhibitory neurons in the amygdala, facilitating NMDA-dependent excitatory inputs onto amygdala cells. This could produce an NMDA-dependent LTP of excitatory influences on amygdala cells. The resulting increase in amygdala cell excitability could, for example, account for the LTP of the amygdalo–VMH potential. In addition, block of GABA-mediated inhibition in the PAG by FG-7142 might facilitate LTP of excitatory transmission from the amygdala to the PAG.

The purpose of the present studies was to test these hypotheses. If NMDA-mediated LTP is produced by FG-7142 in these circuits, then NMDA-dependent LTP of transmission between the amygdala and the VMH and PAG should be observed. One way to test this hypothesis is to administer FG-7142 with the NMDA receptor blocker, AP7. If behavior and limbic LTP are NMDA receptor-dependent, then both behavioral and neural changes should be blocked. This was done in the present study.

Which pathways are critical for the behavioral change is not yet certain. They should be the ones in which LTP persists as long as the behavioral change. It is already known that left amygdala–VMH LTP is not critical for persisting behavioral change following FG-7142. LTP in this pathway fades, while increased defensiveness persists (Adamec, 1993a). Recent data in rodents suggest that the amygdala of the right hemisphere may be more important for aversive learning (Coleman and McGaugh, 1993), and kindling-induced increased defensiveness (Adamec and Morgan, 1994). So it is hypothesized that changes in the amygdala projections to the VMH and PAG in the right hemisphere may be more important in mediating lasting behavioral changes. For this reason, activity over these pathways in both hemispheres was monitored. Moreover, transient reversal of lasting increases in defensiveness should be accompanied by transient decreases in LTP in those pathways that are critical for the behavioral change. Transient reversal of increased defensiveness was induced in the present study with the benzodiazepine receptor blocker, Flumazenil. Flumazenil drug-dependently reduces the increase in defensive response produced in cats by FG-7142 (Adamec, 1991a, 1993a), and by partial kindling (Adamec, 1993b). It also drug-dependently reduces amygdala–VMH LTP in partially kindled cats (Adamec, 1993b). It was hypothesized that Flumazenil would reduce LTP only in those pathways which are critical for the behavioral change.

METHODS

Subjects
Subjects were seven adult male cats weighing between 3.5 and 5 kg. The cats were housed individually in large cages and maintained on a normal day-night cycle of 14 hr light, 10 hr darkness. Two to three cats lived in one room. Cats could see and smell each other, but could not interact.

Surgery and Electrode Implantation
Surgery. Seven electrodes were stereotaxically implanted in each cat using sterile technique while the cats were under combined xylazine (2.2 mg/kg) and sodium pentobarbital (15 mg/kg) anesthesia. Stimulating electrodes were placed in medial basolateral amygdala (BM) (area BM of Krettek and Price, 1978) of the left and right hemispheres. Recording electrodes were placed in the ventromedial hypothalamus (VMH), and dorsolateral periacqueductal gray (PAG) of both hemispheres. In the left hemisphere a stimulating electrode was placed in the ventral amygdalofugal pathway (VAF), which carries the output fibers of the amygdala to the VMH (Dreifuss et al., 1968; Dreifuss, 1972)

Stimulating electrodes were made of two teflon-coated platinum-10% iridium wires (0.18 mm in diameter) for bipolar stimulation. Stimulating electrodes were fixed in a 23 gauge needle and separated by 0.5 mm in the vertical plane. A stainless steel recording electrode (0.18 mm in diameter) was positioned between the two stimulating electrodes so that EEG could be recorded from the amygdala. Recording electrodes were stainless steel microelectrodes (0.5 to 1.5 megohms impedance at 1000 Hz).

Electrode coordinates were taken from the Atlas of Jasper and Ajmone-Marsan (1952). They were: BM — AP (anterior-posterior plane anterior to interaural zero) 11.0 mm; L (lateral to the midline) 9.0 mm; VMH— AP 11.0 mm; L 1.3 mm; VAF — AP 11.0 mm; L 5.0 mm;

PAG — AP 2.5 mm; L 1.0 mm. Vertical coordinates were adjusted during surgery to maximize the evoked potentials.

Electrode placements were optimized during surgery by stimulating and recording evoked potentials. Electrode positions were adjusted to maximize amygdalo–VMH, VAF–VMH, and amygdalo–PAG field potentials (Figure 1).

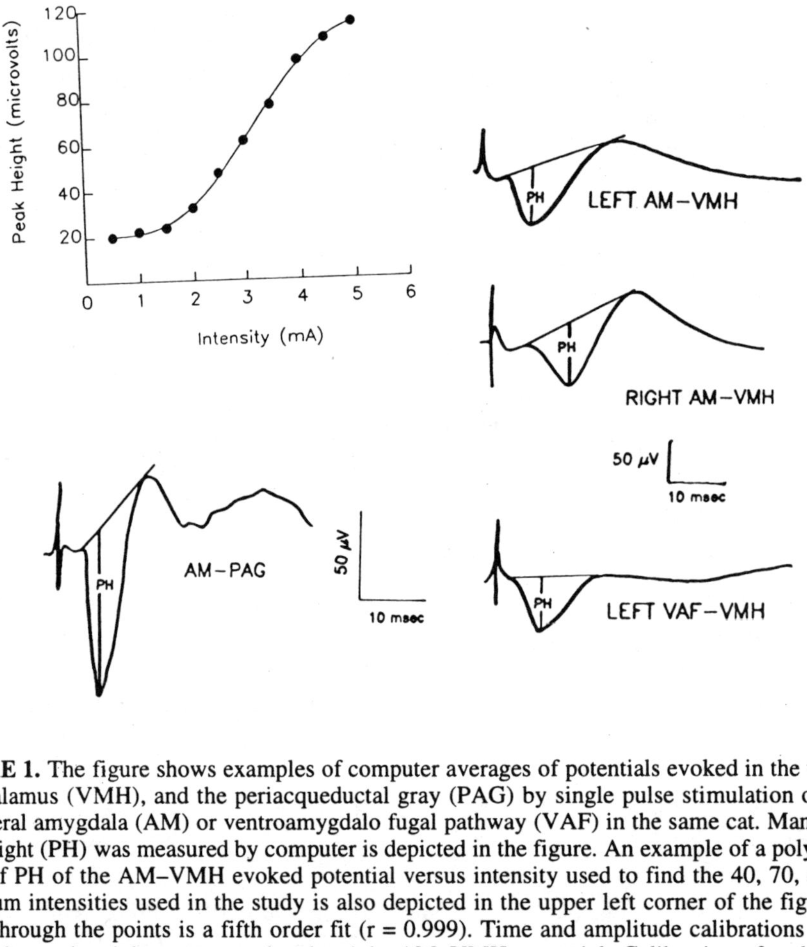

FIGURE 1. The figure shows examples of computer averages of potentials evoked in the ventromedial hypothalamus (VMH), and the periacqueductal gray (PAG) by single pulse stimulation of the medial basolateral amygdala (AM) or ventroamygdalo fugal pathway (VAF) in the same cat. Manner in which peak height (PH) was measured by computer is depicted in the figure. An example of a polynomial fit to a plot of PH of the AM–VMH evoked potential versus intensity used to find the 40, 70, and 100% of maximum intensities used in the study is also depicted in the upper left corner of the figure. The line drawn through the points is a fifth order fit (r = 0.999). Time and amplitude calibrations for the three potentials on the right appear under the right AM–VMH potential. Calibrations for the AM–PAG potential appear to right of the AM–PAG computer average.

Electrode leads were attached to male amphenol pins. These pins were pressure fitted into plastic connectors, and the whole assembly imbedded in a dental acrylic crown. The crown was anchored to the skull with three stainless steel bolts. A fourth bolt was screwed into the skull over the frontal cortex and served as a reference. Electrode positions were verified at the end of the experiment using histological techniques published elsewhere (Adamec, 1991b).

Cats were allowed a four-week postsurgical recovery period before behavioral and physiological testing began.

Behavioral and Drug Testing of Experimental Cats

The sequence of experimental events is outlined in Table 1. Behavioral tests involved exposing the cats to rats. Male hooded rats (140g) were used as prey for the cats. Rats were housed individually in plastic cages and kept on a 12 hr light-dark cycle for three days prior to use. In this time they were handled five times.

TABLE 1. Test Sequence

Test	Test typc	Physiological test	Days(1)	Days(2)	Post*
1	Adapt to lab and test chember	—	1–30		
2	Pre OP 1 Rat Test	—	31		
3	Pre OP 2 Rat Test	—	32		
4	Readapt to test chembers	—	57–64		
5	Post OP 1 Rat Test	—	65	1	
6	Post OP 2 Rat Test	—	66	2	
7		Intensity series	68–69		
8	Jugular cannulation		71	7	
9	Postjugular Rat Test	Intensity series	74	10	
10	i.v. injection of saline	Intensity series	76	12	
	i.v. injection of AP7 (5 mg/kg)	(Both 10 min after)	76	12	
11	i.v. injection of vehicle for AP7 + FG		78	13	
12	Rat Test 2 days after vehicle	Intensity series evoke potentials	80	15	
13	i.v. injection of AP7 followed by FG-7142	Monitor EEG in all pathways	81	16	
14	Rat Test 2 days after AP7 + FG-7142	Intensity series	83	17	
15	Rat Test 6 days after AP7 + FG-7142	Intensity series	87	21	
16	i.v. injection of FG-7142	Monitor EEG in all pathways	94	28	
17	Rat Test 2 days after FG-7142	Intensity series	96	30	2
18	Rat Test 6 days after FG-7142	Intensity series	102	34	6
19	Rat Test 12 days after FG-7142	Intensity series	108	40	12
20	i.v. Flumazenil vehicle Rat Test 17 min after	Evoke potentials 10 min after	114	48	20
21	i.v. Flumazenil, Rat Test 17 min after	Evoke potentials 10 min after	115	49	21
22	Rat Test 2 days after Flumazenil	Intensity series	117	51	23
23	Final Rat Test	Intensity series	134	68	40

*Days (1) are days from the arrival of the cats in the laboratory. Days (2) are days of the experiment following electrode implantation. These are the "Days" in the figures. Post refers to days after the injection of FG-7142. When rat testing and physiological testing occurred on the same day, behavior testing followed physiological testing by ten minutes.

Tests with rats were five minutes in length. The cat was placed in the testing environment first, then the rat was introduced into the room. Cats investigated the rats, and in some cases mildly attacked them. Cats did not injure the rats.

Behavioral testing was done in an enclosed testing chamber between 12:30 pm and 1:00 pm, after the initial daily feeding at 9 am. Behavior was videotaped, and analyzed later. Video recording equipment was located outside the testing chamber, as were the experimenters (see Adamec et al., 1980 for details).

All drugs were injected intravenously (i.v.) through a vascular access port cannula chronically implanted in the left jugular vein (implantation procedure described below). Injections were done with the cat in the physiological recording box (also described below). This is a large electrically shielded box with a narrow opening in the top through which passed the electrode leads connecting the cat to recording equipment. A sterile i.v. line was also passed through this opening and connected the vascular access port to a pump which injected the drugs at a uniform rate (1 ml/min). The box also had a door with a plexiglass front. A video camera was positioned in front of the box door. Experimenters and recording equipment were behind the box. In this way injections could be administered without handling the cat, and the cats' behavior could be monitored on a remote video monitor. Cats were supplied with a sterilized drape on which to lay in the box. Cats were well adapted to the box, and lay down quietly in it during testing.

Preliminary baseline testing (Pre–Post OP, Tests 2, 3, 5, 6, Table 1) involved two tests with rats. The first two tests took place on two consecutive days, four weeks after the arrival of the cats in the laboratory. Cats had been adapted to the testing room for ten days (1/2 hour per day) before testing. The second two tests took place four weeks after surgery (Post OP 1 and 2, Table 1).

Then the left jugular vein was cannulated using sterile technique while the cats were under combined xylazine (2.2 mg/kg) and sodium pentobarbital (10 mg/kg) anesthesia. The cannula (polysulfone plastic) was inserted into the vein and led under the skin to an exit just behind the electrode crown. The cannula terminated in a vascular access port (Access Technologies) which was attached to the crown. These cannulas were soft, pliable, and nonthromobogenic, so they were tolerated well by the cats. Cannulas were flushed daily thereafter with heparinized sterile saline (20 iu/ml, 1 ml flush).

Three days after the jugular cannulation, another rat test took place (Test 9, Table 1). Two days later, cats were injected with vehicle (saline, i.v.) followed in half an hour by AP7 dissolved in saline (5 mg/kg, 25 mg/ml PH adjusted with 1 N NaOH) (Test 10, Table 1). The electrophysiological data collected in this test are currently under analysis. No rat test was done on this day. Two days later (Test 11, Table 1) cats were injected i.v. with saline (vehicle for AP7) followed by a sterile saline–Tween 80 suspension (vehicle for FG-7142). Tween 80 vehicle was prepared by sonicating 2 drops of Tween 80 in 10 ml of saline. Both vehicles were sterile. Volumes of vehicle were adjusted to match the volumes to be injected when AP7 + FG-7142 were injected on Test 13 (Table 1). No rat test was done on this day. Two days after the vehicle injections, another rat test was done (Test 12, Table 1). On the next day, cats were injected with AP7 (5 mg/kg dissolved in saline, 25 mg/ml). After 10 min, a second injection of FG-7142 (10 mg/kg suspended in Tween 80 vehicle, 18 mg/ml) was given i.v. EEG from all brain areas was monitored by computer for seizure activity for 30 min after the injection of FG-7142. Cats were also observed by remote video camera for 20 min after the FG-7142 injection, and returned to their home cages. No rat test was done on this day.

Rat tests were done to determine if there were any changes in behavior following the combined injection of AP7 and FG-7142 two and six days after the injection (Tests 14 and 15, Table 1).

Seven days later cats were injected with FG-7142 (10 mg/kg suspended in Tween 80, 18 mg/kg) (Test 16, Table 1). EEG and behavior was monitored for 30 min after the injection as described above. No rat tests were done on this day. Response to rats was tested 2, 6, 12, and 20 days after the injection (Tests 17–20, Table 1). The rat test on 20 days after FG-7142 was done 17 min after an i.v. injection of sterile vehicle for Flumazenil. This is the same vehicle as the one used for FG-7142. Seventeen minutes was chosen as the test interval after injection, because previous work has shown that Flumazenil affects cat behavior for at least 20 min after injection (Adamec, 1991a). Moreover, the half-life of Flumazenil (in rodents at least) is 16 min (Lister et al., 1984). On the next day, cats were given an equivolume injection of vehicle containing Flumazenil (10 mg/kg, 18 mg/ml). Response to a rat was tested 17 min after injection (Test 21, Table 1).

The last two behavioral tests took place 2 and 19 days after Flumazenil (Tests 22 and 23 Table 1).

Behavioral Measures — General Considerations

Behavior was videotaped and later analyzed in slow motion. A variety of measures of defensive response to rats and approach-attack on rats were taken.

Defensive behaviors in the predatory tests. The measure of defensive response to rats was approach-withdrawal (Withdrawal). Withdrawal was defined as the interval between a rapid withdrawal of head or body from the rat and another approach to the rat (to within one foot if the cat moved away). Withdrawal times were summed and averaged to produce the withdrawal score. This measure reflects the average "willingness" of the cats to approach rats after a threatening defensive response of the rat to approach.

Cat withdrawal is dependent on prey self-defense (Adamec et al., 1980). Therefore rodent self-defense was recorded. Rat defense was measured as the frequency of: 1) active defense (e.g., upright boxing posture, vocalization); 2) passive defense (e.g., crouching and freezing);

and 3) escape [fleeing or jumping, following the postural analysis of Grant and Mackintosh (1963)].

When threatened, cats will often seek shelter in a dark corner. This behavior (Hiding) was measured as the latency to enter, and time spent under a canopied shelter located in one corner of the testing room.

A final measure was latency and time spent in a crouched posture with the tail wrapped around the body. This behavior has been described as a self-warming response to threat in a variety of mammalian species (Andrew, 1972).

Approach-predatory attack and activity measures. There were four measures of approach. Two were latency to approach to within one foot of the prey and latency to sniff the rat. The remaining two were time spent near and time spent sniffing the prey. Passive viewing of the prey was also scored as the time spent watching, but not interacting with, the prey.

The most effective predatory attack involves pinning the prey with a forepaw followed by a bite to the nape of the neck (Leyhausen, 1956 and personal observations). Paw pinning (Pinning) most often involves placing a forepaw on the back or rump of the prey to stabilize the prey for a bite. Latency and frequency of pinning was measured. Cats will also grasp the rat with both paws. Latency and frequency of grasping was measured. Latency and time spent biting rats was also scored.

Gentle and hard paw strikes with the forepaws were scored as described by Leyhausen (1956). Gentle paw striking involves either an aborted strike or light striking with insufficient force to move the prey. Hard paw striking is forceful enough to move the prey. Latency and frequency of these pawing attacks was scored.

Attention was paid to the targeting of biting and pawing. Attacks to front (head-neck), middle (body), and rear (rump-tail) were scored separately.

Activity during prey tests was measured to control for possible sedating effects of the drugs. The floor of the testing enclosure was divided by white tape into one foot squares. The squares crossed by the cat during a five-minute prey test were counted. Taped squares were also used to determine when the cat was near the prey (within one foot of the prey).

Physiological Recording

Physiological recording was done in a large electrically shielded box (Adamec, 1991b for more details). Cats were unrestrained during these sessions except for two nine cable (microdot) leads attached to electrode connectors on the head of the cats. These leads connected the cats to stimulating and recording equipment. Cats were well habituated to the testing box before testing began, and were relaxed and quiet during testing. Behavior of the cat was observed on a remote video monitor during brain stimulation. Cats were relaxed and unmoving during brain stimulation throughout the experiment.

All recording was monopolar referenced to the skull screw. Brain activity was preamplified with Grass P-15 AC coupled amplifiers, and amplified again with a Tektronix oscilloscope. Evoked potential data were filtered between 10 Hz and 3 Khz. The outputs of the oscilloscope were led to analog to digital converters of an EMPAC 386 personal computer.

Evoked potential activity was digitized at a rate of 12 kHz. Eighty msec samples of evoked potential activity were taken after single pulses to either the BM, or VAF. Potentials were evoked at a rate of 1/2 sec. Ten evoked potentials were taken, and the potentials averaged by computer.

Brain Stimulation

All stimuli were balanced, bipolar, constant current square wave pulses delivered with a WPI model 600 photon coupled stimulator. Pulse width varied between 100 and 300 μsec.

Physiological Testing

The sequence of physiological testing appears in Table 1. Cats were adapted to the physiological testing chamber for ten days before and ten days after the electrode implants. Physiological tests done on the same days as the behavioral tests were done before the behavioral testing. The time interval between physiological tests and behavioral tests was ten minutes.

Before physiological testing began, an intensity series was done (Test 7, Table 1). Single pulse stimulation of BM or VAF was done with recording in the VMH and PAG. Intensity of stimulation ranged from 0.5 to 5.0 mA peak current in 0.5 mA steps. Ten potentials were evoked at each intensity. Peak height of the average of ten potentials was measured as depicted in Figure 1. A curve was fit by computer (Table Curve program by Jandel) to the plot of peak height versus intensity. One intensity with a peak height of nearly zero was included, followed by intensities with peak heights greater than zero. Polynomial fits of order five to six were good (all $r^2 > 0.90$).

The fitted equations were used to determine the intensity required to evoke a potentials with peak heights of 40, 70, and 100% of the maximum. These intensities was used throughout the remainder of the experiment with a few exceptions. In Table 1, where it says intensity series, the three intensities determined from the curves were used. Where it says Evoke Potentials, a single intensity was used, the one needed to evoke a response at 70% of maximum. This latter procedure was adopted because there was time for sampling only one intensity 10 min following injection of Flumazenil.

Evoked potentials were averaged and stored on disk. Size of the potentials evoked in the VMH and PAG by all stimulating electrodes were measured as peak height taken from the potential average as described above. It should be mentioned that one test, Test 10 (Table 1), was done to monitor the effects of AP7 on evoked potentials in all pathways. These data are currently under analysis, and will not be reported here.

Drug Supply

FG-7142 was purchased from Research Biochemicals. Flumazenil was kindly donated by Hoffman – La Roche (courtesy of Dr. Peter Sorter). Both drugs were suspended in sterile vehicle just before administration.

Statistics

Data were analyzed with analysis of variance using BMDP for PC (Solo program). *A priori* mean contrasts were done using t tests. *Post hoc* tests used Duncan multiple range tests or the Bonferroni procedure (Krauth, 1988). Data were $\log_{10}(x+1)$ or arctangent transformed when necessary to meet test assumptions of normality.

Ethical Approval

The research reported here was reviewed for compliance with the guidelines of the Canadian Council on Animal Care (CCAC), and approved by the President's Animal Care Committee of Memorial University.

RESULTS

Behavioral Changes over Baseline

Response to rats over tests 2, 3, 5, 6, 9, and 12 (Table 1) was analyzed. These tests span pre- and postsurgical testing through the test two days after vehicle injection. They cover a time interval of 80 days. Test effects were examined using one way analysis of variance for repeated measures over tests. In addition, 16 behavioral measures were dropped from this and further analysis because cats showed ceiling (for latency) and floor (for frequency and duration) effects. Ceiling and floor effects were judged when the mean of a latency measure over the six tests did not differ from 300 sec, or the mean of a frequency or duration measure did not differ from 0 respectively (t test of mean against a constant of 300 or 0). These measures are listed in Table 2.

There were no test effects on any behavioral measure. Therefore behavioral response was stable over 80 days of testing. Since Tests 9 and 12 (Table 1) were before and after the injection of AP7 (Test 10, Table 1), the data also indicate no lasting aftereffects of injection of AP7 on behavior.

Effects of AP7 + FG-7142, FG-7142, and Flumazenil on Behavior

Statistical analysis. The effects of injection of AP7 + FG-7142 and later injection of Flumazenil on behavior were assessed with analysis of variance. Test effects were examined using one way analysis of variance for repeated measures over tests. Twelve tests were

TABLE 2. Approach-Attack Measures Showing Floor or Ceiling Effects

Measure	Floor effect	Ceiling effect
Biting rear	yes	yes
Hard paw striking		
Front	yes	yes
Middle	yes	yes
Rear	yes	yes
Pinning		
Front	yes	yes
Grasping		
Front	yes	yes
Middle	yes	yes
Rear	yes	yes

included in the analysis. These were tests 6, 9, 12, 14, 15, 17, 18, 19, 20, 21, 22, and 23 (Table 1). Tests 6–12 were a pre–drug injection baseline, including a test two days after an injection of vehicle (Test 12). Tests 14–15 assessed the after effects of combined injection of AP7 and FG-7142 at two and six days after the injection. Tests 17–19 examined after effects of injection of FG-7142 alone at 2, 6, and 12 days after the injection. Tests 20–22 examined the effects of Flumazenil on behavior. Test 23 assessed the longevity and persistence of recovery of behavior from the effects of Flumazenil.

Effects on approach-attack. There were no changes over tests in any measure of approach to the rat or of attack on the rat. This finding is consistent with previous work showing that a single dose of FG-7142 has no effect on approach attack behavior of cats with mild, nonlethal attacks on rats (Adamec, 1993a).

Effects on rat defense. There were no changes over tests in any measure of rat defense. This finding is consistent with previous work (Adamec, 1991a,1993a).

Effects of FG-7142 alone on defense. Withdrawal from rats was the only measure of defensive response which changed significantly over tests ($F\{11,55\} = 2.42$, $p < 0.02$, $\log_{10}(x)$ transformation). FG-7142 alone increased withdrawal over baseline over the 40 days of the experiment (Figure 2). The increase appeared to grow over days two and six after injection, peaking at day six. This apparent growth, however, is not reflected in a significant difference between days two and six. Withdrawal tended to decrease over days 12, 20, 23, and 40 after the drug, but again withdrawal on these days did not differ from day six after the drug.

Effects of Flumazenil on Withdrawal

Withdrawal from rats was reduced to pre–FG-7142 baseline values following injection of Flumazenil on day 23 after FG-7142 (Figure 2). This reduction in defensive response was

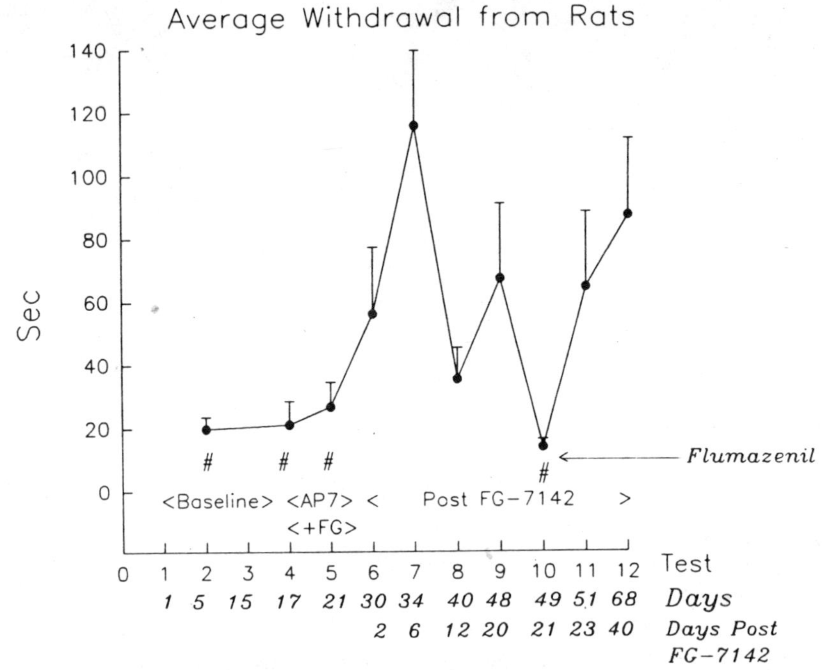

FIGURE 2. Plotted in the figure are mean + SEM of average withdrawal time from rats (in sec). Means are plotted versus test. The three tests prior to injection of AP7 + FG-7142 (Baseline in the figure) are averaged and plotted as one mean since they do not differ. Tests done two and six days after AP7 + FG-7142 (Tests 4 and 5) and after FG-7142 (Post FG-7142, Tests 6–12) are marked in the figure. Means marked with a "#" do not differ from baseline but differ from unmarked means, which do not differ from each other (all $t\{55\} \geq 2.84$, $p < 0.01$, planned comparisons). The mean pointed to with the arrow terminating in the word "Flumazenil" is from the test taken 17 min after an i.v. injection of Flumazenil. Days on the abscissa refer to the Days (2) column of days in Table 1, and are days from the start of the experiment. Days Post FG-7142 are days from the i.v. injection of FG-7142 alone.

drug-dependent. Withdrawal following injection of the Flumazenil vehicle, and 2 and 19 days after Flumazenil, did not differ from withdrawal seen on other days after FG-7142. Withdrawal on these days, however, was greater than withdrawal observed after injection of Flumazenil.

Effects of AP7 + FG-7142 on Withdrawal

Injection of FG-7142 in combination with AP7 had no effect on defensive behavior measured two and six days after injection (Figure 2). Therefore, AP7 blocked the lasting effects of FG-7142 on defensive response.

Effects of AP7 + FG-7142, FG-7142, and Flumazenil on Limbic Physiology

Statistical analysis. Changes in limbic physiology over tests were assessed using a repeated measures analysis of variance design. Factors in the analysis were tests and intensity of stimulation. Recall that for each intensity series done, cats were stimulated at intensities needed to evoke a field potential whose peak height was 40, 70, or 100% of maximum in baseline testing. Data recorded during each of these three intensities were entered as each of three levels of the intensity factor.

Peak heights of potentials evoked in the VMH by amygdala and VAF stimulation, and in the PAG by amygdala stimulation were expressed as a ratio of baseline. Baseline was taken as the peak heights observed at each intensity two days after vehicle injection (Test 12, Table 1).

There were nine tests in this analysis. Predrug baseline was tests 9 and 12 (Table 1), spanning the postjugular cannulation and FG-7142 vehicle tests. Subsequent tests paralleled the tests used in the assessment of behavior with one exception. Two of the tests of the effects of Flumazenil on physiology were not included in this analysis (Tests 20 and 21, Table 1). A separate analysis of the effects of Flumazenil on physiology was done because only one intensity of brain stimulation was used following Flumazenil vehicle and Flumazenil due to time constraints of the procedure. So Tests 9, 12, 14, 15, 17, 18, 19, 22, and 23 (Table 1) were used in the present analysis.

Amygdalo–VMH and VAF–VMH evoked potentials. There was no change in the size of the VAF–VMH potential over all tests. This finding replicates previous reports (Adamec, 1993a). In contrast, the size of the left amygdalo–VMH potential increased over all intensities after FG-7142 (planned t test comparisons, Figure 3). Peak height first rose above baseline six days after injection of FG-7142. There was a rising trend from days two to six after FG-7142, but peak height at day two did not differ from baseline. Peak height returned to baseline values 12 days after FG-7142 and remained at baseline to the end of the experiment. AP7 blocked the transient potentiation of the left amygdalo–VMH potential. Peak height did not differ from baseline at 2 and 6 days after AP7 + FG-7142 (Figure 3).

In a contrast, the size of the right amygdalo–VMH potential increased at two days after FG-7142 over all intensities (Test Effect, $F\{8,81\} = 3.26$, $p < 0.003$, arctangent transformation, Figure 3, no Test by Intensity interaction). Unlike the left amygdalo–VMH potential, potentiation of peak height in the right hemisphere persisted to the end of the experiment (40 days after FG-7142).

As in the left hemisphere, AP7 blocked potentiation of the right amygdalo–VMH potential by FG-7142. Peak height did not differ from baseline at two and six days after AP7 + FG-7142 (Figure 3).

Amygdalo–PAG evoked potentials. The size of the right and left amygdalo–PAG potentials increased at two days after FG-7142 (Test Effects, $F\{8,135\} = 2.81, 2.73$, $p < 0.01$, arctangent transformation, left and right hemisphere respectively, Figure 4, no Test by Intensity interactions). The potential in the right hemisphere remained potentiated throughout the experiment. Potentiation of potentials in the left hemisphere remained elevated to 23 days after injection of FG-7142, but returned to baseline by 40 days after FG-7142.

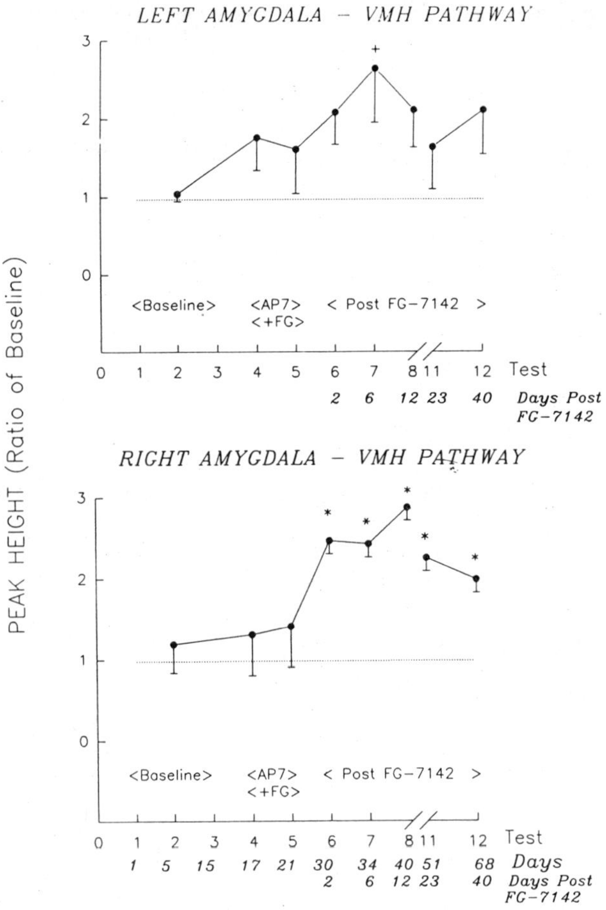

FIGURE 3. Plotted in the figure are mean – SEM of peak height of the VMH potentials evoked in the left and right hemispheres by left and right amygdala stimulation (LEFT, RIGHT AMYGDALA – VMH PATHWAY respectively in the figure). Means are averaged over the three intensities at each test and are plotted versus test. Data are expressed as a ratio of baseline (as defined in the text). The two tests prior to injection of AP7 + FG-7142 (Baseline in the figure) are averaged and plotted as one mean since they do not differ. They are plotted at Test 2 in the figure to align them with Figure 2 for comparison purposes. Tests done two and six days after AP7 + FG-7142 (Tests 4 and 5) and after FG-7142 (Post FG-7142, Tests 6–12) are marked in the figure and are plotted as Tests 4 and 5 to align them with Figure 2 for comparison purposes. A gap appears on the abscissa between test 8 and 11 because the tests of effects of Flumazenil with a single intensity of stimulation fall within this interval. Days on the abscissa refer to the Days (2) column of days in Table 1 and are days from the start of the experiment. Days Post FG-7142 are days from the i.v. injection of FG-7142 alone. In the upper panel, the mean marked with a "+" differs from baseline ($t\{21\} = 1.74$, $p < 0.05$ 1-tailed test of mean against a constant of 1, planned comparisons). Unmarked means do not differ from baseline. In the lower panel, means marked with an "*" differ from baseline. Unmarked means do not differ from baseline or marked means (all $t\{162\} \geq 5.03$, $p < 0.001$, planned comparisons, $p < 0.05$ bonferroni criteria). Data from seven cats are plotted in the upper and lower figures.

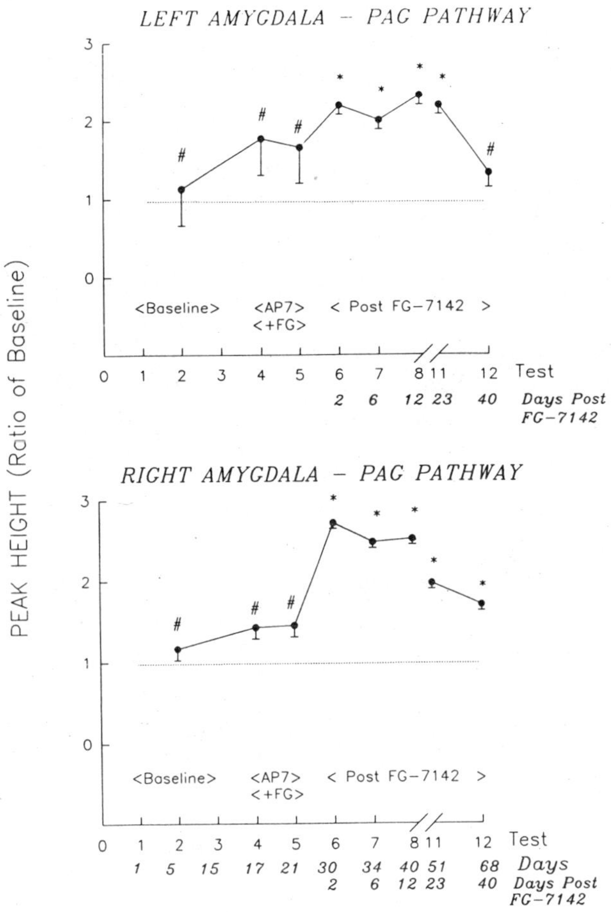

FIGURE 4. Plotted in the figure are mean - SEM of peak height of the PAG potentials evoked in the left and right hemispheres by left and right amygdala stimulation (LEFT, RIGHT AMYGDALA - PAG PATHWAY respectively in the figure). Data are expressed as a ratio of baseline (as defined in the text). Means are averaged over the three intensities at each test and are plotted versus test as described in Figure 3. Means marked similarly do not differ from each other, but differ from means marked differently. In both panels the means marked with a "*" differ from baseline and means marked with a "#" (all $t\{135\} \geq 2.48$, $p < 0.008$ planned comparisons, $p < 0.05$ bonferroni criteria). Data from five cats with correct electrode placements are plotted in the upper and lower figures.

AP7 blocked potentiation of the left and right amygdalo–PAG potentials by FG-7142. Peak heights did not differ from baseline at two and six days after AP7 + FG-7142 (Figure 4).

Baseline limbic physiology. Evoked potentials in all pathways did not change over all tests prior to the injection of FG-7142. Therefore potentials were stable over a retest period of 11 days. This is consistent with previous work in this laboratory, where amygdalo–VMH and VAF–VMH evoked potentials have been found to be stable over periods of two weeks or more. The present findings add to these observations by showing amygdalo–PAG potentials to be stable over time.

Effects of Flumazenil on Limbic Physiology

For this analysis, peak heights were expressed as a ratio of the peak height observed 17 min after an injection of Flumazenil vehicle (Test 20, Table 1). Peak heights observed 17 min after Flumazenil, and two days after Flumazenil, were tested for difference from a constant of 1 (the Flumazenil vehicle baseline). The only evoked potential which was changed by Flumazenil was the right amygdalo–PAG potential. Flumazenil reduced the peak height to 84 % of the Flumazenil vehicle ($t\{4\} = 10.66$, $p < 0.01$, 0.842 ± 0.015 mean ± SEM). This value tended to be larger than baseline values taken prior to FG-7142 ($t\{8\} = 2.06$, $p < 0.08$ 2-tailed test, comparing 0.842 with 0.585 ± 0.124, mean ± SEM of baseline taken prior to FG-7142 at test 9, Table 1, expressed as a ratio of Flumazenil vehicle peak height).

Effects on Body Weight

Cats were weighed before each rat test. Body weights recorded before the rat tests analyzed above were analyzed in the same way as were the behavioral data. Body weight did change over tests (Test Effect, $F\{11,55\} = 4.32$, $p < 0.001$, Figure 5). Body weight began to rise after AP7 + FG and rose again after FG-7142. The trend in the data fit a quadratic function well. Therefore these changes over tests likely reflect a gradual growth, rather than some drug induced increase in body weight.

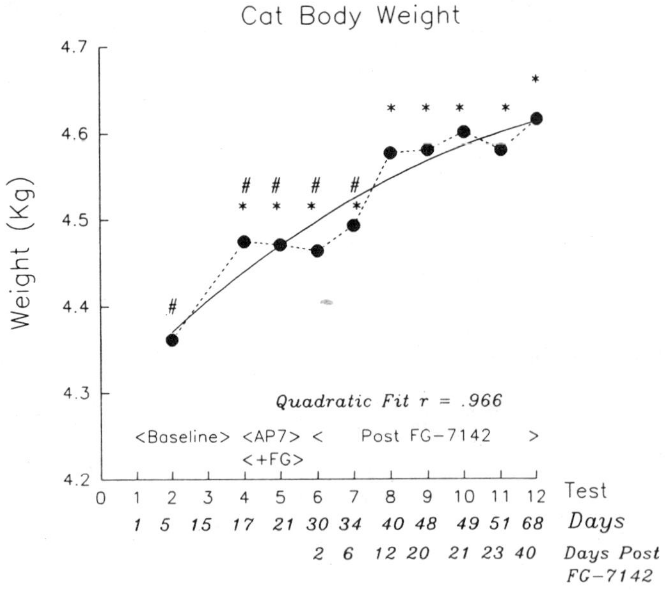

FIGURE 5. Plotted in the figure are mean body weight (in kg) versus tests. The abscissa is as described in Figure 2. Means marked similarly do not differ, but differ from means marked differently. Means marked with two symbols fall between the means marked with each of the symbols (Duncan multiple comparison test, $p < 0.05$). The curve plotted as a solid line is the quadratic fit to the data, which was good ($r = 0.966$ in the figure).

DISCUSSION

Effects of FG-7142 on Defense and Related Limbic Physiology

The findings replicate and extend previous work. In this study, a single injection of FG-7142 lastingly increased defensive response to rats (withdrawal). The cats became more defensive to the same threat, since defensive response of the rats to the cats did not change. Similar increases in defensive response after a single injection of the same dose of FG-7142 have been reported (Adamec, 1991a, 1993a). Moreover, changes in behavior were restricted to defensive response to rats. Approach-attack behavior was unaffected by FG-7142. This finding also replicates previous work (Adamec, 1991a,1993a).

A transient potentiation of the left amygdalo–VMH potential accompanied the increase in defensive response. Potentiation appeared between two and six days after FG-7142 and faded by 12 days after FG-7142. This result is consistent with a previous report of a transient potentiation in the left amygdalo–VMH potential in cats following FG-7142 (Adamec, 1993a).

Potentiation of the amygdalo–VMH potential occurred without changes in the VAF–VMH potential, as has been reported previously (Adamec, 1993a,b). This result suggests that increased amygdalo–VMH transmission following FG-7142 is due to changes within the amygdala, and not in the output pathway (VAF), or in the VMH. This conclusion is partly consistent with the distribution of benzodiazepine receptors in the amygdalo–VMH system. The amygdala in contains high densities of benzodiazepine receptors (Young and Kuhar, 1980). On the other hand, benzodiazepine receptors have also been found in lower, but still high, density in the VMH (Young and Kuhar, 1980).

Unfortunately, there are no data on benzodiazepine receptor distribution in the cat brain. Assuming the data in other species reflect the situation in the cat, then it is puzzling that the lasting excitatory action of FG-7142 should be restricted to the amygdala. A possible explanation is in the NMDA receptor dependence of amygdalo–VMH potentiation. This will be addressed later.

The rise in withdrawal after FG-7142 appeared to be gradual in this study. In previous work, the rise in withdrawal has definitely been gradual. In those studies, withdrawal at two days after FG-7142 fell between baseline and withdrawal observed at six days after FG-7142 (Adamec, 1991a, 1993a). So the pattern observed here is similar but not identical to two previous studies. This gradual change is reminiscent of time-dependent sensitization phenomena reported by Antelman (Yehuda and Antelman, 1983).

In a previous report, it was suggested that initiation of increased defensive response was dependent on a transient potentiation in the left amygdalo–VMH pathway. It was also suggested that changes elsewhere in the brain mediate the lasting change in defensive response to rats (Adamec, 1993a). The present findings bear on the suggestion that changes in pathways other than the left amygdala–VMH path mediate the lasting behavioral change.

Candidates for pathways mediating lasting increases in defense are the left and right amygdalo–PAG pathways and the right amygdalo–VMH path. The patterns of change are different among the paths, suggesting each may contribute to the increase in withdrawal in a different manner.

Evoked activity increased in the left and right amygdalo–PAG and in the right amygdalo–VMH pathways two days after FG-7142, coincident with the increase in withdrawal. So the left and right amygdalo–PAG pathways and the right amygdalo–VMH pathway are implicated in mediating the onset of increased withdrawal. It is of interest that withdrawal tends to peak at six days after FG-7142, when all four pathways show long-term potentiation (LTP). Withdrawal tends to rise (two days after FG-7142) and drop (days 12 and onward after FG-7142) when three of the four pathways (left and right amygdalo–PAG and right amygdalo–VMH) show LTP. Withdrawal remains relatively constant thereafter, despite the fact that left amygdalo–PAG LTP fades by 40 days after FG-7142. This suggests that maintenance of withdrawal does not require LTP of amygdala efferents to either the VMH or PAG in the left hemisphere. The data also suggest, though weakly, that the more pathways showing LTP, the greater the defensive response.

Taken together, these findings suggest that LTP of amygdala efferents to VMH and PAG in the right hemisphere may mediate the lasting increase in withdrawal following FG-7142. The effects of Flumazenil on behavior and physiology suggest further that LTP of the right amygdalo–PAG pathway may be the most critical for lasting increases in withdrawal.

Effects of Flumazenil on Defense and Related Limbic Physiology

Flumazenil reversed the lasting increase in withdrawal in a drug-dependent manner. This finding is consistent with two previous reports. In those studies, Flumazenil partially reversed the increase in withdrawal (Adamec, 1991a, 1993a) whereas here it completely reversed the lasting effects. The only difference between the previous studies and this one is the fact that the cats in this study received two doses of AP7 prior to the dose of FG-7142 which increased defense. Perhaps the AP7 produced some change which made Flumazenil more efficacious.

The effects of Flumazenil on limbic physiology are revealing. One would expect that LTP should be reduced in those pathways that are maintaining the increase in withdrawal. Flumazenil reduced LTP in a drug-dependent manner in only one of the pathways, the right amygdalo–PAG pathway. Given that LTP in this pathway was present from two days after FG-7142, it could be concluded that LTP in the right amygdalo–PAG path mediates the lasting increases in defense. On the other hand, the pattern of change of withdrawal over days 2–12 after FG-7142 suggest LTP in the other pathways might contribute to the magnitude of defensive response. More direct tests of the involvement of these pathways in behavioral change are required to settle this issue.

The lack of effects of Flumazenil on LTP of the right amygdalo–VMH potential are surprising. Previous work has shown that Flumazenil does reduce LTP of the left amygdalo–VMH potential induced by partial kindling (Adamec, 1993b). The effect of Flumazenil was shown to be in the amygdala, and not in the efferents to the VMH, since VAF–VMH LTP was unchanged by Flumazenil in the same cats. Moreover, subsequent work demonstrated that Flumazenil only reduced left amygdalo–VMH potentials when those potentials were showing LTP (Adamec, 1993a). The latter finding is consistent with the present study, since the left amygdalo–VMH potential was neither potentiated at the time of Flumazenil injection, nor was it changed by Flumazenil. In contrast, in the present study, Flumazenil has no effect on a potentiated right amygdalo–VMH potential. Given the evidence that Flumazenil works in the amygdala in the left hemisphere, this finding suggests Flumazenil is without effect within the right amygdala. Together these data suggest some differences exist in the action of Flumazenil in the left and right amygdalas. This finding is novel, and to my knowledge there is no reported evidence of lateralized effects of Flumazenil in limbic structures in animals.

If Flumazenil is not working in the right amygdala, then its site of action on the right amygdalo–PAG potential is in the right PAG. Benzodiazepine receptors to which Flumazenil could bind are present in the PAG (Young and Kuhar, 1980). Moreover, this conclusion is consistent with the evidence that LTP of the amygdalo–PAG pathway is dependent on NMDA receptor-mediated processes within the PAG. This evidence is discussed next.

The Dependence of Changes in Behavior and Physiology on the NMDA Receptor

The data clearly indicate that lasting changes in behavior, and changes in limbic physiology, are NMDA receptor-dependent. The competitive NMDA receptor blocker, AP7, when administered ten min before FG-7142, blocks subsequent behavioral and physiological changes. It is apparent that AP7 is also blocking the acute effects of FG-7142. A common reaction to the injection of FG-7142 alone is panting, which appears at 6 minutes after the injection and fades by 20 minutes. Cats may also crouch and show facially defensive displays (ears retracted) during this time as well. If AP7 is given prior to FG-7142, none of these reactions are seen.

The neuroanatomical and behavioral data of Siegel's group (Lu et al., 1992; Shaikh et al., 1993b) suggest that an excitatory amino acid projection from amygdala to PAG exists. This pathway modulates PAG-elicited defense in an excitatory manner (Shaikh et al., 1993b). Moreover, AP7 reduces the size of the amygalo–PAG potential (unpublished observations). These data, together with the findings of this study, suggest that LTP of the left and right amygdalo–PAG pathways involves NMDA receptor-based processes within the PAG.

The situation in the amygdalo–VMH pathways may be different. Modulation and potentiation of the amygdalo–VMH potential by NMDA receptors may take place within the amygdala. Support for this conclusion is as follows. Systemic AP7 reduces the left and right amygdalo–VMH potentials, but not the left VAF–VMH potential (in preparation). Therefore the NMDA-dependent component of the amygdalo–VMH potential is likely within the amygdala and not in the efferents to the VMH. In addition, LTP of the left amygdalo–VMH potential occurred without changes in the VAF–VMH potential in this and previous studies (Adamec, 1993a).

Taken together, these data suggest that increased amygdalo–VMH transmission following FG-7142 is due to changes within the amygdala, and not in the output pathway (VAF), or in the VMH. LTP in these pathways might then involve some NMDA-dependent process within the amygdala that increases the response of amygdala cells to excitatory inputs. This conclusion is supported by the evidence of NMDA receptor-mediated excitatory processes within the amygdala (Rainnie, 1991a), and the high concentration of benzodiazepine receptors in the amygdala (Young and Kuhar, 1980).

Work in other limbic system pathways has shown that NMDA receptors mediate the initiation, but not maintenance of LTP (Collingridge and Bliss, 1987). It is an open question whether the same is true here. If it is true, then AP7 should not affect the lasting changes in behavior or limbic physiology produced by FG-7142.

Together, all of the data support the hypothesis stated in the introduction. FG-7142 likely engages NMDA receptor-mediated processes by blocking GABA-mediated inhibition. This leads to NMDA-dependent LTP in limbic pathways subserving defensive (fearful) behavior. LTP in these pathways, or possibly only the right amygdalo–PAG pathway, then mediates the lasting increase in defensiveness.

The Dependence of Changes in Behavior on the Benzodiazepine Receptor

It has been shown that the acute effects of FG-7142 on cat behavior are blocked by the benzodiazepine receptor antagonist, Flumazenil (Adamec, 1991a). Preliminary data from this laboratory also suggest that Flumazenil blocks the lasting effects of FG-7142 on behavior. Moreover, Flumazenil has been shown in other studies to reverse the lasting effects of FG-7142 on some behaviors in a drug-dependent manner (Adamec, 1991a, 1993a). In previous work, the reversal of behavioral effects appeared as either a complete or a partial return to baseline values. Those behavioral changes produced by FG-7142 and later reversed by Flumazenil were said to be benzodiazepine receptor-dependent. It was concluded that some lasting change in conformation of the benzodiazepine receptor may underlie behavioral changes. It was suggested that Flumazenil temporarily reverses these changes by altering the conformation of the benzodiazepine receptor.

The findings of this study are consistent with some of those previous results. Lasting effects of FG-7142 on withdrawal were partially reversed by Flumazenil in the Adamec (1991a, 1993a) studies, and completely reversed in the present study. The partial reversal has been interpreted as indicating that more than one mechanism mediates benzodiazepine receptor-dependent behavioral changes (Adamec, 1991a). The only difference between one of the studies showing partial reversal (Adamec, 1993a) and the present study is the administration of the NMDA receptor blocker, AP7, twice prior to the dose of FG-7142 that increased defense. This suggests the possibility that the two doses of AP7 altered some NMDA receptor-dependent process to increase the efficacy of Flumazenil in reversing the increase in defense.

Implications of Neurobehavioral Sensitization by FG-7142 for Chemical Exposure

The studies reviewed have some implications for those attempting to model affective changes following chemical exposure. First, the data suggest it is at least plausible that a single exposure to a chemical that interferes with GABA-mediated inhibition might sensitize limbic circuits, with behavioral consequences. The data also suggest that such a chemical should have particular properties. It should be relatively long lasting. FG-7142 acts for some time [its action has been estimated to last one hour in cats (Ongini, 1983)]. Conversely, if not long lasting, then it may require prolonged exposure or repeated exposure to such chemicals to have behavioral effects. Compounds which have these properties might also be proconvulsant, as is FG-7142 (Jeevanjee et al., 1985). It may also be important that it acts to engage NMDA receptor-mediated processes, as does FG-7142.

Candidates for such compounds found in the environment are insecticides. Two major classes are organophosphate and chlorinated hydrocarbon compounds. Both classes have potent effects on the nervous system of animals and humans (Burchfiel et al., 1976). Single high dose exposures of primates (*Macaca mulatta*) to insecticides such as sarin or dieldrin produce convulsions acutely and lasting changes in EEG without convulsions for up to one year after exposure (Burchfiel et al., 1976). These changes were restricted to the high beta band of the EEG in the temporal lobe and lasted for at least one year after the exposure. Multiple low subconvulsant doses had similar effects on the EEG.

It is of interest that high beta EEG activation in the temporal lobe has been associated with drug induced dysphoria (fear and anxiety) in humans (Stark-Adamec et al., 1982; Adamec et al., 1985). The dysphoria followed intravenous administration of procaine–HCL, which reduces GABA-mediated inhibition in the limbic system of animals (Adamec et al., 1987), and increases PET scan detected activity in the amygdala of humans (Ketter et al., 1993).

These data are of interest because there is a literature that suggests that exposure to insecticides may lead to lasting changes in affect, visuomotor, attention, and language functioning in humans (Savage et al., 1988; Rosenstock et al., 1991). Lasting increases in anxiety have been reported following acute and chronic organophospate insecticide poisoning (Dile and Smith, 1964; Midtling et al., 1985; Savage et. al, 1988; see also Burchfiel et al., 1976 for a review of other studies).

While high doses of insecticides are convulsant, lower doses are not. Yet lower doses do have lasting neurological effects. In this regard it is of interest that several chlorinated hydrocarbon insecticides have been shown experimentally to be proconvulsant in the kindling model (endosulfan, Gilbert, 1992a; lindane, Joy et al., 1980).

Of further interest is the report that convulsions induced in mice by the chlorinated hydrocarbon insecticide, lindane, are antagonized by the noncompetitive NMDA receptor blocker, MK-801. This suggests an activation of NMDA receptor-dependent processes that are highly excitatory.

Finally, a variety of insecticides, including lindane (Joy and Albertson, 1987), have been found to interfere with GABA-mediated inhibition (Abalis et al., 1985, 1986; Eldefrawi et al., 1985; Cole and Casida, 1986). Most studies were done in vitro, but the Joy and Albertson study was in the limbic system (dorsal hippocampus). This area has been shown to be important for some of the behavioral effects of FG-7142 (Lipska et al., 1991).

One might then expect that insecticides would be anxiogenic in animals. Few studies have been done in animals. One study did examine the effects of repeated subconvulsant doses (10 mg/kg) of lindane on behavior in the elevated plus maze, a popular animal model of anxiety. No effects were detected. Clearly, more work with other models is warranted.

One difficulty with ascribing affective disorder to chemical exposure, particularly in putative multiple chemical sensitivity syndrome, is the lack of identification of the offending agent(s). The data I have presented here are only useful if an offending agent can be identified. Moreover, the data presented set certain boundaries on the properties of such offending agents. In my view, if the etiology of behavioral changes following chemical exposure is to be modelled and better understood, then it is essential that criteria for characteristics of potential agents be firmly established.

The data and literature reviewed in this paper speak most directly to possible mechanisms of neurotoxicological effects of specific chemical exposures. In my view, they do not at this time offer an explanation of any of the symptoms of multiple chemical sensitivity disorder.

REFERENCES

ABALIS, I.M., ELDEFRAWI, M.E., and ELDEFRAWI, A.T. (1985). "High-affinity stereospecific binding of cyclodiene insecticides and gamma hexachorlorocyclohexane to gamma–butyric acid receptors of rat brain." Pesticide Biochem. Physiol. 24:95–102.

ABALIS, I.M., ELDEFRAWI, M.E., and ELDEFRAWI, A.T. (1986). "Effects of insecticides on GABA-induced chloride influx into rat brain microsacs." J. Toxicol. Environ. Health 18:13–23.

ADAMEC, R. (1978). "Normal and abnormal limbic system mechanisms of emotive biasing." In: Limbic Mechanisms (Livingston and O. Hornykievicz, ed.). Plenum Press, New York. pp. 405–456.

ADAMEC, R. (1990a). "Does kindling model anything clinically relevant?" Biol. Psychiat. 27:249–279.

ADAMEC, R. (1990b). "Kindling, anxiety and limbic epilepsy: human and animal perspectives." In: Kindling 4 (J. Wada, ed.). Raven Press, New York. pp. 329–341.

ADAMEC, R. (1990c). "Amygdala kindling and anxiety in the rat." NeuroReport 1:255–258.

ADAMEC, R. (1991a). "Acute and lasting effects of FG-7142 on defense and approach-attack behavior in cats: implications for animal model of anxiety which use response suppression." J. Psychopharmacol. 5(1):29–55.

ADAMEC, R. (1991b). "Partial kindling of the ventral hippocampus: Identification of changes in limbic physiology which accompany changes in feline aggression and defense." Physiol. Behav. 49(3):443–453.
ADAMEC, R. (1993a). "Lasting effects of FG-7142 on anxiety, aggression and limbic physiology in the cat." J. Psychopharmacol. 7(3):232–248.
ADAMEC, R. (1993b). "Partial kindling of the ventral hippocampus: The role of the benzodiazepine receptor in changes in limbic physiology which accompany changes in feline defense." Physiol. Behav. 54:531–545.
ADAMEC, R. and MORGAN, H. (1994). "The effect of kindling of different nuclei in the left and right amygdala on anxiety in the rat." Physiol. Behav. 55:1–12.
ADAMEC, R.E., STARK-ADAMEC, C., and LIVINGSTON, K.E. (1980). "The development of predatory aggression and defense in the domestic cat (Felis Catus) I. Effects of early experience on adult patterns of aggression and defense." Behav. Neur. Biol. 30:389–409.
ADAMEC, R.E., STARK-ADAMEC, C., and LIVINGSTON, K.E. (1985). "Basic science and clinical aspects of procaine HC1 as a limbic system excitant." Prog. Neuro-Psychopharmacol. Biol. Psychiat. 9(20):109–120.
ADAMEC, R.E., STARK-ADAMEC, C. and LIVINGSTON, K.E. (1987). "The effects of procaine HC1 on population cellular and evoked response activity within the limbic system of the cat: Evidence for differential excitatory action of procaine in a variety of limbic circuits." Prog. Neuro-Psychopharmacol. Biol. Psychiat. 11:345–364.
ANDREW, R.J. (1972). "The information potentially available in mammal displays." In: Non-Verbal Communication (R.A. Hinde, ed.). Cambridge University Press, London. pp. 204–206.
ASHFORD, N.A. and MILLER, C.S. (1991). Chemical exposures. Low Levels and High Stakes. Van Nostrand Reinhold, New York.
BELL, I.R. (1975). "A kinin model of meditation for food and chemical sensitivities: biobehavioral implications." Ann. Allergy 35:206–215.
BELL, I.R.(1982). "A new medical approach to environmental illness." In: Clinical Ecology. Common Knowledge Press, Bolinas, CA.
BELL, I.R. (1987). "Environmental illness and health: the controversy and challenge of clinical ecology for mind-body health." Advances 4(3):45–55.
BELL, I.R., MILLER, C.S., and SCHWARTZ, G.E. (1992). "An olfactory-limbic model of multiple-chemical sensitivity syndrome: possible relationships to kindling and affective spectrum disorders." Biol. Psychiat. 32(3):218–242.
BLACK, D.W, RATHE, A., and GOLDSTEIN, R.B.(1990). "Environmental illness. A controlled study of 26 subjects with '20th century disease'." J. Am. Med. Assoc. 264:3166–3170.
BOYLES, J.H. (1985). "Chemical sensitivity. Diagnosis and treatment." Otolaryngologic Clin. N. Am. 18:787–795.
BRODSKY, C.M. (1983). "'Allergic to everything': a medical subculture." Psychosom. 24:731–742.
BURCHFIEL, J.L., DUFFY, F.H., and SIM, V.M. (1976). "Persistent effects of sarin and dieldrin upon the primate electroencephalogram." Toxicol. Appl. Pharmacol. 35:365–379.
COLE, L.M. and CASSIDA, J.E. (1986). "Polychlorocycloalkane insecticide-induced convulsions in mice in relation to disruption of GABA regulated chloride ionophore." Life Sci. 39:1855–1862.
COLLINGRIDGE, G.L. and BLISS, T.V.P. (1987). "NMDA receptors — their role in long-term potentiation." Trends Neurosci. 10(7):288–293.
COLEMAN, K. and McGAUGH, J.L. (1993). "The right and left amygdalae are differentially involved in the expression of memory for an aversively motivated task." Soc. Neurosci. Abstr. 19(2):1226.
CRAWLEY, J.N., NINAN, P.T., PICKAR, D., CHROUSOS, M., LINNOILA, M., SKOLNICK, P., and PAUL, S.M. (1985). "Pharmacological Antagonism of carboline induced anxiety in rhesus monkeys." J. Neurosci. 5:485–497.
DILE, J.R. and SMITH, D.W. (1964). "Central nervous system effects of chronic exposure to organophosphate insecticides." Aerospace Med. 35:475–478.
DOROW, R., HOROWSKI, R., PASCHELKE, G., AMIN, M., and BRAESTRUP, C. (1983). "Severe anxiety induced by FG-7142, a β-carboline ligand for benzodiazepine receptors." Lancet 2:98–99.
DREIFUSS, J.J., MURPHY, J.T., and GLOOR, P. (1968). "Contrasting effects of two identified amygdaloid efferent pathways on single hypothalamic neurons." J. Neurophysiol. 31:237–247.

DREIFUSS, J.J. (1972). "Effects of electrical stimulation of the amygdaloid complex on the ventromedial hypothalamus." In: The Neurobiology of the Amygdala, Advances in Neurology (B.E. Eleftheriou, ed.). Vol. 2. Plenum Press, New York. pp. 295–318.

DRUGAN, R.C., MAIER, S.F., SKOLNICK, P., PAUL, S.M., and CRAWLEY, J.N. (1985). "An anxiogenic benzodiazepine receptor ligand induced learning helplessness." Eur. J. Pharmacol. 113:453–457.

ELDEFRAWI, M.E., SHERBY, S.M., ABALIS, I.M., and ELDEFRAWI, A.T. (1985). "Interactions of pyrethroid and cyclodiene insecticides with nicotinic acetylcholine and GABA receptors." Neurotox. 6:47–62.

FILE, S.E. and PELLOW, S. (1986). "The anxiogenic action of FG-7142 in the social interaction test is reversed by chlordiazepoxide and RO-15-1788 but not by CGS 8216." Arch. Int. Pharmacodyn. 271:198–208.

GILBERT, M.E. (1992a). "A characterization of chemical kindling with the pesticide endosulfan." Neurotoxicol. Teratol. 14:151–158.

GILBERT, M.E. (1992b). "Neurotoxicants and limbic kindling." In: The Vulnerable Brain and Environmental Risks, Malnutrition and Hazard Assessment (R.L. Isaacson and K.F. Jensen, ed.). Vol. 1. Plenum Press, New York. pp. 173–192.

GLUSHANKOV, P., VOROBYOV, V., and SREBITSKY, V. (1985). Bull. Exp. Biol. and Med., Moscow. 12:641–768.

GODDARD, G.V., McINTYRE, D.C. and LEECH, C.K. (1969). "A permanent change in brain function resulting from daily electrical stimulation." Exptl. Neurol. 25:295–330.

GRANT, E.C. and MACINTOSH, J.H. (1963). "A comparison of the social postures of some common laboratory rodents." Behavior. 21:246–259.

HIYOSHI, T., MATSUDA, M., and WADA, J.A. (1991). "Centrally induced feline emotional behavior and limbic kindling." Epilepsia 31:259–269.

HUNKELER, W., MOHLER, H., PIERI, L., POLC, P., BONNETTI, E.P., CUMIN, R., SCHAFFNER, R., and HAEFELY, W. (1981). "Selective antagonists of benzodiazepines." Nature 290:51–52.

JASPER, H.H. and AJMOME-MARSAN, C.A. (1952). A Stereotaxic Atlas of the Diencephalon of the Cat. The National Research Council of Canada, Ottawa.

JEEVANJEE, F., LITTLE, H.J., MARTIN, I.L., NICHOLASS, J.M., and NUTT, D.J. (1985). "Is FG-7142 mediated kindling due to changes in efficacy at the benzodiazepine/GABA receptor complex?" Br. J. Pharmacol. 84:186P.

JOY, R.M., STARK, L.G., PETERSON, S.L., BOWYER, J.F., and ALBERTSON, T.E. (1980). "The kindled seizure: Production of and modification by dieldrin in rats." Neurobehav. Toxicol. 2:117–124.

JOY, R.M. and ALBERTSON, T.E. (1987). "Interactions of lindane with synaptically mediated inhibition and facilitation in the dentate gyrus." Neurotoxicol. 8:529–542.

KETTER, T.A., ANDREASON, P.J., GEORGE, M.S., HERCOVITCH, P., and POST, R.M. (1993). "Paralimbic rCBF increases during procaine induced psychosensory and emotional experience." Abstract Soc. Biol. Psychiat.

KRAUTH, J. (1988). Distribution-Free Statistics; an Application-Oriented Approach. Elsevier, Amsterdam. p. 36.

KRETTEK, J.E. and PRICE, J.L. (1978). "Amygdaloid projections to subcortical structures within the basal forebrain and brainstem in the rat and cat." J. Comp. Neurol. 178:225–254.

LEYHAUSEN, P. (1956). "Verhaltstudien an Katzen." Z. Tierpsychol. 2:1–120.

LIPSKA, B.K., JASKIW, G.E., and WEINBERGER, D.R. (1991). "Dorsal hippocampal lesion abolishes the response to FG-7142 in the rat." Pharmacol. Biochem. Behav. 40:169–172.

LISTER, R.G., GREENBLATT, D.J., ABERNETHY, D.R., and FILE, S.E. (1984). "Pharmacokinetic studies on RO-1788, a benzodiazepine receptor ligand, in the brain of the rat." Brain Res. 290:183–186.

LU, C.L., SHAIKH, M.B., and SIEGEL, A. (1992). "Role of NMDA receptors in hypothalamic facilitation of feline defensive rage elicited from the midbrain periaqueductal gray." Brain Res. 581:123–132.

MAEDA, H. (1976). "Effects of psychotropic drugs upon the hypothalamic rage responses in cats." Fol. Psychiat. Neurol. Jap. 30:539–546.

MAEDA, H. and HIRATA, K. (1978). "Two-stage amygdaloid lesions and hypothalamic range: A method useful for detecting functional localization." Physiol. Behav. 8(21):529–530.

MAIER-BODE, H. (1968). "Properties, effects, residues and analytic of the insecticide endosulfan." Residue Rev. 22:1–44.

MIDTLING, J.E., BARNETT, P.G, and COYE, M.C. (1985). "Clinical management of field worker organophosphate poisoning." West J. Med. 142:514–518.

NASHOLD, B.S., JR., WILSON, N.P., and SLAUGHTER, G.S. (1969). "Sensations evoked by stimulation in the midbrain of man." J. Neurosurg. 30:14–24.

NASHOLD, B.S., JR., WILSON, N.P., and SLAUGHTER, G.S. (1974). "The midbrain and pain." In: Advances in Neurology, International Symposium on Pain (J.J. Bonica, ed.). Vol. 4. Raven Press, New York, NY. pp. 191–196.

NIEMINEN, S.A., SIRVIOE, J., TEITTINEN, K., PITKAENEN, A., AIRAKSINEN, M.M., and RIEKKINEN, P. (1992). "Amygdala kindling increases fear-response but did not impair spatial memory in rats." Physiol. Behav. 51:845–849.

ONGINI, E. (1983). "Behavioral and EEG effects of benzodiazepines and their antagonists in the cat." In: Benzodiazepine Recognition Site Ligands: Biochemistry and Pharmacology (G. Biggio and E. Costa, ed.). Raven Press, New York. pp. 211–225.

POST, R.M., KENNEDY, C., SHINOHARA, M., SQUILLACE, K., MIYOAKA, M., SUDA, S. INGVAR, D.H., and SOKOLOFF, L. (1984). "Metabolic and behavioral consequences of lidocaine-kindled seizures." Brain Res. 324:295–303.

RACINE, R.J. (1978). "Kindling: The first decade." Neurosurg. 3(2):234–252.

RAINNIE, D.G., ASPRODINI, E.K., and SHINNICK-GALLAGHER, P. (1991a). "Excitatory transmission in the basolateral amygdala." J. Neurophysiol. 66:986–997.

RAINNIE, D.G., ASPRODINI, E.K., and SHINNICK-GALLAGHER, P. (1991b). "Inhibitory transmission in the basolateral amygdala." J. Neurophysiol. 66:999–1007.

RANDOLF, T.G. (1970). "Domiciliary chemical air pollution in the etiology of ecologic mental illness." Intl. J. Soc. Psychiat. 16: 243–265.

RANDOLF, T.G (1974). "The history of ecological mental illness." In: Annual Review of Allergy, 1973 (C.A. Frazier, ed.). Medical Exam. Pub. Co., Flushing, New York.

REA, W.J. (1977). "Environmentally-triggered small vessel vasculitis." Ann. Allergy 38:245–251.

REA, W.J. (1978). "Environmentally-triggered cardiac disease." Ann. Allergy 40:243–251.

ROSENBERG, J.F., FREEDMAN, M.R., SCHMALING, K.B., and ROSE, C. (1990). "Personality styles of patients asserting environmental illness." J. Occup. Med. 32:678–681.

ROSENSTOCK, L., KEIFER, M., DANIELL, W.E., McCONNELL, R., and CLAYPOOLE, K. (1991). "Chronic central nervous systems effects of acute organophosphate pesticide intoxication." Lancet 338:223–227.

SAVAGE, E.P., KEEFE, T.J., and MOUNCE, L.M. (1988). "Chronic neuropsychological sequelae of acute organophosphate pesticide poisoning." Arch. Environ. Health. 43:38–45.

SHAIKH, M.B., STEINBERG, A., and SIEGEL, A. (1993a). "Evidence that substance P is utilized in medial amygdaloid facilitation of defensive rage behavior in the cat." Brain Res. 625:283–294.

SHAIKH, M.B., SCHUBERT, K., and SIEGEL, A. (1993b). "Excitatory amino acids in basal amygdala facilitate defensive rage in the cat." Soc. Neurosci. Abstr. 19:754.

SCHUBERT, K., SHAIKH, M.B., and SIEGEL, A. (1993). "Excitatory amino acid pathway from the basal amygdala to the midbrain periaqueductal gray defensive rage sites in the cat." Soc. Neurosci. Abstr. 19:754.

SKOLNICK, P. (1983). "New concepts in the neurobiology of anxiety." J. Clin. Psychiat. 44:12–19.

STAUDENMAYER, H. and CAMAZINE, M. (1989) "Sensing type personality, projection, and universal 'allergic' reactivity." J. Psychological Type 18:59–62.

STAUDENMAYER, H. and SELNER, C. (1990) "Neuropsychophysiology during relaxation in generalized, universal 'allergic' reactivity to the environment: a comparison study." J. Psychosom. Res. 34:259–270.

STARK-ADAMEC, C., ADAMEC, R.E., GRAHAM, J.M., BRUUN-MEYER, S.E., PERRIN, R.G., POLLOCK, D. and LIVINGSTON, K.E. (1982). "Analysis of facial displays and verbal report to access subjective state in the non-invasive detection of limbic system activation by procaine hydrochloride." Behav. Brain Res. 4:77–94.

STEWART, D.E. and TASKIN, J (1985). "Psychiatric assessment of patients with '20th century disease' ('total allergy syndrome')." Can. Med. Assoc. J. 133:1001–1006.
TERR, A.I. (1986). "Environmental illness. A clinical review of 50 cases." Arch. Intern. Med. 146:145–149.
WIGSTRON, H. and GUSTAFFSON, B. (1985). "Facilitation of hippocampal long-lasting potentiation by GABA antagonists." Acta Physiol. Scand. 125:159–179.
YEHUDA, R. and ANTELMAN, S.M. (1993). "Criteria for rationally evaluating animal models of post–traumatic stress disorder." Biol. Psychiat. 33:479–486.
YOUNG, W.S. and KUHAR, M.J. (1980). "Radiohistochemical localization of benzodiazepine receptors in rat brain." J. Pharm. Exp. Therap. 212(2):337–346.
ZHANG, S.P., BANDLER, R., and CARRIVE, P. (1990). "Flight and immobility evoked by excitatory amino acid microinjection within distinct parts of the subtentorial midbrain periaqueductal gray of the cat." Brain Res. 520:73–78.

CAVEATS IN THE USE OF THE KINDLING MODEL OF AFFECTIVE DISORDERS

SUSAN R.B. WEISS AND ROBERT M. POST

Biological Psychiatry Branch
National Institute of Mental Health
Bethesda, Maryland

INTRODUCTION

Kindling is, at best, a nonhomologous model for the affective disorders (Post and Weiss, 1989). As such, its utility in the study of chemical hypersensitivity, a syndrome with its own unique profile and ambiguities, must be demonstrated using considerable conceptual and experimental conservatism. In this manuscript, we discuss the strengths and limitations of the kindling model as it has been applied to affective illness. Based on this analysis, we suggest that kindling may be a useful construct to help formulate various questions regarding the longitudinal course of affective disorders (see Table 1), but may ultimately be inappropriate to apply directly.

The kindling model does not meet any of the criteria laid out by McKinney and associates (1969, 1989) for a valid animal model of a psychiatric illness. Specifically, they suggest that an animal model should possess one or more of the following characteristics: 1) Parallelism in the *inducing characteristics*; 2) Some element of *behavioral homology* to the condition being modeled; 3) Some direct correspondences in the *pathophysiological mechanisms* involved (in some models, e.g., hypercortisolemia in depression, these can be examined in isolation, independent of whether or not they are associated with other appropriate behavioral characteristics); 4) Parallelism in the *pharmacology* that would prevent or reverse the behavioral alterations and the syndrome being modeled. Given the fact that kindling does not meet any of these criteria, how might it be a useful construct in the consideration of affective illness? In this paper we will attempt to illustrate how some of the principles and neurobiological mechanisms uncovered in the kindling process may redirect attention to important clinical and neurobiological issues in the affective disorders, and allow a reframing of questions asked and experimental designs used to test specific hypotheses more explicitly.

1. Address all correspondence to: Susan R.B. Weiss, Ph.D., Biological Psychiatry Branch, NIMH Bldg. 10, Room 3N212, 9000 Rockville Pike, Bethesda, MD 20892. Tel.:(301)496-4875. Fax:(301)402-0052.

Toxicology and Industrial Health, Vol. 10, No. 4/5, pp. 421–447

ISSN: 0748-2337

TABLE 1. Phenomena in the Course of Affective Disorders Modeled by Kindling and Behavioral Sensitization

Descriptors	Kindling (K)	Sensitization (S)	Phenomenon
Stressor vulnerability	++	++	Initial stressors early in development may be without effect but predispose to greater reactivity upon rechallenge
Stressor precipitation	++	++	Later stress may precipitate full-blown episode
Conditioning may be involved	—	++	Stressors may become more symbolic
Episode autonomy	++	—	Initially precipitated episodes may occur spontaneously
Cross-sensitization with stimulants	—	++	Comorbidity with drug abuse may work in both directions affective illness ↔ drug abuse
Vulnerability to relapse	++	++	S and K demonstrate long-term increases in responsivity
Episodes may:			
a. become more severe	++	—	S and K both show behavioral evolution in severity or stages
b. show more rapid onsets		++	Hyperactivity and stereotypy show more rapid onsets
Anatomical and biochemical substrates evolve	++	±	K memory-trace evolves from unilateral to bilateral
IEGs involved	++	++	Immediate Early Genes (IEGs) such as c-fos are induced
Alterations in gene expression occur	++	++	IEGs may change gene expression, especially of peptides over long time domains
Change in synaptic microstructure occurs	++	—	Neuronal sprouting and cell loss indicate structural changes
Pharmacology differs as function or stage of versus evolution	++	++	K differs as a function of stage; S differs as a function of development vs. expression

Kindling may be a good model for the development of seizures, since repeated subthreshold stimulations eventually result in the development of a full-blown convulsive response and, with sufficient repetitions, seizures emerge spontaneously in the absence of exogenous electrophysiological stimulation. However, there is much debate over whether kindling, even as a model of clinical epilepsy, is a good representation of processes involved in the unfolding of clinical seizures, and what type of seizure disorder in man might be most closely paralleled by a kindling process.

In this literal sense, then, kindling is *not* a good model for the affective disorders, since seizures are obviously not an end product of the illness (and, in fact, the seizures of ECT are used to treat affective illness). Moreover, in terms of McKinney criterion #1, electrical stimulation of the brain is not a primary inducer of affective illness (although complex partial seizures of the temporal lobe and limbic system are often associated with secondary affective illness). In addition, there is little reason to suspect that pathophysiological mechanisms involved in kindling evolution are identical to those involved in the affective disorders. Furthermore, there are multiple pharmacological disjunctions among drugs that are effective in kindling and affective disorders. Given this sorry state of affairs that clearly invalidates kindling as a traditional or homologous model of affective disorders, what are the potential ways that kindling could nonetheless be useful? Herein we will not only review ways that it may be a useful analogy for considering the etiology and longitudinal course of psychiatric illness, but also highlight the multiple caveats that limit its utility. We suggest that considerable conservatism be employed in interpreting how the principles uncovered in kindling evolution may be applied to the course of affective disorders. This academic exercise on the limitations of the kindling model for affective illness should be of interest in its own right, but, in addition, should be relevant in considering the possible utility of kindling and sensitization models for the area of chemical hypersensitivity.

Electrical Kindling

Electrophysiological kindling is an increase in convulsive responsivity resulting from repeated subthreshold electrical stimulation that culminates in a seizure end point. Following a sufficient number of repetitions, full-blown seizures come to occur spontaneously. Consequently, kindling appears to have three distinct stages of evolution. The first is a *development* phase in which subthreshold stimulation of a distinct region of the brain (e.g., the amygdala) produces focal afterdischarge activity, which increases in duration and spreads to other regions of the brain, eventually resulting in a full-blown major motor seizure. The *completed* phase is characterized by the reliable elicitation of seizures by each stimulation. Finally, seizures become *spontaneous*; that is, they occur in the absence of exogenous stimulation. Many of the major characteristics of kindling are outlined in Table 2 and schematized in Figure 1 (top).

Electrical kindling of the amygdala is associated with the induction of the proto-oncogene c-fos, which is thought to act as a transcription factor and to be part of a cascade of effects involving a series of immediate early genes (IEGs), which subsequently affect DNA and regulate the transcription of late effector genes (LEGs), potentially altering the synthesis of peptides, growth factors, etc. (Figure 9) (Morgan and Curran, 1990). The pattern of c-fos induction can also be used as an indicator of cellular activity at the level of gene transcription (as opposed to deoxyglucose, which is thought to reflect activity in nerve terminals).

During amygdala kindling evolution c-fos is transiently induced unilaterally on the side of electrode placement in the pyriform cortex (with short afterdischarges) or the dentate gyrus of

TABLE 2. Electrical Kindling: Major Characteristics*

1.	Repeated stimulations
	a. Initially subseizure threshold
	b. Intermittent
2.	Local afterdischarges and seizure activity
	a. Increase in amplitude, frequency
	b. Increase in duration
	c. Increase in complexity of wave form
	d. Increase in anatomical spread
3.	Replicable sequence of seizure stages: Behavioral arrest, blinking and masticatory movements, head nodding, opsithotonis, contralateral then bilateral forelimb clonus, rearing and falling
4.	Discharges kindle in quantum jumps
5.	Limbic system kindles more readily than cortex
6.	In kindled animals the history of convulsion development is recapitulated as seizure builds
7.	Transfer effects to secondary sites; kindling facilitated in other sites even after primary site destroyed
8.	Interference: a secondary kindled site interferes with primary site kindling
9.	Kindling is a transsynaptic processs
10.	Relatively permanent change in connectivity; a kindled animal will still have seizures after a one-year seizure-free interval
11.	Seizures may develop spontaneously in chronically kindled animals
12.	Interictal spikes and spontaneous epileptiform potentials develop

*See Goddard et al. (1969) regarding 1, 4–10; Wada and Sato (1974) regarding 2–4, 6; Wada et al. (1974) regarding 10–12; Racine (1978) and Pinel and Rovner (1978a,b).

the hippocampus (with long afterdischarges) during stage 1 and 2 seizures. With the generalization to stage 3 seizures c-fos induction becomes bilateral, and with full-blown stage 4 and 5 seizures a variety of other cortical areas become involved as well (Clark et al., 1991). Thus, there appears to be a spatiotemporal progression of the neuroanatomical substrates involved in the kindled pathway (or putative memory trace), using this marker of transcriptional activation.

Given the evolution of the anatomical substrates of the kindled seizure, it is perhaps not surprising that a differential pharmacological responsivity also exists during the different phases of kindling evolution (Figure 1, bottom). For example, diazepam is an excellent anticonvulsant in the early development and completed stages of amygdala kindling, but is without effect on the spontaneous phase (Pinel, 1983). In contrast, phenytoin shows the opposite profile, in that it is ineffective until animals reach the spontaneous phase of kindling evolution (Pinel, 1983). These data suggest the importance of the stage of seizure (or illness) development as a variable for consideration in terms of appropriate therapeutic interventions.

Pharmacological Kindling

Many of the essential characteristics of electrophysiological kindling can be reproduced with a variety of pharmacological probes. In these instances, instead of subthreshold electrical

PHARMACOLOGICAL RESPONSIVITY AS A FUNCTION OF STAGE OF KINDLING

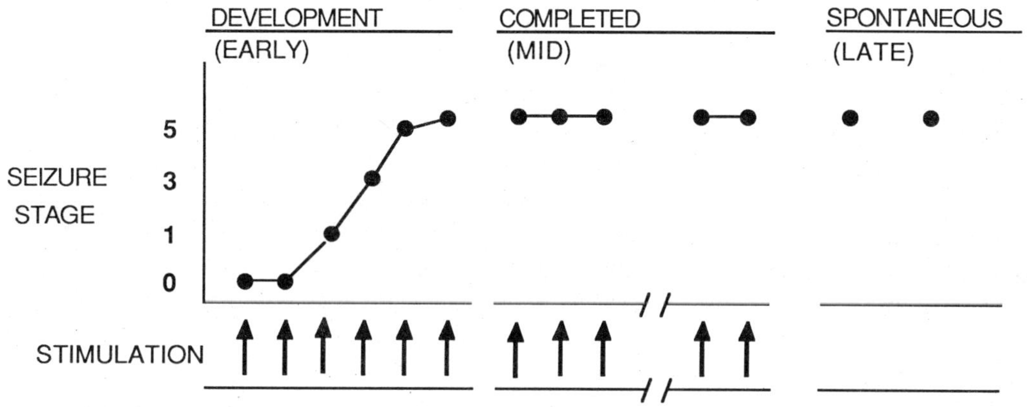

	DEVELOPMENT (EARLY)	COMPLETED (MID)	SPONTANEOUS (LATE)
AMYGDALA KINDLING			
DIAZEPAM	++	++	0
PHENYTOIN	0	±	++
CARBAMAZEPINE	0	++	
CLONIDINE	++	0	
ATROPINE	(++)	0	
ECS	++	++	
VALPROATE	++	++	
OPIATES	0	0	
MK 801*	++	0	
LOCAL ANESTHETIC KINDLING			
DIAZEPAM	++	++	
CARBAMAZEPINE			
(CHRONIC)	++	0	
(INTERMITTENT)	0		

FIGURE 1. Anticonvulsant manipulations are effective in some phases of amygdala kindling, but not others. Phases include: 1) early, development; 2) mid, completed; and 3) late, spontaneous; e.g., carbamazepine is effective on the completed amygdala-kindled seizures, but not on their development. Pinel (1983) documents the efficacy of diazepam on the early and mid, but not late, phases of kindling evolution, whereas phenytoin shows the opposite pattern.

stimulation of the brain being the inducing stimulus, repeated parenteral or intracerebral administration of selective compounds appears to be associated with progressive increases in behavioral and physiological responsivity, culminating in the development of full-blown seizures.

We have observed a kindling-like progression in the development of seizures when cocaine is administered to rats and rhesus monkeys (Post and Contel, 1983), and others have observed this progression in dogs (Tatum and Seevers, 1929; Downs and Eddy, 1932) and humans (Dhuna et al., 1991). In these instances, however, it was unclear whether the seizure end point achieved was based on cocaine's local-anesthetic properties or its psychomotor-stimulant properties. In order to clarify this ambiguity we administered repeated doses of the pure local anesthetic lidocaine, which is equal to cocaine in its local-anesthetic properties but is ineffective on catechol- and indoleamine reuptake, and thus does not produce the motor activating symptoms typically associated with amphetamine, cocaine, or related psychomotor stimulants (Post et al., 1975, 1984a). As illustrated in Figure 2, we found that

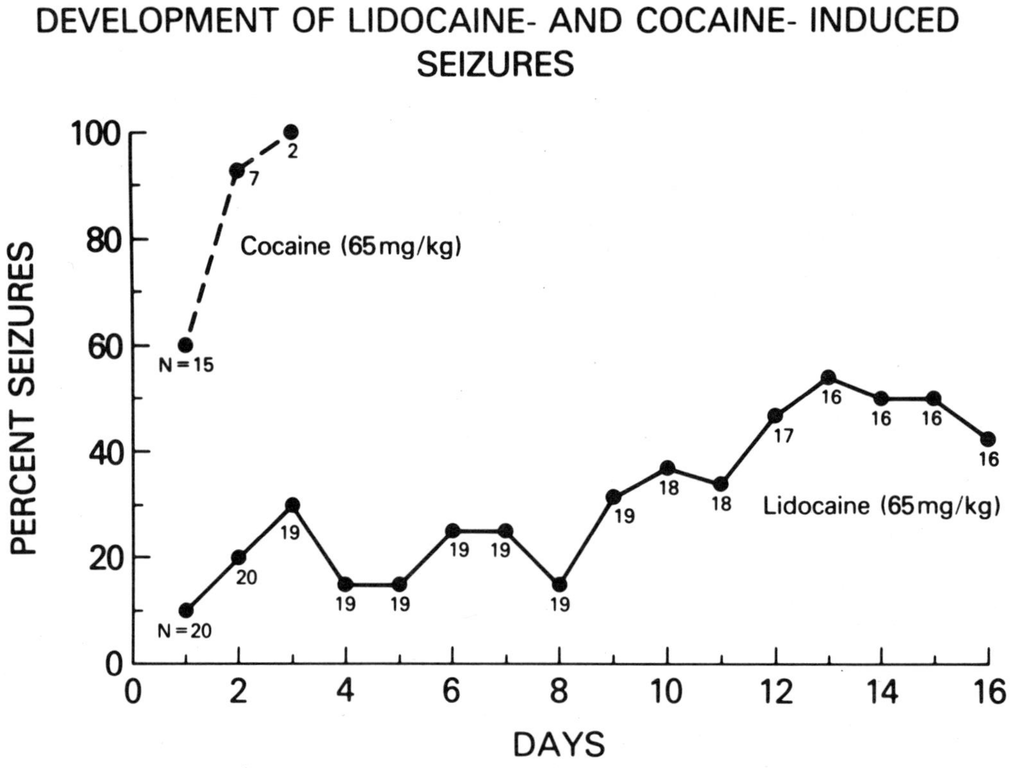

FIGURE 2. The percentage of rats experiencing lidocaine- or cocaine-induced seizures is shown. Rats received a single i.p. injection (65 mg/kg) each day. The study began with 15 or 20 subjects per group; the number remaining is shown below each day's percentage. Despite the equal local anesthetic potency of these drugs, cocaine is much more convulsant than lidocaine on day 1; animals kindle faster to it, and are more likely to die during a seizure episode.

repeated intraperitoneal administration of a dose of lidocaine that was initially subconvulsant came to elicit seizures in approximately 50% of male Sprague-Dawley rats. In most groups of animals, seizures were induced at an average of 1–2 weeks following repeated once-daily i.p. administration of 65 mg/kg. As illustrated in Figure 3, in individual animals seizures were inconsistently evoked from one day to the next, with some animals eventually showing seizure regression despite repeated daily dosing with the drug, whereas others persisted, having increased severity and duration of lidocaine seizures. In a number of instances these animals progressed to the stage of spontaneity and demonstrated seizures on occasions when they were not drug-injected (Post and Contel, 1983; Post and Weiss, 1988, unpublished observations).

The behavioral seizures induced by lidocaine are very similar to those induced by electrical stimulation of the amygdala. Lidocaine-kindled animals show spiking in the amygdala similar to that observed in electrical kindling (Weiss and Post, 1984, unpublished observations). The stages of lidocaine seizure progression also parallel electrical kindling evolution including: (a) a *developmental*, preictal stage; (b) a *completed* stage of full-blown seizures; and then, in some animals, (c) *spontaneity*. Studies of lidocaine-induced seizures using deoxyglucose as a marker of cerebral glucose metabolism demonstrate rather selective limbic involvement (Post et al., 1984a). Seizures either markedly involved amygdala and hippocampus or, in other instances, perirhinal areas and structures in the deep pyriform cortex. Recent data also indicate that lidocaine- and cocaine-induced seizures are associated with the induction of the proto-oncogene c-fos. The pattern of c-fos induction is highly similar to that induced by electrical kindling of the amygdala (Clark et al., 1991). Following the administration of either lidocaine or cocaine at doses that lead to seizures either acutely or following kindling (Clark et al., 1992; Nakajima et al., 1994), c-fos induction is observed in the pyriform cortex and limbic cortices, as well as the dentate gyrus of the hippocampus, the olfactory bulbs, and the striatum. The latter two structures also show enhanced c-fos expression following caffeine-induced seizures in mice (Nakajima et al., 1989). However, with lower doses of cocaine, which do not produce seizures, c-fos is selectively enhanced in the striatum. This occurs via cocaine's ability to block dopamine reuptake and induce c-fos through a D1 receptor mechanism (Graybiel et al., 1990; Young et al., 1991). Nonetheless, the data pertaining to lidocaine administration and cocaine seizures strongly suggest that the kindling-like progression with cocaine is related to this compound's local-anesthetic as opposed to its psychomotor-stimulant properties. They are also convergent with the observations by Ellinwood and Kilbey (1977) that amphetamine administration is not productive of full-blown seizures unless highly unusual circumstances of disulfiram administration are utilized.

Again, as has been demonstrated with electrical kindling, the efficacy of anticonvulsant drugs on local anesthetic kindling also differs according to the stage of kindling evolution. Carbamazepine, which is completely ineffective in preventing the development of amygdala-kindled seizures, can markedly inhibit the development of local anesthetic seizures but not the completed kindled seizures (Weiss et al., 1989b). Carbamazepine's efficacy in this model is also dependent upon chronic administration of the drug (orally, in a diet preparation) and

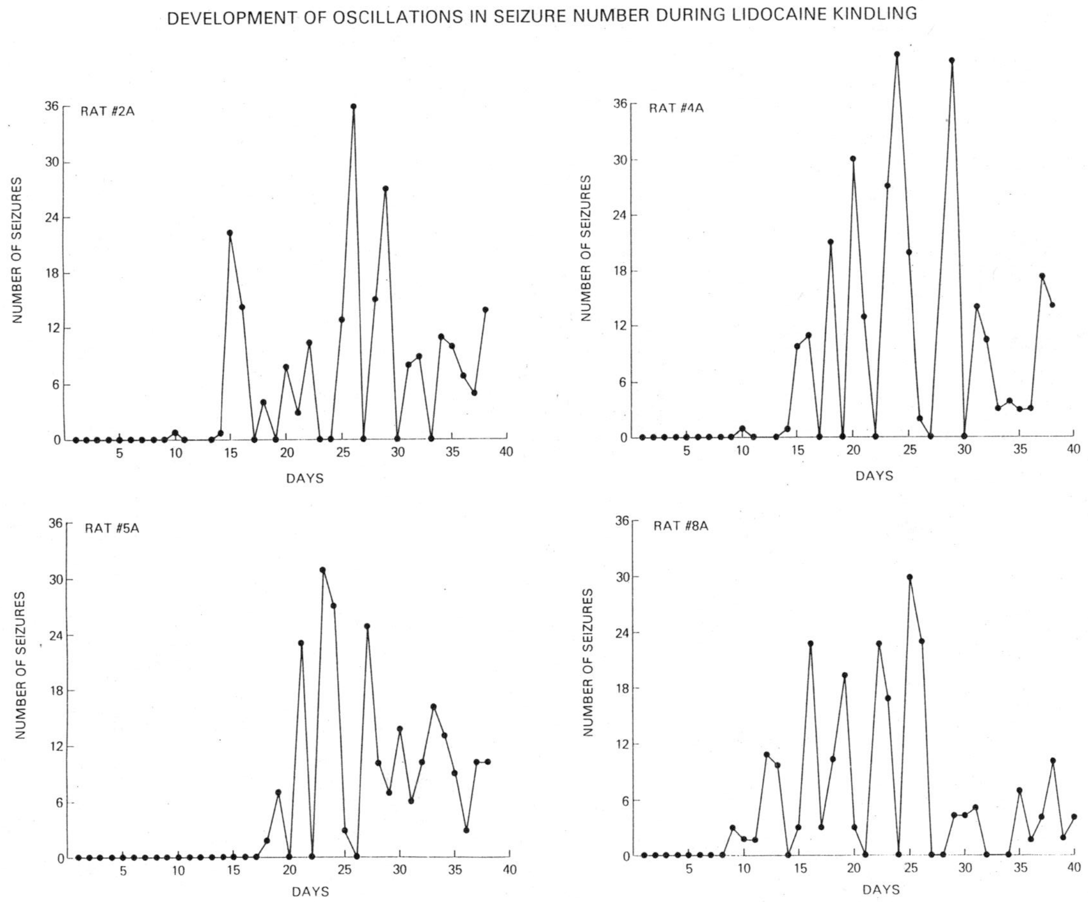

FIGURE 3. This figure displays seizure patterns in four rats receiving the same dose of lidocaine (65 mg/kg). It is worth noting that few seizures were evoked during the first days (1–14). After this time all subjects began to manifest seizures. However., an oscillatory pattern emerged in individual animals. In particular, subject #8A demonstrated spontaneous regression of seizures, regardless of dose amount.

cannot be reproduced with repeated intraperitoneal administration prior to each local anesthetic kindling injection. Thus, carbamazepine's pattern of anticonvulsant efficacy is complex, depending upon the stage of kindling evolution, the nature of the inducing stimulus(local anesthetic versus electrical kindling), and the mode of carbamazepine administration (chronic oral versus repeated intermittent intraperitoneal injection).

Other drugs also appear to produce a kindling-like progression of convulsive responsivity. These include inverse benzodiazepine agonists (e.g., FG-7142 [Little et al., 1984]), picrotoxin (Cain, 1986), cholinergic agonists (Wasterlain et al., 1981), the peripheral-type benzodiene agonist Ro5-4864 (Weiss et al., 1986b), and opiates such as enkephalin or morphine administered directly into the brain (Cain and Corcoran, 1984). When enkephalin is administered into the amygdala but not the ventral hippocampus, it induces a kindling-like progression to full-blown seizures. It is of interest that these seizures can also be triggered by handling stress (Cain and Corcoran, 1984), whereas footshock-induced stress can inhibit amygdala-kindled seizures (Pinel et al., 1973).

Corticotropin-releasing hormone (CRH) is a particularly interesting neuropeptide in relationship to kindling-like progression. Although intracerebroventricular (i.c.v.) administration of CRH produces tonic-clonic seizures of the head, trunk, and forepaws similar to those induced by amygdala or local anesthetic kindling (Ehlers et al., 1983; Weiss et al., 1986a), repeated administration is associated with tolerance to these seizure-inducing effects, rather than a kindling-like progression (Weiss et al., 1986a). Nonetheless, when animals have experienced CRH-induced seizures they show some degree of cross-sensitization to amygdala-kindled seizures. In contrast, animals in which amygdala-kindled seizures have been repeatedly induced electrically are less prone to acute CRH seizures, suggesting that (like repeated administration of CRH itself) electrically kindled seizures are associated with a decreased responsiveness to the seizure-inducing properties of CRH. However, these same kindled animals show CRH-induced increases in aggression, suggesting that substrates mediating these two effects of CRH are differentially regulated.

Cross-sensitization has also been demonstrated between amygdala kindling induced with electrical stimulation and a variety of pharmacological agents, as reviewed by Cain (1986). Taken together, these data add to the evidence described above suggesting a substantial cross-reactivity, in some instances, between the substrates activated by electrical stimulation and those provoked by pharmacological manipulation of endogenous neurotransmitter and neuropeptide substances.

Behavioral Sensitization

Psychomotor stimulant effects of a drug such as cocaine should be distinguished from the drug's properties as a pharmacological kindling agent. Nonlocal anesthetic psychomotor stimulants such as amphetamine, methylphenidate, and even direct dopamine agonists, are capable of inducing behavioral sensitization. The essential aspect of behavioral sensitization is an increased behavioral responsivity (in animals, measured as hyperactivity or stereotypy) in response to repeated administration of the same dose of drug. Many of the characteristic properties of behavioral sensitization are outlined in Table 3. As depicted in Figure 4, there is a matrix of dose and repetition number that affects the magnitude and duration of sensitization. That is, administration of a single low dose of cocaine is insufficient to produce sensitization, whereas a high dose (40 mg/kg i.p. on day 1) produces increased reactivity to a subsequent low-dose injection, an effect that lasts for between five days and a week. Repeated administration of low doses of cocaine (10 mg/kg i.p.) for a period of a week to ten days, however, results in very long-lasting behavioral sensitization, with increased responsivity being observed even many months later.

TABLE 3. Characteristics of Behavioral Sensitization to Psychomotor Stimulants

1.	Shorter latency to maximum response following repeated exposure.
2.	Increased magnitude of behavioral response to same dose.
3.	Selective increase in some behaviors, not in others.
4.	Alterations in responsivity can be long-lasting (week to months).
5.	Intermittent administration produces greater sensitization, as compared with continuous administration.
6.	Effects are dose-related.
7.	Genetic component is evident in degree of sensitization.
8.	Sensitization is not related to alterations in blood levels.
9.	Sensitization shows environmental context dependency; i.e., is conditionable.
10.	Cross-sensitization occurs among many stimulants and dopamine agonists.
11.	Cross-sensitization can occur between stimulants and a variety of stresses (tail-pinch, shock, starvation), but may be stressor-, intensity-, and time-dependent.
12.	Neurotransmitter, receptor, and subreceptor changes postulated.
13.	Hormones and peptides (i.e., vasopressin) modulate and may produce behavioral sensitization.
14.	Dopaminergic substrates in nucleus accumbens (and amygdala for environmental context-dependent component) may be critical to sensitization.

See Ellinwood and Kilbey, 1977; Post and Contel, 1981; Antelman and Chiodo, 1983; Weiss et al., 1989a, Post et al., 1993, 1994 a,b for additional references.

It is of considerable interest that in the paradigms mentioned thus far the sensitization to cocaine is highly context-dependent or conditioned (Post et al., 1981; Weiss et al., 1989a). That is, single or repeated administration of cocaine in the same environmental context as that in which the animals are subsequently tested results in the increased reactivity, whereas administration of similar doses in a different environment does not result in sensitization (Figure 5). The neuroanatomical studies of Pert and Weiss and their colleagues have demonstrated that lesions of the amygdala and nucleus accumbens are able to block this type of conditioned sensitization, though lesions of the dorsal and ventral hippocampus, cerebellum, striatum, and frontal cortex are without effect (Post et al., 1988). Although it has not yet been definitively demonstrated, we surmise that the nucleus accumbens component is necessary for cocaine-induced behavioral sensitization because it mediates the reward and motor program aspects of cocaine-induced locomotion, while the amygdala may be crucial

COCAINE		EFFECT									
Number of Injections	Dose mg/kg i.p.	Behavioral Sensitization	Duration	Activity Context Dep.	Activity Context Indep.	Stereotypy Context Dep.	Stereotypy Context Indep.	Saline Conditioning	Sensitization Neuroleptic Independent	Seizure Kindling	Death
↑↑↑	COC_{65}									++	++
x10 days	COC_{160} subcut. (K. Gale)	++			++		++		++		
↑↑↑↑↑↑↑↑↑↑	COC_{10}	++	months	++	0			++			
↑↑↑	COC_{40}	+		0	++	++	++	±			
↑	COC_{40}	++	days	++	0	0	0	0	0		
↑ ↑ ↑	COC_{20}	0	0	0	0	0	+				
↑ ↑ ↑	COC_{10}	0	0	0	0	0	0				
↑	COC_{10}	0	0								

FIGURE 4. Size and number of arrows indicates dose of cocaine and number of administrations. COC = cocaine and dose in mg/kg administered once-daily i.p., except in the study of Gale, in which case the doses were administered subcutaneously. 0 = no effect; ± = equivocal; + = moderate effect; ++ = marked or definite effect. Effects of cocaine increase and are more persistent with dose and number of repetitions. The highest doses may be associated with seizures, kindling, and lethality. Thus, effects shift from behavioral sensitization to kindling as the dose is increased.

because of the environmental context information that is related to the associative aspects ofthe sensitization. With repeated high doses of cocaine a context-independent form of sensitization is observed; i.e., animals pretreated with cocaine in any environment show increased reactivity to subsequent challenge (even though animals pretreated and tested in the same environment show more persistent degrees of sensitization). Whereas the three-day high-dose paradigm is capable of engendering context-independent sensitization as well as the context-dependent component, amygdala lesions are no longer able to block the conditioned sensitization in the repeated-dose paradigms (Weiss et al., 1990, unpublished observation). Presumably, other substrates outside of the amygdala begin to code the context-dependent

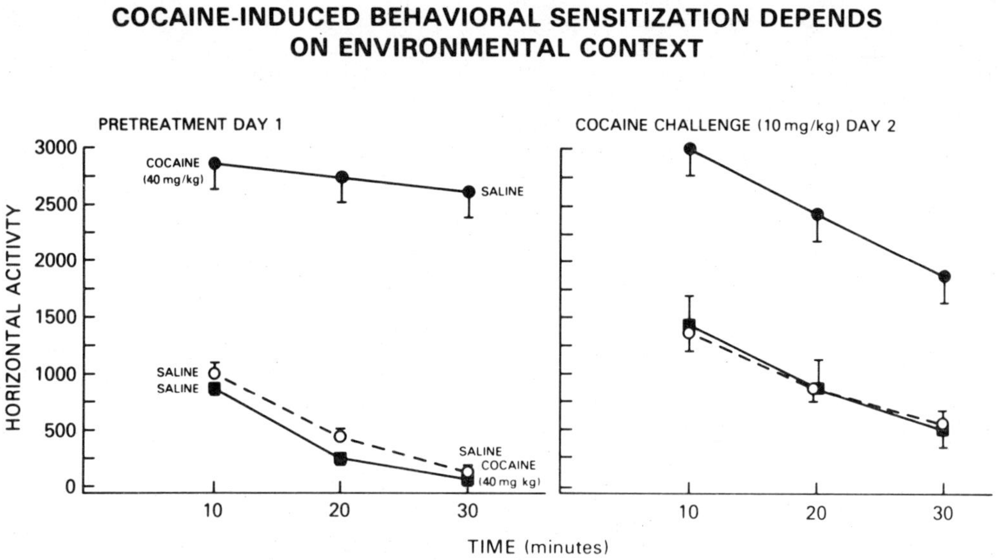

FIGURE 5. Left panel: Horizontal locomotor activity is illustrated on the ordinate, and time on the abscissa, for three groups of rats (n = 10/group) receiving the following day-1 pretreatments: cocaine in the test cage and saline in the home cage (●); saline in the test cage and cocaine in the home cage (■; saline in both cages (○). Right panel: Horizontal activity following a cocaine challenge (10 mg/kg) in all three groups is illustrated. Only the rats that received cocaine in the test apparatus on day 1 showed sensitization to cocaine on day 2.

information, making the amygdala no longer critical for the repeated-administration sensitization paradigm.

Fontana and associates (1993) have demonstrated that there is increased dopamine overflow in the n. accumbens in the two-day conditioned sensitization paradigm. The increased dopamine overflow occurs selectively in animals showing the conditioned increases in hyperactivity, but not in animals similarly pretreated with cocaine in a different environment so that there is no behavioral sensitization. The increase in dopamine overflow also occurs in comparison with saline pretreated controls. These increases in the n. accumbens are selective and do not occur in the striatum or the amygdala. These data suggest that even the conditioned components of cocaine-induced behavioral sensitization have a differential neurobiology that can be observed on a biochemical basis.

As noted previously, cocaine (and amphetamine) can induce changes in gene transcription by activating IEGs such as c-fos and zif/268 (Graybiel et al., 1990; Cole et al., 1992; Moratella et al., 1992). Brown et al. (1992) have demonstrated that conditioned cues associated with cocaine administration can increase c-fos in selected areas of the brain as well. These include the amygdala, cingulate cortex, lateral septum, claustrum, lateral habenula, and the paraventricular nucleus of the thalamus. Thus, regionally selective activation of transcription factors has also been shown to occur with conditioning.

There are interesting pharmacological dissociations in drugs that are able to block the development versus the expression of sensitization. For example, neuroleptics block sensitization when administered during pretreatment phases (Weiss et al., 1989a). The same doses of neuroleptics, however, are unable to block the conditioned component of sensitization when administered prior to the day-2 challenge. These data suggest that dopaminergic mechanisms are necessary to the induction but not the expression of sensitization and that different neurobiological mechanisms may come into play. The data are of interest in terms of similar disjunctions in a variety of other models of learning and memory, including long-term potentiation and kindling, in which drugs effective in the initiation phase are not necessarily able to block the expression phase (as previously mentioned).

Implications for Affective Illness: Utility of the Models and Caveats

It is in the attention to the longitudinal evaluation and potential dependence on sequentially different neuroanatomical and pharmacological substrates that the sensitization/kindling model may be useful in conceptualizing various aspects of the course of manic-depressive illness. Like other models of learning and memory, either the repeated inducing stimuli (typically psychosocial stressors in affective illness, as opposed to electrical or pharmacological stimulation of the brain in kindling) or the resultant episodes themselves (i.e., hyperactivity in sensitization or seizures in kindling, and extended periods of hypo- and hyperactivity in the bipolar mood disorders) result in long-lasting alterations in subsequent behavioral responsivity. Thus, the nonhomologous models of sensitization and kindling organize data in a longitudinal fashion that allows questions of the impact of prior stresses and episodes on the course of illness to be formulated. Prior to considering these models, almost all of the neurobiological theorizing pertaining to affective illness involved either perturbations during acute episodes or consideration of genetic vulnerability. Consequently, these models refocus on longitudinal progression of the illness (its waxing and waning course) and can help formulate testable hypotheses which are useful in many ways, as outlined below.

1. The kindling and sensitization models allow a conceptual reformulation suggesting that experiential (i.e., both psychosocial stresses and affective episodes themselves) as well as genetic factors may leave an impact on the long-term course of illness at the level of alterations in gene expression (Post, 1992).

From data regarding illness patterning and progression in the prepsychopharmacological era (Kraepelin, 1921), one can ask whether "episodes beget episodes." Kraepelin and many other investigators indicate that the modal pattern of affective illness is one of cycle acceleration, with longer healthy intervals between initial episodes and shorter intervals between subsequent episodes, and that this overall pattern of cycle acceleration (Figure 6) occurs despite a marked variation in illness frequency and pattern. This variation is all the more apparent in the current psychopharmacological era, in which effective treatments may markedly affect the natural course of illness, leading to remissions with the use of mood stabilizers (lithium, carbamazepine, and valproate) (Post et al., 1992) and, at times, to exacerbations in the form of increased cycle acceleration with the use of unimodal antidepressant compounds (Wehr and Goodwin, 1979, 1987; Altshuler et al., 1994, unpublished manuscript; Rouillon et al., 1992) (although this latter issue is not without controversy — see Angst, 1992 and Coryell et al., 1992).

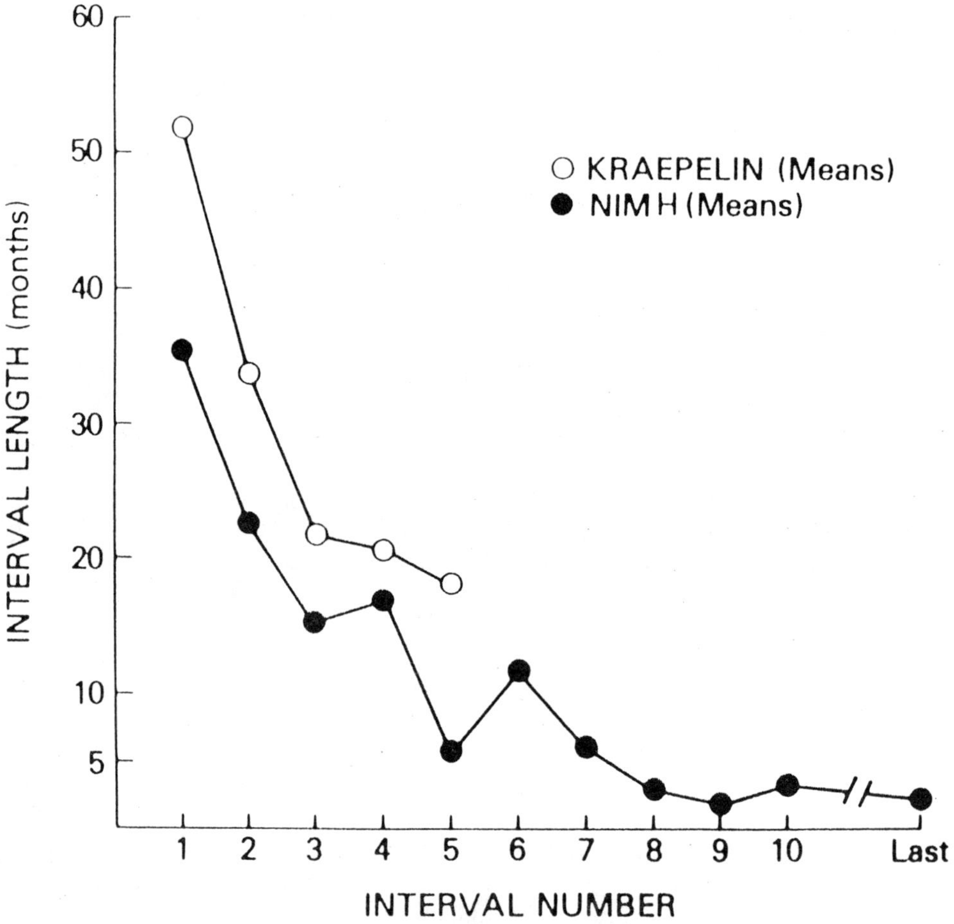

FIGURE 6. This figure depicts the decrements in healthy intervals in recurrent affective illness. Notice that the length of time between the first and second intervals is longer than that of subsequent intervals.

Given this general pattern of cycle acceleration (Figure 6), however, one is then in a position to ask whether this is attributable to aging or some other illness-independent factor, or if repeated episodes themselves convey part of this vulnerability. Part of this formulation can be directly tested by instituting adequate and early prophylaxis and seeing whether, in this group of patients, there is less likelihood of cycle acceleration compared with another group randomized only to acute interventions (Figure 7).

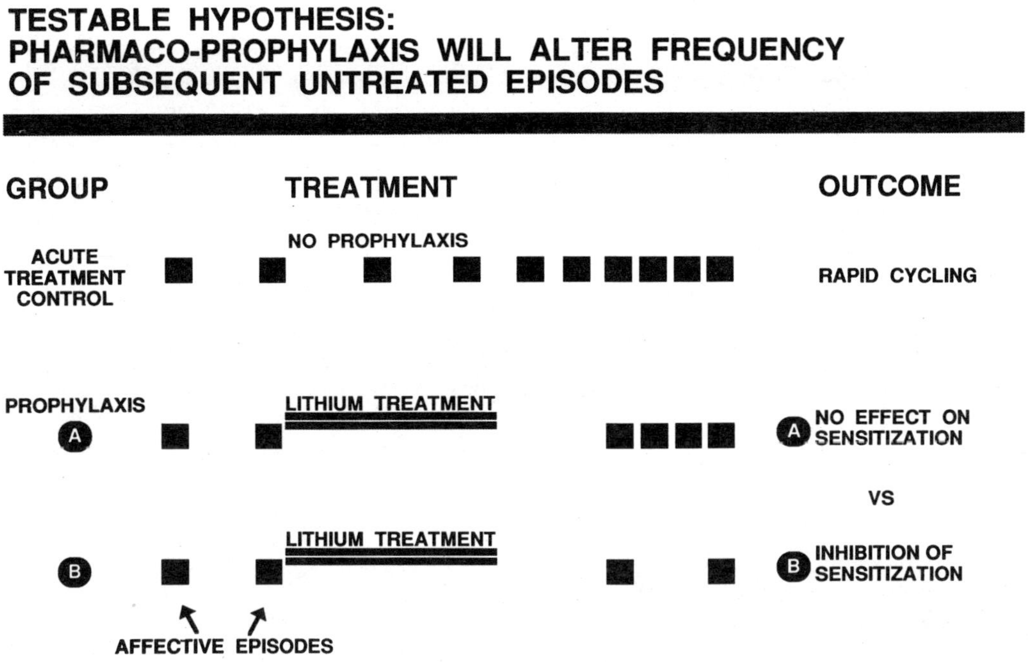

FIGURE 7. Different outcomes of effective lithium prophylaxis on the course of illness are illustrated in rows A and B. Squares represent affective episodes which, if untreated, may result in rapid cycling (top row). If prevention of episodes prevents sensitization (line B) then patients who were randomized to lithium prophylaxis should show less rapid illness progression than those left untreated or only acutely dosed.

2. A second use of the kindling model is the provision of a neurobiological basis for the idea that early life events, losses, and stressors may provide a nidus and pathway for the precipitation of future full-blown episodes by psychosocial stresses. Again, this formulation is only assisted by the kindling model in the sense that repeated electrical stimulation of the brain can initially appear to be without behavioral or physiological effect, but can produce a full-blown seizure with subsequent repetition of the same subthreshold stimulus.

Given this possibility for an episode of a seizure disorder, could this kind of long-term vulnerability and progression occur in response to repeated stressor experiences? The impact of early life events has long been a tenet of psychoanalytic and neural developmental concepts. A modicum of empirical data support a role for these kinds of early life experiences in the development of subsequent depressions.

3. Much more robust data, however, indicate that psychosocial stresses are involved in the initial episodes of unipolar and bipolar affective illness, but, with occurrences and progression, episodes begin to occur more autonomously, unassociated with psychosocial stresses.

It is in this transition from precipitated to spontaneous episodes that the kindling model is of particular conceptual interest. Although the neurobiology of such a transition has not yet been adequately delineated, the model helps focus on neurobiological principles involved in how the experience of repeated precipitated episodes could lead to spontaneous ones. The search for potential mechanisms involved could take place: a) in both the nonhomologous model of kindling, and b) the more homologous models of the affective disorders, such as repeated separations between mothers and offspring in rodents and primates, as well as in the illness itself.

> a) For example, Clark et al. (1993, unpublished data) have observed that in spontaneously kindled seizures c-fos is induced unilaterally, but on the side opposite the kindled electrode. Does spontaneity involve the development and spontaneous triggering of a mirror focus?
>
> b) After repeated primate social separations, animals begin to look depressed upon the appearance of the investigator, even in the absence of further separations (McKinney et al., 1984). Evidence can be marshalled that supports that such a transition in responsivity to more remote conditioned stimuli does, in fact, take place in the recurrent affective illnesses (Post, 1992).

4. A fourth series of clarifications provided by the sensitization and kindling models involves the nature (quality and quantity) of inducing stimuli sufficient to precipitate affective episodes. In this regard "intermittency" of stimulation appears to be a key property in both models. Chronic stimulation of the amygdala does not result in kindling, nor does stimulation at two-minute intervals (Goddard et al., 1969). As the intervals increase toward 15 minutes, one hour, or optimally, 24 hours, the development of full-blown kindled seizures is facilitated. Does such a principle hold for the affective disorders (or for the putative progression of the chemical hypersensitivity syndrome)? In pharmacological approaches, chronic continuous drug exposure is less likely to produce either behavioral sensitization or kindled seizure end points than are repeated intermittent exposures.

5. In the sensitization models, dose and number of repetitions are crucial factors in determining the nature, magnitude, and duration of the sensitization (Figure 4). Do similar "rules" hold for stress sensitization in animal models, the affective disorders, PTSD, and even, hypothetically, the role of stress sensitization in chemical hypersensitivity?

6. Is there cross-sensitization from drugs to stress? That is, in behavioral sensitization to the psychomotor stimulants cocaine or amphetamine, is there also increased reactivity to stress-induced changes in motor activity, and vice versa (Table 4)? These data provide a background for considering how stresses could increase or decrease vulnerability to exogenous or endogenous chemical sensitization, or vice versa (Post et al., 1994a). Based on the data in Table 4, one could postulate that once someone has been sensitized by psychomotor stimulants, stresses could engage and occupy some of the same pathways. In this fashion,

stimulant abuse could propel pathophysiological mechanisms involved in affective illness to full-blown episode end points, and affective episodes could enhance stressor reactivity.

TABLE 4. Cross-Sensitization between Stressors and Stimulant Drug Effects

Stressor	Drug	Reference
Tail pinch --------------------->	Amphetamine	Christoph et al., 1978; Antelman et al., 1980; Antelman and Chiodo, 1983;
Food deprivation ------------>	Amphetamine	Yuwiler, 1976; Antelman and Eichler, 1979; Antelman and Chiodo, 1983
Foot shock ------------------->	Amphetamine	Antelman and Chiodo, 1983; Herman et al., 1984; Leyton and Stewart, 1990
--------------------->	Cocaine	Sorg and Kalivas, 1993
Immobilization (restraint) --->	Amphetamine	Antelman and Chiodo, 1983; Anisman et al., 1985; Hahn et al., 1986
Immobilization and shock --->	Amphetamine	Hahn et al., 1986
Tail pinch <-------------------	Amphetamine	Antelman and Nelson (p.c.)
ECT <-------------------	Amphetamine	Chiodo and Antelman, 1980; Castaneda et al., 1988
Foot shock <-------------------	Amphetamine	Robinson et al., 1987
<-------------------	Cocaine	Kalivas and Duffy, 1989

In this case the caveats are most crucial because some types of stresses do not show sensitization; rather, tolerance is observed. Footshock stress has been demonstrated to decrease amygdala-kindled seizure duration (Pinel, 1973) but potentiate the evolution of these seizures (Grahnstedt and Ellertsen, 1984). In addition, in some instances, repeated administration of stress hormones such as i.c.v. CRH results in a tolerance to some end points (seizures and aggression), while at the same time increasing the likelihood of others; e.g., amygdala-kindled seizure development. This emphasizes that in response to a given manipulation some end points may show adaptation by decremental responses (tolerance), while other elements may show sensitization or persistent increased responsivity. In addition, the effects of repeated stresses and repeated electrical stimulation of the brain (in the kindling paradigm) may produce opposite effects on a number of neurobiological parameters, most recently including the induction of neurotrophic factor mRNAs, with kindling increasing BDNF and decreasing NT_3 and stress producing the opposite effect (Smith et al., 1994 and 1994 unpublished observation).

Consequently, whereas sensitization and kindling can provide organizing principles for considering what characteristics of the inducing stimuli are critical to the production of increased responsivity over time, it remains crucial to assess whether a given neuropsychiatric syndrome is actually enhanced or inhibited by a given set of inducing stimuli, i.e., whether the sensitization and kindling analogies can in fact be usefully applied. The variation in the qualitative and quantitative aspects of the inducing stimuli may make all the difference in the final outcome of the syndrome evolution, and determine whether it is enhanced or diminished, based on prior exposures.

7. A particularly robust example of the importance of historical factors involves the role of conditioning in sensitization. As noted above, the context-dependent or conditioned component of cocaine sensitization is particularly robust, raising interesting issues of whether conditioned cues in the environment can, over time, evoke relapse in the addictive disorders, behavioral alterations in animals, or psychiatric disturbances in man. To the extent that conditioning is relevant to tolerance development, conditioned compensatory reactions (Siegel, 1985) can be evoked by conditioned cues and can explain the paradoxical appearance of opposite behavioral manifestations to the same inducing cue, depending on the type of paradigm utilized. This analysis could lead to the consideration of how either depression or mania may be precipitated by the same environmental stimulus (Post et al., 1984b). It becomes interesting to ask whether, in the putative chemical sensitivity models, all the observed symptoms are directly related to chemical exposures or whether some of them could be reinvoked by conditioned cues and, at least in part, show environmental context-dependence similar to that observed for the psychomotor stimulants (Weiss et al., 1989a).

8. Do the neural substrates involved in the affective disorders evolve over time, in terms of their neurobiochemistry and anatomy, similarly to the evolution observed in sensitization and kindling? With the recent advent of a variety of neural imaging techniques, this hypothetical question may be specifically studied and answered within the next several years.

9. In a parallel fashion, does the psychopharmacology of the affective disorders change as a function of stage of evolution (Figure 8), similar to that observed in amygdala kindling (Figure 1)? Here again the caveats are particularly crucial. The utility of this model is not in the suggestion that the specific drugs effective in different stages of amygdala kindling are also effective in different stages of affective disorders. However, the principle might pertain, and existing data sets should be examined to determine if the data support such a notion. Appropriate prospective assessments of this possibility can then be studied as well. As mentioned earlier, carbamazepine blocks the early development stage of cocaine- and lidocaine-induced pharmacological kindling, but is unable to block this initial development stage of amygdala kindling. Conversely, the drug is highly effective in the completed stages of amygdala kindling, but is largely ineffective on completed lidocaine- or cocaine-kindled seizures. These more precise drug dissociations highlight the idea that a given drug may be differentially effective not only as a function of stage of kindling evolution, but also as a function of the type of paradigm utilized.

As such, it would appear to be an egregious error to consider that the effectiveness of a given pharmacological agent in a given clinical syndrome in some way implicates a kindling-like process. Carbamazepine is effective in the treatment of paroxysmal pain syndromes and in some forms of epilepsy, even when a kindling process does not underlie their evolution.

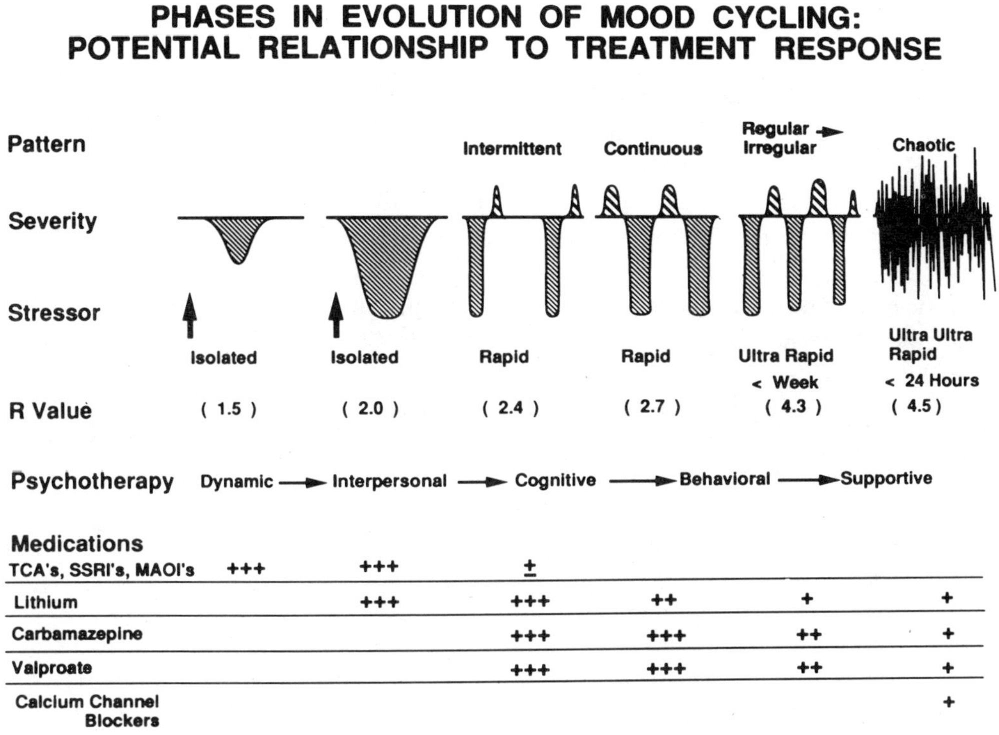

FIGURE 8. Schematic illustration of the progression of bipolar illness from stress-precipitated intermittent episodes of depression to autonomous rapid cycling, and finally to chaotic ultradian cycling. As the illness evolves to greater severity, optimal treatment strategies may also be changing, reflecting alterations in the underlying pathophysiology of the disease. TCA, tricyclic antidepressant; SSRI, specific seratonin reuptake inhibitor; MAOI, monoamine oxidase inhibitor

Similarly, carbamazepine is equally effective in treatment of rage and aggression syndromes in patients with or without EEG abnormalities (Luchins 1983, 1984). Carbamazepine is also effective in various phases of the affective disorders in patients without evidence of a seizure disorder on EEG evaluation or through more indirect measures (Post et al., 1991). One cannot surmise, therefore, that the efficacy of a given anticonvulsant agent is evidence for a kindled pathophysiology.

At the same time, it may be useful to consider the principle of pharmacological disjunctions as a function of stage of illness evolution (derived from the kindling model), in an effort to see if it is helpful in targeting appropriate therapeutics. For example, carbamazepine and phenytoin are not effective in preventing the development of amygdala-kindled seizure evolution, and there is an increasing amount of data to suggest that they are also ineffective in prophylaxis of post-traumatic epilepsy in humans. As a result, to the extent that a kindling-like evolution process is responsible for the late development of post-traumatic epilepsy, one might postulate that valproate and phenobarbital (which are effective in the initial developmental phase of kindling) might be more effective, and that clinical trials should be carried out to test this more specifically.

If lithium is less effective in patients with rapid cycling, is this a function of cycling frequency or (late) stage of evolution of the illness, and would other drugs be more helpful? As carbamazepine and valproate (Post, 1990) are effective in some lithium-refractory rapid cyclers, are they equally effective in early stages of the illness as well? The calcium channel blocker nimodipine is effective in some patients with ultra-ultra rapid (ultradian) cycling frequencies. Is it also effective earlier in the course of illness? Does its profile of effects in ultradian cycling implicate additional dysregulation of calcium in this phase of illness (see Figure 8)?

Taken together, these models suggest that affective illness may be an evolving process with an evolving neurochemistry, neuroanatomy, and neuropsychopharmacology, as well as suggesting differential requirements for different types of psychotherapies as a function of stage of illness (Post, 1992). These considerations may help to explain some of the biological variance in illness measures as a function of stage of syndrome evolution.

Final Caveats for Affective Illness

Kindling as a nonhomologous model of affective disorders opens new vistas for consideration of the longitudinal development of the illness and the potential types of long-term mechanisms involved. In a nonhomologous model, however, there is also the direct implication that although many of the principles of kindling may be appropriately applied to the development of the affective disorders, the precise mechanisms involved in amygdala kindling are not likely to be identical to those of affective disorders. One can still ask where the convergences and divergences are.

Just as the site of electrical kindling and the type of pharmacological kindling have vastly different implications for the final output of the syndrome, so too might stressors and chemical exposures have differential outcomes. Elsewhere we have discussed how the type of stressor may be crucial in differential precipitation of the affective disorders versus PTSD (Post, 1992; Post et al., 1994b). In the latter instance threat of loss of life and limb are most likely involved in an event resulting in PTSD (almost by definition), whereas more typically, actual or threatened separations and losses tend to be productive of an affective episode. Nonetheless, with these caveats in mind, the way the long-term kindled "memory trace" is maintained for both types of stressors may have certain analogies with those in the affective disorders. Specifically, to the extent that learning and memory mechanisms share fundamental principles across differential paradigms, some processes and fundamental mechanisms may occur in parallel.

With regard to this, we have postulated that a complex spatiotemporal cascade of events at a cellular level is likely to underlie the stressor memory trace (Post, 1992; Post et al., 1992, 1993, 1994b) just as is likely to be the case in amygdala kindling. In these instances, we postulate that there are not only acute changes in synaptic responsivity, but also longer-term changes in the synapse and in the microstructure of the brain (axonal and dendritic sprouting and retraction, as well as cell death induced by a cascade of effects on gene expression; Figure 9). These changes initially involve the induction of a series of IEGs, followed by sequential waves of LEG alterations which encode changes in neuropeptides and growth factors, as well as inducing a variety of other neurotransmitter receptor and structural changes (Figure 9). To the extent that these elements of the sensitization and kindling formulations are appropriate to a given syndrome, such as the affective disorders (Post, 1992), panic disorder (Post et al., 1993), or PTSD (Post et al., 1994b), then a systematic search for acute and long-lasting changes may be better organized with an awareness of the complexities of events that appear to underlie kindling evolution, even at this very early stage of understanding its pathophysiological mechanisms. Inherent in this view is the notion that the neurobiology of the syndrome may change as a function of its stage of evolution. As such, treatment interventions may also change. While models are often used for their simplicity relative to the real syndrome being studied, in these latter instances the sensitization and kindling formulations help indicate the potential complexities of processes involved and the precision that may be required in formulating a neurochemical hypothesis.

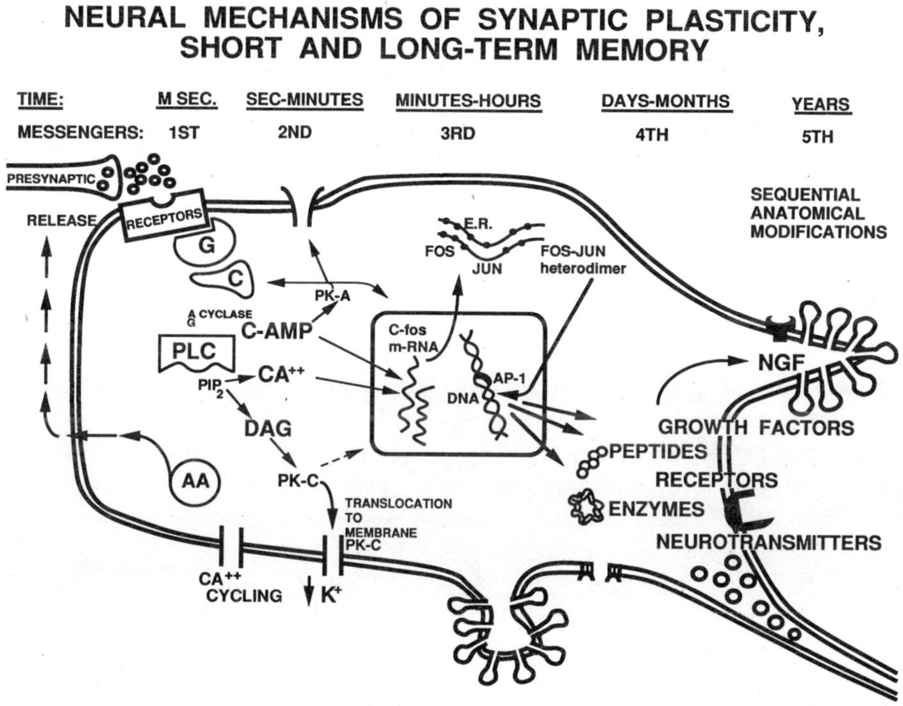

FIGURE 9. Neural mechanisms of short- and long-term synaptic changes following c-fos induction. PLC = phospholipase c; PIP_2 = phosphatidyl inositol 4,5-biphosphate; AA = arachidonic acid; DAG = diacylglycerol; PK-C = protein kinase C; AP-1 = activator protein 1 (binding site on DNA); E.R. = endoplasmic reticulum; PK-A = protein kinase A; NGF = nerve growth factor.

Caveats in the Kindling Analogy for the Chemical Hypersensitivity Syndromes

As is the case in the affective disorders, kindling is also a potential nonhomologous model of chemical hypersensitivity, as seizures are not the recognized end point. Nonetheless, one might ask whether kindling to a nonconvulsive end point might still be applicable to some chemical hypersensitivity syndromes. In this regard it is important to recognize that kindling does not occur unless a certain magnitude of effect is achieved; i.e., stimulation current needs to be sufficient to produce an afterdischarge (Racine, 1978). However, repeated stimulation below the afterdischarge threshold, in some instances, is capable of eventually producing an afterdischarge such that even this caveat may not be overriding. Moreover, Adamec (1991) has shown that prekindling stimulation (i.e., not sufficient to produce a seizure) is nonetheless capable of producing profound and long-lasting behavioral alterations in experimental animals. Cats that are habitual rat killers can be converted to more placid nonkiller animals with amygdala prekindling. Likewise, cocaine in high doses may cause local spiking in the amygdala and olfactory bulb of animals, as well as in man (Stevens et al., 1969), and thus a prekindling stimulation may be achieved on the basis of cocaine's local anesthetic effects (Post et al., 1975, 1987) in combination with its more well-known dopaminergic-mediated, psychomotor-stimulant properties. Nonetheless, in each of these instances, some critical threshold for magnitude of inducing stimulation is required in order to induce a sensitization or kindling-like effect on behavior or preconvulsant or convulsant end points (see Figure 4). In this regard, the low-level exposures often implicated in chemical hypersensitivity inductions (even though they may be subchronic to intermittent) may not be sufficient to engage sensitization and kindling-like mechanisms.

Thus it would appear of central importance to use the kindling and sensitization models for testing hypotheses about the nature of chemical hypersensitivity and to evaluate their potential underlying mechanisms in more appropriate chemical exposure models and in affected patients. Although most organisms show a clear capability for one-trial psychomotor-stimulant sensitization (Weiss et al., 1989a) (even a single administration of a pharmacological dose of CRH induces a kindling-like process) or even stress sensitization (Antelman, 1988; Antelman and Caggiula, 1990) (which we know can result from severe enough trauma in PTSD), single exposures, by definition, almost invalidate a classical or pharmacological kindling process or mechanism as being of etiopathological significance.

If there are repeated chemical exposures, do they in fact produce an increased behavioral response or is there down-regulation and adaptation, which would be more likely if exposures are low-level and more chronic? If there is a sensitivity to one set of chemical agents, to what extent does this show cross-sensitivity to other agents or to stressors (and of which type)? Even though Antelman and Caggiula (1990) have suggested that single exposures to a variety of drug types can induce long-lasting alterations in responsivity, which appear to show progressive alterations over time, this should be considered as an alternative to a kindling-like process, as it can be induced by a single exposure. Even in these instances critical questions arise about the generality of the cross-reactivity induced by the single dose exposure, and whether low-dose exposures are sufficient to produce long-lasting behavioral changes. Further preclinical studies are also needed to explore the common and differential mechanisms and principles underlying time-dependent sensitization, repeated sensitization, the role of conditioning mechanisms, and kindling, as well as whether each of these does or does not show cross-sensitization to stress. Thus, it would appear useful to ask to what extent kindling-like phenomena contribute to the syndrome, and to what extent other pathophysiological mechanisms need to be invoked, as in the case of single exposures.

Finally, it might even be useful to consider using analogies from the divergent models of single-dose time-dependent sensitization, repeated-dose sensitization, and pharmacological and electrophysiological kindling to give greater diagnostic clarity and precision to the multiple chemical sensitivity syndromes. That is, given the heterogeneity of inducing circumstances (with different intensities, durations, and numbers of exposures to many different agents) perhaps the notion of differences in presumptive mechanisms involved could be used to subdivide the chemical hypersensitivity syndromes and make definitions more explicit. In this fashion some clinical syndromes could be excluded, based on the characteristics of the inducing agents or the failure of symptoms to meet criteria of the classical syndrome. This type of clarification and subcategorization may ultimately lead not only to a better understanding of potential relationships or lack thereof of kindling and sensitization phenomena to chemical hypersensitivity phenomena, but also to a greater precision in the types of pathophysiological mechanisms involved and, ultimately, to better treatment approaches.

A modern example may be useful in this regard. Rubinow and associates (1984, 1985) found that a very high percentage of patients arriving at their clinic with a self-referred or presumptive diagnosis of premenstrual syndrome (now called late luteal dysphoric disorder) did not, in fact, have a pattern consistent with that illness when they were followed prospectively. It should be emphasized that many of these non-PMS patients were highly symptomatic and in considerable distress, but they had other affective and related psychiatric syndromes which were not temporally closely linked to the menses on a consistent basis, as required by the definition. Once the diagnostic heterogeneity of the syndrome was thus clarified, much more precise and valid data could be garnered regarding pathophysiological mechanisms and treatment response (Rubinow et al., 1985). PTSD, as well, has struggled with precision of its definitions, eventually arriving at ones that are fairly easily agreed upon despit a panoply of potential inducing circumstances, with either single or multiple exposures. It would appear that once these phenomenological and diagnostic considerations are further clarified in the multiple chemical sensitivity syndrome, potential analogies (or lack thereof) to time-dependent sensitization, classic repeated pharmacological sensitization, and pharmacological and electrophysiological kindling, can be most usefully elucidated. These differential models, moreover, may even help in the development of such a classification schema.

REFERENCES

ADAMEC, R.E. (1991) "Partial kindling of the ventral hippocampus: identification of changes in limbic physiology which accompany changes in feline aggression and defense." Physiol. Behav. 49:443–453.

ANGST, J. (1992). "Course of mood disorders: A challenge to psychopharmacology." Clin. Neuropharmacol. 15:444A–445A.

ANISMAN, H., HAHN B., HOFFMAN D., and ZACHARKO, R.M. (1985). "Stressor invoked exacerbation of amphetamine-elicited perseveration." Pharmacol. Biochem. Behav. 23:173–183.

ANTELMAN, S.M. (1988). "Stressor-induced sensitization to subsequent stress: Implications for the development and treatment of clinical disorders." In: Sensitization in the Nervous System (P.W. Kalivas and C.D. Barnes, ed.). Telford Press, Caldwell, N.J. pp. 227–254.

ANTELMAN, S.M. and CAGGIULA, A.R. (1990). "The effects of acute exposure to stressors and drugs grow with the passage of time." In: Psychobiology of Stress (A. Oliverio and S. Puglisi-Allegra, ed.). Kluwer, Dordrecht, Germany. pp. 161–170.

ANTELMAN, S.M. and CHIODO, L.A. (1983) "Amphetamine as a stressor." In: Stimulants: Neurochemical, Behavioral, and Clinical Perspectives (I. Creese, ed.). Raven Press, New York. pp. 269–299.

ANTELMAN, S.M. and EICHLER, A.J. (1979) "Persistent effects of stress on dopamine-related behaviors: Clinical implications." In: Catecholamines: Basic and Clinical Frontiers (E. Usdin, I.J. Kopin, and J. Barchas, ed.). Pergamon Press, New York. pp. 1759–1761.

ANTELMAN, S.M., EICHLER, A.J., BLACK, C.A., and KOCAN, D. (1980) "Interchangeability of stress and amphetamine in sensitization." Science 207:329–331.

BROWN, E.E., ROBERTSON, G.S., and FIBIGER, H.C. (1992). "Evidence for conditional neuronal activation following exposure to a cocaine-paired environment: Role of forebrain limbic structures." J. Neurosci. 12:4112–4121.

CAIN, D.P. (1986). "The transfer phenomenon in kindling." In: Kindling 3 (J.A. Wada, ed.). Raven Press, New York. pp. 231–248.

CAIN, D.P. and CORCORAN, M.E. (1984). "Intracerebral beta-endorphin, met-enkephalin and morphine: Kindling of seizures and handling-induced potentiation of epileptiform effects." Life Sci. 34:2535–2542.

CASTANEDA, E., BECKER, J.B., and ROBINSON, T.E. (1988). "The long-term effects of repeated amphetamine treatment *in vivo* on amphetamine, KCl and electrical stimulation evoked striatal dopamine release *in vitro*." Life Sci. 42:2447–2456.

CHIODO, L.A. and ANTELMAN, S.M. (1980) "Electroconvulsive shock: progressive dopamine autoreceptor subsensitivity independent of repeated treatment." Science 210:799–801.

CHRISTOPH, G.R., KUHN, D.M., and JACOBS, B.L. (1978). "Dopamine agonist pretreatment alters LSD's electrophysiological action from dopamine agonist to antagonist." Life Sci. 23:2099–2110.

CLARK, M., POST, R.M., WEISS, S.R.B., CAIN, C.J., and NAKAJIMA, T. (1991). "Regional expression of c-fos mRNA in rat brain during the evolution of amygdala-kindled seizures." Mol. Brain Res. 11:55–64.

CLARK, M., POST, R.M., WEISS, S.R.B., and NAKAJIMA, T. (1992). "Expression of c-fos mRNA in acute and kindled cocaine seizures in rats." Brain Res. 592:101–106.

COLE, A.J., BHAT, R.V., PATT, C., WORLEY, P.F., and BARABAN, J.M. (1992). "D1 dopamine receptor activation of multiple transcription factor genes in rat striatum." J. Neurochem. 58:1420–1426.

CORYELL, W., ENDICOTT, J., and KELLER, M. (1992). "Rapidly cycling affective disorder: Demographics, diagnosis, family history, and course." Arch. Gen. Psychiatry 49:126–131.

DHUNA, A., PASCUAL-LEONE, A., LANGENDORF, F., and ANDERSON, D.C. (1991). "Epileptogenic properties of cocaine in humans." Neurotoxicology 12:621–626.

DOWNS, A.W. and EDDY, N.B. (1932). "The effect of repeated doses of cocaine on the dog." J. Pharmacol. Exp. Ther. 46:195–198.

EHLERS, C.L., HENRIKSEN, S.J., WANG, M., RIVIER, J., VALE, W., and BLOOM, F.E. (1983). "Corticotropin releasing factor produces increases in brain excitability and convulsive seizures in rats." Brain Res. 278:332–336.

ELLINWOOD, E.H. and KILBEY, M.M. (1977) "Chronic stimulant intoxication models of psychosis." In: Animal Models in Psychiatry and Neurology (I. Hanin and E. Usdin, ed.). Pergamon Press, New York. pp. 61–74.

FONTANA, D.J., POST, R.M., and PERT, A. (1993). "Conditioned increase in mesolimbic dopamine overflow by stimuli associated with cocaine." Brain Res. 629:31–39.

GODDARD, G.V., MCINTYRE, D.C., and LEECH, C.K. (1969). "A permanent change in brain function resulting from daily electrical stimulation." Exp. Neurol. 25:295–330.

GRAHNSTEDT, S. and ELLERTSEN, B. (1984). "Brain state and kindled seizures." Exp. Neurol. 84:596–605.

GRAYBIEL, A.M., MORATALLA, R., and ROBERTSON, H.A. (1990). "Amphetamine and cocaine induce drug-specific activation of the c-fos gene in striosome-matrix compartments and limbic subdivisions of the striatum." Proc. Natl. Acad. Sci. 87:6912–6916.

HAHN, B., ZACHARKO, R.M., and ANISMAN, H. (1986). "Alterations of amphetamine elicited perseveration and locomotor excitation following acute and repeated stressor application." Pharmacol. Biochem. Behav. 25:29–33.

HERMAN, J.P., STINUS, L., and LEMOAL, M. (1984). Repeated stress increases locomotor response to amphetamine." Psychopharmacology 84:431–435.

KALIVAS, P.W. and DUFFY, P. (1989). "Similar effects of daily cocaine and stress on mesocorticolimbic dopamine neurotransmission in the rat." Biol. Psychiatry 25:913–928.

KREAPELIN, E. (1921). Manic Depressive Insanity and Paranoia. (G.M. Robertson, ed.) E.S. Livingstone, Edinburgh. pp. 179–181.

LEYTON, M. and STEWART, J. (1990). "Preexposure to foot-shock sensitizes the locomotor response to subsequent systemic morphine and intra-nucleus accumbens amphetamine." Pharmacol. Biochem. Behav. 37:303–310.

LITTLE, H.J., NUTT, D.J., and TAYLOR, S.C. (1984). "Chronic effects of benzodiazepine receptor ligand FG 7142: Proconvulsant properties and kindling." Br. J. Pharmacol. 83:951–958.

LUCHINS, D.J. (1983). "Carbamazepine for the violent psychiatric patient." Lancet 1:766.

LUCHINS, D.J. (1984). "Carbamazepine in violent non-epileptic schizophrenia. " Psychopharmacol. Bull. 20:569–571.

MCKINNEY, W.T. (1989). "Basis of development of animal models in psychiatry: An overview." In: Animal Models of Depression (G.F. Koob, C.L. Ehlers, and D.J. Kupfer, ed.). Birkhauser, Boston. pp. 3–17.

MCKINNEY, W.T., MORAN, E.C., and KRAEMER, G.W. (1984). "Separation in nonhuman primates as a model for human depression: Neurobiological implications." In: Neurobiology of Mood Disorders (R.M. Post and J.C. Ballenger, ed.). Williams & Wilkins, Baltimore. pp. 393–406.

MCKINNEY, JR., W.T. and BUNNEY, JR., W.E., (1969). "Animal model of depression. I. Review of evidence: Implications for research." Arch. Gen. Psychiatry 21:240–248.

MORATALLA, R., ROBERTSON, H.A., and GRAYBIEL, A.M. (1992). "Dynamic regulation of NGFI-A (zif268, egr1) gene expression in the striatum." J. Neurosci. 12:2609–2622.

MORGAN, J.I. and CURRAN, T. (1990). "Inducible proto-oncogenes of the nervous system: Their contribution to transcription factors and neuroplasticity." Prog. Brain Res. 86:287–294.

NAKAJIMA, T., DAVAL, J.L., MORGAN, P.F., POST, R.M., and MARANGOS, P.J. (1989). "Adenosinergic modulation of caffeine-induced c-fos mRNA expression in mouse brain." Brain Res. 501:307–314.

NAKAJIMA, T., WEISS, S.R.B., CLARK, M., CAIN, C.J., NIERENBERG. J., MIYATAKE, T., and POST, R.M. (1994). "C-fos mRNA induction in mouse brain following local anesthetic seizures induced by lidocaine and cocaine." Brain Res. Submitted.

PINEL, J.P.J. (1983). "Effects of diazepam and diphenylhydantoin on elicited and spontaneous seizures in kindled rats: A double dissociation." Pharmacol. Biochem. Behav. 18:61–63.

PINEL, J.P.J., PHILLIPS, A.G., and MACNEILL, B. (1973) "Blockage of highly-stable kindled seizures in rats by antecedent footshock." Epilepsia 14:29–37.

PINEL, J.P.J. and ROVNER, L.I. (1978a) "Electrode placement and kindling-induced experimental epilepsy." Exp. Neurol. 58:335–346.

PINEL, J.P.J. and ROVNER, L.I. (1978b). "Experimental epileptogenesis: Kindling-induced epilepsy in rats." Exp. Neurol. 58:190–202.

POST, R.M. (1990). "Alternatives to lithium for bipolar affective illness." In: Review of Psychiatry (A. Tasman, S.M. Goldfinger, and C.A. Kaufmann, ed.). Vol 9. American Psychiatric Press, Washington D.C. pp. 170–202.

POST, R.M. (1992). "Transduction of psychosocial stress into the neurobiology of recurrent affective disorder." Am. J. Psychiatry 149:999–1010.

POST, R.M., ALTSHULER, L.L., KETTER, T.A., DENICOFF, K., and WEISS, S.R.B. (1991). "Antiepileptic drugs in affective illness: Clinical and theoretical implications." In: Advances in Neurology: Neurobehavioral Problems in Epilepsy (D.B. Smith, D.M. Treiman, and M.R. Trimble, ed.). Vol 55. Raven Press, New York. pp. 239–277.

POST, R.M. and CONTEL, N.R. (1981). "Cocaine-induced behavioral sensitization: A model for recurrent manic illness." In: Biological Psychiatry (C. Perris, G. Struwe, and B. Jansson, ed.). Elsevier, Amsterdam. pp. 746–749.

POST, R.M. and CONTEL, N.R. (1983). "Human and animal studies of cocaine: Implications for development of behavioral pathology." In: Stimulants: Neurochemical, Behavioral, and Clinical Perspectives (I. Creese, ed.). Raven Press, New York. pp. 169–203.

POST, R.M., KENNEDY, C., SHINOHARA, M., SQUILLACE, K., MIYAOKA, M., SUDA, S., INGVAR, D.H., and SOKOLOFF, L. (1984a). "Metabolic and behavioral consequences of lidocaine-kindled seizures." Brain Res. 324:295–303.

POST, R.M., KOPANDA, R.T., and LEE, A. (1975). "Progressive behavioral changes during chronic lidocaine administration: relationship to kindling." Life Sci. 17:943–950.

POST, R.M., LOCKFELD, A., SQUILLACE, K.M., and CONTEL, N.R. (1981). "Drug-environment interaction: Context dependency of cocaine-induced behavioral sensitization." Life Sci. 28:755–760.

POST, R.M., RUBINOW, D.R., and BALLENGER, J.C. (1984b). "Conditioning, sensitization, and kindling: Implications for the course of affective illness." In: Neurobiology of Mood Disorders (R.M. Post and J.C. Ballenger, ed.). Williams and Wilkins, Baltimore. pp. 432–466.

POST, R.M. and WEISS, S.R.B. (1989). "Nonhomologous animal models of affective illness: Clinical relevance of sensitization and kindling." In: Animal Models of Depression (G. Koob, C. Ehlers, and D.J. Kupfer, ed.). Birkhauser, Boston. pp. 30–54.

POST, R.M., WEISS, S.R.B., and CHUANG, D.-M. (1992). "Mechanisms of action of anticonvulsants in affective disorders: Comparisons with lithium." J. Clin. Psychopharm. 12:23S–35S.

POST, R.M., WEISS, S.R.B., and PERT, A. (1988). "Cocaine-induced behavioral sensitization and kindling: Implications for the emergence of psychopathology and seizures." In: Mesocorticolimbic Dopamine System (P.W. Kalivas and C.B. Nemeroff, ed.). NY Acad. Sci., New York. pp. 292–308.

POST, R.M., WEISS, S.R.B., PERT, A., and UHDE T.W. (1987). "Chronic cocaine administration: Sensitization and kindling effects." In: Cocaine: Clinical and Biobehavioral aspects (A. Raskin and S. Fisher, ed.). Oxford University Press, New York. pp. 109–173.

POST, R.M., WEISS, S.R.B., ROSEN, J., SMITH, M., THOMAS, N., and PERT, A. (1994a). "Co-morbidity of cocaine abuse and affective disorders." Am. J. Psychiatry. Submitted.

POST, R.M., WEISS, S.R.B., and SMITH, M. (1994b). "Sensitization and kindling: Implications for the evolving substrate of PTSD." In: Neurobiology and Clinical Consequences of Stress: From Normal Adaptation to PTSD (M.J. Friedman, D.S. Charney, and A.Y. Deutch, ed.). Raven Press, New York. In press.

POST, R.M., WEISS, S.R.B., UHDE, T.W., CLARK, M., and ROSEN JB. (1993). "Implications of cocaine kindling, induction of the proto-oncogene c-fos and contingent tolerance." In: Biology of Anxiety Disorders (R. Hoehn-Saric, ed.). APA Press, Washington D.C. pp. 121–175.

RACINE, R. (1978) "Kindling: The first decade." Neurosurgery 3:234-252.

ROBINSON, T.E., BECKER, J.B., YOUNG, E.A., AKIL, H., and CASTANEDA, E. (1987). "The effects of footshock stress on regional brain dopamine metabolism and pituitary β-endorphin release in rats previously sensitized to amphetamine." Neuropharmacology 26:679–691.

ROUILLON, F., LEJOYEUX, M., and FILTEAU, M.J. (1992). "Unwanted effects of long-term treatment." In: Long-Term Treatment of Depression (S.A. Montgomery and F. Rouillon, ed.). John Wiley and Sons, pp. 81–111.

RUBINOW, D.R. and ROY-BYRNE, P.P. (1984). "Premenstrual syndromes: Overview from a methodologic perspective." Am. J. Psychiatry 141:163–172.

RUBINOW, D.R., ROY-BYRNE, P.P., and HOBAN, M.C. (1985). "Menstrually-related mood disorders: Methodological and conceptual issues." In: Premenstrual Tension and Dysmenorrhea (M.Y. Dawood, J.L. Mcguire, and L.M. Demers, ed.). Urban and Schwarzenberg, Baltimore. pp. 27–40.

SIEGEL, S. (1985) "Drug-anticipatory responses in animals." In: Placebo: Clinical Phenomena and New Insights (L. White, B. Tursky, and B. Schwartz, ed.). Guilford Press, New York. pp. 288–305.

SMITH, M.A., MAKINO, M.A., KVETNANSKY, R., and POST, R.M. (1994) "Stress alters the expression of brain-derived neurotropic factor and neurotrophin-c mRNAs in the hippocampus." J. Neurosci. In press.

SORG, B.A. and KALIVAS, P.W. (1993). "Effects of cocaine and footshock stress on extracellular dopamine levels in the medial prefrontal cortex." Neuroscience 53:695–703.

STEVENS, J.R., MARK, V.H., ERWIN, F., PACHECO, P., and SUEMATSU, K. (1969). "Deep temporal stimulation in man." Arch. Neurol. 21:157–167.

TATUM, A.L. and SEEVERS, M.H. (1929). "Experimental cocaine addiction." J. Pharmacol. Exp. Ther. 36:401–410.

WADA, J.A. and SATO, M. (1974). "Generalized convulsive seizures induced by daily electrical stimulation of the amygdala in cats." Neurology 24:565–574.

WADA, J.A., SATO, M., and CORCORAN, M.E. (1974). "Persistent seizure susceptibility and recurrent spontaneous seizures in kindled cats." Epilepsia 15:465–478.

WASTERLAIN, C.G., MASUOKA. D., and JONEC, V. (1981). "Chemical kindling: A study of synaptic pharmacology." In: Kindling 2 (J.A. Wada, ed.). Raven Press, New York. pp. 315–329.

WEHR, T.A. and GOODWIN, F.K. (1979). "Rapid cycling in manic-depressives induced by tricyclic antidepressants." Arch. Gen. Psychiatry 36:555–559.

WEHR, T.A. and GOODWIN, F.K. (1987). "Do antidepressants cause mania?" Psychopharmacol. Bull. 23:61–65.

WEISS, S.R., POST, R.M., GOLD, P.W., CHROUSOS, G., SULLIVAN, T.L., WALKER, D., and PERT, A. (1986a) "CRF-induced seizures and behavior: Interaction with amygdala kindling." Brain Res. 372:345–351.

WEISS, S.R.B., POST, R.M., PERT, A., WOODWARD, R., and MURMAN, D. (1989a). "Context-dependent cocaine sensitization: Differential effect of haloperidol on development versus expression." Pharmacol. Biochem. Behav. 34:655–661.

WEISS, S.R.B., POST, R.M., SZELE, F., WOODWARD, R., and NIERENBERG, J. (1989b). "Chronic carbamazepine inhibits the development of local anesthetic seizures kindled by cocaine and lidocaine." Brain Res. 497:72–79.

WEISS, S.R.B., POST, R.P., MARANGOS, P.J., and PATEL, J. (1986b). "'Peripheral-type' benzodiazepines: behavioral effects and relationship to the anti-convulsant carbamazepine." In: Kindling 3 (J.A. Wada, ed.). Raven Press, New York. pp. 375–392.

YOUNG, S.T., PORRINO, L.J., and IADAROLA, M.J. (1991). "Cocaine induces striatal c-fos immunoreactive proteins via dopaminergic D1 receptors." Proc. Natl. Acad. Sci. 88:1291–1295.

YUWILER, A. (1976). "Stress, anxiety and endocrine function." In: Biological Foundations of Psychiatry (R.G. Grenell and S. Gaby, ed.). Raven Press, New York. pp. 889–944.

NEUROBIOLOGICAL SENSITIZATION MODELS OF POST–TRAUMATIC STRESS DISORDER: THEIR POSSIBLE RELEVANCE TO MULTIPLE CHEMICAL SENSITIVITY SYNDROME

MATTHEW J. FRIEDMAN

National Center for PTSD, VAM, and ROC
White River Junction, Vermont
and
Departments of Psychiatry and Pharmacology
Dartmouth Medical School
Hanover, New Hampshire

Neurological sensitization has been proposed as a model for post–traumatic stress disorder (PTSD) (Lipper et al., 1986; van der Kolk, 1987; Friedman, 1988; Post et al., 1988, 1994; Charney et al., 1993). Laboratory paradigms in which repeated exposure to a discrete stimulus is associated with progressive intensification of a neurophysiologic, behavioral, or pharmacologic response has many parallels with the sequence of events that precipitates PTSD.

Investigators with other clinical interests have also been attracted to sensitization models. Specifically, Bell and associates (1992) have proposed that olfactory-limbic kindling is a very good model for understanding the etiology of multiple chemical sensitivity (MCS) syndrome. A number of articles in this volume have addressed the goodness-of-fit between this model and MCS.

My major assignment is to review laboratory data and clinical observations pertinent to sensitization models of PTSD. I will show that although there are intriguing parallels between the two phenomena, one must have great respect for the complexity and polymorphism of both sensitization and PTSD before grasping for simplistic theoretical conclusions. Secondly, I will address the following question; if both PTSD and MCS can be understood as sensitization phenomena, are PTSD patients at greater risk to develop MCS and vice versa? This article is divided into four sections: a) a description of three distinct sensitization phenomena; b) a description of the symptoms of PTSD; c) a review of the applicability of sensitization models to the clinical phenomenology of PTSD; and d) a review of the hypothesis that PTSD patients might be more vulnerable to MCS.

1. Address all correspondence to: Dr. Matthew J. Friedman, National Center for PTSD, VAM & ROC, Hartland Road, White River Junction, VT 05009. Tel.:(802)296-5132. Fax:(802)296-5135.

Toxicology and Industrial Health, Vol. 10, No. 4/5, pp. 449–462

ISSN: 0748-2337

NEUROBIOLOGICAL SENSITIZATION

It is essential to state at the outset that there are at least three sensitization phenomena that have been proposed as suitable models for PTSD: behavioral sensitization, time dependent sensitization, and kindling (Charney et al., 1993; Yehuda and Antelman, 1993; Post et al., 1994). A thorough description of each of these phenomena is far beyond the scope of this article and the reader is referred to excellent reviews of this topic (Antelman, 1988; Charney et al., 1993; Post et al., 1988, 1994). Behavioral sensitization can be produced by repeated exposure to a noxious environmental stimulus or a psychomotor stimulant such as cocaine. With repeated exposure, responses are enhanced progressively so that a stimulus that initially produced little, if any, response can subsequently produce profound effects. The outcome variable is frequently behavioral, such as locomotor activity (see Post et al., 1988) but it can also be pharmacological, such as dopamine or norepinephrine release in discrete brain areas (Anisman and Sklar, 1978; Kalivas and Duffy, 1989). Although behavioral sensitization has been studied most extensively in adrenergic and dopaminergic systems, a large number of brain structures and neurochemical systems have been implicated (Charney et al., 1993). As noted by Post et al. (1988), these are characteristics of behavioral sensitization: 1) shorter latency of response; 2) increased magnitude of response; 3) sensitized responsivity persists for weeks or months; 4) effects are dose related; 5) intermittent stimulus administration produces greater sensitization, compared with continuous administration; 6) genetic factors may influence sensitization; 7) sensitization shows environmental context dependency, i.e., it is conditionable; 8) cross-sensitization occurs with many psychomotor stimulants and dopamine agonists; 9) cross-sensitization occurs between stimulants and a variety of stressors (such as tail-pinch, shock, and starvation).

Time is a crucial factor. Behavioral sensitization does not appear immediately after exposure to the sensitizing stimulus, but only after the passage of a sufficient interval of time. The most extreme example that proves the point is the phenomenon discovered and named by Antelman (1988) as time-dependent sensitization (TDS). In TDS, behavioral sensitization can be detected following exposure to only one sensitizing stimulus provided that sufficient time has elapsed between initial exposure and subsequent testing. For purposes of discussion, I will consider TDS a special subtype of behavioral sensitization in which only one, rather than a sequence, of stimulus presentations is needed to produce a sensitized response. This distinction will be useful when we discuss the applicability of these models to PTSD.

The final sensitization model is kindling. Here the endpoint is seizure activity rather than a behavioral response, and the abnormalities preceding actual seizure induction are neurophysiologic such as increased spike magnitude. First described by Goddard et al. (1969), kindling is produced by repeated electrophysiological stimulation of a given brain area (such as the amygdala) on an intermittent basis. The stimulus is kept constant and produces no observable effects during the initial presentation. With repeated stimulation, animals exhibit a progressively widespread electrophysiological effect and marked behavioral effects until the previously ineffective stimulus precipitates a full-fledged major motor seizure. Post and associates (1988) have produced a similar sequence of events using pharmacological rather than electrophysiological stimuli. A variety of diverse pharmacological agents can kindle seizures, including local anesthetics (lidocaine), stimulants (cocaine), cholinergic agonists (physostigmine), GABA antagonists (bicuculline), benzodiazepine inverse agonists (FG-7142), peripheral-type benzodiazepine ligands (Ro5-4864), endogenous opiates (beta-endorphin, enkephalin), and corticotropin-releasing hormone (CRH). It should be noted parenthetically that many of those substances are thought to have profound effects on the major neurobiologic systems involved in the organism's response to stress (see Friedman et al., 1995). Other important characteristics of kindling (see Post et al., 1988) are: 1) the limbic system kindles more readily than the cortex; 2) no toxic or neuropathological changes are evident: kindling is a transsynaptic process; 3) kindling persists: animals will continue to have seizures after a one-year seizure-free interval; 4) interictal spikes and spontaneous epileptiform potentials develop; 5) seizures may develop spontaneously in chronically kindled animals; and 6) the efficacy of pharmacological interventions differs as a function of the stage of kindling (i.e., developing, completed, spontaneous).

Before ending this section, it is important to note that there is some overlap between these three mechanisms of sensitization. Most notably, cocaine can produce behavioral sensitization, TDS, or kindling, depending on the experimental protocol in which it is administered. Furthermore, there may be overlapping synaptic, transsynaptic, and genomic mechanisms by which each is produced. (See Post et al., 1995, for a full discussion of these issues.) Finally, TDS can be considered a special subtype of behavioral sensitization. There are also important differences. TDS alone can produce sensitization after only one exposure to the stimulus. Only kindling can result in spontaneous effects after stimulus-induced seizures have become well established. These facts all have important implications for the applicability of sensitization models to PTSD.

POST–TRAUMATIC STRESS DISORDER

Post–Traumatic Stress Disorder (PTSD) is a psychiatric syndrome that develops when people are exposed to catastrophic stressors such as war, torture, rape, genocide, nuclear attack, natural or industrial disasters, and airplane or motor vehicle accidents. Such events are considered qualitatively and quantitatively different from the painful stressors that constitute the normal vicissitudes of life such as divorce, failure, rejection, serious illness, financial reverses and the like. Diagnostic criteria for PTSD were first proposed in 1980 in the American Psychiatric Association's Diagnostic and Statistical Manual of Mental Disorders, Third Edition (DSM-III). They have been revised twice, most rcently in the DSM-IV (American Psychiatric Association, 1994). As shown in Table 1, DSM-IV criteria for PTSD consist of exposure to a traumatic event, intrusive recollections of the event, trauma-related avoidant/numbing symptoms, hyperarousal symptoms, a minimum duration of disturbance and significant functional impairment.

The prominence of traumatic stress as an etiological factor distinguishes PTSD from most other psychiatric disorders. In their review of animal literature pertinent to PTSD, Foa and colleagues (1992) have proposed that, in contrast to a normal stressor, a traumatic stressor is one which is both uncontrollable and unpredictable. Indeed, the PTSD syndrome is based on an explicit stimulus-response causal relationship between traumatic exposure and subsequent psychopathology. This is best illustrated by the "B," or intrusive recollection, cluster of symptoms. For individuals with PTSD, the traumatic experience is an overwhelming event that retains its power to evoke panic, terror, dread, grief, or despair, as manifested in daytime fantasies, traumatic nightmares, and psychotic reenactments known as PTSD flashbacks. Traumamimetic stimuli that trigger recollections of the original event have the power to evoke mental images, emotional responses, and psychological reactions associated with the trauma. Researchers taking advantage of this phenomenon can reproduce PTSD symptoms in the laboratory by exposing affected individuals to auditory or visual traumamimetic stimuli (Keane et al., 1987). Indeed, as shown in Table 1, intrusive recollections may occur in response to such stimulation (symptoms B_4 and B_5) or they may occur spontaneously (symptoms B_1-B_3). PTSD patients often feel that their intrusive recollections have a life of their own and that they appear with or without exposure to traumamimetic stimuli. In this regard they resemble the relentless bombardment of obsessional thoughts experienced by individuals with obsessive-compulsive disorder.

The "C," or avoidant/numbing, criterion includes symptoms that reflect behavioral, cognitive, or emotional strategies by which PTSD patients attempt to avoid exposing themselves to disturbing traumamimetic stimuli, or if exposed, strategies by which they attempt to minimize the impact of such exposure. PTSD patients often feel powerless to master trauma-related thoughts, feelings, and actions that they experience as unwanted and intolerable.

The "D," or hyperarousal, criterion includes symptoms seen in other anxiety disorders such as insomnia and irritability, as well as symptoms such as hypervigilence and startle which are more specific to PTSD. The hypervigilence in PTSD may sometimes be so extreme that it appears to be paranoid behavior. The startle response has a unique neurobiological substrate and may be one of the most pathognomonic PTSD symptoms. Animal and human studies with fear-potentiated startle paradigms have suggested that this may be a very useful animal model for PTSD (Davis, 1990). Other criteria concerning duration of symptoms and functional impairment are as specified in Table 1.

APPLICABILITY OF SENSITIZATION MODELS TO PTSD

Post and Kopanda (1976; Post, 1977) were the first to propose that a kindling mechanism might produce a psychiatric disorder. Extrapolating from research with cocaine, they suggested that neuroanatomic structures in the limbic system might become increasingly sensitized following repeated stimulation with cocaine-like drugs, thereby producing

Table 1. DSM-IV Criteria for PTSD

A. The person has been exposed to a traumatic event in which both of the following have been present:
   1. the person has experienced, witnessed, or been confronted with an event or events that involve actual or threatened death or serious injury, or a threat to the physical integrity of oneself or others.
   2. the person's response involved intense fear, helplessness, or horror. Note: in children, it may be expressed instead by disorganized or agitated behavior

B. The traumatic event is persistently reexperienced in at least one of the following ways:
   1. recurrent and intrusive distressing recollections of the event, including images, thoughts, or perceptions. Note: in young children, repetitive play may occur in which themes or aspects of the trauma are expressed
   2. recurrent distressing dreams of the event. Note: in children, there may be frightening dreams without recognizable content
   3. acting or feeling as if the traumatic event were recurring (includes a sense of reliving the experience, illusions, hallucinations, and dissociative flashback episodes, including those that occur upon awakening or when in toxicated). Note: in young children, trauma-specific reenactment may occur
   4. intense psychological distress at exposure to internal or external cues that symbolize or resemble an aspect of the traumatic event
   5. physiologic reactivity upon exposure to internal or external cues that symbolize or resemble an aspect of the traumatic event

C. Persistent avoidance of stimuli associated with the trauma and numbing of general responsiveness (not present before the trauma), as indicated by at least three of the following:
   1. efforts to avoid thoughts, feelings, or conversations associated with the trauma
   2. efforts to avoid activities, places, or people that arouse recollections of the trauma
   3. inability to recall an important aspect of the trauma
   4. markedly diminshed interest or participation in significant activities
   5. feeling of detachment or estrangement from others
   6. restricted range of affect (e.g., unable to have loving feeelings)
   7. sense of a foreshortened future (e.g., does not expect to have a career, marriage, children, or a normal life span)

D. Persistent symptoms of increasing arousal (not present before the trauma), indicated by at least two of the following:
   1. difficulty falling or staying asleep
   2. irritability or outbursts of anger
   3. difficuly concentrating
   4. hypervigilance
   5. exaggerated startle response

E. Duration of the disturbance (symptoms in B, C, and D) is more than one month.

F. The disturbance causes clinically significant distress or impairment in social, occupational, or other important areas of functioning.

   Specify if: **Acute:** if duration of symptoms is less than three months
   Chronic: if duration of symptoms is three months or more

   Specify if: **With delayed onset:** onset of symptoms at least six months after the stressor

behavioral abnormalities resulting in psychopathology. These observations were extended to consider a similar paradigm in which limbic stimulation by endogenous dopamine (rather than by exogenous cocaine) might, under certain conditions, produce a psychotic state that could be reactivated subsequently by an excessive burst of dopaminergic stimulation. This model was applied to recurrent psychotic disorders and generated great excitement when it was shown that carbamazepine, an anti-convulsant with an anti-kindling action, could produce symptom reduction in bipolar manic patients who had previously failed to respond to lithium treatment.

Several investigators (Lipper et al., 1986; van der Kolk, 1987; Friedman, 1988; Yehuda and Antelman, 1993, Post et al., 1995) have suggested that a similar mechanism might occur in PTSD and have therefore proposed a sensitization model for this disorder. Specifically, they hypothesize that chronic central catecholaminergic arousal in PTSD, mediated by input to the locus ceruleus, sensitizes limbic nuclei, thereby producing a stable neurobiological abnormality. Table 2 lists fourteen characteristics of PTSD and shows how consistent they are with each of the three sensitization models discussed previously.

TABLE 2. Similarities between Characteristics of PTSD and Those of Behavioral Sensitization, Time-Dependent Sensitization, and Kindling

Characteristis of PTSD	Behavioral Sensitization	Time-Dependent Sensitization	Kindling
1. May develop after a single exposure to a traumatic or sensitizing stimulus		X	
2. Develops after repeated exposure to a sequence of traumatic/sensitizing stimuli	X		X
3. Increased amplitude and duration of response	X	X	X
4. Increased responsivity is persistent	X	X	X
5. Effects are dose-related	X	X	X
6. Effects are behavioral	X	X	
7. Prior traumatization/sensitization enhances subsequent response	X	X	X
8. Responses initially elicited by stimulation may become spontaneous over time			X
9. Context independent			X
10. Context dependent and conditionable	X	X	
11. Cross-sensitization with stress	X	X	
12. Cross-sensitization with psychomotor stimulants	X	X	X
13. Mediation by limbic system	X	X	X
14. Efficacy of anti-kindling agents			X

First, PTSD may develop after a single exposure to an overwhelming event that involves actual or threatened death or serious injury, or a threat to the physical integrity of oneself or others. Only TDS provides a neurobiologic model for this clinical scenario. What most clinicians believe to be a more common sequence of events is the second characteristic, the development of PTSD following repeated exposure to traumatic stimuli. Examples of this include repeated physical or sexual assault in the context of child abuse, domestic violence, or political incarceration, repeated exposure to war trauma as a combatant or survivor, and repeated exposure to privation, violence, or atrocities as a political refugee during a genocidal war. Behavioral sensitization and kindling are obviously better suited to model this factor. Admittedly, one cannot draw this parallel too tightly, since it is much more difficult to define a single traumatic event than it is to define the parameters of a single stimulus used in a behavioral sensitization or kindling experiment. However, one might argue that the psychological representation of such a traumatic event (i.e., an intrusive recollection such as a traumamimetic nightmare) might itself serve as a sensitizing stimulus that continues to promote the sensitization process.

Third, patients with PTSD appear to exhibit increased amplitude and duration of response following subsequent exposure to traumatic stimuli. Stimuli with which they could previously cope evoke a more intense and sometimes intolerable response subsequently. An example is Solomon and colleagues' (1987) observation that some Israeli veterans who had apparently coped with the trauma of the 1967 Yom Kippur War became distraught and incapacitated by PTSD symptoms when redeployed for combat duty during the 1982 war in Lebanon.

Fourth, PTSD symptoms are persistent, and when untreated the syndrome may last for decades or a lifetime (Archibald and Tuddenham, 1965; Schnurr, 1991). Fifth, there appears to be a dose-response relationship between the severity (intensity and frequency) of traumatic exposure and the risk of developing PTSD (Pynoos et al., 1987; McFarlane, 1988; Kulka et al., 1990). Both persistence and dose-response characteristics are included in all three sensitization models.

Sixth, PTSD is a psychiatric disorder with behavioral symptoms and it does not include major motor seizures as one of its diagnostic criteria. Strictly speaking, we must certainly rule out kindling as a suitable model for PTSD from this perspective. Post and associates (1988, 1995) have argued, however, that although we should only use the term kindling to describe processes that result in seizures, it is important to recognize that there are kindling-like processes that stop short of full-blown classical kindling. They involve potentially relevant electrophysiological and behavioral abnormalities.

Seventh, prior traumatization and sensitization enhances sensitivity to subsequent exposure. By definition, this characteristic applies to all three sensitization models. A number of studies suggest that adults who have been exposed to sexual or physical abuse during childhood are more likely to develop PTSD following exposure to trauma than traumatized adults who were not abused during childhood (Davidson, 1993).

Eighth, responses initially elicited by stimulation may become spontaneous over time. PTSD patients may have spontaneous recollections of the traumatic event that are not triggered by environmental stimuli. Ninth, intrusive recollections, traumatic nightmares, and PTSD flashbacks sometimes appear to have a life of their own that is completely independent of the situational context in which the patient finds him or herself. My clinical impression is that context-independent spontaneous symptoms are most often found in patients with the most severe cases of PTSD. Post et al. (1988, 1995) have argued that kindling may be the best model for patients who exhibit such a clinical pattern since neither behavioral sensitization nor TDS can account for context independent or spontaneous symptoms.

Tenth, PTSD is often context-specific and conditionable. Indeed, classic behavioral models of fear conditioning and failure of extinction have been proposed as models for PTSD (see Charney et al., 1993). Keane et al. (1985) have reinvoked Mowrer's two-factor theory of classical fear conditioning and operant escape conditioning as a suitable learning model for PTSD. Likewise, Foa et al. (1992) have reviewed inescapable vs. escapable stress paradigms as they relate to PTSD. Finally, cognitive-behavioral treatment of rape victims with PTSD that is predicated on such conditioning models has proven effective (see review by Foa et al., 1995). Behavioral sensitization and TDS but not kindling are pertinent in this regard.

Eleventh, psychological factors such as stress and context can exacerbate both behavioral sensitization, TDS, and PTSD. Several studies have shown cross-sensitization between different classes of sensitizing and laboratory-induced stress. This is consistent with the well known clinical observation that PTSD patients often become more symptomatic when forced to deal with ordinary (nontraumamimetic) environmental stressors. Relatively mild family, social, vocational, or other challenges can exacerbate PTSD symptoms in patients who had been well stabilized before being forced to confront such situations.

Twelfth, psychomotor stimulants acting on catecholaminergic systems can elicit PTSD symptoms as well as behavioral sensitization and TDS effects (Charney et al., 1993; Post et al., 1988, 1995). A large literature has emerged regarding catecholaminergic — both adrenergic and dopaminergic — mechanisms in PTSD (Ende et al., 1990; Friedman, 1991; Jensen et al., 1991; Charney et al., 1993; Murburg, 1994). Perhaps the most dramatic and convincing example is that administration of yohimbine, a centrally acting alpha-2 adrenergic antagonist that disinhibits central adrenergic mechanisms, can elicit panic attacks and trauma-related flashbacks in Vietnam veterans with PTSD, but not in control subjects (Southwick et al., 1993). It is important to remember in this regard that psychostimulants such as cocaine can induce kindling as well as behavioral sensitization and TDS.

Thirteenth, limbic structures appear to mediate the development and persistence of PTSD symptoms as well as the effects of behavioral sensitization, TDS, or kindling. As shown in Table 1, the DSM-IV definition of a traumatic event includes the subjective emotional response of the exposed individual. Therefore, exposure to a catastrophic stressor is only traumatic if the subjective response involved "intense fear, helplessness, or horror."

Finally, anticonvulsant drugs that reverse the effects of kindling have reportedly produced symptom reduction in PTSD patients. There are three published open trials in which antikindling agents produced reduction of symptoms in Vietnam veterans with PTSD. In one study, carbamazepine reduced the intensity and frequency of traumatic nightmares, flashbacks, and intrusive recollections (Lipper et al., 1986). In a second report, carbamazepine reduced impulsivity, irritability, and violent behavior in PTSD patients (Wolfe et al., 1988). Finally, valproate treatment resulted in significant improvement in both hyperarousal and avoidant/numbing symptoms among Vietnam veterans with PTSD (Fesler, 1991).

Taken together, these fourteen observations suggest that sensitization models may have heuristic value in furthering our understanding of PTSD. They are hardly conclusive, but they do suggest a number of laboratory experiments that might strengthen our confidence in this conceptual approach. For example, what will happen when a PTSD laboratory paradigm such as inescapable stress is augmented by a behavioral sensitization or kindling paradigm? Will animals exposed to both conditions be more affected than those exposed to either paradigm alone? Will there be cross-sensitization? Will the end result be context-dependent or context-interdependent? Can the process be accelerated by certain drugs (such as psychostimulants and kindling agents) and prevented by others (such as antiadrenergic drugs or antikindling agents?) Will the augmenting effects of behavioral sensitization, TDS, and kindling be comparable in this regard, or is one process more likely to augment the effects of inescapable stress than the others? Clearly, data from experiments of this nature would give us a better basis for evaluating the attractiveness of sensitization models of PTSD.

SOME QUESTIONS ABOUT THE MODEL

Although there are a number of ways in which a sensitization model is consistent with some of the known phenomenology and pathophysiology of PTSD, the goodness-of-fit between model and syndrome is not consistent in all respects.

First, there are many varied experiential roads to PTSD. In some cases the trauma can be a single terrifying event such as a brutal rape or physical assault, while in other cases traumatic exposure may consist of chronic violence measured in months or years, such as incarceration in a Nazi death camp or captivity and torture as a political prisoner. In other words, the magnitude and frequency of traumatic events that lead to PTSD may vary greatly from one patient to another. Furthermore, in addition to such quantitative differences, there may also be critical qualitative differences in the neurobiological processing of different traumatic experiences (e.g., between rape and motor vehicle accidents).

Second, any model of PTSD must address the fact that PTSD is not a unidirectional process in which patients become progressively worse over time. Indeed some patients improve and even recover subsequently. It is possible that number, frequency, and intensity of prior traumatic exposures may hold the key to predicting PTSD recovery or chronicity, but that hypothesis needs to be confirmed empirically.

Third, any model of PTSD must address the fact that traumatic exposure is only a necessary but not a sufficient condition for the later development of PTSD. Clearly, there are individual differences in vulnerability to developing PTSD. This criticism is probably the easiest to handle since, extrapolating from Post et al. (1995), the sensitization model predicts that traumatic exposure initially induces immediate early gene expression which in turn triggers a cascade of subsequent intermediate and late gene expression involved in coding long-lasting changes in synaptic excitability and microstructure. Individual differences in this regard might explain relative vulnerability or resistance to developing PTSD.

ARE PTSD PATIENTS AT GREATER RISK FOR MCS?

Proponents of the MCS syndrome propose that this is a chronic polysymptomatic condition that affects individuals who become sensitized to low levels of common indoor and outdoor environmental chemicals such as pesticides and solvents (Bell et al., 1992). MCS is a rare, poorly understood, and controversial disorder that is often associated with psychiatric conditions. Since most people exposed to low-level chemical stimulation do not develop MCS symptoms, it appears that this disorder results only when susceptible individuals encounter such low-level stimulation. Therefore, it is also important to identify those factors that might make certain individuals more vulnerable than others to develop MCS.

There are two reasons why PTSD might be a risk factor for MCS. First of all, PTSD patients may be more susceptible to a wide variety of medical problems than non-PTSD patients. Secondly, a sensitization model predicts that PTSD patients will be especially vulnerable to MCS.

In a recent review of the trauma and health literature, Friedman and Schnurr (1995) concluded that PTSD patients are at greater risk for adverse health outcomes. This was a consistent finding, whether such health outcomes were indicated by self-reports of somatic symptoms, by medical problems confirmed by a physician's examination, by increased utilization of inpatient or outpatient medical resources, or by mortality data. Friedman and Schnurr suggested that adverse health outcomes among PTSD patients might be explained by pathophysiological and behavioral abnormalities associated with PTSD. The former include dysregulated physiological, neurotransmitter, endocrine, and immunologic systems that are essential to maintain health. The latter include smoking, drinking, immoderate dietary habits, sexual promiscuity, and other behaviors that increase the risk of adverse health outcomes. From this perspective, MCS is one of many medical problems for which PTSD patients are at greater risk than others.

A sensitization model of PTSD proposes a mechanism by which PTSD patients might be specifically vulnerable to MCS. Bell and associates (1992) have suggested that MCS itself might be best understood as a sensitized endpoint following exposure to low-level environmental chemicals such as pesticides and solvents. In their comprehensive review of this literature, they cite numerous experiments in which repeated exposure to previously undetected olfactory stimuli produced a progressive sensitivity and intolerance of such stimulation. Furthermore, Bell et al. propose that the time pattern of MCS initiation resembles that of kindling or partial kindling. As in other sensitization phenomena, individuals with MCS appear to exhibit cross-sensitization to other stimuli that cause limbic kindling. In contrast to MCS patients whose sensitization is presumably caused by environmental chemicals, it is hypothesized that PTSD patients are sensitized by exposure to traumatic stimuli. Therefore, it is proposed that two very different classes of stimuli, environmental chemicals on the one hand, and psychological trauma-related stimuli on the other, both access the same final common pathway through which they cause limbic sensitization. It is further proposed that under these conditions, there would be cross-sensitization between chemical and traumamimetic stimulation. It follows from the above argument, if sensitization is indeed an important mechanism for both MCS and PTSD, that PTSD patients should be especially susceptible to develop this disorder.

HOW CAN WE TEST THIS HYPOTHESIS?

First we need more conclusive evidence that sensitization models are applicable to both PTSD and MCS. Secondly, we need to know whether behavioral sensitization, TDS, or kindling is the best model in this regard. On the other hand, this might not be the right question to ask given the complexity and polymorphism of both sensitization and clinical phenomena. A better question to ask might be whether certain expressions of PTSD and MCS are better understood as TDS (in the case of a single exposure) as kindling (in the case of symptoms that are spontaneous rather than stimulus-induced) or as behavioral sensitization. In any case, much more basic research is needed.

Assuming satisfactory data supporting sensitization models of both PTSD and MCS, it would be useful to test for cross-sensitization. For example, what will happen following concurrent exposure to a PTSD laboratory paradigm (such as inescapable stress) and a chemical stimulus? Is the sensitization produced by simultaneous exposure to both stimuli greater than the sensitization produced by exposure to either the stressful or chemical stimulus alone? Are these changes context dependent or independent? Can these effects be modified by drugs affecting catecholaminergic systems or kindling mechanisms?

Some less conclusive but suggestive clinical studies might also be undertaken that would be consistent with the hypothesis of sensitization in both PTSD and MCS. If, as I have proposed, there is cross-sensitization in both PTSD and MCS, so that both chemical and traumamimetic stimuli access the same final pathway, it follows that PTSD patients should be at greater risk for MCS and vice versa. A cohort in which individuals were at risk to develop PTSD, MCS, or both disorders would provide an excellent opportunity to test this prediction. Such a cohort has been described by Wolfe et al., (1993, 1994). It consists of American participants in the

Persian Gulf War who have already been assessed and followed longitudinally for PTSD symptoms. A sensitization model of both disorders would predict that, given equal exposure to environmental chemicals, Persian Gulf returnees with PTSD are more likely to develop MCS than those without PTSD.

Another suggestive clinical study might involve neuropsychological testing of PTSD and MCS patients. A number of neuropsychological tests for attention, concentration, and memory have been utilized in the assessment of MCS patients (Bell, et al., 1992; Fiedler et al., 1992; Simon et al., 1993). As proposed by Wolfe (personal communication), it would be of great interest to compare a neuropsychological performance of MCS patients with and without PTSD, PTSD patients without MCS, psychiatric controls, and normal subjects to see whether there are any similarities in cognitive patterns or deficits between these different groups. Similarities between MCS and PTSD patients would also suggest the possibility of a final common pathway for sensitization in both disorders.

Finally, assuming the applicability of a kindling (rather than a behavioral sensitization or TDS) model to both MCS and PTSD, MCS patients would be expected to exhibit symptomatic improvement when treated with antikindling agents such as carbamazepine or valproate. Furthermore, it would be predicted that these drugs should ameliorate symptoms of both PTSD and MCS in patients who are simultaneously afflicted with both disorders. It should be noted, however, that MCS patients often cannot tolerate therapeutic doses of many medications. Therefore, it remains to be seen whether this experiment is feasible.

SUMMARY

The implications of three sensitization models of PTSD have been reviewed. It appears that such a conceptual approach may have great heuristic value in furthering our understanding of PTSD. We must be cautious. The goodness-of-fit between PTSD and sensitization varies with the clinical phenomenology of the patient in question. In general, behavioral sensitization seems to be the best overall model. For PTSD following a single exposure to trauma, time-dependent sensitization seems best and for PTSD with spontaneously occurring intrusive recollections, kindling may be the best.

If MCS can also be understood as a disorder resulting from neurobiological sensitization, it is proposed that there is cross-sensitization between psychological (traumamimetic) stimuli that elicit PTSD symptoms and environmental chemical stimuli that elicit MCS. Assuming that this is indeed the case, it is further proposed that PTSD patients are at greater risk to develop MCS and vice versa.

Several experiments are proposed to test these speculations.

REFERENCES

AMERICAN PSYCHIATRIC ASSOCIATION (1980). Committee on Nomenclature and Statistics. Diagnostic and Statistical Manual of Mental Disorders, Third Edition. American Psychiatric Association, Washington, D.C.

AMERICAN PSYCHIATRIC ASSOCIATION (1994). Diagnostic and Statistical Manual of Mental Disorders, Fourth Edition. American Psychiatric Association, Washington, DC.

ANISMAN, H. and SKLAR, L.S. (1978). "Catecholamine depletion upon reexposure to stress: Mediation of the escape deficits produced by inescapable shock." J. Comp. Physiol. Psychol. 93:610–625.

ANTELMAN, S.M. (1988). "Time dependent sensitization as the cornerstone for a new approach to pharmacotherapy: Drugs as foreign or stressful stimuli. Drug Dev. Res. 14:1–30.

ARCHIBALD, H.C. and TUDDENHAM, R.D. (1965). "Persistent stress reaction after combat: A 20-year follow-up." Arch. General Psychiatry 12:475–781.

BELL, I.R., MILLER, C.S., and SCHWARTZ, G.E. (1992). "An olfactory-limbic model of multiple chemical sensitivity syndrome: Possible relationships to kindling and affective spectrum disorders." Biol. Psychiatry 32:218–242.

CHARNEY, D.S., DEUTCH, A.Y., KRYSTAL, J.H., SOUTHWICK, S.M., and DAVIS, M. (1993). "Psychobiologic mechanisms of post–traumatic stress disorder." Arch. General Psychiatry 50:294–305.

DAVIDSON, J. (1993). "Issues in the diagnosis of post–traumatic stress disorder." In: Review of Psychiatry (J.M. Oldham, M.B. Riba, and A. Tasman, ed.). Volume 12. American Psychiatric Press, Inc., Washington, D.C.

DAVIS, M. (1990). "Animal models of anxiety based on classical conditioning: The conditioned emotional response (CER) and the fear potentiated startle effect." Pharmacol. Therapeut. 47:147–165.

ENDE, N., GERTNER, S.B., and SOCHA, B. (1990). "Unexpected changes in urinary catecholamines and vanillylmandelic acid following rape assault." Horm. Behav. 24(1):62–70.

FESLER, F.A. (1991). "Valproate in combat-related post–traumatic stress disorder." J. Clin. Psychiatry 52(9):361–364.

FIEDLER, N., MACCIA, C., and KIPEN, H. (1992). "Evaluation of chemically sensitive patients." J. Occup. Med. 34:529–538.

FOA, E.B., ZENBERG, R., and ROTHBAUM, B.O. (1992). "Uncontrollability and unpredictability of post–traumatic stress disorder: An animal model." Psychol. Bull. 112:218–238.

FOA, E.B., ROTHBAUM, B.O., and MOLNAR, C. (1995). "Cognitive-behavioral therapy of PTSD." In: Neurobiological and Clinical Consequences of Stress: From Normal Adaptation to PTSD (M.J. Friedman, D.S. Charney, and A.Y. Deutch, ed.). Raven Press, New York. In press.

FRIEDMAN, M.J. (1988). "Towards rational pharmacotherapy for post–traumatic stress disorder." Am. J. Psychiatry 145:281–285.

FRIEDMAN, M.J. (1991). "Biological approaches to the diagnosis and treatment of post–traumatic stress disorder." J. Traumatic Stress 4:67–91.

FRIEDMAN, M.J., CHARNEY, D.S., and DEUTCH, A.Y. (1995). Neurobiological and Clinical Consequences of Stress: From Normal Adaptation to PTSD. Raven Press, New York. In press.

FRIEDMAN, M.J. and SCHNURR, P.P. (1995). "The relationship between trauma, PTSD, and physical health." In: Neurobiological and Clinical Consequences of Stress: From Normal Adaption to PTSD (M.J. Friedman, D.S. Charney, and A.Y. Deutch, ed.). Raven Press, New York. In press.

GODDARD, G.V., McINTYRE, D.C., and LEECH, C.K. (1969). "A permenant change in brain function resulting from daily electrical stimulation." Exp. Neurol. 25:295–330.

JENSEN, J.B., PEASE, J.J., TEN BENSEL, R., and GARFINKEL, B.D. (1991). "Growth hormone response patterns in sexually or physically abused boys." J. Am. Acad. Child Adol. Psychiatry 30(5):784–790.

KALIVAS, D.W. and DUFFY, P. (1989). "Similar effects of daily cocaine and stress on mesocorticolimbic dopamine transmission in the rat." Biol. Psychiatry 25:913–928.

KEANE, T.M., WOLFE, J., and TAYLOR, K.I. (1987). "Post–traumatic stress disorder. Evidence for diagnostic validity and methods of psychological assessment." J. Clin. Psychol. 43:32–43.

KEANE, T.M., ZIMERING, R.T., and CADDELL, J.M. (1985). "A behavioral formulation of post–traumatic stress disorder in Vietnam veterans." Behav. Ther. 8:9–12.

KULKA, R.A., SCHLENGER, W.E., FAIRBANK, J.A., HOUGH, R.P., JORDAN, B.K., MARMAR, C.R., and WEISS, D.S. (1990). Trauma and the Vietnam War Generation. Brunner/Mazel, New York.

LIPPER, S., DAVIDSON, J.R.T., GRADY, T.A., EDINGAR, J.D., HAMMETT, E.B., MAHORNEY, S.L., and CAVENAR, J.O., (1986). "Preliminary study of carbamazepine in post–traumatic stress disorder." Psychosomatics 27:849–854.

McFARLANE, A.C. (1988). "The longitudinal cause of post–traumatic morbidity: The range of outcomes and their predictors." J. Nerv. Mental Dis. 176:30–39.

MURBURG, M.M., ed. (1994). Catecholamine Function in Post–Traumatic Stress Disorder. American Psychiatric Press, Washington, D.C.

POST, R.M. (1977). "Clinical implications of a cocaine-kindling model of psychosis." In: Clinical Neuropharmacology (H.L. Klawans, ed.). Vol. 2. Raven Press, New York. pp. 25–42.

POST, R.M., and KOPANDA, R.T. (1976). "Cocaine, kindling and psychosis." Am. J. Psychiatry 133:627–634.

POST, R.M., WEISS, S.R.B., and PERT, A. (1988). "Implications of behavioral sensitization and kindling for stress-induced behavioral change." In: Mechanisms of Physical and Emotional Stress (G.P. Chrousos, D.L. Loriaux, and P.H. Gold, ed.). Plenum Press, New York. pp. 441–464.

POST, R.M., WEISS, S.R.B., and SMITH, M.A. (1995). "Sensitization and kindling: Implications for the evolving neural substrate of PTSD." In: Neurobiological and Clinical Consequences of Stress: From Normal Adaptation to PTSD (M.J. Friedman, D.S. Charney, and A.Y. Deutch, ed.). Raven Press, New York. In press.

PYNOOS, R., FREDERICK, C., NADER, K., ARROYO, W., STEINBERG, A., ETH, S., NUNEZ, F., and FAIRBANKS, L. (1987). "Life threat and post–traumatic stress in school-age children." Arch. General Psychiatry 44:1057–1063.

SCHNURR, P.P. (1991). "PTSD and combat-related psychiatric symptoms in older veterans." PTSD Res. Q. 2:1–6.

SIMON, G.E., DANIELL, W., STOCKBRIDGE, H., CLAYPOULE, K., and ROSENSTOCK, L. (1993). "Immunologic, psychological and neuropsychological factors in multiple chemical sensitivity." Ann. Internal Med. 19:97–103.

SOLOMON, Z., GARB, R., BLEICH, A., and GRUPPER, D. (1987). "Reactivation of post–traumatic stress disorder." Am. J. Psychiatry 144:51–55.

SOUTHWICK, S.M., KRYSTAL, J.H., MORGAN, C.A., JOHNSON, D., NAGY, L.M., NICOLAOU, A., HENINGER, G.R., and CHARNEY, D.S. (1993). "Abnormal noradrenergic function in post–traumatic stress disorder." Arch. Gen. Psychiatry 50:266–274.

VAN DER KOLK, B.A. (1987). "The drug treatment of post–traumatic stress disorder." J. Affective Disorders 13:203–213.

WOLF, M.E., ALAVI, A., and MOSNAIM, A.D. (1988). "Post–traumatic stress disorder in Vietnam veterans clinical and EEG findings: possible therapeutic effects of carbamazepine." Biol. Psychiatry 23:642–644.

WOLFE, J. Personal communication, conversation, March 18, 1994.

WOLFE, J., BROWN, P., and KELLEY, J. (1993a). "Reassessing war stress: Exposure and the Gulf War." J. Social Issues 49:15–31.

WOLFE, J., KEANE, T.M., and YOUNG, B. (1994). "From soldier to civilian: Acute adjustment patterns of returned Persian Gulf veterans." In: Those Left Behind and Those Who Returned: Psychological Response to War in Families, Children and Soldiers (R.J. Ursano and A.E. Norwood, ed.). American Psychiatric Press, Washington, D.C. In press.

YEHUDA, R. and ANTELMAN, S.M. (1993). "Criteria for rationally evaluating animal models of post–traumatic stress disorder." Biol. Psychiatry 33:479–486.

DRUG SENSITIZATION, SUBSTANCE ABUSE, AND CHEMICAL SENSITIVITY

DAVID B. NEWLIN

Addiction Research Center
National Institute on Drug Abuse
Baltimore, Maryland

Substance abuse, involving drugs such as cocaine, heroin, alcohol, marijuana, nicotine, barbiturates, etc., is by far by the most prevalent psychiatric disorder. Much has been learned about the abuse of these substances that may be useful to consider in designing and analyzing research concerning multiple chemical sensitivities (MCS). We review the central role of sensitization in this literature, including its definition, measurement, and expression in animals and human volunteers. Common factors among abused drugs, including sensitization, are discussed. Finally, empirical studies are delineated or proposed to test some of the notions presented in this paper.

INTRODUCTION

Research on substance abuse, which studies the mechanisms controlling the use of alcohol and other abused drugs, has shown explosive growth over the last two decades. The result has been that information concerning the properties of abused drugs in animals, normal human volunteers, and substance abusing individuals has become plentiful. An explosion of information of this magnitude requires integration of theory and empirical data to organize the field and to conceptualize the ways in which new information can be used to better understand and address old problems. In this case, the enduring problems of addiction to drugs by many individuals in our society and the effects of exposure to toxic substances represent grist for the conceptual mill.

A second purpose for review and conceptualization of data in this field is that the information obtained in relation to substance abuse may be useful in addressing important issues in other fields of inquiry. The purpose of this chapter is to focus on the problem of sensitization to

1. Address all correspondence to: David B. Newlin, Ph.D., Addiction Research Center, National Institute on Drug Abuse, 4940 Eastern Ave., P.O. Box 5180, Baltimore, MD 21224. Tel.:(410)550-1411. Fax:(410)550-1438.

2. Key Words: alcohol, cocaine, common factors, mesolimbic dopamine, multiple chemical sensitivity, nicotine, sensitization, substance abuse, tobacco smoke.

Toxicology and Industrial Health, Vol. 10, No. 4/5, pp. 463–480

ISSN: 0748-2337

abused drugs in order to develop research hypotheses and experimental designs that may be useful in the study of exposure to toxic chemicals.

The chapter begins with a rather extensive examination of definitional issues concerning drug sensitization phenomena. "Sensitization" is a word that has become ubiquitous in the substance abuse literature, and is often used with vague or overlapping meanings. It is hoped that sound definitions will provide a solid foundation for subsequent conceptual work concerning both substance abuse and exposure to toxic chemicals.

The discussion then moves to issues concerning the functional role of sensitization both for the individual organism that is exposed to drugs or toxic chemicals and for the evolutionary function that the particular response to these elements may serve. The chapter emphasizes risk for alcoholism as a particularly rich area of inquiry in which to apply these principles. Finally, it turns to chemical sensitivity as an area that has not been as thoroughly studied as substance abuse, and which may benefit from information contained in the substance abuse literature.

This chapter focuses upon sensitization as a particularly important phenomenon in the substance abuse literature. However, this emphasis is not meant to imply that sensitization is the only relevant mechanism of addiction or even the most important. This becomes particularly relevant in relation to exposure to toxic substances. I certainly do not assume that sensitization is a unique phenomenon that can account for all substance abuse or problems from chemical exposure.

Before turning to definitional issues, it may be useful to examine the incidence and prevalence of substance abuse problems. The Epidemiologic Catchmen Area study in the early 1980s (Robins et al., 1984) demonstrated that approximately one in four American men met formal diagnostic criteria for alcohol abuse or alcohol dependence at some time in their lives. The comparable rates in women were significantly lower, or about 1/3 those in men. More recent data from the National Comorbidity Study (Kessler et al., 1994) demonstrated lifetime prevalence of a substance abuse problem (alcohol or other drug abuse or dependence) on the order of 34% of American men and about half that figure in women. Twelve-month prevalence was approximately half that of lifetime prevalence.

These results may be alarming to those not familiar with the substance abuse field. Abuse and dependence on drugs and alcohol are by far the most prevalent psychiatric disorders; in fact, the mental health professions would be completely overwhelmed if all individuals meeting these diagnostic criteria applied for treatment.

A second type of statistic concerning substance abuse is equally relevant. Relapse rates, or the percentage of individuals who either continue to abuse a substance despite treatment or who return to abuse after a period of abstinence, are extremely high. For most substances, they range from 80% to 90%, even with treatment (Surgeon General's Report, 1988). However, it must be remembered that among those individuals who are able to maintain abstinence for long periods of time (i.e., years), there is often a history of many failed attempts at quitting prior to success.

These results indicate that: 1) substance abuse is a particularly tenacious problem, 2) the goals of treatment are often to prevent relapse rather than simply to stop using the substance in the short term (e.g., 30-day in-patient stay), and 3) successful treatment may come after many failed attempts.

DEFINITION

A broad definition of sensitization is that it is a process inferred when the response to identical repeated stimuli grows with each presentation of the stimulus. Conversely, sensitization is inferred when the response remains unchanged to a stimulus that diminishes in intensity with each presentation.

When applied to drugs, sensitization is inferred from responses that grow in magnitude to repeated administrations of the drug at the same dose, or it may be inferred from responses to a drug that remain the same despite decreasing drug doses (Gilman et al., 1991). In practice, sensitization paradigms usually involve the same dose of a drug given repeatedly rather than decreasing doses because the interpretation of drug sensitization is more straightforward in the former case. Drug sensitization is often defined as the opposite of drug tolerance. However, this does not imply that the mechanisms of sensitization and tolerance are parallel or mirror images of each other.

"Initial sensitization" refers to a larger than expected response to a drug in a naive animal, or one that has never before been exposed to the drug. "Acute sensitization" is inferred when the response to the initial effects of a drug (or those during the rising blood drug curve) is larger than expected on the basis of the response to the drug when the curve is at its peak. Note that these (initial and acute sensitization) are the only types of sensitization that are not defined in relation to repeated drug doses.

"Chronic sensitization" refers to the classic definition (above) in which the response grows with repeated administrations of the same dose of drug. This is the form of sensitization that is most common in the animal literature on abused substances, perhaps because it is the form that is most readily operationalized.

Both acute and chronic sensitization may be "context-dependent" or "context-independent" (Baker and Tiffany, 1985). Contextual dependency refers to experimental paradigms in which it is possible to separate the degree to which sensitization is specific to the environment (i.e., context) in which the drug has been given repeatedly from the degree to which sensitization in nonspecific or expressed in a variety of different contexts. This distinction has proven very important in understanding mechanisms of sensitization. For example, context-dependency has been interpreted as evidence that sensitization is learned or acquired as a result of experience with the drug. Context-interdependent sensitization has been interpreted as a characteristic only of the drug-organism interaction, with little or no environmental modulation (Baker and Tiffany, 1985).

Experimental paradigms for studying sensitization require choosing dependent measures that have the potential to grow with repeated administrations of the drug. For example, one might decide to measure locomotor activation in an activity box as the measure of sensitization to cocaine. The animal may be observed to show greater and greater locomotor activation to each subsequent dose of cocaine. Dependent measures often used in the animal and human literatures on sensitization may be physiological (e.g., dopamine efflux in nucleus accumbens of freely moving, conscious rats measured by microdialysis), behavioral (e.g., stereotypic, repetitive movements to high doses of stimulant drugs), or subjective (e.g., report of "euphoria" on a standardized questionnaire in humans). It is important to recognize that sensitization can occur at different rates with different dependent measures. In fact, it is common for sensitization to occur in some measures at the same time that tolerance is developing in other measures. This phenomenon can lead to the conclusion that sensitization is specific to certain measures in the same sense that it may be specific to certain environments or contexts. In other words, there is no necessary assumption that sensitization is parallel across different measures or that different measures operate together in relation to time-dependent phenomena.

It is also important to recognize that sensitization is *not* a property of the drug itself, or even an invariant property of the organism. Specifically, sensitization refers to a drug-organism *interaction* that occurs over time in a given environment. One must be relatively specific in discussing sensitization, or risk the incorrect assumption that it is some kind of unitary, invariant, and inflexible characteristic of a drug or of an organism (Pavlov, 1927).

For example, we might imagine that the response to repeated exposures to cigarette smoke may be very different among human volunteers who are themselves heavy smokers, compared to those who are virulent antismokers. Sensitization could potentially occur in both groups, although it may be misleading to assume that the same process was involved in both cases, even though the physical stimuli were the same. Another example might be a study of the effects of intravenous morphine in terminal cancer patients vs. heroin abusers. Again, the physical stimulus is the same, but the process may be very different.

A common problem with animal research in this area is that one dependent measure is often recorded by itself and taken as evidence of "organismic" sensitization (Newlin and Shoaib, 1994). For example, the animal researcher may record hyperthermia in response to intraperitonial morphine and come to the conclusion that rats sensitize to morphine. However, if the researcher had recorded analgesia in a tail-flick apparatus in the same experiment, he or she might just as easily have concluded that rats tolerate to morphine (Baker and Tiffany, 1985). There is a real sense in which neither conclusion is correct.

A further definitional problem with sensitization should not be overlooked. Because sensitization refers to a function that grows in magnitude over time, it must have inherent limits. This is analogous to positive feedback loops in control theory that are explosive and are not realizable in nature (Di Stephano, 1967). The function cannot grow endlessly, but must be limited by asymptotic stabilization of the process or by the subsequent development of tolerance. This limitation may be either inherent to the underlying mechanism of sensitization, or may be produced by a second, different process that develops over time.

It is important to recognize that sensitization is a description of results rather than a statement of the mechanism of a process. It is possible that any number of different functional mechanisms could produce sensitization. Description of results does not replace specification of mechanism.

Finally, methodology in psychopharmacology (Goldstein et al., 1968) suggests a robust but simple definition of sensitization. Psychopharmacology emphasizes that most drug effects are dependent upon dose. This simply means that the drug effect is different with doses that are different. The drug effect/dose curve represents the relationship between the magnitude of the dose compared to the magnitude of the drug response. This relationship is very often sigmoidal, but usually log-linear within the central range of doses (Goldstein et al., 1968). An idealized representation of a typical dose-effect curve is illustrated in Figure 1a.

Following a sensitization period in which a number of doses of an individual drug are given in a consistent temporal pattern, we might then test the magnitude of chronic sensitization using a broad range (e.g., logarithmic) of different dosages. When we then plot the dose against the obtained drug effect, chronic sensitization will represent a significant shift to the left in the dose-response curve relative to untreated or control subjects. Conversely, chronic tolerance will be indicated by a significant shift to the right in the dose-effect curve. Shifts in the dose-effect curve are illustrated in idealized form in Figure 1b. This is the preferred experimental definition of sensitization, although it cannot always be accomplished in all experiments for various reasons.

SENSITIZATION AND "REWARD"

In 1984, Stewart et al. suggested that sensitization processes are associated with "reward." Specifically, they argued that the positive euphorigenic effects of drugs tend to sensitize, while the aversive, dysphorigenic effects tend to show tolerance. This seminal proposal placed the research emphasis in the animal literature on sensitization phenomena rather than tolerance, while the latter had dominated the field prior to this paper. A second impetus to focus on sensitization phenomena was Wise and Bozarth's (1987) hypothesis that locomotor activation was isomorphic with reward processes. They noted that virtually all abused drugs produce some degree of locomotor activation in animals, particularly at low doses. Moreover,

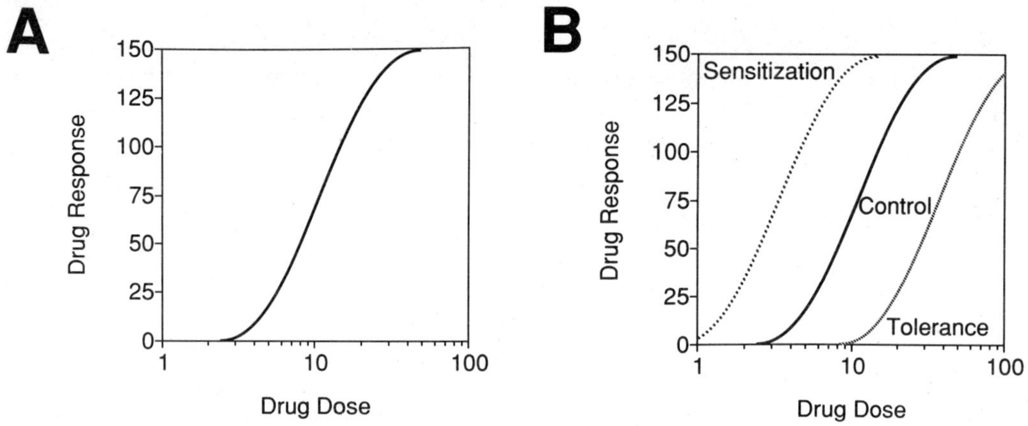

FIGURE 1. 1a) Idealized representation of a common dose-effect curve. Note that it is sigmoidal in shape, and that the X-axis is the logarithm of dose, and the Y-axis is the response in linear units.1b) Idealized representation of chronic tolerance and sensitization as shifts to the roght and left, respectively, in the dose-effect curve relative to controls.

this psychomotor activation tended to show sensitization rather than tolerance, particularly with stimulant drugs.

The term "reward" has several meanings in the drug abuse literature. Drug reward was originally defined in relation to instrumental self-administration of abused drugs such as morphine or cocaine (Johanson and Schuster, 1981). For example, an animal might be instrumented with an intravenous line to deliver unsignalled infusions of morphine; an operant response, such as a bar press, is then immediately followed by infusions. Reinforcement in this case refers to the increasing likelihood that the animal will press the bar when this is "rewarded" by infusions of low doses of morphine.

More recently (Mucha et al., 1981), reward has been defined in relation to place conditioning. In this procedure, an animal is repeatedly given drug in one chamber and saline in a different chamber. The animal is then given the "choice" as to whether it spends more time in the chamber previously associated with drug or the chamber associated with saline. This apparatus has a physical connection between two dissimilar chambers. Note that the animal is not under the influence of the drug at the time of testing. This is an important difference between the conditioned place preference procedure and instrumental conditioning in which the drug infusions take place at or near the time of testing.

Newlin (1992, 1994b) has emphasized that the conditioned place preference paradigm is, in fact, a particularly awkward form of auto-shaping. The place is the conditioned stimulus, the drug response is the unconditioned response, and the behavioral choice of the animal is the skeletal conditioned response. Newlin (1992) proposed that auto-shaping (also called sign-tracking) may provide a model of drug craving that has many attractive theoretical and experimental properties.

Finally, in humans, reward processes are inferred primarily from positive self-report of enhanced mood or euphoria on standardized questionnaires that are administered under the influence of the drug. This inference is based on the assumption that positive affective responses to drugs are related to substance abuse. No such inference is made with drug self-administration, although it can be made in conditioned place preference studies in which the animal chooses stimuli previously associated with the drug.

Newlin and Shoaib (1994) reviewed the drug conditioning literature in relation to the hypotheses advanced by Stewart et al. (1984) and Wise and Bozarth (1987). They came to slightly different, or at least, more specific conclusions than Stewart et al. (1984). Newlin and Shoaib (1994) noted that activating drug effects, such as locomotor activation, hyperthermia, hyperalgesia, seizures, etc., tended to be associated with drug isodirectional conditioned responses and sensitization. Conversely, deactivating drug effects, such as sedation, hypothermia, analgesia, etc. show drug-opposite conditioned responses and tolerance. Note that in both cases, the conditioned response is activating. Therefore, Newlin and Shoaib concluded that the conditioned response, when manifest, was always activating to abused drugs.

Our review (Newlin and Shoaib, 1992) generally supported the observations of Stewart et al. (1984) and Wise and Bozarth (1987). Our conclusions were more specific than those originally proposed as hypotheses by Stewart et al. (1984), and more parsimonious than Eikelboom and Stewart's (1982) homeostatic analysis concerning the direction of conditioned responses. Moreover, when considering Wise and Bozarth's (1987) psychostimulant theory of addiction, we see that activation, sensitization, and conditioned responses all appear to suggest "reward" processes, while deactivation and tolerance suggest "aversion" processes.

Note that tolerance to deactivating effects could be inadvertently interpreted as sensitization to activating effects. For example, tolerance to sedation produced by morphine might appear to be sensitization to activation. This illustrates the point that the direction of the overall effect of abused drugs is toward activation, both conditioned and unconditioned. In other words, animals and humans tolerate the aversive effects of abused drugs, and sensitize to their rewarding effects (Stewart et al., 1984), and this is expressed in animals in terms of greater conditioned and unconditioned activation.

It is important to emphasize that these dynamics correspond to drug effects (activating vs. deactivating) rather than to drugs themselves. "Stimulant" drugs always have some deactivating effects, and "sedative" drugs have some activating effects, such as locomotor activation (Wise and Bozarth, 1987). The power of these analyses is that classification of drug effects rather than drug classes reveals orderly relationships among measurable characteristics. However, interpretation of these relationships is more perilous than simple description of the nature of their underlying order.

COMMON FACTORS

One approach to interpretation that has proven effective is the analysis of "common factors" of abused drugs. A very wide variety of substances are abused. It is difficult to imagine a more diverse group of drugs than those that are voluntarily consumed by humans and animals despite long-term deleterious consequences, immediate negative effects (such as vomiting, anxiety, dizziness, overdose death, etc.) and, in the case of humans, potentially serious social and legal sanctions. These drugs differ markedly in the receptors to which they bind (Jaffe, 1991), the specific receptors that have been identified as mediating their reinforcing effects (e.g., Ritz et al., 1987), and the endogenous ligands that the drugs appear to mimic (Jaffe, 1991). In fact, these marked differences extend to many, if not most, of their physiological, behavioral, and subjective effects (Newlin, 1994).

In contrast, a common factors approach emphasizes the ways in which abused substances are similar rather than different. The rationale is that these common factors may account in some way for their liability for abuse. Moreover, common characteristics may indicate underlying etiological mechanisms, or physiological substrates of the processes that control drug-taking behavior. It seems unlikely that the mechanisms of abuse and addiction are totally different for each abused substance, although this cannot be ruled out at the present time. It seems far more likely that there are at least some common mechanisms that most, if not all, abused chemicals share. Moreover, these common factors are in all likelihood directly relevant to the etiology of their abuse.

Several common factors have been known for some time. These include drug self-administration by animals, sensitivity to Pavlovian conditioning, conditioned place preference conditioning, and in humans, production of subjective effects that are often described as euphoria (Newlin, 1992). More recent observations indicate that virtually all abused drugs are "craved" by human addicts, produce relatively slow (e.g., 10 Hz) synchronized activity in the electroencephalogram when euphoria is being unobtrusively reported (Lukas et al., 1986), and produce some degree of locomotor activation in animals, particularly at low doses (Wise and Bozarth, 1987).

The common factor that has been considered the most important is that virtually all drugs of abuse increase dopamine efflux in the nucleus accumbens of conscious, freely moving animals (Di Chiari and Imperato, 1988). This is normally measured by microdialysis, in which a small probe is inserted in the brain with stereotaxic localization of the nucleus accumbens. Extracellular brain fluid is collected by the probe, usually averaged over 10- or 20-minute periods, so that extracellular dopamine can be measured in the dialysate.

The importance of this common factor is that it suggests a common mesolimbic, dopaminergic mechanism for drug-seeking behavior (Di Chiari and Imperato, 1988). Although most researchers would consider this last statement a vast oversimplification, there is some agreement that this common factor is more or less directly related to the biological substrate of drug self-administration behavior.

Although originally interpreted as a relatively direct measure of a "reward center" of the brain (Di Chiari and Imperato, 1988) that is important in substance abuse, more recent theorizing has emphasized a broader role for dopaminergic neurotransmission in this cortico-mesolimbic system in motivated behavior (Kalivas and Samson, 1992; Newlin 1994a,b). For example, there is good evidence that mesolimbic dopamine is enhanced not just by natural reinforcers, such as food and sex, and exogenous reinforcers, such as cocaine and morphine, but also by stimuli that have been used to "stress" laboratory animals. This latter observation is not consistent with a simple reward circuitry interpretation of the mesolimbic dopamine system, but quite consistent with broader notions of a complex motivational system involved in survival and reproductive functions. For example, one would not expect a stressor, such as tail-pinch, to produce "reward." However, one would expect a stressor to activate motivational systems that control survival functions.

Recently, Cabib and Puglisi-Allegra (1994) reported that dopamine outflow is increased by controllable/escapable stressful stimuli, but decreased by uncontrollable/inescapable stressors. This finding is again consistent with the notion that the mesolimbic dopamine system controls survival and reproductive motivation rather than being a simple reward center (Newlin 1994a,b). Escapable stress would be expected to enhance active motivational processes, while inescapable stress would be expected to reduce motivation. It would be difficult to argue that an escapable stressor would be in some way "rewarding."

Two additional common factors have been identified. First, evidence has accrued over the last decade that abused substances reduce global glucose utilization in the brain (Newlin, 1992). These results have been provided by positron emotion tomography (PET), in which a radioactive substance such as ^{18}FDG is injected in the bloodstream and sensors arrayed around the head record photon emissions. This allows reconstruction of the three-dimensional location of the uptake by the brain of the radioactive substance, which should be closely related to brain activity. Therefore, this evidence indicates that a *reduction* in the gross metabolic activity of the brain is produced by most, if not all, abused drugs (London et al., 1990; Newlin, 1992).

This is a more or less surprising finding. One might have assumed that psychostimulant drugs such as d-amphetamine or cocaine would increase rather than decrease brain metabolism because they have such excitatory effects on mood and behavior. However, this is not the case. It might also seem surprising that a drug such as alcohol would have qualitatively the same effect on brain metabolism as cocaine. This is the power of a common factor. It lends credence to the notion that this global decrease in brain metabolism is mechanistically related to other common factors, such as euphoria.

Finally, this author has studied cardiovascular responses to abused and nonabused drugs, and found a common cardiovascular pattern to drugs of abuse (Newlin, 1992, 1994b). Heart rate is normally controlled by a balance of excitatory sympathetic and inhibitory parasympathetic innervation (Berne and Levy, 1977). We have found that almost all abused drugs that we have studied, including cocaine, marijuana, nicotine, alcohol, morphine, pentobarbital, and methylphenidate all increase heart rate and decrease vagal tone index, a measure of parasympathetic inhibitory influences on the heart (Newlin, 1994b). When parasympathetic inhibition is released, such as from an abused drug, then heart rate increases, in some cases dramatically (Dellinger et al., 1987). The mean heart rate increases from cocaine and marijuana are on the order of 25–35 beats per minute, and the decreases in vagal tone index are from 2 to 2.5 log units (Newlin, 1994b).

We have not found this cardiovascular pattern with aversive drugs. Naloxone given to oploid-dependent individuals increased heart rate, an effect consistent with increased sympathetic tone because there was no change in vagal tone index (Newlin et al., 1992). Similar results were found in a limited number of subjects that received the benzodiazepine antagonist flumazenil after three days on alprazolam, and four subjects who had dysphorigenic responses to mCPP, a serotonergic probe (Newlin et al., 1992). Therefore, the cardiovascular pattern noted above of increased heart rate and decreased vagal tone index appeared to be specific to abused drugs (Newlin, 1994b).

We view this unique cardiovascular pattern to be of central (i.e., brain) origin, and mechanistically related to another common factor to abused drugs: increased dopamine efflux in the nucleus accumbens (Newlin, 1994b). It would appear that this common cardiovascular pattern more or less directly reflects the operation of the dopaminergic mesolimbic motivational system. We do not know at this time what the neuroanatomical or neurophysiological connection between mesolimbic and vagotonic systems might be, although we certainly predict that this connection exists. We can even speculate on some of the dynamics of this system. For example, we would expect that pharmacological blockade or lesioning of the mesolimbic system would prevent the cardiovascular pattern from being observed, but that blocking the vagotonic center would have no effect on reward processes. This would place this unique cardiovascular pattern in the same general category as psychomotor stimulant activation (measured by locomotor activity) in the sense that it is a marker rather than mechanism of the motivational system (Wise and Bozarth, 1987).

SENSITIZATION AS A COMMON FACTOR

We might propose that the tendency toward chronic sensitization to rewarding effects of abused drugs is also a common factor. A remarkable paper by Lett (1989) would seem to support this proposition. She found that multiple pre-exposures to d-amphetamine, cocaine, or morphine *increased* conditioned place preference for the drug itself, as well as preference for other drugs. Therefore, preference appeared to sensitize with multiple exposures rather than show tolerance, and it also showed cross-sensitization rather than cross-tolerance.

FUNCTIONAL ROLE OF SENSITIZATION

What function does sensitization serve? Why does it serve that role? These questions appear somewhat imponderable, but they have important implications for how we understand sensitization phenomena.

We might note that one important effect of sensitization is amplification. A response to stimulus that is at first small and unremarkable can become amplified through sensitization processes to the point that the response to the same stimulus become overwhelming with repeated exposures. The organism is constantly bombarded with a myriad of stimuli in many different sensory and interoceptive (internal) modalities. Somehow, it must separate those stimuli that are in some way "important" from those that are trivial. One way to do this is to amplify some repeated stimuli through sensitization, and to minimize others through tolerance. When it comes to drugs, we might suppose that the stimuli are entirely interoceptive, or perceived through systems for responding to internal sensations rather than through the usual sensory systems. However, this is rarely the case. Most drugs are administered or cued by some obvious procedure that is perfectly associated with the internal stimuli. For example, if we repeatedly inject an animal with morphine, the animal might readily perceive the association between injection and drug effect. Note that now we are dealing with two stimuli, one internal and one external: the internal registration of drug effect and the administration ritual. Therefore, it is important to keep in mind that drugs are complex stimuli.

When we speak of sensitization to a drug stimulus, we usually mean that some measurable response is enhanced with repeated administrations of the drug. In other words, the response is amplified. We might ask why the organism would evolve in such a way that it amplifies some drug responses, but not others. Recent research has raised the very important question of whether underlying physiological substrates, such as mesolimbic dopamine, sensitize (Kalivas and Samson, 1992) at the same time that measures such as analgesia or hypothermia show tolerance (Newlin and Shoaib, 1994). This question has not been answered in any definitive way, although we might expect (or predict) that mesolimbic dopamine will sensitize to most drugs of abuse.

Let us assume for the moment that mesolimbic dopamine sensitizes. We might then argue that the organism has evolved in such a manner that the mesolimbic dopamine, which we have suggested (above) may be isomorphic with activation of a motivational system directed to survival and reproductive functions, is amplified when internally activated. It certainly appears sensible that a motivational system sensitizes. However, it is clear in most cases that drugs such as cocaine and alcohol serve no useful function to the organism, and are deleterious in most cases.

We might speculate that the normal function of the mesolimbic system is to activate motivational systems controlling basic survival and reproductive functions, and that abused drugs (but not aversive drugs) artificially mimic exogenous stimuli that would normally stimulate those very functions. These stimuli include environmental stressors, food, sexual stimuli, etc. (Salamone, 1994). In effect, we are proposing that abused drugs are short-circuiting a motivational system that normally would be amplified (i.e., sensitized) by these stimuli. The activation of the motivational system by abused drugs leads to processes such as sensitization, that are appropriate to the system (in an evolutionary sense) but not to the drugs themselves.

This is the central paradox of drug abuse. Why do people and animals so persistently and ineluctably approach and consume drugs that are clearly deleterious to them? We might speculate, based on the analysis above, that these drugs are approached and consumed because they activate a motivational system that is essential for survival and reproduction. It is paradoxical only because the drugs are obviously deleterious to survival and reproductive functions, at the same time that they artificially activate these very systems.

Note that this (admittedly speculative) argument would make no sense if we assumed, as most drug abuse researchers do, that drugs simply activate a reward system. If this were the case, then animals would determinedly pursue any stimulus that was in some sense rewarding, would sensitize to that reward, and would become so directed to reward that they would fall victim to any deleterious correlate of a rewarding stimulus. For example, for predators (such as humans), the acquisition of meat is a very dangerous occupation in the wild, although highly rewarding. It would be very disadvantageous, indeed, to become preoccupied with food to the point of minimizing all the clear dangers associated with acquiring it. However, it would not be disadvantageous in the same way if we assumed that the anticipation of food activated basic survival and reproductive motivational systems (as argued above). In fact, this would appear to be highly advantageous to the organism.

SENSITIZATION AND CHEMICAL SENSITIVITY

The central points of this paper are: 1) it is important to conceptualize the function of sensitization to certain exogenous stimuli, 2) a common factors approach may facilitate this process, and 3) there are some ways in which MCS and drug abuse are polar opposites. We have argued above that animals move toward abused drugs because they artificially activate basic survival and reproductive systems. Conversely, we might argue that in MCS patients move away from aversive chemicals because they are represented cognitively and affectively as threats to basic survival and reproduction. We might expect that a similar, but opposite, process is in play. Sensitization or amplification processes make this movement away from chemicals aggressively motivated and overwhelming to the patient. We might further propose that these chemicals, or stimuli (such as odors, sights, etc.) that are highly correlated with the chemicals, short-circuit normal processes (i.e., activation of survival and reproduction functions) that evolved to handle environmental threats.

This analysis would suggest that two empirical functions are predicted in MCS patients: 1) sensitization either to chemicals or stimuli that are highly predictive of those chemicals, and 2) a short-circuiting process in which basic motivational systems are activated inappropriately.

AUTO-SHAPING

Auto-shaping is an experimental procedure in which animals show a conditioned skeletal response in a Pavlovian procedure (Brown and Jenkins, 1968; Hearst and Jenkins, 1974; MacIntosh, 1983; Newlin, 1992). It is often referred to as "sign-tracking" because the animal demonstrates directed movements toward a positive (but arbitrary) predictor of a reinforcer such as food, and directed movements away from a positive predictor of punishment, such as electric shock. It is Pavlovian conditioning because there is no contingency between the animal's behavior and the conditioned stimulus or unconditioned stimulus. In other words, the directed skeletal behavior appears superfluous because it has no effect on the availability or presentation of the reinforcer, nor does it serve any other obvious purpose in the experimental procedure. Instead, this behavior is understood as a highly species-specific response that maintains behavior in the wild. It keeps the animal oriented toward positive signs and away from negative signs in the wild. Sign-tracking has been observed in human subjects as well (Newman et al., 1980).

Newlin (1992) has applied this model to craving for abused drugs. In this approach, we might understand the excessive orientation toward drugs and drug stimuli that we refer to as "craving" in terms of a preoccupation with positive signs of drug effects. Therefore, a stimulus such as the smell of tobacco smoke or the sight of another smoker lighting up a cigarette elicits through Pavlovian conditioning a subjective state and directed skeletal movements toward the conditioned stimulus. There is now little doubt that drug effects are highly questionable in Pavlovian paradigm (Siegel, 1975, 1981; Baker and Tiffany, 1985), although researchers rarely measure skeletal behavior in these conditioning paradigms.

It was argued above that conditioned place preference is an example of a very awkward form of auto-shaping [which was noted by Much et al. (1981) in their original study of place conditioning]. Other than this paradigm, there have been attempts to apply auto-shaping with drugs as the unconditioned stimuli only very recently (Krank and Squarey, 1984; Carroll and Lac, 1993). Even in these cases, the procedures were designed more to facilitate drug self-administration rather than to develop an auto-shaping model of addiction or craving.

Auto-shaping brings together many of the common factors that we have outlined above. These include Pavlovian conditioning, craving, positive subjective effects, sensitization, conditioned sensitization, and possibly activation of the mesolimbic dopamine system in anticipation of abused drugs (or elicited by conditioned drug stimuli).

The application of this model to MCS is straightforward. The powerful aversion to chemical stimuli, and perhaps signs of chemicals, in MCS patients reflects the "moving away," or directed skeletal activity *away from odors* (conditioned stimuli) associated with toxic chemicals. This illustrates the notion that MCS and substance abuse may reflect polar opposites. In substance abuse, the addict habitually moves toward drugs and drug stimuli through an auto-shaping process. In MCS, the patient habitually and robustly moves away from chemicals and odors of chemicals through negative auto-shaping.

EMPIRICAL RESEARCH

Although we might assume that sensitization to drugs is a relatively ubiquitous finding in human volunteers, it has actually been tested rarely. Newlin and Thomson (1991) studied college-aged men at high risk for alcoholism by virtue of having a biological father who was alcoholic. Although evidence indicates that this elevated risk is genetic in origin, observed differences between sons of alcoholics and sons of nonalcoholics may be due to genetic or environmental influences, or an interaction of both (Newlin and Thomson, 1990). These high-risk individuals were compared to sons of nonalcoholics, who were presumably at low risk for developing alcoholism.

There were four sessions in the laboratory for all subjects, three in which subjects drank 0.5 g ethanol/kg body weight, and a final session in which subjects drank the mixer, but with no alcohol in it. This last session was a placebo session intended to assess conditioning effects in the two groups. The dependent measures were autonomic and locomotor.

We found that sons of alcoholics demonstrated chronic sensitization to alcohol in the first three sessions, when sons of nonalcoholics tended to show tolerance. There was a significant tendency for the high-risk group to show greater locomotor activity (measured by a stabilometer under the subjects' chairs), an effect also noted in animals that show more rapid acquisition of drug self-administration. The sensitization was observed late in each session, after the effect of oral alcohol was measurable in the subjects' physiology. These processes are illustrated in Figure 2. Note that these differences were in the rising blood alcohol curve in which excitatory, euphoriant effects are most clear in human volunteers. Unfortunately, we did not record during the falling blood alcohol curve in which depressant, aversive effects of alcohol are most prominent.

Newlin and Thomson (1990) reviewed evidence in the human literature that sons of alcoholics tend to show acute sensitization, or a greater than expected response, in the rising blood alcohol curve; in the falling blood alcohol curve, these same high-risk subjects tend to show acute tolerance. The former effect was clear in Newlin and Thomson's (1991) study, while Schukit's program of research has tended to show acute tolerance in the falling curve in the high-risk group (Newlin and Thomson, 1990).

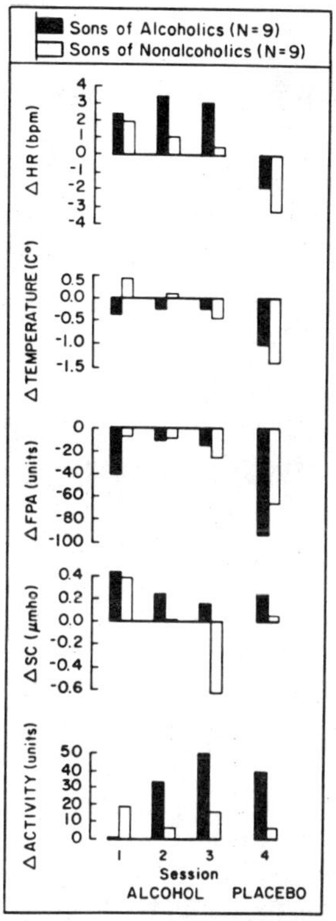

FIGURE 2. Development of chronic sensitization in high-risk (sons of alcoholics) individuals compared to low-risk subjects (sons of nonalcoholics). Note that the pharmacological effects of alcohol were at the end of each session following absorption of alcohol. (Used with permission of the authors, Newlin and Thomson, 1991; in the public domain.)

We (Newlin and Thomson, 1990) interpreted these relationships in motivational terms. Specifically, we proposed that the high-risk subjects accentuate the positive effects of alcohol in the rising curve when euphoria and stimulant effects are most prominent, and attenuate the aversive effects that predominate in the falling curve (such as headache, depression, anxiety, etc.). Therefore, this interesting relationship may suggest that what is being transmitted in high-risk subjects is a response to alcohol that motivates greater drinking and escalation of drinking over time.

We might propose a relatively simply empirical study to test some of the conceptualizations in a laboratory analog of MCS. Consider two groups of human subjects, all of whom are exsmokers: Group I reports that they find the smell of tobacco smoke very appealing, and wish that they could return to smoking but refuse to do so (for obvious reasons). Group II are also exsmokers, but they are virulent in their distaste for the smell of tobacco smoke. These two groups are then exposed individually to tobacco smoke in room air (calibrated in parts per million) in a laboratory environment on three to five occasions. This exposure to tobacco smoke is always presented in association with a distinctive, but arbitrary sensory cue, such as white noise or red light. Half of Group I and half of Group II are control subjects that smell a distinctive but affectively neutral odor during each session. We might record self-report of mood and affective responses to the smell, autonomic measures, hormonal measures from in-dwelling catheter, and locomotor activity continuously during each session. A final session is presented in which there is no smell of tobacco smoke, but the distinctive sensory cue is presented. For half the subjects in each group, this cue signals tobacco smoke, for the other half, it signals the neutral odor. Positron emission tomography scans could be used in the last session in which smoke is presented in order to assess chronic sensitization in response to the tobacco smoke. The design of this study is illustrated in Table 1.

TABLE 1. Design of an Study in Humans of Conditioned Sensitization to Tobacco Smoke

Session	1	2	3	4	5
Group I: Ex-Smokers	Cig	Cig	Cig	Cig	CUE
Group I: Ex-Smokers	Air	Air	Air	Air	CUE (meaningless)
Group II: Anti-smokers	Cig	Cig	Cig	Cig	CUE
Group II: Anti-smokers	Air	Air	Air	Air	CUE (meaningless)

Key: Cig = cigarette smoke exposure, quantified in ppm; CUE = distinctive, but arbitrary cue associated with cigarette smoke; Air = room air with affectively neutral odor.
Dependent measures: 1)Subjective; 2)Autonomic; 3)Hormonal; 4) PET sessions 1 and 4.

This design, which would not be particularly difficult to perform, could be used to assess the meaning and clinical significance of chronic sensitization to a specific low-level exposure to a chemical. At the same time, the design permits the assessment of conditioning factors in the last session in which the cue is presented without the chemical exposure. We will let the reader make predictions of the results that would be obtained in this study.

SUMMARY AND CONCLUSIONS

Sensitization is a process that occurs in individual response systems. It is highly conditionable, particularly with odor stimuli, a characteristic that has important implications for the development of MCS. There are some ways in which substance abuse, in which animals and people become habitually oriented *toward* chemicals, and MCS, in which people become strongly oriented *away from* chemicals, are polar opposites. Auto-shaping is proposed as a process that could describe both functions, although in opposite ways. Consideration of the biological significance of the response to chemicals in terms of basic survival and reproductive functions leads to new ideas about how substance abuse and MCS may develop in humans. Finally, empirical studies of sensitization processes in human volunteers are reviewed or proposed.

ACKNOWLEDGMENTS

The author is indebted to Iris Bell, M.D., Ph.D. for the many fascinating conversations that led indirectly to this manuscript. He would also like to thank Roy Pickens, Ph.D. for commenting on an earlier draft of the paper. The views expressed in this manuscript are entirely those of the author, and do not reflect those of the National Institute on Drug Abuse or the U.S. Government.

REFERENCES

BAKER, T.B. and TIFFANY, S.T. (1985). "Morphine tolerance as habitutation." Psychol. Rev. 92:78–108.

BERNE, R.M. and LEVY, P. (1977). Cardiovascular Physiology. Third ed. Mosby, St. Louis.

BROWN, J.S. and JENKINS, H.M. (1968). "Auto-shaping of the pigeon's key peck." J. Exp. Anal. Behav. 11:1–8.

CABIB, S. and PUGLISI-ALLEGRA, S. (1994). "Opposite responses of mesolimbic dopamine system to controllable and uncontrollable aversive experiences." J. Exp. Anal. Behav. 11:1–8.

CARROLL, M.E. and LAC, S.T. (1993). "Autoshaping IV cocaine self-administration in rats: Effects of nondrug alternative reinforcers on acquisition." Psychopharmacol. 110:5–11.

DI CHIARA, G. and IMPERATO, A. (1988). "Preferential stimulation of dopamine release in the nucleus accumbens by opiates, alcohol, and barbituates: Studies with transcerebral dialysis in freely moving rats." Ann. New York Acad. Sci. 473:367–381.

DI STEPHANO, J.J., STUBBERUD, A.R., and WILLIAMS, I.J. (1967). Feedback and Control Systems. McGraw-Hill, New York.

EIKELBOOM, R. and STEWART, J. (1982). "The conditioning of drug-induced physiological responses." Psychological Rev. 89:506–528.

GILMAN, A.R., RAU, T.W., and NIES, A.S., ed. Goodman and Gilman's The Pharmacological Basis of Therapeutics, Eighth ed. Pergamon Press, New York.

GOLDSTEIN, A., ARANOW, L., and KALMAN, S.M. (1968). Principles of Drug Action: The Basis of Pharmacology. Harper and Row, New York.

HEARST, E. and JENKINS, H.M. (1974). Sign Tracking: The Stimulus-Reinforcer Relation and Directed Action. Monograph of the Psychonomic Society, Austin, TX.

JAFFE, J.H. (1990). "Drug addiction and drug abuse." In: Goodman and Gilman's The Pharmacological basis of Therapeutics. Eighth ed. (A.G. Gilman, T.W. Rall, and Nies, A.S., ed.). Pergamon Press, New York. pp. 357–365.

JOHANSON, C.E. and SCHUSTER, C.R. (1981). "Animal models of drug self-administration." In: Advances in Substance Abuse: Behavioral and Biological Research, Vol. 11. JAI Press, Greenwich, CT.

KALIVAS, P. and SAMSON, H.H. (1992). "The neurobiology of drug and alcohol addiction." Ann. New York Acad. Sci.654

KESSLER, R.C., McGONAGLE, K.A., ZHAO, S., et al. (1994). "Lifetime and 12-month prevalence of DSM-III-R psychiatric disorders in the United States: Results from the national Comorbidity Survey." Arch. Gen. Psychiatry 51:8–19.

KRANK, M.D. and SQUAREY, K. (1994). "Directed incentive effects of local signals for unsweetened ethanol." Alcoholism: Clin. Exp. Res. 18:517.

LETT, B.T. (1989). "Repeated exposures intensify rather than diminish the rewarding effects of amphetamine, morphine, and cocaine." Psychopharmacol. 98:357–365.

LUKAS, S.E., MENDELSON, J.H., BENEDIKT, R.A., and JONES, B. (1986). "EEG alpha activity increases during transient episodes of ethanol-induced euphoria." Pharmacol. Biochem. Behav. 25:889–895.

MACKINTOSH, N.J. (1983). Conditioning and Associative Learning. Oxford University Press, New York.

MUCHA, R.F., VAN DER KOOY, D., and O'SHAUGHNESSY, M. (1981). "Drug reinforcement studied by the use of place conditioning in rats." Brain Res. 243:91–105.

NEWLIN, D.B. (1992). "A comparison of drug conditioning and craving for alcohol and cocaine." In: Recent Developments in Alcoholism, Vol 10, Alcohol and Cocaine (M. Galanter, ed.). Plenum Press, New York.

NEWLIN, D.B. (1994a). "In the belly of the beast: Toward a motivational view of addiction." Contemp. Psychol. 39: 511–512.

NEWLIN, D.B. (1994b). "Common factors in substance abuse: Effect of cocaine on vagal tone." Drug and Alcohol Dependence. Submitted.

NEWLIN, D.B. and THOMSON, J.B. (1990) "Alcohol challenge with sons of alcoholics: a critical review and analysis." Psychological Bull. 108:383–402.

NEWLIN, D.B. and THOMSON, J.B. (1991). "Chronic tolerance and sensitization to alcohol in sons of alcoholics." Alcoholism: Clin. Exp. Res. 15:399–405.

NEWLIN, D.B. and SHOAIB, M. (1994). "Pavlovian conditioning of activating and de-activating effects of abused drugs: A review and conceptualization." In preparation.

NEWLIN, D.B., BYRNE, E.A., and PORGES, S.W. (1990). "Vagal mediation of the effect of alcohol on heart rate." Alcoholism: Clin. Exp. Res. 14:421–424.

NEWLIN, D.B., PRETORIUS, M.B., WONG, C.J. et al. (1991). "Acute intravenous cocaine decreases cardiac vagal tone." Paper presented at the Annual Meeting of the Committee on Problems of Drug Dependence. Palm Beach, Fl, June.

NEWLIN, D.B., PRETORIUS, M.B., WONG, C.J. and DAX, E.M. (1991). "Acute marijuana smoking reduces vagal tone." NIDA Research Monograph, Vol. 105. Committee on Problems of Drug Dependence. U.S. Government Printing Office, Washington, D.C. 585–566.

NEWLIN, D.B., WAITS, R.P., WONG, C.J., PRETORIUS, M.B., and FROEHLICH, J.C. (1994). "Methylphenidate produces vagally mediated tachycardia in human subjects." In preparation.

NEWLIN, D.B., WONG, C.J., and CHESKIN, L.J. (1992). "Cardiovascular responses to naloxone challenge in opiate-dependent individuals." Pharmacol. Biol. Behav. 43:357–360.

NEWLIN, D.B., WONG, C.J., WAITS, R.P., et al. (1992). "Tachycardia from intravenous nicotine is mediated vagally in human subjects." Paper presented at the Annual Meeting of the Committee on Problems of Drug Dependency, Keystone, CO.

NEWMAN, J., WOLFF, W.T., and HEARST, E. (1980). "The feature-positive effect in adult human subjects." J. Exp. Psychol.: Human Learning and Memory 6:630–650.

PAVLOV, I.P. (1927). In: Conditioned Reflexes. (G. Anrep, ed.). Oxford University Press, London.

RITZ, M.C., LAMB, R.J., GOLDBERG, S.R. et al. (1987). "Cocaine receptors on dopamine transporters are related to self-administration of cocaine." Science 237:1219–1223.

ROBINS, L.N., HELZER, J.E. WISSMAN, M.M., et al. (1984). "Lifetime prevalence of specific psychiatric disorders in three sites" Arch. Gen. Psychiatry 41:949–958.

SALAMONE, J.D. (1994). "The involvement of nucleus accumbens dopamine in appetitive and aversive motivation." Behav. Brain Res. 61:117–133.

SIEGEL, S. (1975). "Evidence from rats that morphine tolerance is a learned response." J. Comp. Physiological Psychol. 89:498–506.

SIEGEL, S. (1981). "Classical conditioning, drug tolerance, and drug dependence." In: Research Advances in Alcohol and Drug Problems, Vol. 7 (Y. Isreal, F.B. Glaser, and H. Kalant, ed.). Plenum Press, New York.

STEWART, J., DE WIT, H., and EIKELBOOM, R. (1984). "Role of unconditioned and conditioned drug effects in the self-administration of opiates and stimulants." Psychological Rev. 91:251–268.

SURGEON GENERAL'S REPORT (1988). "Nicotine addiction: The health consequences of smoking." U.S. Department of Health and Human Services, Rockville, MD.

WISE, R.A. and BOZARTH, M.A. (1987). "A psychomotor stimulant theory of addiction." Psychological Rev. 94:469–492.

WONG, C.J., NEWLIN, D.B., PRETORIUS, M.B., BETTER, W.B., and PICKWORTH, W.B. (1991). "Acute effects of various abused drugs on heart rate and vagal tone: A common factors approach." Paper presented at the Annual Meeting of the Committee on problems of Drug Abuse, Palm Beach, Florida, June.

RESPONSE PANEL #2

HOWARD KIPEN, IRIS BELL, CLAUDIA MILLER, RICHARD LETZ, AND NORMAN ROSENTHAL

Dr. Rosenthal: I'm actually feeling rather excited by the various talks that we've had, because I think one of the issues in Multiple Chemical Sensitivity appears to be, "does it exist at all?" Is this a real condition, or is it a bunch of malingerers out to claim compensation, get out of work, shirk duties? Based on my own clinical experience with sound people who have become crippled following very discreet and clear cut exposures, I do believe it is a valid entity. But as scientists we want to be able to rely on more than just belief. Validity, as I will mention, rests on a number of factors: face validity, descriptor validity, predictability, but above all, construct validity. Is there some kind of way, whereby we can understand, in terms of processes that come from other areas of science, what might be going on in this condition? I think that the speakers have provided us with tantalizing clues to the possibility of construct validity. We have seen so many different kinds of evidence for biological responses to low-dose noxious stimuli, whether it be cellular discharges, or behavioral responses, that occur when animals or people are exposed again on subsequent occasions. What is exciting to me about the various talks is that there is really so much science out there. Even though it isn't taken from people who actually had MCS, it shows how the brain can actually respond to low repetitive dosages or challenges of various kinds.

Dr. Letz: I think there has probably been a great deal of work in animal learning over the past 15 years that is applicable. The few times that I've heard conditioning models mentioned, they are discussed in a strictly Pavlovian sense. I think those might be more profitably exploited. But, perhaps classical conditioning is not the most applicable model. A lot of work has been done in avoidance conditioning, and avoidance is a large part of the after effects of the behaviors of MCS patients. That should be looked into much more. Also, I haven't heard anyone relate MCS to a very special type of classical conditioning, taste or odor aversion. There is great literature, beginning 15 years ago, that indicates there are very special parameters related to that type of learning. For example, the delay between the presentation of the conditioned stimulus and the unconditioned stimulus. I think that literature should be looked into and applied to this important problem much more than it has been today.

Dr. Miller: The drug abuse literature and the PTSD literature have some very interesting parallels to observations in chemical sensitivity. I just wanted to highlight a few of those. With addiction, the patients often report addictions to certain foods. Interestingly, they may be driving down a road in heavy traffic, or have some other chemical exposures, and they say the cravings get worse. I don't know what it means, but they often report that cravings are triggered by specific exposures.

Toxicology and Industrial Health, Vol. 10, No. 4/5, pp. 481–486

ISSN: 0748-2337

I have some difficulty equating being in a sick building with some kind of traumatic event like being in a war zone, or with a sexual abuse situation. One the one hand, I think it was said that you don't think losing a spouse from a divorce would qualify as a PTSD-type stimulus, but I have trouble understanding why, then, being in a sick building, when people certainly have other job options, would qualify. I don't think they would view that as an inescapable kind of stressor.

You talked about the stress of giving animals saline injections. I don't know if I can equate that with some of the stressors that some of these patients experience with this particular problem. I think it is difficult to make those kind of parallels. We have to be very careful about how we draw parallels between animal models and what we're seeing clinically.

I see PTSD patients and the substance abuse patients as being the flip side of the chemical sensitivity patients. The PTSD patients and the substance abuse patients often use large amounts of caffeine, alcohol, and they smoke. The chemical sensitivity population, on the other hand, mostly avoids smoking, caffeine, and alcohol. There is a difference in the two groups.

If the chemical sensitivity patients were to adopt the lifestyle of these substance-abusing patients, or many of the PTSD patients who use drugs heavily, then they might also experience very severe symptoms. They find they have to avoid these things, just to feel normal. It makes me wonder what would happen if you took all these substance-abusing people, and you removed them from all of those substances; what would they be like at baseline?

Some other conditions that are increasing in the population as a whole are fatigue, depression, asthma, and migraine headaches. These are conditions for which patients with chemical sensitivity have often seen doctors. Most of the patients, in fact, carry a double diagnosis of chronic fatigue syndrome. The label they've gotten depends upon the practitioner that they've gone to see.

A couple of other analogies with PTSD: the patients who develop chemical sensitivity report an increase in startle response, and they complain about problems with cold and heat intolerance.

So both PTSD and chemical sensitivity patients complain about physical stimuli, and increase startle kinds of responses. I think that there are some parallels there that may suggest, again, a common, perhaps limbic origin, a kindling type of origin.

I'd be interested in hearing from the panel about what individual differences may predispose to PTSD and other things that may predispose to some of these disorders we have talked about. Maybe that would shed some light about who in the population may be more sensitive to chemical exposure, as well.

Dr. Friedman: If I gave anyone the impression that I thought being in a sick building was traumatic, as it is vigorously defined, please accept my apologies. However, when I listened to the three very eloquent patients talk this morning about not being able to quit a job, or having to live in a tent, and having to go to the bathroom even knowing how bad the consequences would be, I really thought that these were major events in their lives, and I thought the psychological meaning of those did in some ways seem uncontrollable and unpredictable.

I recognize that these events are not traumatic, but they are stressful. I believe that when the dust settles on traumatic syndromes, we're going to be talking about a spectrum of disorders, not a yes-no type of choice. Not the forced choice that we clinicians are forced to make at this particular time in history.

Secondly, what's interesting in the Persian Gulf cohort that Jessica Wolfe has been following is that people with PTSD report adverse health complaints since their return. Almost 93% of the women and 75% of the men, defined as having PTSD by very strict criteria, most frequently report headache, fatigue, and musculoskeletal pain. Again, that's a correlational analysis, but it just means to me that we ought to look at these people more carefully.

Thirdly, I think that individual differences are key. Because one of the things that we do know about PTSD is that everyone doesn't get it. Take a hundred people, put them next to a Mt. Pinatubo, or a war zone, or something else; even in the case of rape victims, or political torture victims, where 95% will have it immediately, by 4, 6, 12 months, there will be recovery, or at least there will be symptom amelioration at that time. Everyone is not equally sensitive, or vulnerable. I think vulnerable would be a better term, given the general thrust of this conference today, rather than sensitive.

But why, we don't know. Certainly, there is a dose response relationship in terms of exposure. Certainly, prior exposures, in terms of physical or sexual abuse, or other kinds of priming doses of trauma, play a role. Epidemiological studies on Vietnam Veterans suggest that some of the buffering things are education, age, the kind of recovery environment, etc.

Your comments about food, I thought were very interesting from two perspectives. One is that, from studies with both Israeli veterans and with traumatized women, behavior around food seems to be very variable. Looking at weights, one finds tremendous variability. Either people are obese or ectomorphic. One finds that incest survivors often present clinically as having eating disorders; as anoretic bulemics.

I think the last thing I want to comment on is about addiction, and the use of psychoactive substances. I don't have a good answer for your question. I think it is a very complicated one. I think what we do know is that when a PTSD patient goes through an alcohol or other kind of drug program to clean themselves out, they often have a resurgence of the symptoms. I think there is a risk benefit ratio between giving up substances that may not be good for you in the short or long term, and what the consequences of doing that laudatory activity might be in terms of abstinence.

Dr. Kipen: I wonder whether we are studying a group of sick individuals who are defined by some bright line separating them from the not-sick among us, or whether we are really trying to look at a phenomenon that is kind of a continuum. The extent to which we choose to study the sicker among us will determine some of what we end up uncovering.

My second thought is whether we seek to uncover an underlying biology to explain the behavior. Is that going to lead to therapies and interventions that are effective and that will be able to define chemical sensitivities, much as we are beginning to define depression or bipolar disorder?

Finally, I find myself thinking about the relationship between so-called conditioning influences, chemical influences, and stress influences. Maybe taking a retrospective approach, comparing cases and controls in various paradigms, may allow us to really tease apart the important contributing factors. Perhaps we should consider designing prospective studies defining people who are not manifestly chemical sensitive, and seeing what happens as they go through various influences in their lives.

Dr. Bell: I think many of the important comments about what's been said today have already been made, but I wanted to amplify certain particular points. One of them being the notion of host vulnerability factors and individual differences.

In certain elements of psychology, we rely on looking at the host and looking specifically at unique factors in the host that increase vulnerability to particular phenomena. There are some distinctive ways in which different disciplines approach the questions we are talking about. Part of what I am impressed by today is that the individual difference approach may be especially important in this research endeavor.

The other point that I thought Dr. Friedman emphasized very nicely is the notion of the potential for synergistic effects. We certainly talk about mixtures of environmental chemicals as potential inducers or triggers of chemical sensitivity, and he was again expanding the notion into the possibility that one might have both psychological stressors and environmental chemical stressors, simultaneously.

We are not necessarily just talking about one making vulnerability to the other more likely, but we're also talking about the synergistic interaction between the two, setting off a process which may change a person's responsivity to the environment in a variety of ways.

I would like to answer a couple of questions people came up with in the course of the discussions about this area. We do have some preliminary data, similar to what has been reported in at least one study of PTSD, in which we looked at elderly depressives who had low-dose dexamethasone suppression tests.

We found that the people who said they were more reactive to various chemical odors tended to not only be more depressed, but there tended to be greater suppression of their cortisone response to the low-dose dexamethasone challenge.

This is in distinction to what a standard major depressive might show. It's just a unique feature of PTSD, and it may not offer a specific marker, but, given the emphasis by some of the speakers on the possible requirement for corticosteroid and some of the animal work, it might point to the importance of studying the HA axis in this population. We have a very non-specific kind of marker, so it wouldn't be giving us ultimate answers, but it might be a very important place to start looking.

Finally, I would like to address this question to members of the panel: If we learn that it is possible to sensitize human subjects in an acute exposure episode such as a chemical spill, and we have an individual who enters an emergency room for treatment following an identifiable exposure event, given what is known about the potential for blocking the initial sensitization, do they have any pharmacologic suggestions about where we might start looking, from an experimental point of view?

Dr. Friedman: I'd like to say two things. First of all, I'd like to thank Dr. Bell for raising the issue of dexamethasone suppression, because this is potentially a very valuable and easily applied clinical test to distinguish what appears to be depression phenomenologically, but biologically is very different. I would be willing to bet that many of the so-called depressed people who have been diagnosed with multiple chemical sensitivity might find, with a good psychiatric evaluation, that they really have a trauma-related disorder that has depression as a co-morbidity factor.

Secondly, some of you may recall the big spill in the Sacramento River a couple of years ago. Some isothiocyanate, hydrogen sulfide, some really bad stuff, spilled. What was really interesting was that some patients subsequently reported that they had a very bad smell at the time of the spill. My guess is that they would probably meet criteria for what you would call cacosmia. Those people might very well have multiple chemical sensitivity, although I don't think there has been any evaluations of that sort that this group would feel worthwhile.

I certainly am not happy with the PTSD evaluations that were done. It would seem that we should consider some kind of prospective longitudinal studies, as you suggest. I think before we do any kind of preventive pharmacologic therapy we have to know what the longitudinal course is likely to be, who is at greater risk, and who isn't.

Dr. Rosenthal: I think that Dr. Friedman's idea about sexual abuse is a very compelling analogy, because we psychiatrists often see people who have been abused previously get into, for example, abusive marriages, and once there, feel very stuck. When I hear stories of people stuck in their jobs in sick buildings there are analogous stories, of supervisors saying, "come on, get your job done, there is nothing wrong". And there they are, with their Tylenol, and their aspirins, and I think, obviously, what Dr. Weiss was saying, is that early intervention is critical. Because the further you go along, the more disabling the condition is. At this point, we don't have drug interventions, but we can try to empower our patients to get out, before they have lost their short-term memory.

PSYCHIATRIC SYMPTOMS IN MULTIPLE CHEMICAL SENSITIVITY

GREGORY E. SIMON

Center for Health Studies, Group Health Cooperative of Puget Sound
and
Department of Psychiatry and Behavioral Sciences
University of Washington
Seattle, Washington

Neuropsychiatric symptoms are among the most prominent manifestations of generalized chemical sensitivity. Patients, clinicians, and researchers are in agreement that symptoms such as depression, irritability, and mood instability are prominent among the distressing and disabling symptoms occurring in response to low-level chemical exposure. Beyond that point, however, agreement is difficult. The pathophysiology and clinical management of these symptoms remain quite controversial.

This paper will review available data on the prevalence and form of psychiatric symptoms among those suffering from multiple chemical sensitivity. Various models explaining the relationship of psychiatric symptoms to chemical sensitivity will be discussed. Finally, the implications of these models for clinical management and future research will be reviewed.

REVIEW OF PRIOR RESEARCH

Several reports over the last decade have described the form and frequency of psychiatric symptoms in chemically sensitive patients. Published reports vary widely in procedures for subject selection and assessment of psychiatric morbidity. Greater weight should be given to studies which attempt to include representative patient samples and incorporate systematic assessment. Relevant findings are summarized below.

In 1983, Brodsky described eight patients seeking worker's compensation for multiple chemical sensitivity (Brodsky, 1983). These patients were seen during an independent

1. Address all correspondence to: Gregory Simon, Center for Health Studies, 1730 Minor Ave., #1600, Seattle, WA 98101-1448. Tel.:(206)287-2979. Fax:(206)287-2871.

2. Abbreviations: MCS, multiple chemical sensitivity.

3. Key Words: chemical sensitivity, envioronmental illness, pathophysiology, psychiatric.

Toxicology and Industrial Health, Vol. 10, No. 4/5, pp. 487–496

ISSN: 0748-2337

medical evaluation performed by the author. He reported that "most have a history of overt psychiatric symptoms" and "long history of recurring physical complaints unsupported by objective physical findings." No further details were reported.

A report from the Yale Occupational Medicine Program by Schottenfeld and Cullen described 21 patients "disabled by medically unexplained symptoms following occupational exposure to toxic substances" (Schottenfeld and Cullen, 1985). Chart review and clinical assessment found that seven of these 21 patients satisfied diagnostic criteria for post–traumatic stress disorder. At least three patients also satisfied criteria for major depressive episode.

In 1985, Stewart and Raskin described 18 patients referred to a university psychiatric consultation-liaison service because of environmental illness. According to clinical assessment, three patients suffered from psychotic disorders and seven from depressive or anxiety disorders. Seven patients were diagnosed as suffering from somatoform disorders.

A report by Dager and colleagues (1987) described three patients with histories of acute chemical overexposure who later developed recurrent symptomatic attacks when reexposed to low levels of similar substances. The authors considered these cases of panic disorder precipitated by chemical exposure. Two of three patients had panic responses to infusions of sodium lactate, a biological "probe" for panic disorder. All three patients responded to medication usually used for treatment of panic disorder (tricyclic antidepressants).

Simon and colleagues reported on 37 patients involved in an outbreak of illness following low-level solvent exposure in a plastics manufacturing plant (Simon et al., 1990). Of approximately 400 exposed workers, these 37 had become sufficiently ill to file worker's compensation claims and complete an assessment by an independent medical evaluation board. Evaluation included a structured psychiatric diagnostic interview (the Diagnostic Interview Schedule) and various standardized self-report questionnaires. Approximately 75% of subjects reported symptoms consistent with current anxiety or depressive disorders. Means scores on the anxiety, depression, and somatization scales of the Hopkins Symptom Checklist-90 were at least two standard deviations above population norms. When workers were classified according to the presence of generalized chemical sensitivity, the 13 workers developing generalized sensitivity reported (in retrospect) a much higher prevalence of anxiety, depression, and somatization symptoms preceding chemical exposure than the 23 workers whose symptoms were confined to the workplace. The authors concluded that prior psychological morbidity or a tendency toward somatization may influence the development of generalized chemical sensitivity following chemical exposure.

Black and colleagues compared psychiatric morbidity among 26 patients with environmental illness with that of 13 age- and sex-matched controls from a community sample (Black et al., 1990). Subjects with environmental illness were recruited from support groups, occupational medicine clinics, psychiatric clinics, and local health food stores. Evaluation included a structured psychiatric diagnostic interview (the Diagnostic Interview Schedule).

Approximately 40% of cases met diagnostic criteria for depressive disorder (compared to 13% of controls, and 43% of cases were found to have anxiety disorders (compared to 17% of controls). The authors concluded that chemically sensitive patients "may have one or more commonly recognized psychiatric disorders that could explain some or all of their symptoms."

In 1992, Fiedler and colleagues described eight patients referred to a university occupational health clinic for assessment of multiple chemical sensitivity (Fiedler et al., 1992). One patient with a "significant psychiatric history" was excluded. Assessment included a semistructured psychiatric assessment (the Structured Clinical Interview for DSM-IIIR). Half of the sample were found to have current depressive disorders, but none reported symptoms of psychiatric disorder prior to the onset of chemical sensitivity.

In a follow-up to the cohort study described above, Simon and colleagues compared 41 chemically sensitive patients to a comparison group of 34 patients with chronic musculoskeletal injuries (primarily back injuries and repetitive motion syndromes) (Simon et al., 1993). Chemical sensitivity subjects were systematically selected from the practice of an allergist specializing in care of chemical sensitivity. Assessment included a structured psychiatric diagnostic interview (the Diagnostic Interview Schedule) and self-report measures of psychiatric symptoms. Current anxiety or depressive disorders were found among 44% of chemical sensitivity patients but only 15% of controls. Chemical sensitivity patients also reported markedly more symptoms from the anxiety, depression, and somatization subscales of the Symptom Checklist-90. When premorbid psychiatric symptoms were assessed, chemical sensitivity patients did not report a higher prevalence of prior psychiatric morbidity. A significant minority of chemical sensitivity patients scored within the normal range on all measures of psychiatric symptoms.

Many of the studies above have significant methodological weaknesses. Chemical sensitivity patients were often recruited from referral clinics or from disability claimants. As discussed below, such patients would be expected to report higher level of psychopathology. Control samples were often absent or inappropriate. In some studies, unstructured clinical assessments allow significant potential for bias.

Despite these limitations, all published reports find higher rates of anxiety and depressive symptoms among chemical sensitivity patients than expected by chance. Overall, approximately half of cases satisfy diagnostic criteria for formal anxiety or depressive disorders. These prevalence rates are significantly higher than those of patients with most comparably severe medical illnesses. While some studies find high rates of preexisting psychopathology, this finding is not universal. Most studies find that psychiatric symptoms are common in chemical sensitivity, but not universal.

THEORETICAL MODELS

The data described above are consistent with a number of different models of psychiatric symptoms in chemical sensitivity. While not mutually exclusive, these models emphasize different aspects of the relationship between low-level chemical exposure and symptoms of anxiety and/or depression. Four such models are described below:

"Masked" Psychiatric Disorder
This model explains chemical sensitivity symptoms as somatic expressions of an underlying psychiatric disorder — typically an anxiety or depressive disorder (Black et al., 1990). According to this theory, patients with apparent chemical sensitivity are either unable or unwilling to acknowledge psychological factors that underlie physical symptoms and therefore "convert" this psychological distress or conflict into more socially acceptable physical symptoms. Given current concern about environmental contamination and low-level chemical exposure, chemical sensitivity is a convenient "somatic idiom" for the expression of psychological distress. This model has its roots in the early psychoanalytic descriptions of conversion hysteria. In those accounts, unacceptable psychological conflicts or impulses were less disturbing when expressed as dramatic physical symptoms. A more contemporary example of such a model is the attribution of many chronic pain complaints to "masked depression." Data supporting this model include the increased prevalence of diagnosable psychiatric disorders among chemically sensitive patients, as well as the finding that psychiatric disorders often appear to predate the onset of chemical sensitivity.

This model's clearest drawback is its immediate discounting of patients' primary symptoms and concerns. Chemical sensitivity symptoms are given weight only as indicators of underlying psychological distress. Treatment focuses on acknowledgment of the "real" problem with the expectation that physical symptoms will then fade away. Not surprisingly, chemically sensitive patients confronted with this approach feel discounted and misunderstood. This discounting of physical symptoms is not supported by more modern research. The relationship between depression and chronic pain, for example, is clearly bidirectional (Coderre et al., 1993). Pain can cause depression and withdrawal just as depression can increase sensitivity to painful stimuli. Such a bidirectional relationship is more consistent with the amplification model discussed below.

Despite the above failings, this model does focus attention on how cultural beliefs influence our understanding of both physical and psychological distress. Certain explanations or attributions may be culturally favored, and cultural preferences may vary across place and time. The epidemiology of chemical sensitivity suggests that cultural factors may significantly influence the likelihood that symptoms will be attributed to low-level chemical exposure. The demographic and geographic groups with highest apparent rates of chemical sensitivity are not usually those with highest rates of solvent or pesticide exposure. The apparently higher prevalence of chemical sensitivity among white-collar workers in more economically developed countries suggests that higher levels of knowledge or concern about low-level chemical exposure increase the likelihood of developing generalized sensitivity (Mooser, 1987; Cullen et al., 1992). It is equally possible, however, that cultural factors affect the recognition or diagnosis of chemical sensitivity rather than its incidence.

Classical Conditioning

This model explains reactions to low-level chemical exposure as conditioned physiologic responses (Bolla-Wilson et al., 1988). Such a model fits best for patients who develop chemical sensitivity after a discrete episode of over-exposure. Physical symptoms occurring with that initial exposure are linked to accompanying sensations (such as odors or tastes) and those physiologic responses recur whenever the sensory cue is encountered. This learning process is both involuntary and unconscious. Phobic anxiety (such as agoraphobia) is one example of such a conditioning process. Conditioning mechanisms are not limited to strictly psychological phenomena. Animal studies have demonstrated conditioning of such clearly "physical" processes as immune response (MacQueen et al., 1989).

Symptom Amplification

This model explains chemical sensitivity as a tendency to amplify the noxious sensations associated with low-level chemical exposure (Barsky, 1979). Extreme sensitivity to low-level chemical exposure is viewed as a special case of the general phenomenon of symptom sensitivity. Chemically sensitive patients may be more aware of or more disturbed by the noxious odors or irritant symptoms of low-level chemical exposure. Extensive research in the area of symptom perception has demonstrated that sensitivity to noxious physical sensations is a relatively stable individual characteristic which can be acutely influenced by psychological stress (Barsky et al., 1988; Pennebaker, 1982). According to this model, chemically sensitive patients may amplify symptoms associated with low-level exposure or amplify normal ambient somatosensory phenomena and falsely attribute those sensations to chemical exposure.

Neurotoxicity

This model explains the psychological symptoms of chemically sensitive patients as evidence of neurotoxicity due to low-level chemical exposure (Bell et al., 1992). Consequently, psychiatric symptoms are seen as consequences of chemical sensitivity and have no etiologic role. The neuropsychiatric symptoms of chronic solvent exposure serve are one example of such a process. In that syndrome, psychiatric symptoms such as depression, irritability, or mood instability may precede any objective neuropsychological deficit.

As noted earlier, these models are not mutually exclusive. Toxic injury may interact with prior psychological vulnerability to produce the syndrome of generalized sensitivity. Borrowing from epidemiologic vocabulary, a tendency toward symptom amplification could be viewed as an effect modifier of mild neurotoxic injury (i.e., among patients with mild neurotoxicity, those with preexisting tendencies toward symptom amplification will become more symptomatic). It is then an exactly equivalent statement to say that toxic injury is an effect modifier of preexisting symptom sensitivity (i.e., among those with an average level of symptom sensitivity, development of MCS will depend on level of exposure to solvents or toxins). Whether psychological or neurotoxic factors appear to predominate depends on the population observed and the observers' point of view. In a population with equal levels of chemical exposure, individual differences in vulnerability will be the primary determinant of chemical sensitivity symptoms. Psychological distress or a tendency to symptom amplification may be central components of vulnerability. In a population with equal levels of prior vulnerability, development of chemical sensitivity will be determined solely by intensity of exposure. The study of plastics workers described above fits the first scenario. In that sample of workers with similar levels of exposure, prior psychological vulnerability (reported retrospectively) was a strong correlate of developing chemical sensitivity.

Individual vulnerability and toxic injury may also interact in determining the functional or behavioral outcomes of low-level chemical exposure. Given identical injuries, psychological factors such as depression have a powerful influence on the likelihood of seeking professional treatment or becoming disabled. This finding should not be interpreted as evidence that disabling injury is somehow imagined or factitious. Even among patients with well-defined medical illnesses or mechanical injuries, psychosocial factors exert a strong influence on disability and help-seeking (VonKorff et al., 1992; Simon, 1992). This relationship does, however, have important implications for interpreting research on psychological factors in chemical sensitivity. Samples of patients seeking treatment or claiming disability are likely to include those with highest levels of preexisting psychological vulnerability.

The neurotoxic and conditioning models do suggest differing mechanisms of sensitization. A neurotoxic model based on kindling mechanisms implies continuing injury with each episode of low-level exposure. Structural neuronal changes caused by repeated episodes of toxicity result in a progressively lower threshold for neuronal injury. A behavioral model leads to a quite different explanation for the same set of surface events. Repeated symptomatic episodes lead to progressively greater sensitivity through a learning or conditioning mechanism. No lasting structural injury occurs. The data available so far do not clearly distinguish between these two models. Electrophysiological or functional imaging findings reported so far are not inconsistent with psychological mechanisms. For example, similar functional abnormalities are demonstrable during the course of panic attacks (Reiman et al., 1984). While both the neurotoxic and conditioning mechanisms may be active simultaneously, the distinction between them has important treatment implications.

CLINICAL IMPLICATIONS

The various models described above have quite different implications for the clinical management of chemically sensitive patients. The neurotoxic or kindling model strongly supports the avoidance regimens endorsed by most clinical ecologists and environmental physicians. Any episode of exposure which causes symptoms may also contribute to further neuronal damage. Repeated episodes of injury will lower the threshold for future sensitivity and increase the likelihood that symptoms will become autonomous. The behavioral conditioning model leads to a nearly opposite conclusion. Avoidance will reinforce the learned association between noxious sensations and physical symptoms. Symptom control and gradually increasing exposure are the cornerstones of treatment. This conflict appears, for the moment, unresolvable. One doctor's cure is clearly another person's poison.

RESEARCH IMPLICATIONS

Further research is not needed to establish the increased prevalence of psychiatric symptoms in chemically sensitive patients. The studies described above are unanimous in finding that patients with generalized chemical sensitivity report a greater number of neuropsychiatric symptoms than do nonpatients or patients with unrelated disorders. Further research assessing the prevalence of psychopathology among the chemically sensitive will not clarify the relationship of chemical sensitivity to psychiatric symptoms and may only serve to inflame an already controversial area. Unfortunately, prior research on psychiatric symptoms has been interpreted as evidence that chemical sensitivity is "all in your head."

Future research should focus on the mechanism of sensitization to low-level exposure. As discussed above, the available data are consistent with both behavioral models and neurotoxic models. These two models lead the clinical management of chemically sensitive patients in nearly opposite directions. Understanding mechanisms of sensitization will require different research focus and methods. While the studies described above used psychiatric epidemiologic methods to evaluate the form and frequency of psychiatric symptoms, the next generation of studies will focus on electrophysiology and functional imaging.

While masking or deception experiments may seem best suited to distinguishing behavioral from neurotoxic mechanisms, such methods have significant drawbacks. First, masking olfactory sensation may not necessarily isolate conditioned responses. While conscious sensory cues are the most common model for conditioning experiments, conditioning may be mediated by other mechanisms. Consequently, persistence of reactions after masking obvious sensory cues does not exclude a conditioning mechanism. Conversely, absence of more dramatic arousal-type reactions after masking does not preclude simultaneous toxicity. It is conceivable that arousal-type symptoms (e.g., sweating, palpitations) may be conditioned while simultaneously occurring symptoms reflect neurotoxicity. Second, masking experiments will only perpetuate the polarization and mistrust that have plagued research in this area. Implicit in the design of such experiments is the assumption that chemically sensitive patients must be tricked or double-crossed. Such methods carry a message (intended or not) that chemical sensitivity symptoms may well be factitious or malingered. While blinded challenges may appear the purest method for dissecting toxic reactions from conditioning, such studies may reinforce the destructive controversy about whether chemical sensitivity symptoms are "real."

Research on neurobiologic mechanisms of chemical sensitivity should avoid the errors of earlier research on immunologic mechanisms. First, uncontrolled case series should be used only for generation of hypotheses. Any definitive statements about neurobiologic changes in chemically sensitive patients should be supported by comparison of systematically selected patients with appropriately selected control groups. Second, findings described by investigators with commercial interests in diagnosis of chemical sensitivity should be independently confirmed. Finally, planning of future studies should include advance consultation with researchers of varying views regarding appropriate methods. Advance agreement regarding appropriate methods will increase the acceptability and impact of subsequent results. Criticisms of study methods are more helpful and less subject to bias when made before results are known.

Neurobiologic research should not lose sight of epidemiologic findings. Neurotoxic or kindling models must somehow account for the apparent dramatic variation in susceptibility to chemical sensitivity. Only a small portion of those with histories of solvent or pesticide exposure develop generalized sensitivity and risk shows no clear relationship to level of exposure. While some who report sensitivity to low-level exposure recall identifiable inciting episodes of overexposure, many describe an insidious onset with no unusual exposure episodes. Kindling models would appear to imply a progressive course. While few data are available on the natural history of chemical sensitivity, some cases appear to resolve spontaneously.

BARRIERS TO USEFUL RESEARCH

Current policies for compensating liability and disability claims are a definite impediment to research on neuropsychiatric aspects of chemical sensitivity. In many cases, toxic mechanisms of injury are compensable while behavioral mechanisms are not. Given these policies, most patients and treating clinicians have strong incentives to emphasize neurotoxicity as the explanation for symptoms of anxiety and depression. Employers, insurers, and industry have equally strong incentives to emphasize behavioral mechanisms or preexisting psychopathology in order to reduce costs of compensation. Those with vested interest may overinterpret research data on psychiatric symptoms in chemical sensitivity as proving or disproving particular theories of pathophysiology. This political environment leaves little room for impartial research.

Discussion about neuropsychiatric symptoms in chemical sensitivity has also been tainted by the stigma attached to psychiatric illnesses. Data on the prevalence of depressive and anxiety symptoms among chemically sensitive patients has sometimes been met with protests that chemically sensitive patients are not "a bunch of crazies." Some of the current enthusiasm for research on kindling and neurotoxicity may reflect attempts to avoid this psychiatric stigma. Neurobiologic abnormalities may be seen as proof that chemically sensitive patients aren't "just imagining things." Ironically, this research probably brings our understanding of chemical sensitivity closer to current models of anxiety and depressive disorders. Modern genetic and neurobiologic research is rapidly blurring the distinction between "organic" and "functional" mental disorders. Whatever the results of research on the neurobiology of chemical sensitivity, patients, clinicians, and researchers would do well to acknowledge and avoid prejudicial attitudes toward psychiatric conditions. Better science and better patient care will result.

SUMMARY

Psychiatric symptoms are among the most common and disabling manifestations of multiple chemical sensitivity. Despite some methodological limitations, published research invariably finds elevated rates of anxiety and depressive symptoms among chemically sensitive patients. Available data, however, are consistent with a number of possible mechanisms explaining psychiatric morbidity in MCS. These models lead to quite different conclusions about the clinical management of chemically sensitive patients and the risks of continued low-level chemical exposure. Future research should focus on isolating and evaluating neurotoxic and behavioral mechanisms. Unfortunately, research to date has been too focused on whether MCS is a "real" or legitimate clinical entity. Persistent stigma against patients with psychiatric conditions, as well as current liability and compensation policies, are serious barriers to useful research.

REFERENCES

BARSKY, A.J. (1979). "Patients who amplify bodily sensations." Ann. Int. Med. 91:63–70.

BARSKY, A.J., GOODSON, J.D., LANE, R.S., and CLEARY, P.D. (1988). "The amplification of somatic symptoms." Psychosom. Med. 50:510–519.

BELL, I., MILLER, C.S., and SCHWARTZ, G.E. (1992). "An olfactory-limbic model of multiple chemical sensitivity syndrome: possible relationships to kindling and affective spectrum disorders." Biol. Psychiat. 32:218–42.

BLACK, D., RATHE, A., and GOLDSTEIN, R.B. (1990). "Environmental Illness: A controlled study of 26 subjects with '20th Century Disease'." J. Am. Med. Assoc. 264(24):3166–3170.

BOLLA-WILSON, K., WILSON, R.J., and BLEECKER , M.L. (1988). "Conditioning of physical symptoms after neurotoxic exposure." J. Occup. Med. 30(9):684–686.

BRODSKY, C.M. (1983). "'Allergic to everything': A medical subculture." Psychosomatics 24:731–742.

CODERRE, T., KATZ, J., VACCARINE, A.L., and MELZACK, R. (1993). "Contribution of central neuroplasticity to pathological pain: Review of clinical and experimental evidence." Pain 52:259–285.

CULLEN, M., PACE, P.E., and REDLICH, C.A. (1992). "The experience of the Yale occupational and environmental medicine clinics with Multiple Chemical Sensitivities, 1986–1991." Toxicol. Ind. Health 8:15–19.

DAGER, S., HOLLAND, J.P., COWLEY, D.S., and DUNNER, D.L. (1987). "Panic disorder precipitated by exposure to organic solvents in the work place." Am. J. Psychiat. 144:1056–1058.

FIEDLER, N., MACCIA, C., and KIPEN, H. (1992). "Evaluation of chemically sensitive patients." J. Occup. Med. 34(5):529–538.

MACQUEEN, G., MARSHALL, J., PERDUE, M., SIEGEL, S., and BIENENSTOCK, J. (1989). "Pavlovian conditioning of rat mucosal mast cells to secrete rat mast cell protease." Science 243:83–85.

MOOSER, S. (1987). "The epidemiology of multiple chemical sensitivities." Occup. Med. State Art Rev. 2:663–669.

PENNEBAKER, J.W. (1982). The Psychology of Physical Symptoms. Springer-Verlag, New York.

REIMAN, E., RAICHLE, M.E., BUTLER, F.K., HERSCOVITCH, P., and ROBINS, E. (1984). "A focal brain abnormality in panic disorder, a severe form of anxiety." Nature 310:683–685.

SCHOTTENFELD, R.S. and CULLEN, M.R. (1985). "Occupation-induced post–traumatic stress disorder." Am. J. Psychiat. 142:198–202.

SIMON, G. (1992). "Psychiatric disorder and functional somatic symptoms as predictors of health care use." Psychiat. Med. 10(3):49–60.

SIMON, G., KATON, W.J., and SPARKS, P.J. (1990). "Allergic to life: Psychological factors in environmental illness." Am. J. Psychiat. 147:901–906.

SIMON, G., VONKORFF, M., WAGNER, E.H., and BARLOW, W. (1993). "Patterns of antidepressant use in community practice." Gen. Hosp. Psychiat. 15:399–408.

STEWART, D.E. and RASKIN, J. (1985). "Psychiatric assessment of patients with '20th century Disease.'" Can. Med. Assoc. J. 133:1001–1006.

VONKORFF, M., ORMEL, J., KATON, W.J., and LIN, E.H.B. (1992). "Disability and depression among high utilizers of health care." Arch. Gen. Psychiat. 49:91–100.

PSYCHOLOGICAL BASES OF SYMPTOM REPORTING: PERCEPTUAL AND EMOTIONAL ASPECTS OF CHEMICAL SENSITIVITY

JAMES W. PENNEBAKER

Southern Methodist University
Dallas, Texas

The reporting of physical symptoms is influenced to a large degree by psychological processes. Individuals are more likely to notice subtle sensations in environments lacking in stimulation than those demanding external attention. The beliefs or schemas that people hold dictate where and how they attend to their bodies as well. These normal perceptual processes help explain why people are often poor at accurately detecting internal physiological activity. Several individual differences are also related to the symptom reporting process. Females are more likely to base their symptom reports on external situational cues than are males. In addition, individuals with chronic anxiety — those high in Negative Affectivity (NA) — report more symptoms than those low in NA. Finally, individuals who have had traumatic experiences, either in childhood or within 1–6 months prior to a major symptom reporting episode, tend to be high symptom reporters. Several recommendations are made to help researchers and clinicians distinguish between psychological or perceptual factors with presumed biological effects. One implication of this work is that MCS and allied syndromes should be viewed as both a mental and a physical health problem.

INTRODUCTION

The relatively new phenomenon of chemical sensitivity or multiple chemical sensitivity (MCS) is characterized by individuals' experiencing a variety of unpleasant and often ambiguous physical symptoms in response to relatively low dosages of chemicals in the environment. Cases of chemical sensitivity have been reported among individuals who appear to be hypersensitive to a variety of measurable and/or perceived chemicals. Some of the

1. Address all correspondence to: James W. Pennebaker, Department of Psychology, Southern Methodist University, Dallas, Texas 75275 (e-mail address: h7dr0001@vm.cis.smu.edu). Tel.:(214)768-3727. Fax:(214)768-2187. I am indebted to David Watson for his input in several sections of the paper. Preparation of this paper was made possible by National Science Foundation grant BNS90-21518.

Toxicology and Industrial Health, Vol. 10, No. 4/5, pp. 497–511

ISSN: 0748-2337

perplexing facets of this phenomenon include the failure to find reliable dose-response relations among presumed toxins and symptoms, difficulty in identifying immune or other biological responses associated with the reported symptoms, and the growth of an alternative health care system to deal with people complaining of MCS (Bell, 1994; Miller, 1994).

At the heart of the MCS controversy is the question "Is it real?" Rather than directly answering the reality question, the purpose of this chapter is to explore how symptoms in general are perceived and reported. For example, there is little doubt that the physical symptoms of both well-understood disorders such as influenza and less-understood problems such as MCS are subjectively distressing. Despite their possible physiological bases, the symptoms of both influenza and of MCS indicate that psychological and perceptual factors are related to both the etiology and possibly the treatment of their symptoms. Virtually all diseases — whether biologically validated or not — have physical symptoms that are influenced by psychological processes. In the first part of this chapter, physical symptoms and sensations are shown to be normally perceived in the natural environment. From this work, parallels between perceptual processes, mass psychogenic illness, and MCS are drawn. Finally, some methods are proposed by which to isolate psychological, perceptual, and social variables in understanding MCS.

THE PERCEPTUAL PROCESS: THE ACTIVE VERSUS PASSIVE PERCEIVER

The most common symptoms of MCS include fatigue, difficulty concentrating, racing heart, shortness of breath, anxiety, headache, upset stomach, dizziness, and muscle tension (Miller, 1994). Indeed, these sensations are also among the most common among healthy samples of normal individuals (Pennebaker, 1982), those diagnosed with somatization disorders (e.g., Robbins and Kirmayer, 1991), and even those diagnosed with hypertension (Robinson, 1969; Pennebaker and Watson, 1988) and diabetes (Cox et al., 1985; Pennebaker et al., 1985). Similarly, most of these physical symptoms are also typically associated with diagnoses of clinical depression (APA, 1987). Within the personality literature, these constellations of symptoms are the basis of Negative Affectivity, or NA (Watson and Pennebaker, 1989; Pennebaker and Watson, 1991).

Although people suffering from MCS report a variety of symptoms that do not have a clear biological basis, the prevailing opinion is that their reports are subjectively real. That is, when individuals report symptoms and sensations, they subjectively experience significant bodily activity. The question that immediately comes to mind, then, concerns the perceptual information that influences how individuals attend to and interpret the body's ambiguous signals. Research addressing this issue has evolved considerably over the last century.

The laboratories of Wilhelm Wundt and Gustav Fechner yielded the first scientific investigations relevant to symptom perception. In developing the science of psychophysics, Wundt and others demonstrated a one-to-one correspondence between external stimuli and the perception of stimuli (Boring, 1950). That is, when all extraneous variables were held constant, perceptions of light, sound, touch, and other sensory dimensions were mathematically related to the sources of the percepts.

J.J. Gibson (1966, 1979) was one of the first researchers to question the psychophysics tradition within perceptual psychology. The crux of his argument was that organisms rely on information from a variety of sources in order to perceive even the simplest stimulus arrays. Further, this information was processed by virtually all of the senses simultaneously in order to create a stable percept. Accordingly, organisms were not passive recipients of information. Rather, they actively searched the environment in order to understand it better and, consequently, to behave appropriately and efficiently.

Consider, for example, how individuals perceive the brightness of the moon at night. Of course they rely heavily on their eyes and their visual systems in general. But the perception of the brightness of an object depends on much more than the stimulation of a portion of rods and cones within the retina. Visually, perception of brightness will also vary depending on the background of the sky, other visible objects, and degree of adaptation to darkness. Beyond the visual system, the perception of brightness will also be affected by our beliefs about or needs concerning the object. For example, if we live in a primitive culture and believe that moonlight exerts an evil influence, we may over- or underestimate its brightness. Similarly, an adolescent couple trying to walk undetected in their neighborhood at night may overestimate the moon's brightness. Conversely, their curious neighbors may underestimate its brightness.

More relevant to the present discussion is the role of nonvisual information that can affect visual perception. The apparent brightness of an object can be altered if the perceiver experiences intense pain, loses vestibular cues (and a corresponding sense of balance), becomes paralyzed, or even hears a loud noise. The point that modern day perceptual researchers emphasize is that we simultaneously rely on all of our perceptual systems in making sense of any visual array.

The Role of Attention

How do individuals first notice and attend to internal physical sensations as opposed to external visual cues? Given that individuals can process only a finite amount of information at any given time, we have proposed that internal sensory and external environmental cues compete for attention (Pennebaker, 1982). As the number and salience of external cues increases, attention to internal stimuli will necessarily decrease, and vice versa. When the environment lacks meaningful external information (such as when people engage in boring or tedious tasks), attention will tend to focus more internally, thereby causing an increase in symptom reporting. Thus, people should perceive and report more symptoms in unstimulating environments than in interesting ones.

A great deal of research now supports this competition of cues model. Various experiments demonstrate that people report higher levels of fatigue (Fillingham and Fine, 1986), increased heart palpitations, and even cough more in boring situations than in stimulating ones. Manipulations that heighten self-attention also increase physical symptom reporting (e.g., Mechanic, 1980). Indeed, epidemiological studies indicate that symptom reports are elevated when individuals live alone or in rural environments, or work in undemanding or unstimulating settings (e.g., NCHS, 1980). Conversely, increased focus of attention to the body heightens symptom reports. It is noteworthy, however, that these increased symptom reports are unrelated to accuracy of perceiving physiological change (e.g., Pennebaker and Watson, 1988).

The Selective Search for Information

Another line of research is based on the assumption that organisms actively search their environment for information that will enable them to behave more adaptively (e.g., Neisser, 1976). This scanning is not random but, rather, is guided by beliefs or mental sets that direct the ways in which information is sought and ultimately found. This principle is also relevant to the formation of health complaints. That is, health-related beliefs influence how people attend to and interpret bodily sensations (e.g., Skelton and Croyle, 1991).

Dramatic examples of the power of health beliefs, or schemas, can be seen in cases of "medical student's disease" and mass psychogenic illness. Regarding the former, approximately 70% of first year medical students report symptoms of the diseases that they are studying (Woods et al., 1978). The students, who are undoubtedly under stress from sleep deprivation, exams, or other reasons, can detect a number of subtle bodily sensations that probably reflect heightened autonomic activity. When they read about various obscure illnesses associated with ambiguous symptoms, the students now scrutinize their bodies particularly closely. Their disease beliefs or schemas direct the ways they attend to their bodies and interpret their symptoms.

Schemas and selective search also play an important role in cases of mass psychogenic illness, or MPI (Colligan, et al., 1982). In MPI, large groups of individuals who typically work together report a related set of physical symptoms that have no clear organic basis. MPI usually develops when one person in a setting becomes overtly sick and displays observable symptoms such as vomiting or fainting. These symptoms affect the belief processes of others in the setting — especially friends — who consequently experience similar, but less dramatic symptoms such as feelings of nausea or dizziness. It is also important to appreciate that cases of MPI are most likely to occur in settings where people are anxious or tense. Many of the most famous cases have occurred during the height of the production season, in companies with poor worker-management relations, loud and unpredictable noise, and jobs that restrict people from talking to one another. Because of several striking similarities between MPI and MCS, we will return to this topic.

The perceptual approach is important in clarifying how normal individuals typically detect and report symptoms in the real world. Obviously, the symptom-reporting process can also be exaggerated or dampened by rewards or punishments or other forms of primary and secondary gains (e.g., Mechanic, 1978). Most of our research on symptom reporting has been conducted on individuals who would not be classified as suffering from MCS or any of the somatoform disorders. Nevertheless, it is easy to see how extreme variants of these same processes could lead to the greatly magnified symptom reporting that characterizes many cases of MCS.

GENDER, PERSONALITY, AND LIFE EXPERIENCES THAT PREDISPOSE PEOPLE TO SYMPTOM REPORTING

Some people tend to report physical symptoms at higher rates than others. By the same token, cases of MCS do not appear to be randomly distributed within the population nor within groups of people presumably exposed to the same doses of chemicals. As will be discussed below, there is evidence to suggest that women, people high in Negative Affectivity, and individuals who have experienced a traumatic life event either in childhood or recently are prone to report a relatively high degree of a variety of symptoms and sensations.

Gender Differences in Symptom Reporting and Information Processing

An interesting outgrowth of our research on symptom perception has been the emergence of a consistent sex difference in how individuals notice, define, and react to symptoms. Specifically, women are particularly sensitive to external environmental cues and men to internal physiological cues in defining their symptoms. This conclusion is based on laboratory and field studies that have attempted to learn how accurate individuals are at perceiving specific physiological activity.

Across virtually all highly controlled laboratory studies, men are consistently better able than women to detect heart rate, stomach activity, blood pressure, and blood glucose levels. In naturalistic field studies, however, both women and men are equally good at estimating blood pressure, blood glucose, and various autonomic channels (for general review, see Roberts and Pennebaker, in press).

In a particularly strong test of this, Cox and his colleagues (1985) asked 19 diabetics who had had experience monitoring their blood glucose levels to participate in a two-phase study in the hospital and, later, at home. In the hospital phase, participants' glucose levels were directly manipulated over the course of a day. That is, a machine that simulated the activity of the pancreas took each diabetic on a blood glucose "roller coaster ride" over the 8-hour experiment. Once every 10 minutes, subjects estimated their glucose levels. Overall, the correlation between actual and estimated blood glucose was +.42 for men and +.13 for women. Later, the same subjects estimated and measured their glucose levels several times each day for two weeks at home during their normal days. At home, where a variety of situational cues were present, the correlation between actual and estimated glucose levels were +.58 for men and +.69 for women.

As discussed elsewhere (Pennebaker and Roberts, 1992), situational cues — such as time of day, room temperature, and so forth — are usually redundant with internal physiological cues. In other words, we can make fairly accurate educated guesses about people's physiological activity if we know the settings they are in. What is interesting, however, is that men and women differentially rely on internal versus external cues in defining the symptoms that they are feeling.

The use of differential perceptual strategies by men versus women may explain the large gender differences in somatization disorders and, perhaps, in cases of MCS. Given that women are especially sensitive to situational cues, their symptom-reporting patterns should reflect the settings that they are perceiving as stressful and potentially toxic. Men, however, will tend to ignore the settings and focus on their physiological cues. Symptom reporting, then, will tend to mirror situational fluctuations in women and physiological changes in men.

Interestingly, these gender effects are consistent with cases of MPI. Across cultures (e.g., Phoon, 1982) and over the centuries (e.g., Sirois, 1982), women have been found to be far more prone to MPI episodes than to men. It could be argued that MPI, like MCS, is a phenomenon wherein a salient smell or label emerges to help explain the existence of naturally occurring symptoms of stress or illness. The more prominent the situational label, then, the more likely that females may use it to help define their internal state. This could be both adaptive and maladaptive. If, in fact, low levels of chemical exposure are directly causing subtle adverse biological activity, females would be more likely to detect it than males. On the other hand, if the situational cues are benign and detectable biological changes are resulting from chemical exposure, males may be slightly more likely to detect them. Note that the gender differences in the perceptual bases of symptom reporting has been found in generally healthy individuals. Further, the effects are moderately strong but not overwhelming. At this point, the findings and explanations are intriguing and worthy of future study.

Negative Affectivity and Physical Symptoms

Within the last few years, researchers have begun to examine the role that personality variables play in the formation and reporting of physical symptoms. Much of this research has been based on a mood-related disposition called "Negative Affectivity" or trait NA (Watson and Clark, 1984). Trait NA is essentially identical to several other dispositional constructs such as neuroticism (Costa and McRae, 1987), trait anxiety (Taylor, 1953), pessimism versus optimism (Scheier and Carver, 1987), and so on. Trait NA reflects pervasive individual differences in negative mood and self-concept. High NA individuals experience consistently higher levels of distress and dissatisfaction over time and across different situations. High NA subjects are also more introspective and tend to dwell differentially on their failures and shortcomings. They tend to be negativistic, focusing on the negative aspects of themselves and others.

Several NA scales are available in the personality literature (e.g., Taylor, 1953; Byrne, 1961; Watson et al., 1988). A growing number of studies indicate that trait NA is highly correlated with virtually all measures of symptom reporting. Across several samples, using different NA scales and various measures of symptom reporting, NA markers typically correlate in the .30 to .50 range with symptom reports, with a mean coefficient of approximately r = .40. Interestingly, high NA individuals consistently report all types of sensations and physical symptoms to a greater degree than do low-NA individuals — even though high and low-NA subjects do not differ noticeably on various objective health markers (Costa and McRae, 1987; Watson and Pennebaker, 1989).

The reliable link between NA and symptom reporting, in the absence of any NA-related differences in objective health status, indicates that symptom reports are strongly affected by the NA trait. High NA individuals appear to be hypervigilant about their bodies and have a lower threshold for noticing and reporting subtle bodily sensations. Because of their generally pessimistic view of the world, they are also most likely to worry about the implications of their perceived symptoms. Indeed, these high NA individuals would also be the ones to worry more about chemicals, pollens, physicians, and the environment. In short, high NA subjects would seem to be at higher risk for MCS.

The possibility of an NA-MCS link cannot be discounted. First, consistent with other aspects of the trait, high NA individuals will be more likely than others to report symptoms across all situations and over long time intervals. Because of this, excessive symptom reporting is not only influenced by transient situational stressors, but also reflects a stable personality trait. Second, reliance on symptom reports without a concurrent measure of NA can lead to a distorted view of the meaning and significance of these symptoms. Third, no matter what the final conclusions surrounding MCS, researchers and clinicians should be alert to the role of NA as a nuisance factor that must be assessed when trying to evaluate and treat reports of symptoms.

As discussed above, symptom reports are influenced both by situational cues and by broad dispositional differences in distress and complaining. Closely allied with this evidence is recent work suggesting that the tendency to report symptoms and negative affect is strongly heritable (e.g., Rose et al., 1988; Tellegen et al., 1988). The genetic argument reflects common assumptions about the phenotypic bases of physiological functioning, as well as recent findings concerning the inheritance of perceptual and emotional styles.

The awareness and reporting of physical symptoms depends, to a large degree, on the way information is processed in different parts of the brain. The somatosensory cortex, for example, fundamentally affects how individuals perceive sensations in their bodies. Beyond basic perception, the ability to report symptoms is dependent on the proper functioning of the language centers in the temporal and parietal lobes (e.g., Luria, 1980). Further, it is well documented that central nervous system structure and function are, in turn, strongly genetically determined. Monozygotic twins, for example, have remarkably similar — albeit not identical — cortical structures, neurotransmitter activity, EEG and autonomic nervous system activation compared with DZ twins (Lykken, 1982). In short, the brain's biological hardware that underlies symptom perception clearly has a heritable basis.

Of particular importance is a series of discoveries pointing to the genetic bases of personality dispositions and their associated cognitive-perceptual-behavioral styles. Of greatest relevance are the findings of Tellegen and his colleagues at the University of Minnesota (1988) who have conducted a large-scale examination of the heritability of trait NA. The Minnesota researchers investigated over 400 pairs of identical and fraternal twins who were reared either together or apart. Overall, an estimated 55% of the variance on trait NA could be attributed to genetic factors, whereas only 2% was due to the shared familial environment. The remaining 43% of the variance is presumably attributable to measurement error and idiosyncratic situational influences that were not assessed by the researchers. A number of other large-scale studies have now reported similar findings in regard to the heritability of trait NA (Pederson et al., 1988; Rose et al., 1988).

Unfortunately, the Minnesota project did not directly examine the heritability of symptom reporting. Nevertheless, the heritability of the NA trait strongly suggests that individuals' proclivity to report symptoms and sensations has a similar genetic basis. Given the currently available data, however, it is impossible to point with any certainty to the biological and/or psychological mechanisms underlying the genetics of somatization. Perhaps the most promising hypothesis links attentional vigilance with specific physiological substrates.

Jeffrey Gray (1982) points to the importance of inhibitory centers within the brain (such as those in the septum and hippocampus) as influencing — both directly and indirectly — individual differences in trait NA. When these inhibitory centers are activated, individuals become hypervigilant about the presence of novel stimuli in the environment. Gray believes that high NA subjects, whom he calls trait anxious, have overactive inhibitory centers which result in their being characterologically hypervigilant. This hypervigilance probably affects symptom reporting in two ways. First, high NA subjects should be more attentive to subtle sensations in their bodies. Second, because their scanning is fraught with anxiety and uncertainty, high NAs may be more likely to interpret minor symptoms and sensations as painful or pathological. Interestingly, Barsky and Klerman (1983) also argue that hypervigilance, selective attention, and the tendency to view somatic sensations as ominous are all important elements in the amplification of symptoms. Thus, the perceptual style of high NAs may be largely responsible for their enhanced somatic complaining.

Traumatic Experiences and Symptom Reporting

Although situational and inherited dispositional factors can be powerful predictors of symptom reporting, certain broad life experiences also appear to influence symptom awareness — often in pathological ways. Symptom reports increase after traumatic experiences. This is seen in studies where large groups of traumatized individuals are followed for weeks, months, or years after such traumas as rape, death of spouse, or other trauma (Pennebaker, 1989, 1990, 1993; Pennebaker et al., 1988). Similarly, studies that have focused on large groups of individuals diagnosed with various somatoform disorders typically report trauma rates significantly above those of individuals either without somatoform disorders or other problems. What is it about a trauma that appears to exacerbate symptom reports?

For the last few years, my students and I have been examining this question from several perspectives. One important feature of traumas that is linked to self-reports of health and illness (and even health-related behaviors) is that those traumas that are not openly discussed with others are more problematic than those that are talked about. Across various large scale surveys, for example, we have consistently found that individuals who report having one or more traumas at any point in their lives about which they did not talk reported having significantly higher rates of reported minor health problems (headaches, upset stomach, racing heart) as well as serious diagnoses (high blood pressure, cancer, ulcers) — see Pennebaker and Susman (1988). Indeed, these symptom and illness rates were significantly higher than for subjects who had experienced the same types of traumas but who did talk about the events.

Why, then, would undisclosed traumas maximally elevate symptom reports? At least four overlapping hypotheses exist:

Long-term stress responses resulting from the trauma. One possibility is that these traumas are simply more biologically stressful and, in some way, result in adverse autonomic and immune function changes which are accurately detected by the perceivers. This may be true with many cases. However, closer inspection of people suffering from various somatoform disorders would indicate that greater accuracy of bodily state due to higher ANS activity is, at best, a partial explanation.

Symptom reporting as a distraction. People who have had traumas in their lives and who have not been able to resolve them may adopt whatever cognitive strategies they can to avoid thinking about these events. That is, they are always in search of distracters. Of all types of information, bodily cues are always available. By focusing on symptoms and sensations, individuals may be able to avoid addressing the overwhelming thoughts of emotional upheavals.

Traumatic thought suppression. A related hypothesis assumes that individuals who actively avoid trying to think about their traumas consistently work to block out trauma-relevant thoughts and emotions. In reality, aspects of the trauma are continuously processed on both a conscious and nonconscious level. When a dimension of the trauma pops into the individual's thoughts, he or she can suppress the thought fairly quickly. Of prime importance, the brief appearance of the thought together with the work of trying to suppress it, results in an emotion and autonomic response. From the individual's perspective, however, the bodily changes associated with the emotional response are not immediately interpretable since trauma-relevant thoughts continue to be suppressed. In other words, the person experiences an emotion without a perceptible eliciting event. Unable to truly define the bodily state as an emotion, the person's only recourse is to label the emotional changes as their components: physical symptoms (cf., Wegner, 1992; Wegner et al., 1990).

Secondary gain. The reporting of symptoms is ultimately a social act. By telling others of one's symptoms, the person is seeking help in reducing the symptoms, seeking more information about the causes or consequences of the symptoms, and, in some cases, searching for acknowledgment, attention, or other forms of reward from others. Reporting of symptoms following a stressful event has previously been found to bring stressed families closer together (e.g., Minuchin et al., 1975) and allow the symptom reporters a way by which to escape from other stressful situations such as school or work (e.g., Mechanic, 1978).

In many ways, it is beyond the goals of this paper to explore the many links between stressful life experiences and symptom reporting. Suffice it to say that in cases where ambiguous symptoms are reported at a high rate without a clear environmental or biological correlate, it would behoove the clinician or researcher to explore the patient's earlier traumatic life experiences.

DISTINGUISHING THE PSYCHOGENIC CORRELATES OF MCS

Over the last 20 years, we have witnessed the growth and popularization of several disorders that have helped people label and understand their ambiguous symptoms: iron-poor blood, hypoglycemia, chronic fatigue, mitral valve prolapse, Epstein-Barr, Seasonal Affective Disorder, and others. Similarly, a multitude of environmental factors have also been blamed for a comparable group of ambiguous symptoms: ionization of the air, video display terminals, refined sugar, animal fat, food additives, etc. Are these disorders and environmental toxins "real"? I suspect that the answer is sometimes yes. Among a group of 100 people who experience the standard symptoms of fatigue, anxiety, racing heart, dizziness, upset stomach, and headache, a small percentage probably have iron-poor blood, low blood glucose, or spend too much time in front of a video display terminal. Similarly, some of these 100 people will be hyperresponsive to low dosages of chemicals. In fact, a few may be slowly being poisoned to death by the chemicals they fear while their physicians dismiss their symptoms as psychogenic.

But here is the dilemma: although virtually all of our 100 people will be subjectively feeling the sensations they report, perhaps only a minority will exhibit biological activity that can explain the symptoms. Further, the presence of biological markers will typically be secondary to the power of perceptual and emotional factors in the symptom reporting process. Is it possible to disentangle truly dangerous low-level chemical exposure from psychological factors in explaining the symptoms of MCS? Using single session survey or experimental designs, this is an almost impossible task — something akin to proving the null hypothesis. On a case by case basis, however, where individuals are repeatedly exposed to potential toxins and placebos in a double-blind fashion, we may be able to distinguish the individual symptom reporting patterns in responses to substances. Note that this strategy may help to determine the dose-response gradient of toxins to symptoms for a particular individual in a controlled laboratory setting. The degree to which this knowledge generalizes to symptom reports in the real world may still be unknown since, after all, symptoms are subject to the various distortions inherent in the perceptual process.

A useful strategy for clinicians and researchers would be to acknowledge that symptoms are based on a multitude of factors. In the case of MCS, certain psychological variables may be particularly potent in driving the symptom reporting process whether or not toxic chemicals exist. The following checklist may be helpful in identifying the degree to which cases of MCS are being influenced by perceptual, emotional, or other psychological factors:

I. Perceptual Factors

A. Boring or tedious environment. People are more likely to pay attention to and to amplify bodily sensations in situations lacking in stimulation.

B. Situations fraught with tension or anxiety. Poor worker-management relations, interpersonal conflict at home, and other forms of stress or pressure can exacerbate physical symptoms. People in settings where they are unable to directly address the causes of their tension may be forced to look elsewhere to define their symptoms. For example, a person who has a specialized job with an abusive boss may not be able to acknowledge the tremendous tension brought about by his/her boss. An alternative label for this tension might be illness or MCS.

C. Isolation at work or home. Cases of MPI and, perhaps MCS, are likely to occur in places where individuals are unable or not allowed to talk with others. This can occur in places with high levels of ambient noise (e.g., factories) or in places where talking is discouraged (e.g., libraries). Lack of interpersonal communication allows people more time to ponder their bodily sensations, may exacerbate tension or anxiety, and may fail to allow for the normal social comparison that occurs when people try to understand their sensations (e.g., Singer et al., 1982).

D. An appropriate trigger or causal attribution. A new or unique smell, knowledge of chemical use, or recent exposure in the media of a disorder such as MCS can help people organize their symptoms into a coherent explanation. A particularly powerful trigger can be the overt and unexplained sickness or death of a coworker or family member.

E. The social spread of the disorder. Within the workforce, psychogenic diseases have been found to spread along friendship lines rather than by proximity.

F. Secondary gain. Symptoms of all disorders are much more likely to be reported if the person receives some kind of reward from others. The reward could be attention or alleviation of responsibilities at work or in the home. Research within a family systems framework suggests that somatization disorders (and perhaps MCS) can help to control the dynamics within the family.

II. Individual Differences

A. Gender. Females are more likely to suffer from episodes of MPI and, by extension, MCS. This may be partially due to the fact that males and females rely on different sources of information to interpret bodily state.

B. Negative Affectivity. Individuals with a history of reporting negative moods, thoughts, and symptoms may be especially prone to MCS. The ambiguous symptoms of MCS are strikingly similar to those of depression.

C. Traumatic experiences in childhood. People report far more physical symptoms if they have had a traumatic experience in childhood. Further, those who have not disclosed this trauma to others are more likely to report a variety of health and symptom problems throughout the life cycle.

D. Recent traumatic experiences. Psychological upheavals (death of family member, divorce) in the six months preceding the reporting of MCS may also contribute to the symptoms being reported.

IMPLICATIONS FOR TREATMENT

One of the concerns of the Public Health Service and many in the scientific and medical communities is that people with MCS and allied syndromes are turning to nontraditional healers to allay their distress. Given our current medical and psychiatric systems, I think this is both an understandable and, in some cases, a desirable trend. As outlined in this paper, it is reasonable to assume that many of the symptom reports of an unknown percentage of MCS sufferers are the result of psychological processes. Physicians and even patients who fail to appreciate these psychological dimensions will become quite frustrated in attempting to isolate subtle chemicals that may not exist or explain the symptoms themselves.

When dealing with real human beings suffering from vague symptom complaints, a strict medical model will neither explain the symptoms nor be valuable in treating them. The health professional should be encouraged to explore alternative sources of stress, situational factors, and even personality domains in attempting to better understand the symptoms themselves. Note that this strategy can also be helpful in trying to isolate possible toxic chemicals as well.

To the degree that medical professionals are unable or unwilling to explore the psychological domains of MCS symptoms, it may be appropriate for their patients to seek nontraditional healers to reduce their distress. Any healer who helps the individual to acknowledge perceptual and emotional dimensions of symptom reporting will probably be helpful in improving patients' physical and psychological health. The value of nontraditional healers may be limited, however. If truly toxic chemicals are causing the ambiguous symptoms, the nontraditional healers may ultimately be harmful. Those nontraditional healers who claim to represent the medical model who offer questionable treatments at high prices may, in some ways, be mildly beneficial. Unlike traditional healers, these individuals may be listening to their patients, offering treatments comparable to placebos, and giving the patients a sense of control — all psychological ingredients of healing (cf. Frank and Frank, 1991).

Ultimately, MCS is both a mental and a physical health problem. If there are, in fact, subtle chemicals poisoning the sufferers' bodies or their immune systems are failing in some as-yet-to-be-discovered way, the biological changes are highly stressful. Any mental health interventions that can positively affect this process would be beneficial. Alternatively, if a significant percentage of MCS cases are shown to be overwhelmingly psychological in their etiology, the stress of the symptoms will also impact sufferers' physical health. In short, all conceivable outcomes associated with MCS suggest a treatment wherein mental and physical health components are coordinated.

REFERENCES

AMERICAN PSYCHIATRIC ASSOCIATION (1987). Diagnostic and Statistical Manual of Mental Disorders. (3rd edition, revised). American Psychiatric Association Press, Washington, D.C.

BARSKY, A.J. and KLERMAN, G.L. (1983). "Overview: Hypochondriasis, bodily complaints, and somatic styles." Am. J. Psychiat. 140:273–283.

BELL, I.R. (1994). "Neuropsychiatric aspects of sensitivity to low-level chemicals: A neural sensitization model." Toxicol. Ind. Hlth. 277–312.

BORING, E.G. (1950). A History of Experimental Psychology. Appleton-Century, Crofts, New York.

BYRNE, D. (1961). "The Repression-Sensitization scale: Rationale, reliability, and validity." J. Personal. 29:334–349.

COLLIGAN, M.J., PENNEBAKER, J.W., and MURPHY, L.R., eds. (1982). Mass Psychogenic Illness: A Social Psychological Analysis. Erlbaum, Hillsdale, NJ.

COSTA, P. T., JR. and MCCRAE, R. R. (1987). "Neuroticism, somatic complaints, and disease: Is the bark worse than the bite?" J. Personal. 55:299–316.

COX, D.J., CLARKE, W.L., GONDER-FREDERICK, L.A., POHL, S., HOOVER, C., SNYDER, A, ZIMBELMAN, L., CARTER, W.R., BOBBITT, S., and PENNEBAKER, J.W. (1985). "Accuracy of perceiving blood glucose in IDDM." Diabetes Care 8:529–535.

FILLINGIM, R.B. and FINE, M.A. (1986). "The effects of internal versus external information processing on symptom perception in an exercise setting." Health Psych. 5:115–123.

FRANK, J.D. and FRANK, J.B. (1991). Persuasion and Healing. (3rd edition). Johns Hopkins University Press, Baltimore.

GIBSON, J.J. (1966). The Senses Considered as Perceptual Systems. Houghton-Mifflin, Boston.

GIBSON, J. J. (1979). The Ecological Approach to Visual Perception. Houghton-Mifflin, Boston.

GRAY, J.A. (1982). The Neuropsychology of Anxiety: An Enquiry into the Functions of the Septo-Hippocampal System. Oxford University Press, New York.

LEHMAN, D.R., ELLARD, J.H., and WORTMAN, C,B, (1986). "Social support for the bereaved: Recipients' and providers' perspectives on what is helpful." J. Consult. Clin. Psych. 54:438–446.

LURIA, A.R. (1980). Higher Cortical Functions in Man. Basic Books, New York.

LYKKEN, D.T. (1982). "Research with twins: The concept of emergenesis." Psychophys. 19:361–373.

MECHANIC, D. (1980). "The experience and reporting of common physical complaints." J. Health Soc. Behav. 21:146–155.

MECHANIC, D. (1978). Medical Sociology (2nd edition). Freeman, New York.

MILLER, C.S. (1994). "Chemical sensitivity: History and phenomenology." Toxicol. Ind. Hlth. 4/5: 253–276.

MINUCHIN, S., BAKER, L., ROSMAN, B.L., LIEBMAN, R., MILMAN, L., and TODD, T.C. (1975). "A conceptual model of psychosomatic illness in children." Arch. Gen. Psychiat. 32:1031–1038.

NATIONAL CENTER FOR HEALTH STATISTICS (1980). Basic Data on Depressive Symptomatology, Series 11, Number 216. USGPO, Washington, D.C.
NEISSER, U. (1976). Cognition and Reality. Freeman, New York.
PEDERSON, N.L., PLOMIN, R., McCLEARN, G.E., and FRIBERG, L. (1988). "Neuroticism, extraversion, and related traints in adult twins reared apart and reared together." J. Personal. Soc. Psych. 55:950–957.
PENNEBAKER, J.W. (1982). The Psychology of Physical Symptoms. Springer-Verlag, New York.
PENNEBAKER, J.W. (1989). "Confession, inhibition and disease." In: Advances in Experimental Social Psychology (L. Berkowitz, ed.). vol. 22. .Academic Press, New York. pp. 211–244.
PENNEBAKER, J.W. (1990). Opening Up: The Healing Power of Confiding in Others. Morrow, New York.
PENNEBAKER, J.W. (1993). "Social mechanisms of constraint." In: Handbook of Mental Control (D.M. Wegner and J.W. Pennebaker, ed.). Prentice Hall, Englewood-Cliffs, NJ.
PENNEBAKER, J. W., GONDER-FREDERICK, L., COX, D. J., and HOOVER, C. W. (1985). "The perception of general vs specific visceral activity and the regulation of health-related behavior." In: Advances in Behavioral Medicine (E.S. Katkin and S.B. Manuck, ed.). vol. 1. JAI Press, San Francisco. pp. 165–198.
PENNEBAKER, J. W. and WATSON, D. (1988). "Blood pressure estimation and beliefs among normotensives and hypertensives." Health Psych. 7:309–328.
PENNEBAKER, J.W., KIECOLT-GLASER, J., and GLASER, R. (1988). "Disclosure of traumas and immune function: health implications for psychotherapy." J. Consult. Clin. Psych. 56:239–245.
PENNEBAKER, J.W. and ROBERTS, T.A. (1992). "Toward a his and hers theory of emotion: Gender differences in visceral perception." J. Soc. Clin. Psych. 11:199–212.
PENNEBAKER, J.W. and SUSMAN, J.R. (1988). "Disclosure of traumas and psychosomatic processes." Soc. Sci. Med. 26:327–332.
PENNEBAKER, J.W. and WATSON, D. (1991). "The psychology of somatic sympto." In: Current Concepts of Somatization (L.J. Kirmayer and J.M. Robbins, ed.). American Psychiatric Association, Washington, D.C.
PHOON, W.H. (1982). "Outbreaks of mass hysteria at workplaces in Singapore: Some patterns and modes of presentation." In: Mass Psychogenic Illness (M. Colligan, J.W. Pennebaker, and L. Murphy, ed.). Erlbaum, Hillsdale, NJ. pp. 21–32.
ROBBINS, J.M. and KIRMAYER, L.J. (1991). "Cognitive and social factors in somatization." In: Current Concepts of Somatization (L.J. Kirmayer and J.M. Robbins, ed.). American Psychiatric Press, Washington, D.C. pp. 107–142.
ROBERTS, T.A. and PENNEBAKER, J.W. (1994). "Women's and men's strategies in perceiving internal state." In: Advances in Experimental Social Psychology (M. Zanna, ed.). vol. 28. Academic, New York.
ROBINSON, J.O. ((1969). "Symptoms and the discovery of high blood pressure." J. Psychosomatic Res. 13:157–161.
ROSE, R.J., KOSKENVUO, M., KAPRIO, J., SARNA, S., and LANGINVAINIO, H. (1988). "Shared genes, shared experiences, and similarity of personality: Data from 14,288 adult Finnish co-twins." J. Personal. Soc. Psych. 54:161–171.
SCHEIER, M.F. and CARVER, C.S. (1987). "Dispositional optimism and physical well-being: The influence of generalized outcome expectancies on health." J. Personal. 55:169–210.
SINGER, J.E., BAUM, C.S., BAUM, A., and THEW, B.D. (1982). "Mass psychogenic illness: The case for social comparison." In: Mass Psychogenic Illness (M. Colligan, J.W. Pennebaker, and L. Murphy, ed). Erlbaum, Hillsdale, NJ. pp. 155–170.
SIRIOS, F. (1982). "Perspectives on epidemic hysteria." In: Mass Psychogenic Illness (M. Colligan, J.W. Pennebaker, and L. Murphy, ed.). Erlbaum, Hillsdale, NJ . pp. 217–236.
SKELTON, J.A. and CROYLE, R. T., eds. (1991). Mental Representation in Health and Illness. Springer-Verlag, New York.
TAYLOR, J.A. (1953) "A personality scale of manifest anxiety." J. Abn. Soc. Psych. 48:285–290.
TELLEGEN, A., LYKKEN, D.T., BOUCHARD, T.J., JR., WILCOX, K.J., SEGAL, N.L., and RICH, S. (1988). "Personality similarity in twins reared apart and together." J. Personal. Soc. Psych. 54:1031–1039.

WATSON, D. and CLARK, L. A. (1984). "Negative Affectivity: The disposition to experience aversive emotional states." Psych. Bull. 96:465–490.

WATSON, D., CLARK, L.A., and TELLEGEN, A. (1988). "Development and validation of brief measures of Positive and Negative Affect: The PANAS Scales." J. Personal. Soc. Psych. 54:1063–1070.

WATSON, D. and PENNEBAKER, J. W. (1989). "Health complaints, stress, and distress: Exploring the central role of negative affectivity." Psych. Rev. 96:234–254.

WEGNER, D.M. (1992). "You can't always think what you want: Problems in the suppression of unwanted thoughts." In: Advances in Experimental Social Psychology (M. Zanna, ed.). vol. 25. Academic Press, San Diego. pp. 193–225.

WEGNER, D.M., SHORTT, J.W., BLAKE, A.W., and PAGE, M.S. (1990). "The suppression of exciting thoughts." J. Personal. Soc. Psych. 58:409–418.

WOODS, S., NATTERSON, J., and SILVERMAN, J. (1978). "Medical students' disease: Hypochondriasis in medical education." J. Med. Ed. 41:785–790.

NEUROPSYCHIATRIC STATUS OF PATIENTS WITH CHRONIC FATIGUE SYNDROME: AN OVERVIEW

JOHN DELUCA,[*‡] SUSAN K. JOHNSON,[*†‡] AND BENJAMIN H. NATELSON[§]

[*]Kessler Institute for Rehabilitation
Department of Research
West Orange, New Jersey

[†]Kessler Institute for Rehabilitation
Department of Psychology
West Orange, New Jersey

[‡]UMDNJ — New Jersey Medical School
Department of Physical Medicine and Rehabilitation
Newark, New Jersey

[§]UMDNJ — New Jersey Medical School
Department of Neurosciences
Newark, New Jersey

Chronic fatigue syndrome (CFS) is an illness that results in debilitating fatigue as well as rheumatological, infectious, and neuropsychiatric symptoms. The present paper is a brief overview of the neuropsychological and psychiatric research on CFS. Studies from our laboratory contrasting CFS with patients with multiple sclerosis, depression, and healthy controls are detailed. Our hypothesis of the neuropsychological impairments in CFS is discussed.

INTRODUCTION

Chronic fatigue syndrome (CFS) is a controversial illness that has recently become the focus of research and clinical interest. The symptom profiles of multiple chemical sensitivity (MCS) and CFS are similar, although there are different triggers for onset, i.e., a chemical exposure in MCS and a viral illness in CFS. Furthermore, in joint research on the two syndromes it has been found that a third of CFS patients report a general sensitivity to chemicals. In addition, the number of environmental substances to which they have become symptomatic is comparable to individuals with MCS (Nancy Fiedler, personal communication). As with

1. Address all correspondence to: John DeLuca, Ph.D., Kessler Institute for Rehabilitation, 1199 Pleasant Valley Way, West Orange, NJ 07052. Tel.:(201)243-6974. Fax:(201)325-5583.
2. Key Words: Cognition, Fatigue, MRI, Personality, Somatization.

Toxicology and Industrial Health, Vol. 10, No. 4/5, pp. 513–522

ISSN: 0748-2337

MCS, arguments exist as to whether CFS is an "organic" or emotional disorder (Jones et al., 1985; Manu et al., 1989; Stewart, 1990). Despite the polarization of the argument, it appears most likely that CFS represents a complex interaction of physiological and psychological disturbances. The purpose of the present paper is to provide a brief overview of the neuropsychiatric aspects of CFS. It is not intended to be an exhaustive review.

DEFINING CFS

The Centers for Disease Control (Holmes et al., 1988) has developed diagnostic criteria for CFS (see Table 1). The two "major" criteria require that the patient experience severe fatigue for a minimum of six months that reduces activity 50% below premorbid levels, and that the fatigue is not explained by the presence of other medical or psychiatric illness. In addition to debilitating fatigue, the CDC criteria also list "minor criteria," which include rheumatological, infectious, and neuropsychiatric symptoms (see Table 1). In conjunction with the major criteria, the patient must have at least 8 of the 14 symptoms and signs which make up the minor criteria in order to be diagnosed with CFS. Recently, modifications have been made to the original criteria to clarify inclusion and exclusion conditions to the CFS case definition (Schleuderberg et al., 1992).

TABLE 1. CDC Criteria for the Diagnosis of CFS (Adapted from Holmes et al., 1988)

Major Criteria:

1) Chronic or recurring fatigue that reduces activity by at least 50% for six months or more.
2) Fatigue is not explained by the presence of other evident medical or psychiatric illness.

Minor Criteria:

At least six symptoms plus two signs, or at least eight symptoms from below.

Symptoms

- Mild fever or chills
- Sore throat
- Painful lymph nodes in the anterior or posterior cervical or axillary distribution
- Unexplained generalized muscle weakness
- Muscle discomfort or myalgia
- Prolonged generalized fatigue after previously tolerated levels of physical activity
- Generalized headaches
- Migratory arthralgia without swelling or redness
- Neuropsychologic complaints
- Sleep disturbance
- Main symptom complex developing over a few hours to a few days

Physical signs

- Low-grade fever
- Nonexudative pharyngitis
- Palpable or tender anterior or posterior, cervical or axillary lymph nodes

NEUROPSYCHOLOGY OF CFS

Many CFS patients describe their cognitive difficulties as one of the more disabling and troubling symptoms (Abbey and Garfinkfel, 1991) of their illness. Symptoms typically include problems with attention, concentration, and memory. Despite these complaints, it is only recently that cognitive dysfunction has been examined systematically. The early studies suffered from serious methodological difficulties (e.g., absence of a control group) and thus are of little practical use (Marrie et al., 1987; Millon et al., 1989; Altay et al., 1990). More recent studies have yielded somewhat conflicting results. Studies have found impairments in verbal memory (Riccio et al., 1992; Sandman et al., 1993), selective impairments in visual memory (Grafman et al., 1993), and slowed information processing (Deluca et al., 1993). Other studies have found no cognitive impairments (Johnson et al., 1994), or impairments that can be accounted for by depression (Krupp et al., 1994).

In our preliminary work with CFS patients, we noticed that despite significant cognitive complaints, few if any impairments were observed on formal neuropsychological evaluations. However, patients tended to show relative impairment on complex tasks requiring simultaneous processing of multiple elements of information. Thus, in our initial study (DeLuca et al., 1993), we compared CFS subjects with subjects with multiple sclerosis (MS), a population known to show impaired information processing efficiency, and with healthy controls. MS subjects were also an ideal contrast group because they share many of the same features as CFS subjects,s including fatigue, cognitive complaints, weakness, and depression. Subjects were tested on the Paced Auditory Serial Addition Test (PASAT), a complex auditory processing task (Gronwall, 1977). The 12 CFS (diagnosed in accordance with CDC criteria), 11 clinically definite MS (Poser et al., 1983), and 11 healthy controls were matched for age, education, and verbal intelligence. Results showed that performance on the PASAT was significantly below the control group for both the CFS and MS groups. However, the CFS and MS subjects did not differ significantly from each other (Figure 1).

In contrast to information processing, the three groups did not differ from each other on the Similarities Test of verbal abstract reasoning (Wechsler, 1981). Mean depression on the Beck Depression Inventory (BDI) (Beck, 1961) did not differ between the CFS and MS groups, but was significantly higher relative to controls, as expected. Nevertheless, cognitive performance was not significantly correlated with depression. We hypothesized that CFS subjects have difficulty in the simultaneous processing of complex cognitive information, and that this primary problem may lie at the root of other cognitive complaints (e.g., memory).

To test this hypothesis of selective cognitive dysfunction in CFS, we compared 22 CFS subjects with persistent cognitive complaints to 21 age- and education-matched healthy controls on a verbal memory task (California Verbal Learning Test: CVLT) (Johnson et al., 1994). No significant differences were observed between groups on any measure of verbal memory, including acquisition, recall (immediate or delayed, free or cued) and recognition.

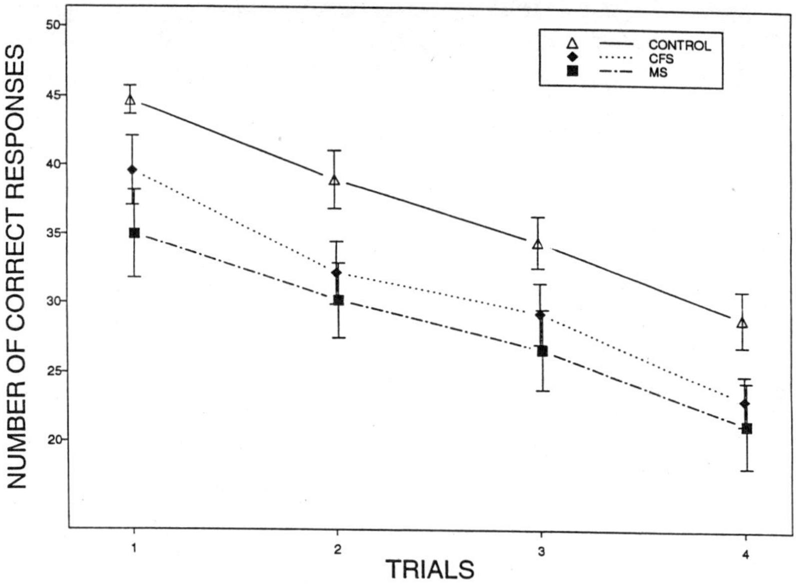

FIGURE 1. Paced Auditory Serial Addition Test performance in CFS, MS, and healthy controls (Mean ± SEM). (Reprinted from DeLuca et al., 1993, with permission. Copyright 1993, American Medical Association).

These results were interpreted as support for the hypothesis of selective impairment in the efficiency of information processing in CFS. We concluded that complaints of impaired memory may be secondary to deficient information processing and encoding of material rather than impaired storage and/or retrieval.

As a result of an NIH funded research project, we have recently completed a major study examining a broad range of cognitive functions in CFS (DeLuca et al., 1994). Given the current controversy over whether cognitive symptoms in CFS are due solely to depression (Krupp et al., 1994), the addition of a depressed group is critical since individuals with CFS often manifest significant symptoms of depression. Thus, 26 CFS subjects were compared to age-, education-, and verbal intelligence-matched subjects with clinically definite but "mild" MS (n = 12), depression (DEP) (n = 14), and healthy controls (n = 20). DEP subjects had to be diagnosed with either major depression or dysthymia according to the DSM III-R criteria based on a structured clinical interview QDIS-III-R (Quick Diagnostic Interview Schedule III-R, Robins et al., 1991). Subjects were tested on a broad range of neuropsychological functions including: attention/concentration, intellectual functions, concept formation, and verbal and visual memory. Depression, anxiety, and subjective cognitive complaints were assessed using self-report instruments.

Once again, CFS subjects were significantly impaired on the PASAT relative to controls, confirming our earlier findings (DeLuca et al., 1993). The MS and DEP groups did not differ from each other, the CFS group or controls on the PASAT. There was a suggestion that impaired information processing in CFS was selective to the auditory domain, leaving visual processing relatively intact. No significant group differences were observed on tests of intellectual ability, conceptual reasoning, or visual memory. The CFS group showed relatively intact verbal memory ability, although there were a few "hints" of reduced performance relative to controls. Verbal memory was assessed using the California Verbal Learning Test. The CFS group showed significantly more difficulty than controls in learning the verbal list learning task. This is in contrast to our earlier study (Johnson et al., 1994), which did not find any impairments on the CVLT. However, an examination of verbal recall in the present study as a percentage of the amount of information learned, revealed no differences between the CFS and control groups. We hypothesized that the reduced acquisition or learning on the CVLT in the CFS group was due to impaired information processing efficiency, and that the storage and retrieval of encoded information did not differ from controls. In marked contrast to the objective neuropsychological data, the CFS group subjectively expressed significantly higher levels of impairment in the areas of attention, concentration, memory, and reasoning relative to the MS, depressed, or healthy control groups. Both the CFS and DEP groups showed significantly higher scores on the State Anxiety Inventory (Speilberger et al., 1983) than either the MS or healthy controls, and higher scores on the BDI than healthies. An overall impairment rating was computed using the objective test scores based on distance below the healthy group's mean scores. This rating showed the CFS and DEP groups were significantly different from the healthies in overall neuropsychological performance. Although these data reveal that the DEP and CFS groups are closer than the other groups on cognitive and psychological measures, depression is not an adequate explanation for the cognitive difficulties of the CFS group. When the CFS group was split based on high and low depression levels, there was no difference on PASAT scores, the same was found when they were split according to level of anxiety. Furthermore, only the CFS group was significantly lower than the controls on the PASAT.

Natelson et al. (1993) compared brain MRI scans of 52 CFS subjects with 52 age- and gender-matched controls with a history of mild head injury or headache. Fourteen of the 52 CFS subjects (27%) versus 1 of the 52 controls (2%) revealed abnormalities ($p < .0002$). Seven CFS subjects showed multiple small areas of abnormal increased T2 signal in cerebral white matter (primarily in the corona radiata), five revealed ventricular or sulcal enlargement, one showed a single area of hemosiderin or calcification in the pons, while another had a small periventricular lesion. The one control patient had a solitary pontine lesion. Upon follow-up, three of the original 14 abnormal CFS subjects now carried other diagnoses; two were "possible" MS and the third, possible Bechet's disease. However, after dropping these subjects from the original analysis, differences between CFS and controls in the frequency of abnormal MRI scans remain statistically significant ($p < .008$).

Thus, the Natelson et al. study extends a previous study by Buchwald et al. (1992) showing abnormal MRI scans of the brain at least in a subpopulation of CFS subjects. One could hypothesize that the CFS subjects with the most severe cognitive impairments may be those with abnormal MRI scans. However, no such study exists to date which addresses this issue.

In summary, CFS patients display cognitive impairments in complex information processing efficiency, while higher-order cognitive functions are largely intact. This is in marked contrast to the CFS subjects stated degree and type of cognitive impairments. We hypothesize that perceived memory impairments among the CFS population are secondary to reduced information acquisition (resulting from inefficient information processing) rather than to impaired storage or retrieval of the memory trace. Although some CFS subjects show abnormal MRI scans of the brain, the relationship between structural brain pathology and cognitive impairment is unknown.

PSYCHIATRIC AND PERSONALITY ISSUES IN CFS

A number of studies have assessed the psychiatric status of chronic fatigue patients and found relatively high rates of lifetime and current psychiatric disorder. Prevalence of lifetime psychiatric disorder has ranged from 24.5% (Hickie et al., 1990) to 86% (Katon et al., 1991), while rates of psychiatric disorder comorbid with fatigue onset have ranged from 66% (Manu et al., 1988) to 45% (Katon et al., 1991). In studies comparing chronic fatigue patients to other chronic illness groups, the rates of psychiatric diagnoses have been consistently higher than for patients with rheumatoid arthritis (Katon et al., 1991) and muscle disease (Wood et al., 1991). The problem with most of these studies, however, is that they have generally not used patients who met the CDC criteria for CFS.

In an attempt to clarify the psychiatric aspects of CFS, studies in our laboratory have examined patients who meet CDC criteria and whose psychiatric disorder (if any) clearly began *after* the onset of CFS. We have assessed subjects' psychiatric status using the QDIS III-R structured interview which is based on DSM III-R diagnostic criteria. CFS subjects were compared to age- and education-matched groups of MS, DEP and healthies. In a recently completed study (Johnson et al., 1995), we focused on the complex problem of diagnosing somatization disorder (SD) in CFS by reanalyzing the SD section of the DIS using different criteria sets. Both the CFS and DEP subjects endorsed a higher percentage of SD symptoms than either the MS or healthies, but changing the coding of SD symptoms from psychiatric to physical under different sets of criteria dramatically affected the rates of diagnosing SD in the CFS group. Thus, if CDC criteria symptoms are coded as physical, only 2.4% of CFS subjects met the strict DSM II-R criteria for SD. This study also diagnosed one-third of the CFS subjects with major depressive disorder concurrent with CFS onset, which is similar to the rates reported by others. However, the CFS group had less major depression and fewer other Axis I disorders than the depressed group.

Several studies have assessed personality in CFS. Millon et al. (1989) used the self report Millon Clinical Multiaxial Inventory (MCMI) on 24 CFS patients and found elevations on measures of histrionic, schizoid, avoidant, narcissistic, and aggressive/sadistic personality patterns. Blakely et al. (1991) compared CFS to chronic pain and healthy subjects using the Minnesota Multiphasic Personality Inventory and found elevations on hypochondriasis, hysteria, and depression in CFS and chronic pain subjects, patterns often elevated in the medically ill. But the CFS groups showed a marked elevation on psychasthenia, schizophrenia, and anxiety compared to chronic pain subjects, which was interpreted as resulting from the high emotionality and "bizarre somatic complaints" of CFS patients. Pepper et al. (1993) examined Axis I and Axis II disorders in subjects with CFS, MS, and major depression using the Structured Clinical Interview for DSM-III-R (SCID) and the MCMI-II. They found that the CFS group had less depression and fewer personality disorders than the depressed group. Also, the CFS group did not differ from the MS group regarding personality disorders, but had more frequent current depression.

Our laboratory has also examined aspects of personality in CFS that may contribute to a greater understanding of this disorder. We have investigated personality factors in CFS in addition to formal personality disorders. Only preliminary results are available from this ongoing study. Using the same comparison groups mentioned above, we have administered the Neuroticism Scale from the NEO Inventory and the Personality Disorders Questionnaire-Revised (PDQ-R, Hyler and Ryder, 1987). This study has revealed that the DEP subjects show the highest levels of personality disturbance and neuroticism, with the CFS and MS groups showing parallel performance which is significantly higher than healthies but significantly lower than the depressed group. This suggests that living with a chronic illness may have a negative effect on some aspects of personality.

In summary, our studies have largely confirmed and extended the findings of other studies examining psychiatric diagnoses and personality disturbance in CFS. Regarding Axis I disorders, the CFS group had less pathology than the depressed group, but more current depression than the MS group. In the area of personality disturbance, the CFS and MS group show similar patterns, while the depressed group showed significantly greater personality pathology. Compared to the DEP group, the overall level of psychopathology is less severe among CFS subjects.

DISCUSSION

Despite significant subjective complaints of cognitive dysfunction, CFS subjects appear to show only "subtle" impairments on objective tests. Cognitive impairment appears to be specific to complex information processing ability, which likely influences verbal memory performance. However, reduced cognitive efficiency in CFS was not correlated with degree of depression or anxiety. Simple attention, memory, and intellectual ability appear to be intact in CFS patients. An impairment index of cognitive performance which takes into account the entire neuropsychological profile suggests that the CFS group in overall performance looked more like the depressed group than either the MS or healthy control groups. While the presence of structural CNS impairment was significantly elevated in CFS relative to controls, the relationship between abnormal brain MRI and cognitive performance remains an area for future study.

Clearly, a subset of patients with CFS have significant psychiatric disturbance, particularly current depressive disorder. However, in contrast to the neuropsychological studies, the CFS group looked more like the MS group than the depressed group with respect to the frequency of both Axis I psychopathology and, particularly, the personality disorders (Axis II) and traits.

Taken together, the neuropsychological and psychiatric studies suggest that while depression and anxiety are prominent features of CFS, CFS is not simply an abnormal manifestation of formal depressive, anxiety, or somatization disorders. The severity of psychopathology in the depressed group far exceeded what was observed for the CFS group on both interview and self-report measures.

While some investigators may view CFS as either a psychiatric disorder or a manifestation of an "organic" medical disease, there is no clear evidence to show that either of these extreme positions is correct. While it is possible (or even perhaps likely) that acute onset of CFS symptoms may be associated with immune, CNS and/or neuroendocrine dysfunction, a significant number of CFS patients meet clear diagnosis for psychiatric disorders. The heterogeneity of CFS leaves plenty of room to parcel out subgroups within this population. However, questions such as whether symptoms are "organic versus psychologic" are futile and are of little help to the clinician, researcher, or patient. We prefer an empirical approach of examining the specific nature of patient complaints (e.g., examine actual cognitive impairments in the face of stated problems) so that an overall profile can be utilized in treatment of the "whole" patient.

ACKNOWLEDGMENTS

Supported in part by grants AI-32247 from NIH, H133P10002-92 from NIH, and H129J10006 from the Department of Education, Rehabilitation Services Administration.

REFERENCES

ALTAY, H.T., TONER, B.B., BROOKER, H.B., ABBEY, S., and SALT, I.E. (1990). "The neuropsychological dimensions of neuromyasthenia (chronic fatigue syndrome): a preliminary report." Intl. J. Psychiat. Med. 20:141–149.

BECK, A.T., WARD, C.H., MENDELSON, M., MOCK, J., and ERBAUGH, J. (1961). "An inventory for measuring depression." Arch. Gen. Psychiat. 4:561–571.

BLAKELY, A.A., HOWARD, R.C., SOSICH, R.M., MURDOCH, J.C., MENKES, D.B., and SPEARS, G.F.S. (1991). "Psychiatric symptoms, personality and ways of coping in chronic fatigue syndrome." Psychol. Med. 21:347–362.

BUCHWALD, D.P.R., CHENEY, D.I., PETERSON, B., HENRY, S.B., WORMSLEY, A., GEIGER, D.V., ABLASHI, Z., SALAHUDDIN, C., SAXINGER, R., BIDDLE, R., KIKINIS, R.A., JOLESZ, F.A., FOLKS, T., BALACHANDRAN, N., PETER, J.B., GALLO, R.C., and KOMAROFF, A.L. (1992). "A chronic illness characterized by fatigue, neurologic and immunologic disorders and active herpesvirus type 6 infection." Ann. Int. Med. 116:103–113.

DELUCA, J., JOHNSON, S.K., and NATELSON B.H. (1993). "Information processing efficiency in chronic fatigue syndrome and multiple sclerosis." Arch. Neurol. 50:301–304.

DELUCA, J., JOHNSON, S.K., BELDOWICZ, D., and NATELSON B.H. (1994). "Neuropsychologic impairments in chronic fatigue syndrome, multiple sclerosis and depression." J. Neur. Neurosurg. Psychiat. In press.

GRAFMAN, J., SCHWARTZ, V., DALE, J.K., SCHEFFERS, M., HOUSER, C., and STRAUS S.E. (1993). "Analysis of neuropsychological functioning in patients with chronic fatigue syndrome." J. Neurol. Neurosurg. Psychiat. 56:684–689.

GRONWALL, D. (1977). "Paced auditory serial addition task: A measure of recovery from concussion." Perceptual Motor Skills 44:367–373.

HICKIE, I., LLOYD, A., WAKEFIELD, D., and PARKER G. (1990). "The psychiatric status of patients with the chronic fatigue syndrome." Brit. J. Psychiat. 156:534–540.

HOLMES, G.P., KAPLAN, J.E., GANTZ, N.M., KOMAROFF, A.L., SCHONBERGER, L.B., STRAUSS, S.E., JONES, J.F., DUBOIS, R.E., RUNDLES, C.C., PAHWA, S., TOSATO, G., ZEGANS, L.S., PURTILO, D.T., BROWN, N., SCHOOLEY, R.T., and BRUS, I. (1988). "Chronic fatigue syndrome: a working case definition." Ann. Int. Med. 108:387–389.

HYLER, S.E. and REIDER, R.O. (1987). "PDQ-R: Personality Disorders Questionnaire-Revised." New York State Psychiatric Institute, New York.

JOHNSON, S.K., DELUCA, J., FIEDLER, N. and NATELSON, B.H. (1994). "Cognitive functioning in chronic fatigue syndrome." Clin. Inf. Dis. 18(Suppl 1):S84–S85.

JOHNSON, S.K., DELUCA, J., and NATELSON, B.H. (1995). "Assessing somatization disorder in the chronic fatigue syndrome." Psychosomatic Med. Submitted.

JONES, J.F., RAY, G., MINNICH, L.L., HICKS, M.J., KIBLER, R., and LUCAS, D.O. (1985). "Evidence for active Epstein-Barr virus in patients with persistent, unexplained illnesses: Elevated anti–early antigen antibodies." Ann. Int. Med. 102:1–7.

KATON, W.J., BUCHWALD, D.S., SIMON, G.E., RUSSO, J.E., and MEASE, P.J. (1991). "Psychiatric illness in patients with chronic fatigue and those with rheumatoid arthritis." J. Gen. Med. 6:277–285.

KRUPP, L.B., SLIWINSKI, M., MASUR, D.M., FRIEDBERG, F., and COYLE, P.K. (1994). "Cognitive functioning and depression in patients with chronic fatigue syndrome and multiple sclerosis." Arch. Neur. 51:705–710.

MANU, P., MATTHEWS, D.A., and LANE, T.J. (1988). "The mental health of patients with a chief complaint of chronic fatigue." Arch. Intern. Med. 148:2213–2217.

MANU, P., MATTHEWS, D.A., LANE, T.J., TENNEN, H., HESSELBROCK, V., MENDOLA, R., and AFFLECK, G. (1989). "Depression among patients with a chief complaint of chronic fatigue." J. Aff. Dis. 17:165–172.

MARRIE, T.J., ROSS, L., MONTAGUE, T.J., and DOAN, B. (1987). "Post–viral fatigue syndrome." Clin. Ecol. 1:5–10.

MILLON, C., SALVATO, F., BLANEY, N., MORGAN, R., MANTERO-ATIENZA, E., KLIMAS, N., and FLETCHER, M. (1989). "A psychological assessment of chronic fatigue syndrome/chronic Epstein-Barr virus patients." Psychol. Health. 3:131–141.

NATELSON, B.H., COHEN, J.M., BRASSLOFF, I., and LEE, H.J. (1993). "A controlled study of brain magnetic resonance imaging in patients with the chronic fatigue syndrome." J. Neurol. Sci. 120:213–317.

PEPPER, C.M., KRUPP, L.B., FRIEDBERG, F., DOSCHER, C., and COYLE, P. (1993). "A comparison of neuropsychiatric characteristics in chronic fatigue syndrome, multiple sclerosis, and major depression." J. Neuropsychi. Clin. Neurosci. 5:200–205.

POSER, C.M., PATY, D.W., SCHEINBERG, L., McDONALD, W.I., DAVIS, F.A., and TOUTELLOTTE, W.W. (1983). "New diagnostic criteria for multiple sclerosis: Guidelines for research protocols." Ann. Neurol. 13:227–231.

RICCIO, M., THOMPSON, C., WILSON, B., MORGAN, D.J., and LARIENT A.F. (1992). "Neuropsychological and psychiatric abnormalities in myalgic encephalomyelitis: A preliminary report." Br. J. Clin. Psychol. 31:111–120.

ROBINS, L., COTTLER, L., and KEATING, S. (1991) "Diagnostic Interview Schedule (DIS): Version III Revised." Dept. of Psychiatry, Washington University School of Medicine, St. Louis, Mo.

SANDMAN, C.A., BARRON, J.L., NACKOUL, K., GOLDSTEIN, J., and FIDLER, F. (1993). "Memory deficits associated with chronic fatigue immune dysfunction syndrome." Biol. Psychiat. 33:618–623.

SCHLUEDERBERG, A., STRAUSS, S.E., PETERSON, P., BLUMENTHAL, S., KOMAROFF, A.L., SPRING, S.B., LANDAY, A., and BUCHWALD, D. (1992). "Chronic fatigue syndrome: Definition and medical outcome assessment." Ann. Int. Med. 117:325–331.

SPIELBERGER, C.D., GORSUCH, R., and LUSHENE, R. (1983). "The state trait anxiety inventory (STAI) test manual." Consulting Psychologists Press, Palo Alto, CA.

STEWART, D.E. (1990). "Emotional disorders misdiagnosed as physical illness: Environmental hypersensitivity, candidiasis hypersensitivity, and chronic fatigue syndrome." Intl. J. Ment. Health, 19(3):56–68.

WECHSLER, D. (1981). Wechsler Adult Intelligence Scale — Revised, Manual. Psychological Corporation, New York.

WOOD, G.C., BENTALL, R.P., GOPFERT, M., and EDWARDS, R.H.T. (1991). "A comparative psychiatric assessment of patients with chronic fatigue syndrome and muscle disease." Psychol. Med. 21:619–28.

QUESTIONS AND ANSWERS #3

MATTHEW FRIEDMAN, THEODORE SIMON, JOHN DELUCA, AND JAMES PENNEBAKER

Cynthia Wilson: Dr. Friedman said that he was talking of MCS responding to drugs. In the Pacific Northwest, they're having really good success controlling some of our neurological symptoms with dextromethorphan hydrobromide and other neuropeptides. I don't know if that fits into this theory.

Dr. Friedman: It's always nice having someone agree with my theories. I must confess that I'm not familiar enough with those drugs, but perhaps Dr. Weiss would know.

Cynthia Wilson: I know two of the drugs' names. One is dextromethorphan hydrobromide and PHF is the other. There is also MK 801.

Dr. Janette Sherman, Western Michigan University, practicing in Alexandria, VA: If my pharmacology serves me right, dextromethorphan is one of the morphine opioid classification of drugs, and it probably does work, because she probably feels better.

This has been a very interesting conference, and I'd like to make a couple of comments. When people speak of low-level exposure to pesticides, one has to remember that the half-life of chemicals such as Dursban is at least 60 days, so there's no such thing as a low level. It's a continuing exposure. Also, I thought the discussion of the neurotoxicity of the organophosphates was very interesting, because they were designed to be neurotoxic, and if we're finding brain lesions, I think it's because they were designed to cause a certain brain lesion.

I find that the multiple chemical sensitivity problems, and the fatigue syndrome, is a relatively new disease for those of us who have been practicing for a number of years. I did not see these diseases more than ten years ago. I think these are new diseases, brought by the widespread use of toxic chemicals, indoors, where people spend a great deal of time. If they are occurring among the more educated population, it is because the affluent are able to have pest control, or have office buildings that are being sprayed. They are therefore the people who are being more exposed. I think we are truly having an epidemic.

Steve McFadden, independent researcher: I just want to comment about MK-801. It's used a lot in nerve gas research for antidotes. My impression was it is related to phenocylidene, angel dust. So, be careful in experimenting on these things.

Toxicology and Industrial Health, Vol. 10, No. 4/5, pp. 523–535

ISSN: 0748-2337

Grace Zeim, from Baltimore: I had a question for Dr. Simon, on his Boeing-funded study.

There were a couple of problems with the study design, and I wondered if you would comment: I know Dr. Baker felt that about half of the patients from his patient population that were in your MCS group did not have MCS.

The second problem was my concern about using a musculoskeletal population as a control group for immune studies. Since they could have chronic inflammation, you would be comparing inflammation and chemical sensitivity.

Thirdly, your musculoskeletal control group is probably a group with increased pain, and on pain medications. There is a concern about comparing the neuropsychologic performance of a group with chronic pain, on pain medications, with people with chemical sensitivity.

Dr. Simon: I can't resist the opportunity to respond to your comment about the funding for our study.

The study is the second study I referred to, which was a case/control study between chemically sensitive patients. The funding for that study came from a variety of sources, some from the Washington State Department of Health, some from the Department of Environmental Health, University of Washington. About half the funding came from a fund which is administered by a joint labor-management commission at the Boeing Company, which is obviously very interested in this question.

It is a commission which includes representatives from management at Boeing as well as representatives from the International Association of Machinists, Boeing's largest union. It was created specifically to be a supposedly impartial source of funding for research on chemical sensitivity. We felt that this source of funding was about as pure as one could come up with, since we had to go before people who had widely varying opinions. They had to approve our protocol, and agree that they thought it was a reasonable way to address these questions.

I should also say, that of the funding for this study, about 95% of our budget was directly transferred to Antibody Assay Laboratories in California, which performed the immunologic tests. All of our time was donated. Bill Daniell and Dr. Stockbridge worked for free. I was supported by a grant from ATSDR. The funding from Boeing certainly didn't go into any of our pockets, it was directly transferred to the immunology lab. In terms of your question about appropriate selection of cases and controls: the first question is an important one, I think, not just about our study, but about these studies in general. Because there is no accepted case definition of Multiple Chemical Sensitivity, it is unfortunately quite easy in this area to study a population of people, and after the fact have people say, "Well, those weren't people who have real MCS because the results weren't how I wished they had come out." Certainly this is an area crying out for a case definition. I can say that our selected population were people who were given a diagnosis of chemical sensitivity by their physician. They had a duration of symptoms that lasted at least three months, and on a standardized questionnaire they endorsed having symptoms in response to multiple unrelated exposures, and endorsed symptoms in three or more organ systems, including the CNS. Now, I know the allergist we worked with did say later, "Well, I don't think some of those patients had real chemical sensitivities." Although he did, I'm hesitant to argue with him since he's not here.

We actually posed this exact question to our collaborator: "Well, would you be interested in categorizing these subjects, according to your clinical judgment, in terms of who had the true chemical sensitivity syndrome and who were the masquerading CS patients, who would not be presumed to have immunologic abnormalities?" Although we requested he do that for several months, and actually held up the publication of the manuscript hoping we might be able to get that data, it never happened. Once we published the manuscript and people see the results, it's quite easy to criticize our sample selection.

In terms of the selection of comparison subjects, this is very problematic in terms of selecting the appropriate companion groups, especially when you're comparing multiple different measures. Obviously, there's probably an ideal comparison group for looking at immunologic measures, there's probably an ideal comparison group for looking at neuropsychologic measures, and an ideal group for looking at psychiatric symptoms. We obviously couldn't afford three comparison groups. So we had to give it our best shot, and this was what we thought was the most reasonable example.

In terms of your concerns about the immunologic system, none of these patients had systemic illness. All of them had musculoskeletal injuries. These were the result of primarily work-related injuries. There were a few people who had back injuries as a result of automobile accidents, and others who had Carpal Tunnel Syndrome, and others Repetitive Motion Syndrome. I suppose we could say there is some immunologic basis for those injuries (chronic inflammation was one), but we selected them because we believed here were people who had a clearly defined etiology for their injuries. In terms of your concern about medication, we didn't include anybody who was taking opioanalgesics for exactly that reason. We didn't include anybody who was taking medications. We didn't have to exclude anybody for that purpose, since these people were at this point fairly chronic in their courses.

From the Audience: What about other analgesics?

Dr. Simon: Some of them were using aspirin or nonsteroidal anti-inflammatory drugs, but none were using opioanalgesics.

Dr. Ziem: One other discrepancy I noted: I sent immune studies to the same laboratory and I opened the envelopes on my first 50 patients myself, so it's not a coding situation, in my practice at least. I've been interested to see that about 75 to 80% of my patients have increased CD26 cells, but you show a significant decrease of CD26 cells using the same laboratory, presumably the same type of patient population. To me this suggests the possibility of a coding error with CD26 cells.

Dr. Simon: One thing we didn't publish in our paper, which may be coming out in the paper Bill Daniell is writing, is that we sent, unbeknownst to the immunologic laboratory we used, pair-blinded specimens on ten patients. Consecutive vacutainer tubes from the same person's arm were given different coding numbers. The reliability on most of their measures is no better than chance. Immunologic measures in the same patient drawn into consecutive vacutainer tubes are not statistically significant when related to each other.

Susan Clark, Environmental Advocacy League: I have a question for Dr. Simon: You spoke of comparably severe medical illnesses in your talk, and I'm wondering what those comparably severe illnesses would be?

Dr. Simon: In the study I referred to, we compared chemically sensitive patients to people with musculoskeletal injuries. I'm basing that really on a number of other studies we've done in various other patient populations using similar measures. I think comparison of chemically sensitive patients to community controls is inappropriate. Patients must be compared with patients, and if we were to look at patient populations with any kind of medical illness, they would have elevated rates of anxiety and depressive disorders, reflecting both the influence of medical illness on one's mood, and anxiety and depressive disorders on the likelihood of receiving treatment. When two people have equal severity of medical illnesses, the one who is more distressed is more likely to see the doctor. So patients must be compared with patients.

Even when one compares chemically sensitive patients to other kinds of patients, the rates of anxiety and depressive disorders seem higher than one might expect. For instance, I can refer you to literature on rheumatoid arthritis, diabetes, people who have fairly significant chronic medical conditions. Patients with chemical sensitivity appear to have higher rates. Now what that tells us is that chemical sensitivity is primarily a brain disease, as opposed to some of these other conditions which probably have their roots in other organ systems. I want to be careful to say that I am not maintaining that that says that chemical sensitivity is solely an anxiety or depressive disorder, but that it certainly steers us away from immunologic abnormalities and toward central nervous system abnormalities. I think all the evidence we've heard here is pointing us in that direction.

Susan Clark: I might be a good case study. I would be willing to share some personal information. I have nearly died of the following conditions: nephritis, pneumonia, angina, and pelvis fractures from a car accident. I've survived all of that, and by the way, I don't get depression, I don't get fatigue, and I don't get PMS, so I consider myself, in many ways, very lucky. Perhaps I am strange as an MCS victim as well. I am cacosmic and moderately to severely so. Just consider that I feel the pain that I experience, for example, from these particular kinds of migraines that come from reactions to low levels of chemicals. Sometimes

I am not even aware I'm being exposed, and find out only after the fact. The headaches are far, far more painful than those I have experienced from those other medical conditions.

Rosalind Taratoot: I come from 13 years of pesticide research with the Federal Government and I've been on disability with MCS for 10 years. My comment is to Dr. Simon. When I first got ill, 10 out of 52 employees in my lab were diagnosed with the same problems that I had. We got no medical or emotional support. Our anxiety went through the roof. We were diagnosed and actually showed autoimmune disorders, autoantibodies, very symptomatic. OSHA and NIOSH documented that we were indeed ill. But the psychosomatic-type diagnosis came very readily.

I worked with a clinical psychologist. I did not give her my medical diagnosis first, and I was, indeed like many others, diagnosed with psychological problems and somaticization disorders. When I gave her my medical history, it turned out her husband was a chemical engineer at Georgia Tech. He wrote her a note saying, "This lady can indeed be suffering from the psychological disorders she describes due to the solvents and pesticides she was exposed to." The lady looks at me and says, "I've never dealt with anybody like you, what do I do with you?" And I told her, "Teach me to cope with the chronic illness so I can get through the anger and get on with my life." And she has done so.

I was told by ten different specialists that I would never leave my home again, that I would be like the person in the bubble. I'm here today, this is the first trip I've taken alone in ten years.

Michael Gray, Benson, Arizona: Dr. Simon: In relation to the ten split specimens that you sent to Antibody Assay Labs, was I correct in hearing you say that you got absolutely no statistical correlation specimen to specimen? Have you turned that information over to the agencies that regulate their performance?

Dr. Simon: No. Actually, you will notice in our paper the fact that we did not name the lab we worked with, nor did we name the allergist from whose practice we were treating patients. Our intention here is not to be inflammatory. Our intention here is not to question the practice. I certainly wouldn't say that on the basis of ten specimens that a laboratory should have its license yanked or anything like that.

Dr. Gray: But don't you think it should be investigated?

Dr. Simon: I'm not an immunologist. For me to make some comment about the quality of their immunologic work . . . I mean, I'm a psychiatrist.

Bill Meggs, North Carolina and New York: Is MCS a disease of the brain? Maybe not, maybe so, but we've heard one objective physical parameter that has been shown in a controlled study to be statistically significantly different in the MCS patients from the controls, and that was nasal resistance. Dr. Doty, Dr. Bascomb, and her collaborators from this city, the University of Maryland, have all shown that patients reporting sensitivity to environmental tobacco smoke also develop an increase in nasal resistance upon exposure to this irritant. It may go beyond just a brain problem, and airway inflammation may be very important in this syndrome. Certainly we need further empirical studies to sort out of these different mechanisms. The fact that we are now talking about mechanisms, and have models, will greatly advance empirical science.

Bob Adamec, Memorial University of Newfoundland, Canada: I have a question for Dr. DeLuca. I was wondering whether antidepressants were efficacious in Chronic Fatigue Syndrome symptomatology?

Dr. DeLuca: That's a good question. Most people do get treated with antidepressants. Typically I would say it doesn't improve their fatigue or cognitive status. They may feel a little better in terms of their mood. Some people do say they start feeling a little better, but certainly nobody goes back to work, or nobody starts jogging again. I would say it might give some patients the sense they feel a little better, but I don't find that they sense they are totally relieved of their symptoms.

Dr. Adamec: Has a broad spectrum of antidepressants been tried?

Dr. DeLuca: I really don't know. There are certainly not any studies that I know of [of antidepressant use in CFS].

I think CFS is a good population to compare and contrast with MCS, but I think there are a lot of good populations. Tomorrow morning Dr. Fiedler is going to talk about some of the studies she's done in comparing MCS with CFS. I agree with Dr. Simon that control groups should not be healthy individuals, they should be other patient populations. I think CFS is a good one, as well as other medical problems such as arthritis or other populations.

Dr. Simon: I didn't think I should let it pass that the efficacy of antidepressant medicines in either CFS or MCS is somehow established. I think in both of those areas we are dealing within the realm of anecdote. I think there are anecdotal reports of both great benefit and terrible consequences of antidepressant use.

Nicholas Ashford, MIT: Lets' take stock of where we are. We can focus on patient groups and our experience with listening and talking to patients, hundreds and hundreds and hundreds of them. We understand from the patients' testimony what a traumatic experience it is to go through the medical establishment, cost problems and so forth. As this problem evolves, I think the ratio of psychological to physical problems becomes larger. That being the case, I would echo Howard Kipen's comments earlier that we either follow natural experiments or do retrospective studies. We have heard today that both psychological and physical problems may kindle. If kindling is an important problem, it may lead to chemical sensitivity. I'd be very interested to see, Dr. Simon for example, what the neuropsychological tests look like at the beginning of the laying of new carpet and what they look like three months later. I think you will find a very different picture than the obvious psychological sequelae of physical problems. There's a problem dealing at the patient end of this business. It's important to help patients, but if we want to get at the root of this issue, I believe you have got to start earlier. Is there any way your particular views on these issues could be useful in the context of natural experiments rather than the patient referral?

Dr. Simon: I think your comment is a very good one, and I'm not sure I have any answer for it. I certainly agree with you that when looking at people that have fairly established illnesses, you often get very confusing answers about the etiology of those illnesses. There is certainly a lot of information from biological models about compensatory changes and what happens after the fact, and how you really can't tell much.

I'm not aware of any published evidence that certain kinds of symptoms predominate in the early course of these things, and other kinds of symptoms later. I am sort of skeptical about retrospective reports of anything. I can say that, being a person who published a report about retrospective reports of psychological symptoms in patients with chemical sensitivity. I'm now extremely doubtful about methods of asking people about the kinds of symptoms they had several years ago, or several months ago. I think to try to reconstruct such histories and draw any conclusions from them is really on very shaky ground.

Charles Hinshaw, Wichita Kansas: Dextromethorphan, I believe, is a nonnarcotic, anticough medicine available over the counter, sometimes in pure form, but it does have an action on the CNS. Secondly, this is back at Dr. Simon: I would like to support one of Dr. Ziem's observations, that in the patients I study in my office there are frequently abnormalities obtained from the same laboratory in a very high percentage of the patients. If you study those patients for immunologic disorders, such as inhalant allergies and so forth, there's a high percent of positives.

In reference to what Dr. Ashford said, I think he's right. If there's a flaw in your study, and perhaps it's just a bias or an unintentional sampling bias, maybe you should do another study with different patient selections such as Dr. Ashford suggested.

Dr. Simon: I wouldn't want to say that the results we had from the laboratory we used were by their standards normal. The vast majority of the chemical sensitivity patients were, by the standards of the laboratory, abnormal. The problem is that the vast majority of patients with back injuries and repetitive motion syndromes were also, by the standards of the laboratory, abnormal. They were all abnormal, but not in a consistent direction, or on the same tests, just all abnormal. The interpretation from the laboratory that these were consistent with multiple chemical sensitivities did not vary in frequency between the patients who were chemically sensitive and our comparison group. You may well be right; we heard today that you really need to look at people at a certain stage of their illness or people who have been in some ways prepared to have these tests done. I can't argue with that. Given the money and time, that would be a very good idea.

From the Audience: Speaking about the laboratory from the stand point of a pathologist colleague and not from the stand point of chemical sensitivity, I think he would be interested in knowing about the dissimilarity between your paired samples and also the similarity between the controls and the test subjects. Supposedly these laboratories are regulated and these tests are run under control conditions, so that they can distinguish abnormal from normal. I would encourage you to relay that information to the laboratory.

Dr. Simon: We have.

Jacob Berkson: I have MCS, and it's real, and it was preventable. If I had only known back in March 1988 that a termiticide, known as Dursban, had a risk to it, and that people all over the country had become ill following an application of it, and these people subsequently developed these symptoms which devastated their lives. We need disclosure and labeling. When the public doesn't know, the public is defrauded. I was injured by carpet fumes more than five years ago, and I went through this whole process of hearing, "Gee, we never heard about this before." The same thing, the same process as with the termiticide.

This isn't publicized enough. It's real. I don't know how many anecdotes you need before you understand that this is real. I got a 1000-page handbook from Johns Hopkins Hospital, of the leading illnesses of people over 55. This book was written by the leading authorities in the country, and many of the hundred illnesses described there have no known origins, but they're treating them. They even have clinics that treat these illnesses that they don't know the causes of, but that they accept as real. I don't know if I'm one out of a million, or one out of one hundred thousand. We don't like being a negligible risk. We're the canaries. We're telling you that what has happened to us is going to happen to you, or your children or your grandchildren, because the air you breath is the same air we breath. We breath polluted air and you're breathing it, and sooner or later you're going to get it. So let's get down to work and find a definition.

Dr. Simon: I feel it is important for me to respond to something you said, because Multiple Chemical Sensitivity is real. I don't think there was anyone up in front of the room today who would deny that. I think that everyone would agree that it real and that it is an important problem, and that it's terribly disabling. My concern is that I must somehow sound that way. When I think about why I sound that way, it seems that when you start talking about anxiety and depression and words like that, people read, no matter what you say, they read in big letters — LIE, FAKE, NOT REAL. I think this reflects a variety of things, perhaps a general societal prejudice against people with psychiatric problems.

The second issue that is very unfortunate is that insurers, industry, and states, in their workers' compensation evaluations, have used psychiatric diagnoses as epithets with which to deny claims that may well be legitimate. People are told, "You're just depressed," or, "You're just anxious," as if that meant you don't have a real problem. "It's not our fault and stop bothering us." I think the problem is an unfortunate use of those terms and that has contributed to some of this misunderstanding. I certainly believe these problems are real. My comments about immunologic mechanisms, I hope, will not be interpreted as meaning that I don't believe these problems are real or important. My concern is that an undue influence paid to immunologic mechanisms has been barking up the wrong tree and is really not helpful to people who suffer from this problem, and that treatments based on those mechanisms may be ineffective. I'm encouraged by this conference, and should give some credit to Drs. Bell and Miller, who, I think, have caused a real shift in this area over the last two or three years in bringing a stronger scientific focus toward neuropsychiatric mechanisms for explaining chemical sensitivity symptoms. They have managed to do something I unfortunately have not been able to do, which is to raise some of those questions without sounding like I'm saying it's not real.

Albert Donnay, environmental health engineer and public health activist: I want to second the call for more epidemiologic research that is focused on populations at risk, because now there is plenty of evidence to suggest that we know what some of those populations are: farm workers, carpet layers, etc. I'm concerned about the large effort now being focused on the Gulf Veterans who have come forward for exams or compensation. I want to offer an anecdote from the Phil Donohue show. Some of you may be aware that two weeks ago Senator Reigel and a few Gulf veterans were on that program. Sen. Reigel got 1200 calls, about half from veterans, half solidarity calls. We should compare that with the response to Dr. Doris Rapp's appearance on that program in 1987/8, when she spoke about what she had seen in people exposed to toxins in carpets, especially in schools. Phil Donohue got 144,000 letters, his highest response for any program.

Stephen McFadden, independent research advocate: Time-dependent sensitization, in my understanding, should be looked at as a neuronal-threshold autoregulation mechanism. In a nerve, closely spaced stimuli result in down regulation of neuronal output, or tolerance, whereas widely spaced input results in up regulation of output, or sensitization.

It would seem that brain damage is a fundamental characteristic of poisoning by excitatory neurotoxins in mammalian systems. We need better neurotoxicity screening tests. With 65,000 chemicals commonly used in commerce, we only have basic data on about 650. And there is very little neurotox data. My impression is that EPA/ORD has about $5 million to study neurotoxicity, which is really very little, compared to about a half billion of Federal toxicology research. We really need better tests to screen new chemicals for these things. Ethylparathion hit the market in 1948. In 1991, we finally got proposed pesticide toxicity guidelines, and they have yet to be finalized. We're really running blind.

Richard Conrad, biochemist: Psychological symptoms are a result, and not a cause, in both CFS and chemical sensitivities. Remember that. I think the kindling model is excellent, as are the other talks on the neurophysiological. But don't emphasize the neuropsychological as a cause. It's a result.

Norman Rosenthal: Just a comment or two for Dr. Simon. As a fellow scientist and psychiatrist, I share with you some of the difficulties in tackling this problem, and appreciate the efforts you've taken to do your study. Although Dr. Ashford's point that a retrospective study has its limitations is well taken, we're stuck with that, in more or less all areas of psychiatry. Maybe MCS presents a unique opportunity for being able to follow things prospectively.

I'd like to shift, though, from the scientific to the psychiatric, in terms of where I think we're running into some trouble. Saying to the person with MCS, "You're not crazy," is actually perceived as a very therapeutic intervention. These MCS people, in their dealings with the outside world, are being treated as though they're crazy, all the time. They're feeling like they're losing it. The ability to recognize that these are all understandable consequences of something that happened to you is a very organizing and reassuring, and actually a very honest, intervention. I think they find that realistically helpful.

I think they need to be dealt with differently from the manic depressive, and the panic anxiety, in that they have been legitimately injured. The injury has to be acknowledged. It is a necessary part of the therapeutic process. It's a little bit analogous to early childhood sexual abuse, which for many years was treated by the analyst who said, "Well, I understand your suffering is real, but we don't know if it really happened." But it really does matter. The injury has to be acknowledged, that it actually happened, that it's real, and that their anger is legitimate. And I think that has not really come across, and that is where the anger is coming from.

Dr. Simon: I certainly hear both things that you say. Some minor qualification, though: Words such as anxiety and depression have often been used in a very destructive way, by insurers or compensation evaluation boards. Because of that, they have taken on a really negative and pejorative meaning, and have come to mean, "You're crazy, you're lying, you're faking." I agree with you, I think it is terribly important to acknowledge this to people who have this problem. I certainly don't believe they're lying, or faking, or making it all up, or that they're flakes, or whatever. I firmly believe they are people who have a real problem related to chemical exposure.

I agree with you how important it is to acknowledge that. I have a hard time leaving the other people out of it, because they have brain diseases too. I think most of what we've heard today is a broader understanding of chemical sensitivity, much closer to a modern understanding of neuropsychiatric conditions. Unfortunately, a modern understanding of neuropsychiatric conditions is miles removed from many public perceptions, such as insurance companies and compensation evaluation boards, and so on.

I agree with you also that it is important to acknowledge that this injury is real. I certainly hope that nothing I say is interpreted otherwise. I have no doubt that these people have very real, disabling conditions, which are related to chemical exposure. My concern about some of the previous work has been that focusing on, say, an immunologic injury, or some other kind of injury, has led to a premature jump to accept those results. We say, "Here's proof that it's real," but that leads us down a blind alley.

Helen Mayberg, University of Texas: I just want to make a clarifying comment. In the neurological literature there are many examples: stroke, post-trauma, Parkinson's disease, Huntington's disease, where there are clear, identifiable pathologies. The clinical phenomenology may be weakness or tremor; each disease has its own array of clinical constellations, which help to define it. But what's been very clear over the last ten years, is that many very different etiologic diseases have associated with them a clear organic mood disorder, or anxiety disorder. What that's helped us to understand is something about the localization of mood disorder itself.

Depression is not outside your brain. Depression is in your brain, as are many of these other things. Maybe the mood disorder, and anxiety disorder, clearly in higher incidence, tell us something about where in the brain some of these problems are, and that basic science work overlaps with that. I think that these things are all focusing on where in the brain we should be looking, and not that one is independent from another.

Dr. Simon: All mood disorders are organic. The brain is a chemical organ, and one's experience has a neurochemical substrate. I think that is completely undeniable. What differs and what we call organic mood disorders, and what we call other mood disorders, is just our ability to accurately describe or detect the abnormality.

Dr. Pennebaker: There's a tension here, in terms of how a psychiatrist interprets a real illness, and how patients interpret what the psychiatrist says is a real illness. I think there is some doublespeaking going on here.

Maybe Dr. Simon can clarify, when he talks about real illness related to an exposure, is he talking about depression, or anxiety? Is that illness? Or is that a consequence of an organic factor, or damage, which results in those symptoms? There's a difference. Patients are very sensitive to that shade of difference. They tend to think that psychiatric diagnoses are pejorative.

Dr. Simon: I think there is no shade of difference. People develop symptoms in response to chemical sensitivity, and very often, those symptoms are neuropsychiatric symptoms. Those symptoms must have an organic basis. The brain does not work by mysticism. It works by chemicals. So those symptoms must be organically based. I don't see how anyone could logically deny that.

Continuing episodes of low-level exposure clearly provoke symptoms. I think that is clearly true and undesirable. The question is, by what mechanisms are these systems provoked, and is it possible, through any means, to uncouple the destructive effects of continued exposure, from the exposure?

The effects of the exposure are mediated by the direct pharmacologic or toxic mechanism. For example, if exposure to organic solvents is the direct cause of the symptoms, that would be very difficult to uncouple, barring some pharmacologic magic we don't currently have. If, however, there are a number of intervening steps, between that exposure and the ultimate destructive consequences, that leaves the door open for both behavioral and pharmacologic means for uncoupling.

To me, the real problem in this area is a completely false distinction between organic and nonorganic conditions. A completely false distinction between the belief that, if chemical sensitivity causes these symptoms through a one-step mechanism vs. a three-step mechanism, then it's real or not real.

Dr. Pennebaker: I appreciate what you're saying, and I certainly understand the different views in this area. I just think that patients, again, are very sensitive to that nuance. They are telling us that their symptoms occur directly as a result of chemical exposure. Other people here express different views, that perhaps it could be related to some other kind of beliefs, or other kinds of conditioning, or PTSD, or something else, that intervenes in those steps.

Dr. Simon: Beliefs are chemical events. They must be. That's how they persist. Maybe I'm taking this to the point of the ridiculous, but to say that there are chemical events and there are beliefs, maintains some distinction, which I think in the end will fall.

Dr. Pennebaker: My good friends in biological psychiatry tell me everything is chemical, and I understand what you are saying. But patients see a difference between what they think as an organically induced, chemically induced kind of depression or anxiety, and the other kind that you're talking about. Yes, there is some change to brain chemistry, but really it bears no relationship to that exposure directly, in a toxic sort of sense.

Dr. DeLuca: I'd just like to make a comment. I'm not that familiar with MCS, but I'm familiar with CFS. I can tell you that a lot of these same issues, a lot of these same questions and concerns, occur in the population with CFS.

I think a lot of what happens in the patient group is also what's pervasive in the health care community. Health care professionals don't believe that this is a real illness. You hear them (patients) going from physician to physician, who say, "You're just depressed." Many of the people with CFS are, in fact, depressed, but they are confusing the fact of causality.

I think the fact is that there is a lot of depression, a lot of anxiety, but it is not necessarily causative of the disease. I think that distinction has to be clear in CFS, MCS, mild head injury, and a lot of other things.

From the Audience: More specifically, I think what we're talking about are two things — stigma and validation.

After the disorder of PTSD became recognized, two very interesting things happened: First of all, the difference between PTSD, depression, and anxiety disorder, in the mind of many of my patients and the patients seen by my colleagues, is that when you have determined that the disease, or the disorder, is caused by an agent outside of yourself, it becomes a very different cognitive frame.

With depression and anxiety disorders, there is something wrong with me. We talk about a depressive diathesis, etc. The concept of PTSD is that, "There but for the grace of God, go I." Anybody in the wrong place at the wrong time could get this problem. I think the same thing is true of MCS.

Many of us in the PTSD field are seeing people who have carried certain diagnoses for years, if not decades, pleading for us to change the diagnoses to PTSD. PTSD has become respectable, even honorable, under certain circumstances. I truly believe that if MCS develops recognition, if patients can receive the validation from their clinicians that, "It's not your fault that you were allergic to this carpet cleaner," that itself is going to be very, very therapeutic.

THE UTILITY OF PSYCHOPHYSIOLOGICAL MEASURES IN ASSESSING THE CORRELATES AND CONSEQUENCES OF ORGANIC SOLVENT EXPOSURE

LISA A. MORROW, STUART R. STEINHAUER, AND CHRISTOPHER M. RYAN

Western Psychiatric Institute and Clinic
Department of Psychiatry
University of Pittsburgh School of Medicine
Pittsburgh, Pennsylvania

SOLVENT NEUROTOXICITY

By definition, organic solvents are chemical agents that have the ability to dissolve non–water-soluble substances. They are found most frequently in paint products, dyes, glues, and cleaning agents, and are prominent in the manufacture of plastics, textiles, and agricultural products. Estimates suggest that at least nine million people in the United States are routinely exposed to organic solvents in the workplace (National Institute for Occupational Safety and Health, 1987).

Reports of adverse behavioral effects associated with neurotoxic exposure have been prevalent in the occupational and medical literature for over a century. In this country, Francis Braceland was one of the first investigators to describe the neurobehavioral consequences of carbon disulfide exposure in workers employed in the viscose rayon industry (Braceland, 1942). The hallmark symptoms that he noted — headaches, fatigue, insomnia, confusion, amnesia, changes in personality, depression, irritability — describe the primary symptoms of solvent neurotoxicity. Because solvents are highly lipophilic, the neuropsychiatric symptoms have been attributed to a depression of central nervous system (CNS) activity. In fact, chemicals of this class are often used as anesthetics.

The somatic and neuropsychiatric consequences of exposure to chemicals at levels below threshold limit values (TLVs) have been the subject of several consensus workshops (Baker and Fine, 1986; Cranmer and Goldberg, 1986). A typology describing the neurobehavioral symptoms associated with solvent neurotoxicity, which reflects both the severity and reversibility of symptoms, has been proposed. The mildest form, *Type 1,* is characterized by

1. Address all correspondence to: Lisa A. Morrow, Ph.D., Neurobehavioral Toxicology, Western Psychiatric Institute and Clinic, 3811 O'Hara St., Pittsburgh, PA 15213. Tel.:(412)624-4626. Fax:(412)624-2413.

2. Key Words: Event-Related Potentials, Heart Rate, Kindling, Pupil Diameter, Solvents.

Toxicology and Industrial Health, Vol. 10, No. 4/5, pp. 537–544

ISSN: 0748-2337

symptoms including decreased concentration, fatigue, loss of initiative, and memory loss. There is no evidence of impairment on formal neurobehavioral measures, and symptoms are thought to be reversible if exposure is terminated. *Type 2A/2B* is characterized by more severe symptomatology that may or may not be reversible. *Type 2A* refers to a change in mood or personality, such as increased lability, irritability, depression, and poor impulse control, whereas impairments on neuropsychological evaluation constitute a marker for *Type 2B.* Typically, psychomotor slowing, memory impairment, decreased attention, and poor learning can be documented. Subtle neurological signs may also be present. The most serious impairment, *Type 3,* follows years of voluntary abuse of solvents (e.g., glue sniffing) or long-term chronic occupational exposure with repeated overexposures. There is a generalized decrease in intellectual function, neuroradiological evidence of structural brain damage, and little likelihood of reversibility.

It is unfortunate that this tripartite typology has not been widely embraced by researchers. Nevertheless, one could argue that most of these individuals would meet current criteria for Type 2A/2B solvent neurotoxicity. On formal neuropsychological evaluation, exposed workers typically perform more poorly than nonexposed control subjects on measures of reaction time, learning and memory, abstract reasoning, visuospatial ability, and perceptuomotor speed (Gregersen et al., 1984; Ekberg et al., 1986; Ryan et al., 1988; Morrow et al., 1990b; Morrow et al., 1991; Morrow et al., 1992a; Morrow, 1994). In addition, self-report measures of psychological symptomatology reveal more somatic complaints: fatigue, tension, irritability, and mood changes (Hane et al., 1977; Lindstrom, 1980; Morrow et al., 1989; Morrow et al., 1993).

Although this symptomatology is consistent with solvent-induced central nervous system damage, there is still skepticism about the extent of actual solvent-induced cognitive and emotional dysfunction. Baker and Fine (1986), among others, have argued that a strong dose-response relationship is necessary to support the hypothesis that solvent exposure produces cognitive and behavioral impairment. Unfortunately, while many studies of exposed and nonexposed workers have found evidence of neuropsychological and psychological dysfunction, a dose-response relationship is often lacking (a situation not unique to solvents, but found with virtually any neurotoxic substance). That is, increased duration or dose of solvent exposure is not always associated with increased psychological and cognitive impairment. Because solvents have such a short half-life — 24 to 48 hours — it is rare that measurements are available in order to quantify solvent "body burden." If blood or urine samples are not available, and air sampling measures from the workplace have not been obtained, then estimates of duration, frequency, and amount (i.e., dose) of exposure must rely on the workers' recollection. Furthermore, accurate assessment of neurotoxicity may be confounded by other factors, such as job stress or previous alcohol use. Other organismic variables such as liver dysfunction (Tarter et al., 1988) or a history of multiple peak exposures (Morrow et al., 1991) may increase the risk of neuropsychological impairment. In addition, serious methodological limitations (e.g., failure to equate experimental and control groups on demographic variables) may restrict the generalizability of many earlier studies.

FUNCTIONAL IMAGING AND PSYCHOPHYSIOLOGICAL STUDIES

Because quantifiable changes to the central nervous system (CNS), as measured by neurological examination, computerized tomography, or magnetic resonance imaging, are often absent following exposure, researchers have begun to utilize functional imaging and psychophysiological measures to document the extent to which the central nervous system may be altered following solvent exposure.

Several recent reports have found evidence of decreased neurofunctional capacity in both cortical and subcortical areas following solvent exposure. Blood flow measurements in solvent-exposed workers show the lowest flow to the frontotemporal areas (Hagstadius et al., 1989). Studies utilizing single photon emission computerized tomography (SPECT) imaging procedures in persons with neurotoxic exposure demonstrated significantly lower perfusion in frontal and temporal cortex and in the basal ganglia and thalamus (Callender et al., 1993). Positron emission tomography (PET) demonstrated reduced glucose uptake to frontal and medial temporal cortex, as well as in the basal ganglia, hippocampus, amygdala, putamen, and thalamus, in a worker with a single peak solvent exposure (Morrow et al., 1990a). In this latter study, there was an association between deficits seen on neuropsychological tests and reduced glucose uptake in certain brain areas. This underscores the notion that even relatively brief exposures can eventuate in significant neurological and neuropsychological decrements.

The use of psychophysiological measures [e.g., event-related potentials (ERPs), cardiac and pupil reactivity] to assess persons with toxic exposure has been limited, but recent findings suggest these procedures may be particularly sensitive to chemical exposure. It has been argued that the assessment of ERPs may be more sensitive to CNS dysfunction than cognitive measures (Newton et al., 1989). That is, a higher level of impairment may be required before deficits are detected on the cognitive tests, but subclinical impairments may be ascertained with ERPs. In one of the first studies measuring ERPs in solvent-exposed persons, amplitude and latency values were within normal ranges, but changes in the amplitudes of the N100 and P300 components during attentional switching tasks were significantly smaller than those of controls (El Massioui et al., 1986). The latency of the P300 was delayed in a small group of persons tested after exposure to hydrogen sulfide (Wasch et al., 1989).

Work in our lab assessing ERP waveforms and information processing changes has demonstrated evidence for disruptions of both neurophysiological function and cognitive processing in individuals with a history of organic solvent exposure (Morrow et al., 1992b). Data on ERP measures were compared for solvent-exposed subjects, nonexposed controls, and psychiatric controls. Measures of visual sensitivity were determined with a Continuous Performance Test (CPT) and discrimination of a target presented with distractors provided a probe of iconic memory [Span of Apprehension (SOA)]. Two "oddball" paradigms — Counting and Choice Reaction-Time (RT) — were employed to assess ERPs (Steinhauer and Zubin, 1982). In the Counting task, subjects silently counted the occurrence of infrequent high-pitched tones, but not low-pitched tones. In the Choice RT task, subjects pressed a button following each high tone, and a different button following each low tone. For all subjects, *the same order of tasks is purposely employed:* Counting first, Choice RT second. This is done because it is emphasized that only target tones are overtly task relevant for the initial Counting task, while all tones become relevant for the Choice RT task.

Results showed significant decrements on the CPT and poorer iconic memory for both solvent-exposed subjects and psychiatric patients, compared to normal controls. While the exposed subjects did not show evidence of significantly reduced P300 amplitudes, there was evidence of greatly increased N250 and P300 latencies across tasks and conditions when compared to both controls and schizophrenic patients. The average P300 latency for the exposed group was over 400 msec, while the normal and psychiatric groups had latency values of 344 msec and 350 msec, respectively. The morphology of the waveforms that we have observed in the solvent-exposed subjects was similar to that seen in patients who have evidence of CNS disease, epilepsy, and Alzheimer's disease (Goodin et al., 1979; Nelson et al., 1991). Solvent-exposed subjects did show larger amplitudes of the early waveforms (N100, P200), suggesting a heightened sensitivity to the sensory effects of stimuli. In addition, when we compared the two oddball paradigms — Counting and Choice RT — there was a tendency for the exposed subjects, like normal subjects, to have longer P300 latencies for the Choice RT task — a more cognitively challenging task. We also found a rank order correlation between length of exposure and P300 latency (rho = 0.50). These data indicate that ERPs may be sensitive to subclinical impairment. Of the patients we have assessed on the ERPs, four had cognitive test results that were within normal limits but showed substantial increases in P300 latency (over 400 msec). All of these patients reported irritability and mood changes which have been hypothesized to precede evidence of clinically significant cognitive changes (Parkinson et al., 1990).

In addition to the ERP measures, data are also available for autonomic nervous system function. Both cardiac and pupillary reactivity were measured in exposed subjects and nonexposed controls (Morrow and Steinhauer, 1994). Autonomic functions such as heart rate and pupil size provide objective measures of mental activity and are sensitive to changes in information processing. Both are highly correlated with the ability to initiate and maintain attention. For this study we looked at cardiac and pupil reactivity during the Counting and Choice RT conditions in the oddball paradigm and found atypical responding in the solvent-exposed subjects. While control subjects show a decline in heart rate going from the Counting to the Choice RT task — the normal habituation response, exposed persons show an increase in heart rate. Likewise, initial pupil diameter was similar for both groups, but only the control subjects exhibited habituation across the two tasks. This increase in heart rate and pupil diameter demonstrated for exposed subjects suggests an increase in sympathetic arousal. We have suggested that this failure to habituate may be a response to a cognitively challenging event. That is, the Choice RT condition requires more demands on cognitive processing since the subject must attend to each stimulus, make a decision about each stimulus, and select the appropriate response. The Counting task only requires attention to the rare tones and no attention is demanded when the frequent tones are presented. Indeed, many of our subjects spontaneously report that the Choice RT condition is more difficult, and in looking at error rates, the average error rate is higher for the Choice RT condition.

To determine the extent to which these measures change over time, we reevaluated eight of the exposed persons on both ERP and autonomic functions. None of these persons worked with solvents in the interim. For the ERP measures, there was evidence of mild improvement in one subject: P300 latency improved from 440 msec to 384 msec on the Counting task and 416 msec to 392 msec on the Choice RT task. It should be noted that the latencies at the follow-up testing are still more than two standard deviations slower than controls. The other seven subjects showed virtually no change, or an *increase*, in P300 latency.

Surprising findings were noted for the autonomic measures. Six of the eight subjects demonstrated a *more severe* alteration of autonomic reactivity at follow-up. For example, one person initially showed a normal cardiac habituation response — an average heart rate of 77 beats per minute (bpm) on the Counting task and 76 bpm on the Choice RT task — but no habituation of pupil response (7.09 cm to 7.16 cm). This initial testing was completed two weeks after a peak exposure to aromatic solvents. After 33 months, a second evaluation was done. Both pupil and heart rate failed to show habituation: the average heart rate increased from the Counting to the Choice RT — 83 and 86 bpm, respectively. Pupil diameter also showed a dramatic increase across tasks — 7.10 cm on the Counting task to 7.65 cm on the Choice RT task. The P300 latency of this subject of 384 msec and 384 msec for Counting and Choice RT at first testing was virtually unchanged at the second testing — 400 msec and 384, respectively. This same atypical response pattern — increasing sympathetic arousal when comparing the initial and follow-up testings — was noted in five other exposed subjects, the majority having a history of peak exposure. Only two of the eight subjects had a normal habituation response at both test sessions 1 and 2.

While it is somewhat premature to speculate about the mechanisms responsible for the deficits observed in exposed patients, several hypotheses have been developed. Jennings (1992) has postulated a midbrain inhibitory circuit, consisting of limbic structures that regulate autonomic requirements (e.g., slowing of heart rate). Recently, Bell and her colleagues have hypothesized that changes in neurophysiology, cognition, and affect following chemical exposure may result from kindling of limbic structures (Bell et al., 1992). Kindling refers to the sensitization of limbic neurons (including amygdala, olfactory bulb, piriform cortex, and hippocampus) to repeated intermittent stressors. Kindling develops from repeated intermittent exposures, either mild or marked, that initially do not elicit neuronal excitability. Over time, however, excitability of the neurons is altered so that subsequent low-level stressors may serve as triggers. It is this mechanism that "serves as an amplifier for reactivity to low level chemical exposures and as an initial common pathway for a range of clinical phenomenology, including cognitive and affective dysfunctions" (Bell et al., 1992, p. 221). Pathways regulated by the limbic system would thus be altered. Thresholds are lowered such that subsequent stressors, which were initially benign, result in neuronal excitability. Autonomic pathways regulated by limbic structures would also be modified. If this inhibitory system were altered via limbic kindling, there would be a state of decreased attention, and disruptions of arousal. Exposure to organic solvents via the olfactory pathways (which are the primary access route to the CNS and particularly susceptible to kindling) may affect limbic nuclei such that subsequent stressors (e.g., low-level chemical or psychological stressors) occurring intermittently over time, could modify neuronal sensitization. This would lead to an increasing susceptibility to a cognitively challenging event, as well as an alteration of arousal and attention. Consistent with our model are our own pilot data showing, in a small sample of exposed patients, a failure to habituate that seems to worsen over time. Future studies to test the kindling model should focus on a challenge paradigm (e.g., exposing subjects to low-level chemicals) to determine psychophysiological alterations as a direct response to the stressor.

SOLVENTS AND MULTIPLE CHEMICAL SENSITIVITIES

Our work to date demonstrates the value of psychophysiological measures in studying individuals with significant solvent exposures. We are convinced that this approach may also be quite useful in studying individuals with multiple chemical sensitivities (MCS). If, as has been suggested by several investigators, some type of limbic system dysfunction underlies the development of neuropsychiatric symptoms in both groups of subjects, then psychophysiological evaluations of the sort outlined above might be particularly useful in identifying brain dysfunction in individuals suspected of having MCS. Unfortunately, before this endeavor can proceed further, investigators will have to establish working diagnostic definition(s) of MCS. Our own work in solvent neurotoxicity has used the typology developed at several consensus conferences to aid in the diagnosis and classification of individuals with a history of solvent exposure. A similar typology, perhaps established in the same way — at a consensus conference — could certainly facilitate the identification of subtypes of MCS individuals. It is our expectation that those individuals who meet criteria for even moderate MCS (equivalent to Type 2A/2B solvent neurotoxicity) will manifest atypical anomalies that are similar to those found in solvent-exposed adults.

CONTINUED APPLICATION OF PSYCHOPHYSIOLOGICAL TECHNIQUES

As may be evident from the foregoing discussion, the study of psychophysiological parameters during cognitive activation provides a variety of tools for the study of chemically exposed individuals. Moreover, these noninvasive techniques can be repeated over short time periods. Beyond the ability to identify group differences, these measures have a unique ability to supplement individual data on neuropsychological function. For example, P300 latency provides a clear indication of central nervous system integrity. P300 amplitude, the extent of pupillary dilation, and changes in phasic cardiac processes provide direct assessment of information utilization. Tonic autonomic measures, such as overall pupillary diameter and heart rate, provide an assessment of generalized autonomic arousal and maintenance of attention, and appear to be increased in subjects having difficulty attending to demands of specific tasks. Thus, not only are they useful in the research domain, but they are applicable to the clinical evaluation of patients in whom there are suspected changes in cognitive function following neurotoxic exposure.

REFERENCES

BAKER, E. and FINE, L. (1986). "Solvent neurotoxicity: the current evidence." J. Occup. Med. 28:126–129.

BELL, I., MILLER, C., and SCHWARTZ, G. (1992). "An olfactory-limbic model of multiple chemical sensitivity syndrome: possible relationships to kindling and affective spectrum disorders." Biol. Psychiat. 32:218–242.

BRACELAND, F. (1942). "Mental symptoms following carbon disulphide absorption and intoxication." Ann. Intern. Med. 16:246–261.

CALLENDER, T., MORROW, L. SUBRAMANIAN, K., DUHON, D., and RISTOV, M. (1993). "Three-dimensional brain metabolic imaging in patients with toxic encephalopathy." Environ. Res. 60:295–319.

CRANMER, J. and GOLDBERG, L., ed. (1986). Proceedings of the workshop on neurobehavioral effects of solvents. Intox Press, Roland, AR.

EKBERG, K., BARREGARD, L., HAGBERG, S., and SALLSTEN, G. (1986). "Chronic and acute effects of solvents on central nervous system functions in floorlayers." Br. J. Ind. Med. 43:101–106.

EL MASSIOUI, F., LESEVRE, N., and FOURNIER, L. (1986). "An event-related potential assessment of attention impairment after occupational exposure to organic solvents." In: Evoked Potentials III: The Third International Evoked Potentials Symposium (C. Barber and T. Blum, ed.). Butterworths, Boston. pp. 346–350.

GOODIN, D., SQUIRES, K., and STARR, A. (1979). "Long latency components of the auditory evoked potential in dementia." Brain 101:635–648.

GREGERSEN, P., ANGELSO, B., NIELSEN, T., NORGAARD, B., and ULDAL, C. (1984). "Neurotoxic effects of organic solvents in exposed workers: an occupational, neuropsychological, and neurological investigation." Am. J. Ind. Med. 5:201–225.

HAGSTADIUS, S., ORBAEK, P., RISBERG, J., and LINDGREN, M. (1989). "Regional cerebral blood flow at the time of diagnosis of chronic toxic encephalopathy induced by organic-solvent exposure and after cessation of exposure." Scand. J. Work Environ. Health 15:130–135.

HANE, M., AXELSON, O., BLUME, J., HOGSTEDT, C., SUNDELL, L., and YDREBORG, B. (1977). "Psychological function changes among house painters." Scand. J. Work Environ. Health 3:91–99.

JENNINGS, J.R. (1992). "Is it important that the mind is in the body? Inhibition and the heart." Psychophysiology 29:369–383.

LINDSTROM, K. (1980). "Changes in psychological performances of solvent-poisoned and solvent-exposed workers." Am. J. Ind. Med. 1:69–84.

MORROW, L. (1994). "Cueing attention: disruptions following organic solvent exposure." Neuropsych.

MORROW, L., RYAN, C., GOLDSTEIN, G., AND HODGSON, M. (1989). "A distinct pattern of personality disturbance following exposure to organic solvents." J. Occup. Med. 31:743–746.

MORROW, L., CALLENDER, T., LOTTENBERG, S., BUCHSBAUM, M., HODGSON, M., and ROBIN, N. (1990a). "PET and neurobehavioral evidence of tetrabromoethane encephalopathy." J. Neuropsychiat. Clin. Neurosci. 2:431–435.

MORROW, L., RYAN, C., HODGSON, M., and ROBIN, N. (1990b). "Alterations in cognitive and psychological functioning after organic solvent exposure." J. Occup. Med. 32:444–450.

MORROW, L., RYAN, C., HODGSON, M., and ROBIN, N. (1991). "Risk factors associated with persistence of neuropsychological deficits in persons with organic solvent exposure." J. Nerv. Ment. Dis. 179:540–545.

MORROW, L., ROBIN, N., HODGSON, M., and KAMIS, H. (1992a). "Assessment of attention and memory efficiency in persons with solvent neurotoxicity." Neuropsychologia 30:911–922.

MORROW, L., STEINHAUER, S., ROBIN, N., and HODGSON, M. (1992b). "Delay in P300 latency in patients with organic solvent exposure." Arch. Neurol. 49:315–320.

MORROW, L., KAMIS, H., and HODGSON, M. (1993). "Psychiatric symptomatology in persons with organic solvent exposure." J. Con. Clin. Psych. 61:171–174.
MORROW, L., and STEINHAUER, S. (1994). "Alterations in heart rate and pupillary response in persons with organic solvent exposure." Unpublished manuscript.
NELSON, C., COLLINS, P., and TORRES, F. (1991). "P300 brain activity in seizure patients preceding temporal lobectomy." Arch. Neurol. 48:141–147.
NEWTON, M., BARRETT, G., CALLANAN, M., and TOWELL, A. (1989). "Cognitive event-related potentials in multiple-sclerosis." Brain 112:1637–1660.
NATIONAL INSTITUTE FOR OCCUPATIONAL SAFETY AND HEALTH (1987). "Organic solvent neurotoxicity." NIOSH Publication No. 87-104. U.S. Government Printing Office, Washington, D.C.
PARKINSON, D., BROMET, E., COHEN, S., DUNN, L., DEW, M.A., RYAN, C., and SCHWARTZ. J. (1990). "Health effects of long-term solvent exposure among women in blue-collar occupations." Am. J. Ind. Med. 17:661–675.
RYAN, C., MORROW, L., and HODGSON, M. (1988). "Cacosmia and neurobehavioral dysfunction associated with occupational exposure to mixtures of organic solvents." Am. J. Psychiat. 145:1422–1445.
STEINHAUER, S. and ZUBIN, J. (1982). "Vulnerability to schizophrenia: information processing in the pupil and event-related potential." In: Biological Markers in Psychiatry and Neurology (E. Usdin and I. Hanin, ed.). Pergamon Press, Elmsford, NY. pp. 371–385.
TARTER, R., VAN THIEL, D., and EDWARDS, K., ed. (1988). Medical Neuropsychology: The Impact of Disease on Behavior. Plenum Press, New York.
WASCH, H., ESTRIN, W., YIP, P., BOWLER, R., and CONE, J. (1989). "Prolongation of the P300 latency associated with hydrogen sulfide exposure." Arch. Neurol. 46:902–904.

NEUROPSYCHOLOGY AND PSYCHOLOGY OF MCS

NANCY FIEDLER,* HOWARD KIPEN,* JOHN DELUCA,† KATHIE KELLY-MCNEIL,* AND BENJAMIN NATELSON‡

*UMDNJ — Robert Wood Johnson Medical School
Environmental and Occupational Health Sciences Institute
Piscataway, New Jersey

†Kessler Institute
West Orange, New Jersey

‡Veterans Administration Medical Center
East Orange, New Jersey

Neurological symptoms are frequently reported by patients with multiple chemical sensitivities (MCS). Methods to compare the psychiatric, personality, and neuropsychological function of patients with MCS, chronic fatigue syndrome (CFS), and normal controls are described. Increased rates of Axis I psychiatric diagnoses are observed in the literature for MCS and CFS subjects relative to controls. Findings on the MMPI-2 and the Toronto Alexithymia Scale reveal profiles consistent with the tendency to report somatic rather than emotional symptoms in response to stress. However, many of the reported somatic symptoms also coincide with those found in neurologic disorders. The overall neuropsychological profile for MCS subjects does not reflect cognitive impairment. Relative to normal controls, the only difference in neuropsychological performance observed is reduced recognition of nontarget designs on a visual memory task. More fruitful areas for future psychological research will include measurement of the interaction between behavioral response styles and attentional processes in cognition, as well as observations under controlled challenge conditions.

1. Address all correspondence to: Nancy Fiedler, UMDNJ — Robert Wood Johnson Medical School, Environmental & Occupational Health Sciences Institute, 681 Frelinghuysen Road, Room 210, Piscataway, New Jersey 08855. Tel.:(908)932-0190. Fax:(908)932-0127.

2. Abbreviations: CFS, Chronic Fatigue Syndrome; CVMT, Continuous Visual Memory Test; DIS, Diagnostic Interview Survey; MCS, Multiple Chemical Sensitivity; MMPI, Minnesota Multiphasic Personality Inventory; SCID, Structured Clinical Interview for the Diagnostic and Statistical Manual.

3. Key Words: Chronic Fatigue Syndrome, Multiple Chemical Sensitivity, Neuropsychology, Psychiatric.

Toxicology and Industrial Health, Vol. 10, No. 4/5, pp. 545–554

ISSN: 0748-2337

INTRODUCTION

Neurotoxicants such as pesticides and solvents are ubiquitous in the environment and cause acute, as well as chronic, emotional and cognitive symptoms (e.g., Anger and Johnson, 1992). Patients exposed to neurotoxicants often complain of poor memory, difficulty in concentration, and changes in mood (Hartman, 1988). Standardized neuropsychological tests can quantify these symptoms. For example, these tests have documented cognitive impairments of memory and concentration due to chronic solvent exposure (NIOSH, 1987).

Exposure standards have been established to protect the health of workers using neurotoxicants. However, the concentration and duration of exposure that may result in permanent cognitive and emotional symptoms is not clearly defined and probably interacts with host factors. Recently, we have seen a group of patients with multiple nervous system and other symptoms which they attribute to environmental chemical exposures well below accepted exposure standards. A subgroup of such individuals are clinically defined to have multiple chemical sensitivities (MCS).

MCS patients frequently complain of cognitive and emotional symptoms, yet few data are available to quantify these complaints (e.g., concentration and memory). The purpose of this component of our research is to evaluate objectively the psychiatric and neuropsychological function of MCS patients and to compare them to patients with similar health complaints not attributed to chemicals (i.e., chronic fatigue syndrome (CFS)) and to normal controls.

Based on a description of their symptom profile, MCS patients resemble patients with psychiatric diagnoses such as post–traumatic stress (Schottenfeld and Cullen, 1986) or somatoform disorders (e.g., Brodsky, 1983; Stewart and Raskin, 1985). These similarities have lead some investigators to hypothesize that MCS can be explained by preexisting psychiatric disorders (e.g., Brodsky, 1983). Due to these observations, several recent studies investigated the premorbid psychiatric status of MCS patients. For example, as compared to musculoskeletal injury patients (Simon et al., 1993) and community controls (Black et al., 1990), a significantly greater proportion of MCS subjects had premorbid psychiatric disorders. However, a significant proportion of MCS subjects also did not qualify for a preexisting psychiatric diagnosis (53% in Simon et al., 1993; 35% in Black et al., 1990). Fiedler et al. (1992a) reported in a case series that MCS subjects, chosen according to the more specific criteria proposed by Cullen (1987), infrequently had a history of psychiatric disorder prior to the onset of MCS. The discrepancy in these investigations may be due, in part, to differences in the criteria used to select MCS patients.

Depending on the criteria employed, a broad range of patients may be considered chemically sensitive. For example, from the most inclusive perspective, clinical ecologists propose chemical sensitivity as an early manifestation of diseases involving any system (e.g. cardiovascular, renal) (Rea, 1992). Based on this conceptualization, patients diagnosed with traditional diseases are also said to suffer from chemical sensitivity. While not as broad, previous investigators (e.g., Black et al., 1990; Simon et al., 1993) included patients under the care of clinical ecologists who were diagnosed with systemic candidiasis and chemical hypersensitivity. In part, the goal of these studies was to characterize the psychiatric, neuropsychologic, and immunologic status of the "typical" chemical sensitivity patient from a clinical ecology practice. In contrast, the goal of studies from our laboratory (Fiedler et al., 1992a; Fiedler et al., unpublished[1]) is to characterize patients who meet criteria proposed to define MCS as a primary problem generally in the absence of other known medical and psychiatric diagnoses. This allows a clear separation of findings related to chemical sensitivity from findings associated with known medical or psychiatric diagnoses.

Patients with chronic fatigue also frequently complain of poor attention, impaired concentration and memory, as well as symptoms indicative of psychiatric illness (e.g., depression) (Katon and Russo, 1992). As with MCS patients, little evidence is available to suggest a consistent underlying organic basis for their symptoms. Nevertheless, data from some standardized neuropsychological tests are available to substantiate cognitive symptoms (e.g., DeLuca et al., 1993). It has also been suggested that chronic fatigue and MCS are modern variations of somatoform disorders (Stewart and Raskin, 1985). However, chronic fatigue patients do not attribute their symptoms to chemical exposures. Rather, the onset of their symptoms typically follows a flu-like illness and exacerbations of symptoms are attributed to exertion. Due to the symptomatic similarities of CFS and MCS, another goal of our studies has been to use standardized psychiatric and neuropsychological instruments to compare MCS and CFS patient groups.

Although the neuropsychological and psychiatric symptoms reported by MCS patients are often similar to those reported by patients with ongoing solvent exposures, the exposure history of these two groups appears to be different. In contrast to MCS, neuropsychological impairment among solvent-exposed patients typically occurs after considerably greater exposure duration (e.g., 7 to 10 years) and involves clearly higher concentrations of solvent exposure. The neuropsychological tests used in our laboratory overlap those used to evaluate the long-term effects of solvent exposure. Therefore, the pattern of performance in MCS and CFS patients can be compared to those seen among workers with chronic occupational solvent exposure.

RATIONALE FOR SUBJECT SELECTION CRITERIA

Based on the clinical similarities between MCS and CFS, a direct objective comparison of these groups could help resolve discrepant hypotheses about the conditions. For example, some investigators have suggested that the symptomatology of CFS is the somatic form of depression (Holmes et al., 1988). On the other hand, the symptoms of MCS have been

---

[1]FIEDLER, N., KIPEN, H., DELUCA, J., KELLY-MCNEIL, K., and NATELSON, B. "A controlled comparison of Multiple Chemical Sensitivity and Chronic Fatigue."

compared to anxiety-based disorders such as phobias (Bolla-Wilson et al., 1988). In addition, healthy normal controls provide a nonclinical comparison standard.

MCS subjects entering our studies meet the following criteria suggested by Cullen (1987): 1) symptoms acquired in relation to an initial identifiable environmental exposure(s); 2) symptoms currently involve more than one organ system (e.g., respiratory and nervous system); 3) symptoms recur and abate in response to exposure to chemical/substances; 4) symptoms are elicited by exposures to very low levels of chemicals of diverse structural classes (e.g., solvents, pesticides); 5) other medical conditions do not account for the symptoms. Subjects with a history of psychiatric hospitalizations or the following premorbid psychiatric conditions are excluded: psychoses, manic depression, alcoholism or drug abuse. Also, subjects currently involved in a legal case related to their exposure are not included.

Chronic fatigue subjects (CFS) evaluated for our studies meet the revised Centers for Disease Control criteria for CFS (Schluederberg et al., 1992). Severe fatigue that reduces activity 50% below premorbid levels for a minimum of six months is the "major" criterion. This fatigue cannot be explained by other medical or psychiatric illness. Minor criteria include at least 8 of 14 rheumatological, infectious, and neuropsychiatric symptoms (e.g., myalgia, sore throat, poor concentration).

Normal controls are recruited from a general internal medicine clinic population or from newspaper advertisements. Subjects with any of the following disorders are excluded: neurologic disease or brain injury, toxic exposure, stroke or cardiovascular disease, liver or kidney disease, serious gastrointestinal disorders (e.g., colitis), and major psychiatric conditions described previously.

RATIONALE AND SELECTION OF NEUROPSYCHOLOGIC AND PSYCHIATRIC MEASURES

A battery of neuropsychological tests that include traditional neuropsychological tests and performance based tests are used to assess cognitive function. Traditional neuropsychological tests have been used to evaluate dysfunction due to brain injury (Lezak, 1983). Since MCS and CFS patients report cognitive symptoms of permanent dysfunction, traditional tests reflecting neuropsychological impairment are being used as a part of the neuropsychological battery. On the other hand, performance based tests such as reaction time have been used as subtle indicators of changes in performance due to such environmental factors as stress or drugs (Wittenborn, 1987). In addition to general cognitive dysfunction, MCS patients report relatively subtle performance changes under exposure conditions. While MCS patients are accommodated as much as possible during testing, our testing rooms are not free of exposures reported to cause symptoms. Therefore, if subtle exposure and symptoms occur, the performance tests would be most sensitive in evaluating these effects.

Tests of concentration include computerized Simple Reaction Time (dominant and nondominant hand) and Continuous Performance Test (Baker et al., 1985), Digit Span (Wechsler, 1981), and the Stroop Color Word Test (Trenerry et al., 1989). Tests of visuomotor skills are Digit Symbol (Wechsler, 1981), computerized Hand-Eye Coordination (Baker et al., 1985), and Grooved Pegboard, dominant and nondominant hand (Trites, 1981). Tests of memory are the California Verbal Learning Test (Delis et al., 1987), the Continuous Visual Memory Test (Trahan and Larrabee, 1988), and Visual Reproduction I and II (Wechsler, 1987). The Metamemory Questionnaire (Mateer et al., 1987) is administered to assess subjects' perception of memory function.

To examine the current psychiatric status of subjects, the Structured Clinical Interview for the Diagnostic and Statistical Manual, 3rd Edition Revised (Spitzer et al., 1990) and the Minnesota Multiphasic Personality Inventory-2 are administered (Hathaway and McKinley, 1989). In addition, the somatization section of the Diagnostic Interview Survey, III-A (Robins and Helzer, 1985) is administered by a nurse to ascertain the rate of medically explained and unexplained symptoms. These instruments assessed traditional psychiatric diagnostic categories. Results from these measures provide an objective comparison of the rates of psychiatric disorders between MCS and CFS. In addition to the traditional clinical scales, the MMPI-2 provides content scales based on single symptom dimensions such as depression and health concerns. The latter allows separation of somatic (health concerns) from emotional symptoms (anxiety). The Toronto Alexithymia Scale is administered to assess the tendency to express feelings in somatic rather than verbal or symbolic terms (Taylor et al., 1988). Performance on this scale has been associated with psychosomatic illness and somatization (Taylor et al., 1988).

PSYCHIATRIC AND PERSONALITY STATUS OF MCS PATIENTS

Our studies as well as those from other investigators (e.g., Simon et al., 1993), evaluating Axis I psychiatric diagnoses with structured psychiatric interviews, suggest that MCS subjects have significantly higher rates of psychiatric diagnoses than age-, gender-, and education-matched controls.

The diagnoses occurring most frequently among MCS subjects are depression and anxiety disorders. For example, our preliminary results with MCS patients suggests a rate of current depression at 17% (Fiedler et al., 1992a). These rates are lower than the rates of current depression (29%) and anxiety (34%) reported in the most recent study by Simon et al. (1993). MCS and CFS subjects have a significantly higher overall rate of current psychiatric diagnoses than various control groups (e.g., normal controls, musculoskeletal injury), a significant proportion of subjects from our studies as well as other laboratories (e.g., Simon et al., 1993) have no current psychiatric diagnosis. The data do not support that anxiety based disorders are more prevalent among MCS patients. Also, a significant proportion of subjects from our studies (e.g., 83%) (Fiedler et al., 1992a) as well as other investigators (e.g., Simon et al., 1993) have no current psychiatric diagnosis.

For several reasons, traditional Axis I psychiatric diagnoses may not provide the most useful information regarding the psychological attributes that contribute to this disorder. First, thus far, the data has not shown MCS to be fully explained by premorbid or current psychiatric disorders. Even though considerable overlap exists between the symptoms of somatoform disorders and MCS, a majority of these patients do not meet the criteria for somatization disorder as currently outlined in DSM-III-R. Second, the cause and effect relationship between MCS and psychiatric disorders cannot be addressed among individuals who are already ill with MCS at the time of testing. That is, it can be effectively argued that psychiatric disorders such as depression or anxiety may have been a consequence of MCS.

Instead of focusing on the dichotomous diagnostic categories within psychiatry, use of measures that assess personality traits associated with physical symptom reporting may be of greater heuristic value. For example, our studies with the MMPI-2 reveal that the group profile for MCS subjects is consistent with somatoform disorders (Fiedler et al., 1992a). That is, the peak scale elevations are on scales of hypochondriasis and hysteria. When the items from these scales are analyzed based on their content, the primary symptoms expressed by MCS subjects load on a scale of Health Concerns. This scale includes somatic items that are psychologically based as well as somatic items endorsed by neurologically impaired populations such as closed head injury and multiple sclerosis (Meyerink et al., 1988; Alfano et al., 1990).

Other studies have shown that MCS subjects also scored significantly higher than controls on scales associated with somatization such as the Barsky Amplification Scale (Simon et al., 1990), and the Illness Behavior Questionnaire (Black et al., 1993). The Toronto Alexithymia Scale, a measure reflective of difficulty with verbal and symbolic expression of emotion, was selected in our laboratory to assess the personality construct hypothesized to be associated with post–traumatic syndromes and psychosomatic disease (Taylor et al., 1988). That is, MCS subjects are hypothesized to express symptoms of distress in somatic rather than emotional terms.

In summary, traditional, Axis I psychiatric diagnoses have not adequately described the majority of MCS subjects. However, several studies support that MCS subjects report a mixture of somatic symptoms that reflect both neurological and psychosomatic processes.

NEUROPSYCHOLOGICAL STATUS OF MCS PATIENTS

To date, few studies have systematically evaluated the neuropsychological symptoms reported by MCS patients and no studies have evaluated cognitive performance under controlled exposure conditions. Simon et al. (1993) administered traditional measures of neuropsychological performance including tests of memory, concentration, and visuo-motor speed. A few significant differences were noted between MCS and musculoskeletal injury controls on immediate verbal recall. However, the authors concluded that when the effects of psychological distress (e.g., anxiety and depression) were controlled, differences in cognitive function were no longer apparent.

In our laboratory, applying the standardized neuropsychological tests described previously, we also detected few overall differences in performance between MCS subjects and normal controls. Relative to normative standards, we observed no differences on tests of concentration, which included Simple Reaction Time, Continuous Performance, Stroop Color-Word Task, and Digit Span (Fiedler et al., 1992a,b). We also detected no differences relative to normative standards on tests of visuomotor coordination such as Digit Symbol, Hand-eye Coordination, and Grooved Pegboard (Fiedler et al., 1992a,b). Preliminary comparisons with age, gender, and education matched CFS, and normal controls revealed no significant differences in performance on these measures.

We observed more mixed results on tests of verbal and visual memory. Preliminary data using the California Verbal Learning test revealed no significant overall differences in performance between MCS and normal controls. However, MCS subjects performed significantly more poorly on the initial trial of the word lists (Fiedler et al., 1992a). That is, the ability of MCS subjects to encode the list after one presentation was significantly lower for both lists presented within this task. After repetition of the word list (five learning trials), however, the overall recall as well as the slope of the learning curve for the MCS subjects was comparable to normal controls. This preliminary result is similar to the findings cited above by Simon et al. (1993). That is, initial verbal learning trials were reduced relative to controls. However, after a delay of 30 minutes, no differences in retention of the material were noted.

Preliminary data comparing MCS to normal controls on a complex signal detection task of visual memory, i.e., Continuous Visual Memory, reveals that MCS subjects commit significantly more False Alarms. That is, in signal detection terms, they have difficulty distinguishing target from nontarget designs or noise from the signal.

Thus far, performance on neuropsychological testing has not revealed evidence of cognitive impairment among MCS subjects. Even among the more sensitive tests of performance such as Reaction Time and Continuous Performance, the preliminary data from our laboratory do not show significant reductions in performance relative to normal controls. However, caution is indicated since only two laboratories to date have reported neuropsychological evaluation of MCS patients and none of these evaluations have been conducted under controlled exposure conditions.

SUMMARY

The most consistent finding in our studies as well as those of other investigators (e.g., Simon et al., 1993) is that MCS subjects report numerous somatic symptoms associated with Axis I psychiatric disorders such as depression, and with neurologic disorders such as multiple sclerosis (Alfano et al., 1990; Gass, 1991). The challenge from a psychological perspective is to develop objective methods that will help us understand the conditions under which these symptoms are produced, their neurobiological basis, and their impact on cognitive performance.

Thus far, our research with MCS subjects has been conducted in relatively uncontrolled exposure conditions. Subjects travel to our clinical facility, and while we make every effort to minimize exposures, our facility is not free of substances to which MCS subjects report reactions (e.g., cleaning products). Despite these uncontrolled exposure conditions, with one exception, the overall neuropsychological test results do not reveal cognitive deficits. Thus, the neuropsychological performance pattern of MCS subjects has not reflected reductions in performance that resemble other groups reporting cognitive deficits due to neurotoxicant exposures (e.g., solvents) (Ryan et al., 1988) nor have they reflected permanent brain injury.

On tests associated with personality attributes of somatic concerns, MCS subjects have shown a consistent pattern of increased focus on bodily sensations. This behavioral style may reflect difficulty with filtering out irrelevant or subclinical stimuli. For example, James et al. (1987) reported that based on evoked response potential, somatization disorder patients had difficulty separating relevant from irrelevant stimuli. Our preliminary data from the CVMT, a behavioral task in which MCS subjects had to separate relevant from irrelevant stimuli, provides some impetus for further exploration of this behavioral response style.

In summary, our studies suggest that future research should focus on attentional processes and on personality factors that may contribute to symptomatology. Dichotomous psychiatric diagnoses and traditional neuropsychological test batteries are not likely to enhance our understanding of the psychological or neurobiological mechanisms of MCS.

REFERENCES

ALFANO, D., FINLAYSON, M., STEARNS, G., and NEILSON, P. (1990). "The MMPI and neurologic dysfunction: profile configuration and analysis." Clin. Neuropsych. 4:69–79.

ANGER, W.K. and JOHNSON, B.L. (1992). "Human behavioral neurotoxicology: workplace and community assessments." In: Environmental and Occupational Medicine (W.N. Rom, ed.). Little Brown, Boston. pp. 573–592.

BAKER, E., LETZ, R., and FIDLER, A. (1985). "A computer-administered neurobehavioral evaluation system for occupational and environmental epidemiology." J. Occup. Med. 27:206–212.

BLACK, E.W., RATHE, A., and GOLDSTEIN, R.B. (1990). "Environmental illness: a controlled study of 26 subjects with '20th century disease'." J. Am. Med. Assoc. 264:3166–3170.

BLACK, D.W., RATHE, A., and GOLDSTEIN R.B. (1993). "Measures of distress in 26 'environmentally ill' subjects." Psychosom. 34:131–138.

BOLLA-WILSON, K., WILSON, R.J., and BLEECKER, M.L. (1988). "Conditioning of physical symptoms after neurotoxic exposure." J. Occup. Med. 30:684–686.

BRODSKY, C.M. (1983). "Psychological factors contributing to somatoform diseases attributed to the workplace." J. Occup. Med. 25:459–464.

CULLEN, M.R. (1987). Workers with Multiple Chemical Sensitivities. Hanley and Belfus, Philadelphia. pp. 655–662.

DELIS, D.C., KRAMER, J.H., KAPLAN, E., and OBER, B.A. (1987). California Verbal Learning Test Manual. The Psychological Corporation, San Antonio, Texas.

DeLUCA, J., JOHNSON, S., and NATELSON, B. (1993). "Information processing efficiency in chronic fatigue syndrome and multiple sclerosis." Arch. Neurol. 50:301–304.

FIEDLER, N., MACCIA, C., and KIPEN, H. (1992a). "Evaluation of chemically sensitive patients." J. Occup. Med. 34:529–538.

FIEDLER, N., KIPEN, H., and KELLY-McNEIL, K. (1992b). "Neuropsychological and psychiatric profile of patients with multiple chemical sensitivities." Clin. Neuropsych. 6:347.

GASS, C.S. (1991). "MMPI-2 interpretation and closed head injury: a correction factor." Psychol. Assess.: J. Consult. Clin. Psychol. 3:27–31.

HARTMAN, D.E. (1988). Neuropsychological Toxicology: Identification and Assessment of Human Neurotoxic Syndromes. Cook County Hospital, University of Illinois Medical Center. Pergamon Press, Chicago.

HATHAWAY, S.R. and McKINLEY, J.C. (1989). MMPI-2. University of Minnesota Press, Minneapolis, MN.

HOLMES, G.P., KAPLAN, J.E., NELSON, M., GANTZ, N.M., KOMAROFF, A.L., SCHONBERGER, L.B., STRAUSS, S.E., JONES, J.F., BUBOIS, R.E., CUNNINGHAM-RUNDLES, C., PAHWA, S., TOSATO, G., ZEGANS, L.S., PURTILO, D.T., BROWN, N., SCHOOLEY, R.T., and BRUX, I. (1988). "Chronic fatigue syndrome: a working case definition." Ann. Int. Med. 106:387–389.

JAMES, L., SINGER, A., ZURYNSKI, Y., GORDON, E., KRAIUHIN, C., HARRIS, A., HOWSON, A., and MEARES, R. (1987). "Evoked response potentials and regional cerebral blood flow in somatization disorder." 47:190–196.

KATON, W. and RUSSO, J. (1992). "Chronic fatigue syndrome criteria. A critique of the requirement for multiple physical complaints." Arch. Int. Med. 152:1604–1609.

LEZAK, M.D. (1983.) Neuropsychological Assessment. Oxford University Press, New York.

MATEER, C.A., SOHLBERG, M.M., and CRINEAN, J. (1987). "Focus on clinical research: perceptions of memory function in individuals with closed-head injury." J. Head Trauma Rehab. 2:74–84.

MEYERINK, L.H., REITAN, R.M., and SELZ, M. (1988). "The validity of the MMPI with multiple sclerosis patients." J. Clin. Psychol. 44:764–769.

NATIONAL INSTITUTE FOR OCCUPATIONAL SAFETY AND HEALTH (NIOSH) (1987). Current Intelligence Bulletin 48: "Organic solvent neurotoxicity". U.S. Department of Health and Human Services, Centers for Disease Control.

REA, W.J. (1992). Chemical Sensitivity. Lewis Publishers, Boca Raton, Florida.

ROBINS, L.N. and HELZER, J.E. (1985). "Diagnostic Interview Schedule (DIS) Version III-A." Washington University School of Medicine. St. Louis, MO.

RYAN, C.M., MORROW, L.A., and HODGSON, M. (1988). "Cacosmia and neurobehavioral dysfunction associated with occupational exposure to mixtures of organic solvents." Am. J. Psychiat. 145:1442–1445.

SCHLUEDERBERG, A., STRAUS, S., PETERSON, P., BLUMENTHAL, S., KOMAROFF, A.L., SPRING, S.B. LANDAY, A., and BUCKWALD, D. (1992). "Chronic fatigue syndrome research: definition and medical outcome assessment." Ann. Int. Med. 117:325–331.

SCHOTTENFELD, R.S. and CULLEN, M.R. (1986). "Recognition of occupation-induced post–traumatic stress disorders." J. Occup. Med. 28:365–369.

SIMON, G.E., KATON, W., and SPARKS, P.J. (1990). "Allergic to life: psychological factors in environmental illness." Am. J. Psychiat. 147:901–906.

SIMON, G.E., DANIELL, W., STOCKBRIDGE, H., CLAYPOOLE, K., and ROSENSTOCK, L. (1993). "Immunologic, psychological, and neuropsychological factors in multiple chemical sensitivity: A controlled study." Ann. Int. Med. 19:97–103.

SPITZER, R.L., WILLIAMS, J.B.W., GIBBON, M., and FIRST, M.B. (1990). User's Guide for the Structured Clinical Interview for DSM-III-R. American Psychiatric Press, Washington, D.C.

STEWART, D.E. and RASKIN, J (1985). "Psychiatric assessment of patients with '20th century disease' (Total Allergy Syndrome)." Can. Med. Assoc. J. 133:1001–1006.

TAYLOR, G.J., BAGBY, R.M., RYAN, D.P., PARKER, J.D.A., DOODY, K.F., and KEEFE, P. (1988). "Criterion validity of the Toronto alexithymia scale." Psychosom. Med. 50:500–509.

TRAHAN, D.E. and LARRABEE, G.J. (1988). Continuous Visual Memory Test. Psychological Assessment Resources, Odessa, FL.
TRENERRY, M.R., CROSSON, B., DEBOE, J., and LEBER, W.R. (1989). Stroop Neuropsychological Screening Test Manual. Psychological Assessment Resources, Odessa, FL.
TRITES, R.L. (1981). Neuropsychological Test Manual. Technolab, Montreal.
WECHSLER, D. (1981). Wechsler Adult Intelligence Scale — Revised. The Psychological Corporation, New York.
WECHSLER, D. (1987). Wechsler Memory Scale — Revised Manual. The Psychological Corporation, San Antonio.
WITTENBORN, J.R. (1987). "Psychomotor tests in psychopharmacology." In: Human Psychopharmacology: Materials and Methods (I. Hindmarch and P.D. Stonier, ed.). vol. 1. Wiley, Chichester.

RESPONSE PANEL #3

HOWARD KIPEN, IRIS BELL, CLAUDIA MILLER, RICHARD LETZ, AND NORMAN ROSENTHAL

Dr. Kipen: I'd like to reemphasize some of the things that came out in Dr. Pennebaker's talk, and some additional data regarding Dr. Fiedler's presentation of our environmental questionnaire of 122 substances.

There were different responses, by gender, on that test, not just for people with MCS, but for anybody who took it. We've given it to over 1000 people, some with diagnoses, many without. Females respond significantly higher to the number of substances with the given symptoms, within any diagnostic group. The other thing regarding the environmental questionnaire data, is that, whether you have symptoms, it seems, is not always a great predictor of whether or not you will present yourself for medical care or attention. Trying to focus on who is sensitive, as opposed to who is sick, might help us.

The group that is high in reporting symptoms from substance exposure, is asthmatics. That's not a big surprise, but if you go back into the asthma literature, you find that it's kind of underplayed. If you go back a little further, when people tried to look at exposing asthmatics to various substances (in certain, not very extreme paradigms) they found decreases in lung function. It's not clear whether they were getting that because of a direct bronchial stimulation from the substance, which in this case was cologne, or whether it's mediated through the olfactory system.

I would like to ask a question of Dr. Steinhauer: Can you talk a little bit more about how the solvent-exposed people got into your study, and how they differed in terms of exposure and complaints that led them to be involved?

Dr. Steinhauer: One of the reasons we are doing the painter study is to try to get a group of people who are homogeneous in exposure. Unless there is some major incident, it's impossible. All of the patients for the Sullivan study were people who were referred because they had been seen by at least one occupational physician, and very often more than one. What that meant is that we were seeing a group that was highly selected for having a preponderance of problems, and that also would explain why very few of the people we saw would have had normal responses.

They've had a variety of different exposures. Many were single exposures. They range from 15 seconds to a room filled with kerosene fumes. But they also included individuals who had received low-level exposures over many years.

Toxicology and Industrial Health, Vol. 10, No. 4/5, pp. 555–560

ISSN: 0748-2337

Dr. Ryan (in response to a question by Dr. Bell): We've been doing neuropsychological evaluations on chemically exposed workers in Pittsburgh over the last ten years. There is no doubt in my mind that people who have a significant exposure, such as a solvent exposure, show clear evidence of neuropsychological impairments, when compared to groups of nonexposed people, or when you look at them clinically.

But, I've been very distressed at my inability to find evidence of significant neuropsychological impairments in people with MCS. Dr. Fiedler's paper almost supports that view. She's given a variety of standard, clinically psychometrically sound measures, and on only one test has she found evidence of impairments. This is a very challenging test, with scores that can be predicted best by levels of health concerns.

We have a real problem with MCS patients. We are not able to document the types of complaints that these individuals are reporting. People report "a significant memory problem," or, "I have difficulty concentrating," and we are not able to find evidence for that on the neurologic measures.

So we have some problems. Either the neuropsychological measures are insensitive, or we're dealing with disorders that are not traditional neuropsychological disorders. Now, it's easy to say that there's a problem with the neuropsychological measures. But, my patients who have closed head injuries or who have all the other kinds of medical disorders who also report those types of symptoms, do show impairments on the neuropsychological tests.

Dr. Bell: I wanted to add that I have also looked at some similar tests, with the CBLT, and the Continuous Visual Memory Test in the cacosmic students.

We had the same results. These are students, all about sophomores in college, so they're quite matched for education and socioeconomic status. There were very subtle findings, especially in the delayed recognition component of the Continuous Visual Memory Test. It was a small change; I wouldn't consider it clinically significant, and it was a set of findings we did at baseline, without the chemical challenges.

Cullen's definition talks a great deal about symptom flares during exposure, and symptom remissions during avoidance. All of the proposed definitions of MCS bring up that specific point, and we ignore it, because it is so hard to do the exposure challenges. One other point that I wanted to make is the notion of peak exposure. That is, the notion of a particular single episode, such as a toxic spill, that may require acute medical treatment. Dr. Morrow has pointed out that the people with the persisting neuropsychological deficits tend to have these histories of peak exposures, and that appears to be a somewhat better predictor than cumulative exposure to the chemical.

This, again, goes along with the sensitization notion, in the sense that a single exposure could set off a process that could have persisting implications. Dr. Rosenstocks's paper on pesticide-exposed workers which followed them up two years later, has somewhat similar findings. I think the Morrow paper was averaging 16 months later. We're talking about a fairly substantial, prolonged evidence of deficits, that was not dependent upon litigation or disability status, although the people who had these deficits tended to be disabled in a persistent manner. Finally, Dr. Fiedler has just reported that 21% percent of the people who appear to have MCS do not have any identifiable event or onset. I'd like to reemphasize our own data, showing that we found 15% of the population, recruited without reference to the environment, would report chemical illness.

When you ask people the question a different way, and say "Do you consider yourself especially sensitive to certain chemicals?" you get an answer of "yes" from about 30%, both in our surveys, and Dr. Wallace's survey in the EPA building. This is potentially a different group of people who may not, as a group, have histories of peak exposures, or have histories of identifiable exposures, and yet they may have some elements of this illness.

Dr. Miller: I want to amplify again the need to do exposure challenges to evoke symptoms, and make an analogy to some other areas. This may not be something we classically do in the neurological field, but we certainly do if we're dealing with a patient who has a very subtle reactive airway disease. We've all used methacholine challenges to evoke that kind of response, if there's nothing clinically visible on a standard pulmonary function testing. Another analogy, perhaps, might be panic attacks. Looking at a patient at baseline, they may look perfectly normal, but if you give them a lactate challenge, or a $C0_2$ challenge that is of the appropriate concentration, you can evoke that kind of a panic attack. There is certainly precedent for using exposure challenges to evoke the state that you are trying to observe, and that gives you the opportunity, then, to look at markers, pre- and postchallenge. I think that may be very critical.

I wanted to refer back to some of the speakers yesterday. Dr. Simon has done some interesting work, looking at immunological changes. I really appreciate the fact that people are trying to do this carefully. People are doing split samples, and blinded testing; I think that's a very important contribution. The immunological area, I think, is still very ill-defined. I think everyone would acknowledge that there may turn out to be some relevant markers we are not currently looking for.

Dr. Pennebaker talked about his own personal experience, a very interesting one, where he was giving a lecture, and when someone commented on smelling some gas, he surveyed himself for all of his own personal symptoms, and how we all do this, introspectively. I think he said something like 70% of medical students have this sort of syndrome, where they read their diseases into themselves. But I wonder if there is a practical way of differentiating between symptoms that are being over-read by the organism vs. symptoms that have truly been increased by some kind of exposure? A striking difference between his own experience and what we're hearing from patients is that his symptoms went away. These people's symptoms are persistent. Are they persisting because of over-reading, and how might we differentiate?

I'm interested in the similarities between the chronic fatigue syndrome and chemical sensitivity that Dr. Fiedler and Dr. DeLuca highlighted in the last two days. I've had the opportunity to talk to some chronic fatigue conferences, where there have been, like this one, many patients present in the audience. I have asked them about intolerance for alcohol, for tobacco, for a variety of cleaning chemicals, different drug responses, adverse drug responses they've had in the past; and find that it is an incredibly high percentage, almost all of the people in one conference that I attended, who said that they had all of these problems.

I would like to ask Dr. Steinhauer whether the patients he was seeing, and doing the evoked potential studies on, had overt neurological symptoms, or whether they showed softer signs, more like the chemical sensitivity patients? Has he looked at either chemical sensitivity patients or patients who would fit that definition, or Gulf War veterans?

Dr. Steinhauer: The impetus for beginning the psychophysiological study was to determine whether we could detect subtle signs of central nervous system disturbance. What it really has provided is fairly good converging evidence of deficits in both places. It also gives us a direct measure of central nervous system function, which is different from the inferred data that we get looking at neuropsychological tests.

Most of the people that we see are being referred because of the inability to find clear neurological disturbance. It is there, in some of the cases. Some of them do show such things as peripheral neuropathy, but very often there have been no signs. It is typical that we see this latency prolongation in persons who have normal CT scans. There are very few who are going to have SPECT or PET or MRI scans done, though even on MRI we don't see large structural abnormalities.

From the Audience: Dr. Pennebaker, can you distinguish people who are simply amplifying their symptoms from those whose symptoms are coming from toxic sources?

Dr. Pennebaker: That's a terribly complicated question. With enough people, you can statistically pull them apart, but on a case-by-case basis, I don't think it's possible.

Dr. Letz: I wanted to pick up on a theme that came up earlier, that perhaps there is a limitation of clinical neuropsychological testing in this situation. Some patients, not just MCS patients, have memory complaints, and it can be terribly difficult to establish a memory problem, in part because it is not strictly a memory problem, but rather an "attentional" problem in that they are not able to attend to the stimuli. They may for brief periods of time be able to pull it together enough, in the time that we have to test them, to perform adequately on the tests that we have. That does not mean that they don't then go home and in their normal course of events have memory difficulties, so that may be a limitation we have to keep in mind. Another point that arose during that discussion was: Since neuropsychological tests in general are effort-dependent, those studies reporting few neuropsychological deficits in MCS patients are actually a powerful argument that MCS patients are giving full effort in performing the tests. So the thesis that people are trying to fake the tests or are not performing up to their capacity is a hard one to maintain, given these experimental results.

I wanted to spend just a moment to try to discuss some lessons from the epidemiologic studies of solvent-exposed workers and other occupational neurotoxic-exposed subjects. There has been an evolution of the way in which we study such exposure-response relationships. We have gone from what is generally the case in the MCS area, of studying small numbers of subjects in a case-control fashion, to identifying subjects on the basis of their complaints, to performing studies on subjects on the basis of their exposures. These cross-sectional studies also have their limitations, but in general those studies have been of much larger sample sizes so we've seen an increase in the average occupational sample siz;e from the 10 to 30 range, to the 100 to 200 to 300. What we find is that some of the findings in the case-control studies are not corroborated in the cross-sectional studies.

Another trend in occupational studies is a greater emphasis on exposure assessment to actually perform within the larger studies, i.e., exposure-response analyses. Again, those types of analyses have not always corroborated the initial case-control results.

What these trends indicate to me is that there is tremendous heterogeneity of response to exactly the same exposures. This is an area in the occupational sphere that is incredibly understudied, and I think it is an even more important issue to study in the MCS realm. I don't have any answers, but I think it is the direction we need to go.

Dr. Rosenthal: I'd like to underscore some of the comments of the other speakers. I think the recurrent theme is: Is the condition real or is it fictitious? When we get these negative neuropsychological tests, is the problem that our tests are deficient, or is there really nothing to test? I think this is going on at some level all the time.

We may learn something more from the neuropsychological testing of depressed patients, as opposed to closed head injuries. Maybe there is something to be gained from Herb Weingarten's work on effortful vs. noneffortful memory, which shows that depressed people have good memory when it's just information that's coming in, but when they actually have to attend and put energy into retaining and processing, then they can't come back with the material quite as easily.

That moves on to what Drs. Bell and Miller were saying about provocative tests. I think that Dr. Fiedler's study was a really excellent and interesting study, but the fact is that it was baseline, and we're all sensitive, caring clinicians: so we make sure that our work space is going to be hospitable to our patients, that we haven't got smelly fumes, and we assure that it's well-ventilated; and thus, in a sense, though we are trying to be kind, we are actually biasing ourself against positive findings.

A small anecdote from my own practice, with my own patient whom you saw yesterday: He would come in regularly, and I knew he had had Dursban exposure, so I would be sure everything was well-ventilated. One time they came and sprayed the waiting room for ants, and I went in there and smelled and I couldn't smell a thing. I thought, "All right, we'll see is this real (see, I was still suspicious), I won't say anything." I walked out into the waiting room. His nose was running; his eyes were running. He said, "I've got to get out of here." After that time we never met inside the house, we always met out on the deck. I can tell you, that on that particular occasion, he would not have been cognitively intact. For the rest he was quite coherent. We saw with Cynthia Wilson yesterday that her IQ moves around, depending on what's in the room. I think that Dr. Miller's point is very well taken that we need to have very refined, specific blind challenges. We know we're going to need a lot of money before we can do that, and that's going to happen soon. I hope it's going to be a gradual, cumulative process and that this conference is going to be part of that process. In the meanwhile, we might just take one test room and have it nicely ventilated and take the other one and put a piece of cheap carpet down, and just do it in a kind of crude way, until all this good money rolls in and we can do it with sophistication.

Dr. Miller: I'd like to add that we are an occupational center and we do try to keep the window open. We are starting to do olfactory threshold testing. We are trying not to do it in a biased way, but simply telling people, at the beginning of threshold testing, that if you have any symptoms, to tell us at once. At very small levels of exposure, we are seeing a significant increase in symptoms among our MCS patients compared to normal controls. At super threshold levels we are seeing an increase in trigeminal symptoms that we wouldn't expect.

NEUROSPECT FINDINGS IN PATIENTS EXPOSED TO NEUROTOXIC CHEMICALS

GUNNAR HEUSER, ISMAEL MENA, AND FRANCISCA ALAMOS

Division of Nuclear Medicine
Harbor-UCLA Medical Center
Torrance, California

Exposures to neurotoxic chemicals such as pesticides, glues, solvents, etc. are known to induce neurologic and psychiatric symptomatology. We report on 41 patients — 16 young patients (6 males, 10 females, age 34 ± 8 yrs.) and 25 elderly patients (9 males, 16 females, age 55 ± 7 yrs). Fifteen of them were exposed to pesticides, and 29 to solvents. They were studied with quantitative and qualitative analysis of regional cerebral bood flow (rCBF), performed with 30 mCi of Xe-133 by inhalation, followed by 30 mCi of Tc-HMPAO given intravenously. Imaging was performed with a brain dedicated system, distribution of rCBF was assessed with automatic ROI definition, and HMPAO was normalized to maximal pixel activity in the brain. Results of Xe rCBF are expressed as mean and S.D. in ml/min/100g, and HMPAO as mean and S.D. uptake per ROI, and compared with age-matched controls — 10 young and 20 elderly individuals.

Neurotoxics

HMPAO% Uptake	Young	Elderly
R. Orbital frontal		
R. Dorsal frontal	.70 **	.66*
R. Temporal		.64*
R. Parietal	.66**	.66**

* $p < 0.05$
** $p < 0.001$

We conclude that patients exposed to chemicals present with diminished CBF, worse in the right hemisphere, with random presentation of areas of hypoperfusion, more prevalent in the dorsal frontal and parietal lobes. These

1. Address all correspondence to: Ismael Mena, M.D., Chief, Division of Nuclear Medicine, Lac Harbor UCLA Medical Center, 1000 W. Carson St.-BX#23, PO Box 2910, Torrance, CA 90509-2910. Tel.:(310)222-2842. Fax:(310)328-7288.

2. Abbreviations: MRI, magnetic resonance imaging; rCBF, regional cerebral blood flow; SPECT, single photon emission computed tomography.

3. Key Words: blood flow, brain, pesticides, solvents, SPECT.

Toxicology and Industrial Health, Vol. 10, No. 4/5, pp. 561–571

ISSN: 0748-2337

findings are significantly different from observations in patients with chronic fatigue and depression, suggesting primary cortical effect, possibly due to a vasculitis process.

INTRODUCTION

Exposure to neurotoxic chemicals occurs often in our environment. Intermittent contact with solvents and pesticides can even be as frequent as a daily occurrence. This pattern can be intensified if there is an accidental exposure (Dean et al., 1985; Heuser, 1992; Heuser et al., 1992).

Many exposed individuals present with chronic ill health, memory loss, attention-span loss, emotional instability, headaches, depression, fatigue, impaired coordination, tremor, insomnia, and dysesthesias (Baker et al., 1985).

Because of the predominantly neurologic and psychiatric symptomatology, we are reporting on results of functional brain imaging assessing cerebral blood flow by NeuroSPECT. It is hypothesized that selective diminution of rCBF should correlate quantitatively with symptomatology (Hagstaduis et al., 1989; Callender et al., 1993; Read et al., 1993).

NeuroSPECT has been determined quantitatively by ^{133}Xe inhalation and qualitatively with high resolution HMPAO ^{99m}Tc (Ceretec TM) imaging.

MATERIALS AND METHODS

Patient Population
We report on 41 patients — 16 young individuals (6 males and 10 females, age 34 ± 8 years) and 25 elderly individuals (9 males and 16 females, age 55 ± 7 years). Fifteen of them were exposed to pesticides and 26 to solvents.

Patient selection. Right-handed adults were selected for the study if they satisfied the following criteria: 1) Exposure to toxin was well documented; 2) Patients and their significant others volunteered complaints of intermittent or continuous impairment of cerebral function (memory impairment, emotional liability, etc.) ever since exposure; 3) Absence of diagnosis such as diabetes, hypertension; and 4) Negative history of head injury, alcoholism, and excessive drug intake.

As a rule, patients had been exposed to more than one chemical. For instance, most pesticides are propelled with solvents that also contain a high percentage of inert material that may also be potentially toxic to humans. New carpet and related materials (glue, backing) contain many potentially toxic chemicals, including solvents and pesticides. While exposure was always well documented in our cases, we were at times unable to pinpoint individual chemicals and therefore had to accept the fact that our patient had been exposed to several chemicals. We also realized that interaction between some chemicals could be assumed to take place before as well as after the chemicals entered the body.

Our patient population had been exposed any time between six months to four years prior to the SPECT study and had continued to have persistent complaints since exposure.

Control population. These patients were compared for ^{133}Xe rCBF with 40 elderly controls, age 66.1 ± 8.8 years, for Tc-HMPAO with 20 controls, age 65.1 ± 8.9 years, and for ^{133}Xe rCBF and HMPAO with 10 young individuals, age 33 ±6.8 years (Mena et al., 1993).

Brain Imaging

Two methods were applied consecutively for the assessment of cerebral blood flow: 1) ^{133}Xe rCBF for absolute quantification of cerebral blood flow, followed by 2) high-resolution ^{99m}Tc-HMPAO for qualitative analysis of brain perfusion imaging. All brain perfusion measurements were performed with the subjects lying supine with the head positioned carefully, verified by means of three beams of light projecting over the orbitomeatal line and the vertical midline. This achieved parallel transaxial cuts to the orbitomeatal line and correction for lateral tilting of the head. Environmental stimuli such as light and noise were controlled, particularly at the time of HMPAO intravenous injection (Holman et al., 1992).

^{133}Xe rCBF Imaging. ^{133}Xe rCBF measurements were performed with a high-sensitivity brain-dedicated system that provided high count-rate dynamic studies (Headtome 2 SET 031, Shimadzu, Kyoto, Japan). The sensitivity of the system was 31,000 counts per second per microcurie per ml of ^{133}Xe. The Headtome had three individual rings of 64 NaI detectors, each arrayed in a 42 cm. diameter circle. Studies were performed by inhalation of 30 mCi (1100 MBq) of I- ^{133}Xe. These measurements were based on the method described by Fick (1870), relating to the clearance of a freely diffusible tracer substance. Kety (1951) later applied this principle for the measurement of cerebral blood flow. The washout of ^{133}Xe (133Xe), an inert lipophilic gas and its exponential constant, are an exclusive function of cerebral blood flow. A fraction of inhaled ^{133}Xe passes through the alveolar wall and enters the pulmonary circulation. In the brain, it diffuses through the blood brain barrier and distributes in the extracellular space, exchanging freely with the blood compartment. A period of two minutes of inhalation for equilibrium is followed by a rapid brain clearance once the patient begins to breathe room air. This blood brain clearance is directly related to regional perfusion. Namely, if perfusion is diminished, blood clearance is slow; the opposite is also true — if blood flow is increased, the blood clearance is accelerated. Sequential one-minute images were gathered simultaneously for three transaxial slices during the study period. During the acquisition, the patient breathed through a mouthpiece with the nose clipped during the three acquisition phases: 1) Background acquisition, performed for one minute; 2) Inhalation or equilibrium of one minute, during which ^{133}Xe mixed with 30% oxygen was administered in a closed system at a concentration of 30 mCi per liter (1100 MBq/l) ^{133}Xe gas (DuPont, Billerica, MA); and 3) Washout phase of five to six minutes, during which the patient breathed room air and exhaled into a charcoal activated trap to retain

the exhaled Xenon gas. End tidal CO^2 and respiratory rate were monitored continuously (Hewlett Packard 78354A, Palo Alto, CA) in order to document and avoid hyperventilation.

A sequence of six tomograms, one every minute, were reconstructed using the backprojection algorithms with Butterworth and Ramachandran filtering (cutoff 20 mm, order = 8), on a 32 x 32 pixel matrix. Images were obtained simultaneously at three different levels, namely, two, six, and ten cm above the orbitomeatal line. An air curve was constructed from the end tidal concentration of ^{133}Xe that served as the input function for an iterative convoluted procedure, using a least square criteria for goodness of fit. The algorithm performed a pixel by pixel analysis and generated a family of output curves. Each pixel had a time versus activity curve from the concentration of ^{133}Xe in the sequence of pictures. Each curve had a simple rising component for the first minute of the inhalation period, followed by an exponential decrease (the washout period). rCBF was calculated from the sequence of pictures method, applying a single exponential component model on the washout segment using the equation

$$F = \text{logn}2 \text{ x } \lambda/\text{T}1/2 \text{ EXP}$$

where:
F = flow in ml/min/100g
logn2 = 0.693
λ = mean partition coefficient between blood and brain = 1.5 ml/100g, and
T1/2 EXP = negative slope for the brain exponential function expressed in minutes.

SPECT images delimit gray from white matter and therefore produce minimal overlap curves, in marked contrast to multiprobe systems that image simultaneously and overlap white and gray matter, requiring correction of slow and fast washout components.

Three transaxial images were obtained at two, six, and ten cm simultaneously above the orbitomeatal line, with a thickness of two cm. Blind areas existed at four cm and eight cm above the reference plane. The images were extrapolated to a 128 x 128 matrix and displayed using a color scale coded to ml/100g/min.

^{99m}Tc-HMPAO Imaging. After completing the ^{133}Xe study, we used ^{99m}Tc-DL-hexmethylpropylene-amineoxime (HMPAO Exametazime Ceretec™ Amersham) as a cerebral perfusion high resolution imaging tracer. It is a lipophilic molecule that crosses the blood brain barrier in its DL optical form and has a high first-pass extraction. The brain uptake ranged from 3.5 to 7% of the injected dose, and equilibrium was reached within two minutes. Only 15% of the amount taken up by the brain could diffuse back. After this period, the brain concentration diminished slowly at a rate of not greater than 1%/hour. Once in the brain, ^{99m}Tc-HMPAO converts to a secondary compound in hydrophilic form and is therefore trapped so it cannot diffuse back across the blood brain barrier. Because this phenomenon occurs outside of the brain as well, entrance into the brain of fresh Tc-HMPAO was

completed in this two-minute period. This means that the images gathered later corresponded to a window of time no greater than two minutes after the intravenous injection.

The patient's antecubital vein was cannulated 10 minutes prior to injection and a dose of 30 mCi (1100 MBq) was administered intravenously. All patients were maintained in a low ambient light and noise environment during the intravenous injection and brain uptake phase (two minutes) after injection. After one hour, in order to allow for background washout, imaging of HMPAO distribution in the cerebral cortex was performed with the same brain dedicated imaging device, using a fan beam high resolution collimator. A set of three 1.6 cm thick transaxial cuts were gathered over 10 minutes. These images were separated by 3.2 cm blind areas filled by two sequential precise 1.6 cm imaging bed shifts, and acquisitions during two additional 10-minute periods were performed. A final fourth 1.6 cm shift completed the acquisition and generated transaxial cuts at the level of the convexity of the brain, which were necessary for completion of the top of the brain image. The total number of counts of this series was 20 million, which is approximately six times more than what is obtained with a rotating single head camera. Reconstruction was done with the backprojection reconstruction algorithm using the Butterworth and Ramachandran filters. We applied Chang's attenuation correction, and the spatial resolution was 8 mm in the brain cortex and 9.6 mm at the level of the basal ganglia. Twelve contiguous transaxial images were displayed with a color scale based on regional tracer uptake that defined the normal to abnormal cortical interfaces at 60% of maximal activity in the brain with a 95% confidence level.

Quantification of rCBF

Both ^{133}Xe rCBF and ^{99m}Tc-HMPAO data sets were further quantified as follows: ^{133}Xe rCBF images obtained at two and six cm above the orbitomeatal line were analyzed. All images were displayed using the same color scale that defines the outer edge of the cortex at the level where rCBF drops to 45 ml/min/100g. Fourteen regions of interest (ROIs) were manually placed over the cortex in each slice, and all ROIs were the same size, with a diameter of 1.9 cm. The outer edge of the ROIs was set at the interphase of 45 ml/min/100g, or its projection. For ^{99m}Tc-HMPAO, five consecutive transaxial images were analyzed. The most caudal image was at the level of the inferior temporal lobes, extending cephalad to the right parietal lobe. Twelve proportional and irregular ROIs were defined by a program that automatically detected cortical boundaries. The algorithm automatically created brain cortical circumferential profiles from an ellipse which had its center over the basal ganglia. The external cortical edge was defined at the point of maximal difference between two adjacent points and the internal cortical edge was defined at 30% of the distance from the external boundaries to the center of the ellipse. The mean counts per pixel in each region were normalized to the maximum observed in the brain and expressed as a percent. We refer to this value as HMPAO uptake. All values are reported as mean and standard deviation. rCBF and HMPAO uptake were compared with control subjects age matched. In addition, the within-group Manova test was used to assess consistency within groups.

Normal Controls
133Xe rCBF. Studies performed in 43 elderly normals demonstrate a homogeneous distribution of blood flow in two cm-thick cuts at two and six cm above the orbitomeatal line (OM). Mean rCBF fluctuated in basal state between 40 and 45 ml/min/100g in both hemispheres. In 19 young normal controls (age 18–45), mean rCBF fluctuated between 50 and 60 ml/min/100g in both hemispheres.

HMPAO uptake. In 19 normal individuals, HMPAO uptake was normalized to the maximal rCBF in the brain, which fluctuated between 67% in the anterior temporal lobes and 76% in the occipital cortex. The uptake was thus uniformly distributed throughout the cortex.

EXPOSURE TO NEUROTOXICS

rCBF Xe-133
Figure 1 (Color Plate 1) demonstrates in one patient (male, age 58) a marked diminution of cerebral blood flow in the right dorsal frontal lobe, both parietal lobes, and left temporal lobe, with blood flow fluctuating between 24 and 34 ml/min/100g. Maximal perfusion is observed in the visual cortex and anterior cingulate gyrus at 64 ml/min/100g, while the remainder of the gray matter blood flow fluctuates between 34 and 54 ml/min/100g which is mildly diminished. Mean rCBF values in the elderly group is 40 ± 7; mean rCBf in the young group is 45 ± 7, p < 0.02 ml/min/100g (Figures 2 and 3).

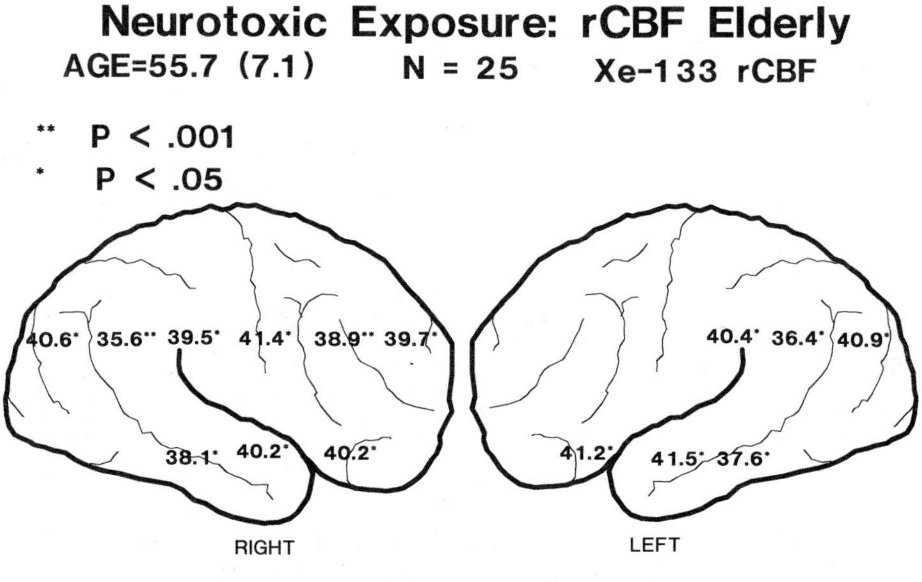

FIGURE 2. rCBF demonstrates a moderate diminution throughout the areas tested in the brain, with lateralization to the right hemisphere.

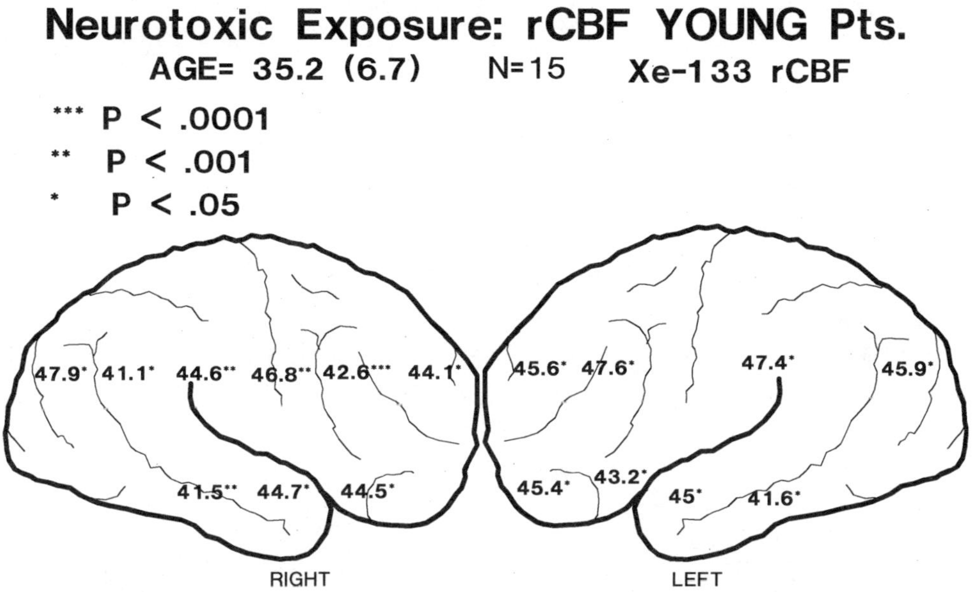

FIGURE 3. In young patients, regional cerebral blood flow (rCBF) is diminished bilaterally.

The same individual demonstrated, with a confidence value of 95, a marked diminution of cerebral blood flow in both temporal lobes, extensively in the right frontal, right dorsal, and right parietal lobe (Figure 4 — Color Plate 1). There was hypoperfusion with marked thinning of the functional cortical gray matter in the dorsal aspects of frontal and parietal lobes, with a scalloping presentation. In 25 elderly patients (Figure 5), there was a marked diminution of rCBF in the dorsal aspects of frontal and parietal lobes and dorsal right temporal lobe, with uptakes fluctuating between 64–70%, with a $p < 0.001$, fluctuating to 0.05. In the left hemisphere, there was diminished uptake of HMPAO in the dorsal occipital lobe. In the young group of patients (age 34 years) (Figure 6), there is diminished uptake in the dorsal frontal lobe and dorsal parietal lobe in both hemispheres, without evidence of a strong lateralization as demonstrated in the elderly population. There was also diminution of perfusion in the left temporal lobe, with diminished uptakes and p values between 0.001 and 0.05.

Statistical analysis among patients exposed to solvents versus pesticides produced borderline differences, with solvents showing more diminution of blood flow in the left temporal lobe than pesticides, with a $p < 0.05$ (Figure 7).

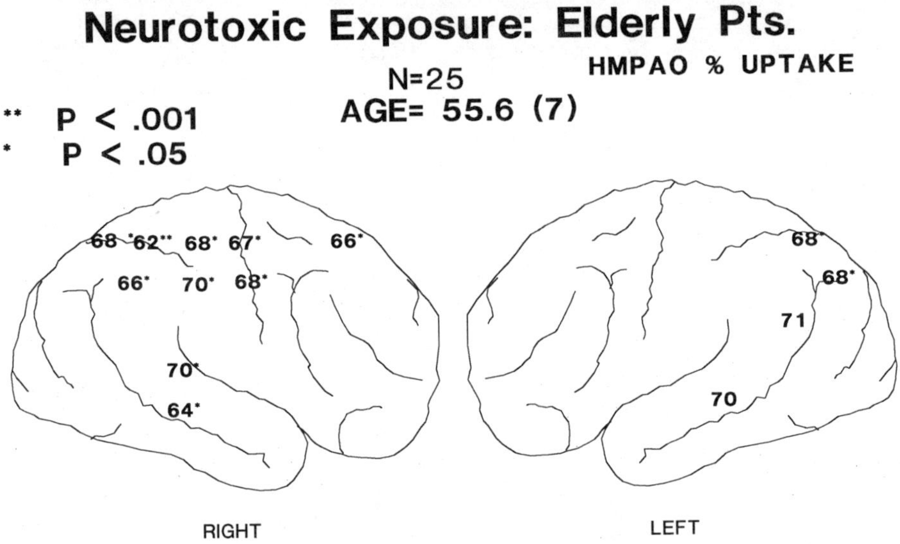

FIGURE 5. In 25 elderly patients, there was significant diminution of HMPAO uptake, denoting hypoperfusion extensively in the right hemsiphere, dorsal frontal lobe, parietal lobe, and dorsal temporal lobe. In the left hemisphere, there was hypoperfusion in the occipital lobe. Age = mean years ± (S.D.)

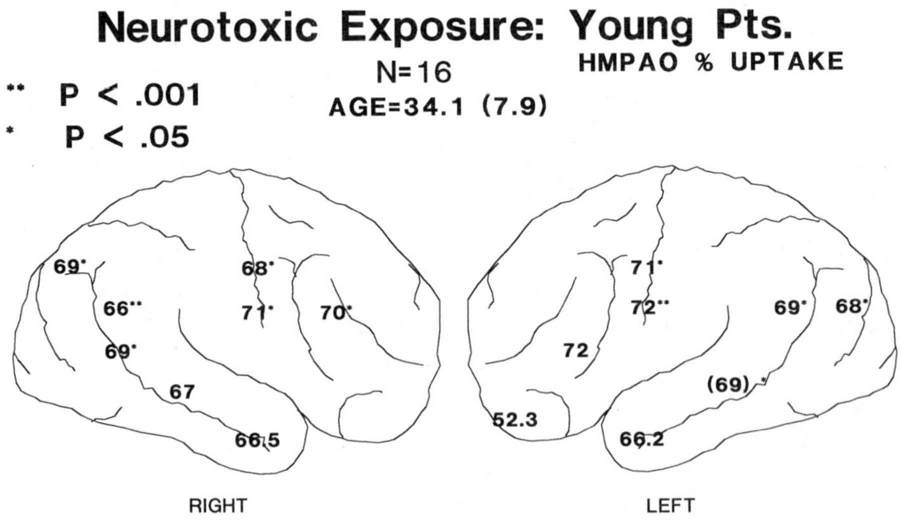

FIGURE 6. In 16 young patients, there was bilateral dorsal frontal, parietal and dorsal temporal hypoperfusion. There was no evidence of a significant lateralization phenomenon.

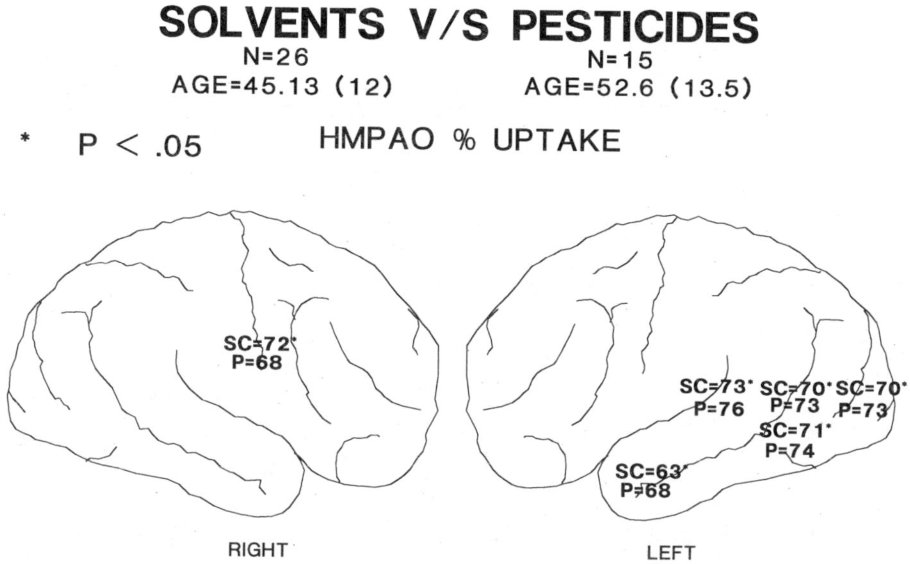

FIGURE 7. Comparison of cerebral perfusion by HMPAO uptake demonstrates more marked impairment of cerebral perfusion in patients exposed to solvents (SC) versus pesticides (P). These findings are more evident in the left hemisphere in the temporal lobes.

DISCUSSION

Our studies show that significant impairment of brain function may last years after exposure to neurotoxic chemicals has ceased. In 1991 and again in 1992, we suggested the use of SPECT as a sensitive and potent indicator of CNS function impairment after neurotoxic exposure (Heuser, 1992; Heuser et al., 1992; Heuser et al., 1993). In 1993, Callender et al. reported a series of patients who had abnormal SPECT scans after occupational neurotoxic exposure. While their SPECT scans were evaluated by visual inspection done by several radiologists, our results are based on both visual inspection and quantitative visual analysis. Also, our patient population was statistically compared to the normal control population. Our studies included many patients whose exposure had not been in an occupational setting.

Most patients had MRI brain scans, the majority of which were nondiagnostic. Therefore, even severe functional impairment as assessed by SPECT is not usually associated with cell death, demyelinization, atrophy, or other anatomical changes as assessed by MRI (Holman and Devous, 1992; Simon, 1992). We believe that neurotoxic damage leaves a diagnostic signature that is dissimilar to other diseases and conditions such as depression, Alzheimer's disease, multiple infarct dementia, CVA, head injury, obsessive compulsive disorders, and others (Rubin et al., 1992; Miller, 1992).

Quantitative cerebral blood flow demonstrates a modest diminution of global cerebral blood flow, while regional quantitative analysis documents significant focal clustered impairment of blood flow in the dorsal aspects of frontal and parietal lobes. There is lateralization of these abnormalities to the right hemisphere in the elderly, but not in young patients. There is also focal hypoperfusion in both temporal lobes in the young, and in the right temporal lobe in the elderly.

These findings contrast significantly with observations in late-life depression and late-life chronic fatigue syndrome (Heuser et al., 1993). In these conditions, there is a more severe global diminution of cerebral blood flow, and the localization of areas of abnormalities is centered in the orbitofrontal lobe, with predominance in the right hemisphere. There is also involvement of the temporal lobe. Only in young patients with neurotoxic exposure have we observed significant temporal hypoperfusion. Thus the findings in depression and chronic fatigue syndrome may correspond to a diaschisic phenomenon originating in the limbic system, while neurotoxic chemicals may have a direct effect on the cortex.

This cortical focal hypoperfusion presents in a scalloped pattern of perfusion, consisting of random thinning of cortical gray matter — a pattern of perfusion shared with cocaine abuse (Holman et al., 1992; Mena et al., 1992; Strickland et al., 1993), lupus, and early HIV infection of the brain. Its exact nature will be investigated by means of coregistration of neuroSPECT with magnetic resonance imaging (MRI) (Darcourt et al., 1993). Thus far, it appears to point to a vasculitis mechanism.

The possible reversibility of these lesions has not been investigated and, if present, would shed light on the mechanism of damage. Cellular death would be ruled out.

Refincment of analytical tools, such as circumferential profile analysis to better document both progression of disease and/or natural improvement after pharmacological intervention, is being accomplished at our laboratory.

Exposure to neurotoxic substances, solvents, or pesticides appears to significantly impair cerebral function as assessed by neuroSPECT.

ACKNOWLEDGMENT

The authors wish to thank Ms. Linda Blakely for typing of the manuscript and preparation of graphic material.

REFERENCES

BAKER, E.L., SMITH, T.J., and LANDRIGAN, P.J. (1985). "The neurotoxicity of industrial solvents: A review of the literature." Am. J. Ind. Med. 8:207–217.

CALLENDER, T.J., MORROW, L., SUBRAMANIAN, K., DUHON, D., and RISTOVV, M. (1993). "Three-dimensional brain metabolic imaging in patients with toxic encephalopathy." Env. Res. 60:295–319.

DARCOURT, J., CHANG, L., McBRIDE, D., CAUVIN, J.-C., MILLER, B., and MENA, I. (1993). "Comparison of Tl-201 and Tc-^{99m}-Sestamibi SPECT for brain tumor detection." Clin. Nucl. Med. 18(10):925.

DEAN, J., LUSTER, M.I., and MUNSON, A.E. (1985). Immunotoxicology and Immunopharmacology. Raven Press, New York.

FICK, A. (1870). Ueber edie Messung des Blutquantums in den Herzventriklen." S.B. Phys. Med. Ges. Wurzburg 2:16.

HAGSTADUIS, S., ORBAEK, P., RISBERG, J., and LINDGREN, M. (1989). "Regional cerebral blood flow at the time of diagnosis of chronic toxic encephalopathy induced by organic solvent exposure and after the cessation of exposure." Scan. J. Work. Environ. Health 15:130–135.

HEUSER, G. (1992). "Diagnostic markers in clinical immunotoxicology and neurotoxicology." (Editorial) Intl. J. Occup. Med. Tox. 1(4):v–x.

HEUSER, G., MENA, I., GOLDSTEIN, J., THOMAS, C., and ALAMOS, F. (1993). "NeuroSPECT findings in patients exposed to neurotoxic chemicals." Clin. Nucl. Med 18(10):923.

HEUSER, G., WOJDANI, A., and HEUSER, S. (1992). "Diagnostic markers of multiple chemical sensitivity." In: Markers in Immunotoxicology. National Academy Press, Washington, D.C. pp. 117–138.

HOLMAN, B.L., GARADA, B., JOHNSON, K., MENDELSON, J., HALLGRING, E., TEOH, S.K., WORTH, J., and NAVIA, B., (1992). "A Comparison of brain perfusion SPECT in cocaine abuse and AIDS dementia complex." J. Nucl. Med. 33:1312–1315.

KETY, S.S. (1951) "Theory and applications of the exchange of inert gas at the lungs and tissues." Pharmacol. Rev. 3:1–41.

HOLMAN, B.L. and DEVOUS, M.D. 1992. "Functional brain SPECT: The Emergence of a powerful clinical method." J. Nucl. Med. 33:1888–1904.

MENA, F., MENA, I., ALAMOS, F., SOLE, C., and NEUBAUER, S. (1993). "Normal data base of SPECT ^{99m}Tc-HMPAO brain perfusion and ^{133}Xe rCBF in children, young adults and elderly." Clin. Nucl. Med. 18:(10):924.

MENA, I., GIOMBETTI, R.J., MILLER, B.L., GARRETT, K., VILLANUEVA-MEYER, J., MODY, C., and GOLDBERG, M.A. (1992). Cerebral Blood Flow Changes with Acute Cocaine Intoxication: Clinical Correlations with SPECT, CT, and MRI. "Proceedings of Imaging Techniques in Medications Development: Preclinical and Clinical Aspects" (the NIDA Technical Review Meeting) Keystone, Colorado June 18–19. In press.

MILLER, B.L., CHANG, L., MENA, I., BOONE, K., and LESSER, I.M. (1992). "Progressive right frontotemporal degeneration: Clinical, neuropsycholgical and SPECT characteristics." Dementia 4:204–213.

READ, S.L., MILLER, B., MENA, I., KIM, R., and ITABASHI H. (1993.) "SPECT in dementia: Clinical and pathologic correlation." J. Am. Ger. Soc. Submitted.

RUBIN, R.T., VILLANUEVA-MEYER, J., ANANTH, J., TRAJMAR, P.G., and MENA, I. (1992). "Regional ^{133}Xe cerebral blood flow and cerebral ^{99}mTc-HMPAO uptake in unmedicated patients with obsessive compulsive disorder and matched normal control subjects: Determination by high-resolution single-photon emission computed tomography." Arch. Gen. Psych. 49:695–702.

SIMON, T.R. (1992). "Breast implants and organic solvent exposure can be associated with abnormal cerebral SPECT studies in clinically impaired patients." 78th Scientific Assembly and Annual Meeting of RSNA, Nov. 29–Dec. 4.

STRICKLAND, T., MENA, I., VILLANUEVA-MEYER, J., MILLER, B., CUMMINGS, J., MEHRINGER, C.M., SATZ, P., and MYERS, H. (1993). "Cerebral perfusion and neuropsychological consequences of chronic cocaine use." J. Neuropsych. Clin. Neurosci. 5:419–427.

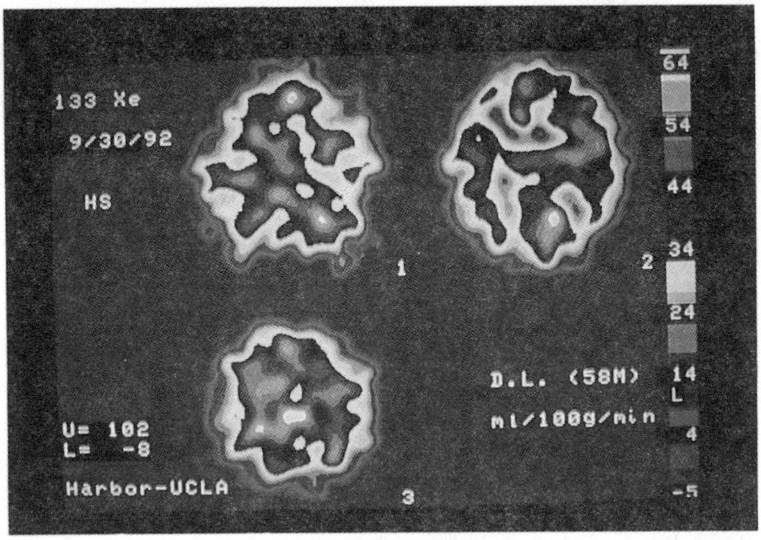

FIGURE 1. ^{133}Xe regional cerebral blood flow (rCBF) in a 58-year-old male. Transaxial cuts demonstrate in color coded display to ml/minute/100g right dorsal frontal and both parietal and left temporal lobe with markedly diminished cerebral blood flow fluctuating between 24 and 34 ml/min/100g.

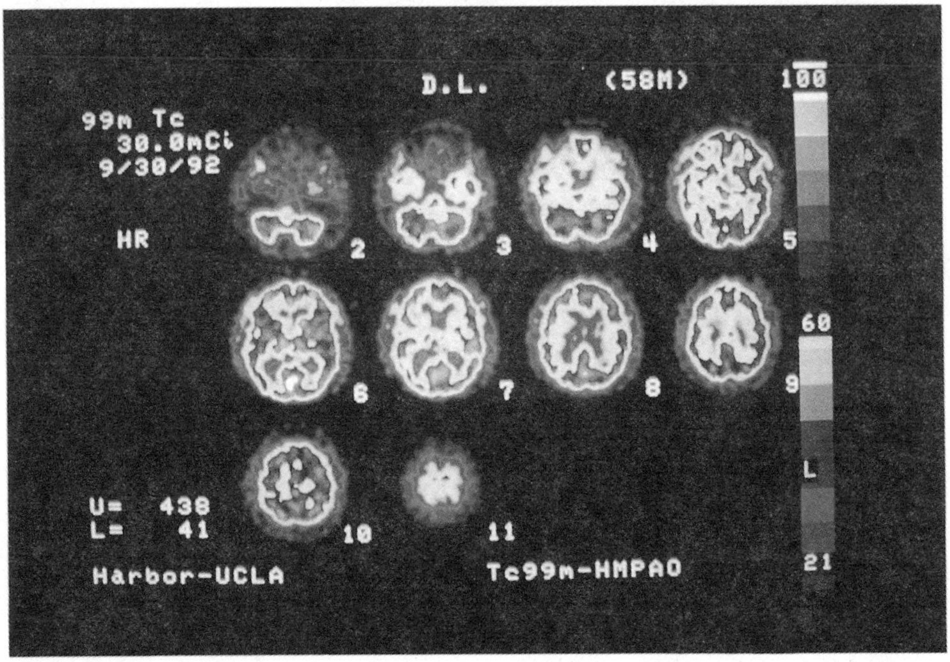

FIGURE 4. Figure 4 denotes the distribution of cerebral blood flow in a series of transaxial cuts. The color scale is thresholded with a 95% confidence level, denoting normal perfusion in red and white and abnormal in yellow, green, and blue. There is marked thinning of the functional cortical gray matter in the dorsal aspects of frontal and parietal lobes, with a scalloping presentation.

Color Plate 1. As referenced in Heuser et al., pp. 566, 567.

SINGLE PHOTON EMISSION COMPUTED TOMOGRAPHY OF THE BRAIN IN PATIENTS WITH CHEMICAL SENSITIVITIES

THEODORE R. SIMON,* DAVID C. HICKEY,* CYNTHIA E. FINCHER,† ALFRED R. JOHNSON,‡ GERALD H. ROSS,‡ AND WILLIAM J. REA‡

***Advanced Metabolic Imaging**
North Dallas
Dallas, Texas

†University of North Texas
Denton, Texas

‡Environmental Health Center
Dallas, Texas

Chemical sensitivities display a recurrent pattern on scintigraphic examinations of the brain. The pattern can include mismatching between early and late imaging, multiple hot and cold foci distributed throughout the cortex without regard to lobar distribution (salt and pepper pattern), temporal asymmetries, and sometimes increased activity in the basal ganglia.

This study used Desert Shield/Desert Storm veterans who present with abnormal neurological and psychological symptoms as a model to exhibit abnormalities by brain scintigraphy. These are typical of those seen in patients with documented exposure to neurotoxic compounds who develop a clinical syndrome that has been termed "chemical sensitivity."

Exposure to cocaine, alcohol, and other substances of abuse can result in abnormal scintigrams of the brain using tracers such as [technetium-99m]hexamethylpropyleneoxime. This study used techniques combining regional cerebral blood flow data with delayed distributional data after the intracellular conversion of the tracer into a hydrophilic molecule.

In addition to delayed image abnormalities, a mismatch occurs in the regional activity between the two image sets of the veterans. This degree of mismatch was not seen in control subjects who were screened for avoidance of neurotoxic

1. Address all correspondence to: Theodore R. Simon, M.D., 4429 Southern Avenue, Dallas, TX 75205. Tel.:(214)661-4710. Fax:(214)661-4762.
2. Abbreviations: DS: Desert Storm/Desert Shield Veterans.
3. Key Words: Brain Scintigraphy, Neurotoxicity, SPECT.

Toxicology and Industrial Health, Vol. 10, No. 4/5, pp. 573–577

ISSN: 0748-2337

agents. Patterns identified from examinations performed on patients with known exposure to petroleum distillates, pesticides and other materials linked with neurotoxicity were identified in some veterans of the Desert Shield/Desert Storm operation. A single case of repeated examinations on a veteran showed a reversion of these patterns toward normal after therapy. This reversion followed independent assessments of clinical improvement.

INTRODUCTION

Various groups have shown that exposure to cocaine, alcohol, and other substances of abuse can result in abnormal scintigrams of the brain using tracers such as Tc-99m HMPaO (Tumeh et al., 1990; Holman et al., 1991; Woods et al., 1991; Erbas et al., 1992). This study sought to confirm that work and add information relating to regional cerebral blood flow. This combined approach imaged the tracer twice. First, a fast gamma camera system was used with a controlled intravenous injection procedure to obtain flow data. Later, the tracer was measured after being trapped in neurons by a mechanism correlating with a glutathione-mediated process through which the lipophilic tracer that enters the neurons is converted to a hydrophilic molecule and trapped in the cell (Andersen, 1989). In addition to the well described functional image abnormalities, a mismatch occurred in the regional activity between the flow and functional image sets of patients with a history of chemical abuse. This degree of mismatch was not seen in control subjects who underwent careful screening to exclude candidates with exposure to neurotoxic chemicals. As these observations were extended to other groups, a substance abuser who became psychotic after sniffing organic solvents was tested and found to exhibit the same pattern. This examination lead to an assessment of environmental exposures to organic solvents, petroleum distillates, pesticides, and other materials linked with neurotoxicity. Various similar patterns are seen in patients who have sustained such exposure. This study sought to determine if those patterns also occur in veterans of the Desert Shield/Desert Storm conflict. In addition, this study provided an opportunity to determine the stability of the pattern in an individual undergoing therapy.

METHODS

Six Desert Shield/Desert Storm veterans (DS) who were referred for brain scintigraphy as part of a comprehensive clinical evaluation were examined. Scintigraphs were compared to images obtained from normal subjects who were matched by gender and approximate age. Normal controls were taken from a group determined to be free of exposure to neurotoxic agents who had given informed consent to participate in a study approved by the appropriate Investigational Review Board.

Subjects were injected with approximately 925 MBq (25 mCi) of [technetium-99m]hexamethylpropyleneoxime over two minutes using a microprocessor controlled Harvard Pump. Fifteen seconds after initiation of the injection, a two minute single photon emission computed tomographic (SPECT) acquisition was begun, using a three headed

SPECT system (Triad) with image binning at six degree intervals. Eighteen minutes thereafter, imaging at three degree intervals was performed for approximately 33 minutes. Both image sets were filtered and backprojected onto three planes orthogonal to the canthomeatal plane.

The images were then qualitatively evaluated by two experienced nuclear medicine physicians using a standardized questionnaire (Figure 1). One patient consented to a repeat examination after therapy. During therapy, his symptoms of memory loss and fatigue had greatly remitted. This second examination was compared to that patient's earlier examination to determine whether any scintigraphic changes had developed.

Tc99m-HMPaO RESULTS

Desert Storm/Desert Shield Veteran Study

Patient Name.________________ Date of Examination __/__/9__

Overall impression? ☐ Severely abnormal ☐ Moderately ☐ Mildly Abnormal ☐ Normal

Is cross-cerebellar diaschisis present? ☐ None ☐ Left ☐ Right

Diversion to soft tissue? ☐ No ☐ Yes

Please rate the lobar perfusion using a scale from 1 to 5.
Severe-C; Moderate-B; Mild-A
1 = Significantly Hypoperfused (>50% Decrease)
2 = Hypoperfused (<50% Decrease)
3 = Normal
4 = Hyperperfused (<50% Increase)
5 = Significantly Hyperperfused (>50% Increase)

LEFT		RIGHT
☐	Frontal Lobe	☐
☐	Temporal Lobe	☐
☐	Parietal Lobe	☐
☐	Occipital Lobe	☐
☐	Subcortical Region	☐
☐	(Basal Ganglia, Thalamus)	☐

Please rate focal findings using a scale from 1 to 6.
Severe-C; Moderate-B; Mild-A
1 = Multiple hot foci
2 = Single hot focus
3 = Normal
4 = Single cold focus
5 = Multiple cold foci
6 = hot and cold foci

LEFT		RIGHT
☐	Frontal Lobe	☐
☐	Temporal Lobe	☐
☐	Parietal Lobe	☐
☐	Occipital Lobe	☐
☐	Subcortical Region	☐
☐	(Basal Ganglia, Thalamus)	☐

Please rate the general uptake in the flow phase compared to that seen in the functional phase using a scale from 1 to 5.
1 = Flow much more than function
2 = Flow slightly more than function
3 = Matched
4 = Function slightly more than flow
5 = Function much more than flow

Reader Initials: TRS☐ DCH☐ **Date:** __/____/9_

FIGURE 1. Standardized questionnaire used for qualitative evaluations of brain scintigrams.

RESULTS

All six DS subjects and six controls were males. Ages ranged from 23 to 54 years in the DS group (averaging 36.6) and from 25 to 46 (averaging 35.5) in the control group.

All six DS subjects were subjectively rated as abnormal (three as severe, two as moderate, and one as mild). Diversion of activity to the soft tissues was noted in three of the DS patients (50%) and was absent in the controls. Lobar perfusion discrepancies were present in three of the DS patients (50%) but only one control (16%). The degree of difference was much greater in the DS group (Table 1). Five of the DS veterans (83%) showed widespread focal abnormalities that were much more prominent than the three variant sites identified in the same control subject who had discrepant lobar findings. Significant mismatches between the flow and functional phases were identified in 4 DS patients (67%). A single case of slight mismatch was identified in the control group. All mismatched cases showed more activity in the functional phase than in the flow phase.

TABLE 1. Scintigraphic Findings

Subject	Impression	Soft Tissue Diversion	Lobar Discrepancies	Focal Findings	Phase Matching
DS1	Severe	No	2	Many	Severe mismatch
DS2	Severe	Yes	No	Many	No
DS3	Severe	Yes	1	Many	Severe mismatch
DS4	Mild	No	No	No	No
DS5	Moderate	No	2	Many	Severe mismatch
DS6	Moderate	Yes	No	Many	Severe mismatch
C1	Normal	No	2	Three	No
C2	Normal	No	No	No	Mild mismatch
C3	Normal	No	No	No	No
C4	Normal	No	No	No	No
C5	Normal	No	No	No	No
C6	Normal	No	No	No	No

DISCUSSION

This limited study shows differences between the DS group and normal controls in terms of regional findings and overall flow and functional comparisons. The pattern encountered is similar to that seen in patients with known exposure to widely recognized neurotoxins including petroleum distillates and pesticides. This pattern is, as yet, nonspecific for the particular neurotoxic agent. However, other testing and a history of known exposures can help clinically in isolating the etiologic agents.

REFERENCES

ANDERSEN, A.R. (1989). "^{99M}Tc-D,L-Hexamethylene-propyleneamine Oxime (^{99m}Tc-HMPAO): Basic kinetic studies of a tracer of cerebral blood flow." Cerebrovasc. Brain Met. Rev. 1:288–318.

ERBAS, B., BEKDIK, C., ERENGI, G., ENUNLUT, T., AYTAC, S., KUMBASCAR, H., and DOGAN, Y. (1992). "Regional cerebral blood flow changes in chronic alcoholism using Tc-99m HMPAO SPECT: Comparison on CT parameters." Clin. Nuc. Med. 17:123–127.

HOLMAN, B.L., ARVALHO, P.A., MENDELSON, J., TEOH, S.K., NARDIN, R., HALLGRING, E., HEBBEN, N., and JOHNSON, K.A. (1991). "Brain perfusion is abnormal in cocaine-dependent polydrug users: a study using Technetium-99m-HMPAO and ASPECT." J. Nucl. Med. 32:1206–1210.

TUMEH, S.S., NAGEL, J.S., ENGLISH, R.J., MOORE, M., and HOLMAN, B.L. (1990). "Cerebral abnormalities in cocaine abusers: demonstration by SPECT perfusion brain scintigraphy: work in progress." Radiol. 176:821–824.

WOODS, S.W., PEARSALL, H.R., SEIBYL, J.0., and HOFFER, P.B. (1991). "The Quinn essay: single photon emission computed tomography in neuropathic disorders." In: Yearbook of Nuclear Medicine (P.B. Hoffer, ed.). Mosby Year Book, Chicago. pp. xiii–xlvi.

EEG AND ERP STUDIES OF LOW-LEVEL ODOR EXPOSURE IN NORMAL SUBJECTS

TYLER S. LORIG

Washington and Lee University
Lexington, Virginia

Evidence from four EEG/ERP experiments is presented. Findings of these experiments consistently demonstrate the ability of low-level and undetected odors to alter neurophysiology. Behavioral and cognitive effects are also described. These data may have applicability to theories of MCS since they illustrate the possibility that low-level or undetected odors affect central nervous system activity and may precipitate or cue MCS symptoms.

INTRODUCTION

Human olfactory ability is extraordinary and paradoxical. It is extraordinary for the fact that, as mammals, we have a large and sophisticated rhinencephalon which is the hallmark of this class of animals. The paradox arises because despite this neuroanatomical substrate associated with the sense of smell, humans care little about odors. Numerous experiments have demonstrated the breadth and variety of errors humans make when labeling odors (Schab, 1991 for review).

Humans often ignore odors and underestimate the importance of this modality. For instance, in an experiment designed to assess "sensory conditions" on the ratings of neutral slides of "computer art," Lorig and Thompson (cited by Lorig, 1992) found that only three of 93 subjects believed that odor was the sensory condition being manipulated, in spite of the use of easily detectable and contextually unusual odors (lavender and vanilla odors in a psychology laboratory). In contrast to this seeming lack of olfactory sophistication, fragrance manufacturers thrive, with annual sales of many billions of dollars, odors influence seating choices (Kirk-Smith and Boothe, 1987), cause menstrual synchrony in human females (McClintock, 1971), and affect the success of job applicants (Baron, 1983). Anosmic

1. Address all correspondence to: Tyler S. Lorig, Ph.D., Washington and Lee University, Lexington, VA 24450. Tel.:(703)463-8839. E-mail: tlorig@nose.wlu.edu.

2. Abreviations: electroencephalograph, EEG; event-related potential, ERP; multiple chemical sensitivity, MCS.

3. Key Words: EEG, ERP, olfaction, MCS.

Toxicology and Industrial Health, Vol. 10, No. 4/5, pp. 579–586

ISSN: 0748-2337

individuals suffer dietary disturbances and also report changes in sexual behavior (Tennen et al., 1991). This amalgam of facts suggests that odors have a pervasive influence on human behavior even though humans report caring little about smells.

The human lack of interest in and awareness of odors is especially interesting from an evolutionary perspective. Consider the fact that each human has an unique odor "signature," making tracking by bloodhounds possible (Labows and Preti, 1992). Foods and poisons often have distinctive smells and the odors of vaginal secretions vary as a function of menstrual cycle (Doty et al., 1975). Human illnesses also carry distinctive odors. Additionally, people are often alarmed by the smell of smoke, but such anxiety is context dependent. Smoke from a wood fire may be pleasant on a cold winter day when a hearth is nearby but that same odor may be quite alarming on the fourteenth floor of an apartment building in mid-summer. This is a rich chemosensory world and it seems unlikely that during the course of human evolution, we would have extinguished our responsiveness to such a potentially valuable source of information.

Those skeptical of odor's role might argue that vision has largely supplanted the need for olfactory information processing. Certainly much of the world is redundant in regard to olfactory and visual information. Rotted fruit both looks and smells different from ripe fruit. Similarly, humans both look and smell different from one another. Still, unlike other primate species, there are no visible signs of ovulation in human females. Stoddart (1990) has even suggested that perfume use may have developed as a masking agent for odors related to the menstrual cycle. Like odors related to ovulation, chemosenosry stimuli may provide the earliest cue that an individual is becoming ill. Such information should be especially salient in the survival of a species; a view contrary to the neuroanatomist Brodal (1981) who states "the sense of smell is of relatively minor importance in the normal life of civilized man (p. 640)."

The foregoing discussion is made to suggest that olfaction may be somewhat more important and complicated than is typically assumed. Similarly, the role of olfaction in multiple chemical sensitivities (MCS) may also be subtle and complicated. It is possible that toxic chemicals in the environment may kindle MCS by acting directly and indirectly on the limbic system (Bell et al., 1992). It is also quite possible that MCS may be an extreme example of olfactory conditioning. To further complicate matters, such conditioning can take place in the absence of any overt olfactory cues. For instance, Kirk-Smith et al. (1983) found that undetectable odors could be conditioned to produce anxiety in college students. This study appears to be the first well-controlled illustration of odor effects which occur in the absence of subjective awareness. Our laboratory has made similar findings using EEG and ERPs.

Experiment 1

In a study designed to assess the sensitivity of EEG as a measure of odor effects on humans, Lorig and Schwartz (1988) recorded EEG from four scalp sites (F7, F8, T5 and T6) as subjects smelled several odors (commercial perfume products). The odors were perceptually very similar and subjective reports indicated no differences in a number of perceptual categories. Many subjects were convinced that the same odor was being repeated. Results of the EEG analyses were not so similar. A complex interaction was found between the scalp distribution of theta activity determined by period analysis (Johnson, 1980) and odor. The point of interest here is that, even although perceptually similar, neurophysiologically these odors were quite unique. These findings suggested the possibility that undetected chemical differences in the odors altered EEG but not perception.

Experiment 2

To further evaluate the hypothesis that undetected chemicals could alter EEG, a second experiment was conducted using several concentrations of lavender and spiced apple odors (Lorig et al., 1990). Of greatest interest were the responses of subjects not smelling the lowest concentrations of these odors. Only those subjects reporting no detection of the odors and those subjects free of false alarms were included in the analysis. EEG was recorded from four sites as in Experiment 1. Period analysis of these data revealed that the undetected lavender and spiced apple odors produced different scalp distributions with greater beta activity occurring posteriorly during lavender administration (see Figure 1). Theta activity also tended to reflect similar differences, but failed to reach statistical significance ($p = 0.08$).

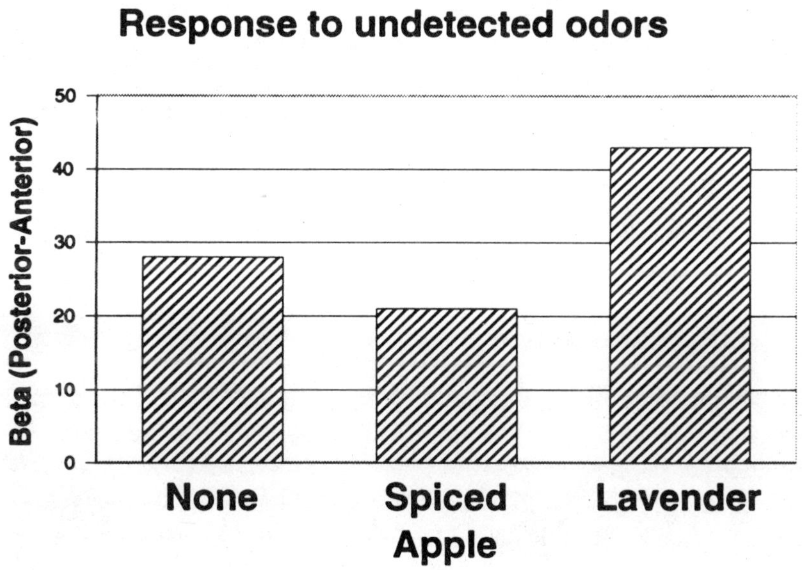

FIGURE 1. Changes in beta (13–30 Hz) activity during administration of undetected odors.

Experiment 3

While Experiment 2 was provocative, it did not fully evaluate the psychophysical responses to the odors since thresholds were not determined individually. Additionally, the findings of Experiment 2, while significant, offered no index of any behavioral or cognitive effects of low-concentration odor administration. In Experiment 3, twelve subjects participated as EEG was recorded during administration of high, medium, and low concentrations of Galaxolide, a musk-like odor. In preliminary testing, both the high and medium concentrations were detected. The low concentration was not detected by subjects in a stairstep detection test run prior to EEG data collection. Following this psychophysical evaluation, subjects were exposed to all concentrations of Galaxolide and a no-odor control, which were presented in randomized order. EEG data were collected from nine electrode sites. Subjects were told that "on some trials, no odors would be presented" in order to reduce odor searching (Lorig and Schwartz, 1988). Following EEG testing, the subjects were again exposed to the low concentration of Galaxolide and no-odor while they performed a visual search task (see Figure 2). This task required subjects to locate the target stimulus in the letter matrix using a joystick. Latency to correct detection data were collected.

E

K	F	G	L	T	N	F	A	S	Q	W	M	N	F	Z	P	Y	N	I	C
L	V	J	R	P	D	M	N	Q	K	X	S	E	L	O	I	R	B	H	U
M	R	T	U	V	G	J	H	K	D	L	K	I	Z	X	W	S	L	I	V
G	F	D	H	J	K	A	S	W	P	O	U	N	B	H	Y	R	C	D	W
I	L	K	J	U	Y	N	D	C	X	V	T	Y	N	V	A	H	L	K	J
Z	S	W	C	F	R	V	T	G	V	B	Y	N	U	J	I	L	H	T	W
P	M	I	Q	L	S	K	F	H	V	R	U	M	O	V	S	K	I	Y	U
I	N	K	W	S	X	R	D	C	Y	H	M	I	K	L	P	U	R	Q	Z
O	I	N	K	M	G	Y	T	D	T	N	P	I	Q	J	U	R	W	C	P
V	U	J	T	G	B	B	N	O	L	S	A	W	P	L	U	T	N	M	S

Results from the EEG portion of this experiment indicated that alpha activity was greater in the no-odor condition than during administration of an undetected concentration of Galaxolide (see Figure 3, top panel). Theta, again, tended toward significance ($p = 0.08$). Perhaps more interesting than the EEG findings were the behavioral effects. Subjects took twice as long to correctly isolate the target when exposed to the undetected concentration of Galaxolide (see Figure 3, lower panel).

These findings are especially provocative since, in addition to clear electrophysiological differences between undetected odor exposure and no-odor conditions, there were even more dramatic behavioral effects. Such data lead to the conclusion that odor chemicals can exert behaviorally relevant CNS effects even when undetected.

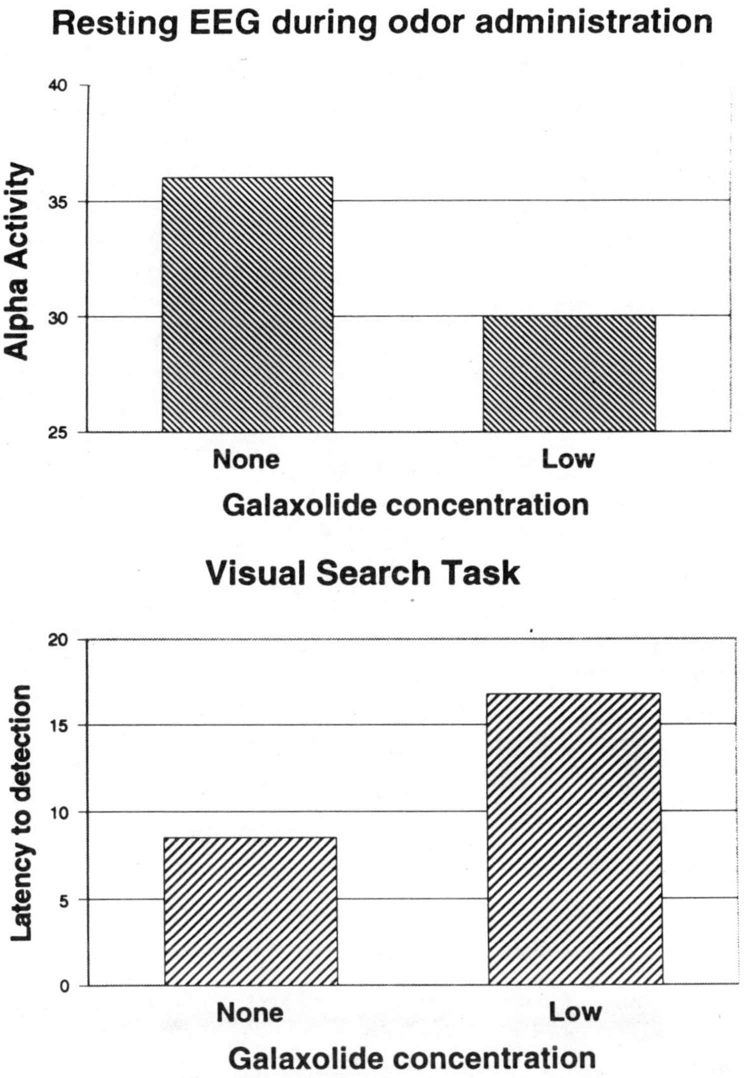

FIGURE 3. Alpha (8–12 Hz) activity during administration of undetected levels of Galaxolide (top panel) and during visual search (bottom panel).

Experiment 4

Experiment 3 was surprising in some regards. Increases in stimulation typically led to EEG arousal and behavioral improvement. While increased EEG arousal was apparent at central electrode sites, behavior was impaired by low-level odor administration. Experiment 4 was performed to evaluate the nature of any temporary cognitive impairment induced by the odor administration. Low-level odor exposure could interfere with cognition in several ways. Odors could, for instance, be distracting. They might also interfere with the process of "context updating" necessary for adequate stimulus evaluation (Donchin et al., 1978). This process is associated with the P300 response evoked during the "oddball" paradigm and may have its neurophysiological origin in hippocampus and entorhinal cortex.

Eight subjects participated in the study. Like the previous experiment, the same three levels of Galaxolide were tested. After stairstep testing to evaluate subjects' thresholds, participants were asked to count rarely occurring auditory stimuli (low-pitched tones presented in a series of tones) while odors were presented via a lightweight oxygen mask (Lorig et al., 1991). EEG recording was synchronized with auditory stimulus presentation thus permitting averaging of the EEG. These "event-related potentials" (ERPs) indicated odor-related differences in two portions of the waveforms. The positive-going waveform at approximately 200 msec (P200) was largest during low-level odor exposure (see Figure 4, lower panel) and P300 was also larger during odor exposure (see Figure 4, top panel). Since P200 has been found to be

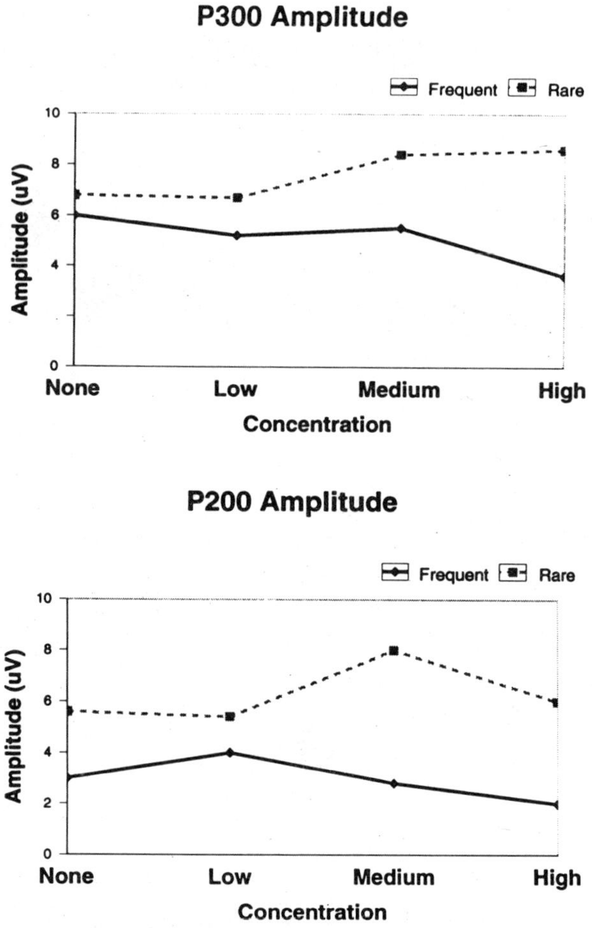

FIGURE 4. P300 (top) and P200 (bottom) amplitude in an auditory oddball paradigm as a function of Galaxolide adminstration.

enhanced during tasks where attention is divided between two channels (Hansen and Hillyard, 1983), these data may be interpreted as suggesting that low-level Galaxolide administration divides attention. The resulting inattention to the auditory channel alters the subjective probability that the rare tone will occur and leads to a larger P300. In effect, the lack of attention to the tones leads to greater "surprise" when the target tone is presented. The relationship between subjective probability, surprise, and P300 amplitude is well established (Donchin et al., 1978).

CONCLUSIONS

The experiments described here provide clear evidence that undetected odors alter neurophysiology and behavior. Furthermore, they suggest that odors may exert these effects by distracting subjects even when the odor is not detected. Other mechanisms of action may also be at work but have not been evaluated. The implications of these data for theories of MCS are interesting. It is possible that MCS patients may respond to undetected odors quite differently than unaffected subjects because of kindling or conditioning. Odors may induce or trigger MCS episodes and may also be undetected by experimenters and clinicians. MCS patients who appear to "spontaneously" exhibit symptoms could lead diagnosticians to misattribute the disorder to hysteria when a chemical cue undetected by clinician and patient may be the trigger. If nothing else, these data should alert those scientists evaluating MCS patients to be less trusting of their own noses and the noses of their patients.

REFERENCES

BARON, R.A. (1983). "The sweet smell of success? The impact of pleasant scents on the evaluations of job applicants." J. Appl. Psych. 68:709–713.

BELL, I.R., MILLER, C.S., and SCHWARTZ, G.E. (1992). "An olfactory-limbic model of multiple chemical sensitivity syndrome: Possible relationship to kindling and affective spectrum disorders." Biol. Psychiat. 32:218–242.

BRODAL, A. (1981). Neurological Anatomy. Oxford University Press, New York. p. 608.

DONCHIN, E., RITTER, W., and McCALLUM, W.C. (1978). "Cognitive psychophysiology: The endogenous components of the ERP." In: Event-Related Brain Potentials in Man (E. Callaway, P. Tueting, and S. Koslow, ed.). Academic Press, New York.

DOTY, R.L., FORD, M., PRETI, G., and HUGGINS, G.R. (1975). "Changes in the intensity and pleasantness of human vaginal odors during the menstrual cycle." Science 190:1316–1318.

HANSEN, J.C. and HILLYARD, S.A. (1983). "Endogenous brain potentials associated with selective auditory attention." Electroenceph. Clin. Neurophys. 49:277–290.

JOHNSON, L. (1980). "Measurement, quantification, and analysis of cortical activity." In: Techniques in Psychophysiology (I. Martin, and P. Venables, ed.). John Wiley & Sons, New York. pp. 329–356.

KIRK-SMITH, M., VANTOLLER, C., and DODD, G. (1983). "Unconscious odour conditioning in human subjects." Biol. Psych. 17:221–231.

KIRK-SMITH, M.D. and BOOTH, D.A. (1987). "Chemoreception and human behaviour: Experimental analysis of the social effects of fragrances." Chem. Senses 12:159–166.

LABOWS, J.N. and PRETI, G. (1992). "Human semiochemicals." In: Fragrance: The Psychology and Biology of Perfume (S. VanToller and G. Dodd, ed.). Elsevier, New York. pp. 69–88.

LORIG, T.S. (1992). "Cognitive and 'non-cognitive' effects of odor exposure: electrophysiological and behavioral evidence." In: Fragrance: The Psychology and Biology of Perfume (S. VanToller, and G. Dodd, ed.). Elsevier, New York. pp. 161–173.

LORIG, T.S., HERMAN, K.B., SCHWARTZ, G.E., and CAIN, W.S. (1990). "EEG activity during administration of low concentration odors." Bull. Psychonomic Soc. 28:405–408.

LORIG, T.S., HUFFMAN, E., DEMARTINO, A., and DEMARCO, J. (1991). "The effects of low concentration odors on EEG and behavior." J. Psychophys. 5:69–77.

LORIG, T.S. and SCHWARTZ, G.E. (1988). "Brain and odor I: Alteration of human EEG by odor administration." Psychobiol. 16:281–284.

MCCLINTOCK, M.N. (1971). "Menstrual synchrony and suppression." Nature 299:244–245.

SCHAB, F.R. (1991). "Odor memory: Taking stock." Psych. Bull. 109:242–251.

STODDART, D.M. (1990). The Scented Ape: The Biology and Culture of Human Odor. Cambridge University Press, Cambridge.

TENNEN, H., AFFLECK, G., and MENDOLA, R. (1991). "Coping with smell and taste disorders." In: Smell and Taste in Health and Disease. (T. Getchell, R. Doty, L. Bartoschuk, and J. Snow, Jr., ed.). Raven Press, New York. pp. 787–804.

OLFACTORY (CHEMOSENSORY) EVENT-RELATED POTENTIALS

GERD KOBAL AND THOMAS HUMMEL

Department of Experimental and Clinical Pharmacology and Toxicology
University of Erlangen-Nürnberg
Erlangen, Germany

Chemosensory event-related potentials (CSERPs) have found their way into a number of fields of research where they help to determine the function of both the trigeminal and the olfactory system. The investigation of chemosensory deficits in patients with Parkinson's or Alzheimer's disease is only one of the typical applications. It can be assumed that scientists involved in research on patients with multiple chemical sensitivities will also benefit from having access to objective data covering different aspects of the sense of smell.

INTRODUCTION

Many clinicians realize that objective assessment of the sense of smell is critical for (1) establishing the validity of a patient's complaint, (2) characterizing the specific nature of the chemosensory problem, (3) accurately monitoring medical or surgical interventions, (4) detecting malingering, (5) counseling a patient about how to best compensate for the disorder, and (6) assigning disability compensation.

A variety of techniques have been used in an effort to electrophysiologically assess olfactory sensations. The first recording of an event-related potential (ERP) evoked by chemical stimulation was obtained in 1890 by Fleischl von Marxow, who presented ammonia to the nasal mucosa of a rabbit. This stimulation resulted in a negative electrical deflection on the surface of the brain. Following pioneering animal experiments (Hosoya and Yoshida, 1937; Ottoson, 1956) Osterhammel et al., (1969) were the first to record the human electro-olfactogram from the olfactory epithelium. ERPs in response to odorants were obtained in humans by Finkenzeller (1965), Allison and Goff (1967), and Hughes and Andy (1979).

1. Address all correspondence to: Dr. Thomas Hummel, Dept. of Experimental and Clinical Pharmacology and Toxicology, Universitätsstr. 22, 91054 Erlangen, GERMANY. Tel.:+49-9131-852277. Fax:+49-9131-856898.
2. Key Words: Chemical Stimulation, Chemosensory Event-Related Potential, Olfactory nerve, trigeminal nerve.

Toxicology and Industrial Health, Vol. 10, No. 4/5, pp. 587–596

ISSN: 0748-2337

MATERIALS AND METHODS

To record chemosensory event-related potentials (CSERP) considerable sophistication in stimulus presentation is needed. Small fluctuations in stimulus concentration can increase the variability of latencies which, in turn, may distort or even cancel CSERP components. For the same reason, temporal presentation of the stimuli must be as precise as possible. To guarantee the simultaneous excitation of a large number of cortical neurons, onset of stimulation must be steep (≤ 50 msec). The concomitant excitation of other sensors such as mechano- or thermosensors must be avoided.

Recently, stimulators meeting the above requirements have been developed (Kobal, 1981, 1985). Monomodal chemical stimulation is possible by embedding odorant pulses into a constantly flowing air stream (140 ml/sec; 36.5°C; 80% humidity). Odorant switching is performed by means of a vacuum (Figure 1). Stimulation is performed using preestablished concentrations *(odor + dilution)*; during interstimulus intervals only nonodorous air *(control)* is applied to the nasal cavity. This ensures that subjects receive no tactile, thermal, or acoustic cues to provide extraneous information about the timing of stimulus presentations (for details, see Kobal, 1981, 1985, 1987; Kobal and Hummel, 1988, 1992).

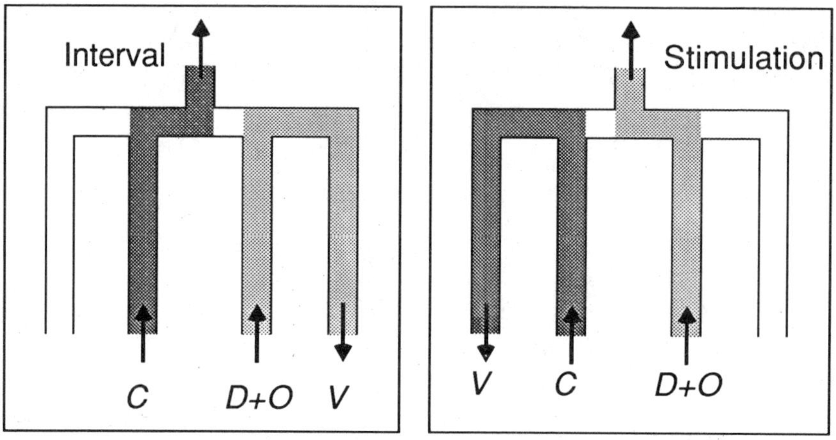

FIGURE 1. Schematic drawing of switching device of a chemical stimulator (**C**ontrol air stream, **D**ilution + **O**dorant, **V**acuum). When the odorant is switched on or off, the subject will not notice turbulences or changes in flow rate or pressure. The carrier gas (air) is thermostabilized (36.5 °C) and humidified to 80% relative humidity.

RESULTS AND CONCLUSIONS

The Human Electro-olfactogram (EOG)

The EOG is considered to reflect summated generator potentials from olfactory receptor cells (Ottoson, 1956). It is recorded from the surface of the olfactory mucosa by means of macroelectrodes. Since any foreign material in the nose may provoke both sneezing and mucous discharge, recording of EOGs from the human nasal mucosa is difficult.

Kobal (1981) performed an EOG study using the stimulants amyl acetate, hydrogen sulfide (H_2S), and eugenol. Responses to all odorants exhibited characteristic negative potentials. Both amplitudes and latencies of the EOG were dependent upon stimulus concentration. When H_2S-stimuli of longer durations were applied, temporal integration was observed over a ten-second period. The same phenomenon occurred when pairs of stimuli were presented. Although the response to the second stimulus had a smaller amplitude, it was superimposed upon the remnant of the preceding one (Figure 2). This finding supports the view that desensitization following repetitive stimulation is predominantly due to central nervous system habituation (Zwaardemaker, 1927) rather than peripheral adaptation.

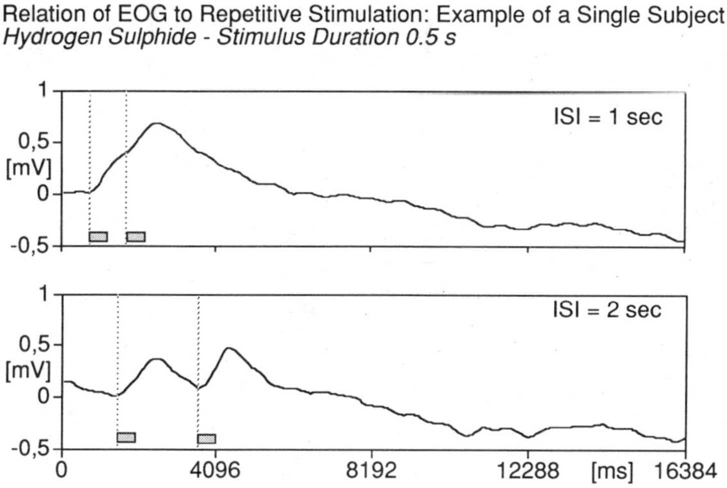

FIGURE 2. Electro-olfactograms (recordings from the olfactory epithelium; example of a single subject) after repetitive olfactory stimulation with H2S at an interstimulus interval (ISI) of 1 and 2 sec, respectively. Stimulus duration was 500 msec as indicated by the filled rectangular boxes. At an ISI of 1 sec responses are superimposed on each other. For the second response no major decrease of amplitudes could be observed when compared to the response to the first stimulus.

While recording EOG from the olfactory mucosa, Kobal (1981) additionally recorded from the respiratory epithelium. Here, it was possible to register a negativity after stimulation with odorants which also produced trigeminal activity. This negative mucosal potential (NMP) was not correlated with the subjects' olfactory sensations but rather it was related to the painful sensations elicited by the stimuli. In experimental animals these responses can be eliminated after systemic administration of capsaicin (Thürauf et al., 1991), indicating the NMP's relationship to excitation of nociceptive C- or Adelta-fibers. Additionally, in human subjects, the NMP was shown to be independent of autonomic reflexes (Thürauf et al., 1993). Thus, it appears to represent peripheral events closely related to the activation of nociceptors of the trigeminal nerve (Kobal, 1985).

Chemosensory Event-related Potentials (CSERPs) — General Considerations
The first human ERPs in response to chemical stimulation were obtained by Finkenzeller (1965) and Allison and Goff (1967). However, Smith et al. (1971) contested the olfactory nature of these responses because they were unable to obtain them in patients without trigeminal sensitivity. More than a decade passed before Kobal and Hummel (1988) reported that CSERPs could be obtained using the odorant vanillin (Doty et al., 1978). Importantly, vanillin responses were not obtained in anosmic patients (Figure 3) while the nonodorous trigeminal stimulant CO_2 elicited responses.

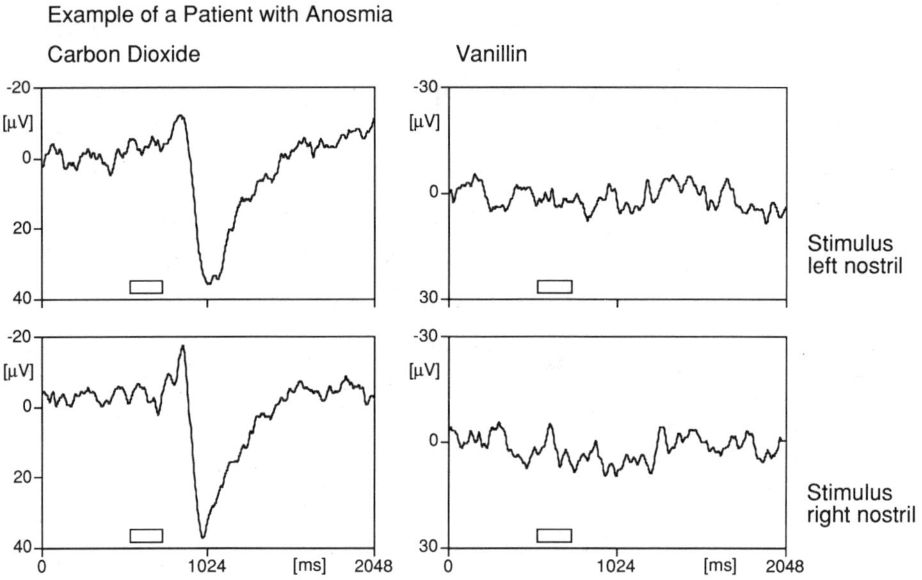

FIGURE 3. CSERP in a 22 year old woman with anosmia (recording position Pz). No responses could be observed after olfactory stimulation with vanillin. In contrast, the patient clearly responded to trigeminal stimulation with CO_2.

Coding of Stimulus Quality
Attempts to experimentally determine differences in the shape of CSERP waveforms in relation to stimulus quality have not been successful. No difference could be found between the ERPs' shape when they were elicited either by activation of the olfactory or the trigeminal system (Kobal and Hummel, 1988). Thus, it is highly unlikely that different CSERP amplitudes result from the activity of different sensory systems. The differential responses observed by Herberhold (1975) and Tonoike et al. (1990) were due rather to the extent of afferent fiber activation.

Cortical Sources of the CSERP
The location of the cortical generators of CSERPs is unclear. One approach to localizing their sources is to analyze the electroencephalogram. Unfortunately, surface potential distributions must be interpreted very cautiously. Realistic models must consider that (1) ERP wave forms

are usually not produced by a single intracranial generator, and (2) such waves can be compatible with a variety of intracerebral charge configurations (Vaughan and Arezzo, 1988). Therefore, in order to characterize the underlying processes in more detail, research is needed using multichannel EEG recordings in combination with current source density techniques (Mitzdorf, 1985; Scherg and von Cramon, 1986; Zhang and Jewett, 1993).

In contrast to the EEG, magnetoencephalography (event-related magnetic field, ERMF) allows for better localization of cortical activity (Hari et al., 1982; Okada, 1985; Williamson and Kaufman, 1987). Importantly, ERMFs are not influenced by physical properties of the conductive media (Hari and Lounasmaa, 1989).

In the area of chemically induced sensations, only one study was performed on recording of ERMFs (Huttenen et al., 1986). CO_2 stimuli at concentrations which produced pain were applied to the nasal mucosa (Kobal, 1985). Simultaneous recordings were made from three channels, and responses were measured from 15 to 87 locations on the scalp covering an area of 60-200 cm^2. ERMF waves had maximal deflections at 350–400 msec after stimulus onset coinciding with the CSERP negative potential recorded from the vertex. The equivalent dipoles were found within a 2.5 x 2.5 cm^2 area in the secondary somatosensory projection area (SII). According to Chudler et al. (1985), this area is closely connected with the processing of nociceptive information. This observation is in line with the view that CO_2 specifically activates trigeminal nocisensors (Thürauf et al., 1991; Steen et al., 1992). Results of more recent experiments using odorants with little or no trigeminal activity suggest that the generators of the olfactory event-related magnetic field are located in the temporal lobe.

Topographical Distribution of the CSERP

Topographical ERP mappings have permitted the differentiation between olfactory and somatosensory (trigeminal) responses. Hummel and Kobal (1992) found a different topographical distribution of CSERPs after stimulation with vanillin or acetaldehyde compared to stimulation with sulfur dioxide (SO_2) or ammonia (NH_4). When acetaldehyde and vanillin were used, amplitudes were largest at the parietal site Pz and the central site Cz, and smallest at lateral frontal positions. When SO_2 and NH_4 were used, amplitudes obtained at Cz were significantly larger than those obtained at any other recording position, indicating a more central distribution for trigeminal stimuli (Figure 4). These results have been confirmed by a number of other studies (Hummel et al., 1992; Kobal et al., 1992; Livermore et al., 1992).

Clinical Applications

An important application of OERPs is the diagnosis of olfactory dysfunction, as well as the detection of malingering. The test procedure, standardized by Kobal and Hummel (1992), includes the recording of responses to CO_2 (52% v/v), H_2S (0.8 ppm) and vanillin (2.0 ppm). All three stimulants are applied 15 times, both to the left and the right nostrils (interstimulus interval 30–40 sec, stimulus duration 200 msec). EEG is recorded from 6 positions (Fz, Cz, Pz, C3, C4, and Fp2; referenced to A1+A2); EEG-records of 2048 msec duration are digitized (sampling frequency 250 Hz; bandpass 0.2–30 Hz) and averaged. In all anosmic patients tested so far, no OERPs could be detected after stimulation with vanillin or H_2S (Kobal, 1994). However, after trigeminal stimulation with CO_2 CSSERPs were obtained in all of these cases (Figure 3).

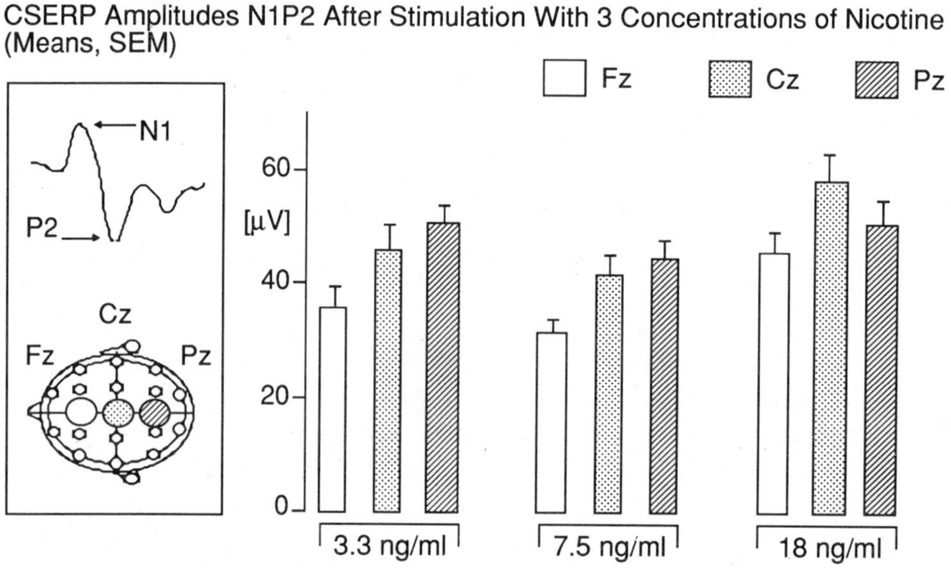

FIGURE 4. Measurement of CSERP amplitudes N1P2 (means, standard errors, n =9) at midline recording positions Fz, Cz, and Pz. Responses are shown to three concentrations of nicotine, which produces odorous (3.3 ng/ml), burning (7.5 ng/ml), and stinging (18 ng/ml) sensations in a concentration-related manner. At low concentrations, the distribution of ERP amplitudes revealed an "olfactory" pattern; i.e., largest amplitudes were obtained at the parietal recording site. Following stimulation with the strongest stimulus the distribution exhibited a "trigeminal" pattern with largest amplitudes at central sites; i.e., this particular pattern emerged only when the concentration of nicotine was above recognition threshold for stinging sensation.

OERP recording may prove helpful in the diagnosis of neurological diseases associated with changes in olfactory function, e.g., Alzheimer's Disease (Doty et al., 1987), Parkinson's Disease (Doty et al., 1988), multiple sclerosis (Hawkes et al., 1994), and progressive supranuclear palsy (Doty et al., 1994). In a recent study, CSERP in response to CO_2, vanillin and H2S were examined in 8 patients with Parkinson's Disease (PD) and 9 controls (Hummel et al., 1993). While there was no difference between the two groups regarding trigeminal stimulation, significantly longer latencies for olfactory responses were found in PD patients.

Recently, a study was published on patients with temporal lobe epilepsy (Hummel et al., 1994). Twenty-two patients were investigated, 12 of whom suffered from epilepsy with a left-sided focus while the remaining 10 patients had a right-sided focus. In both groups, prolonged CSERP latencies were found after left-sided stimulation with CO_2, compared to stimulation of the right nostril. In contrast, a different pattern emerged for olfactory stimuli. After right-sided olfactory stimulation, latencies were prolonged in patients with right-sided epileptic foci. Similarly, when the left nostril was stimulated in patients with a left-sided focus, CSERP latencies were prolonged (Figure 5). These data suggest that the neocortical processing of olfactory, but not trigeminally-mediated, information is affected by functional lesions of the temporal lobe.

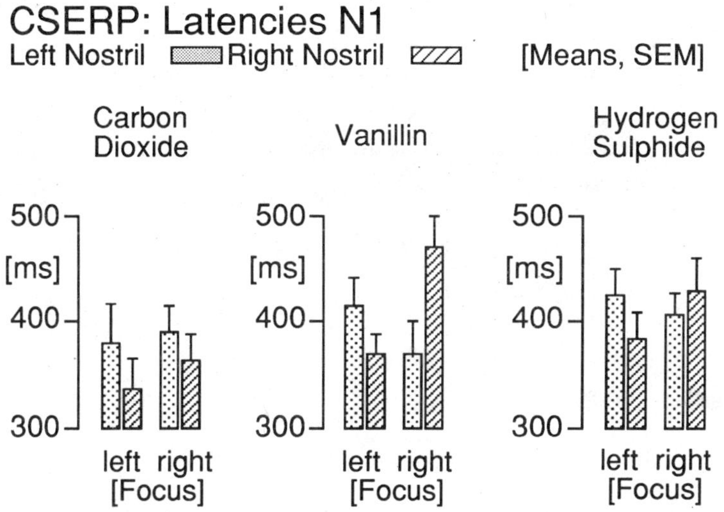

FIGURE 5. Means and standard errors of CSERP latency N1 for epileptic patients with left- (n = 12) or right-sided foci (n = 10) at recording position Pz after stimulation with CO_2, vanillin, and H_2S. While mean latencies of CSERP in response to CO_2 were always longer after stimulation of the left nostril, a different pattern emerged for the two olfactory stimuli (vanillin and H_2S). In patients with a left-sided focus latencies were prolonged after left-sided stimulation; in patients with a right-sided focus prolonged latencies were found after stimulation of the right nostril.

Olfactory-Induced Emotions

There are indications of ERP correlates of odor-induced emotions. When chemical stimulants were presented to each side of the nose, differences between ERPs emerged in relation to the stimulated side (Kobal et al., 1992). Responses to CO_2, menthol, and H_2S showed both significantly shorter latencies and smaller amplitudes after stimulation of the left nostril. In contrast, when vanillin was used as a stimulant, CSERP latencies were shorter and amplitudes were smaller after stimulation of the right nostril. Since the four stimulants did not produce statistically significant differences in intensity estimates after stimulation of the right or the left side, this finding could not be explained by differences in perceived intensities.

The major difference between vanillin and CO_2, menthol, and H_2S related to the stimulants' perceived degree of pleasantness, respectively. That is, while vanillin was perceived as pleasant by all subjects, all other stimulants were associated with more or less negative experiences. CO_2 was always perceived as painful; menthol was perceived as both pleasant and unpleasant (e.g., cooling, slightly painful sensations); H_2S reminded all subjects of rotten eggs. It was hypothesized that the ERP differences caused by the lateralized stimulation was related to the way the two hemispheres process emotional information. This is in line with previous studies on hemispheric differences in emotional information processing (Dimond et al., 1976; Davidson, 1984). Confirming this interpretation, hedonic-related latency shifts were also observed when the odorants phenyl ethylalcohol and acetaldehyde were used as stimuli (Kobal et al., 1989; Hummel and Kobal, 1994).

To summarize, considerable promise is being realized in electrophysiological responses to odorants. Such measures may be of value in the exploration of the physiological underpinnings of odor-related hedonic and cognitive processes. In addition, CSERPs are likely to be of value in the understanding of diseases such as multiple chemical sensitivities where exposure to odors may play a crucial role.

ACKNOWLEDGMENTS

This work was supported by the Marohn-Stiftung, Germany. We would like to thank Dr. Frank Gullotta, Virginia Commonwealth University, Richmond, Virginia, for helpful suggestions during preparation of the manuscript.

REFERENCES

ALLISON, T. and GOFF, W.R. (1967). "Human cerebral evoked responses to odorous stimuli." Electroenceph. Clin. Neurophysiol. 23:558–560.

CHUDLER, E.H., DONG, W.K., and KAWAKAMI, Y. (1985). "Tooth pulp evoked potentials in the monkey: cortical surface and intracortical distribution." Pain 22:221–223.

DAVIDSON, J.D. (1984). "Affect, cognition, and hemispheric specialization." In: Emotions, Cognition, and Behavior (C.E. Izard, J. Kagan, and R.B. Zajonc, ed.). Cambridge University Press, Cambridge, UK. pp. 320–365.

DIMOND, S.J., FARRINGTON, L., and JOHNSON, P. (1976). "Differing emotional response from right and left hemispheres." Nature 261:690–692.

DOTY, R.L., BRUGGER, W.E., JURS, P.C., ORNSDORFF, M.A., SNYDER, P.J., and LOWRY, L.D. (1978). "Intranasal trigeminal stimulation from odorous volatiles: Psychometric responses from anosmics and normal humans." Physiol. Behav. 20:175–185.

DOTY, R.L., DEEMS, D., and STELLER, S. (1988). "Olfactory dysfunction in Parkinson's Disease: A general deficit unrelated to neurologic signs, disease stage, or disease duration." Neurology 38:1237–1244.

DOTY, R.L., REYES, P., and GREGOR, T. (1987). "Presence of both odor identification and detection deficits in Alzheimer's Disease." Brain Res. Bull. 18:597–600.

DOTY, R.L., SMITH, R., McKEOWN, D., and RAJ, J. (1994). "A comparision of tests of olfactory function: Factor analysis and reliability studies." Percept. Psychophys. In press.

FINKENZELLER, P. (1965). "Gemittelte EEG-Potentiale bei olfactorischer Reizung." Pflügers Arch. 292:76–85.

FLEISCHL VON MARXOW, E. (1890). "Mittheilung, betreffend die Physiologie der Hirnrinde." Zbl. Physiol. 4:s37–s4.

GEVINS, A.S. and REMOND, A. (1987). Handbook of Electroencephalography and Clinical Neurophysiology, Vol 1: Methods of Analysis of Brain Electrical and Magnetic Signals. Elsevier, Amsterdam.

HARI, R. and LOUNASMAA, O.V. (1989). "Recording and interpretation of cerebral magnetic fields." Science 244:432–436.

HARI, R., KAUKORANTA, E., REINIKAINEN, K., HUOPAINIEMI, T., and MAUNO, J. (1982). "Neuromagnetic localization of cortical activity evoked by painful dental stimulation in man." Neurosci. Lett. 42:77–82.

HAWKES, C.H., SHEPHARD, B.C., and KOBAL, G. (1994). "Assessment of olfaction in multiple sclerosis: evidence of dysfunction by olfactory evoked response and identification tests." Brain. In press.

HERBERHOLD, C. (1975). "Funktionsprüfungen und Störungen des Geruchssinnes." Arch. Oto-Rhino-Laryng. 210:67–164.

HOSOYA, Y. and YOSHIDA, H. (1937). "Über die bioelektrischen Erscheinungen an der Riechschleimhaut" J. Med. Sci. III Biophys. 5:22.

HUGHES, J.R. and ANDY, O.J. (1979). "The human amygdala. I. Electrophysiological responses to odorants." Electroencephalogr. Clin. Neurophysiol. 46:428–443.

HUMMEL, T. and KOBAL, G. (1992). "Differences in human evoked potentials related to olfactory or trigeminal chemosensory activition." Electroenceph. Clin. Neurophysiol. 84:84–89.

HUMMEL, T. and KOBAL, G. (1994). "Chemosensory event-related potentials: effects of dichotomous stimulation with eugenol and dipyridyl." In: Olfaction and Taste XI (K. Kurihara, N. Suzuki, and H. Ogawa, ed.). Springer, Berlin. pp. 659–663.

HUMMEL, T., KOBAL, G., and MOKRUSCH, T. (1993). "Chemosensory evoked potentials in patients wirh Parkinson's Disease." In: New Developments in Event-Related Potentials (H.-J. Heinze, T. Münte, and G.R. Mangun, ed.). Birkhäuser Verlag, Boston. pp. 275–281.

HUMMEL, T., LIVERMORE, A., HUMMEL, C., and KOBAL, G. (1992). "Chemosensory event-related potentials: relation to olfactory and painful sensations elicited by nicotine." Electroenceph. Clin. Neurophysiol. 84:192–195.

HUMMEL, T., PAULI, E., STEFAN, H., KETTENMANN, B., SCHÜLER, P., and KOBAL, G. (1994). "Chemosensory event-related potentials in patients with temporal lobe epilepsy." Epilepsia. In press.

HUTTUNEN, H., KOBAL, G., KAUKORONTA, E., and HARI, R. (1986). "Cortical responses to painful CO2-stimulation of nasal mucosa: a magnetencephalographic study in man." Electroenceph. Clin. Neurophysiol. 64:347–349.

KOBAL, G. (1981). Elektrophysiologische Untersuchungen des menschlichen Geruchssinnes. Thieme, Stuttgart.

KOBAL, G. (1985). "Pain-related electrical potentials of the human nasal mucosa elicited bychemical stimulation." Pain 22:151–163.

KOBAL, G. (1987). Process for measuring sensory qualities and apparatus therefore. United States Patent Number 4,681,121.

KOBAL, G. (1994). "Olfactory event-related potentials in the diagnosis of smell disorders." In: Olfaction and Taste XI (K. Kurihara, N. Suzuki, and H. Ogawa, eds.). Springer, Berlin. pp. 626–630.

KOBAL, G. and HUMMEL, C. (1988). "Cerebral chemosensory evoked potentials elicited by chemical stimulation of the human olfactory and respiratory nasal mucosa." Electroenceph. Clin. Neurophysiol. 71:241–250.

KOBAL, G. and HUMMEL, T. (1992). "Olfactory evoked potentials in humans." In: Smell and Taste in Health and Disease (T.V. Getchell, R.L. Doty, and J.B. Snow, Jr., eds.). Raven Press, New York. pp. 255–275.

KOBAL, G., HUMMEL, T., and KETTENMANN, B. (1994). "Chemosensory event-related potentials in temporal lobe epilepsy and first recordings of olfactory event-related magentic fields." Chem. Senses. In press.

KOBAL, G., HUMMEL, T., and PAULI, E. (1989). "Correlates of hedonic estimates in the olfactory evoked potential." Chem. Senses 14:718.

KOBAL, G., HUMMEL, T., and VAN TOLLER, S. (1992). "Differences in chemosensory evoked potentials to olfactory and somatosensory chemical stimuli presented to left and right nostrils." Chem. Senses 17:233–244.

LIVERMORE, A., HUMMEL, T., and KOBAL, G. (1992). "Chemosensory evoked potentials in the investigation of interactions between the olfactory and the somatosensory (trigeminal) systems." Electroenceph. Clin. Neurophysiol. 83:201–210.

MITZDORF, U. (1985). "Current source-density method and application in cat cerebral cortex: Investigation of evoked potentials and EEG phenomena." Physiol. Rev. 65:37–100.
OKADA, Y. (1985). "Discrimination of localized and distributed current dipole sources and localized single and multiple sources." In: Biomagnetism: Applications and Theory (H. Weinberg, G. Stroink, and T. Katila, ed.s). Pergamon Press, New York. pp. 266–272.
OSTERHAMMEL, P., TERKILDSEN, K., and ZILSDORFF, K. (1969). "Electro-olfactograms in man." J. Laryngol. 83:731–733.
OTTOSON, D. (1956). "Analysis of the electrical activity of the olfactory epithelium." Acta Physiol. Scand. 35(Suppl. 122):1–83.
SCHERG, M. and VON CRAMON, D. (1986). "Evoked dipole source potentials of the human auditory cortex." Electroenceph. Clin. Neurophysiol. 65:344–360.
SMITH, D.B., ALLISON, T., GOFF, W.R., and PRINCITATO, J.J. (1971). "Human odorant evoked responses: effects of trigeminal or olfactory deficit." Electroenceph. Clin. Neurophysiol. 30:313–317.
STEEN, K.H., REEH, P.W., ANTON, F., and HANDWERKER, H.O. (1992). "Protons selectively induce lasting excitation and sensitization of nociceptors in rat skin." J. Neurosci. 12:86–95.
THÜRAUF, N., FRIEDEL, I., HUMMEL, C., and KOBAL, G. (1991). "The mucosal potentials elicited by noxious chemical stimuli: is it a peripheral nociceptive event?" Neurosci. Lett. 128:297–300.
THÜRAUF, N., HUMMEL, T., KETTENMANN, B., and KOBAL, G. (1993). "Nociceptive and reflexive responses recorded from the human nasal mucosa." Brain Res. 629:293–299.
TONOIKE, M., SETA, N., MAETANI, T., KOIZUKA, I., and TAKEBAYASHI, M. (1990). "Measurements of olfactory evoked potentials and event related potentials using odorant stimuli." Ann. Intl. Conf. IEEE Eng. Med. Biol. Soc. 12:912–913.
VAUGHAN, H.G. and AREZZO, J.C. (1988). "The neural basis of event-related potentials." In: Human Event-Related Potentials (T.W. Picton, ed.). Elsevier, Amsterdam. pp. 45–96.
WILLIAMSON, S.J. and KAUFMAN, L. (1987). "Analysis of neuromagnetic signals." In: Handbook of Electroencephalography and Clinical Neurophysiology, Volume 1: Methods of Brain Electrical and Magnetical Signals (A.S. Gevins and A.A. Remond, eds.). Elsevier, Amsterdam. pp. 405–448.
ZHANG, Z. and JEWETT, D.L. (1993). "Insidious errors in dipole localization parameters at a single time-point due to model misspecification of the number of shells." Electroenceph. Clin. Neurophysiol. 88:1–11.
ZWAARDEMAKER, H. (1927). "The sense of smell." Acta Oto-Laryngologica 11:3–15.

RESPONSE PANEL #4

HELEN MAYBERG, HOWARD KIPEN, IRIS BELL, CLAUDIA MILLER, RICHARD LETZ, AND NORMAN ROSENTHAL

NOTE: The following comments by Dr. Helen Mayberg were part of a response panel. However, because of the great interest in her comments, Dr. Mayberg was asked to develop a short overview on the topic of imaging. Those prepared comments are included as the Appendix.

Dr. Mayberg: I'm a neurologist, but my expertise is in positron emissions tomography and single photon emissions tomographies, so I'll restrict my comments to those two. I want to make a couple of clarifying comments, because I thought there were a few things said that I think may be disturbing, particularly to the patients.

These techniques do not measure holes in your brain. "Holes in the brain" is a euphemism; changes in flow, function, or metabolism cause changes in the behavior of neurons that can be measured with the detection of radioactive decay. I'm sure Dr. Simon didn't mean to imply that people would have holes in their head.

When one makes measurements of blood flow, it is either absolute blood flow in the case of Dr. Mena with his xenon, or relative blood flow in the case of the seratech tracer, which is a commercially available tracer available in most nuclear medicine departments. As we look at these kinds of studies we have to look at what the application is in the general community, and to use the work from the research arm to define how we proceed. We need to have very standardized ways in which these procedures are done. What wasn't given time in the discussion by either presenter was the fact that these methods are very labor intensive, are very dependent on the condition of the subject at the time, and that the methods themselves undergo tremendous scrutiny and testing and validation. Whenever any change is made in the conditions under which the patient is studied or the time at which a measurement is made, those studies need to be validated irrespective of patient groups and controls, so that the method and its reliability and standardization are well known.

The second issue is one that neither of the two presentations went into, the history of how these studies are done. I have learned over the last two days that this group is still arguing about the definition of the disease. With functional brain imaging, what has generally gone on, as in most kinds of pathological states, is that the pathology of the brain is known. With Alzheimer's Disease we know about plaques and tangles, with Temporal Lobe Epilepsy we know about cell dropout in the hippocampus, and that we have known pathology. Even if we can't see it structurally, we know where we are looking from pathology studies. We don't have

Toxicology and Industrial Health, Vol. 10, No. 4/5, pp. 597–603

ISSN: 0748-2337

this in this entity. So the next thing we have to do is have very carefully defined clinical criteria. If you have heterogeneity, these imaging scans are not going to parse out your patient subgroups for you. It's been very clear in studies of dementia; that has been a very difficult issue. But by very carefully defining subgroups, studying groups with hypothesis-driven techniques, asking a question, testing a question with well-validated methods, we should look first in groups to see whether there are differences between controls and patients, and then, secondarily, determining the sensitivity to look for in a given individual to see a difference. I point you again to the history of functional imaging in well-defined diseases like Temporal Lobe Epilepsy. There, you know before you start where the abnormality is due to depth electrodes recording from the temporal lobe. They are only sensitive in 70% to 80% of the cases, depending on how good your machine is. So even in well-known diseases sensitivities are argued about, and at best they are 70–80%. So when you start to see 90–95% sensitivity, that is a red flag for me to really question the sensitivity or method in this group.

The last issue I want to cover is the need for proper controls. It's very clear that in many studies of functional imaging, for other reasons, that the behavioral state of the subject at the time of the experiment is vitally important. We start in well-characterized diseases by making measurements at rest. We can start by looking at rest, and looking for patterns of abnormality, but the caveat is that not seeing anything doesn't mean that, with provocation, something might not be detected. Controls are really essential. It's very clear from experiments in normals that creating an emotional state can have profound effects on patterns of cerebral blood flow and metabolism, and that we have to have very carefully controlled studies or the data are really uninterpretable.

I would offer one approach: People with clear, comparable exposures who complain are compared with people with comparable exposures who don't have problems, because then you have what's also very clear from functional imaging, that abnormalities may be present with no behavioral correlate. The brain has a number of ways in which it compensates, so that people with strokes with clear lesions in their head, loss of brain tissue, with metabolic or flow concomitants, may have absolute resolution of all their neurological problems. The scan abnormality may have no meaning. We need to know: there may be subthreshold exposures that have effects on brain metabolism or blood flow patterns that do not have clinical correlates and we need to compare that potential scan pattern to patients who clearly have symptoms from the similar exposure.

The notion of a challenge experiment or treatment experiment is very important, but things have to be similar across subjects. I know its difficult, due to the variable conditions, but we have to decide what is the right time to make the imaging measurement, and what is the right dose.

I leave you with one food-for-thought: the ethics of how one does challenge studies troubles me. The idea of challenging someone without very carefully defined pre-imaging studies is extremely unethical, in my view.

Dr. Simon: Dr. Mayberg is right; the state of the patient is critical. We isolate these studies to a particular room. We habituate the patients to the room, and we give them a mental test and a mental task to perform while they are doing the test to try to make the mental states of the patients as comparable as we can.

The next issue she raised is also very important. In other studies that we've done, for example the ones with breast implants, solvent exposures, and CFS, our general results when we get to a testable group is in the 80–90% range. So Dr. Mayberg is right that on any individual test this is not a definitive, pathognomonic pattern. Rather, it is a pattern that shows up in more of these than in controls.

The third issue that I'd like to raise is the issue of challenge. What is the correct dose, and what are the ethics of challenge? It is my belief that we limit challenges to patients that have had challenges by their clinical physicians, and the doses we use are exactly the same as used before.

Dr. Mena: I agree with Dr. Mayberg that it is extremely important to define, first, the sensitivity of our procedures, and second, the definition of our normal controls. We've been working for more than six years in the area of neuroSPECT, and the definition of our criteria for sensitivity has been recently corroborated by autopsy studies in Alzheimer's Disease. The criteria I described to you this morning were used in 13 of 14 patients where we made the diagnosis of Alzheimer's Disease on neuroSPECT. Eventually the other patient went to autopsy, and we were right concerning the pattern of profusion and the nuclear medicine diagnosis.

Second, the definition of normals is very critical. Particularly important for our young normals is that they have not been exposed to substance abuse. This is very difficult to get, particularly from the population in Los Angeles. At our institution, we were not able to get one single person that withstood the rigor of the inventory of questions we ask, so we had to go to the Mormon Church and all our young normals are Mormons.

Dr. Mayberg: I want to leave the audience with the idea that the three of us are colleagues of sorts. We all have strong beliefs about the power of functional imaging, and I do not want to give the audience the impression that we are adversaries in a methodological sense. We all come with our inherent biases. My particular area of interest is in affective disorder, so I was very intrigued by the overlap of the comments. Another issue is the specificity of findings we may see. This multifocal pattern may indeed be common to the multitude of patients who have a variety of exposures, and reflect some final common pathway, not necessarily of the disease itself, but of the epiphenomena and the constellation of clinical symptoms. I will caution you that, looking at localization and behavior relationships with functional imaging, we haven't been as smart as we might, based on lesion data, which we all rely on when someone has a left hemisphere stroke and language disturbance. When they have aphasia, we're not surprised, because there's a long history of association of those lesions with that deficit. It's been shown that people with aphasia don't have isolated abnormalities that correlate with their structural lesions, but actually have widespread abnormalities that cover the entire language system. When people have tried to look at more elaborate behaviors from people who have problems, it's less well-localizing. Like an Alzheimer's patient who has trouble with visual/spacial processing, there is a correlation with more right-sided abnormalities. But when one tries to go from, "If I have a scan abnormality and it's more on the left side, can I predict that the person has a language disturbance?" that has been unsuccessful.

The other comment I want to make is about this multifocal pattern. It is very important for the audience to realize that the multifocal pattern is described in a number of etiologically distinct neurologic and neuropsychiatric entities, so that besides drug addiction, it's seen in AIDS Dementia Complex, multi-infarct dementia, and chronic cocaine users, and the pattern is persistent. Again, these are etiologically extremely diverse. The MCS complaints I've heard did not sound like the pattern in these cases. Just because the patterns are similar, that does not mean they are clinically alike in any way, but it shows us that the pictures can only go so far.

I would contend, based on a lot of the other data that were presented, that an alternative approach may be to look at things other than flow with SPECT, or metabolism with PET. The advances in MR spectroscopy measurements are one example.

From the Audience: As a layman, I can't help but look at the pre- and postchallenge studies and say, "Gee, I can see a difference and I'm not a neuroradiologist." I wonder how one might conduct challenge studies and what might be criteria for a believable change in a particular individual? I'd like to hear all the panelists say what are some optimal study designs.

Dr. Simon: We are looking at a complex pattern and certainly hot and cold foci are a part of that pattern, but the mismatch is possibly the cardinal sign of the pattern. In terms of design, one has to come to some kind of a conclusion as to what is an ethical challenge. Apparently, there might be some serious question about that.

Another important thing is to be able to pull in patients, like we did with the controls, with a priori criteria. You can't do that under the current funding climate, or at least we have not been able to. But, in an ideal funding climate, one would define the patient population group and then be able to draw patients from there, rather than taking patients one is able to test and trying to match them up with what group they come from. Of course, I'd like to see us expanding the normal control group. We were only able to do a limited group, from age 22 to 45, because that's all we had resources for. Clearly, Dr. Mena's work and that of others show that children are different centographically than adults, so one would want to expand the population so that one can have a control group, while being cognizant that these control groups are going to be highly technique-dependent. One will have to be very careful when one does assemble that group, so that they are assembled using the same technique that the test will actually be performed in.

Dr. Bell: I have several comments to make in response to some of the points that have been raised in the discussion, and I'd also like to put some perspective on the EEG data that have been presented.

In response to Dr. Kipen's last point, about different routes of administration: There is a limited amount of work in animals, looking at intraperitoneal injection of herbicide. It has been found, with some of these substances, that they are subsequently found to be concentrated in the nasal mucosa. There is a P450 detoxification enzyme system in the mucosal region, and apparently materials are transported there. Now whether they then end up in the limbic system per se, or whether they have any influence on brain function by that route, is certainly not established. But I think it is very interesting.

In response to Dr. Mayberg's comment about the effects of the challenges, I would agree completely. We certainly don't want to expose people to gratuitous studies that may present any dangers. At the same time, I think we've emphasized the need for challenges, and I think we have to find some balance between these two considerations. Obviously, we don't want to find out, 25 years from now, that we have exposed people, and have the same kind of scandals that are coming out now about the radiation testing that was done on various veterans and other people. There are some data in the sensitization literature to guide us about what doses and schedules of exposure will produce more lasting sensitization, as opposed to less permanent sensitization.

When we're talking about variable levels that are, in fact, the equivalent to standing in a mall, I don't see that we're adding tremendous risk, from an ethical point of view, to what that individual would have by simply trying to go outside the home. If we were to take the people who were trying to stay away from all of those exposures, and never go near the mall, and test them, then I think we have separate issues. We may have some subject selection problems that are very significant, because the sickest people may be the ones that we would ethically be the most concerned about.

In connection with that, I would like to remind you, and it is mentioned in more detail in my review paper, that there are some findings showing beta activity changes in chemically sensitive patients. I know Drs. Staudenmayer and Selner have published a paper suggesting that MCS patients have higher levels of parietal data, in a nonspecific way, without chemical challenges. They concluded that since psychiatric patients also showed the same pattern, that these patients must be basically identical with psychiatric patients. I think that Dr. Mayberg's point, that this is a very nonspecific finding, is very well taken.

Dr. Rosenthal: Just a quick point about the ethics of challenges. I agree with Dr. Bell that if we're going to document the illness, we are going to have to find some paradigm. After all, a lot of the MCS folks are already experiencing these reactions in their ordinary lives, and so the challenge does not have to be great for them.

But that part of the ethics of challenges, or any experimentation, is the issue of informed consent. I think it is really critical to bring the patients in as collaborators, even in areas where we may not know the ultimate effect. We can just state that, honestly. "We're going to give you these little challenges, and, do they make a difference in the long run? We don't really know. It's a risk we're taking, and a risk, that if you're going to join us in this enterprise to try and understand and treat your problem, you're going to have to share that risk with us, and sign on the dotted line."

Dr. Janette Sherman, from Western Michigan University, practicing in Alexandria, VA: Several of the people discussed the fact that about 20+% of the people could not identify an initiating event. We need to understand that most pesticides, and certainly Dursban LO (LO means Low Odor), is produced and marketed. Many people cannot detect it at all. We have to understand that masking chemicals are used widely, and, many of the fancy odors that we are now exposed to will also mask an odor of pesticides.

It's difficult to determine differences between solvent-exposed persons and pesticide-exposed persons. Unfortunately, most pesticides are delivered in a solvent base.

Stephen McFadden, independent research advocate: Dr. Simon, in the challenged SPECT scan, is there a decrease in the peripheral blood flow in the whole body, or just the brain?

Dr. Simon: For these challenges, we only have a limited area we can look at, so we have chosen the brain. So, I can't speak to a general reaction.

Dr. McFadden: Is there an infrared peripheral thing you can look at if the blood flow perusal profusion in the finger, like hemoglobin, or something like that?

Dr. Simon: That's really a good idea. I think I'll try that. It would be very easy for us to incorporate in our protocol. I really thank you for the idea.

Jim Rapase, from EPA in Washington: I'm on loan to OSHA right now, and I wanted to point out that, within the next week, there will be a Federal Register notice, by OSHA, on indoor air quality. A lot of the researchers and patients, who might be attending this conference, might well want to get a copy of the Federal Register notice. It's about 350 pages. There is a substantial piece in it about multiple chemical sensitivity, and sick buildings, and some of you may want to submit some information to OSHA concerning your researches or your experiences in buildings.

Dr. Simon: I'd like to respond to a couple of things. When we did the the chronic fatigue study, we had four groups of people. We had people that had chronic fatigue syndrome by

CDC criteria, chronic fatigue syndrome that didn't meet CDC criteria, but their doctor thought they had it, affective disorders, and normals. In that group, by the way, the chronic fatigue, by CDC criteria, were the strongest on the temporals, and CFS without CDC were still in that general range, and the others were not in that area. We had temporal symmetries in all the groups. What we're really trying to do is characterize from a spectrographic standpoint.

The real question was: What is the proper control for multiple chemical sensitivities? I was careful not to use the word in my talk, because I'm still trying to understand what it really means. I think that one of the first things that people interested in this area need to do is give us something that we can really work with; an operational definition that we can know what a control is. Until that is done, I don't know how to assemble a control group.

Dr. Mayberg: Dr. Mena has obviously tried to make comparisons, on depression, chronic fatigue, and MCS. What's really fascinating to me, based on my own work in depression, is that this wasn't an MCS multifocal pattern, plus the depression pattern. I think that's something that's not consistent with a big body of published data on primary and secondary depressions. There is no data that is really consistent on what happens in depressed people, when they're feeling depressed, vs. when they're not depressed.

I think Dr. Simon and Dr. Mena have got to get their data published, and we need to establish reproducible results so that those without very elaborate equipment can understand where we're going, before we try to start doing this on individual patients.

Dr. Rebecca Bascombe, Director, Environmental Research Facility, University of Maryland: We've been doing studies in environmental tobacco smoke sensitivities. In thinking about that work, there are some lessons that can be applied to this discussion. I support Dr. Bell's comment that we can often learn a lot by studying mild forms of a disorder. With asthma, for example, we do not perform inhalation challenges on people with the most severe forms of asthma, and yet there have been tremendous advances in understanding the mediators of the inflammation associated with antigen challenge by taking individuals who have a mild enough form of the condition that the challenge would cause a reversible change.

What about the people who are very severely affected? In that case we can think about other clinical conditions, where we do things to people that cause tremendous disruption. For example, electrophysiologic testing is where you induce the cessation of the heart beat or where you induce a life-threatening arrhythmia. There is ethical precedent for doing potentially dangerous things to people when the clinical condition is considered by them to be sufficiently severe.

LOW-LEVEL CHEMICAL SENSITIVITY: A PERSPECTIVE FROM BEHAVIORAL TOXICOLOGY

BERNARD WEISS

University of Rochester School of Medicine and Dentistry
Rochester, New York

Low-level chemical sensitivity is hardly a new issue in environmental toxicology. It is, in fact, the focus of risk assessment. The risk assessment process is designed explicitly to estimate the health threats posed by low exposure levels, typically by extrapolating from high experimental or environmental levels. The conventional risk assessment structure, however, was designed primarily around cancer. It is only awkwardly applicable to neurobehavioral toxicants because of the multiplicity of endpoints that have to be considered in evaluating neurotoxicity. At the same time, neurotoxic risk assessment maintains certain advantages over cancer risk assessment because of diminished uncertainties over dose extrapolation. It does not have to depart as far from the range of observable data.

The main problem with extending the risk assessment model to issues such as Multiple Chemical Sensitivity (MCS) and Sick Building Syndrome (SBS) is the absence of a specific chemical whose concentration can be measured and then manipulated. A prototypical agent, however, such as a volatile organic solvent, might be selected and studied. Beyond the choice of agent, however, is the question of which behavioral criteria are likely to yield the most useful information. Although neuropsychological test batteries provide one source of data, they typically are administered in a setting other than the one allegedly provoking the syndrome. A different approach invokes what might be called a

1. Adress all correspondence to: Dr. Bernard Weiss, Department of Environmental Medicine, University of Rochester Medical Center, Rochester, NY 14642. Tel.:(716)275-1736. Fax:(716)256-2591. Internet:weiss@envmed.rochester.edu.

2. Abbreviations: ADD, Attention Deficit Disorder; ADI, Acceptable Daily Intake; CFS, Chronic Fatigue Syndrome; FDA, Food and Drug Administration; MCS, Multiple Chemical Sensitivity Syndrome; MMPI, Minnesota Multiphasic Personality Inventory; SBS, Sick Building Syndrome; VOCs, Volatile Organic Solvents.

3. Key Words: behavioral toxicology, experimental design, food additives, organic solvents, performance.

4. Preparation supported in part by grants ES01247 and ES05433 from the National Institute of Environmental Health Sciences and grant DA07737 from the National Insitute on Drug Abuse.

Toxicology and Industrial Health, Vol. 10, No. 4/5, pp. 605–617

ISSN: 0748-2337

miniature work situation. Here, a test subject is evaluated in a setting that emphasizes sustained performance testing in the presence of target chemicals.

Experimental design is another factor to be considered. Two features are especially critical. The most sensitive design, at least for the current stage of knowledge, would probably emphasize consistency of response, and would choose as subjects individuals who claim to be afflicted with low-level sensitivity. Consistency in a single individual may be more informative than significance tests in a large sample. In addition, consistency as a criterion helps overcome the problem that, in any such sample, only a minor proportion of the subjects mcy truly exhibit such sensitivity. At a later stage, a broader range of subjects might be targeted. Research on behavioral disorders evoked by food additives illustrates the importance of such questions. It also demonstrates that the methods currently used to assess the potential toxicity of many substances, including food additives, typically ignore subtle, and often sensitive, neurobehavioral measures.

Did we really need a Congressional mandate to take up the question of low-level chemical sensitivity? Those of us who inhabit the terrain of the environmental health sciences have already elevated the issue to a doctrine. We call it risk assessment. From the outside, risk assessment appears to be a rigorous quantitative discipline. Would I be divulging secrets if I told you, that from the inside, it may sound like incantations about low-level chemical sensitivity? An example: you construct a statistical model, based on extreme doses in animals, to project a risk probability in humans not to exceed one in a million. That is the rule proclaimed by federal regulatory agencies to set exposure standards for presumed carcinogens. Low-level could hardly be more extreme.

The conventional risk assessment framework (National Research Council, 1983) fits the subject of this workshop rather awkwardly, and cannot smoothly adapt to the much broader arena of neurobehavioral toxicology or even to the universe of systemic, or non-carcinogenic toxicants. Cancer, despite its acknowledged biological complexities, is still unidimensional. It became the prototype for health risk assessment because it moves smoothly along the accepted chain of decisions framed by the National Research Council: hazard identification, dose-response assessment, exposure assessment, and risk characterization. Neurobehavioral toxicity ramifies in so many directions that it cannot be crammed into such a neat progression. It can be made to fit only by enlarging the conventional framework to include two key questions. First, which criteria will you accept as evidence of adverse effects? Pathology? Motor activity in rodents? Complex learned behaviors? If you rely on human data, do you base your decision on psychological test scores? Would you accept subjective complaints? Clinical inference? Second, how do you define exposure? Total cumulative dose? Mean dose? Intermittent peaks? Acute episodes? To embrace these complications you need many more boxes in your decision flow chart.

The single greatest advantage neurotoxic risk assessment maintains over cancer risk assessment is, ironically, its less stringent risk standards. Such elasticity derives from the assumption that non-carcinogenic chemicals, unlike carcinogens, exert adverse effects only after they exceed some exposure threshold. Regulators do not flinch at a probability of adverse effects far in excess of 10^{-6} for neurobehavioral toxicants. They are not inclined to worry about 250 people in a population of 250 million who may respond to low-level contamination with complaints of impaired memory. This more relaxed attitude allows us to avoid the monumental uncertainties built into cancer risk assessment due to its need to extrapolate from high experimental exposures to low environmental exposures. We often can narrow the uncertainties by restricting our extrapolations to the vicinity of observable data. That privilege is granted us only if we provide compelling evidence of the utility and significance of our measures.

Such a privilege has not been granted to the labels of Multiple Chemical Sensitivity Syndrome (MCS), Sick Building Syndrome (SBS), or even to Chronic Fatigue Syndrome (CFS). The evidence is not forceful enough. It is viewed as too vague, subjective, and impressionistic. A recent Medical Grand Rounds at the University of Rochester showed me how much these entities are shrouded in skepticism by the medical community. The response of the audience paralleled that of the American Medical Association's Council on Scientific Affairs in its recent report (1992) on MCS, SBS, and CFS. The question posed to this conference is how to conduct a quest for evidence that will be taken seriously by biomedical scientists and by clinicians.

As disagreeable and disappointing as it may be to patient advocates, patient testimony is not sufficient. Only evidence from the laboratory or from controlled clinical trials can dispatch the ambiguities and move the skeptics. Wary suspicion is a symptom of a healthy science. If it were not, many members of the American Physical Society would hold stock in Cold Fusion corporations. Here is the conundrum. Almost always, when we perform a risk assessment, we do so in the context of a particular chemical. The process is not structured to first identify the hazard, then the chemical. So how do you gather evidence?

I find it instructive to think about how to conduct a laboratory experiment on MCS or SBS. The first problem, of course, is how to select the chemical challenge. Hudnell et al. (1992) and Otto et al. (1992), in a pair of papers from the U.S. EPA, solved the problem by assembling a mixture of 22 volatile organic solvents (VOCs). Volatile chemicals are suitable choices because we have to assume that, especially for SBS, the primary exposure medium is inhalation. These investigators exposed their subjects to only one concentration and assessed their responses by administering a battery of neuropsychological tests plus subjective ratings. Although they saw increases or no changes in ratings of fatigue, mental confusion, drowsiness, and eye and throat irritation, ratings of odor intensity decreased, indicating adaptation of olfactory receptors. They detected no changes in test performance. Practice effects due to repeated behavioral testing, they noted, might have been a confounding factor.

Although these two studies illustrate one reasonable strategy for bringing MCS and SBS into an experimental framework, other strategies ought to be essayed because they yield different kinds of answers. Two variations ought to be weighed. One involves the choice of agent or agents. A blend of VOCs is difficult to manipulate experimentally because it presents monitoring problems and because dose-response functions are difficult to achieve. Altering the concentration of the mixture would modify the relative concentrations of the components because each would be characterized by a different dose-response slope. A more efficient approach might be to select a prototype VOC and to systematically vary its concentration.

The second variation lies in how sensitivity is measured. The EPA investigators used a battery of 13 neurobehavioral tests designed to probe different functional endpoints. Test batteries serve a special role in neuropsychological diagnosis. They are designed primarily to sample a broad range of functions. But, precisely because of their design, they may provide only relatively superficial assessments of particular functions. They may be inadequate reflections of how an individual performs in other environments, or under unusual circumstances. Even sleep-deprived subjects may perform competently when aroused to execute brief tests from a battery, but they then drift into the stuporous state that prevailed before testing.

One alternative to the test battery approach is to monitor performance continuously, especially during prolonged experimental sessions. The literature on performance enhancing drugs offers some suggestions. Monotonous tasks, such as searching for deviations from background on a visual display, show the steepest degradation in performance over time. They are also the tasks most responsive to pharmacological intervention by agents such as the amphetamines (Weiss and Laties, 1962). Complex, tedious tasks are also the most sensitive to sleep loss (Naitoh and Townsend, 1970).

Research conducted to ascertain the behavioral effects of bacterial infections yields further support to the usefulness of complex performance indices (Alluisi et al., 1973). The investigators asked how well sustained performance could be maintained during incapacitation with an infectious disease. They defined sustained performance as the continuous performance of tasks or jobs for four or more hours per day during several days, weeks, or months. Tularemia was induced by exposing subjects to an aerosol containing F. tularensis. Sandfly fever was induced by inoculation with human serum containing sandfly fever virus.

Instead of relying on a test battery in which the individual components are presented in sequence, the experimenters made use of what they termed a synthetic work situation. The subjects faced a panel (Figure 1) designed to present six concurrent tasks. Three of these measured vigilance or watch-keeping functions, and were relatively passive, requiring subject responses only when a designated signal or stimulus was presented. The other three tasks measured memory functions, discriminative functions, and procedural functions. A test day occupied eight hours, during which the basic 2-hour test schedule cycled through four repetitions.

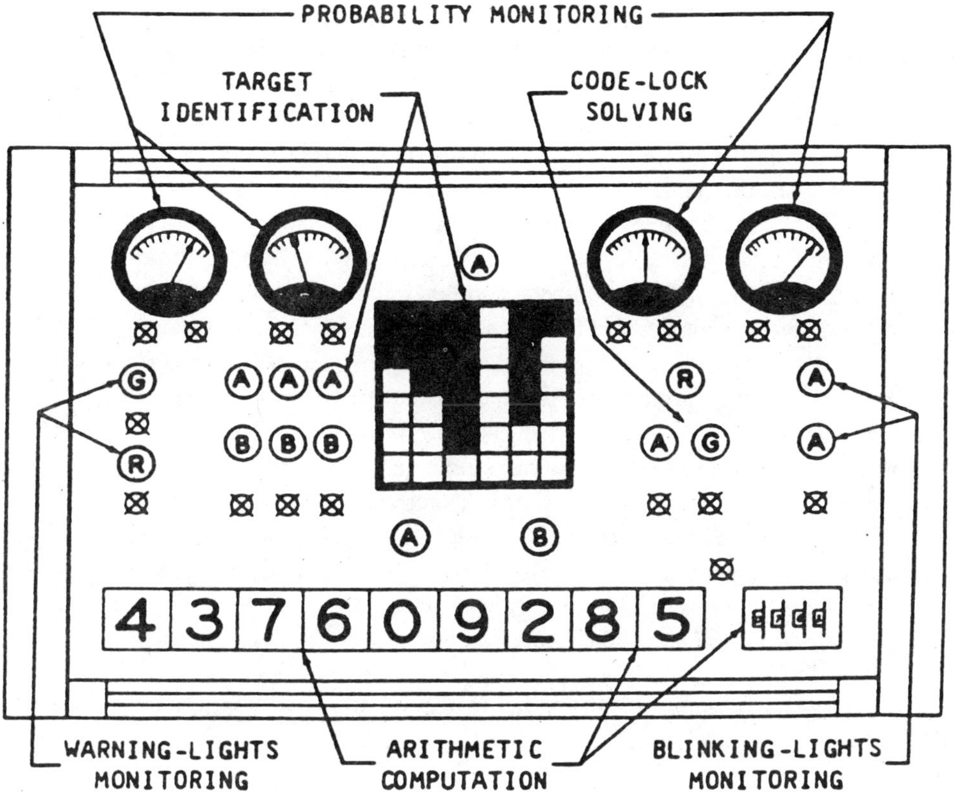

FIGURE 1. Diagram showing the subject's view of the multiple task performance battery. The subject's tasks include: monitoring of blinking lights; monitoring of warning lights; monitoring of probability; arithmetic computations; solving of code locks; and identification of targets (from Alluisi et al., 1973).

As might be expected, decrements in performance tended to reflect the clinical indices of illness such as elevated temperature and pulse rate. Even after clinical recovery, however, performance remained below baseline levels. When performance was aligned with the day of maximum temperature to provide consistency among subjects, restoration of normal body temperature occurred about two days later and performance decrements persisted until the end of the experiment five days later. Furthermore, the performance decrements occurred with all the subtasks, indicating a generalized rather than narrow impact of infection.

Several psychological tests were also included in the assessment. These included the Minnesota Multiphasic Personality Inventory (MMPI), a projective test, and an adjective check list to measure mood. Test scores did not discriminate consistently between control and experimental subjects nor were they correlated with performance.

Data such as these suggest that one useful experimental technique for investigating the impact of low-level chemical exposure might be to select a prototypical agent, such as a volatile organic solvent, and to measure sustained performance on a complex multiple test system over a period lasting at least several hours. Different ambient concentrations of the solvent would yield at least a rudimentary dose-response function. Such a situation corresponds more closely to that confronting SBS individuals than being tested in a physician's or neuropsychologist's office. Moreover, such a situation is not as easy as a conventional single test to deceive. With multiple tasks, degradation of performance is not uniform across tasks in motivated subjects who are subjected to interventions such as sleep loss, infections, or drugs. Some tasks are more sensitive than others, and this information can be used to detect feigned performance disruptions that do not fit such a pattern.

Dependence upon such a prototypical substance is not free of problems either. Solvents such as toluene are detectable by smell even at low concentrations. This property might be used to advantage, however. One type of session would expose the subject to continuous toluene at some specified level. Another could expose the subject to the toluene for only a few minutes. Because of the speed of olfactory adaptation, it would prove difficult for most subjects to differentiate the two situations by depending on odor, so at least brief and prolonged exposures could be compared. Additional anesthetic agents (VOCs are all anesthetics at high levels) such as nitrous oxide, which is odorless, could also serve as controls.

Both epidemiological and experimental investigations suggest an additional approach or an additional source of data. Wayne et al. (1967) sought to measure the impact of atmospheric ozone concentration on athletic performance, a natural question in their Southern California setting. Because high-school cross-country runners typically improve through the competitive season, they correlated expected performance with oxidant concentration at the time of the events and found longer than predicted times. The authors did not view these results as a reflection of physiological impairment. Instead, they offered what I term a behavioral explanation: that these results might be ". . . more related to lack of maximum effort due to increasing discomfort than directly to physiologic capacity."

Similar explanations have been advanced for the results of experimental studies with ozone in which subjects cease exercising despite no other evidence of diminished pulmonary or physiological dysfunction. Animal studies provoke similar explanations. Rats reduce the rate at which they press a lever for access to a running wheel, which usually is a potent reward, in the presence of concentrations of ozone (0.08 ppm) below those specified by the Clean Air Act. I interpret such behavior (Weiss, 1989) as avoidance of the aversive consequences of exercising in the presence of ozone. It should be feasible to arrange analogous performance situations, or to combine them with the kind of monitoring tasks displayed in Figure 1, to sharpen our ability to detect adverse low-level exposure effects. Ozone and some other air pollutants offer additional experimental opportunities as well. Ozone is an odorless deep-lung irritant, and could serve as a test or control agent for situations in which odor is one of the stimulus properties under investigation.

Experimental design is not simply a matter of choosing exposure levels and response criteria. Species extrapolation is a major issue in risk assessment. It is not irrelevant, in another form, in pursuing questions about low-level chemical sensitivity. Hudnell et al. (1992) and Otto et al. (1992) tested 66 healthy young males in their experiment. Although such a sample may be an appropriate population for a preliminary study, most observers would express some doubts about the extending the results to the SBS problem. An appropriate experimental sample would consist of individuals who claim to be afflicted with the syndrome. And an appropriate experimental design would emphasize consistency of results in a small sample rather than statistical tests with a large sample.

A question related to the topic of this conference erupted in 1975 with the publication of a book asserting that certain dietary constituents, particularly food additives, could provoke the signs of hyperactivity (now called Attention Deficit Disorder) in sensitive children (Feingold, 1975). That assertion, known as the Feingold hypothesis, proposed the use of elimination diets as probes for sensitivity, and, if the results were positive, as countermeasures. Feingold singled out synthetic food dyes and artificial flavors for special attention, in part because of their lack of nutritional value. His claims provoked enormous public interest and eventually several experimental studies. I conducted one of them (Weiss et al., 1980), and the experimental design we chose for examining the Feingold hypothesis conveys more than a superficial message for the way the current low-level question is framed.

Feingold never claimed that all children were sensitive. His estimates, given at various times, asserted that perhaps 20% to 50% of hyperactive children might be responders. Our experiment took both a wide and narrow view of the hypothesis. It was narrow in two senses. First, because we decided to focus on whether the hypothesis could be confirmed at all, we chose as our subjects children whose parents believed they had benefitted from the Feingold elimination diet. Our strategy was to determine if consistent responses could be evoked in the individual children by selected food additives. Second, we decided to focus on food dyes because a blend of those approved by FDA could be managed and because several other studies had also focused on dyes. It adopted a wide point of view by excluding children who had been diagnosed as hyperactive. Instead, our subjects consisted simply of children whose parents believed them to respond adversely to certain foods. We excluded clinically hyperactive children because of the risk assessment implications of the Feingold hypothesis. Food additives are not tested for behavioral toxicity, so that positive findings in our sample would suggest that many more children would be subject to adverse behavioral effects than only those with the clinical syndrome. For this reason, we deliberately chose to underestimate the potential impact on clinically hyperactive children.

My collaborators (from the Kaiser system in the San Francisco area and from the Department of Nutritional Sciences at the University of California) and I recruited the parents of 22 children between three and seven years of age as participants. All the families were enrolled in the Kaiser system during the study. The parents actually became collaborators; over the 11-week duration of the study, they devoted themselves to the daily duties listed in Table 1.

Before the study began, each participating parent selected a list of ten items, from a pool derived from several child behavior inventories, that described their child's response to deviations from the elimination diet. These items, plus a standardized hyperkinesis scale, provided the bulk of the behavioral data.

TABLE 1. Parents' Daily Routine: Food Color Study

- Record sleep data
- Note infractions, medications
- Telephone report to staff
- Give soda pop at specified time
- Transfer pop labels to forms
- Put on wrist counter, record scores
- Perform 15-minute behavior counts
- Rate target behavior for total day
- Complete Conners form
- Complete diet inventory

Each day, at a prescribed time, the child consumed a bottle of soft drink especially prepared for the study. On 8 of these days, the drink contained a blend of food colors whose quantities we calculated as the mean daily intake for California children in Kaiser pediatric clinics. On all the other days, the drink contained caramel coloring and cranberry extract. The control and challenge drinks were indistinguishable either by color or taste. We devoted considerable effort to insuring such a placebo-controlled experiment because, in this kind of a study, any distinctive cue could trigger biased evaluations by the parents. We even made certain that the stains made on white napkins by the experimental and control drinks could not be discriminated; both left a similar red residue. A set of bottles was delivered weekly by our staff, and the identification codes were kept in a safe until the end of the experiment.

Two features of the design should be stressed. One is the selection of behavioral endpoints. In essence, each subject represented an individual experiment in what is called a single-subject design. This feature was crucial for the second aspect we deemed important: consistency of response, which we evaluated by the repeated-measures approach we adopted. We regarded replicability in a small sample as a more critical test of the hypothesis than single observations in a large sample.

We found two responders. One, a three-year-old boy, was a mild responder. The other, a 34-month-old girl, was a remarkable responder. Her results, described in Table 2, indicate that on challenge days she produced scores on five of the seven negative target behaviors that greatly exceeded her scores on control days. (The p-values were calculated by Randomization tests, which are extremely well-suited for single-subject designs.)

TABLE 2. Subject 73: Target Behaviors — Effects of Food Dye Challenge

• Short attention span	• $p < 0.0001$
• Acts as if driven by motor	• $p < 0.0001$
• Runs away	• $p < 0.0006$
• Throws and breaks things	• $p < 0.0003$
• Whines	• $p < 0.03$
• Clings to mother	
• Asks for help inappropriately	
• Affectionate	
• Friendly and outgoing	
• Kind to animals	

These findings do not stand alone. They are buttressed by other studies, each of which has found some segment of its sample to be responsive to the total elimination diet or to color challenge (reviewed in Weiss, 1982). A later, and more elaborate study by Kaplan et al. (1989), in which the investigators controlled the total diet, affirmed and extended these results. In general, those studies that concentrated on clinically hyperactive or ADD children tended to find a greater proportion of responders than our own.

The conclusions to be drawn from these data appear in Table 3. The table compares, for the food dyes used in Weiss et al. (1980), the Acceptable Daily Intake (ADI) prescribed by FDA with the amounts we and others have found to provoke adverse behavioral effects in children. The ADI is obtained on the basis of a two-year feeding study in rodents. The highest dose producing no detectable pathology is divided by 100 (a factor of 10 for species differences and a factor of 10 for individual differences). This value is considered to present no human health hazard. ADIs calculated in this way are about 50–60 times greater than those used in our study. It indicates the gap between traditional measures of safety and the variables that need to be assessed to confront the issue of low-level sensitivity.

TABLE 3. Food Dyes: Doses (mg/day) Eliciting Behavioral Responses vs. FDA Asseptable Daily Intake

Color	Behavior*	ADI
Yellow 5	9.07	300
Yellow 6	10.70	300
Red 40	13.80	420
Red 3	0.57	150
Blue 1	0.80	200
Blue 2	0.15	37
Green 3	0.11	150

*Based on Weiss et al., 1980.

This literature also holds significant implications for how we might go about investigating the validity of low-level chemical sensitivity syndromes. First, experimental studies should attempt to include subjects who claim to be victims of the syndrome. Healthy young men are appropriate if you wish to generalize to the population of healthy young men. Moreover, it is palpably unreasonable to assume that, in any such study, all of the subjects who believe themselves to be suffering from the particular syndrome will truly belong to such a group. Consider the statistical problems of dealing with a sample only a proportion of which is actually sensitive to a particular challenge or treatment.

Figure 2A portrays a population whose members consist of 30% responders and 70% nonresponders who cannot be distinguished beforehand. Figure 2B shows the impact on the two subpopulations after a challenge that shifts the responder distribution by one standard deviation. Figure 2C reveals what an investigator would see, namely, a minor and probably nonsignificant shift in the mean. Figure 2D plots the results of random samples of 15 drawn from the before and after treatment distributions. The overlap is daunting. Figure 3 discloses the number of subjects that would have to be studied to achieve p-values of 0.01, 90% of the time for various combinations of subpopulations and challenge-provoked shifts in the means.

What do this example and the food additive experiment tell us about investigative strategies? I interpret them as supporting designs in which the experimenters adopt repeated-measures designs that focus on response consistency. It will be more difficult to recruit participants for such studies, but I'm convinced the yield will be greater. I'm also convinced that a more natural testing environment, such as a synthetic workplace setting, is to be preferred to testing as part of an office visit. And, even if a synthetic workplace setting is not practical, why not standardize a set of criteria for the individual subject, much as we did for the children? Organized in the form of a diary, and even instrumented with an unobtrusive hand-held computer, such an approach could produce an enormous volume of useful information.

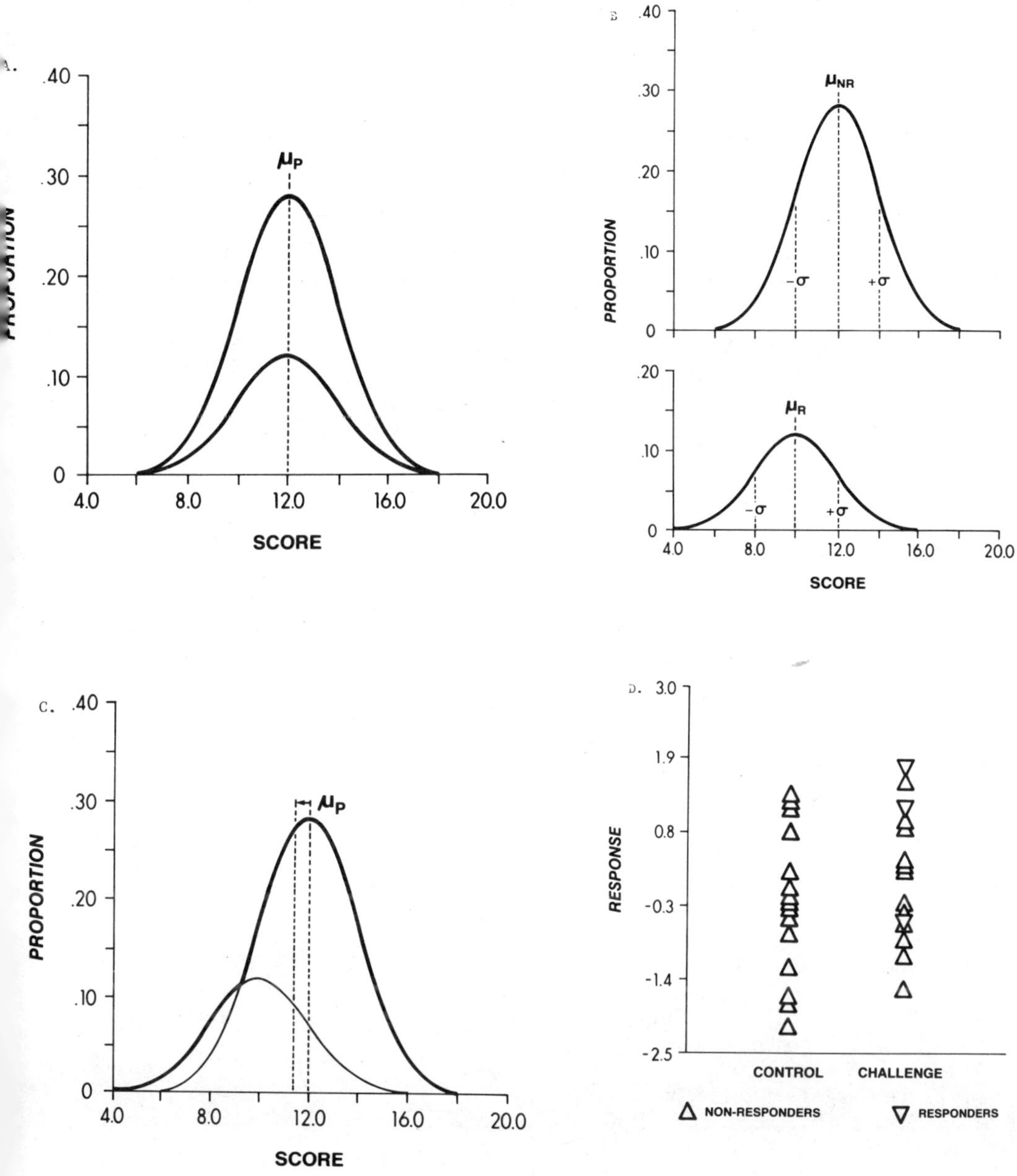

FIGURE 2. Confounding of susceptible and nonsusceptible subpopulations in a sample. 2A. Population distribution before chemical challenge showing 70% nonresponders and 30% responders. 2B. After chemical challenge, the scores of the susceptible portion of the sample are shifted by one standard deviation. 2C. The mean of the total distribution, containing both responders and non-responders, is shifted by a relatively minor amount. 2D. Randomly selected subjects from the hypothetical distribution. In a sample as small as 15 subjects, the contribution of the responders seems negligible.

PERCENTAGE OF RESPONDERS	RESPONSE AMPLITUDE ($\frac{d}{\sigma}$) 1.0	2.0	3.0
10	2102	~550	265
20	~550	~150	65
30	265	65	30
40	~150	~37	16
50	90	23	12
100	23	7	4

FIGURE 3. Cell entries represent the number of subjects, for each combination of response shift (given as standard deviation) and proportion of responders in the sample, to achieve a p-value of 0.01, 90% of the time.

REFERENCES

ALLUISI, E.A., BEISEL, W. R., BARTELLONI, P.J., and COATES, G.D. (1973). "Behavioral effects of Tularemia and Sandfly fever in man." J. Infect. Dis. 128:710–717.

AMERICAN MEDICAL ASSOCIATION, COUNCIL ON SCIENTIFIC AFFAIRS (1992). "Clinical Ecology." J. Am. Med. Assoc. 268:3465–2467.

FEINGOLD, B. F. (1975). Why Your Child Is Hyperactive. Random House, New York.

HUDNELL, H.K., OTTO, D.A., HOUSE, D.E., and MOLHAVE, L. (1992). "Exposure of humans to a volatile organic mixture. II. Sensory." Arch. Environ. Health 47:31–38.

KAPLAN, B.J., McNICOL, J.R., CONTE, A., and MOGHADAM, H.K. (1989). "Dietary replacement in preschool-aged hyperactive boys." Pediatrics 83:7–17.

NAITOH, P. and TOWNSEND, R.E. (1970). "The role of sleep deprivation research in human factors." Human Factors 12:575–585.

NATIONAL RESEARCH COUNCIL (1983). Risk Assessment in the Federal Government: Managing the Process. National Academy Press, Washington, D.C.

OTTO, D.A., HUDNELL, H.K., HOUSE, D.E., MOLHAVE, L., and COUNTS, W. (1992). "Exposure of humans to a volatile organic mixture. I. Behavioral assessment." Arch. Environ. Health 47:23–30.

WAYNE, W.S., WEHRLE, P.F., and CARROL, R.E. (1967). "Oxidant air pollution and athletic performance." J. Am. Med. Assoc. 199:151–154.

WEISS, B. (1982). "Food additives and environmental chemicals as sources of childhood behavior disorders." J. Am. Acad. Child Psychiat. 21:144–152.

WEISS, B. (1989). "Behavior as an endpoint for inhaled toxicants." In: Concepts in Inhalation Toxicology (R. McClelland and R. Henderson, ed.). Hemisphere Publishing Co., Washington, D.C. pp. 493–513.

WEISS, B. and LATIES, V.G. (1962). "Enhancement of human performance by caffeine and the amphetamines." Pharm. Rev. 14:1–36.

WEISS, B., WILLIAMS, J.H., MARGEN, S., ABRAMS, B., CAAN, B., CITRON, L.J., COX, C,. McKIBBEN, J., OGAR, D., and SCHULTZ, S. (1980). "Behavioral responses to artificial food colors." Science 207:1487–1488.

PATIENT COMMENTS PANEL

Mary Lamielle, National Center for Environmental Health Strategies: I was interested in the idea of studying sites where carpet is installed or where pesticide applications are taking place. What obligation is there going to be on the part of the researchers if they see somebody being affected in those situations? We do want to see who is getting sick, and how they get sick, but don't we have some moral and ethical obligations there?

The second thing I want to add, in part with regard to Dr. Simon's presentation yesterday, is that we deal with a lot of folks who are in the initial stages of getting sick. Someone might call after only a week or so in a newly carpeted building and report that they are starting to feel symptoms. What I see is that folks who get out of a situation early, before they have become chronically ill or disabled, don't end up in the same anxiety state or have the major anxiety symptomatology that seems to develop in people who become chronically ill.

And third, a question: When we get to blind studies and double-blind studies, what kinds of chemicals or exposures are we talking about? What can be blinded?

Dr. Rosenthal: It's clear we don't really have the proper data base to develop an informed scientific opinion, but that's the case with most of what we do in life. When I advise patients, I think, "What would I advise if it were my own son?" I would say, "Get the heck out of there, as soon as possible, if possible." Obviously, if you are going to fall into bankruptcy, you have to weigh the circumstances. When we see the disabled people around us, and we think it may be possible to prevent some of this from happening, and we see somebody developing cacosmia and becoming sick, I don't think we can afford to wait for the scientific data base to be in place to make our recommendations. If it was my kid, I'd say, "Get out," and I'd do everything I could to get him out, and if it was my patient I would do the same.

Dr. Miller: One substance that might be used at a subolfactory level is galaxalide, a musk-like fragrance you can use at low concentrations. It influences EEG activity even in normals, and affects the response to a visual tracking paradigm, so it decreases the ability of a person to perform on a visual tracking test; another is propylene glycol. You have to be aware that different individuals may respond to different substances. You will want to try both an open challenge and a blinded challenge to be sure you are getting the effect.

I want to address the ethical consequences. I feel obligated to let patients know what has happened to some people in certain exposure circumstances, and that other people have been fine in those same circumstances. I can't predict for any individual what will happen. I can provide information. If they seek help from some of the patient support groups, I think that's

Toxicology and Industrial Health, Vol. 10, No. 4/5, pp. 619–621

ISSN: 0748-2337

one way as a physician that you can keep from having to make the decision. We're working with an illness that has no known markers and no case definition, and that seems pretty shaky to some people, and I understand that.

Dr. Bell: I wanted to add a methodological comment about the choice of substances. We have looked at galaxalide as well as butanol in our chemically sensitive students, and the biggest effect was actually with the galaxalide. It was a very minor effect with the butanol. Butanol is a common alcohol in indoor air. There is some need for preliminary research for what are the target chemicals. In his work, Dr. Simon identified Decane as one of the irritants in sick building syndrome.

Jacob Berkson: I want to emphasize that prevention is the key. Prevention of pollution will save a lot of people from being ill. Of course, we can't prevent all the pollution, so we have to try to prevent exposure to the pollution. We do that mostly by avoidance.

Bill Hersey, president of the union representing professionals at EPA Headquarters where we had this problem of toxic carpet: First, from the victims' standpoint, it's important to translate what's happened here into victim assistance as rapidly as possible. Our people have had a very difficult time getting workmen's compensation and getting adequate medical treatment, and the Union has heard from many victims across the country with the same problems. Secondly, I would urge the utmost caution about using a new carpeting exercise as some sort of prospective epidemiology study. Our experience was very bad in that regard, people got very sick, some very suddenly, some over a very long period of time, and it could conceivably be that they were involved in some sort of time-sensitization process. The carpet industry has undergone some dramatic changes since 1987–1988, when we had our problem, so before you invest a lot of money in a study, be sure you have a carpet that will induce illness.

Susan Clark, Environmental Advocacy League: A few linguistic points: First, the word "patient." Most of us exist outside the clinical environment most of the time, so the use of the term "patient" is inaccurate. It is often used as a demeaning term. For example, someone earlier thought patients might think we actually had holes in our brains. I think that represents some misunderstanding and perhaps even some classism. Next, I'd like to say that multiple chemical sensitivity is an unfortunate term that we are perhaps stuck with, but it merely takes what is a symptom and elevates it to the level of a title or diagnosis. The reporting of chemical reactivity, which is the reporting of a symptom, is not a self-diagnosis. It is on the level of reporting a headache or a stomachache. I would like to thank Iris Bell for pointing out the linguistic problems with the word cacosmia, that odor perception is not necessary for reaction. I think that as victims are better listened to those kind of linguistic problems will be solved before they begin. Thank you for the discussion of medical ethics. Above all, we want to do no harm and we want to first pursue physiological causes before assuming psychogenesis. That means we have a certain side we need to err on, and that is the side of avoiding toxic exposure, and promoting prevention.

Finally, I want to ask whether the supposedly high incidence of affective disorders in victims may not actually be unstable over time. That is, they may vary in proportion to the number of good research studies that appear per year, or the number of researchers who speak at conferences with integrity and who have read their studies before they get up to the podium, or who have seen patients in a clinical setting before they presume to speak about the subject; and also the validation of disability benefits, and so on.

Steven A. McFadden, Independent Research Advocates: The federal toxicologists at NIEHS and EPA Health Effects Research Lab are insufficiently funded to do the research necessary to protect human health.

With respect to federal toxicology policy, consider pesticide policy. Under FIFRA, EPA can only regulate pesticides to the extent necessary to prevent unreasonable adverse affects on human health and the environment. If the clinical significance of EEG changes from pesticide poisoning was, say, Attention Deficit Disorder, there would certainly be societal risks. Yet, until the effects are documented, EPA cannot regulate. Without toxicity tests on the chemicals up front, with little epidemiology, and little funding for research on MCS, there are no documented effects, so EPA cannot regulate.

I also wanted to bring up a quote by Sheldon Samuels: "Failure to consider sensitive subpopulations is social, as well as natural selection against those individuals." It seems to me that the kindling hypothesis of MCS is visionary. We want to thank Iris Bell and Claudia Miller and others, for that point of view. But kindling would seem to be an effect that is generic to at least mammalian nervous systems, and probably more than that. This is an end point which is also significant to the more general population. It's time to put a research agenda on the table in order to deal with it and improve this situation. If we figure out what research can be done, we can figure out how to fund it.

Rosalind Taratoot: My comment is that I interface with a lot of other support groups, such as the Lupus Foundation, Learning Disability, Chronic Fatigue, Arthritis Foundation, etc. There seems to be a great increase in all of these autoimmune-type dysfunctions in people, and they are occurring at younger and younger ages. This obviously is a concern to me, because I have children and grandchildren. I'd like to see us develop multidisciplinary studies. I've heard many good papers presented. From my personal poison-exposure history, I could fit into almost any research category discussed at this meeting. I would like to see people work together to come up with a coordinated, concerted effort to reach the answers. I don't think any one paper or one explanation is the answer.

MULTIPLE CHEMICAL SENSITIVITY: LESSONS FROM SEASONAL AFFECTIVE DISORDER

NORMAN E. ROSENTHAL

Section on Outpatient Psychiatry Clinical Psychobiology Branch
National Institutes for Mental Health
Bethesda, Maryland

In multiple chemical sensitivity (MCS) we see a condition in its early phases of research and treatment. As yet no case definition has been agreed upon by all workers and researchers. It is not codified in any official manual of medical or psychiatric conditions, and some clinicians and researchers question whether it is even a legitimate medical condition. All of these circumstances existed approximately fifteen years ago with respect to another clinical condition, seasonal affective disorder (SAD) (Rosenthal et al., 1984). Since then, however, the existence of SAD has been widely accepted, criteria for its diagnosis have been developed, and it has been included (albeit in the form of an adjectival modifier, "seasonal pattern") in the standard manual for psychiatric diagnoses, the DSM-III-R (American Psychiatric Association [APA], 1987) and now, the DSM-IV (APA, 1994).

The two conditions, MCS and SAD, share three important features. First, they are both diagnoses made largely on the basis of patient reports, corroborated to some extent by collateral observers or objective evidence of illness patterns. Second, they are both conditions in which symptoms are reportedly elicited in vulnerable individuals by exposure to specific environmental stressors — chemicals in the case of MCS and light deprivation in the case of SAD. Finally, both have been recognized for some time and have moved relatively recently into the public eye and indeed, the awareness of clinicians. SAD has certainly been recognized in one form or another for hundreds of years (Rosenthal, 1993b), MCS at least for decades (Miller, 1994). It would not be surprising, however, to find case descriptions of MCS going back hundreds of years if the appropriate literature were properly combed.

Despite their similarities, the two conditions differ in certain important respects. In the case of SAD, a specific treatment, namely light therapy, was recognized at about the same time as the syndrome was first described; as yet there is no established treatment for MCS. Second, seasonal rhythms are part of the natural world and there is a rich literature on naturalistic and experimental studies of seasonal rhythms in animals on which researchers have been able to

1. Address all correspondence to: Norman E. Rosenthal, M.D., Chief, Section on Outpatient Psychiatry Clinical Psychobiology Branch, NIMH, 9000 Rockville Pike, Bethesda, MD 20892.

Toxicology and Industrial Health, Vol. 10, No. 4/5, pp. 623–632

ISSN: 0748-2337

draw. Although there are basic studies that clinical researchers can refer to in an attempt to understand the pathogenesis of MCS, these seem at present somewhat removed for theclinical phenomenon of MCS. Finally, in the case of SAD, no one can be blamed or held liable; darkness is a feature of our natural landscape. This is not so in the case of MCS, where someone can be held liable for the injury or injuries. The financial implications of such liability could be interpreted as a reason for the victim of such an injury to exaggerate its impact. Alternatively it could be construed as a reason for those with a financial interest at stake to minimize the severity of a patient's syndromes, dismiss them as being "psychological" in nature, or even dismiss the existence of MCS altogether. Given the high stakes involved, it should come as no surprise that the heated controversy about MCS is greater than one might expect to surround a topic of purely academic interest.

Bearing in mind both the similarities in the two conditions and their differences, it might be a useful exercise to evaluate what we have learned about SAD in the past fifteen years that may be of relevance in our approach to patients with MCS.

THE VALIDATION OF MCS AS A SYNDROME

For centuries, psychiatrists have been confronted with having to classify symptom clusters of unknown etiology, and guidelines have been established for validating such clusters (Spitzer and Williams, 1980). Four different types of validity have been recognized: (1) face validity; (2) descriptive validity; (3) predictive validity; and (4) construct validity. These types of validity have been applied to the diagnosis of SAD by Spitzer and Williams (1989) and can similarly be applied to MCS.

Face Validity

Face validity is related to the degree to which experts in the field consider a syndrome to be a distinct clinical entity. Despite great disagreement as to the underlying causes of MCS, no-one disputes the existence of patients who regard themselves as disabled and hypersensitive to chemicals as a result of prior chemical exposure. In an analogous fashion, there was general agreement that patients with SAD reported that they felt worse in the winter, though there was some question as to whether this was indeed a legitimate subcategory of depression. Skepticism about the legitimacy of SAD was perhaps articulated most eloquently by Mrosovsky (1989), who suggested that SAD patients might be seeing "chipmunks in the sky," a nonexistent pattern created in their imagination. In an analogous way, some researchers have suggested that patients with MCS, while perhaps suffering from genuine distress, might misattribute this to innocuous chemicals (Staudenmayer and Selner, 1990; Simon et al., 1993; Bell et al., 1994a,b). Time has borne out the legitimacy of SAD as a valid condition, which is now recognized by the mainstream of medical establishment (Rosenthal, 1993a). It has yet to be determined whether MCS is also a valid condition worthy of recognition.

Descriptive Validity

Descriptive validity refers to the extent to which defining features of an illness are unique to that illness. As can be seen from the article by Miller (1994) in the present monograph, many symptoms of MCS are not unique to that illness. It could be argued, however, that a history of chemical exposure, accompanied by subsequent hypersensitivity to multiple chemical stimuli, *is* somewhat specific. Together with the other less specific symptoms of the condition, these features would appear to provide a reasonable degree of descriptive validity even though, as Miller and others would have noted, a clear-cut history of a single chemical exposure is not always present and subjective hypersensitivity is not unique to MCS patients. Nevertheless, it would seem that a description of MCS could be found and criteria developed that would be as specific, reliable, and easy to administer as those of many other disorders where diagnosis is based primarily or exclusively on the clinical picture.

Descriptive validity is not an all-or-none phenomenon and even if some cases cannot be diagnosed with confidence, that does not invalidate the concept of MCS. The field will have to decide the degree to which an individual needs to exhibit features of the condition in order to meet diagnostic criteria. As detailed by Miller (1994) in her accompanying article, criteria for diagnosis of MCS have been suggested. It is important for the field to agree on a set of criteria not only for the purpose of research and treatment, but also for reasons of general acceptance of the condition by those outside the inner sanctum of experts and by non-medical establishments, such as insurance companies and the courts. The initial criteria of Rosenthal et al. (1984) for SAD have been widely used by researchers and have provided a basis for comparison of clinical populations. The establishment of a widely accepted set of criteria for MCS would serve a similarly useful function and should be regarded as a priority for those working in the field.

A review of the currently available criteria for MCS suggests that they could benefit from theinclusion of severity criteria — items referring to duration and repercussion of the condition. Such criteria exist, for example, in most classifications of major depression, which require that an individual has suffered the symptom of the depression for at least two weeks, with significant functional consequences of the condition (Spitzer et al., 1978; APA, 1987). Another important element in the diagnosis of psychiatric disorders is the historical course of the condition. The diagnosis of SAD, for example, relies heavily on prior course, in that depressions are required to have occurred in a specific relationship to the seasons. A famous example where course provided a critical diagnostic distinction is that of Kraepelin's division of the major psychotic illnesses into "dementia precox" (schizophrenia) and "manic-depressive insanity" (bipolar disorder) on the basis of the deteriorating course of the former and the relapsing and remitting course of the latter (Kraepelin, 1921). Current diagnostic criteria for MCS do not specify the course of the condition and probably should. It would seem that a person with good premorbid functioning who suffered severe functional deterioration with symptoms of MCS following a chemical exposure would provide a more clear-cut diagnostic example than someone whose functioning has always been impaired.

That is not to say that the latter individual could not develop MCS; merely that it would be harder to diagnose such a case.

Predictive Validity

The course of a condition is relevant in the next type of validity to be considered — predictive validity. Predictive validity refers not only to one's capacity to make predictions about the natural history of a diagnostic entity, but also about the likelihood of its response to specific treatments. To my knowledge, no studies have been published to date on the long-term course of MCS. No specific effective treatments have been developed for the condition, although avoidance of noxious stimuli has been suggested by some to be helpful (Ashford and Miller, 1989). In the case of SAD, multiple studies of light treatment have been undertaken (Terman, et al. 1989; Oren and Rosenthal, 1992; Rosenthal, 1993a) and these have been regarded as very helpful in validating SAD as a syndrome (Spitzer and Williams, 1989) Furthermore, long-term studies have been performed and the persistence of seasonal symptoms over time has been documented (Schwartz, unpublished observation). Studies of clinical course and treatment response would clearly be helpful in furthering our understanding of MCS as a diagnostic entity.

Construct Validity

The final type of validation of a diagnostic entity is construct validity. This involves understanding the pathophysiological basis of a condition. Evidence of causation or biological abnormalities of relevance to the hypothesized basis of a condition provide construct validation of a syndrome. In the case of SAD, the concept of seasonality is inherent in the symptoms of the condition, their presumed biological underpinnings, and the use of light therapy to treat the condition. Researchers in the field of MCS are grappling with trying to understand how a chemical injury to the brain could be related to subsequent hypersensitivity to exposure to multiple chemical stimuli, often in the absence of demonstrable neuropsychological damage.

In an accompanying paper in this monograph, Bell (1994) has noted different paradigms suggested by various researchers for understanding MCS. These can be divided into two main groups — biological mechanisms and psychological mechanisms. Biological mechanisms include time dependent sensitization (Antelman, 1988) and kindling (Bell et al., 1992), while psychological mechanisms include misattribution (Staudenmayer and Selner, 1990; Simon et al., 1993; Terr, 1993) and projection of psychological conflicts (Staudenmayer and Camazine, 1989). A parallel dichotomy existed in attempts by clinicians to understand SAD long before it was formally described (Rosenthal, 1993b). Thus, in 1946 two different clinicians working on opposite sides of the Atlantic both had dealings with patients who, in retrospect, suffered from SAD. Their interpretations of their patients' symptoms and the treatments they administered were, however, quite different, reflecting their different conceptualization of their patients' conditions.

A German physician, Dr. Helmut Marx (1946), observed four men who became depressed following exposure to the dark winter days of an Arctic winter. Reasoning that light deprivation might be the cause of their symptoms, he exposed them to bright light and reported good results. In the same year, Dr. George Frumkes (1946), a U.S. psychiatrist, described SAD in a young man. His clinical description was precise enough to allow us to diagnose the patient's problem easily in retrospect. His explanation for the man's problem, however, was heavily influenced by Freudian analytic theories that were popular at the time. According to Dr. Frumkes, the patient's depressions

> began about the time he learned that masturbation was not a unique sin of his; when there was a decrease in the intense, conscious feeling of guilt. The depressions represented a redistribution of the punishment in the psychic economy . . . Masturbation was for him an unconscious infantile sexual striving for his mother, and the associated hostile impulses connected with this drive.

Frumkes also suggested that the depression might have represented "memorial observances of the births of his brothers and sisters." He treated the patient with conventional psychoanalysis and reported that "as treatment progressed, the depressions diminished in regularity and intensity."

Such a psychodynamic formulation of SAD might seem quaint nowadays. Nevertheless, in the past decade, we have seen psychodynamic explanations of similar vintage applied to patients with MCS. Staudenmayer and Camazine (1989), for example, observe that patients with MCS, whom they term "Universal Reactors" are "the latest individuals in a long list of those who historically manifested psychosomatic illness." They note that "in his personality theory, Carl G. Jung presented characteristics of the sensing type which describe Universal Reactors."

> What comes from inside seems to him morbid and suspect. He always reduces his thoughts and feelings to objective causes, to influences emanating from objects, quite unperturbed by the most glaring violations of logic. Once he can get back to tangible reality in any form, he can breathe again. In this respect he is surprisingly credulous. He will unhesitatingly connect psychogenic symptoms with a drop in the barometer, while on the other hand the existence of a psychic conflict seems to him morbid imagination (Jung, 1971)."

These researchers go on to suggest that "Universal Reactors project their problems onto the environment 'since they substitute for an internal instinctual danger an external perceptual one.' (Freud, 1936)."

The problem with such psychodynamic formulations is that they are easier to construct than to test. In fact, they are almost impossible to test. Even in studies that show patients with MCS to have a tendency to projective defenses, such as Staudenmayer and Camazine, have reported, who can say whether these tendencies are a cause, a result, or an epiphenomenon of MCS?

Elsewhere, Staudenmayer and Selner (1990) emphasize the importance of "discriminating verifiable medical disease from psychosomatic illness." This statement begs the question as to what constitutes verification. In many medical conditions, diagnosis is easily verified. This is not the case with psychiatric conditions. Many, if not most, patients with schizophrenia, mood disorders, obsessive compulsive disorders, and anxiety disorders will show no specific findings on a multitude of objective tests. Yet their conditions are deemed legitimate by most of the medical profession, perhaps because of their time-honored status. It does not seem reasonable, though, to hold conditions such as SAD or MCS to a different standard simply because they have been described more recently.

Various authors have undertaken biological investigations in MCS patients but after finding no specific abnormalities, have turned to psychological explanations (Staudenmayer, 1990; Simon, 1993; Terr, 1993). I would suggest that such turning away from biological mechanisms may be premature. Perhaps it would be more prudent to question the status of our technology and be suitably modest about our current ability to rule out a pathophysiological basis for our patients' conditions. Elsewhere in this monograph, Bell (1994) has reviewed possible reasons for a failure to find specific abnormalities in patients with MCS, and she and Miller (1994) have suggested strategies to overcome some of the shortcomings of earlier studies, specifically the use of challenge studies following a period of withdrawal from the putative toxic substances.

In summary, it would seem that tough testable constructs have been advanced to help validate MCS as a condition. Future research efforts should be directed towards such validation.

THE PSYCHOLOGY OF DIAGNOSIS

Patients with MCS believe that their condition was caused by exposure to environmental chemicals and is exacerbated by subsequent exposure. To some professionals such an explanation appears credible, despite the absence of any special tests that validate the history. Other professionals, having evaluated the evidence, doubt that symptoms are the result of true responses to toxins but see them rather in psychological terms. Presumably even this latter group would acknowledge that psychological responses have their basis in neurochemistry, so the two groups might be less polarized than would appear at first glance to be the case.

Yet the psychological impact of these divergent views in a therapist must surely lead to very different responses on the part of the MCS patient. Such patients are likely to see the clinical ecologist as an ally who validates their perceptions of their condition. While such a sense of validation might be therapeutic in its own right, it does not necessarily imply that the treatments used by clinical ecologists are specifically effective. The skeptical therapist, on the other hand, essentially says to the patient "You are mistaken in thinking that your problems are caused by chemicals. They are actually due to your own psychological problems." Such a formulation is likely to be heard as an unempathic "blaming of the victim" and result in disaffection on the part of the patient for whom it is of little consolation to be told, "Even though you are mistaken in your beliefs, I acknowledge that your suffering is real."

Given that we don't know what causes MCS and that our technology may well be inadequate for determining subtle forms of neurophysiological abnormalities in MCS patients, it would seem not only kinder, but also fairer, to entertain seriously that the patient may be correct and that the injury, however subtle and elusive at this time, may nonetheless be real. While different standards of proof may be required for special purposes, such as determination of liability, an open-minded approach in our therapeutic dealings with MCS patients would appear to be appropriately respectful of their suffering and our ignorance.

THE POLITICS OF DIAGNOSIS

Diagnoses are not actual entities, like the earth and the planets, but rather human artifacts, like countries, the boundaries of which are sometimes drawn by consensus but perhaps more often haggled or fought over. In the case of SAD, the diagnosis was fairly readily accepted into DSM-III-R, but only as "seasonal pattern," an adjectival modifier of any recurrent form of mood disorder. The acceptance of a diagnosis by the medical community confers an important status to a medical condition. It makes it easier to seek third party reimbursement, lends weight to claims of disability or liability, and assists in obtaining research grants, though in itself it doesn't guarantee success for any of these ventures.

A good illustration of the politics of diagnosis was apparent in the treatment of a condition that may be closely allied to MCS — the Gulf War Syndrome. As reported by Miller (1994) in this monograph, large numbers of Gulf War veterans have reported symptoms reminiscent of MCS. Though the cause of these symptoms cannot be determined with certainty, it is interesting to note that many of these patients had been exposed to some of the same types of chemicals that have been reported to cause MCS, especially the fumes of burning oil wells and pesticides. The syndrome has been greeted with skepticism by many, including those in the medical profession. After all, there is no appropriate diagnostic category into which it could be classified. Some regard it as a variant of post–traumatic stress disorder, though reports of civilians with similar problems challenge this formulation. Finally, now that a chemical explanation has surfaced, the emphasis is on exposure to exotic chemical weapons rather than familiar domestic chemicals.

Following a great deal of publicity and lobbying, a decision was made to regard the condition as service-related despite the continuing uncertainty about its exact nature and etiology. That decision — appropriate, in my view — is a good example of a direct path from symptoms to compensation without consensus about diagnosis or pathophysiology, or any general agreement as to the construct behind the condition. In principle, however, there is no essential difference between our status of knowledge about the Gulf War Syndrome and MCS, especially in those cases where a clear exposure and subsequent deterioration in function can be documented. So the essential difference between the status of the Gulf War Syndrome and MCS from the standpoint of compensation is a political one. Thus, it seems clear that acceptance of MCS as a legitimate diagnosis, together with all the benefits that attend such an officially recognized condition, will result not only from scientific endeavors, but also from political efforts.

CONCLUSION

As with SAD, the condition of MCS has been well described by multiple clinicians of different theoretical orientations, working in diverse and scattered locations over the past several decades. Severe cases do not appear to be uncommon and the condition may be even more widespread in its less florid forms. A great deal of work needs to be done to do justice to those who suffer from this serious and disabling condition. We need to agree on diagnostic criteria, study its prevalence and expand our research attempts to understand and treat the condition.

In researching the condition, there will always be debate about the role of psychological factors in the reporting of symptoms. We will need clever strategies to circumvent such ambiguity. Blind challenges have already been undertaken to some extent and seem eminently worthwhile, though the question of withdrawal from environmental toxins prior to the challenge has been raised as an important methodological consideration before negative conclusions can be reached (Bell, 1994). One way to exclude the input of the cerebral cortex altogether might be to study the effects of toxic chemicals on nasal mucosal cultures of affected individuals and healthy controls (Wolozin et al., 1993). Animal models specifically tailored to the type of exposures that reportedly sensitize and later trigger the symptoms of MCS would also appear to be a valuable approach.

In trying to research MCS, however, we are in a catch-22 situation. It is difficult to attract research money for a controversial condition and it is difficult to resolve such controversy without the necessary research. The current ATSDR meeting is an excellent step towards publicizing the importance of the problem and bringing together experts in the field. Further steps in the form of obtaining research dollars for first-rate science in the area should follow.

REFERENCES

AMERICAN PSYCHIATRIC ASSOCIATION (1987). Diagnostic and Statistical Manual of Mental Disorders. (3rd ed. rev.) Rosenthal, Washington, D.C.

AMERICAN PSYCHIATRIC ASSOCIATION (1994). Diagnostic and statistical manual of mental disorders. (4th ed. rev.) Rosenthal, Washington, D.C.

ANTELMAN, S.M. (1988). "Time-dependent sensitization as the cornerstone for a new approach to pharmacology: Drugs as foreign/stressful stimuli." Drug Dev. Res 14:1–30.
ASHFORD, N.A. and MILLER, C.S. (1989). Chemical Sensitivity. (A report to the New Jersey State Department of Health).
BELL, I.R. (1994). "Neuropsychiatric aspects of sensitivity to low-level chemicals: A neural sensitization model." In: Proceedings of the Agency for Toxic Substances and Disease Registry Conference on Low-Level Exposure to Chemicals and Neurobiological Sensitivity (F. Mitchell, ed.). ATSDR, Baltimore. Toxicol. Ind. Health 10:277–312.
BELL, I.R., MILLER, C.S. and SCHWARTZ, G.E. (1992). "An olfactory-limbic model of multiple chemical sensitivity syndrome: Possible relationship to kindling and affective spectrum disorders." Biol. Psychiatry 32:218–242.
BELL, I.R., MILLER, C.S., SCHWARTZ, G.E., PETERSON, J.M., and AMEND, D. (1994a). "Neurobehavioral and somatic characteristics of nondisabled young adults with cacosmia." Submitted.
BELL, I.R., SCHWARTZ, G.E., AMEND, D., PETERSON, J.M., and STINI, W.A. (1994b). "Sensitization to early life stress and response to chemical odors in older adults." Biol. Psychiatry. In press.
BLACK, D.W., RATHE, A., and GOLDSTEIN, R. B. (1990). "Environmental Illness." J. Am. Med. Assoc. 264:3166–3170.
BOLLA-WILSON, K., WILSON, R., and BLEECKER, M.L. (1988). "Conditioning of physical symptoms after neurotoxic exposure." J. Occup. Med. 30:684–686.
FREUD, S. (1936). The Problem of Anxiety. (H.A., Turner, trans.). Psychoanalytic Quarterly Press, New York.
FRUMKES, G. (1946). "A depression which recurred annually." Psychoanalytic Q. Press 15:351–364.
JUNG, C. G. (1971). Psychological Types. (R.F.C. Hull, trans.). Princeton University Press, New York.
KRAEPELIN, E. (1921). In: Manic-depressive Illness and Paranoia (G.M. Robertson, E. Livingstone, and S. Livingstone, ed.). E&S Livingstone, Edinburgh.
MARX, H. (1946). "HypophysÑre Insuffizienz," bei Lichtmangel. Klin Wochensch 24/25:18–21.
MILLER, C.S. (1994). "Chemical Sensitivity: History and phenomenology." In: Proceedings of the Agency for Toxic Substances and Disease Registry Conference on Low-Level Exposure to Chemicals and Neurobiological Sensitivity (F. Mitchell, ed.). ATSDR, Baltimore. Toxicol. Ind. Health 10:253–276.
MROSOVSKY, N. (1989). "Seasonal affective disorder, hibernation, and annual cycles in animals: Chipmunks in the sky." In: Seasonal Affective Disorder (N.E. Rosenthal and M.C. Blehar, ed.). Guilford Press, New York. pp. 127–147.
OREN, D.A. and ROSENTHAL, N.E. (1992). "Seasonal affective disorder." In: Handbook of Affective Disorders (E.S. Paykel, ed.). Churchill Livingstone, Edinburgh. pp. 551–567.
ROSENTHAL, N.E. (1993a). "Diagnosis and treatment of seasonal affective disorder." J. Am. Med. Assoc. 270(22):2717–2720.
ROSENTHAL, N.E. (1993b). Winter Blues. Guilford Press, New York.
ROSENTHAL, N.E., SACK, D.A., GILLIN, J.C., LEWY, A.J., GOODWIN, F.K., DAVENPORT, Y., et al. (1984). "Seasonal Affective Disorder: A description of the syndrome and preliminary findings with light therapy." Arch. General Psychiatry 41:72–80.
SIMON, G.E., KATON, W.J., and SPARKS, P.J. (1990). "Allergic to life: Psychological factors in environmental illness." Am. J. Psychiatry 147:901–906.
SIMON, G.E., DANIELL, W., STOCKBRIDGE, H., CLAYPOOLE, K., and ROSENSTOCK, L. (1993). "Immunologic, psychological, and neuropsychological factors in multiple chemical sensitivity. A controlled study." Ann. Internal Med. 19:97–103.
SPITZER, R.L., ENDICOTT, J., and ROBINS, E. (1978). "Research diagnostic criteria: Rationale and reliability." Arch. General Psychiatry 35:773–782.
SPITZER, R.L. and WILLIAMS, J.B.W. (1980). "Classification of mental disorders and DSM-III." In: Comprehensive Textbook of Psychiatry (H. Kaplan, A. Freedman, and B. Sadock, ed.). Williams and Wilkins, Baltimore. pp. 1035–1072.
SPITZER, R.L. and WILLIAMS, J.B.W. (1989). "The validity of SAD." In: Seasonal Affective Disorder (N.E. Rosenthal and M.C. Blehar, ed.). Guilford Press, New York. pp. 79–86.

STAUDENMAYER, H. and CAMAZINE, M. (1989). "Sensing type personality, projection and universal 'allergic' reactivity." J. Psychol. Type 18:59–62.

STAUDENMAYER, H. and SELNER, J.C. (1990). "Neuropsychophysiology during relaxation in generalized, universal 'allergic' reactivity to the environment: A comparison study." J. Psychosomatic Res. 34(3):259–270.

TERMAN, M., TERMAN, J.S., QUITKIN, F.M., McGRATH, P.J., STEWART, J.W., and RAFFERTY, B. (1989). "Light therapy for Seasonal Affective Disorder: A review of efficacy." Neuropsychopharmacology 2(1):1–22.

TERR, A.I. (1993). "Immunological issues in 'Multiple chemical sensitivities'." Regulatory Toxicol. Pharmacol. 18:54–60.

EEG RESPONSES TO LOW-LEVEL CHEMICALS IN NORMALS AND CACOSMICS

GARY E. SCHWARTZ,*†‡ IRIS R. BELL,*‡ ZIYA V. DIKMAN,* MERCEDES FERNANDEZ,* JOHN P. KLINE,* JULIE M. PETERSON,‡ AND KENNETH P. WRIGHT*

***Department of Psychology**

†Department of Neurology

‡ Department of Psychiatry

University of Arizona
Tucson, Arizona

Recent studies from the University of Arizona indicate that normal subjects, both college students and the elderly, can register the presence of low-intensity odors in the electroencephalogram (EEG) in the absence of conscious awareness of the odors. The experimental paradigm involves subjects sniffing pairs of bottles, one containing an odorant (e.g. isoamyl acetate) dissolved in an odorless solvent (water or liquid silicone), the other containing just the solvent, while 19 channels of EEG are continuously recorded. For the low-intensity odor conditions, concentrations are adjusted downward (decreased) until subjects correctly identify the odor bottle at chance (50%). The order of odorants, concentrations, and hand holding the control bottle, are counterbalanced within and across subjects. Three previous experiments found that alpha activity (8–12 hz) decreased in midline and posterior regions when subjects sniffed the low-intensity odors. The most recent study suggests that decreased theta activity (4–8 hz) may reflect sensory registration and decreased alpha activity may reflect perceptual registration. In a just completed experiment involving college students who were selected based on combinations of high and low scores on a scale measuring cacosmia (chemical odor intolerance) and high and low scores on a scale measuring depression, cacosmic subjects (independent of depression) showed greater decreases in low-frequency alpha (8–10 hz) and greater

1. Address all correspondence to: Dr. Gary E. Schwartz, Department of Psychology, University of Arizona, Psychology 312, Tucson, AZ 85721. Tel.:(602)621-5497. Fax:(602)621-9306.

2. Abbreviations: electroencephalogram (EEG).

3. Key words: cacosmia, EEG, low-level chemicals, multiple chemical sensitivity, olfaction, sensitization, subliminal.

Toxicology and Industrial Health, Vol. 10, No. 4/5, pp. 633–643

ISSN: 0748-2337

increases in low-frequency beta (12–16 hz) to the solvent propylene glycol compared to an empty bottle. Topographic EEG mapping to low-intensity odorants may provide a useful tool for investigating possible increased sensitivity to specific chemicals in chemically sensitive individuals.

INTRODUCTION

The olfactory system is an exquisitely sensitive sensory system. Although individuals vary in the minimal concentrations necessary for the conscious registration of a given odor, the olfactory receptors are capable of detecting individual molecules and generating corresponding neural signals. Most chemoreceptors within the body do not generate neural signals that can be experienced consciously (e.g., perceptual awareness has not been reported for carbon dioxide receptors in the cardiovascular system). It is reasonable to hypothesize that the olfactory system may register a broader range of intensities and molecules than is typically experienced consciously. Early research at Yale University suggested that subthreshold concentrations of an odorant could be registered in the EEG (Lorig et al., 1990), and recent studies from Washington and Lee University indicate that behavioral and cognitive effects can be observed for undetected odors as well (Lorig, 1994). Recent research from the University of Arizona (Schwartz et al. 1992a,b,c, 1993, 1994a,b) suggests that subthreshold concentrations of specific odorant molecules can be detected in the EEGs of normal college students and the elderly, and that subjects showing a specific anosmia for androstenone nonetheless register the presence of androstenone in their EEGs.

There are numerous implications of these findings for basic and applied research in the areas of olfaction, the study of conscious and unconscious processes, and chemical sensitivity. One major implication and advantage of using low-intensity odors with populations claiming to be chemically sensitive is that this procedure removes expectancy and potential "placebo" explanations from the findings that are obtained. Low-intensity odors make it possible to avoid using "masking" odors to test for chemical sensitivity. In addition, low-intensity odors may make it possible to distinguish between malingering and nonmalingering patients. Preliminary analyses of an experiment examining EEG responses to low-intensity odors in subjects scoring high and low in cacosmia and high and low in depression (a 2 X 2 design) suggests that quantitative EEG mapping may be a useful tool in research on chemical sensitivity (Schwartz et al., 1994b).

Research in olfaction, especially involving low-intensity odors, has inherent conceptual and methodological problems (Lorig, 1989). Research in chemical sensitivity has even more conceptual and methodological problems (Bell et al., 1992). Consequently, combining these two areas is especially challenging. The research described in this paper, though suggestive and promising, should be viewed as exploratory.

First, we will review studies presented at the 1992 and 1993 Association for Chemoreception Sciences meetings documenting EEG registration of subthreshold odors and of androstenone in androstenone anosmic subjects. These studies set the stage for research on chemical sensitivity, and therefore they are reviewed in some detail. Then we will report the results of analyses conducted to date on two studies that were presented at the 1994 Association for Chemoreception Sciences and the American Psychiatric Association meetings, indicating EEG registration of subthreshold odors in normals and increased EEG registration of the solvent propylene glycol in subjects claiming to be cacosmic.

SUBTHRESHOLD AND ANDROSTENONE ANOSMIA STUDIES

In the first experiment (Schwartz et al., 1992a), 19 channels of EEG were recorded from 54 elderly subjects between the ages of 60 and 70 while they sniffed pairs of flasks for 8 seconds each, presented sequentially, with eyes closed. One of the flasks contained concentrations of isoamyl acetate predissolved in propylene glycol and then diluted in 50 ml water. The concentrations increased sequentially from subthreshold (3 concentrations) through threshold (4 concentrations) to suprathreshold (4 concentrations). The other flask contained only the solvents (the control condition). The order of experimental and control flasks was counterbalanced across subjects. At the end of each trial, subjects guessed which flask contained the odor and rated their confidence and perceived intensity on a scale from 0 to 10. In all the experiments, EEG was amplified and analyzed using the NeuroSearch-24 EEG system. In this experiment, EEG was sampled at 128 hz, spectral analyses were performed, and alpha power (8–12 hz) was displayed in topographic maps. In normal subjects, decreases in alpha typically reflect cortical activation [as do decreases in theta (4–8 hz) and increases in beta (> 12 hz)]. Correct guesses were 92% for the suprathreshold trial and 52% (chance) for the subthreshold trial selected for analysis.

When subjects smelled the suprathreshold concentration, significant EEG alpha frequency (8–12 hz) decreases were observed in anterior, mid central, and posterior regions, whereas when subjects smelled the subthreshold concentration, significant alpha decreases were observed only in the mid central and posterior regions. All significant effects were at least $p < 0.05$. The data were interpreted as supporting the hypothesis that older adults can be sensitive to odors at subconscious levels, and that conscious perception may involve the frontal regions as originally hypothesized by Luria (1973).

In the second experiment (Schwartz et al., 1992b), 19 channels of EEG were recorded from 10 college students while they sniffed pairs of flasks for four seconds each. One of the flasks of each pair contained common odorants used in olfaction research, isoamyl acetate, di-acetyl, or coumarin, dissolved in 50 ml of water. The other flask contained 50 ml water (the control condition). In the first three sessions, thresholds were determined for the three odors. Two subthreshold concentrations, one "gray zone" (at threshold) concentration, and two suprathreshold concentrations were selected per odor per subject. Over the next three sessions, EEG data were collected during the 6 trials of each odor at each concentration. The order of odors, concentrations, and experimental and control flasks was counterbalanced within subjects and across sessions. At the end of each trial, subjects guessed which flask contained the odor (detection), what the odor was (discrimination), and rated their confidence on a scale from 0 to 10. EEG analyses were performed on the highest subthreshold and suprathreshold concentrations. Correct odor detection was 95%, 95%, and 98% for the suprathreshold concentrations of the three odors, and was 43%, 45%, and 51% (chance) for the subthreshold concentrations.

Across all odors, when subjects smelled the suprathreshold concentration, significant EEG alpha (8–12 hz) decreases were observed in anterior, central, and posterior regions, whereas when subjects smelled the subthreshold concentration, significant EEG alpha decreases were observed only in the central region (somewhat right lateralized). These data replicated and extended the initial findings in the elderly.

In a second experiment involving elderly subjects (Schwartz et al., 1992c), the question was raised whether subjects who did not report smelling androstenone (a purported male sex pheromone) would nonetheless show EEG registration of the molecule. Nineteen channels of EEG were recorded from 33 subjects between the ages of 60 and 70 while they sniffed bottles containing different odors for 60 seconds per bottle. Two bottles contained concentrations of androstenone dissolved in silicone. Two bottles contained either liquid silicone or water (control conditions). Four bottles contained essential oils. The order of odor presentations was counterbalanced across subjects. Subjects rated from 0 to 10 their level of relaxation, the intensity and pleasantness of the odor, how masculine or feminine the odor smelled, and the percentage of time (0 to 100%) they were aware of the odor. Osmics rated the intensity of the odor and the percent time they were aware of the odor higher for androstenone than silicone and water (5.4 vs. 1.9; 72% vs. 25%). Anosmics rated the intensity of the androstenone and the percent time they were aware of the androstenone similar to silicone and water (2.0 vs. 1.4; 28% vs. 24%). Thirteen subjects were osmic for androstenone, 20 were selectively anosmic for androstenone. The anosmia was highly selective for androstenone since the ratings of the essential oils were virtually identical comparing the osmic and anosmic groups (7.3 and 7.8; 82% and 88%).

Analysis of the EEG of the subjects anosmic for androstenone compared to the control conditions revealed significant, selective EEG alpha decreases in the right central region when smelling androstenone. Further evidence for central nervous system processing of androstenone in the androstenone anosmic subjects was observed in the mood data. Both osmic and anosmic subjects rated themselves feeling significantly less relaxed during the androstenone than during the control conditions (osmics 6.3 vs. 7.2; anosmics 6.4 vs. 7.1), even though the anosmics did not report smelling the androstenone. The EEG findings for the anosmic androstenone subjects curiously paralleled the EEG findings for the subthreshold odorants.

The subthreshold and androstenone findings were replicated and extended in a subsequent experiment (Schwartz et al., 1993). Nineteen channels of EEG were recorded from 52 college students while they sniffed pairs of bottles. One bottle per pair contained isoamyl acetate or androstenone dissolved in odorless silicone. The other bottle contained silicone (the control). Subjects were prescreened for isoamyl acetate and androstenone detection and invited for EEG testing if they consistently detected isoamyl acetate and either detected androstenone (osmic) or failed to detect androstenone (anosmic). EEG was collected during two two-second sniff periods per bottle for eight trials per odor at two concentrations, one suprathreshold and one subthreshold (subthreshold androstenone was determined by androstenone osmic subjects). The order of odors, concentrations, experimental and control bottles, and hand, was counterbalanced within subjects. After each trial, subjects indicated which bottle contained the odor (detection), rated their confidence of detection and the odor's intensity on scales from 0 to 10. In this experiment, EEG was sampled at 256 hz rather than 128 hz. Correct odor detections for the suprathreshold concentrations were 99% for isoamyl acetate, and 99% for androstenone osmics and 41% for androstenone anosmics. Correct odor detections for the subthreshold concentrations were 48% for isoamyl acetate and 57% for androstenone (50% is chance).

When subjects smelled suprathreshold isoamyl acetate, significant alpha decreases were observed in anterior, central, and posterior regions, whereas when subjects smelled subthreshold isoamyl acetate, significant alpha decreases were observed primarily in the central region to sniff two (post hoc analyses of the first two studies revealed that the largest effects for subthreshold alpha also occurred to sniff two). Significant alpha decreases were also observed to suprathreshold androstenone in both osmic and anosmic subjects in both sniffs, whereas significant EEG alpha decreases were observed to subthreshold androstenone in both groups to sniff one. The data replicated and extended the previous findings suggesting that humans can register odors at subthreshold levels, and that androstenone anosmia may involve cortical inhibition of olfactory perception.

In the most recent study (Schwartz et al., 1994a), 19 channels of EEG were recorded from 86 college students, replicating the previous experimental protocol with three important additions. The first was that the subthreshold odors were administered double-blind rather than single-blind (the previous studies all used single-blind procedures). The second was that the number of subthreshold isoamyl acetate trials was increased from 8 to 16. In addition, EEG was spectral analyzed in 2-hz bands instead of 4-hz bands, and analyses were performed on low-frequency theta (4–6 hz), high-frequency theta (6–8 hz), low-frequency alpha (8–10 hz), and high-frequency alpha (10–12 hz).

Figure 1 (Color Plate 2) displays topographic EEG maps subtracting control bottles from the odor bottles for the subthreshold isoamyl acetate trials (n = 16) for the 86 subjects. Subtraction maps are necessary to remove the effects of hand posture, breathing, attention, and registration of the control solvent from the EEG, leaving just the effects of the registration of the isoamyl acetate. The top four maps reflect sniff one, the bottom four maps reflect sniff two. The four EEG bands are all drawn on the same scale. The color scheme shows that reds and whites reflect increases in theta or alpha, while blues and blacks reflect comparable magnitude decreases in theta or alpha.

It can be seen that for low-frequency theta (4–6 hz), EEG registration of the subliminal odor occurred on the first sniff, especially in the central and posterior regions, more on the right side. For high-frequency theta (6–8 hz), EEG registration of the subliminal odor occurred on both sniffs, again central and more posterior on the right side. For low-frequency alpha (8–10 hz), some evidence for EEG registration was obtained, more central and right sided, somewhat for both sniffs. For high-frequency alpha (10–12 hz), EEG registration was obtained especially for sniff 2, and the effect was central and more anterior on the right side. These effects were all statistically significant, at least $p < 0.05$.

Time and space do not permit a more detailed presentation of these data. However, three additional sets of analyses should be summarized briefly:

1. When the 16 trials for each subject were split into hits and misses, significant EEG effects, especially for theta, were replicated for misses when subjects guessed incorrectly that the odor was in the control bottle (which they did on half the trials). Hence, the EEG effects, especially theta, reflected registration of the low-level odor regardless of cognition or belief.
2. When the control trials were compared in terms of belief (trials where subjects thought the odor was in the control — i.e., false positive trials), belief in the absence of odor showed virtually no effect for EEG theta, and showed some effect for EEG alpha. However, belief showed up more in the anterior (frontal regions), where as odor effects showed up primarily in the central and posterior regions. *Hence, belief can be distinguished from odor effects.*
3. When subjects were split into high,- medium-, and low-performance subgroups based on their overall guessing accuracy, the EEG effects were replicated across subgroups for theta. Alpha effects were strongest in the high performance subgroup. However, the groups were identical in ratings of perceived intensity of the odors and perceived confidence of their guesses! Hence, the high performance group reflected a high "sensitivity" group in the absence of conscious awareness. EEG patterns for high sensitive normal subjects may be relevant to the applications of this paradigm to the study of cacosmics and chemically sensitive individuals.

One final point should be mentioned. When ratings of intensity for hit trials versus miss trials (on a 0 to 10 scale) were analyzed separately for the three performance groups, a subtle yet significant difference in reported intensity was observed for the groups (high 1.98 vs. 1.75; medium 2.20 vs. 1.99; low 1.74 vs. 1.65, $p < 0.03$). Hence, even though subjects guessed at chance level, and their reports of odor intensity were minimal, subtle differences between hit versus miss trials were revealed. Since the EEG effects were observed in the miss trials as well as the hit trials, the EEG results could not be explained by these subtle effects. However, as will be seen below, subtle differences in awareness were also found in subjects perceptions of different chemicals included in the design as "control" stimuli, and some of these subtle perceptions were different between noncacosmic and cacosmic subjects.

PRELIMINARY RESEARCH IN CACOSMICS

Bell et al. (1993a,b,c) have conducted a series of studies indicating that it is possible to identify in normal populations subjects reporting symptoms of illness to common environmental chemical odors. Although subjects scoring high in cacosmia tend to indicate higher levels of depression and anxiety, the correlations are small and account for only a small proportion of the variance in cacosmia scores. Hence, it is possible to find subjects who score high in cacosmia and low in depression, subjects who score low in cacosmia and high in depression, subjects who score high in both cacosmia and depression, and subjects who score low in both cacosmia and depression. We have hypothesized that cacosmia and depression are separate yet interactive processes, and that quantitative EEG analyses of responses to low-level chemicals will differ for cacosmia and depression (Bell et al., 1992).

An experiment was conducted on 66 subjects selected for combinations of high and low cacosmia and high and low depression (Schwartz et al., 1994b). Subjects were initially chosen from the top and bottom 40% for depression using the Symptom Checklist 90 Revised and for identifying three of five chemicals as causing illness "sometimes" or more, or identifying no chemicals as causing illness. These subjects were then screened over the telephone. Cacosmia was reassessed using a five-item true/false scale, depression was reassessed using a short depression subscale, and health was assessed using assorted health history questions. In order to be chosen for the high-cacosmia groups, subjects had to identify at least three chemicals as causing illness. For the low-cacosmia groups, all chemicals had to be identified as not causing illness. To be selected for the depression groups, subjects had to have a total of eight for the three Weinberger questions and the "I expression depression" question (five point scale). To be selected for the nondepressed groups, subject's depression scores had to be less than eight. Students were disqualified if they were smokers, if they used prescription or nonprescription drugs regularly, if they had a cold or flu, stuffy nose, asthma, anaphylactic shock, epilepsy, diabetes, heart, liver or kidney disease, or suicidal feelings. Table 1 shows the mean ages, the number of females and males, and the mean cacosmia scores, for the four groups.

TABLE 1. Mean Age, Sex Ratio, and Cacosmia Scores for Cacosmia (C) and Depression (D) Combinations

Group	Age	Females/Males	Cacosmia
Controls (Low C/Low D)	19.7	9/9	7.22
Cacosmics (High C/Low D)	18.9	10/5	13.21
Depressed (Low C/High D)	18.7	9/6	8.60
Cacosmic Depressed	18.2	16/2	13.33

It can be seen that all four combinations of cacosmia and depression scores were obtained, and that the groups were identical in ages. The cacosmia groups had somewhat more females than the noncacosmia groups.

To increase the duration of exposure for subjects, the paradigm used followed the design in Schwartz et al. (1992c) examining EEG registration of androstenone in androstenone anosmics in the elderly. Subjects smelled bottles containing different chemicals for one minute periods with eyes closed. The bottles were held by a metal stand. The experiment included three "control" bottles (an empty bottle and two bottles containing the solvents used in the study — distilled water and propylene glycol), and four chemical mixtures (low and moderate concentrations of butanol in distilled water, and low and moderate concentrations of galaxolide — a musk odorant — in propylene glycol), randomized across subjects.

After each trial, subjects made a series of ratings using the numbers 0 through 10 regarding the intensity of the odor, the duration of the experience, the pleasantness of the odor, odor identification, and moods. Analyses to date have been completed on the EEG data for the three control bottles for theta (4–8 hz), low-frequency alpha (8–10 hz), high-frequency alpha (10–12 hz) and low-frequency beta (12–16 hz).

Significant odor effects (empty versus distilled water versus propylene glycol) were obtained for each of the four bands, primarily in central and posterior regions. In addition, significant interactions with cacosmia were obtained for low-frequency alpha ($p < 0.05$ for the posterior temporal/parietal sagittal strip, and $p < 0.055$ for the occipital strip) and low-frequency beta ($p<.04$ for the posterior temporal/parietal sagittal strip, and $p < 0.007$ for the occipital strip).

Figure 2 (Color Plate 3) displays distilled water minus empty bottle topographic EEG maps (top) and propylene glycol minus empty bottle topographic EEG maps (bottom) separately for the noncacosmics (left) and cacosmics (right), averaged over depression, for low-frequency alpha. As in the previous studies, subtraction maps were used to remove the effects of hand posture, breathing, attention and the smelling of air per se from the EEG, leaving just the effects of the registration of the solvents. The terms increase and decrease therefore refer to relative changes in EEG as compared to the empty bottle control.

It can be seen that the noncacosmics showed, if anything, relative *increases* in low-frequency alpha (red color), whereas the cacosmics showed clear *decreases* in low-frequency alpha in the posterior regions (black and blue colors). The maps suggest that the decreases in low-frequency alpha (relative to the empty bottle control) were greater for propylene glycol than distilled water (especially in more anterior regions). There were no significant interactions involving depression related to the chemicals in low-frequency alpha.

Figure 3 (Color Plate 3) displays distilled water minus empty bottle topographic EEG maps (top) and propylene glycol minus empty bottle topographic EEG maps (bottom) separately for the noncacosmics (left) and cacosmics (right), averaged over depression, for low-frequency beta. It can be seen that the noncacosmics showed relative *decreases* in low-frequency beta (black and blue) to both distilled water and propylene glycol, whereas the cacosmics showed clear *increases* in low-frequency beta in the posterior regions (red and white colors) to propylene glycol, the chemical solvent. Interestingly, increases in beta have been reported for monkeys and humans exposed to the organophosphate sarin (Burchfiel and Duffy, 1982).

Significant effects for depression by bottle by EEG site were obtained for the anterior frontal sagittal strip ($p < 0.03$) and the temporal/central saggital strip ($p < 0.05$), but these effects did not interact with cacosmia. Moreover, the effects were completely opposite. Nondepressed subjects showed *increases* in low-frequency beta to distilled water and propylene glycol, whereas depressed subjects show *decreases* in low-frequency beta to distilled water and propylene glycol.

These preliminary findings suggest that "normal" subjects who report feeling sick when smelling certain chemicals, regardless of their perceived depression, show heightened EEG reactivity in posterior regions, especially to propylene glycol, as evidenced by decreases in low-frequency alpha and increases in low-frequency beta. These findings can not be explained by depression. Depression did not yield significant effects for low-frequency alpha, and the significant effects observed for low-frequency beta were in more anterior regions and were in the opposite direction (i.e., depressed subjects showed reduced EEG reactivity.

The present study has a number of weaknesses that qualify the interpretation. First, the three bottles were not experienced identically. Whereas 26% and 23% of the subjects reported smelling "something" for the empty and distilled water bottles, 45% of the subjects reported smelling "something" for the propylene glycol solvent ($p < 0.02$). There were no differences between groups in reports of smelling "something." This rating was specifically designed to be vague so that reports of subtle experiences of olfaction would be potentially detected.

For ratings of intensity, a similar pattern was observed. Despite the fact that the ratings of intensity were tiny (using the scale 0 to 10, the average ratings for empty, distilled water, and propylene glycol were 0.7, 0.5, and 1.0), the solvent was rated as more intense ($p < 0.01$). Again, there were no significant differences between groups in reports of intensity.

For ratings of pleasantness, the picture is more complicated. On the average (across the three bottles), cacosmic subjects rated the bottles as slightly less pleasant than the noncacosmic subjects (–0.3 versus +0.4, $p < 0.04$). This effect was primarily due to slight ratings of unpleasantness for the nondepressed cacosmics (–0.8) compared to nondepressed noncacosmics (+0.7). Depressed cacosmics and depressed noncacosmics rated the bottles neutral on the average (0.1 and 0.1). Recall that the rating scale was –10 for unpleasantness to 0 for neutral to +10 for pleasantness. Hence, these group differences are very subtle, though they are reliable.

CONCLUSIONS

The pattern of findings obtained to date suggest that it may be possible to quantify subtle central nervous system effects to low-intensity chemicals in normals and cacosmics, and hence individuals claiming to have multiple chemical sensitivities. The discovery that subtle differences in conscious experience can be revealed through careful assessment points to the care that must be taken in establishing concentrations of odors and assessing subjects guessing performance and subjective experience. Though one would expect that the EEG effects should be greater when patients are actually experiencing symptoms, because these effects would be confounded by the subjective experience the symptoms, the interpretation would be complicated. The use of low-intensity chemicals virtually eliminates this inherent complication.

In the cacosmia study reported above, the one-minute duration paradigm (rather than the paired sniff paradigm) was used in this preliminary study because we wanted subjects to experience the odors for a duration of time. The paired sniff paradigm developed for subthreshold olfactory testing needs to be applied to cacosmic subjects before definitive conclusions about EEG responses in the absence of conscious awareness can be drawn.

The use of computer controlled olfactometers may further improve the precision of the findings and conclusions. It is unclear whether spectral analysis and/or olfactory evoked potentials will prove more sensitive for assessing EEG registration of low-intensity chemicals.

Recall that the three bottles analyzed above in the cacosmia study were originally included in the design as "control" bottles to evaluate primary odors of interest (butanol in water and galaxolide in propylene glycol). Nonetheless, subjects differed in their EEG responses to the "control" bottles in a meaningful way (propylene glycol and water versus just an empty bottle). These data are consistent with the hypothesis that quantitative EEG methods may prove to be a sensitive and specific procedure for evaluating central nervous system sensitivity to different chemicals (propylene glycol and water versus air) at low-intensity levels. By using subthreshold concentrations of odorants without conscious sensory awareness, it should be possible to avoid using "masking" odors in future double-blind studies of chemical sensitivity.

REFERENCES

BELL, I.R., MILLER, C.S., and SCHWARTZ, G.E. (1992). "An olfactory-limbic model of multiple chemical sensitivity syndrome: Possible relationship to kindling and affective spectrum disorders." Biol. Psychiatry 32:218–242.

BELL, I.R., SCHWARTZ, G.E., PETERSON, J.M., and AMEND, D. (1993a). "Self-reported illness from chemical odors in young adults without clinical syndromes or occupational exposures." Arch. Environ. Health 48:6–13.

BELL, I.R., SCHWARTZ, G.E., PETERSON, J.M., and AMEND, D. (1993b). "Symptom and personality profiles of young adults from a college student population with self-reported illness from foods and chemicals." J. Am. Coll. Nutr. 12:693–702.

BELL, I.R., SCHWARTZ, G.E., PETERSON, J.M., AMEND, D., and STINI, W.A. (1993c). "Possible time-dependent sensitization to xenobiotics: self-reported illness from chemical odors, foods, and opiate drugs in an older adult population. " Arch. Environ. Health 48:315–327.

BURCHFIEL, J.L. and DUFFY, F.H. (1982). "Organophosphate neurotoxicity: chronic effects of sarin on the electroencephalogram of monkey and man." Neurobehav. Toxicol. Teratol. 4:77–778.

LORIG, T. S. (1989). "Human EEG and odor response." Prog. Neurobiol. 33:387–398.

LORIG, T. S. (1994). "EEG and ERP studies of low-level odor exposure in normal subjects." Toxicol. Ind. Health. 10:579–586.

LORIG, T.S., HERMAN, K.B., SCHWARTZ, G.E., and CAIN, W.S. (1990). "EEG activity during administration of low concentration odors." Bull. Psychonomic Soc. 28:405–408.

LURIA, A.R. (1973). The Working Brain. Basic Books, New York.

SCHWARTZ, G.E., WRIGHT, K.P., POLAK, E.H., and SCHWARTZ, J.I. (1992a). "Conscious and unconscious odor registration in the elderly." Chem. Senses 17:695.

SCHWARTZ, G.E., WRIGHT, K.P., POLAK, E.H., KLINE, J.P., and DIKMAN, Z. (1992b). "Topographic EEG mapping of conscious and unconscious odors. " Chem. Senses 17:695.

SCHWARTZ, G.E., KLINE, J.P., DIKMAN, Z, WRIGHT, K.P., and POLAK, E.H. (1992c). "EEG registration of androstenone in androstenone anosmic subjects." Chem. Senses 17:694.

SCHWARTZ, G.E., KLINE, J.P., DIKMAN, Z.V., and POLAK, E.H. (1993). "EEG registration of conscious and unconscious concentrations of isoamyl acetate and androstenone." Chem. Senses 18:625.

SCHWARTZ, G.E., KLINE, J.P., DIKMAN, Z.V., FERNANDEZ, M., and POLAK, E.H (1994a). "EEG registration of unconscious concentrations of isoamyl acetate: A double-blind experiment." Chem. Senses. In press

SCHWARTZ, G.E., BELL, I.R., HERRING, A., PETERSON, J.M., and KASZNIAK, A.W. (1994b). Memory and EEG in chemically sensitive young adults. Paper presented at the Annual Meeting of the American Psychiatric Association, Philadelphia, PA, May 21–26, 1994.

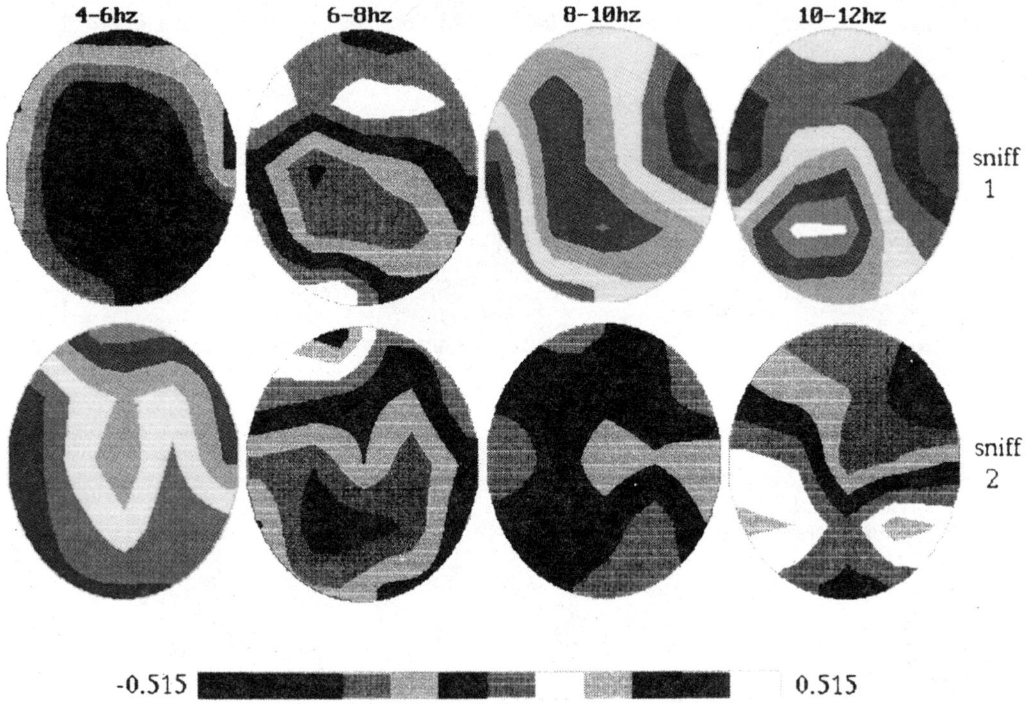

FIGURE 1. Topographic EEG maps subtracting control bottles from the odor bottles for subthreshold isoamyl acetate trials (n = 16) for 86 subjects. The top four maps reflect sniff one (the first sniff for the isoamyl acetate and control trials), the bottom four maps reflect sniff two (the second sniff for the isoamyl acetate and control trials). The four EEG bands (4–6 hz, 6–8 hz, 8–10 hz, 10–12 hz) are all drawn on the same scale. Reds and whites reflect increases in EEG, blues and blacks reflect comparable magnitude decreases in EEG. The top of each map reflects the front of the head, the bottom of each map reflects the back of the head.

Color Plate 2. As referenced in Schwartz et al., p. 637

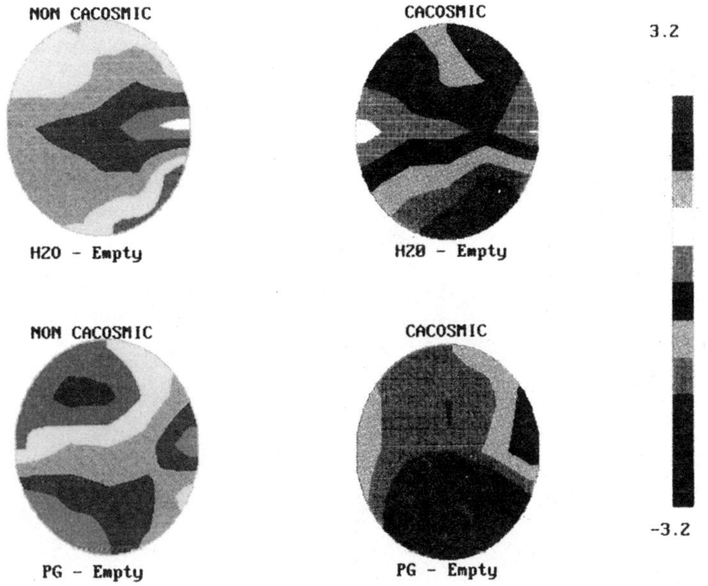

FIGURE 2. Distilled water minus empty bottle topographic EEG maps (top) and propylene glycol minus empty bottle topographic EEG maps (bottom) separately for noncacosmics (left) and cacosmics (right), averaged over depression, for low-frequency alpha (8–10 hz). Reds and whites reflect increases in EEG, blues and blacks decreases in EEG. The top of each map reflects the front of the head, the bottom of each map reflects the back of the head.

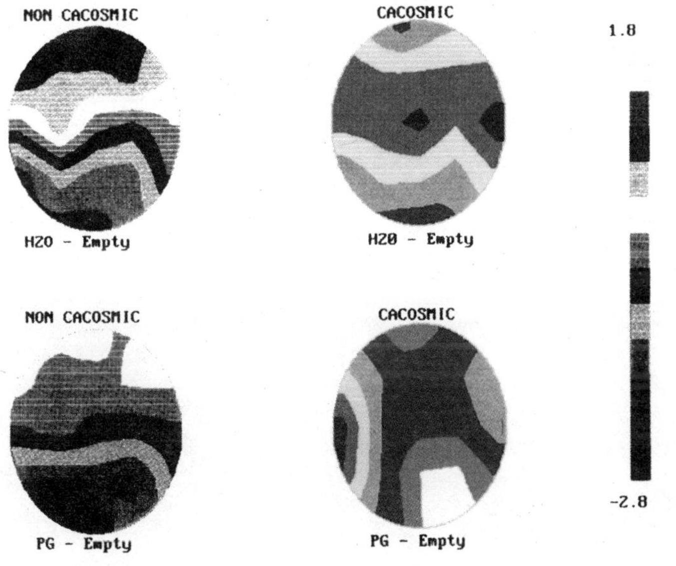

FIGURE 3. Distilled water minus empty bottle topographic EEG maps (top) and propylene glycol minus empty bottle topographic EEG maps (bottom) separately for noncacosmics (left) and cacosmics (right), averaged over depression, for low-frequency beta (12–16 hz). Reds and whites reflect increases in EEG, blues and blacks decreases in EEG. The top of each map reflects the front of the head, the bottom of each map reflects the back of the head.

Color Plate 3. As referenced in Schwartz et al., p. 640

NEUROSCIENCE RESEARCH AGENDA OPTIONS PANEL

NOTE: In order to obtain some perspective on the comments made during the research agenda section, Dr. Howard Kipen, a physician who has published a number of papers on chemical sensitivity, and who was a member of the Planning Committee for the conference, was asked to prepare the following overview. Following his presentation, the comments made during the meeting are presented in edited form.

Dr. Kipen: The panel took the opportunity at the close of the conference to offer final suggestions, recommendations, and guidance on issues related to research which will advance our understanding of low-level chemical sensitivities. Recommendations were offered in sequence by individuals who do not necessarily represent consensus. These concluding recommendations largely aim to supplement rather than repeat the presentations in individual papers, and thus do not emphasize much of the detail on specific neuroscience aspects of chemical sensitivity research. This paper summarizes the comments and arranges them into coherent themes to facilitate their consideration and implementation. The three themes are Definitions, Populations, and Mechanisms, with a broad level of agreement that funding is urgently needed to facilitate exploration of many fertile lines of investigation.

DEFINITIONS

Definitions of chemical sensitivity, although not a primary concern of this meeting, occupied a fair amount of discussion time. A recommendation to establish the definition with attention to its validity (e.g., content, criterion, and predictive) was made in the context of establishing chemical sensitivity as a recognized entity. This was felt to be especially important in the face of strong opposition from those who feel the concept threatens the viability of commercial products. Commentators did not agree that research could only progress with an agreed-upon definition, perhaps declared by a consensus committee, although certainly a few held this position. It was noted that for practical purposes the definition proposed by Cullen and endorsed by the ACOEM had been serviceable for the conduct of clinical research.

POPULATIONS

An alternative point of view proposed that exploration of large populations for salient characteristics of chemical sensitivity, perhaps using sophisticated multivariate methods such as discriminant analysis and cluster analysis, could eventually produce a more valid and representative definition that could bring together a group of experts to engage in discussion. It is likely that chemical sensitivity exists on a continuum, with the extreme end characterized by those who are clinically "ill." By whatever definition is declared or empirically

Toxicology and Industrial Health, Vol. 10, No. 4/5, pp. 645–652

ISSN: 0748-2337

determined, it is necessary to determine population prevalence of the various degrees of chemical sensitivity, with or without comorbid medical or psychiatric conditions. This allows public health planning for research and response to proceed rationally in the face of scarce resources. Analytic epidemiology studies can be undertaken to ascertain risk factors and eventually design prevention strategies. Expertise in population studies and in psychometric and other aspects of questionnaire design should be brought to bear on this problem. Once characteristics on which to base selection of those with and without chemical sensitivity, or at least varying degrees of it, have been determined, mechanistic research becomes more feasible.

MECHANISMS

Suggestions were made to focus on roles for the olfactory system, both as a chemical sense organ and as an important receptor for psychological cues. The importance of controlling for psychological cues which may accompany odor challenges and for perceptual differences which may separate individual responses to the same odor, were emphasized. Some commentators felt the use of clean environments (e.g., environmentally controlled units) was particularly critical to the conduct of any valid challenge study. It was noted that the design and operation of such inhalation facilities was nontrivial and required substantial expertise in aerosol physics and inhalation toxicology.

The immune system, especially its ability to be conditioned by psychological stimuli and the field of psychoneuroimmunology, was proposed as key to investigating the phenomenon of chemical sensitivity. Discussion of more traditional end points, such as neuropsychological tests, emphasized the desirability of global performance tests rather than more typical single functions. Choice of subjects was considered very important. Ideally this would occur after careful population research, not only to improve characterization of the individual with chemical sensitivity, but to improve understanding of how lesser degrees of chemical sensitivity are or are not present among so-called "normals." It was noted that in-depth and repeat testing of selected individuals may often be more desirable than superficial testing of large samples. However, attention must be paid to learning effects and the repeatability of tests.

These concluding remarks should be read and interpreted in the context of the substantial preceding presentations and discussions.

Dr. Bell: I'd like to begin by thanking ATSDR for having the courage for putting on this conference. I have to agree with Dr. Rosenthal: it was probably a turning point in the research in this problem. I hope it will be the beginning of many such conferences, in which we will be able to present direct data on whatever case definition we define.

There are several thoughts that I wanted to emphasize. The call for a case definition is getting very loud. We have to have consensus on what research we are going to do, so that we can start studying roughly the same populations and talking about the phenomenology and ultimately, similar mechanisms. We had talked about the need for prospective research, in termsof event-driven work, where we take advantage of the misfortunes of the individuals who are affected by chemical spills. Clearly, there are ethical considerations in creating such an episode, but there are lesser ones in following up on it. They do offer us an opportunity to look at people longitudinally.

I would like to emphasize the notion of time as an important factor in understanding both the phenomena and the potential mechanism. Because of the expense and complication, we generally prefer to do a one-session study that takes five minutes to accomplish, with one measure of outcome, and have a simple clean answer. I think that none of that is possible. We will, under many different circumstances, and in many different ways, want to study time factors, both longitudinally, and also in short-term studies that may take a few days or weeks to accomplish.

I think some of the neuropsychologists would agree that the difficulty in doing that kind of work is that you must have the ability to do repeated measures of whatever outcomes. They must be comparable to each other. You have to have tests that are not only standardized, but that can be done over time. The test that you gave on Thursday must be the same test, equivalent in difficulty and so on, that you're giving next Tuesday. Many of the clinically oriented tests are limited to a specific, onetime evaluation of the individual. That makes it particularly difficult. So, we do have to take advantage of whatever tools are available to us for doing repeated measures.

People have emphasized the need for within-subject designs. We must look at individuals, compared with themselves in some fashion, during sham and placebo exposures, as well as during active exposures. We've also talked about the need for subthreshold studies using olfactory stimuli, where the possibility of suggestion and placebo effect is lessened by the fact the person is not consciously aware of the presence of the substance. However, as you saw in our data, it's very clear that if the substance is given, the brain registers the presence of that substance — in both normal people and chemically sensitive individuals.

What's even more interesting is that the brain does not register the presence of the chemical when a person simply believes the chemical is there. That's a very important point, because it does say that belief is not the same thing as exposure, at least at that level, and it offers us some hope of teasing apart what's happening under various conditions.

One of the points that Dr. Weiss referred to, maybe somewhat indirectly, that comes to us from the food sensitivity research literature, is that we have to be very careful not to assume a single substance is the problem for all the people with a given diagnosis. One of the things that came up repeatedly in that work, with food, food colorings, sugar, and so on, was that if you designed studies that do not pick responders out by prescreening them, you wind up with very large sample sizes. You also run the risk of missing the effect that is truly there, because you are only using a small number of people who respond to a particular substance that you're presuming is the problem.

That's one of the reasons I believe that in the initial research we do, we are going to have to take a comprehensive approach. There may be chemicals that we can individually pick on that may be worth studying, but we have to be very careful not to bet everything on an expensive experiment, or particular exposure, or particular concentration, until we have done a lot of pilot work.

In terms of patient and subject selection, we want to be very careful not just in who we're picking from the case definition point of view, but also controls. Many people would emphasize the need for having patient groups who are not normals, and I think that is a very important observation. One of the other criticisms that can be made, if you simply pick patients who have other kinds of problems, is that people, so far, have not asked those people if they're chemically sensitive. That was true of Dr. Simon's study, for example. My data show clearly that if you just take a general population of people without health issues, you're going to find a subset who do say, "I have problems with chemical odors." You don't want those in your control group or you're going to have a very messy comparison. You'd better have a group that says they have it and group that says they don't, to start with, and then we refine from there.

Finally, I'd like to emphasize that right now we probably will need multiple outcome measures, because we are talking about a polysymptomatic illness that does affect many different parts of the body and very likely has more than one mechanism. That's going to make it even more complicated from a statistical and experimental design point of view. I'd certainly like to hear from members of the audience who are more experienced than I am in experimental design, on those points.

Dr. Letz: I have been touched by the human stories that I have heard at this conference. They serve to remind us all why it is important to study this problem.

As a scientist who does not work directly in this area, I have two hands, and hopefully, two sides of the brain. On the one hand, I'm encouraged by the ideas and the preliminary data that have been presented at this conference. On the other hand, I'm severely disappointed by the paucity of interpretable data available in the peer-reviewed literature on this problem.

I have a few things I would like to recommend. One is to echo the need for a case definition. Something has to be found that people can live with, so that we can move forward in this field. Once that is done, there is an immediate need for prevalence studies. From these studies, we'll have an idea of the magnitude of the problem. This will help us design further studies. We will be able to identify risk factors. These risk factors will allow us to meet an important need that Mr. Berkson brought up: prevention.

We also need collaborative, interdisciplinary research projects. The first thing we need is well-formed hypotheses that are the bases of these studies. We also need appropriately selected subjects. Those subjects will be appropriately selected by having a good case definition and by using good sampling techniques.

We need adequate sample sizes. Since average sample size of the studies that have been done to date is very small, there is very little statistical power to determine much at all.

We need psychometrically sound symptom questionnaires, made up by psychologists, not by allergists, pathologists, etc. Psychologists have been studying these things for years, and they know how to do this better than others in the health professions.

We also need objectively administered and objectively scored, simple, inexpensive outcomes that can be used in studies with sufficient sample sizes to have power to discover associations. Limited use of expensive techniques is important and will help us in determining mechanisms, but until we find out a lot more about the problem, it's my opinion that money will be wasted.

Dr. Rosenthal: There's one formidable obstacle that we have in the area of MCS that we never had in seasonal affective disorder. We never had a dark industry, watching to prevent us from using lights. I think that there is a formidable set of financial interests stacked against finding out about the dangerous effects of chemicals. I think you need only look at the tobacco companies and the way they have stonewalled, to get some sense of the obstacles that stand in the way of advancing knowledge in this area. If solid things like lung cancers and emphysema could be successfully fought off for so long, imagine how much more difficult it will be with the "flaky" symptoms of sadness, dizziness, and memory loss, and all these other things that nonetheless are so very real to the people who suffer them.

I agree with the previous speaker that each of us has only two hands, and that's why I think some kind of united work, collaborative work, is needed. We're unfortunately in a bit of a catch-22; if MCS doesn't exist, and it's not a real problem, why should we have grants to fund research in it? Without grants, we're not going to advance the science. So we're in this position of having to pull ourselves up by a shoestring. But it can be done, because it is a very widespread and important problem.

I think this has been an important milestone and I'm glad to have been a part of it. I think that MCS patients are a terribly important political force to unite and see that the scientists get the money they need for research. Without that research, progress is not going to occur.

Dr. Weiss: There is a great deal of emphasis, as some of you may know, in current reviews and planned proposals on molecular end points. Because we know nothing else, it is very difficult to propose any molecular outcomes, or endpoints, or systems, for low-level chemical sensitivity.

What I said today was really directed toward research strategies and how we go about trying to uncover whether a particular phenomenon exists. When Feingold first proposed his hypothesis, of course, he was laughed at. After all, the FDA protected us. It was only after we began to look at how the FDA evaluates neurobehavioral toxicity, that we realized that there was no margin of protection, because the technology for assessing it was not practiced.

I can see why Rick Letz would like to see large samples, and simple tests. That may be one component of what has to be a ramified strategy. I think that finding subpopulations, studying individuals intensively in controlled environments, repeatedly, and using not the typical gambit of neuropsychological tests, but more a global performance as an index, could be very useful in trying to define what we're talking about and what has bewildered me throughout this meeting.

Dr. Miller: I want to continue along the line of that Dr. Weiss has just begun and say that chemical sensitivity probably is the study of individual differences, and having large numbers of patients, who fit one particular category, may not be possible.

A very unique approach, which was really developed almost 40 years ago by Dr. Randolph, was the concept of placing patients in a very clean and controlled environment and testing them with single foods, one at a time, to sort out these individual differences. And it's a very simple and very elegant concept. One could also then conduct double-blind, placebo-controlled challenges in that same environment, using salient kinds of challenges for each individual. The more I hear, the more I think that approach may offer some resolutions of questions, about the physiological vs. the psychological nature of this problem. I think it is very difficult, from a scientific perspective, to do it any other way. Science is not about belief, it is about guess and test. You formulate a hypothesis and you test that hypothesis.

There's an issue, however, separate from the science. That is the public health questions that we're confronting right now. I'm certainly reminded of John Snow and the cholera epidemic. He saw people getting ill and noticed that the people lived around the Broad Street area and that there was a pump around that area from which they were drawing their water. The people who seemed to be getting ill were the ones who drank that water. So he ordered the pump handle to be broken, and the epidemic stopped.

It was not until 30 years later that Koch discovered the bacterium that causes cholera. So, Snow knew of no mechanism at that time, and yet there was a very important public health intervention, that could be performed, that saved lives. Still, the science had to proceed. So, I think we need to focus on both science and public health. Pursue the science, but not at the expense of patients and others becoming ill in the future from this problem.

Dr. Kipen: I'm just going to return to a brief list of what I think we need to consider in terms of a research agenda. Since everything I have to say has been mentioned before, I'll be quite brief about it.

I think the idea of doing population studies is really critical. Whether we can all agree on a definition for MCS or not, I think we can agree on its salient features. I think we can look at responsiveness to chemicals in populations and see how prevalent that is according to other aspects of the population, whether they be demographic factors, gender factors, or medical history factors. I think we can agree to disagree on whether people do or don't qualify for a diagnosis. That should not hang us up in terms of looking at the prevalence of some of the really salient features. The overlap with chronic fatigue, and historically with a number of other things, is very compelling. There are other quotations from Marcel Proust which have suggested that even more direct responses to chemicals were present many, many years ago. Trying to see how all of these syndromes are different and similar, in terms of population studies, could be very revealing.

The only thing that I would add is that in looking at olfaction exposure studies, we should also consider other routes of exposure, and how that does or doesn't impact the effect of chemicals on the organism.

Mary Lamielle: I also will be very brief. I just want to say that the last two days have been fascinating for a lot of us. There has been a lot of sharing and clarification of ideas, and a dialogue on this issue, that's really been long overdue. It has been a well-documented conference, with audio and video tapes, and proceedings that will come out in the future.

I want to give a thank-you to the other panel members and our expert panel from Annapolis, from which this panel was derived, and also thank the folks at ATSDR for all their work in moving these issues forward. Also, thanks to Senator Lautenberg and his staff for getting the directed Congressional funds, and also to other legislators who have been pushing these issues.

I think this meeting has shown that chemical sensitivity is a legitimate area for research. It is a key piece in moving things forward. Obviously, funding is the key piece on moving a lot of these issues forward. I think it feels really good to see some of the doors opening, and I hope that means more progress.

Dr. B. Weiss: I'm a little disturbed by the separation that I've heard between what we consider to be the discipline of neurobehavioral toxicology and this syndrome we've been discussing, MCS. It's really a question addressed to toxicology, isn't it? We're talking about adverse effects of chemicals. I think these adverse effects ought to be studied in the same way we study other toxic substances, no matter what the end points are.

The cleaving of the questions of how we study toxicology, or toxicants, and how we study chemical sensitivity, is artificial. That artificiality is one of the reasons there is so much difficulty being faced by people who see themselves as victims of this particular syndrome. It doesn't fit. It should fit, for scientific purposes, and for the welfare of everybody else.

There are a couple of issues that I think should have been addressed in this conference. One has been called psychoneuroimmunology. Connections between the nervous system and the immune system, both of which have their functions expressed in manifold ways, interact in many different ways over many different systems. Remember that psychoneuroimmunology also refers to the fact that immune responses can be learned, that is, there are environmental stimuli that can provoke certain kinds of immune phenomena.

I heard Claudia refer to the need for clean, controlled environments. Of course. But it's not something to be performed in somebody's office, or in a room that allegedly is free of carpets. Inhalation toxicology is a very elaborate discipline. It means a lot of technology. I can assure you that the study we've done would not be possible without having aerosol physicists around, inhalation toxicologists, and a lot of other people who know what they're doing. You can't rely on allergists, because they don't know very much about how to control environments.

Dr. Miller: I'm an allergist, so I feel compelled to respond in some way, although I'm not a typical allergist. You're absolutely right. We need to be much more concerned with low levels of chemicals, and how to design a space to meet those needs.

Of course, we've written extensively of the need to develop an environmental unit, in a hospital kind of setting, where you minimize out-gassing from different materials, and you scrupulously avoid introducing items like pesticides and perfumes into it. It seems that's a tool for uncovering, whether some of these syndromes, whether it's chemical sensitivity or chronic fatigue or asthma or whatever, may be related to environmental factors. It is a tool that we as clinicians need very badly. Right now, when we are faced with a patient having any kind of a chronic illness, which we suspect may have resulted from chemical or environmental exposures, we have no way of sorting that out.

Initially, it is a research effort. If it were something that turned out to be useful, we might use it in other places. It's a way of starting to ferret out this very difficult area, and it would rely upon the knowledge of people like Dr. Weiss, who have so much knowledge in this field; people who know about indoor and outdoor air pollutants, and what things to avoid in order to reduce the exposures in such an environment, and do it in a technically competent way that will produce publishable data.

QUESTIONS AND ANSWERS #4

CLAUDIA MILLER, MATTHEW FRIEDMAN, KATHY REST, AND IRIS BELL

Nicholas Ashford, MIT: I don't think we are going to be able to agree on a case definition. What I see described here is a disease, a problem, a symptom, a syndrome, a condition which evolves over time. If you insist that there be three symptoms before you classify someone with MCS, surely at sometime in his life that person had two symptoms. And if he had two, he had one. I think that when we look at prevalence studies with many symptoms, you're getting a patient who is evolved very far along the way.

I can't prove this, but I'd put money on it, that the longer the process is allowed to continue the more nebulous, the more psychiatric and psychological the consequences become, and the ratio of psychological to objective, quantifiable, physical symptoms gets higher and higher.

I propose that we do what IARC does with carcinogens: Class 1, Class 2, Class 3, or Class 4. That is, we have alternative definitions which mirror the evolution of this disease. Not because we want to compensate people, but because we want to do research on people whose symptoms are objectifiable. If you start with prevalence and CNS, you've got the most difficult part of the problem. You've made it incredibly difficult for yourselves. As I listen to all the talk, and I hear about PTSD, but more particularly CFS and cacosmia, I'm left with the striking opinion that these are just steps along the way to full-blown MCS. If you insist on a snap-shot definition, you're going to get yourselves in real trouble.

I will repeat, what you have to do is start at the front end, not at the back end which is the patient end. Let's find out what caused the incident of people getting sick, like a pesticide spill, moving into a new building, taking new kinds of anesthesia, drugs, or numerous consumer products. Look at the things that start the process rolling, and follow them over time if you want.

Albert Donnay, environmental health engineer, office of Dr. Ziem: I'd like to thank Dr. Rosenthal for the ray of light he has shed on the validation process.

I'm very concerned about getting past this first step of case definition. At the end of this month, the NIH will convene its panel to discuss the health effects of Persian Gulf Veterans, and they're considering a definition of something that they may give an entirely new name, called Gulf War Syndrome; yet the clinicians that have seen these patients all think they look remarkably like MCS. We're facing a potential diversion here of all of our efforts if another term and disease entity is introduced into this mix.

Toxicology and Industrial Health, Vol. 10, No. 4/5, pp. 653–660

ISSN: 0748-2337

With respect to that element of diagnosis, I agree with Dr. Ashford that it is very clear from what we've heard that there is a considerable overlap between CFS and MCS. There also may be elements of overlap with depression and other psychiatric disorders. While I commend the studies that have included cohorts of CFS and depressed patients, it's really essential, in any future studies, that those populations be separately reevaluated before they are entered into a study population to determine that they are not, in fact, MCS patients. Because, as we saw with some of the data that were presented today, CFS patients' results are coming out somewhere between those of MCS and depressed patients. That could well be because 30% of them actually have MCS. They say they react to chemicals. If that is indeed our litmus test for MCS, those people should not be left in populations that are used as comparisons and defined as depressed or CFS.

Larry Plumlee, Bethesda, Maryland: I was very moved by what Dr. Weiss said. You know, we've known for a long time there were some people who weren't protected by TLVs. I certainly recall that in the documentation for the TLVs going back in the 1960s, it talked about the fact that these levels were not sufficient to protect some people. And yet, few have come forward to systematically study these especially sensitive persons until recently. It's almost as if toxicology in our society has become a tool to promote and facilitate the marketing of chemicals, rather than a tool for carefully examining what's really going on in the population. Dr. Randolph's papers were written 40 years ago. I worked hard when I was at EPA to try to interest some of the agencies in studying his findings. So, isn't the problem that there just haven't been enough funds, and, perhaps, courageous investigators, to counter the strong inertia within neurotoxicology around uncovering problems that would make the marketing of chemicals more difficult? Are we now stepping through this resistance and saying we won't take it any longer? I hope so.

Bill Hersey, NFFE Local 2050/EPA: With respect to epidemiology studies, again I would like to caution people about prospective studies for the reasons I mentioned before. Some 600 or so employees at EPA reported one kind of adverse effect or other, which they attributed to carpet exposures. Lance Wallace's statistical analyses that have the data on where the carpet was or reports of carpet odor and hot spots within the building, established a solid statistical link between the presence of carpet in certain workplaces and adverse health reports. But only a relatively small portion of the people who reported adverse effects became multiply chemically sensitive. I'd like to encourage the research community to study those of our workers who got MCS and a group of controls of those who reported some adverse effects, but did not acquire MCS, and look for differences in those two populations.

The last point is the connection between funding and progress, and the fact that there is powerful opposition to considering MCS a valid medical condition. As an advocate for people who have become multiply chemically sensitive, both our own and those on the outside who have contacted the union, and also as one who spent almost 20 years in the chemical industry, I would like to urge all of us to be thinking about creative ways to engage the chemical industry in solving this problem.

Dr. Ruth Allen, epidemiologist, Pesticide Health Effects Division, EPA: I review the health studies you all are bringing us and I have a simple request. In your public presentations and in your written work, would you actually name the chemical that is causing the problem, for the following reasons: If you just say on your study that pesticides were tested, this gives me zero help in reviewing the effects of particular chemicals. It also gives the regulatory community zero help in their efforts at product stewardship, because each of the products has a registration, has an advocacy, and is intentionally used throughout the world.

I would also have you, as a second request, report those episodes of poisonings you know about to one of several sources. We have a pesticide incidence data system within the agency, which Congress requested we set up after the metam sodium spill in the Sacramento River. We have a form to input specific poisoning episodes. We have a team that works under the 6A2 provision of the law that reviews these episodes, and we do use this in our deliberations. We also have a hotline at Texas Tech, The National Pesticide Telecommunication Network and we do monitor those incoming calls and pick out the ones that relate to incidence.

Richard Conrad, biochemist, and chemically sensitive: I've been thinking of forming a group of scientists with MCS. We'd like to take part in designing the research protocol. I think we can give you a lot of clues you might not have been heard before or that will come up, in context, that might provide big chunks of information, where it would have been slow, methodical plodding. Discovery occurs when you come up with an anomaly. There are a lot of people who think they're doing science, when they say, "Well, that doesn't fit," or, "That's not conventional. It's too risky. It doesn't fit in our boxes."

What we're talking about here is not having boxes. About looking for unconventional clues that don't fit, where we have to drop our ego, drop our theories, and say, "I give up. OK, here are two things that don't fit, but they're both right." We've got to have a new idea that links them together, and be open to those things. That's the way progress happens, not by grinding, not by setting up huge protocols, doing all the data, doing all the experiments. It comes through some freethinking, and by really listening to the clues that are already there, in us.

Matt Friedman, from the National Center for PTSD: I've really enjoyed the last two days. I think it's been a very exciting convergence of people from very different experiential and scientific and clinical perspectives. But I've been a bit discomfited in the last hour or so by what I'm feeling is a certain amount of polarization. To use a metaphor from the war in Bosnia, and the way there has been some ethnic cleansing, I feel that there has been some attempts at psychiatric cleansing this afternoon, and I think that's a big mistake.

I recognize that, from the perspective of people who suffer from MCS, some psychiatric terms like psychosomatic, or anxious, have been seen as pejorative, or as invalidating. But, I think for those of us who work in fields of affective disorder, or anxiety disorders, or other disorder, this mind-body dualism has got to stop.

The fact of the matter is that the psychiatric data will not go away, whether you're looking at Persian Gulf vets, or other MCS people. I think it's time for us to look at that seriously and with much objectivity, not to see that as something to be ignored, discarded, or argued against. Because that may be a very much an important part of the syndrome, and, as some of us, myself included, tried to indicate in our remarks, our understanding of both the empirical data, and some of the theoretical constructs of some of these psychiatric disorders, the neurobiology of them, may be very important. Not only in our understanding of MCS, but more importantly, in developing interventions that will work, and can be tested.

Dr. Miller: I think that there are few patients who aren't willing to see a psychologist or psychiatrist, if they thought that would be helpful. Many of them have tried that. Some of them have had such dramatic onsets associated with an exposure, and no prior psychiatric history, and they feel like the reality of their condition has been dismissed by referrals, constantly, to psychiatrists or psychologists.

There is a sensitivity here which, as I know you it haven't lived in this field as long as some of us have, that, does give that flavor to it. It's not that people aren't willing to consider that. Every human condition, whether you have cancer or something else, may not be benefited, or may not have psychosocial or psychiatric kinds of ramifications. I think people are appreciative of that, but they also want to have addressed the possibility that it could be all due to chemical exposure. There is no way to sort those things out. But I'm certain it looks polar.

Dr. Friedman: I'm just saying that the evidence is not in yet. And the ignorance of clinicians, who may have seen these people, is not a good reason to buy our search for etiologies and for treatments.

Dr. Rest: I found these last two days incredibly exciting and exhilarating. I appreciated the fact that so many different people from so many different disciplines, looking at problems from so many different angles, agreed to come and participate in this conference, and not only present what they have learned about their own subject areas, but stretch themselves to think about what this might mean for all of us who are trying to advance the understanding of MCS. I think that the conference has really done that. I think a lot of people feel really energized. I loved going outside during breaks and lunch and when the conference was over at the end of the day, and seeing all kinds of different people gathered and talking. I hope that through this conference and through the collaborative networks that have been established we can move forward. I'm a person who considers herself interdisciplinary and so I do applaud the idea of thinking in a more interdisciplinary way, and pulling in experts from other areas that might be able to help us. I'm especially grateful for the patients who were willing to come here, and especially the three of you who were willing to stand up here and be the first three speakers on the first day without knowing any of us here, and sharing your stories. I think it was a courageous thing to do, and it was very useful to have you here. I thank everyone in the audience who has the problem of multiple chemical sensitivity for their questions that were very informative and pointed and gave people things to think about.

Cynthia Wilson: Looking for physiological causes, before we look for psychological causes, is very important to the victims. We don't want this illness to be tagged psychogenic in origin before physical causes have been ruled out. Until this conference, that's pretty much what has been going on with us. We want fair and adequate research and we want it to start where other research starts, and that is looking for immunological causes, and if they can't be found *then* you can go looking for the psychological causes.

Dr. Bell: As a psychiatrist, although it seems to be an embarrassing thing to admit, I want to remind people that I never would have proposed the model I have proposed if I had not been trained by some excellent biological psychiatrists. They taught me about kindling and sensitization and all the related materials we've been discussing, and prepared me to discuss the bio/psycho/social aspects of this kind of problem. That does not mean that I have always agreed with my training or with my colleagues in psychiatry or psychology, in terms of their interpretations of what goes on in chemical sensitivity. But I think there was an important distinction that was just raised, and I'd like to point out that a good psychiatric clinician knows that when you find the phenomenology of a diagnosis, like major depression, your first obligation is to rule out organic causes. I know the presumption of what Dr. Friedman was talking about in terms of diagnostic categories. Those are, as I indicated earlier, descriptive labels; they give us a way to describe what we are seeing. They don't give us etiology. People mistake this particular labeling as an etiology, and I think the first thing we have to do if we see a descriptive label is, in fact, try to understand its mechanisms, starting with the physiological and certainly including the psychological. Most of the data that have been done in the field of psychology indicate that psychological factors are modifiers when biological factors are in the moderate to mild range. When biological factors are in the strong range they overwhelm the psychological factors. We are dealing with a realm where there may be interaction here, but it does not rule out either side.

Laureen Burton: I'm speaking for myself, and not my agency. As a regulatory person for only 2 1/2 years, and a public health person, and a toxicologist, and a chemist, this area has been very interesting to me, and I think this conference is very interesting.

In the last six months or so that my agency has let me go out and attend these conferences, I see a lot of time spent debating the psychological studies vs. the toxicological studies vs. the neurological. That really is a waste of time for those of us who want to help you.

The important thing is to get a definition. The important thing is to get something concrete that those of us who want to help you regulatorily can latch onto, as a definition, as a way to categorize the people who are sick.

There are, possibly, a handful of things that we're talking about. The chemical industry is not going to go away. As a regulatory person I have to work on both sides. I work with the consumer, and with the industry. It does not help, as one of my previous colleagues at EPA said, when the studies use a "general term." They say a "general" carpet, they say a "general" pesticide, they say a "general" household cleaner. That does not help us help you. We have to know what it is we're dealing with.

There are several carpets. Not all carpets cause the problem. Several of them may, and I have to say may, because the association has not been made.

For the citizens who are not scientists, I remind you, don't believe everything you read to be the gospel. This does not help those of us in the agencies who want to help you. We have a hotline as well, and since I'm not speaking for the agency, I will skip that point, but there is a hotline. EPA has one, OSHA has one, NIOSH has one, CPSC has one, everybody has a hotline. It is important; do not use the government as the enemy. Tell us what your problem is. Your particular call may not get immediate action. But the groupings of calls mean something.

Dr. Hancey: I've been tussling for two days to decide whether I was going to come out of the closet, here at this meeting. I'm the worst combination: I'm a psychiatrist employed by a chemical industry carpet manufacturer. It took a lot of guts to say that. I wanted to build on what's been said by several of the last speakers, which is: it does nobody any good to remain divisive.

I think if one of the major themes that comes out of this conference is that there is momentum for consensus building. The chemical industry is not your natural enemy either. You can believe that if you want to, and if you believe that it's true, then you can act that way, and then we will remain enemies. But what Dr. Hersey was saying is absolutely true. I think there are kinder, gentler carpets, there are kinder, gentler chemical industries, and people within them who can be reached out to. I am such a person. I'm trying to network here. I'm not stalking you.

It's unscientific and unthinking to presume that, out of a thousand MCS patients, none of them will have a primary psychiatric disorder. It's also unscientific to think that all of them have a scientific disorder. What you need to have is an open mind, when you design a research protocol, so that you don't foreclose possibilities, turning your brain off to possibilities. We've got to get away from the dualistic either-or: either you're crazy or it's real. That's a waste of time; it's counterproductive. I would just hope that the consensus-building momentum is what people go home with tonight.

From the Audience: I was wondering, as a researcher in this area, whether you have any influence over setting up a research fund among the chemical industries to provide some means to study this problem a little further, which is in all of our best interests, given the shrinking research dollars on the federal level.

Dr. Hancey: Not speaking officially for my company or anybody else, the concern about the litigation consequences of doing research, from the law side of our corporation, is a major stumbling block. I think that maybe, after looking at some well-conceived research protocols with good design, we scientist types can try and be persuasive to the legal types who are hysterical about it, and see if we can get them down off the chandelier.

Michael Gray, internist, Occupational Medicine, Southeastern Arizona: In the last seven years, I have clinically evaluated 55 patients that have presented to me with a variety of concerns, a number of which have ended up in my mind warranting the diagnosis of MCS. Generally, I have ended up finding that they fit a category of immunotoxicity and neurotoxicity. The way I have worked them up is to do a thorough internal medicine evaluation, which is your standard history, physical, CBC, SMACK, urinalysis, thyroid profile, electrocardiogram with appropriate chest X rays, pulmonary function tests, and a battery of tests through antibody assay labs that looks at their lymphocyte profile, autoimmune profile and status, their chronic viral disease status, and their sensitivity to formaldehyde, volatile organic solvents, isocyanates, and latex. I then have consistently referred them to one or another neuropsychologist to be evaluated in the standard Halstead-Rittan profiling and whatever else the neuropsychologist thinks is appropriate to determine whether there is toxic encephalopathy present. I usually wait until I have their report back before I decide whether my suspicion of probable toxic encephalopathy warrants the diagnosis.

It has been very instructive to follow nine patients, all of whom were employed in an environment which was confirmed by OSHA to have formaldehyde present, in excessive amounts, for a period of a year. They had mixed exposures, however, so the data aren't just purely related to formaldehyde, but it also relates to phenol and xylene. I find that when I look at the pesticide-related individuals, we again are confronted with a mixed exposure because almost all pesticides that are used are solubilized with organic solvents before they are mixed with water, so that we rarely have a pure situation in the real world. I also don't choose or exclude individuals based on whether they are in litigation. I try to take them as they are and evaluate and render an opinion based on what I'm seeing clinically. What has struck me is as follows: 1) I also have about a 75% prevalence of CD26 excessive counts, which signifies activated cellular immunity. I'm finding a significant amount of excessive autoimmune markers that may or may not relate to autoimmune disease, but nonetheless reflect excessive autoimmune activity. I'm also finding a significant prevalence of markers in the formaldehyde-exposed group (all but the individual who was on methotrexate for his formaldehyde-induced psoriasis). All of the rest had elevated IgM markers that were antiformaldehyde, and, in the beginning of the groups' study and follow-up, 100% of them had elevated myelin autoantibodies to peripheral nerve myelin, and 75% had elevated myelin autoantibodies to CNS myelin. I think that these are quite significant findings. They have been sufficient for me, as a clinician not in an academic setting, to render diagnoses in these cases and render opinions about causation based on their history. What has struck me is that not one of them, after going through neuropsych testing, turned out to be significant on the malingering status category. They were all found to be sincere. They were almost all found to have suffered significant degradation of intellectual performance. Several of the formaldehyde group were double- and triple-degreed individuals, whose IQs had dropped about 30 points, according to the neuropsych evaluations that were done. Some of them were tested by more than one neuropsychologist. For whatever it's worth, working in the real world without any resources for "research dollars" it has been possible to evaluate and workup patients productively.

Stuart Steinhauer, University of Pittsburgh: I'd like to say very quickly, that I think there's a general misunderstanding of what many people in neuropsychiatry are doing here and why they are applying their talents. Maybe there's a reason there is no listing for neurotoxicity under psychiatric disorders. It's probably because the people from this field are not worried about whether it is a psychiatric disorder. Rather, they are more worried about applying their approaches and studying brain science and behavioral manifestations to whatever it is that occurs. It doesn't matter whether it is approached by one specialty or the other. As Dr. Letz pointed out, psychologists who have experience in test organizations bring special skills. Psychiatrists who have looked at organization of behavior where there are some known, or suspected, or unknown biological contributions, bring skills to observe and interpret. They work together with neurologists, with other neuroscientists in every area, to bring problem solving to bear, and not to try to come up with a psychiatric diagnosis. That's not the problem, and I don't think that's the concern, nor is there interest in talking about either the neuroscience or neuropsychiatric implications of trying to say this is a pure psychiatric disorder. If there is a problem that requires an individual dealing with difficulties in performance and life and expression of behavior, that's something that's dealt with by a different branch of psychotherapy and treatment, but they all may contribute in different ways.

CRITIQUE: SPECT STUDIES OF MULTIPLE CHEMICAL SENSITIVITY

HELEN MAYBERG

University of Texas Health Science Center
San Antonio, Texas

Since their introduction, studies using positron emission tomography (PET) and single photon emission tomography (SPECT) have concentrated on pattern recognition in groups of patients with well characterized and often pathologically confirmed diseases. A growing literature addresses the sensitivity and specificity of patterns seen in single studies to correctly predict the presence of a particular illness (AAN, 1991; Holman and Devous, 1992, Masdeu et al., 1994). These results have already had a tremendous impact on the diagnostic workup and treatment plan for individual patients with epilepsy (Fisher and Frost, 1991), brain tumors (Coleman et al., 1991), and dementia (Duara, 1990), as well as a growing role in the evaluation of stroke (Baron, 1985), movement disorders (Brooks, 1993), and head trauma (Ichise et al., 1994). Scan abnormalities have also been identified in groups of patients with certain well defined psychiatric diagnoses, including depression (Baxter et al., 1989; Mayberg et al., 1994), panic disorder (Reiman et al., 1984), schizophrenia (Gur et al., 1989; Buchsbaum, 1990), attention deficit (Zametkin et al., 1990), anxiety and obsessive compulsive disorder (Baxter et al., 1988; Rauch et al., 1994), alcoholism (Volkow et al., 1992), and substance abuse (Volkow et al., 1991, Holman et al., 1992), among others. However, the sensitivity and specificity of these patterns have not been determined, and in many cases the group patterns may actually be too subtle to detect in individual patients. While these types of studies remain extremely important for identifying previously unrecognized brain abnormalities and potential disease mechanisms in a variety of neuropsychiatric illnesses, including multiple chemical sensitivity, their utility in the management of individual patients is still far from clear.

An enthusiastic but cautious attitude therefore seems appropriate in evaluating preliminary SPECT scan findings in patients with toxic exposures, chronic fatigue syndrome, and presumed multiple chemical sensitivity. These results must be considered in the context of other SPECT and PET studies performed in a variety of neuropsychiatric disorders.

Studies examining functional imaging patterns in patients with well defined clinical deficits following focal neurological insults have been performed and published. Relationships

1. Address all correspondence to: Helen Mayberg, M.D., University of Texas Health Science Center at San Antonio, 7703 Floyd Drive, San Antonio, TX 78284-6240. Tel.:(210)567-8187. Fax:(210)567-8152.

Toxicology and Industrial Health, Vol. 10, No. 4/5, pp. 661–666

ISSN: 0748-2337

between isolated neurobehavioral findings and regional metabolic and flow patterns in patients with strokes have been described, often confirming expected regional abnormalities in the setting of normal anatomy (Baron, 1989). Patients with aphasia, for instance, show left frontal-temporal-parietal hypometabolism, independent of both the anatomical lesion location and clinical characteristics of their language disturbance (Metter et al., 1989). The reverse — whether or not an existing language deficit can be predicted from a specific scan abnormality — has not been addressed. It is clear that persistence of metabolic or flow defects after stroke or trauma may be accompanied by a normal neurological exam. Whether a functional abnormality correlates better with the initial or the residual neurological deficit has not yet been examined fully. This fundamental information is needed before a given functional imaging scan can be effectively and meaningfully used in evaluating the current, past, or future behavioral deficits of a given patient.

Patterns of regional blood flow and metabolism and their relationship to specific behavioral symptoms in diseases such as dementia have also been studied (Duara, 1990). Asymmetric cortical hypometabolism correlates with clinical deficits in patients with probable Alzheimer's disease and may be seen months prior to the onset of specific behavioral changes (language disturbances with left-hemisphere hypometabolism; neglect and visual-spatial deficits with right-hemisphere hypometabolism) (Haxby et al., 1985; Grady et al., 1986). Similarly, striatal (caudate and putamen) hypometabolism seen in clinically normal, at risk Huntington's disease patients has also been demonstrated (Grafton et al., 1990). These studies suggest the presymptomatic efficacy of PET (and likely SPECT) for identifying functional brain lesions prior to their clinical manifestation. These findings have potential clinical implications in diseases where early or prophylactic treatment is an option. There remains however, an obvious gap in terms of the temporal relationship of clinical deficits to metabolic abnormalities and vice versa.

Use of functional imaging to identify markers of behaviors that have no known pathology has also been explored. Depression, for example, has been studied in some detail (Mayberg, 1993). Despite differences in patient selection, medication status, scan acquisition, and data analysis strategies, frontal (and less consistently temporal) abnormalities in flow and metabolism have been demonstrated in patients with both primary affective disorders and depression associated with specific neurological diseases such as Parkinson's disease, Huntington's disease, and stroke. The reproducibility of these findings might initially suggest that functional measurements are clinically useful in the differential diagnosis of patients with cognitive abnormalities in which the recognition of a mood disturbance may be obscured (i.e., Alzheimer's disease, stroke with aphasia, hypokinetic movement disorders with bradyphrenia). However, the specificity of this pattern for depression has not been determined. Paralimbic hypometabolism (inferior frontal, anterior temporal, cingulate) has been demonstrated in patients with chronic amnesia (Fazio et al., 1992) and Pick's disease (Jagust et al., 1989). Subgroups of schizophrenic patients show bifrontal hypometabolism (Gur et al., 1989; Buchsbaum, 1990), as do patients with progressive supranuclear palsy (Foster et al., 1988) and corticonigral degeneration (Brooks, 1993). Frontal defects have also been described in patients with head trauma (Ichise et al., 1994), AIDS dementia (Holman et al., 1992), and cocaine withdrawal (Volkow et al., 1991). Although it is unlikely that depression would be confused clinically with either a pure amnestic syndrome or florid thought disorder, the similarities in the pattern of regional metabolism in patients with dissimilar behavioral symptoms require further careful study. While use of functional imaging in depression may eventually prove helpful in identifying those patients who might benefit from a trial of antidepressants or other treatments targeting specific symptoms, the experimental evidence for these uses does not yet exist.

Given these data from diseases in which the neuropathology is known or where, at minimum, case definitions are well established, caution is needed in designing and interpreting studies in patients with neuropsychiatric sequelae from presumed multiple chemical sensitivity. Chronic fatigue syndrome, specific neurotoxic exposures including pesticides, solvents, and breast implants, and multiple chemical sensitivity (as best as it can be defined) have all been associated with a wide range of neuropsychiatric complaints, often in the setting of normal standard neurological, psychiatric, neuropsychological, and structuring neuroimaging exams. Because of the reported clinical variability, lack of strict diagnostic criteria, and uncharacterized neuropathology in these syndromes, functional imaging studies of these patient groups are difficult to interpret. The specificity of functional abnormalities for particular symptoms or specific causal agents have also not been evaluated. Although preliminary studies report multifocal defects in patients with these problems (Simon, 1992; Callender et al., 1993; Heuser et al., 1993; Mena et al., 1993), no clear lesion-deficit correlations have been demonstrated. Similarly, studies have yet to correlate specific clinical complaints or objective neurological or neuropsychological deficits with specific functional imaging patterns. While presence of multifocal defects may prove to be a marker of some type of brain insult, the patterns thus far appear nonspecific, similar to those reported in AIDS dementia, cocaine abuse, and multi-infarct dementia.

Controlled studies of well characterized patients selected using standardized clinical criteria are clearly needed. Appropriate comparison groups are also required. Subjects with similar exposure histories but without subjective complaints may be a more credible control population than age-, sex-, educationally, or socioeconomically matched subjects.

Means to regulate the behavioral state of the patient at the time of the imaging study may also be critical. While the standard approach has been to examine patients in a basal resting state, it can be reasonably argued that optimal results will be obtained if patients are "challenged" to reproduce their clinical symptoms. Which of these approaches will most consistently detect abnormalities in these patients is unknown. This issue is not trivial, since it is well documented that mood state, anxiety, attention, and performance of specific cognitive tasks will have significant impact on the pattern of regional blood flow or metabolism measured with both PET and SPECT. Experimental manipulation of a behavioral state is, in fact, the strategy used in functional mapping studies of normal subjects, a major area of imaging research (Posner et al., 1988; Fox, 1989). Studies range from simple mapping of primary modalities such as sensation and vision to exploring complex behaviors whose functional localizations are unknown. The applicability of these "activation" or "challenge" strategies to patient populations are in the early stages of testing and the long-range utility of these types of studies in neurological or psychiatric subjects has not been tested. The potential application of these methods to the planning of a neurosurgical resection in patients with tumors or focal epilepsy (Pardo and Fox, 1993), or to the study of functional recovery following brain injury (Weiller et al., 1992) show great promise. How activation studies are optimally used to unmask episodic symptoms in specific psychiatric populations such as panic or obsessive compulsive disorder is also an active area of ongoing research.

Future applications of these various imaging strategies to the study of multiple chemical sensitivity will require careful review of all available clinical and experimental data. Attention to the specific behavioral state (rest versus task), type of chemical challenge (dose, timing, route of exposure), and means to maintain a blinded presentation are of primary importance. Strategies to interpret the contribution of acute behavioral confounds such as anticipatory anxiety or panic, and the influence or preexisting psychiatric diagnoses such as depression are also necessary. Challenge experiments that can take advantage of the data on subthreshold exposures appear well suited to the study of these subjects. Study designs incorporating multiple trials would, in addition, allow direct evaluation of order and placebo effects. The ethics of provocation studies will, however, need to be carefully considered since existing data suggest that additional exposures, regardless of dose, may have clinical and potentially irreversible consequences. Unambiguous results from experiments designed to specifically examine these complex issues are clearly needed.

Despite the preliminary nature of the reported findings and the limited availability of peer-reviewed papers and published replications, there is already growing interest in and use of functional imaging to support the presence of "brain abnormalities" in persons who have had well-documented or suspected exposures. While the pilot studies presented during this conference appear to provide novel research approaches to study changes in brain function that may accompany or result from exposures in susceptible persons, the introduction of these findings either for clinical purposes or in court is still premature. The sensitivity and specificity of the reported scan abnormalities have not been determined and continued debate remains regarding optimal methods for scan acquisition and data analysis. Furthermore, unbiased critical assessment of the potential risks that additional exposures (i.e., during imaging) may have for patients and plaintiffs who are highly motivated to undergo even potentially high-risk studies is also lacking. Therefore, the use of these scans in court prior to completion of controlled trials and peer-review of results not only may be viewed as unethical, but may also be considered legally insupportable under Rule 702 of the Federal Rules of Evidence (Mayberg, 1992; U.S. Supreme Court, 1993; Thornton, 1994). The clinical utility of these scans is also uncertain and their use, at present, should be considered purely experimental.

In summary, preliminary SPECT findings in patients with toxic exposures and multiple chemical sensitivity are enticing, but require replication. Researchers in the field will need to propose additional testable hypotheses, establish consensus case definitions, identify appropriate control groups, and standardize as well as validate both the scan protocols and methods for qualitative and quantitative image analysis. These issues must be resolved to ensure future interpretable and meaningful results.

REFERENCES

AMERICAN ACADEMY OF NEUROLOGY, THERAPEUTICS, AND ASSESSMENT SUBCOMMITTEE (AAN) (1991). "Assessment: Positron emission tomography." Neurology 41:163–167.

BARON, J.C. (1985). "Positron tomography in cerebral ischemia." Neuroradiology 27:509–516.

BARON, J.C. (1989). "Depression of energy metabolism in distant brain structures: Studies with positron emission tomography in stroke patients." Seminars Neurol. 9:281–285.

BAXTER, L.R., JR, SCHWARTZ, J.M., MAZZIOTTA, J.C., PHELPS, M.E., PAHL, J.J., GUZE, B.H., and FAIRBANKS, L. (1988). "Cerebral glucose metabolic rates in nondepressed patients with obsessive-compulsive disorder." Am. J. Psychiatry 145:1560–1563.

BAXTER, L.R., SCHWARTZ, J.M., PHELPS, M.E., MAZZIOTA, J.C., GUZE, B.H., SELIN, C.E., and SUMIDA, R.M. (1989). "Reduction of prefrontal cortex glucose metabolism common to three types of depression." Arch. Gen. Psychiatry 46:243–250.

BROOKS, D.J. (1993). "Functional imaging of movement disorders." In: Movement Disorders 3 (C.D. Marsden and S. Fahn, ed.). Buttersworth Scientific, Boston. pp. 65–87.

BUCHSBAUM, M.S. (1990) "The frontal lobes, basal ganglia and temporal lobes as sites for schizophenia." Schizophrenia Bull. 16(30):379–391.

CALLENDER, T.J., MORROW, L., SUBRAMANIAN, K., DUHON, D., and RISTOVV, M. (1993). "Three dimensional brain metabolic imaging in patients with toxic encephalopathy." Environ. Res. 60:296–319.

COLEMAN, R.E., HOFFMAN, J.M., and HANSON, M.W. (1991). "Clinical application of PET for the evaluation of brain tumors." J. Nucl. Med. 32:616–622.

DUARA, R., ed. (1990). Positron Emission Tomography in Dementia. Frontiers of Clinical Neuroscience Series, Vol. 10. Wiley-Liss, New York.

FAZIO, F., PERANI, D., GILARDI, M.C., et al. (1992). "Metabolic impairment in human amnesia: A PET study of memory networks." J. Cereb. Blood Flow Metab. 12:353–358.

FISHER, R.S. and FROST, J. J. "Epilepsy." J. Nucl. Med. 32:651–659.

FOSTER, N.L., GILMAN, S., BERENT, S., MORIN, E.M., BROWN, M.B., and KOEPPE, R.A. (1988). "Cerebral hypometabolism in progressive supranuclear palsy studies with PET." Ann. Neurol. 24:399–406.

FOX, P.T. (1989). "Functional brain mapping with PET." Seminars Neurol. 9:323–329.

GRADY, C.L., HAXBY, J.V., SCHLAGETER, N.L., BERG, G., and RAPOPORT, S.I. (1986). "Stability of metabolic and neuropsychological asymmetries in dementia of the Alzheimer's type." Neurology 36:1390–1392.

GRAFTON, S.T., MAZZIOTTA, J.C., PAHL, J.J, ST. GEORGE-HYSLOP, P., HAINES, J.L., GUSELLA, J., HOFFMAN, J.M., BAXTER, L.R., and PHELPS, M.E. (1990). "A comparison of neurological, metabolic, structural, and genetic evaluations in persons at risk for Huntington's Disease." Ann. Neurol. 28:614–621.

GUR, R.E., RESNICK, S.M., and GUR, R.C. (1989). "Laterality and frontality of cerebral blood flow and metabolism in schizophrenia: Relationship to symptom specificity." Psychiatry Res. 27:325–334.

HAXBY, J.V., DUARA, R., GRADY, C.L., CUTLER, N.R., and RAPOPORT, S.I. (1985). "Relations between neuropsychological and cerebral metabolic asymmetries in early Alzheimer's disease." J. Cereb. Blood Flow Metab. 5:193–200.

HEUSER, G., MENA, I., GOLSTEIN, J., THOMAS, C., and ALAMOS, F. (1993). "Neurospect findings in patients exposed to neurotoxic chemicals." Clin. Nucl. Med. 18:923 (abstract).

HOLMAN, B.L. and DEVOUS, M.D. (1992). "Functional brain SPECT: The emergence of a powerful clinical method." J. Nucl. Med. 33:1888–1904.

HOLMAN, B.L., GARADA, B., JOHNSON, K.A., MENDELSON, J., HALLGRINGE, E., TEOH, S.K., WORTH, J., and NAVIA, B. (1992). "Comparison of brain perfusion SPECT in cocaine abuse and AIDS dementia complex." J. Nucl. Med. 33:1312–1315.

ICHISE, M., CHUNG, D.G., WANG, P., WORTZMAN, G., GRAY, B.G., and FRANKS, W. (1994). "Technetium-99m-HMPAO SPECT, CT and MRI in the evaluation of patients with chronic traumatic brain injury: A correlation with neuropsychological performance." J. Nucl. Med. 35:217–226.

JAGUST, W.J., REED, B.R., SEAB, J.P., KRAMER, J.H., and Budinger, J.F. (1989). "Clinical-physiological correlates of Alzheimer's disease and frontal lobe dementia." Am. J. Physiological Imaging 4:89–96.

MASDEU, J.C., BRASS, L.M., HOLMAN, B.L. and KUSHNER, M.J. (1994). "Brain single-photon emission computed tomography." Neurology 44:1970–1977.

MAYBERG, H.S. (1992). "Functional brain scans as evidence in criminal court: An argument for caution." J. Nucl. Med. 33(6):18N–25N.

MAYBERG, H.S. (1993). "Neuro-imaging studies of depression in neurological disease." In: Depression in Neurologic Diseases (S.E. Starkstein and R.G. Robinson, ed.). Hopkins University Press, Baltimore, MD. pp. 186–216.

MAYBERG, H.S., LEWIS, P.J., REGENOLD, W., and WAGNER, H.N. (1994). "Paralimbic hypoperfusion in unipolar depression." J. Nucl. Med. 35(6):929–934.

MENA, I., GOLDSTEIN, J., JOUANNE, E., and LESSER, I. (1993). "Cerebral hypoperfusion in late life chronic fatigue syndrome (CFS) and late life depression." J. Nucl. Med. 34:210P (abstract).

METTER, E.J., KEMPLER, D., JACKSON, C.A., HANSON, W.R., MAZZIOTA, J.C., and PHELPS, M.C. (1989). "Cerebral glucose metabolism in Wernicke's, Brocas's and conduction aphasia." Arch. Neurol. 46:27–34.

PARDO, J.V. and FOX, P.T. (1993). "Preoperative assessment of the localization of language functions: a comparison of CBF PET with intracarotid sodium amytal." Hum. Brain Map 1:57–68.

POSNER, M.I., PETERSEN, S.E., FOX, P.T., and RAICHLE, M.E. (1988). "Localization of cognitive functions in the human brain." Science 240:1627–1631.

RAUCH, S.L., JENIKE, M.A., ALPERT, N.M., et al. (1994). "Regional cerebral blood flow measured during symptom provocation in obsessive-compulsive disorder using oxygen 15–labeled carbon dioxide and positron emission tomography." Arch. Gen. Psychiatry 51:62–70.

REIMAN, E.M., RAICHLE, M.E., BUTLER, F.K., HERSCOVITCH, P., and ROBINS, E. (1984). "A focal brain abnormality in panic disorder, a severe form of anxiety." Nature 310:683.

SIMON, T.R. (1992). "Breast implants and organic solvent exposure can be associated with abnormal cerebral SPECT studies in clinically impaired patients." Radiology 185:234 (abstract).

UNITED STATES SUPREME COURT (1993). Daubert versus Merrell Dow Pharmaceuticals. 113 Supreme Court 2786.

THORNTON, J.I. (1994). "Courts of law v. courts of science: A forensic scientist's reaction to Daubert." Shepard's Expert Sci. Evid. Q. 1:475–486.

VOLKOW, N.D., FOWLER, J.S., WOLF, A.P., HITZEMANN, R., DEWY, S., BENDRIEM, B., ALPERT, R., and HOFF, A. (1991). "Changes in brain glucose metabolism in cocaine dependence and withdrawal." Am. J. Psychiatry 148:621–626.

VOLKOW, N.D., HITZEMANN, R., WANG, G.J., WOLF, A.P., and DEWEY, S.J. (1992). "Decreased brain metabolism in neurologically intact healthy alcoholics." Am. J. Psychiatry 149:1016–1022.

WEILLER, C., CHOLLET, F., FRISTON, K.J., WISE, R.J., and FRACKOWIAK, R.S. (1992). "Functional reorganization of the brain in recovery from striatocapsular infarction in man." Ann. Neurol. 31(5):463–472.

ZAMETKIN, A.J., NORDAHL, T.E., GROSS, M., KING, A.C., SEMPLE, W.E., RUMSEY, J., HAMBURGER, S., and COHEN, R.M. (1990). "Cerebral glucose metabolism in adults with hyperactivity of chioldhood onset." N. Engl. J. Med. 323:1361–1366.

MULTIPLE CHEMICAL SENSITIVITY WORKSHOP

SPEAKERS AND PARTICIPANTS

Robert Adamec, Ph.D.
Department of Psychology
Memorial University of Newfoundland
St. John's, Newfoundland
Canada

Seymour Antelman, Ph.D.
Department of Psychiatry
Western Psychiatric Institute and Clinic
University of Pittsburgh
School of Medicine
Pittsburgh, Pennsylvania

Iris Bell, M.D., Ph.D.
Department of Psychiatry
University of Arizona Health Sciences Center, and
Tucson Veteran Affairs Medical Center
Tucson, Arizona

Jacob Berkson
Hagerstown, Maryland

John DeLuca, Ph.D.
Kessler Institute for Rehabilitation
Department of Research
West Orange, New Jersey

Dr. Richard Doty, Director
Smell and Taste Center
University of Pennsylvania Medical Center
Philadelphia, Pennsylvania

Nancy Fiedler, Ph.D.
UMDNJ — Robert Wood Johnson Medical School
Environmental and Occupational Health Sciences Institute
Piscataway, New Jersey

Matthew J. Friedman, M.D., Ph.D.
Departments of Psychiatry and Pharmacology
Dartmouth Medical School, and
Executive Director of National Center for PTSD, VAM, and ROC
White River Junction, Vermont

Mary E. Gilbert, Ph.D.
ManTech Environmental Technologies
Health Effects Research
U. S. EPA
Research Triangle Park, North Carolina

Charles T. Hinshaw, Jr., M.D.
Environmental Medicine
Wichita, Kansas

Dr. Thomas Hummel
Institut fur Experimentelle und Klinische Pharmakologie und Toxikologie der Friedrich-Alexander-Universitat Erlangen-Nürnberg
Nuremberg, Germany

Helen Keplinger
Bethesda, Maryland

Howard Kipen, M.D.
UMDNJ — Robert Wood Johnson Medical School
Environmental and Occupational Health Sciences Institute
Piscataway, New Jersey

Dr. med. Gerd Kobal
Institut fur Experimentelle und Klinische Pharmakologie und Toxikologie der Friedrich-Alexander-Universitat Erlangen-Nürnberg
Nuremberg, Germany

Mary Lamielle
National Center for Environmental Health Strategies
Voorhees, New Jersey

Richard Letz, Ph.D.
Division of Environmental and Occupational Health
Emory University School of Public Health
Atlanta, Georgia

Tyler S. Lorig, Ph.D.
Department of Psychology
Washington and Lee University
Lexington, Virginia

Helen Mayberg, M.D.
Assistant Professor for Neurology, Psychiatry, and Radiology
Research Imaging Center
The University of Texas Health Science Center at San Antonio
San Antonio, Texas

Ismael Mena, M.D.
Director, Division of Nuclear Medicine
Professor of Radiological Sciences
Harbor-UCLA Medical Center
Torrance, California

Claudia Miller, M.D.
Department of Family Practice
The University of Texas Health Science Center
San Antonio, Texas

Lisa Morrow, Ph.D.
Western Psychiatric Institute and Clinic
Department of Psychiatry
University of Pittsburgh School of Medicine
Pittsburgh, Pennsylvania

David Newlin, Ph.D.
Addiction Research Center
National Institute of Drug Abuse
Baltimore, Maryland

James W. Pennebaker, Ph.D.
Professor of Psychology
Department of Psychology
Southern Methodist University
Dallas, Texas

Kathleen M. Rest, Ph.D., M.P.A.
Occupational Health Program
Department of Family and Community Medicine
University of Massachusetts Medical Center
Worcester, Massachusetts

Norman E. Rosenthal, M.D.
Chief, Environmental Psychiatry Section
Clinical Psychobiologic Branch
National Institutes for Mental Health
Bethesda, Maryland

Christopher Ryan, Ph.D.
Western Psychiatric Institute and Clinic
Department of Psychiatry
University of Pittsburgh School of Medicine
Pittsburgh, Pennsylvania

Gary Schwartz, Ph.D.
Department of Psychology
University of Arizona
Tucson, Arizona

Gregory E. Simon, M.D.
Center for Health Studies, Group Health Cooperative of Puget Sound and,
Department of Psychiatry and Behavioral Science
University of Washington
Seattle, Washington

Theodore Simon, M.D.
Department of Radiology
University of Texas
Dallas, Texas

Barbara Sorg, Ph.D.
Department of Veterinary and
Comparative Anatomy, Pharmacology,
and Physiology
College of Veterinary Medicine
Washington State University
Pullman, Washington

Bernard Weiss, Ph.D.
Department of Environmental Medicine
University of Rochester School of
Medicine and Dentistry
Rochester, New York

Susan Weiss, Ph.D.
Biological Psychiatry Branch
National Institutes for Mental Health
Bethesda, Maryland

Cynthia Wilson
Chemical Injury Information Network
White Sulphur Springs, Montana